THE ANCESTRY OF THE VERTEBRATES

Frontispiece. Torsten Richard Emanuel Gislén (1893–1954), originator of the calcichordate theory (for obituary see Anonymous, 1954).

The ancestry of the vertebrates

R. P. S. Jefferies

Cambridge University Press

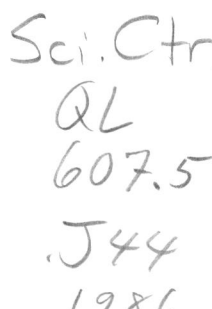
Published by the British Museum (Natural History), Cromwell
Road, London SW7 5BD and the Press
Syndicate of the University of Cambridge,
The Pitt Building, Trumpington Street, Cambridge CB2 1RP
32 East 57th Street, New York, NY 10022, USA
10 Stamford Road, Oakleigh, Melbourne 3166, Australia

First published 1986

Printed in Great Britain by Henry Ling Ltd., at the Dorset Press,
Dorchester, Dorset

British Library Cataloguing in Publication Data

Jefferies, R.P.S.
    The ancestry of the vertebrates.
    1. Vertebrates—Evolution
    I. Title   II. British Museum (Natural History)
    596'.038   QL607.5

    ISBN 0-565-00887-0 hard covers BM(NH) edition
    ISBN 0-521-34266-X hard covers CUP edition USA only

Contents

	Preface	vii
Chapter 1	Aims, history of study, methodology, methods	1
Chapter 2	Metazoan relationships and the hemichordates and echinoderms	17
Chapter 3	Living acraniates–amphioxus and its relatives	55
Chapter 4	The tunicates	87
Chapter 5	Adult anatomy and basic phylogeny of living vertebrates	117
Chapter 6	The embryology of vertebrates and the head problem	163
Chapter 7	Stem chordates–the cornutes	191
Chapter 8	Primitive crown chordates– the mitrates	239
Chapter 9	The phylogeny of the deuterostomes	317
Chapter 10	Some other opinions	345
	Literature cited	359
	Index	371

Preface

This book argues that vertebrates and other living chordates arose from the strange fossils called calcichordates. This view was first proposed in detail by the Swedish zoologist Torsten Gislén (frontispiece) in 1930, but his paper was substantially ignored for more than 30 years. Studies on these fossils carried out at the British Museum (Natural History) since February 1964 show, in my opinion, that Gislén was essentially right. He deserves more honour than he ever received in his life-time.

As a palaeontologist I start with the fossil facts, and these are skeletal. To reconstruct the soft parts I need to use circumstantial argument. In Chapter 8, for instance, I reconstruct the pharynx of mitrates on the basis of many observed skeletal details which all point to one conclusion – that it was tunicate-like. But a zoologist is aware that he could see the pharynx, if the mitrates still lived, merely by cutting the animal open. It is probably this awareness that has led Prof. J. Z. Young, for example, to accuse the calcichordate theory of 'too much interpretation' (1981, p. 69). I deny this, for a complex circumstantial argument can sometimes be almost as certain as a scalpel in its results. However, it must not be judged until all its constituent, mutually supportive reasons have been understood. I expect and hope, therefore, that readers will be sceptical of this book but that they will not judge it until they reach the end. Current orthodoxy concerning the ancestry of vertebrates, which derives from Garstang (1928), is a web of speculation and is believed only because it is orthodoxy (Chapter 10). It is infinitely weaker than detailed interpretation of facts.

The calcichordate theory is not fixed. It has changed in the past and will change in the future, as new evidence comes in. The same is true of all other theories based on evidence.

This book is an interim report, for much remains to be done. Calcichordates have been described, in the published literature, from the Palaeozoic rocks of the following countries: Australia (Tasmania, Victoria); Brazil; Britain; Canada (Ontario); Czechoslovakia; France (Brittany, Montagne Noire, Normandy); Germany (Rhineland); Morocco; New Zealand; South Africa; Spain; Sweden (Gotland); and the United States (Illinois, Maryland, Missouri, Nevada, New York, Ohio, Oklahoma, Utah, Wisconsin). However, it is impossible that this list represents their true distribution. For example, calcichordates must surely exist in the huge palaeozoic outcrops of the USSR and China, but they have been described from neither. Moreover, in most of the published descriptions the animals are regarded as echinoderms. The fossils need

to be placed in their correct phylum as chordates and studied again. There are, in addition, several other peculiar fossils, traditionally placed in the 'carpoid echinoderms' but not classed as 'cornutes' or 'mitrates', which need to be re-investigated, in the suspicion that they also may prove to be chordates. All this, however, is work for the future. I hope this book will persuade some palaeontologists, particularly those at the start of their careers and in whatever country they happen to be, to study these strange fossils so as to throw light on how chordates arose and complete the task that Gislén began. (I hope, also, that such people will communicate with me.) Similarly, specialists in the comparative anatomy, taxonomy and embryology of living chordates should consider the calcichordates. For these fossils show, more clearly than ever before, what the primitive chordate condition was.

I wish to thank my colleagues at the British Museum (Natural History) for providing a friendly yet critical working environment. I think in particular of the members of my twice-daily coffee group – we have cooperatively drunk about six cubic metres of coffee – whom I shall not name in person since they know well enough who they are. I should also like to thank Mr David Lewis for his technical help and unceasing moral support over the years, and Prof. W. T. Dean, formerly of the BM(NH) but now of University College, Cardiff who started the whole thing off, as explained in Chapter 1.

I have been generously helped by colleagues outside the BM(NH) by access to specimens in collections. A palaeontologist well knows how important this access is, particularly when the fossils in question, as in this case, are rare. Here it is a pleasure to thank, in particular: Dr Rudolf Prokop of the National Museum, Prague with whom I have had the satisfaction of collaborating; Dr Ladislav Marek, of the Czechoslovak Academy of Sciences, Prague; Dr Ian Rolfe, of the Hunterian Museum, Glasgow; Mr A. G. Brighton and Dr Barry Rickards, of the Sedgwick Museum, Cambridge; Dr Isles Strachan, of Birmingham University; Prof. Steve Gould of the Museum of Comparative Zoology, Harvard; M. le Docteur Jean Chauvel of Rennes University, France; M. l'Abbé Robert Courtessole and his co-workers of Carcassonne; Messrs R. Feist and J. Mattei of Montpellier; M. Prieur and Mme Sirvan of the University of Lyon; Dr Hermann Jaeger of the Humboldt University, Berlin; Dr Porter Kier of the National Museum of Natural History, Washington, USA; Dr Dennis Kolata of Champaign, Illinois, and many others, unfortunately too numerous to mention.

I am also grateful to Prof. Georges Ubaghs of

the University of Liège, Prof. Ronald Parsley of Tulane University, New Orleans and M. Jean Chauvel of the University of Rennes (who has already been mentioned in another connection). All these gentlemen have remained helpful colleagues despite our total differences in interpretation. Prof. Gerhard Regnell of Lund and Dr Christina Franzén of Uppsala kindly provided the photograph of Gislén (frontispiece) and supplied biographical references concerning him.

I should also like to thank Mr Robert Cross, Mr Chris Owen and the other staff of BM(NH) publications and of Cambridge University Press who edited this book and saw it through the press. Finally I am grateful to Mrs Michèle Bhukhureea, formerly of the Palaeontology Department, BM(NH) who, with astonishing patience, typed a long and almost indecipherable text, and Mrs Harriet Ovenden and Mrs Rita Owen who undertook the laborious task of labelling the diagrams.

Plaster replicas of various calcichordates and wall charts on the subject can be obtained from Stuart A. Baldwin, Fossil Hall, Boars Tye Road, Silver End, Witham, Essex CM8 3QA, England. Copies of an enlarged model of a reconstruction of *Cothurnocystis elizae*, which was made by Mr Frank Munro for museum exhibition, are available by negotiation with the Director, Hunterian Museum, Glasgow University, Glasgow G12 8QQ, Scotland.

There are very few photographs in this book. I have preferred to use drawings, because these are easier to understand . Readers who want photographic evidence for fossil structures must consult my published papers in which photographs are abundant.

R. P. S. JEFFERIES
LONDON, 1983

Chapter 1   Aims, history of study, methodology, methods

In this book I aim to elucidate the ancestry of the vertebrates and of chordates in general. I shall argue that, among living forms, the echinoderms are the closest relatives of the chordates, and that the hemichordates are more distantly related. The crucial evidence for this view comes from a group of curious marine fossils, some 400 to 600 million years old, which I have called calcichordates. The structure of these fossils, however, can be interpreted only by comparison with their living relatives, aided by functional morphology. Much of the book is therefore devoted to describing still-living animals. So as to be understandable and fairly brief I shall concentrate on a few of these living forms out of the thousands belonging to the groups in question. Many of the relevant zoological facts were well known to late-nineteenth-century zoologists, when the problem of vertebrate origins was a focus of research, but have since been generally forgotten. This is why it is necessary to present the zoological background.

When I started on this field of study, from early 1964 onwards, I was prepared to reject classical homologies without much hesitation. However, further evidence and published criticism has often shown that I was mistaken in this. The late-nineteenth-century zoologists were usually right when they asserted that one structure was fundamentally identical (i.e. homologous) to another. But they were often wrong in their detailed conclusions about relationships because their phylogenetic methodology was faulty and because the crucial fossils were either unknown or uninterpreted. Also classical zoologists often disagreed with each other as, for instance, in the controversy about whether the vertebrate head was segmented or not. Often both sides in these quarrels had a large, but different, part of the truth.

In this book I shall not survey all the views that have been presented on the origin of vertebrates (see however Chapter 10). For indeed there are few phyla of Metazoa which have not been presented by somebody for this position. A recent example is the work of Willmer (1975), who has re-argued the case for the nemertines. Methodologically more sophisticated is Løvtrup's recent attempt (1977) to show that the mollusc–arthropod group represents the closest relatives of vertebrates – closer even than amphioxus or the tunicates. I discuss Løvtrup's views in Chapter 10, but here I shall only say that, in annulling the phylum Chordata, he has discounted most of the relevant morphological evidence, including the notochord. Likewise I do not agree with Garstang (1928) that vertebrates arose from a paedogenetic tunicate tadpole, despite the impressive support that he has received (Berrill, 1955; Romer, 1972; Starck, 1963; Bone, 1981; Hennig, 1983).

I use the word 'vertebrate' throughout this book as an exact synonym of 'craniate'. This conforms to present normal practice as reflected in some well known text books such as Weichert (1965), Young (1981). On the other hand Janvier (1981), in an important paper, has restricted the word 'vertebrate' to the group whose living representatives comprise [lampreys + gnathostomes], which is likely to be monophyletic, and uses the word 'craniate' for the wider monophyletic group of [myxinoids + lampreys + gnathostomes] which corresponds to the vertebrates traditionally and in my own usage. Janvier's proposed alteration of usage is apposite because myxinoids, lampreys and gnathostomes all have crania while only lampreys and gnathostomes have vertebrae. I do not adopt it here, however, because the word 'vertebrate' is well established in its wider meaning. To connote the monophyletic group of [lampreys + gnathostomes] I use the term 'myopterygians' proposed earlier by Janvier (1978).

The first person who thought of a possible connection between the calcichordates and chordates was the great British student of echinoderms Francis Bather. He immediately rejected the possibility, however, with the words:

'One can hardly avoid a passing comparison of these elliptical openings [i.e. the gill slits of the calcichordate *Cothurnocystis*] to pharyngeal clefts in Chordata, if only for the sake of warning the speculative morphologist against taking this as a sign of affinity' (1913, p. 417).

The first to disregard Bather's warning was the Japanese zoologist Matsumoto who, in 1929, suggested in passing that the calcichordates were nearer to tunicates than to echinoderms. Much the same conclusion was argued in much greater detail by the Swede Torsten Gislén (frontispiece) in an extremely erudite paper published in 1930. Many of Gislén's conclusions were undoubtedly correct, in my opinion, but his argument was based more on embryos than on fossils, and was almost disregarded. Among those who did mention it was Gregory (1935), who in general terms approved it. He made a comparison, plausible but probably mistaken, between the platy echinoderm-like calcareous skeleton of calcichordates and the bony armour of certain primitive fossil fishes. The same comparison was elaborated by Caster & Eaton (1956) and it was Eaton (1953) who published one of the very few papers where Gislén's views were seriously discussed. Berrill (1955) dismissed Gislén's views in a paragraph, as also did van der Horst (1939, p. 624). Marcus (1958), however, in a general review of the relationships between phyla, suggested that future evidence supporting a connection between calcichordates (carpoids) and chordates would not be surprising. In Germany the great teacher Remane supported Gislén's views (Remane *in* Heberer, 1967; Remane, Storch & Welsch, 1972). Gislén's attempt certainly deserved more attention than it ever got.

Dr Roy Crowson, in the early 1950s, independently reached similar conclusions to those of Gislén's, persuaded, like him, by the asymmetries of living primitive chordates (Crowson, 1982, p. 252).

My own active interest in calcichordates began with two specimens brought into the British Museum (Natural History) by Prof. W. T. Dean in February, 1964, which caused me to re-read Gislén's work. Since then I have published a series of papers on the subject (Jefferies, 1967 to Jefferies, 1984; Jefferies & Prokop, 1972; Jefferies & Lewis, 1978; Jefferies, Lewis & Donovan, in press). These papers do not all present identical opinions, since my interpretations have changed down the years. Criticisms, have been published in Jefferies (1967, 1968a: discussants. Tarlo, Hill, Sellwood); Eaton (1970); Denison (1971); Jollie (1982); Kolata & Jollie (1982); Kolata & Guensburg (1978); Philip (1979, 1981); Bone (1981); Chauvel (1981); Ubaghs (1976, 1978, 1981); and Sprinkle (1983).

The study of calcichordates as fossils, rather than as animals, has been in the hands of systematic palaeontologists who saw them as echinoderms and either denied, or did not consider, any connection with chordates. The first

published description of one (*Ateleocystites huxleyi*) was by the Canadian Elkanah Billings (1858). Most of the known species were first recorded in the great monographs of the late nineteenth century such as that of the American James Hall (1859), on the Palaeozoic rocks of New York, and the Frenchman Joachim Barrande (1887) on the Palaeozoic rocks of Bohemia. The first known English species was *Placocystites forbesianus*, described by the Belgian de Koninck (1869) from the Silurian rocks of the West Midlands. All the species described in the nineteenth century belonged to the more advanced, relatively symmetrical group known as the mitrates. Twentieth-century works describing mitrates include Schuchert (1913), Dehm (1932, 1933), Reed (1925), Thoral (1935), Rennie (1936), Chauvel (1941), Bassler (1943), Caster (1952), Gill & Caster (1960), Ubaghs (1961, 1968, 1970, 1979), Kolata & Guensburg (1978), Philip (1981), Kolata & Jollie (1982) and Caster (1983).

The less advanced, bizarre-looking group of calcichordates known as cornutes were not known until the twentieth century. The first described species was *Ceratocystis perneri*, which was named by Jaekel in (1900), though pieces of it had already been described by Pompeckj (1896). Other cornutes were described, from Scotland, by Bather (1913) in a paper already quoted which was an enormous advance in knowledge. Cornutes have also been described by Thoral (1935), Ubaghs (1963, being the first known occurrence in America), Caster in Ubaghs (1967b), Ubaghs (1969, 1983), Chauvel (1971), Jefferies & Prokop (1972), Sprinkle (1976), Chauvel & Nion (1977). The pace of description seems to be increasing.

The most comprehensive account of the calcichordates, under the name Stylophora, is that of Ubaghs (1967b) with contributions by Caster. Ubaghs interprets these fossils in a completely different way from myself, as explained in Chapters 7 to 10. Indeed, we differ about anterior, posterior, dorsal and ventral. Ubaghs' views have been followed by most later authors but attacked, along with my own, by Philip (1979) and Kolata & Jollie (1982) – see also Ubaghs (1981), Jefferies (1981b) and Chapter 10.

It is historically interesting that MacBride (1909, p. 308) explained the asymmetries of amphioxus as caused by a 'flatfish' stage in its ancestry, resting on one side. This suggestion came originally from his research assistant J. Stafford and was based entirely on embryological evidence, without reference to any fossils.

Methodology is all-important in phylogenetic reconstruction, for without it we are merely writing myths, with no hope of real advance. The works of Hennig (1950, 1966, 1969, 1975, 1981, 1983) started a debate on phylogenetic and classificatory methodology which is still not ended, but which has already led to a great increase in logical rigour. For recent accounts of this debate see Eldredge & Cracraft (1980), Dupuis (1978), Hull (1979), Patterson (1980, 1981), Platnick, Wiley (1981) and Ax (1984). The school of taxonomists that followed Hennig used to be called Hennigian, but has come to be called cladistic (from *clados* = a branch). Here I explain the methodology which I have adopted so that the subsequent argument will be easier to follow. My position is cladistic, but unlike some cladists I think fossils can be crucial. Also I believe that cladistic classification ought to reflect reconstructed phylogeny and *vice versa* (contrast Platnick, 1980; Patterson, 1981).

Hennig (1966) stated that, in order to understand the phylogeny of a group, we should try to place its known members at the end of the branches of a bifurcating phylogenetic tree, according to the likely relationship between the known members. Relationship is here defined strictly in terms of shared common ancestry and is recognized by the possession of shared, derived, homologous character-states, which are present in some members of the group under consideration, but primitively absent in others. This type of phylogenetic tree is now called a cladogram (Mayr, 1974). The nodes of the cladogram, if it is correct and discounting hybridization, will represent the splitting of species in the past history of the group, if none of the species placed on the cladogram was in fact ancestral to the others. A cladogram is therefore an appropriate phylogenetic tree if all the species are alive, do not hybridize and lack a fossil record. When fossils are involved the issues are more complicated.

Given the living species *a*, *b* and *c*, there are three possible fully resolved cladograms (fig. 1.1). Either *b* and *c* are more closely related to each other than they are to *a*; or *a* and *c* are more closely related to each other than they are to *b*; or *a* and *b* are more closely related to each other than to *c*. Hennig's definition of relationship states that two species are more closely related to each other than to a third species if their latest common ancestor is more recent than that of all three species.

This definition recalls the colloquial view of relationship, whereby two sisters are more closely related to each other than to a first cousin, because they share a father and mother rather than two grand-parents.

A cladogram is a topological diagram in that the relative positions of the nodes and segments signify, but nothing else does. Thus, as shown in fig. 1.2, there are four ways of writing the cladogram of fig. 1.1a and the meaning of all four is identical, i.e. *b* and *c* are more closely related to each other than they are to *a*.

When a mother species has split into two daughters, then one daughter species and all its descendants constitutes for Hennig the sister group of the other daughter species and all its descendants. Thus in fig. 1.1a, species *a* is

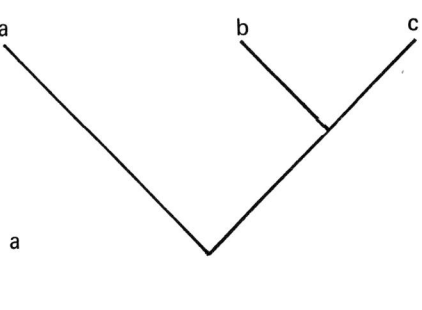

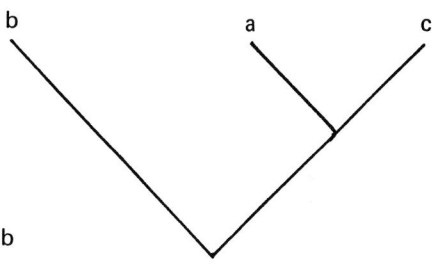

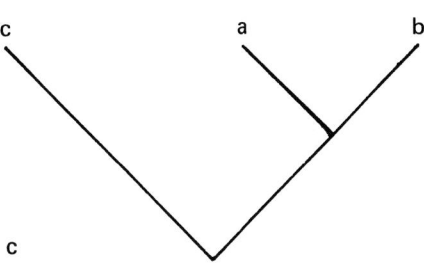

1.1. The three possible fully resolved cladograms for the extant species a, b *and* c.

the sister group of (*b* + *c*), if all other species are left out of consideration. It was important to Hennig's system, as originally proposed, that the mother species should not be taken as identical to either of its daughter species. This stipulation, when applied as it originally was at the species level, has provoked valid criticism (Mayr, 1974) and implies a limitation of the cladistic method, as I shall show later. However, Hennig was certainly right to stress that, given two sister groups with still-living members, neither sister group can be assumed to be identical with the latest common ancestor of both, for both could easily have acquired specialized features independently.

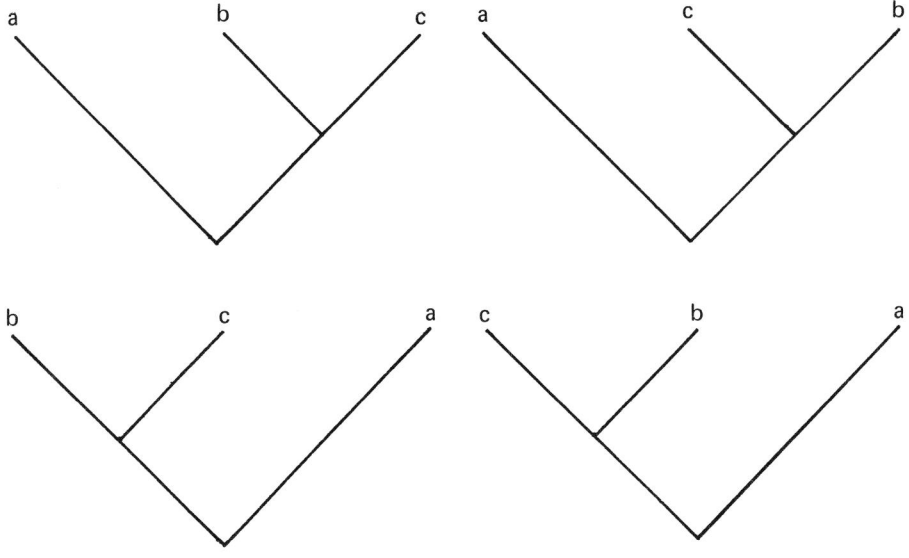

1.2. *Four alternative ways of writing the same cladogram.*

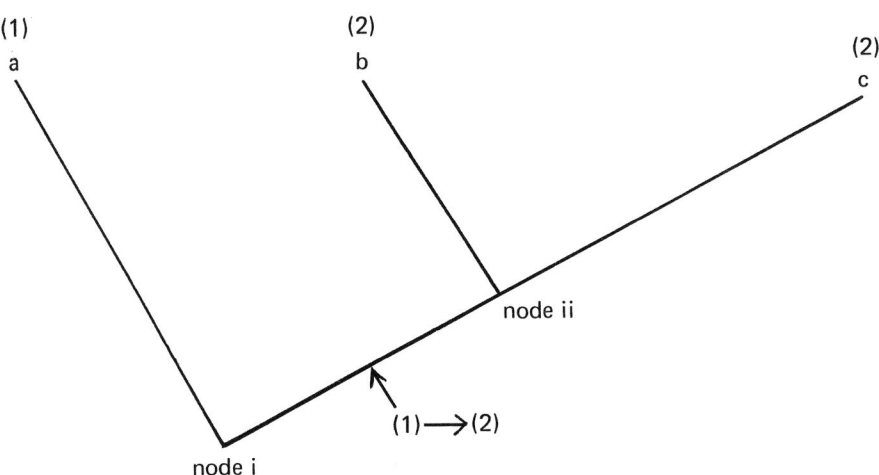

1.3. *The synapomorphy (2), present in b and c but not a, shows that b and c are closer related to each other than to* a.

As already stated, a cladogram is built up on the basis of shared derived homologous character-states. Hennig's term for 'derived' was 'apomorphous', so he called such shared character-states 'synapomorphies'. Thus, in fig. 1.3, suppose there is a character which can exist either in the primitive character-state *1* or the derived character-state *2*, and *2* occurs in species *b* and *c*, whereas *a* has *1*. If so, the most parsimonious and initially most likely assump-

tion is that *b* is closer related to *c* than to *a*. For the cladogram suggests a line of exclusive common ancestry running from node *i* to node *ii* and shared by *b* and *c* but not by *a*, and within this line or segment *1* changed to *2*.

It is important to note that shared possession of the primitive character state *1* would not help in working out relationships. For if *a* and *b* had *1*, but *c* had *2*, then there are three possible fully resolved cladograms, all equally parsimonious and equally likely (fig. 1.4). Hennig's word for primitive was 'plesiomorphous' and he referred to the shared possession of primitive character-states as 'symplesiomorphy'. His realization that symplesiomorphy does not help in reconstructing phylogenies, except indirectly by making synapomorphies possible to recognize, is his central and brilliant insight.

The organisms that existed at the nodes of cladograms can be reconstructed. Thus in fig. 1.3, node *i* would have had character-state *1*, whereas node *ii* would have *2*. A useful graphical trick is that all occurrences of homologous character-states will be continuously connected together by segments of the cladogram.

An example will make the method more concrete. Suppose that, in fig. 1.1a, species *a* is a turtle, *b* a mouse and *c* a man. In that case (*b* + *c*) is the sister group of *a*, within this set of three species, since mouse and man both share a series of derived characteristics (warm blood, hair, milk, live birth, a diaphragm, a single pair of bones in the lower jaw, enucleate red blood corpuscles, etc), which the turtle (*a*) does not have and almost certainly never had. These features evolved within the segment of the cladogram which is the exclusive common ancestry of man and mouse with respect to the turtle. Moreover, the latest common ancestor of mouse and man would have had all these advanced features, whereas the latest common ancestor of turtle, mouse and man would have lacked them.

Three-taxon cladograms cannot always be resolved with such certainty. If hybridization is discounted, the terminal twigs of a cladogram leading to particular still-living species must correspond to the true phylogeny, for they imply merely that the species populations now living result from independent lines that existed before the present day. But recognizing a non-terminal segment of a cladogram depends on what changes have occurred within the segment. If no changes occurred – if the segment was 'empty' – then in principle it is not recognizable by the method of synapomorphy. Thus, in fig. 1.5a, species *b* has the primitive character-state *1*, *a* has *2* (derived) and *c* has *3* (independently derived), and the figure is assumed to represent the true phylogeny. But no evolution occurred in the exclusive common ancestry of *b* and *c* so that the corresponding segment has left no trace.

Such empty non-terminal segments are most likely to exist when the species *a*, *b* and *c* are very closely related. Thus, to take three very similar British birds, it could conceivably be that the Wood Warbler (*Phylloscopus sibilatrix*) is more closely related to the Willow Warbler (*P. trochilus*) than to the Chiff-Chaff (*P. colybita*). But, if so, nothing obvious has happened in the exclusive common ancestry of Wood Warbler and Willow Warbler, the empty segment is not recognizable, and the cladogram,

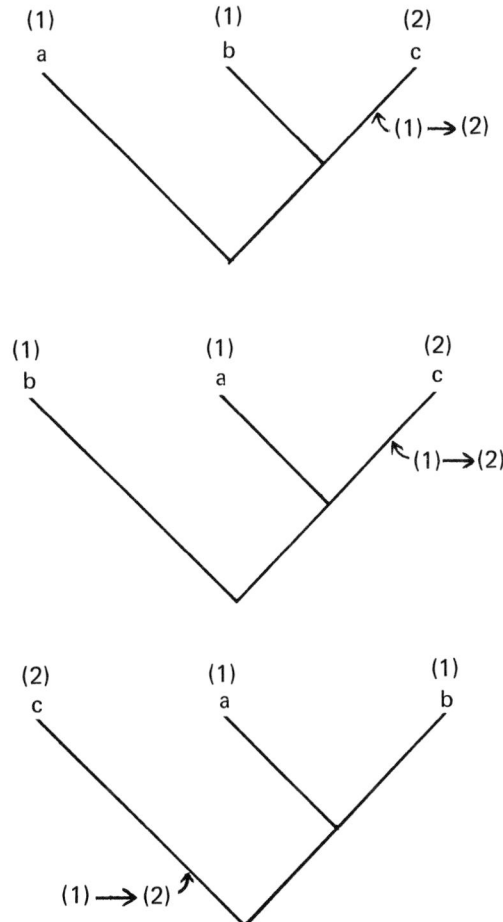

1.4. *Symplesiomorphy does not directly help.*

unless resolved by biostratigraphical or geographical evidence, becomes an unresolvable trichotomy (fig. 1.5c).

Non-terminal segments of a cladogram can therefore be arranged in a spectrum of increasing inherent recognizability. At one extreme the segment is empty and not recognizable by the method of synapomorphy; next comes the segment in which a single simple change has occurred, producing a feature which can easily be lost in descendent species or independently acquired in parallel; next comes a segment with several simple changes; and finally many complex mutually interdependent changes may have occurred in the segment forming a gestalt or gestalten whose recognition is virtually certain, as with the example of turtle, mouse and man. Proceeding along this spectrum the species separated by the segment become less closely related to each other. And observed

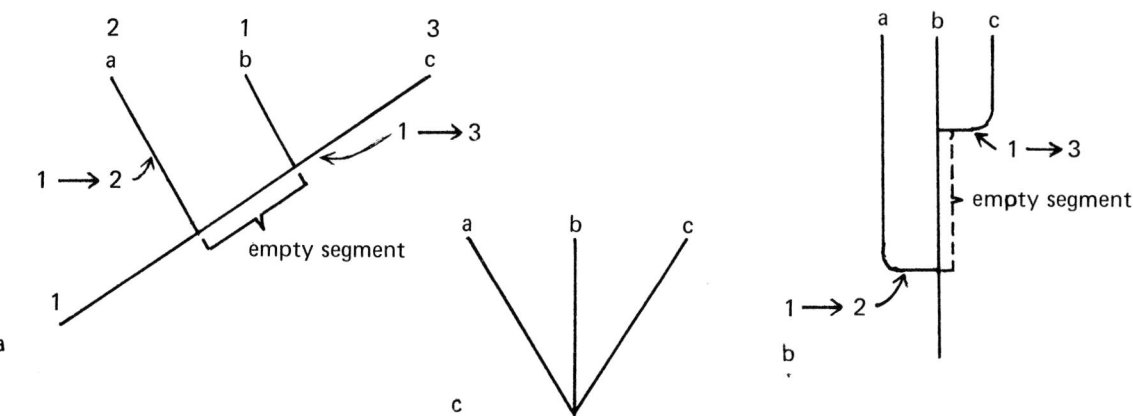

1.5. *a) A cladogram with an empty segment undetectable by synapomorphy; b) the corresponding speciational tree; c) the trichotomous cladogram representing the two unresolved trichotomies of (a).*

stratigraphical sequence, crucial at the beginning for detecting the segment, becomes irrelevant at the end.

The recognition of homology lies behind this spectrum, and calls for discussion. Riedl (1975, 1979) described homology as a special case of biological identicality in which organs are recognized as structurally identical between species. We immediately recognize two occurrences of a complex message as being causally identical, but need several repetitions of a simple message to reach the same conclusion. The occurrence twice of the digit 2 may be accidental; but the sequence 11086165, 11086165 cannot represent an accidental repetition; or rather, assuming that any digit between 0 and 9 could have been chosen with equal probability for each of the eight places, the probability of such repetition being accidental is 1 : 100000000. Even without making a calculation, we assume that these two occurrences of eight digits must depend on identicality i.e. the 'decisions' causing the first occurrence must be the same as the decisions causing the second. We do not feel this with the sequence: 2, 2 whose accidental probability is 1 : 10. Complexity, corresponding to high accidental improbability, is the factor which shows that some resemblances, as between the eye of a dogfish and of a man, are certainly homologous.

The invisibility of empty segments using Hennig's method is important in connection with speciation theory. For it is now believed that new species usually arise as small isolated populations within, or at the edge of the range of, a much larger population of a pre-existing ancestral species (Mayr, 1942, 1974; White, 1978; Eldredge & Gould, 1972; Gould & Eldredge, 1977; Stanley, 1980). The origin of a new species is accompanied by its genetic isolation from the pre-existing species and rather abrupt

evolutionary changes will usually occur in the new species, during or just after the instant of isolation. This is because large populations are stabilized by gene flow, and probably by natural selection of heterozygosity and of non-extreme morphologies; whereas small populations are likely to change quickly, through the founder effect, random drift and the inbreeding that produces homozygotes. In Hennig's terms this means that one daughter species of a split will be identical to the mother, and the other not; or, in traditional terms, the ancestral species will persist alongside the new one. Moreover, it is possible for the persistent species to bud off a second new species at a later time, without changing in the interim. Between the two buddings there will be an empty segment. Speciation theory therefore suggests that empty segments, undetectable by synapomorphy, may be numerous when dealing with closely related species. I return to this matter below.

Cladistics uses its own concept of monophyly. A group is monophyletic (fig. 1.6a) if it contains a stem species, all descendants of that stem species, and by implication nothing else. (A stem species can be defined as the latest common ancestor of all the other members of a group.) A monophyletic group will be genetically continuous in that all its members will be connected by ancestor-descendant links running solely within the group. Also all its members will be more closely related, in Hennig's sense, to all other members, than to any non-members. The members of a monophyletic group will be recognized by the possession of at least one synapomorphy unless this synapomorphy has been secondarily lost. An example of such a group is the birds whose living members are characterized by a whole series of advanced features including bipedality, feathers, no teeth, a horny bill etc.

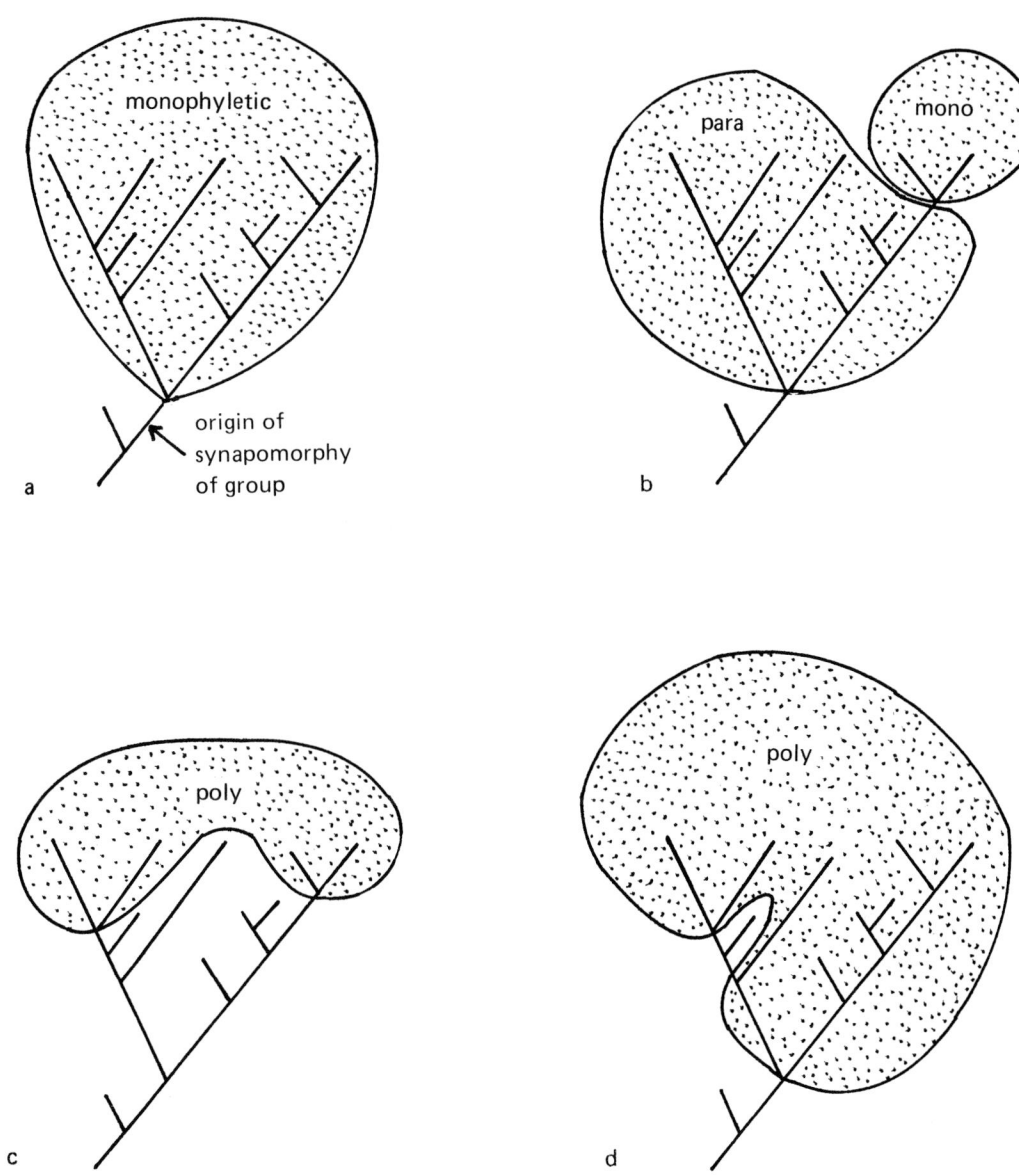

1.6. Mono-, para- and polyphyly.

A group is called paraphyletic (fig. 1.6b) if it includes a stem species and all immediate descendants of that species, but excludes certain advanced forms descended from these descendants and not themselves ancestral to members of the group. Like a monophyletic group, a paraphyletic group will be genetically continuous. Unlike a monophyletic group, however, some members of a paraphyletic group will be more closely related to non-members (within the derived descendent group) than to other members. A paraphyletic group is in principle a primitive portion of a wider monophyletic group. A paraphyletic group will be recognized by possessing the synapomorphies of this wider monophyletic group, and by primitively lacking the synapomorphies of the descendent monophyletic group or groups. An example of a paraphyletic group is the reptiles. These are the primitive portion of the monophyletic group known as amniotes, excluding the birds and mammals. Reptiles have the amniote synapomorphy of an amnion, but lack feathers (unlike birds) and hair (unlike mammals). Within the group of reptiles it is usually thought that the crocodiles are more closely related to birds, in the sense of sharing a more recent common ancestor, than other reptiles are.

A polyphyletic group is not genetically continuous. Usually it excludes its latest common ancestor and the immediate descendants of that ancestor (fig. 1.6c). Conceivably, however, a polyphyletic group could include a stem species and the immediate descendants thereof, but exclude some descendants of these forms which were themselves ancestral to members of the group (fig. 1.6d). A polyphyletic group is recognized on the basis of a convergence. An example would be the group 'flying vertebrates' comprising flying fish, pterodactyls, birds and bats.

All systematists try to exclude polyphyletic groups from their system. Traditional systematists however, find paraphyletic groups as acceptable as monophyletic ones. Cladists try to delimit, at least among still living organisms, only monophyletic groups.

How do we know that one character-state is more primitive than another? Four criteria of primitiveness can be proposed depending on outgroup comparison, embryology, the stratigraphical sequence of fossils, and improvements in the function of a part (adaptational criterion).

The criterion of outgroup comparison says that, if a character-state occurs in a monophyletic group, and also in a wider monophyletic group of which the first group is part, then the character is primitive for members of the smaller group. Thus teeth (as shown in *Archaeopteryx*), are a primitive feature for birds since, although absent in living birds, they are present in crocodiles (the likely living sister group of birds), in amniotes generally, and in gnathostome vertebrates generally. Most well based statements about primitiveness arise from outgroup comparison, rather than from embryology or stratigraphy. They depend crucially, however, on the homology of the character-state in question. When homologies are certain, as is true for the teeth of *Archaeopteryx* compared with those of crocodiles or man, then the results of outgroup comparison are also certain. Other criteria are then unnecessary and can be disregarded if they give a contrary result. Thus it is sometimes thought that the primitiveness of the teeth of *Archaeopteryx* among birds is proven by the antiquity of this fossil (the oldest bird known, being Upper Jurassic or about 150 million years old). This is not so, however, for if tomorrow a toothless bird was discovered in older rocks than *Archaeopteryx*, then nobody would change their opinion on the primitiveness of teeth in birds. Fossils could alter this opinion only by making the homology doubtful, which is incredible. If the relevant homologies are certain, stratigraphical sequence is irrelevant. But equally, if homologies *are* doubtful, stratigraphical sequence can be all-important.

The embryological criterion of primitiveness asserts that a character-state is probably primitive if it appears earlier in ontogeny than an alternative character-state. In my view it is based on Haeckel's law, who wrote (1866, vol. 2, p. 7) that: '. . . ontogeny is nothing else than a short recapitulation of phylogeny'. The reasons why Haeckel's law usually holds have been dealt with by Riedl (1975, 1979). Briefly, the ontogeny of an organism can be seen as a series of branching decisions, of which the later ones depend on some of the earlier ones. Thus the eye lens of a vertebrate, induced in the ectoderm by the optic vesicle, depends on the prior existence in ontogeny of ectoderm and optic vesicle. Early ontogenetic decisions, therefore, usually have a burden of later decisions that depend upon them, and this burden can be estimated numerically simply by counting the number of dependent decisions. Mutations can strike at any stage of ontogeny. If, however, they nullify a decision of high burden (such as the notochord in man), they are more likely to kill the organism than by nullifying one of low burden (such as the fingerprint pattern in man). Decisions of high burden, taken early in the development of an organ, will therefore be difficult to rescind. Evolutionary alterations will usually affect decisions of low burden (which are usually taken late in ontogeny) or else they will be added on to the previous ontogeny without apparently altering it. By being added they will increase the burden of previous decisions and make them still more conservative.

Some authors (Gould, 1977, de Beer, 1958) have distinguished between Haeckel's law and von Baer's law, which states that: 'The general features of a large group of animals appear earlier than the special features' (quoted in Gould, 1977, p. 56). Moreover these authors argue that von Baer was right and Haeckel subtly wrong. To me, however, the two laws state the same thing since by 'general features' von Baer meant features characteristic of a wider monophyletic group, whereas special features were those present in a smaller, included, monophyletic group which is necessarily more recent in origin. Thus in the development of the chicken the notochord (characteristic of chordates) precedes feathers (characteristic of birds). Von Baer and Haeckel were both usually correct and their laws, given the fact of evolution, are logically equivalent. Exceptions to the one will also be exceptions to the other. However, I prefer to use Haeckel's formulation since it speaks directly in terms of phylogeny.

Obviously Haeckel's law must be used with caution, for it is probabilistic and subject to several types of exception. Thus natural selection can sometimes simplify or shorten ontogeny, even when this means altering decisions of high burden as with the caterpillars of butterflies. Also early features that are of low burden are easily altered since such change will not affect later developments. Thus the

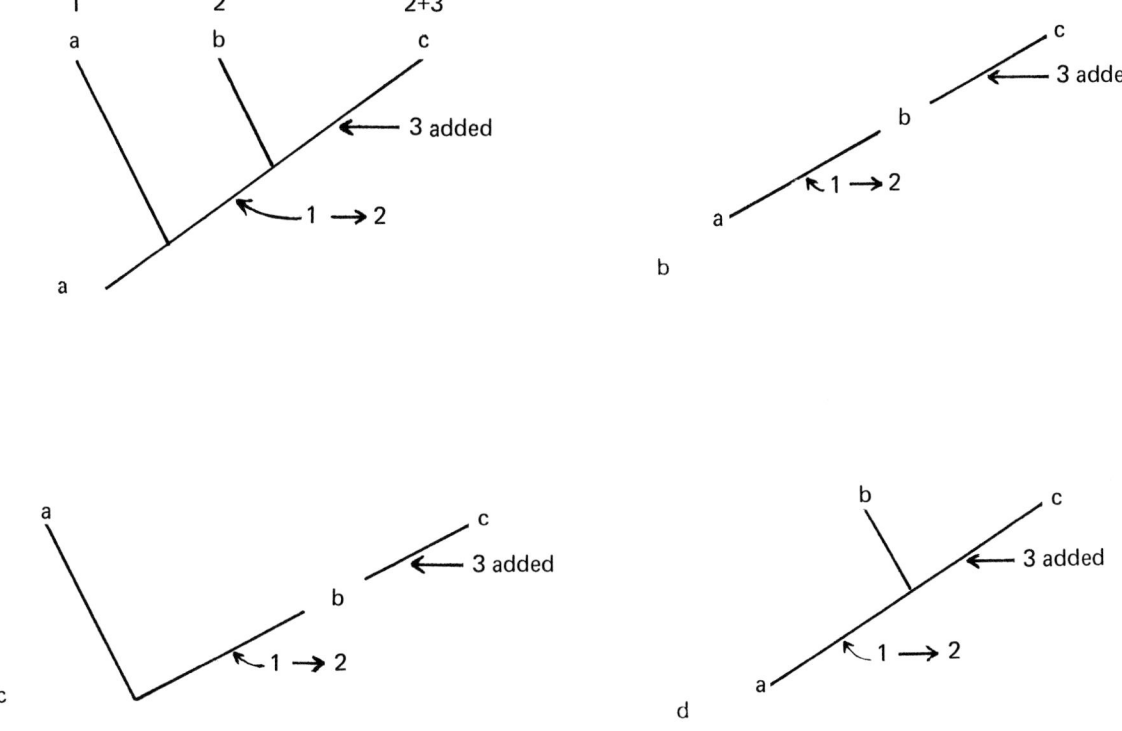

1.7. *A cladogram and its four possible phylogenetic trees. The cladogram of three fossil species (fig. a) may correspond to four actual phylogenies (figs a to d) but, whatever the tree, the cladogram will give, within any one line of descent, the correct temporal sequence for the origin of synapomorphies. The figure (a) represents either a cladogram, with a, b and c arbitrarily placed terminal, or else a phylogeny in which they were in fact terminal.*

amnion was invented as an early ontogenetic adaptation but affects the pre-existing ontogeny very little. Moreover two organs whose development is not interdependent can easily be advanced or retarded relative to each other, since burden is not involved. And the process of paedogenesis (Gould, 1977) can lop terminal stages off ontogeny by accelerating sexual development relative to somatic development, though what ontogeny remains will still be subject to Haeckel's law. Embryology is important in this book since classical students of vertebrate and echinoderm origins used Haeckel's law, and, as I believe, they were right to do so.

The stratigraphical criterion of primitiveness states that a character-state found in early fossils of a monophyletic group is more likely to be primitive than an alternative character-state found in later forms. I believe this criterion is sometimes useful and it is possible to say when. Before discussing it, however, we must consider the whole broad question of fossils in phylogeny.

Suppose that there are three fossil species *a*, *b* and *c*

which are not contemporaneous with each other; furthermore *a* has the primitive character-state 1; *b* and *c* have the derived character-state 2; and *c* has in addition the derived feature 3 that *a* and *b* lack. In that case we can construct a cladogram placing all three species, in the first place by convention, at the ends of its branches (fig. 1.7a). This cladogram will give the correct temporal sequence for the origin of synapomorphies within any one line of descent, but will not always be correct considered as a tree portraying the phylogenetic history of *a*, *b* and *c*. For as such it discounts the possibility that *a* is directly ancestral to *b* and *c*, or that *b* is ancestral to *c* (fig. 1.7, b, c, d). We can only discount these possibilities if the observed stratigraphical sequence makes it unlikely (as when *a* is younger than *b* or *c* which shows that at least the known population of *a* is not ancestral to *b* or *c*); or if the more primitive species have specialized features which disqualify them as ancestors of the more advanced species (as if *a* in text fig. *a* had a derived feature, not shown in the diagram, which *b* and *c* primitively lack); or, more weakly, if fossils are so rare that the chance of one known species being the actual ancestor of another known species is negligible. Considered as a tool, a cladogram is useful in building up a phylogeny for a group of fossil species, because it will place the origin of synapomorphies in correct time sequence. It must be remembered, however, that its terminal branches exist in the first instance by convention only, and that some of them will not have existed in historical fact. The cladogram is therefore an 'expression of nested patterns of synapomorphy' (Eldredge & Cracraft, 1980, p. 53). This difficulty, that terminal branches are in the first place conventional, does not arise with cladograms for still-living species, nor for an assortment of fossils from a single time-plane. It was first emphasized by Harper (1976).

Punctuated equilibria are another complication in using fossils in phylogeny. This term is used to describe the fact that in the fossil record we seldom observe gradual transition from one species to another (Eldredge & Gould, 1972; Gould & Eldredge, 1977; Stanley, 1980). More often a species appears suddenly in a stratigraphical succession, persists unchanged for a time, and then dies out, usually to be replaced by another closely similar species. This behaviour is predictable from modern speciation theory since, as already mentioned, species usually arise by quantum speciation in small, usually peripheral, populations, which would seldom be found fossil. Only later do they sometimes spread over the range of the ancestral species, wiping it out. Two consequences follow from this. Firstly, when the fossil record is sparse, a known single occurrence of a species might come from near the beginning or end of its true time range, without change of morphology. For this reason, the stratigraphical criterion of primitiveness will not apply and phylogeny could only be reconstructed by drawing cladograms on the basis of morphology. Secondly, when the record is good, so that reliable stratigraphical range charts for the fossil members of a group can be drawn up, there will be many empty segments of the type already mentioned. Moreover, the differences between known species will often be slight and of low burden, so that parallelism will be rife and homology difficult to recognize. In such cases it will be impossible to reconstruct phylogeny without consulting stratigraphy. This leads to the curious result that the prospect of working out the correct phylogenetic connections between known fossils can be higher when few are known, or when few are considered, than when many are (fig. 1.8). Hennig's stipulation that both of the daughter species or groups should be seen as different from their mother should not be regarded as a description of speciation, but as a requirement if phylogeny is to be reconstructed on the basis of synapomorphies alone.

The quality of the record can often be judged. It is bad if

genuinely new forms are often being discovered or the range of known forms increased; if most species are known from a single locality or horizon, or even from a single individual; if the morphological gaps between known forms are large and continually being reduced by new finds. The record is good if the opposite conditions hold. For the calcichordates the record is unquestionably bad on the basis of all the criteria mentioned. Cladograms are therefore the correct approach to phylogeny and stratigraphical sequence should be disregarded. Further discussion of the stratigraphical criterion of primitiveness can be found in Fortey & Jefferies (1982).

The adaptational criterion of primitiveness states that, if an organ occurs in two versions, one less and the other better adapted to a function important in the organism's life, then the less adapted version is the more primitive. This criterion can rightly be applied if the functional morphology of an organ is very well understood. In general, however, its use in phylogenetic studies has led to a pullulation of baseless stories.

Hennig's stem-group concept (1969, 1981) is of high value when using fossils to reconstruct the phylogeny of extant groups. Suppose that groups 1 and 2 are two surviving groups which are sister groups of each other among Recent organisms (fig. 1.9). There will then be two obvious delimitations of group 1, one wide and the other narrow. The narrower delimitation contains the latest common ancestor of the living members of group *1* plus all its descendants whether surviving or dead. Hennig referred to the group so delimited as the *group, but I call it the crown group (Jefferies, 1979; Patterson, 1981; Smith, 1984; Paul & Smith, 1984). The wider delimitation is more difficult to define. It comprises all descendants of the latest common ancestor of groups 1 and 2, except members of group 2. It can be called the total group. If the crown group of 1 is subtracted from the total group of 1, then a paraphyletic ancestral group remains which can be called the stem group of 1. Stem groups are paraphyletic and extinct by definition. Every monophyletic group with more than one living member species can in principle be divided, in this fashion, into stem group and crown group. As concerns monophyletic groups containing one surviving species only, all the extinct species of such a group can conveniently be seen as its stem group. 'Stem' and 'crown' suggest an obvious comparison with an apple tree, for example.

Within the stem group it is theoretically important to distinguish the stem lineage ('Stammlinie' of Ax, 1984). This concept, again, is best understood from a diagram. Thus, in fig. 1.9, the stem lineage of group 1 comprises the sequence of direct ancestors and descendants which led from the latest common ancestor of [1 + 2] up to the latest common ancestor of 1, both of these latest common

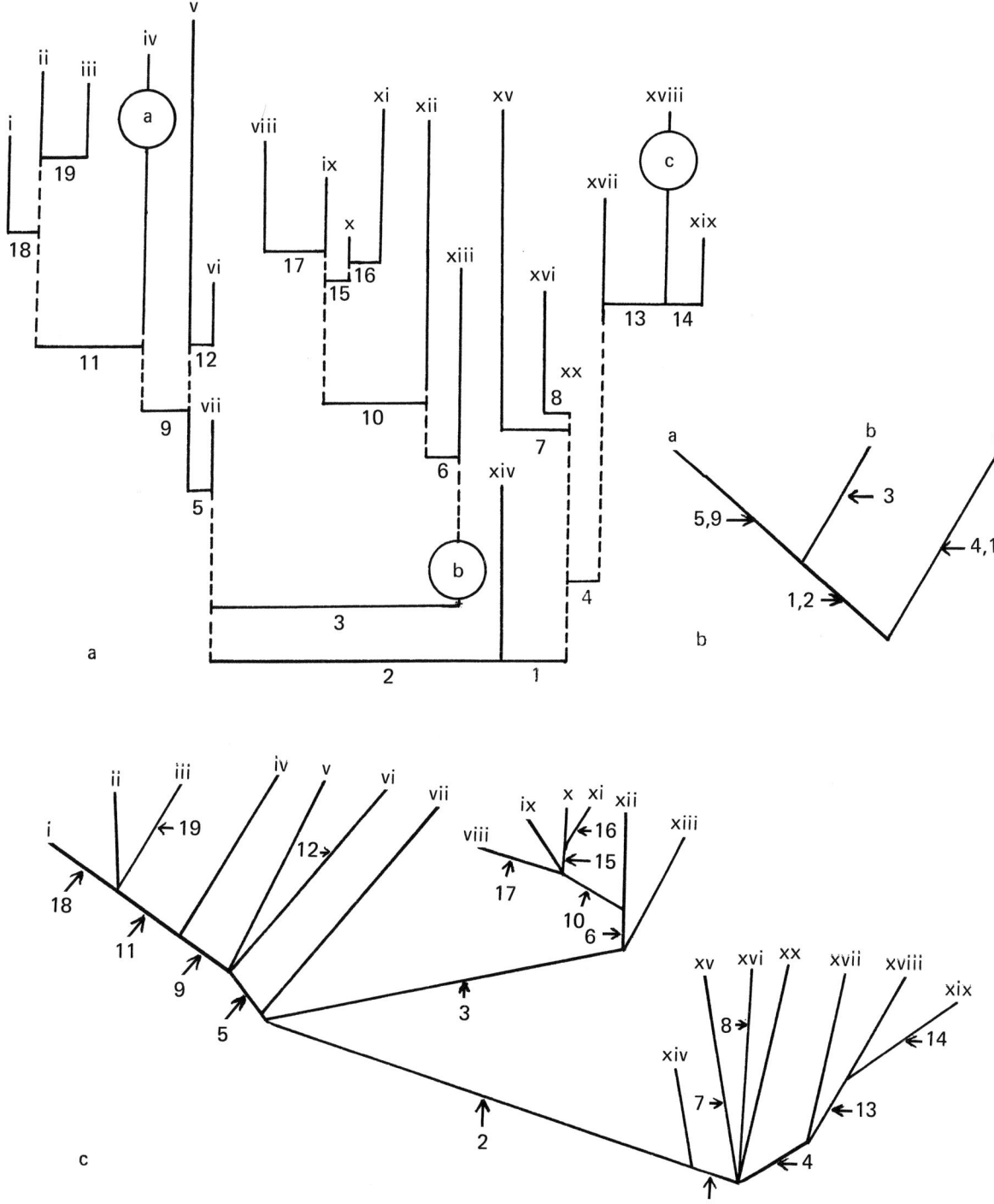

1.8. a) The historically correct speciational tree for a group of twenty fossil species. Empty non-terminal segments are dashed. b) The cladogram for the known fossil populations a, b and c (of species iv, xiii and xviii) within the group, assuming that the primitive morphology of the group is known by outgroup comparison – note the absence of empty segments. c) The best non-stratigraphical cladogram for the whole group – note the large number of unresolved multifurcations and of empty terminal segments.

ancestors being excluded from the stem lineage*. A series of evolutionary changes will have occurred within the stem lineage so that, in course of time, its members increasingly came to resemble the earliest members of the crown group to which they gave rise.

Stem groups may be divided into plesions (short for 'plesiomorphous taxon' Patterson & Rosen, 1977). Plesions are small segments of the stem group arranged in

sequence of increasing relationship to the crown group. A plesion, in my opinion, is best regarded as the smallest such segment that can be recognized i.e. on a practical definition, it comprises all those members of a stem group which, so far as can be discerned, are equally closely related to the crown group. A plesion will therefore consist of a small segment of the stem lineage plus all 'side branches' of that segment – by side branches I mean all those members of the plesion which are not in the stem lineage, and therefore not directly ancestral to more crownward parts of the stem group nor to the crown group.

*Ax (1984) includes the latest common ancestor of the extant members of a group as being the latest member of the stem lineage of the group. My usage differs from his in this respect.

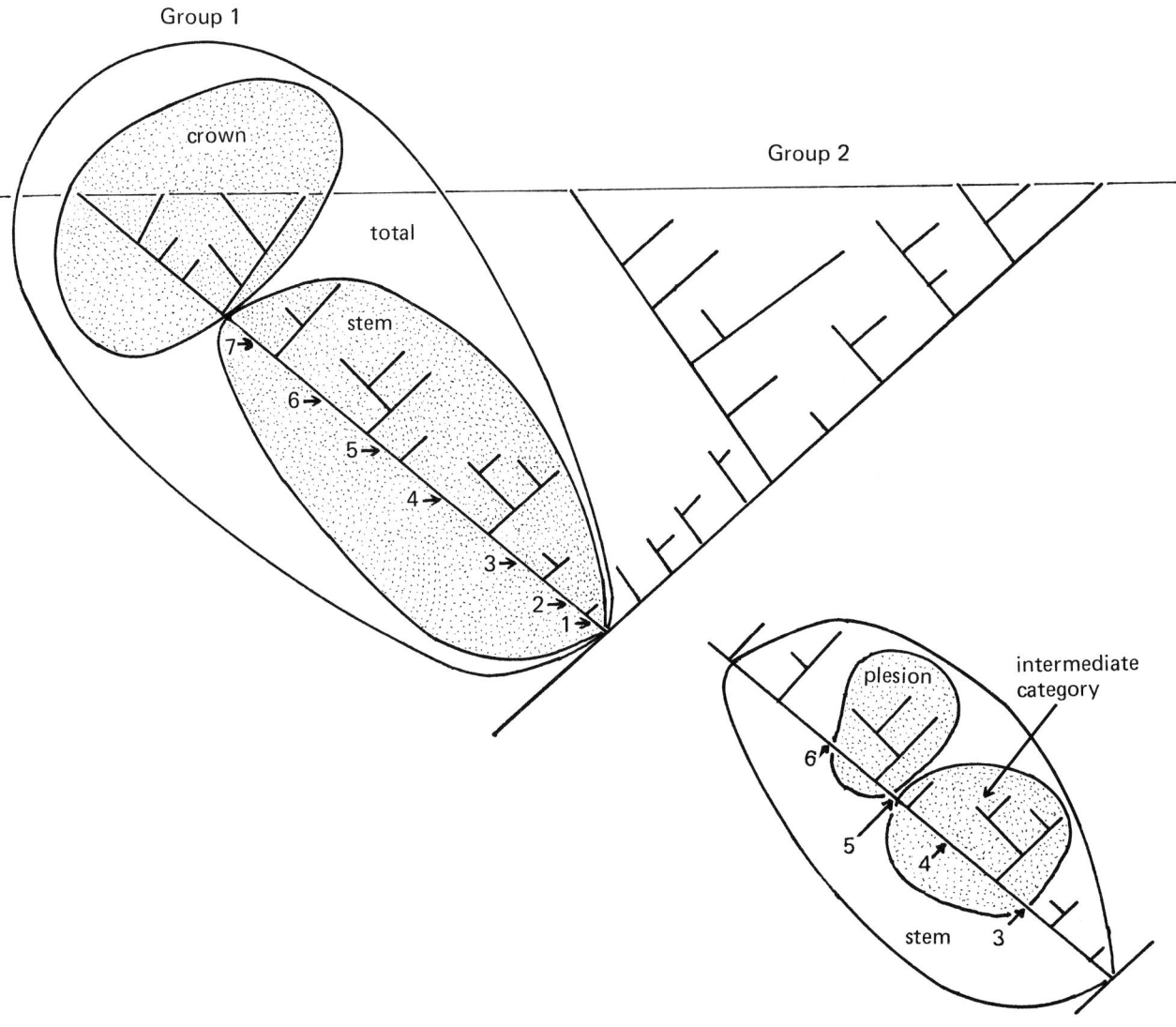

1.9. *Crown group, stem group, total group, plesion and intermediate category. The numbers 1 to 7 refer to evolutionary novelties.*

The term 'more crownward' is here proposed to signify that a fossil or group of fossils is more closely related to the crown group than is another fossil or fossil group. The opposites of 'more crownward' are 'less crownward' to denote position and 'anti-crownward' to denote direction. 'More crownward' is more specific than is 'more advanced' since it implies advance within the stem lineage only. It is also more accurate than 'later' which ought to refer to time only. 'More crownward' does not necessarily, or even usually, mean 'stratigraphically higher'.

The segment of stem lineage included in a plesion will begin with one evolutionary novelty and end with another – ideally it will end with the next evolutionary novelty that actually occurred in the stem lineage. Thus the members of a plesion will normally share with the crown group, or with the most primitive members of the crown group, at least one synapomorphy which less crownward plesions lacked. And they will lack at least one synapomorphy with the crown group which more crownward plesions possessed. Sometimes, however, plesions can be put in sequence by using synapomorphies with more crownward plesions, though these concern features which primitive members of the crown group have lost. For example, as explained in Chapter 7, *Nevadaecystis* is regarded as more crownward, within the chordate stem group, than is *Ceratocystis*. This is partly because *Nevadaecystis* had a strut in common with still more crownward plesions of the chordate stem group such as *Cothurnocystis*; but the strut was absent, by loss, from mitrates and these were primitive members of the chordate crown group. Thus a plesion can be placed in sequence on the basis of synapomorphies with the crown group, or with more crownward members of the stem group. The logical principle of synapomorphy is the same in both cases.

A plesion is paraphyletic if defined as just suggested, because it contains a segment of the stem lineage, and excludes forms which descend from that segment but are not included in the plesion i.e. it excludes more crownward portions of the stem lineage and all descendants of those portions, such as the crown group. On the other hand, as soon as a particular plesion can be shown to be paraphyletic, by demonstrating that some parts of it are more closely related to the crown group than other parts, then the original plesion will split into two plesions, one more crownward than the other. Patterson & Rosen, when they invented plesions, believed them to be monophyletic since they wrote (1977, p. 165) of a plesion as being: 'the plesiomorph sister-group of all those [plesions and monophyletic groups], living and fossil, that succeed it'. I myself formerly assumed that plesions were monophyletic (Jefferies, 1979). Such a definition usually breaks down, however, and with it the concept of plesions as mono-phyletic, when more than one species belongs, so far as discernible, in the same plesion, and it is not possible to say which of these species, if any, is more crownward than the others. The definition proposed here is more practical than Patterson & Rosen's concept but necessarily implies that plesions, like stem groups (or stem lineages for that matter), are paraphyletic.

An intermediate category (Hennig's 'Zwischenkategorie' (1969)) is a grouping of several adjacent plesions together. Such larger paraphyletic groupings may sometimes be convenient to recognize.

The fact that stem groups can be subdivided into plesions, mainly on the basis of synapomorphies with the crown group, without consulting stratigraphy, is important. For it allows us, by using fossils, to reconstruct changes in the stem lineage and thus to work out the morphological and adaptational history of the origins of extant monophyletic groups. Thus *Australopithecus*, who was upright but had a small brain, shows that erect stance preceded intellect in the origin of man. And this conclusion depends solely on the morphology of *Australopithecus*, not at all on his stratigraphical position, for it would still hold if he still lived (though then, of course, *Australopithecus* would not be a plesion).

Ax (1984, in press), in a brilliant book which has greatly influenced me, has proposed that stem groups be abolished and that fossils should be arranged in sequence of plesions entirely by using the stem-lineage concept. He does this because he rightly regards stem groups as paraphyletic and thinks that all paraphyletic groupings should be extinguished. He recognizes the value of sequences of plesions, regarding a plesion as a monophyletic offshoot of the stem lineage. I cannot accept Ax's abolition of the stem group for two reasons. Firstly, although fossils can be shown to belong to a stem group, none can ever be shown to belong to a stem lineage, unless the stratigraphical record is known to be perfect (which it never is known to be). It is sometimes possible to show that a member of a stem group is *not* a member of a stem lineage, either because it is stratigraphically later than more crownward members of the stem group, or else because it shows morphological features which never existed in the stem lineage. But it can never be shown that a fossil *belonged* to a stem lineage – only, at most, that it might have done. Secondly, as to Ax's objection that stem groups are paraphyletic, he is right. But stem lineages are even more paraphyletic since they exclude, not the crown group only, but all side branches within the stem group. Conversely, on a practical definition plesions are also paraphyletic, as I have already shown, but the plesion concept is usable nevertheless. Ax's concept of the stem lineage is important in clarifying the nature and subdivisions of stem groups. But in the actual classification of fossils it can never replace the stem group. Indeed, it is impossible to classify fossils in relation to extant groups without using stem groups (which are paraphyletic), and plesions (which are paraphyletic) – and to recognize plesions it is necessary to reconstruct stem lineages (which are paraphyletic).

I originally set up the calcichordates as a subphylum of the Chordata (Jefferies, 1967). This was before I had realized the power of Hennig's stem-group concept, or of Hennig's thought in general, and now seems unhappy. All known calcichordates can in fact be seen as stem chordates or as stem members of the extant chordate subphyla (whether acraniates, tunicatres or vertebrates). Consequently I now use the word 'calcichordate' only in an informal manner to include all chordates that retain a calcite skeleton of echinoderm type. For subphyla ought to be monophyletic groups with some members still extant.

Hennig (1966) made a number of proposals for defining the categorial rank of monophyletic groups (whether phylum, subphylum, order etc.). The question of rank, in my view, is both arbitrary and secondary, although it causes much quarrelling. Indeed there are good arguments for abolishing categorial rank (Ax, 1984). In what follows I merely use the traditional ranks for the groups that I deal with. Thus acraniates, tunicates and vertebrates, for example, are for me subphyla of the phylum Chordata, although in fact the group [tunicates + vertebrates] is probably the sister group of the acraniates, so that, following Hennig, they ought to be equal in rank.

A linguistic analogy to the use of fossils in phylogenetic reconstruction is helpful. It comes from the methods that scholars use in working out the copying history of manuscripts (Platnick & Cameron, 1977). This study is known as stemmatics or textual analysis and expresses its results in diagrams known as stemmas, which are remarkably like cladograms. Stemmatics is based on the fact that, before the invention of printing, the only way of copying a manuscript was for a scrivener to write it down word for word, which was inaccurate. Each copying produced its own mistakes and these were inherited, and added to, by later copyings. These mistakes are logically equivalent to synapomorphies.

For example, take Chaucer's poem: 'The Dethe of Blaunche the Duchess'. This was written about 1369 on the death by plague of Blanche, wife of Chaucer's patron John of Gaunt, Duke of Lancaster. It is 1333 lines long and there are four 'witnesses', being an early printed edition and three manuscripts. Their interrelationships were worked out by Koch (1881) using several types of 'synapomorphy'. The most reliable of these were the loss of lines of verse, or groups of lines, but there are also instances where alteration has spoiled the sense or the scansion. On this evidence Koch deduced that the three surviving manuscripts were closer related to each other (in the sense of sharing a later common ancestor or 'exemplar') than to the early printed text. One of the manuscripts had the remarkable feature, without biological parallel, that lines 31–96, at first omitted in common with both other manuscripts, had been reinstated in a later hand. Notably enough, alterations arise mainly in the process of copying (if we ignore damage by flood, fire and chaos), just as evolutionary changes mostly arise near the time of speciation. The manuscripts of a poem are therefore very similar to a monophyletic group of species which is evolving by punctuated equilibria. The great usefulness of the stemmatic method, like the cladistic method, is that it can be applied even with a very incomplete record. Chaucer was a popular poet and probably any poem by him would be copied hundreds of times. Most such copies are gone for ever but, if mistakes are introduced in copying and persist over subsequent copyings, we can still work out the interrelationships of the surviving witnesses, even when we have only four of them, and reconstruct the original text. To do so it is not necessary to have contemporary records of when all once-existing copies were made, how long they survived nor which older manuscripts they were made from. Similarly, in reconstructing the phylogeny of a group of fossils we do not need to know all the species that existed, their time ranges, nor which was ancestral to which. Even when the record is meagre we can still draw reliable cladograms.

Functional morphology can be useful in recognizing the homologies of organs preserved in a group of fossils, particularly in a group so bizarre as the calcichordates, for it is logically independent of comparisons with living relatives. The methodology of the functional morphology of fossils has been discussed by Rudwick (1961, 1964). He advocates that, when a palaeontologist has thought of a possible function for a structure in a fossil, he should next conceive a paradigm. This is an imaginary structure which would fulfil the function best, given the materials available to the fossil organism. If the observed structure resembles the paradigm closely, this will suggest that the function was correctly identified. Rudwick points out that analogy is more important than homology in identifying function. Thus, in deducing the function of a pterodactyl's wing, the analogy with a bat's wing is helpful, but comparison with the arm of a crocodile is not, although pterodactyls are probably more closely related to crocodiles than to bats. In my view Rudwick is right to stress the value of analogy in functional comparisons, but wrong to exclude phylogeny completely, since a cladogram may suggest which structures arose simultaneously with which others, and this may be functionally illuminating. The question: 'What was it for?' is valid for biological structures. It is comparable with asking what a machine is for and is based on the fact that organisms are self-conserving systems. It would be

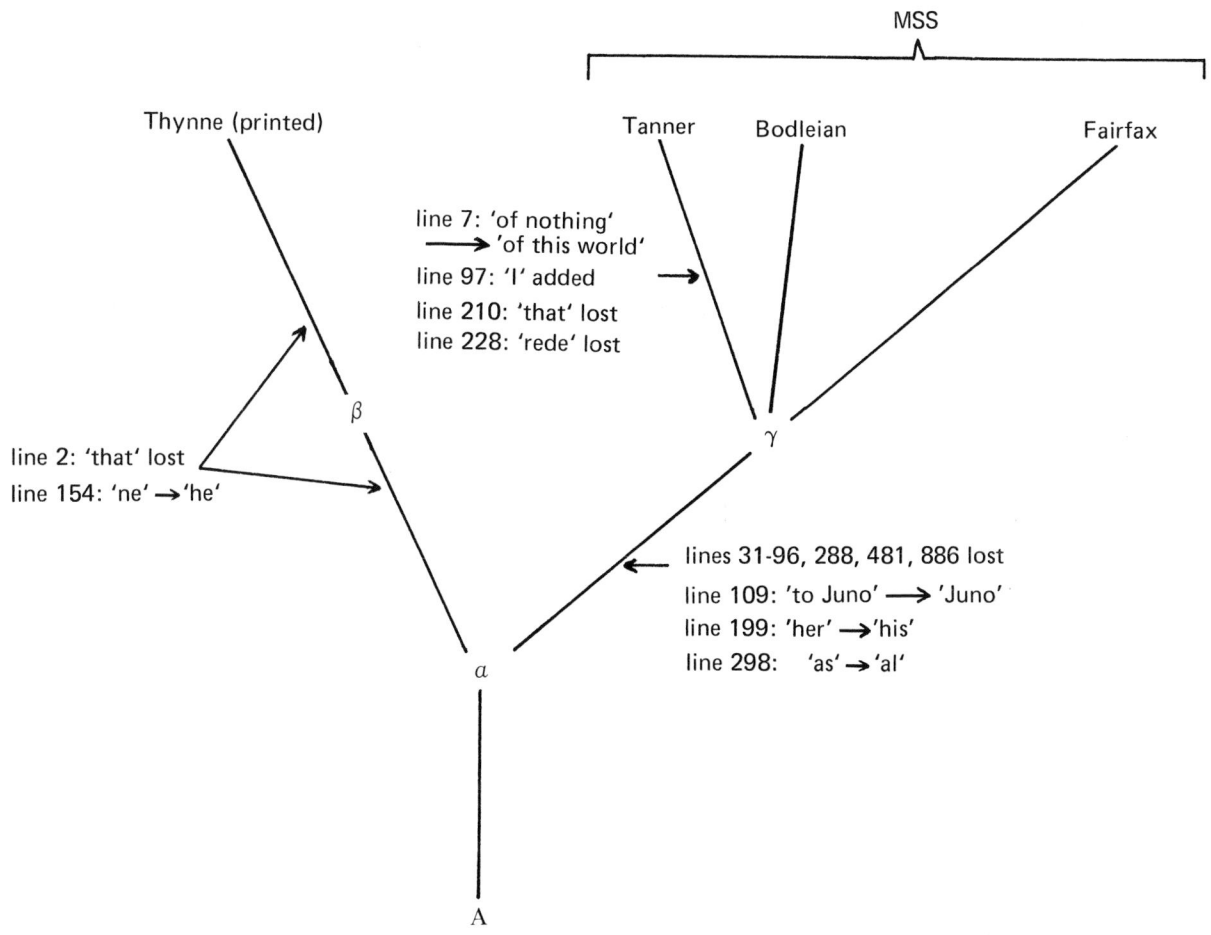

1.10. *Stemma or 'cladogram' of 'The Dethe of Blaunche the Duchess'. A is the presumed original of Chaucer. α, β and γ are deduced ancestral manuscripts. Koch (1881) regarded 'Bodleian' and 'Fairfax' as closer to each other than to 'Tanner', in the sense of sharing a more recent common ancestor, but this grouping was based on primitive resemblance (absence of mistakes) and is illogical.*

wrong to exclude this question in the name of logical rigour.

A final methodological point concerns reciprocal illumination. In solving a crossword, the 'down' answers suggest solutions to the 'across' questions and *vice versa*, and new results confirm or contradict answers already reached. In Hennig's words (1984, p. 51): 'Das Kriterium der Wahrheit ist Vereinbarkeit' – the criterion of truth is coherence (of results). No circular argument is involved, or, at least, not viciously circular. Similarly, with fossils, functional interpretations can suggest homologies with living animals, and these can confirm or contradict the functional interpretations. Løvtrup (1977, p. 21) has stated that: 'The discovery of a new fossil has no impact on classification'. I agree with him that the light cast by

Recent organisms on fossils is greater than the light cast by fossils on Recent organisms. I also agree that phylogenies should be reconstructed primarily on the evidence of still-living forms. I think, however, that in lucky cases, the evidence of fossils can rightly change our views as to how extant groups are related. They will do this, not because of stratigraphical sequence, but by providing otherwise unknown morphologies preserving synapomorphies which have disappeared in living forms.

As to methods, rather than methodology, the calcichordates all had skeletons of echinoderm type and are best studied as uncrushed articulated specimens. Such occur in two states of preservation. Either they are natural moulds, in which case the original skeleton is dissolved away and only the complicated hole remains where once the calcite used to be, or else the original skeleton is preserved. Natural moulds are excellent material for study. A complete idea of the internal and external surfaces of the vanished skeleton can be got by making latex rubber casts of the hole. Such casts show the surface of the stereom in histological detail. Moreover, the mould of the internal surface of the skeleton can be seen as a positive, in rock, of the outer surface of the soft parts, which can therefore be reconstructed. Individual specimens preserved as hollow moulds, however, never reveal an animal completely. It is necessary to puzzle a reconstruction together from several specimens, by making several projections simultaneously on a drawing board. When the original skeleton is preserved the internal structure can be studied by making models in expanded polystyrene on the basis of serial sections. Details of this technique can be found in Jefferies & Lewis (1978).

To sum up, the reconstruction of the phylogeny of a group demands, in the first place, a thorough knowledge of the morphology of extant members. Given this, we can draw cladograms on the basis of synapomorphies. Sometimes we can supplement this information by using fossils. This involves assigning fossils to the stem groups of extant crown groups, and placing them in series of plesions in these stem groups using synapomorphies with the crown groups or with more crownward plesions. Stratigraphy will be of very little help in matters of broad phylogeny. In lucky cases, members of a stem group may show synapomorphies with living groups, or with other stem groups, that extant forms have lost all trace of, and this may aid in working out the relationships of extant forms. To assign fossils correctly to their stem group it is sometimes necessary to reconstruct them in detail, soft parts also when possible, and for this purpose functional morphology can help. The internal coherence and complexity of the resulting picture, and the way it fits all the relevant clues much like a crossword, will confirm that it approaches the truth.

Chapter 2 Metazoan relationships and the hemichordates and echinoderms

In this chapter I shall sketch the classical views on the phylogenetic relationships between the metazoan phyla and then discuss the anatomy, habits and embryology of the hemichordates and echinoderms – two phyla which are generally accepted as close relatives of the chordates. I shall then consider the homologies connecting the hemichordates and echinoderms, in so far as these can be worked out on the basis of still-living forms, and what these homologies imply phylogenetically. For purposes of clarity I shall concentrate on the pterobranch *Cephalodiscus* among hemichordates and on the common starfish *Asterias rubens* and the crinoid *Antedon bifida* among echinoderms, but I shall also touch on other forms.

The earliest phylogeny of the multicellular animals (Metazoa) is obscure. The only evidence comes from embryology (using Haeckel's law or its corollary von Baer's law), or from comparative anatomy using outgroup comparison. It is difficult to decide what relevant features are primitive and what advanced, or how far Haeckel's law can be relied upon. All conclusions must therefore be tentative.

As first pointed out by Haeckel (1874b), most Metazoa (including sponges) pass through a two-layered gastrula stage in their ontogeny. In coelenterates this develops with little change into the two-layered adult and the opening or blastopore becomes the definitive mouth, which combines the functions of mouth and anus. Non-coelenterate Metazoa, best called the Bilateria, have a more complicated fate for the gastrula, since they are three-layered, with mesoderm as well as endoderm and ectoderm. The gastrula of the Bilateria probably recapitulates an adult, two-layered coelenterate ancestor, as Haeckel suggested.

Bilateria can be divided, following Grobben (1908), into Protostomia and Deuterostomia (fig. 2.1). In the Protostomia, the blastopore gives rise to the mouth (or is near the site of the mouth), whereas the anus is a secondary

perforation. In the Deuterostomia the converse is true – the blastopore gives rise to the anus (or is near the site of the anus), whereas the mouth is a secondary perforation. Of these alternatives, the protostomian condition is probably primitive, since protostomians include forms which primitively lack an anus (the Plathelminthomorpha, see Ax, 1984) and therefore cannot derive from the deuterostome condition. This suggests that the 'Protostomia' are probably a paraphyletic group and that the Deuterostomia, conversely, are monophyletic. The pogonophores, though once regarded as deuterostomes, are now also placed in the 'Protostomia', because of features that they share with the annelids (Southwood, 1975). The bryozoans and brachiopods, although both mouth and anus are secondary in them, are likewise placed in the 'Protostomia' because of resemblances to the phoronids (Zimmer, 1973) and these three phyla together are called the Tentaculata, because they carry tentacles.

A group 'Spiralia', with a complicated mode of cleavage and mesoderm arising from a particular single cell (4d), can be recognized within the 'Protostomia'. It includes all the 'Protostomia' with or without an anus, except the tentaculate phyla whose cleavage is always radial (Zimmer, 1973). Since spiral cleavage exists in those Protostomians which primitively lack an anus, it is probably primitive for Bilateria and has been lost by the forms with radial cleavage.

The tentaculate phyla of the 'Protostomia' (fig. 2.2) differ from the other members of that group, not only by lacking spiral cleavage, but also by their trimerous body plan. They are made up, that is to say, of a protosome, a mesosome bearing tentacles, and a metasome or trunk.

The Deuterostomia, defined by the secondary perforation of the mouth and the blastopore becoming the anus (or being near the site of the anus), comprise the Chaetognatha or arrow worms, hemichordates, echinoderms and chordates. The chaetognaths are only

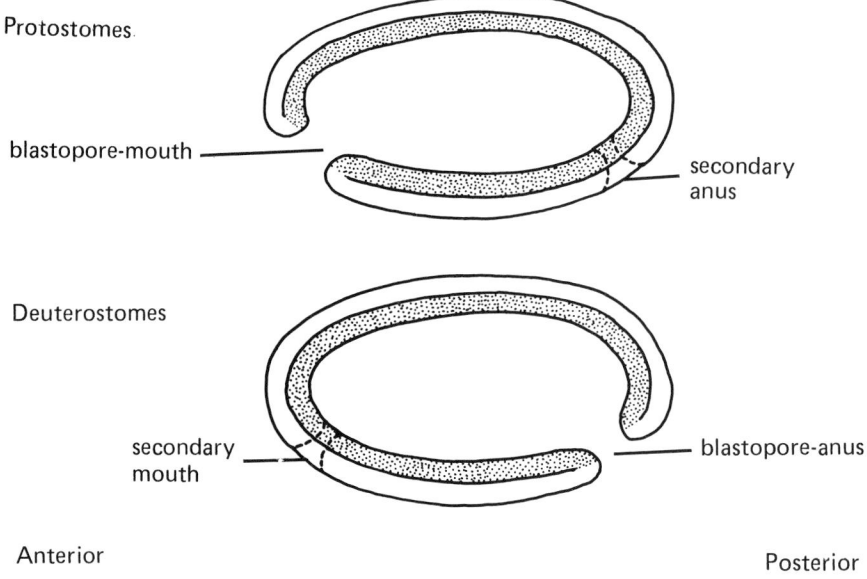

Protostomes

blastopore-mouth

secondary anus

Deuterostomes

secondary mouth

blastopore-anus

Anterior

Posterior

2.1. The embryological distinction between protostomes and deuterostomes.

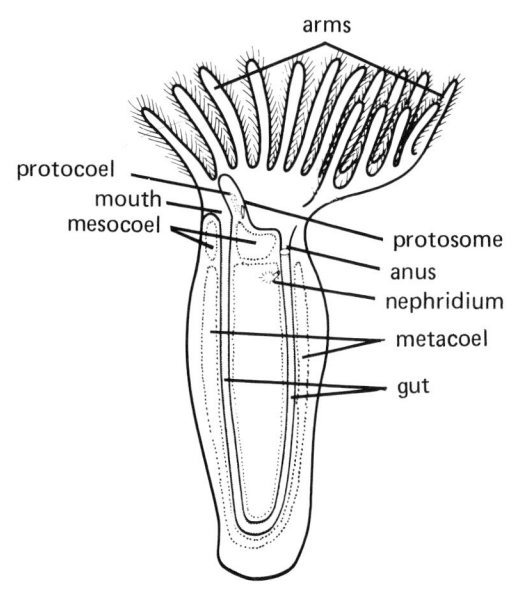

arms

protocoel
mouth
mesocoel

protosome
anus
nephridium
metacoel
gut

2.2. The basic structure of a tentaculate (after Starck, 1978, fig. 23, p. 65).

doubtfully placed in this group, though they have the defining characteristics of the blastoporal anus and secondary mouth (Grobben, 1908); I shall not discuss them further. Among the other three phyla the hemichordates are of immediate interest here since they have a trimerous

body with protosome, mesosome and metasome like the tentaculate protostomes. Zimmer (1980) has used this resemblance to suggest that the tentaculate protostomes, since they also lack spiral cleavage, should be placed among the deuterostomes. This is illogical, however. It is more reasonable to regard all forms with radial cleavage, and primitively with a trimerous body plan, as a monophyletic group of which the Deuterostomia form part. This argument implies that the trimerous condition was primitive among the Deuterostomia, being inherited from trimerous, protostomian ancestors. Among deuterostomes the hemichordates are the only group that show bilaterally symmetrical trimerism directly comparable with that of phoronids, brachiopods and bryozoans. By outgroup comparison, therefore, the hemichordates represent, in these respects, the primitive condition of deuterostomes, and from this condition the echinoderm and chordates would have derived.

The hemichordates are themselves divided into the tentaculate Pterobranchia and the worm-like Enteropneusta of which the widest known is *Balanoglossus*. Pterobranchs probably represent the more primitive condition, being readily comparable with the tentaculate 'Protostomia'. Also the echinoderms, and with them the chordates, can more readily be derived from a pterobranch-like ancestor than from an enteropneust. I shall therefore describe the pterobranchs in some detail. There are three living genera – *Cephalodiscus*, *Atubaria* and *Rhabdopleura*, and I shall concentrate on *Cephalodiscus*.

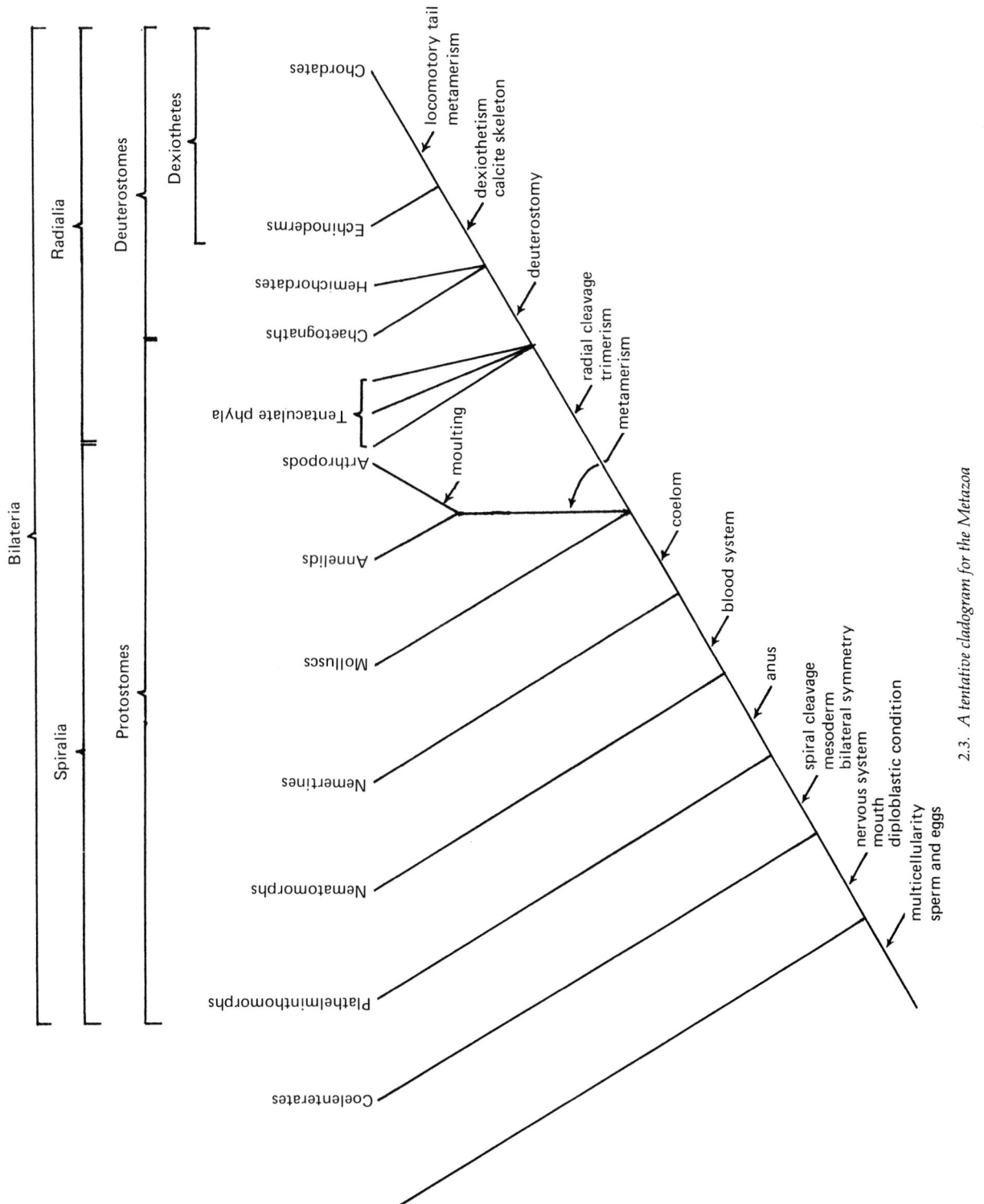

2.3. *A tentative cladogram for the Metazoa*

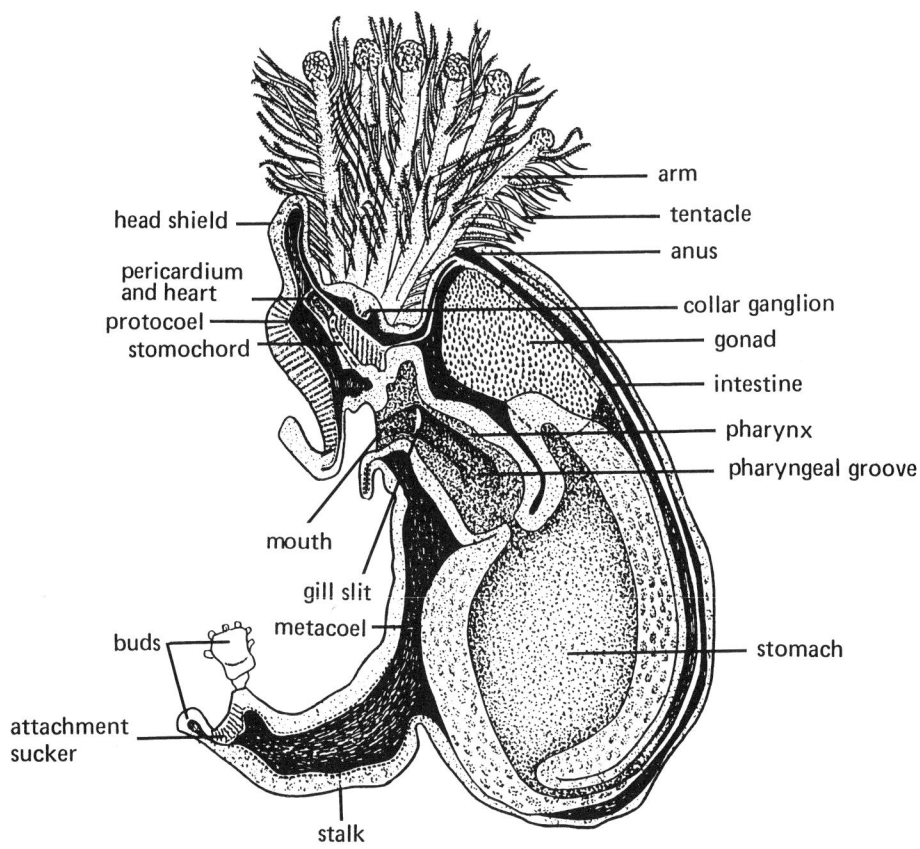

2.4. Cephalodiscus. *A sagittal section of a zooid and the internal organs of the right half of the animal; from Barrington (1965, fig. 29, p. 58) after Schepotieff (1907).*

Cephalodiscus is a marine, bottom-dwelling animal found mainly in the southern hemisphere. It is semi-colonial and inhabits bundles of horny, non-chitinous tubes which it constructs itself and which are usually attached to hard objects on the sea floor.

An individual *Cephalodiscus*, or zooid, as already implied, has a body divided into three parts. The protosome is a plate-like head shield. The mesosome is in the form of a ∩-shaped collar, from which five to nine pairs of arms project upwards on right and left. Each arm carries numerous slender tentacles. The metasome, or trunk, is divided into an anterior swollen part, containing the gut and a pair of gonads, and an extremely extensible stalk. The posterior end of the stalk has an attachment sucker, near which buds are formed that develop into other zooids. The length of a zooid, stalk omitted, is between 5 and 15 mm.

The habits of living zooids of *Cephalodiscus* were observed and described by Andersson (1907) and Gilchrist (1915). Each zooid can fix itself to the inside of its

horny tube by the sucker on the end of the stalk but sometimes there is more than one zooid for each tube. The zooids will come out of the tube and move over the surface of the colony, using the ventral surface of the head shield as a creeping surface, but usually leaving the stalk fixed inside the tube by its terminal sucker or by the head shield of one of its buds (fig. 2.5). The central part of the ventral surface of the head shield has been seen to secrete a sticky mucus (Gilchrist, 1915, p. 236) which is probably used as material in building the horny tube. The zooid can retreat into its tube by contracting its stalk. The whole surface of the zooid is ciliated and the cilia habitually beat anteriorly except for those on the ventral surface of the arms which beat downwards toward the mouth, into which they carry food particles (Gilchrist, 1915, p. 238).

There are five coeloms in each zooid: an unpaired protocoel inside the head shield or protosome; a pair of mesocoels inside the collar or mesosome and extending into the arms; and a pair of metacoels inside the trunk or

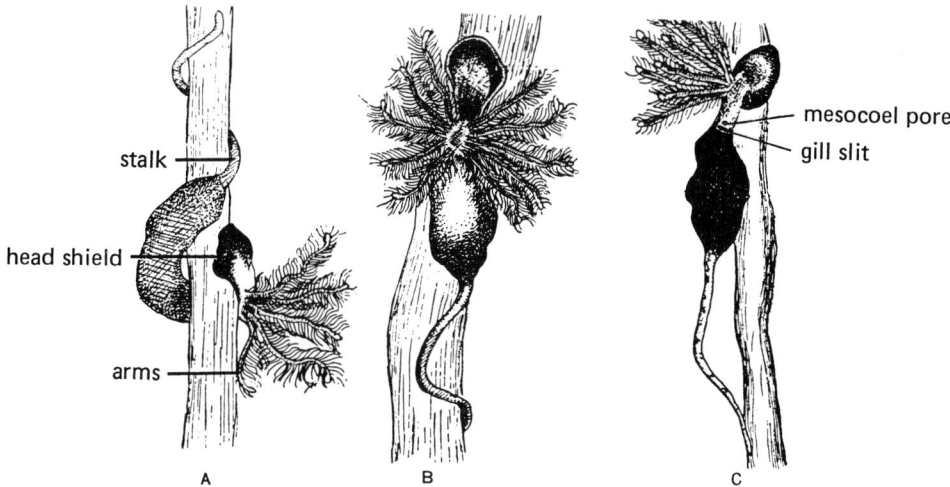

2.5. Cephalodiscus *creeping on its tube using the head shield as a grasping organ and the stalk as a prehensile 'tail'; from Dawydoff (1948) after Andersson (1907).*

metasome. Median mesenteries separate the left metacoel from the right one, and the left mesocoel from the right one. In the stalk the mesentery between left and right metacoels is incomplete. This condition with five coeloms is sometimes misleadingly called 'tricoelomate'.

The body openings number ten. There is a pair of protocoel pores at the posterior end of the protosome, and a pair of mesocoel pores at the posterior end of the mesosome, but the metacoels have no coelomic opening. The mouth is situated ventrally in the mid-line, between the protosome and the mesosome. The anus is in the dorsal mid-line, near the anterior end of the metasome. Just anterior to the anus, also in the metasome, is a pair of gonopores, being the openings of the gonads. And finally, and perhaps the most remarkable feature of *Cephalodiscus*, there is a pair of gill slits situated just behind the mesocoel pores and penetrating the wall of the metasome and of the pharynx.

The alimentary canal is U-shaped. The mouth leads into the pharynx whose internal surface has a pair of pharyngeal grooves in which the gill slits are situated. The ontogeny of the mouth of *Cephalodiscus* has never been described in a sexually produced individual and accounts of its formation in the buds are totally conflicting (van der Horst, 1935–39, p. 512). However, there is probably no buccal cavity of ectodermal origin since the mouth is formed at the surface of the larva in the related genus *Rhabdopleura* (Dilly, 1973). The oesophagus leads into a capacious stomach and this is connected to the anus by the intestine. The whole internal surface of the alimentary canal is ciliated.

The stomochord (figs 2.4, 2.7) is a controversial feature of the gut. It is a cylindrical process, sometimes hollow and sometimes solid, that projects into the protocoel from the dorsal mid-line of the pharynx. This is the structure which Bateson (1885), working on enteropneusts, homologized with the notochord of tunicates and of chordates in general. He was almost certainly mistaken in this for the stomochord has no connection with a motile tail, contrary to the primitive situation for the notochord of chordates. Silén (1954) has plausibly suggested that it may be homologous with the pre-oral gut (Seessel's pouch) of vertebrates. The name 'buccal diverticulum', used by Hyman (1959), is unhappy since it suggests that the stomochord derives from an ectodermal buccal cavity and such, as already mentioned, probably does not exist in *Cephalodiscus*.

The nervous system of *Cephalodiscus* is entirely in the deeper parts of the epidermis, outside the basement membrane. One of its most important parts is the collar ganglion, situated in the dorsal mid-line between left and right groups of arms, which probably coordinates the activities of the arms in feeding. A nerve extends forwards from this ganglion in the dorsal mid-line into the head shield to supply an important nerve plexus in the ventral wall of the shield. Another extends rearwards in the dorsal mid-line to the anus. A pair of nerves pass from the posterior end of the ganglion round the anterior part of the metasome. They meet to form a nerve in the ventral mid-line of the metasome and this passes to the stalk whose epidermis is richly nervous everywhere, especially ventrally. Zoologists have tended to over-emphasize the dorsal elements in the nervous system of hemichordates so as to point a resemblance to vertebrates. In fact, however, there are several concentrations of the nervous layer in *Cephalodiscus*, some of them dorsal, some ventral and some paired. And in any case, as I shall try to show later, the dorsal mid-line of hemichordates is not homologous with that of chordates; rather, it corresponds to the right side of the chordate body.

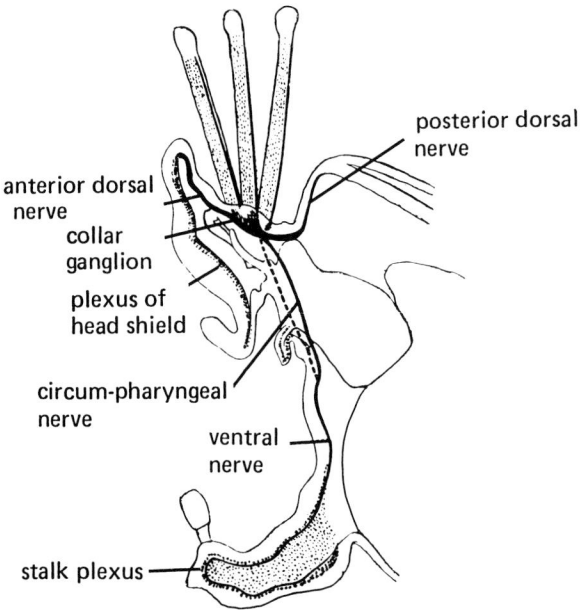

2.6. The nervous system of Cephalodiscus; *from Schepotieff (1907b, pl. 46, fig. 15).*

The blood system (fig. 2.8) consists of a sac known as the heart vesicle inside the protocoel and a system of spaces, mostly tubular, between the limiting membranes of the coeloms. The heart vesicle (figs 2.4, 2.7, 2.8) or pericardial vesicle is a small bag, with no openings, situated at the anterior end of the stomochord. This bag is shaped like a hollow-walled cup and the rearward-facing concavity of the cup contains a central blood sinus. A vessel passes from this central blood sinus rearwards along the ventral surface of the stomochord. When the vessel reaches the dorsal mid-line of the pharynx it splits into a pair of vessels which pass round the pharynx and meet again posterior and ventral to it. There they form a ventral vessel which passes rearwards in the mid-ventral line, in the mesentery between the left and right metacoels, into the stalk. On reaching the end of the stalk it seems to turn dorsally and to continue as a dorsal vessel which passes along the mid-dorsal line of the stalk into the trunk, where it dies out. Blood sinuses on the anterior surfaces of the stomach and of the paired gonads are connected to a vessel in the dorsal mesentery of the trunk which runs to the central sinus of the heart vesicle. The direction of circulation of the blood in this system, if there is a constant direction, is not known.

Excretion is assumed to take place by an organ called the glomerulus (fig. 2.7). This is a mass of thickened epithelium on the folded ventral wall of the vessel ventral to the stomochord. The glomerulus presumably excretes into the protocoel so that excretory products will escape to the outside through the protocoel pores.

The muscles of the protosome or head shield are complicated. They presumably act when the protosome is used in creeping for it is then observed to change shape (Andersson, 1907, p. 15). The muscles of the mesosome

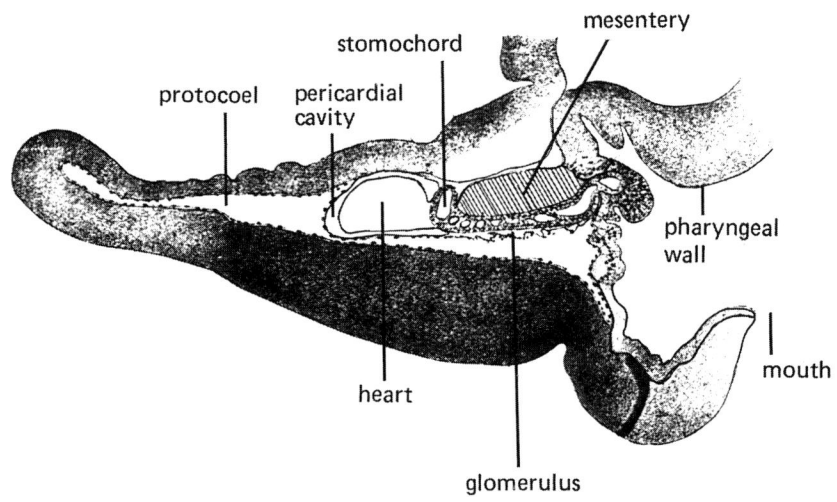

2.7. Cephalodiscus. *Heart, pericardium and associated organs of the protocoel; from Schepotieff (1907b, pl. 46, fig. 15).*

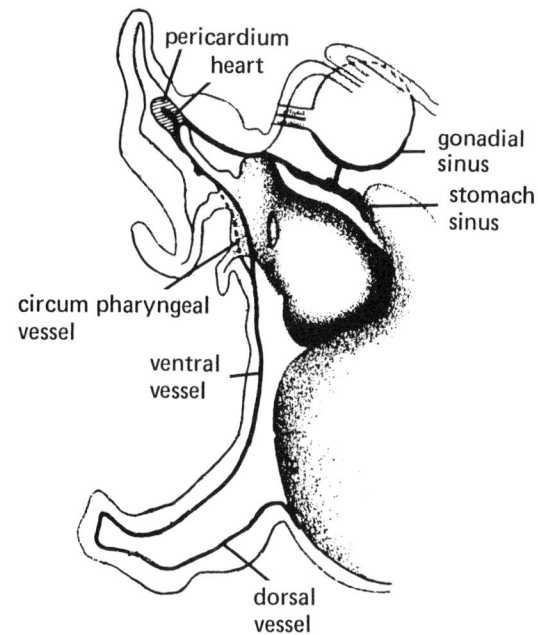

2.8. Cephalodiscus. *The blood vascular system; from Schepotieff (1907b, pl. 47, fig. 12).*

are mainly concentrated round the mouth. The muscles of the metasome are powerful and longitudinal. They reach from the anterior end of the metacoels back to the posterior end of the stalk. These metasomal muscles are used when the zooid suddenly contracts into its tube to escape danger.

Sexual reproduction is by a pair of gonads situated in the metacoels (fig. 2.4). These gonads are connected by gonoducts to the gonopores already mentioned near the anus. According to species, the colonies are sometimes exclusively of one sex, or of both sexes together. The individual zooids are male, female, hermaphrodite or neuter. Fertilization is probably internal, for Andersson (1907) saw sperm within an ovary and Stebbing (1970) came to a similar conclusion for *Rhabdopleura*. How the fertilized eggs are released is unknown.

The embryology of sexually produced individuals of *Cephalodiscus* is not well known. The eggs are large and yolky but cleavage seems to be total producing a blastula with much extracellular yolk in the blastocoel. Gastrulation occurs by invagination. After gastrulation the embryo develops cilia over its whole surface and escapes from the egg membrane. The larva is pear-shaped in outline in dorsal or ventral view, but dorso-ventrally flattened. It has been called a planula. It crawls over the bottom by cilia with the wide end of the pear, which is the anterior end of the animal, always forwards (Gilchrist,

1915, p. 242). There is a visibly glandular concavity near the posterior end of the larva which is probably the anlage of the sucker at the posterior end of the definitive stalk. A much larger glandular area on the antero–ventral surface of the larva is probably the anlage of the ventral surface of the head shield. Antero-dorsally the larva has a thickened plate of ectodermal cells which are probably neurosensory.

The ontogenetic origin of the coeloms is differently described by different authors and perhaps varies with the species (van der Horst, 1935–39, p. 522). Sometimes all five coeloms evaginate separately from the archenteron. Sometimes the protocoel evaginates on its own while the mesocoels and metacoels differentiate from a single common evagination of the archenteron.

The gut is almost straight when the mouth and anus first appear. At a slightly later stage, however, it becomes U-shaped, as in the adult, and at the same time the stalk, stomochord and arm anlagen appear. Gill slits seem to form very late in ontogeny, presumably at about the time when the larva settles permanently on the substrate. The way in which buds of *Cephalodiscus* turn into zooids is complex and variable (van der Horst, 1935–39, p. 503).

The pterobranch *Atubaria* (fig. 2.9) is very similar to *Cephalodiscus* in its individual zooids. All the known individuals were taken in one haul in Sagami Bay, Japan, in 1935. The external features were described by Sato (1936)

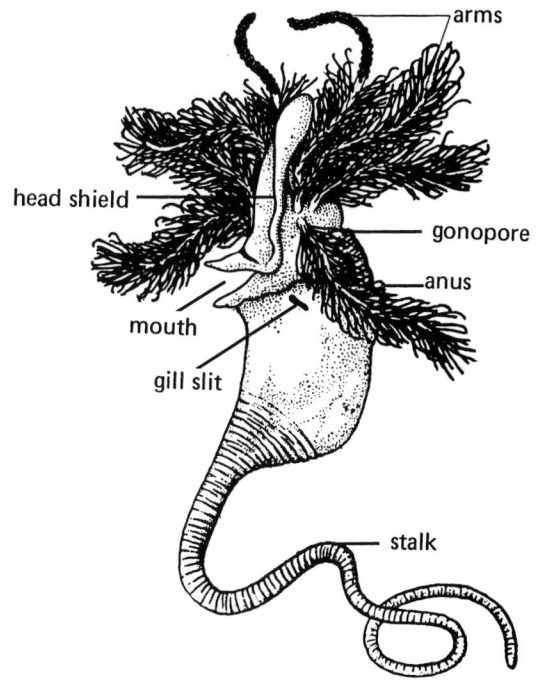

2.9. *External features of* Atubaria; *from Komai (1949, fig. 1, p. 19).*

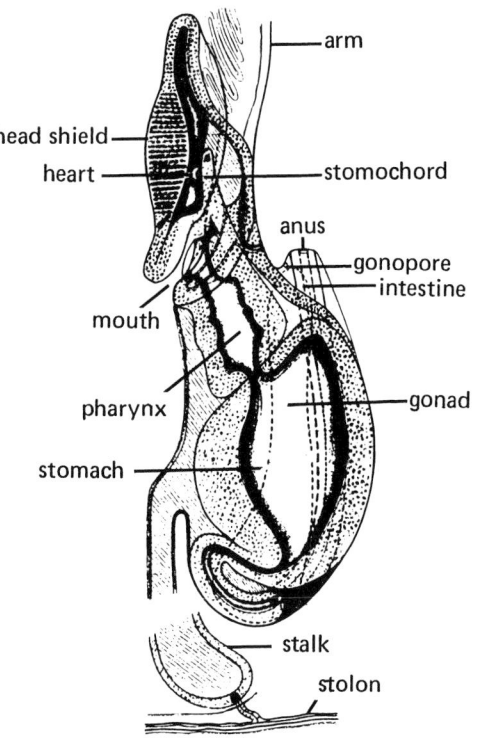

2.10. Rhabdopleura. *Sagittal section of a zooid and the internal organs of the right side; from Schepotieff (1907a, pl. 17, fig. 1a).*

and the internal features by Komai (1949). Unlike all other pterobranchs, *Atubaria* has no horny tubes and is not known to bud. The zooids were found free on a hydroid to which they clung by their stalks. These organs were presumably used in prehensile fashion since they had no sucker.

The pterobranch *Rhabdopleura* (figs 2.10, 2.11) is less similar to *Cephalodiscus*. It is widely distributed in northern and southern hemispheres and has been found alive in British waters off Plymouth (Stebbing, 1970a, b; Stebbing & Dilly, 1972; Dilly, 1972, 1973, 1975a, b). *Rhabdopleura* lives in a branching growth of horny tubes, like *Cephalodiscus* but is more strictly colonial, since the zooids remain throughout life attached to a strand of living tissue, inside the branching tubes, which is known as the stolon. The zooids of *Rhabdopleura* are much smaller than in *Cephalodiscus*, being less than 1 mm long. There are no gill slits. There are a number of asymmetries: thus (1) of the single pair of arms, the right is usually longer than the left; (2) there is only one gonad, situated in the right metacoel and opening right of the mid-line; (3) the mouth is left of the mid-line; (4) the intestine runs to the anus right of the stomach. Having only one gonad, the zooids of *Rhabdopleura* are never hermaphroditic, but the colonies always are. The blood-vascular, nervous and muscular systems are all very like those of *Cephalodiscus*. There is no obvious phylogenetic relationship between the asymmetries of *Rhabdopleura* and those of echinoderms. Cilia on the sides of the tentacles actively create a

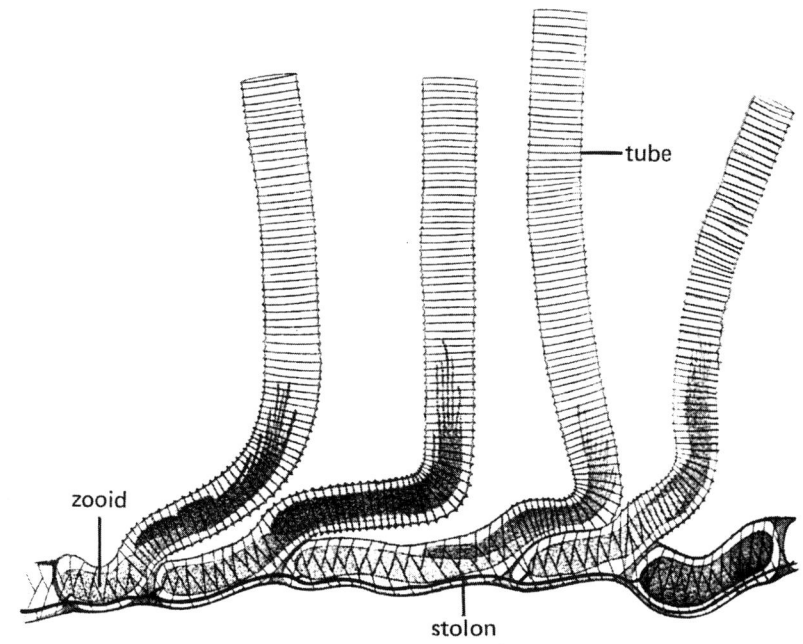

2.11. Rhabdopleura. *The zooids in their tubes attached to the stolon; from Schepotieff (1907a, pl. 22, fig. 2).*

feeding current of water, while other cilia, on the ventral surface of the arms, probably carry food particles to the mouth (Stebbing & Dilly, 1972). Thus *Rhabdopleura* is an active, not a passive, filter feeder. As such it is bilaterally symmetrical, like a bivalve mollusc, or a tunicate, not radially symmetrical in the manner of a passive filter feeder such as a crinoid, a sabellid worm, or a spider with its web.

The fossil record of pterobranchs goes back to the Ordovician. For Kozłowski (1949) described, from the Lower Ordovician of Poland, the horny tubes of *Rhabdopleurites* and *Rhabdopleurides* (both resembling *Rhabdopleura*) and of *Eocephalodiscus* (resembling *Cephalodiscus*). Moreover Kozłowski argued that tubes of the excellent zone fossils known as graptolites were built on the same pattern as *Rhabdopleura* and suggested a relationship. The zoological position of the graptolites, however, is not relevant here. For a recent account see Crowther (1981).

The second main group of hemichordates comprise the worm-like animals known as enteropneusts or acorn worms. *Balanoglossus* and *Saccoglossus* are the two most widely known genera. Unlike *Rhabdopleura* or *Cephalodiscus* the enteropneusts very seldom reproduce by budding (see, however, Packard, 1968) and are not colonial. Many enteropneusts burrow in sand and mud but others lie under stones or in the holdfasts of sea-weeds. There are several reasons why enteropneusts have been better studied than pterobranchs: they live in shallow water, rather than deep; they are much bigger with a length from about 20 mm to about 1·8 m; and they are a more diverse group with about 70 known species and 11 genera.

An enteropneust (fig. 2.12) is regionated into three parts like a pterobranch. These are the protosome or proboscis; the mesosome or collar; and the metasome or trunk. As in pterobranchs the protocoel is unpaired, whereas there is a single pair of mesocoels and of metacoels. The protosome is approximately conical; the mesosome is a short cylinder without arms; and the metasome is a long cylinder terminating at the anus posteriorly and therefore without a stalk, except sometimes in the post-larva.

Body openings comprise: a single protocoel pore (on the left, on the right, or median), or a pair of such pores, penetrating the posterior wall of the protosome; a pair of mesocoel pores in the posterior wall of the mesosome; a left and right series of gill pores situated dorso-laterally in the metasome; sometimes a left and right series of so-called oesophageal pores, which are essentially modified posterior gill pores; a left and right series of almost invisible gonopores, situated between the gill pores and continuing behind them; and the terminal anus already mentioned.

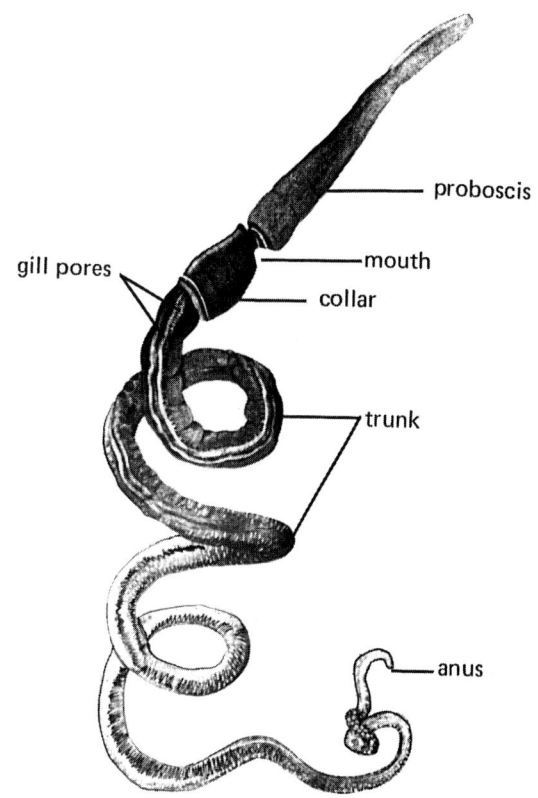

2.12. *External appearance of an enteropneust (the Welsh species* Saccoglossus cambrensis *Brambell & Cole 1939); from Brambell & Cole (1939, pl. 1).*

The gill pores sometimes open into a fold in the surface of the trunk, rather than direct to the outside. They are the external openings of gill slits in the pharyngeal wall (fig. 2.13). These gill slits are always U-shaped in adult enteropneusts, with the concavity of the U facing dorsally. The gill slits are oval when they first appear in ontogeny but a tongue grows down from the dorsal margin towards the ventral side of the slit (fig. 2.14). This tongue has been compared with the tongue bar of a gill slit of amphioxus, which likewise grows down in ontogeny from the dorsal margin of the primordial slit. The comparison is made more appealing by the fact that some enteropneusts, like amphioxus, have strengthened the gill slit by means of thin bridges or trabeculae across it and also because, both in amphioxus and enteropneusts, the gill bars are stiffened with skeletal cartilages (fig. 2.15). However, though the gill slits are probably broadly homologous between amphioxus and the enteropneusts there is no reason to think that the details are so. The U-shape is probably to be regarded in both cases as an independent adaptation to increase the length of the slit, and therefore the number of

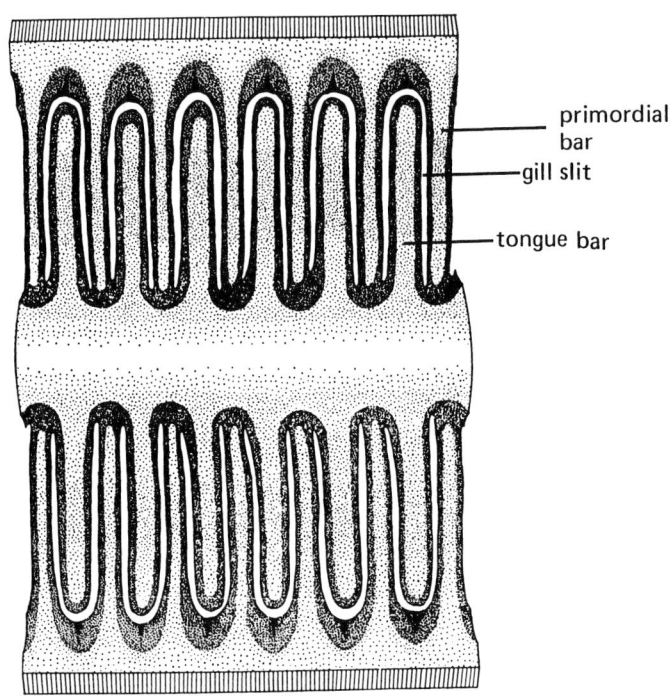

2.13. *Gill slits of an enteropneust (Saccoglossus cambrensis) to show their* **U***-shapes when seen from inside the pharynx; from Brambell & Cole (1939, text fig. 13).*

2.14. *Ontogeny of an enteropneust gill slit (Saccoglossus kowalevskii); from Bateson (1886, pl. 30, fig. 85).*

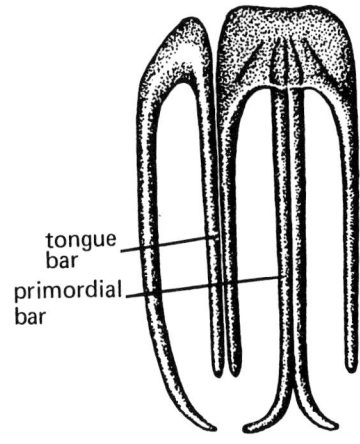

2.15. *Branchial cartilage of an enteropneust (Saccoglossus kowalevskii), from Hyman (1959, fig. 37C, after Spengel, 1893).*

cilia that it carries and its pumping action. The cartilages and trabeculae would in that case be independently acquired adaptations necessary to strengthen this lengthened slit. It is noteworthy that the tongue bars of amphioxus have no coelom, while the primary bars do have such – in enteropneusts the opposite is true (fig. 2.16 and van der Horst, 1939, p. 640).

The alimentary canal of enteropneusts runs nearly straight from the mouth to the anus. The mouth is situated where the protosome joins the mesosome. There is no buccal cavity arising from ectoderm in ontogeny. The most anterior part of the alimentary canal, just inside the mouth, but in front of the gill slits, should therefore be

seen as pre-branchial pharynx. From it the stomochord projects into the protosome as in pterobranchs. Behind the branchial pharynx, whose wall as already mentioned is penetrated by gill slits and 'oesophageal' pores, is an intestine which runs backwards to the anus. The anus has a sphincter muscle. This is almost the only muscle in the wall

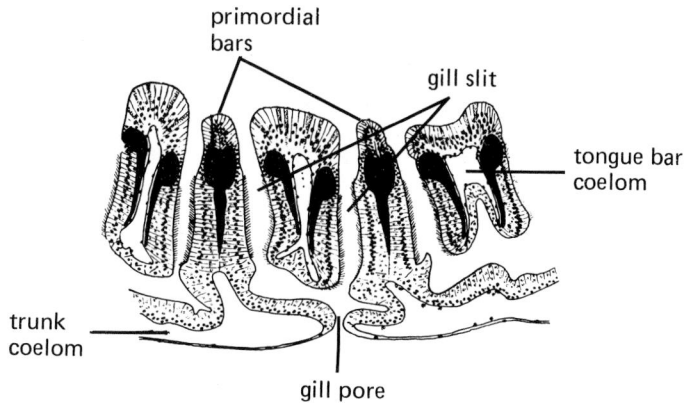

2.16. *Part of a longitudinal section of an enteropneust (*Saccoglossus cambrensis*) to show the presence of coeloms in the tongue bars of the gill slits and their absence in the primordial bars – contrary to amphioxus; from Brambell & Cole (1939, text fig. 12, p. 224).*

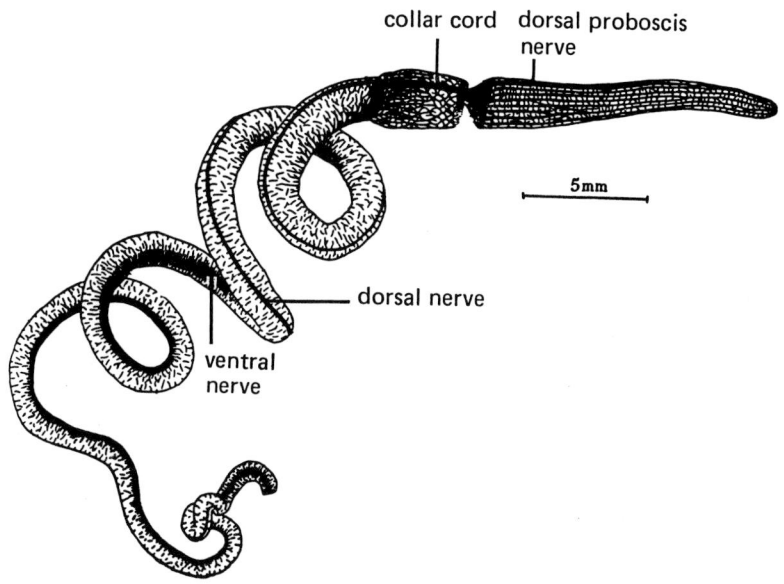

2.17. *Nervous system of an enteropneust (*Saccoglossus cambrensis*); from Knight Jones (1952, fig. 1).*

of the alimentary canal, which is, however, extensively ciliated.

The nervous system of enteropneusts (fig. 2.17) is very like that of *Cephalodiscus* for it consists of a nervous layer in the epidermis, just outside the basement membrane (Knight-Jones, 1952). This nervous layer is thickened in places to form mainly longitudinal tracts. The most striking such tract is the so-called collar cord, situated near the dorsal mid-line of the mesosome, but internal to the mesocoels. Its position corresponds well to the mesosomal

ganglia of pterobranchs except that it is formed by actual invagination of the mid-dorsal collar epithelium. The collar cord passes rearwards into a dorsal nerve along the dorsal mid-line of the metasome and forwards into a dorsal nerve along the mid-line of the proboscis. Right and left paired nerves connect the posterior end of the collar cord with a very large nerve that follows the mid-ventral line of the metosome. This ventral nerve is much stouter than the mid-dorsal nerve. The collar cord and associated dorsal nerve tracts are sometimes held to be homologous with

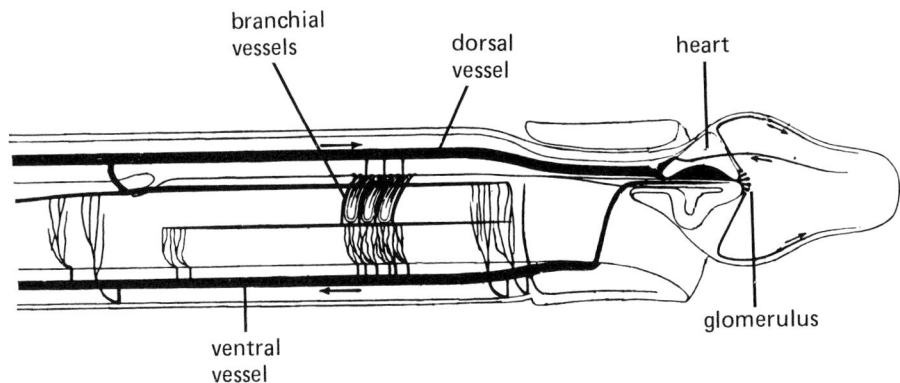

2.18. *Blood system of a diagrammatic enteropneust; from van der Horst (1939, fig. 1).*

the dorsal nerve cord of chordates, as first proposed by Bateson (1884, 1886). This seems unlikely since, as I argue later and have mentioned already, the mid-dorsal line of a hemichordate is probably homologous with the right edge of a chordate body. In addition, however, the suggested homology ignores the ventral nerve tracts on enteropneusts which, as in pterobranchs, are much stouter than the dorsal ones.

The blood system (fig. 2.18) is also like that of *Cephalodiscus*. There is a pulsatile heart vesicle just dorsal to the stomochord and it partly encloses a central sinus. Blood is said to flow into this central sinus along a dorsal vessel and to flow rearwards along a vessel in the ventral mid-line, but published observations on the direction of flow are sparse (Kowalevsky, 1866).

Excretion, as in pterobranchs, is assumed to take place by a glomerulus associated with the stomochord. Excretory products would leave the body through the protocoel pore or pores.

The muscular system is also very similar to that of pterobranchs. Circular muscles are lightly represented in the proboscis and trunk but almost absent elsewhere. By far the greater part of the muscles are longitudinal.

Reproduction is always sexual, except for budding in the South African *Balanoglossus capensis* Gilchrist and the Australasian *B. australiensis* (see Packard, 1968). The sexes are separate. Development can be either indirect, with a long planktonic and plankton-feeding larval stage, or 'direct' with a brief, benthonic, yolk-feeding larva. Cleavage is complete and gastrulation is by invagination. The blastopore closes and marks the posterior end of the larva. The gastrula becomes ciliated and escapes from the egg membrane. In indirect development it then turns into the tornaria larva (fig. 2.19). This is shaped like a very short sausage; it has a tuft of cilia emerging from one end (the future anterior end of the animal); at the opposite,

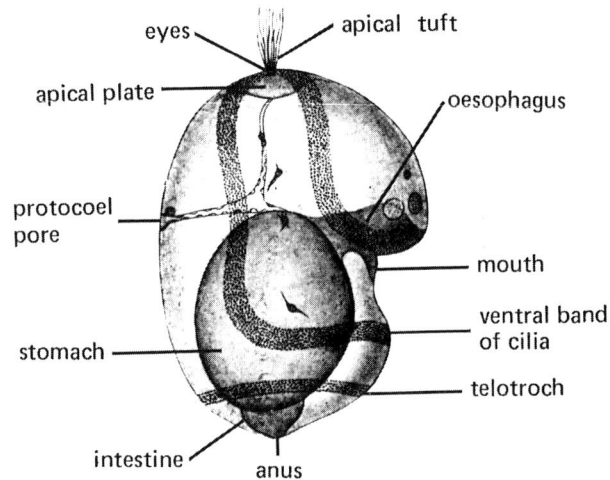

2.19. *Tornaria larva of an enteropneust* (Balanoglossus clavigerus); *from Stiasny (1914, pl. 5, fig. 17).*

posterior end is a ring of cilia called the telotroch; situated between the tuft and the telotroch there is another band of cilia, on the future ventral side of the animal, which follows a complicated bilaterally symmetrical course. The tornaria has a mouth, an anus and a gut which is divided into 'oesophagus' (pharynx), stomach and intestine. At the base of the apical tuft of cilia there is a pair of eyes and a nervous region called the apical plate. Inside, the future protocoel has evaginated from the archenteron and sends an extension to the dorsal surface which opens at a pore – the future protocoel pore. Tornarias are very similar to the auricularia larvae of starfishes and, significantly, were thought to be echinoderm for many years until Metschnikoff (1881) discovered that they belonged to enteropneusts. The most obvious difference is that echinoderm larvae do not have the ring-shaped telotroch.

The origin of the coeloms varies greatly. Sometimes all five coeloms arise from the archenteron independently. Sometimes there is a single anterior evagination from the archenteron; the anterior part of this evagination produces the protocoel and gives rise to a pair of pouches posteriorly which differentiate to form the mesocoels and metacoels of both sides. Sometimes an anterior evagination produces the protocoel only while a single pair of solid outgrowths from the archenteron produce the mesocoels and metacoels. Sometimes the protocoel forms as an archenteric evagination while the mesocoels and metacoels arise as splits in the mesenchyme. The variability of origin in ontogeny, despite the obvious homologies of the various coeloms, is surprising and contradicts Haeckel's law. The heart vesicle seems sometimes to be of ectodermal origin, while sometimes it develops as a space in the mesenchyme. The protocoel sends out a tube to the surface, usually on the left side only, to form the protocoel pore. A pair of similar tubes grow out from the mesocoel to form the mesocoel pores. The collar nerve cord forms by invagination in the dorsal mid-line of the collar. The larval gut, mouth and anus convert gradually into the definitive gut, mouth and anus. The stomochord forms at a late stage as an outpouching from the pre-branchial pharynx and the gill slits likewise arise at a late stage. In *Saccoglossus* the late larva develops an adhesive and contractile tail or stalk ventral and posterior to the anus (fig. 2.20). This is probably homologous with the stalk of

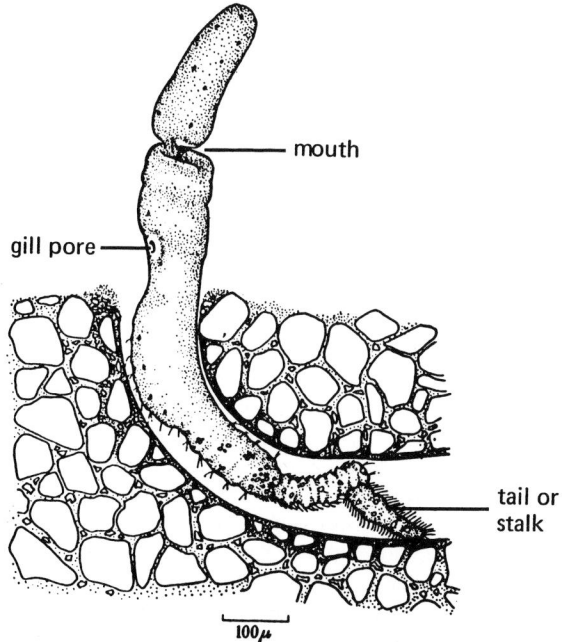

mouth

gill pore

tail or stalk

100μ

2.20. The late larva of an enteropneust showing the tail (Saccoglossus horsti); from Burdon-Jones (1952, fig. 24, p. 583).

pterobranchs. It disappears before the larva becomes adult.

Burrowing, in the forms that burrow, is almost entirely by muscular peristaltic waves in the proboscis. This is reminiscent of the active role that the head shield plays in the creeping of pterobranchs. In feeding, mucus is produced by the proboscis and swept back into the mouth as a mucous rope with ingested food particles. In addition, burrowing enteropneusts ingest sand. A stream of water leaves the pharynx through the gill slits, driven by the lateral cilia of the slits. This stream helps to draw the feeding mucus into the pharynx. Movement of the gut contents is largely by means of the cilia of the alimentary lining.

The phylum Echinodermata is divided into two subphyla (Smith, 1984): (1) the eleutherozoans which are free-living and usually and primitively have the mouth facing down towards the substrate – this group includes the asteroids or starfishes, the ophiuroids or brittle stars, the echinoids or sea urchins, and the holothuroids or sea cucumbers; (2) the pelmatozoans which are primitively attached in the adult and have the mouth facing upwards, away from the substrate – this group includes only the crinoids (sea-lilies and feather-stars) among living animals. As a representative eleutherozoan, and also in most respects as a representative echinoderm, I take the asteroid *Asterias rubens*, the common starfish of European waters.

Asterias rubens is abundant in the shallow sea, especially among boulders. Very similar species exist off North America (*A. vulgaris, A. forbesi*). The anatomy of *Asterias rubens* was well described at an elementary level by Chadwick (1923) and its embryology worked out in two masterly papers by Gemmill (1914, 1915). In external form, *Asterias rubens* is the familiar five-pointed star, usually about 120 mm across (figs 2.21, 2.22). It has five tapering arms attached to a central disc which is usually about 20 mm across. The animal has an upper, aboral and a lower, oral surface. The disc of the upper surface carries a prominent circular plate, the madreporite, near where two arms meet, and also a much less obvious, indeed almost invisible, anus. On the lower or oral surface, in the centre of the disc, there is a circular mouth which is surrounded by a plateless area called the peristome. Five ambulacral grooves, one along each arm, converge towards the mouth. Their name comes from Latin *ambulacrum* (= avenue) and refers to the fact that each groove has on either side a zig-zag row of muscular organs called tube feet, like trees on each side of an avenue.

A skeleton is present in *Asterias* and gives the surface a warty appearance everywhere except on the peristome and ambulacral grooves. The wartiness is caused by solid plates in the body wall which are formed of the mineral calcite (a form of $CaCO_3$). Each plate is crystallographically a single crystal of calcite and takes the form of a

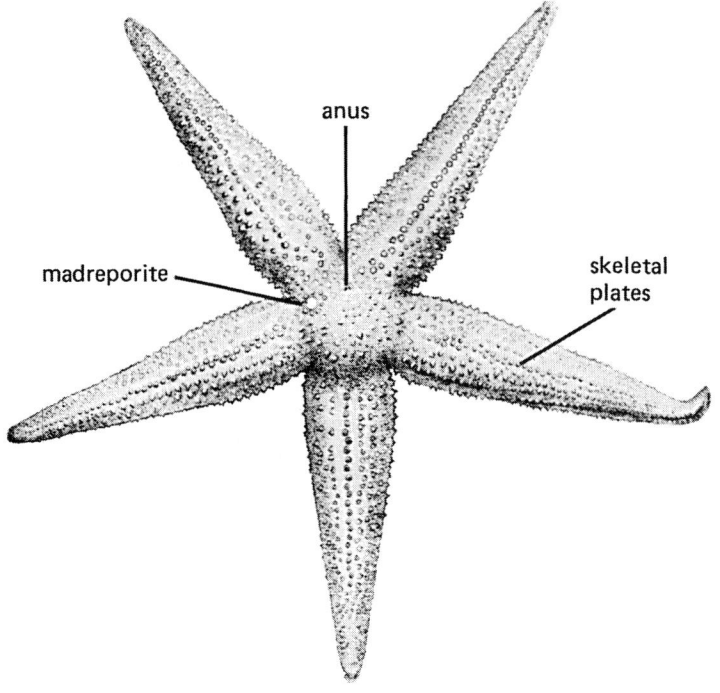

2.21. Asterias rubens. *Aboral or upper surface; from Chadwick (1923, pl. 1, fig. 1).*

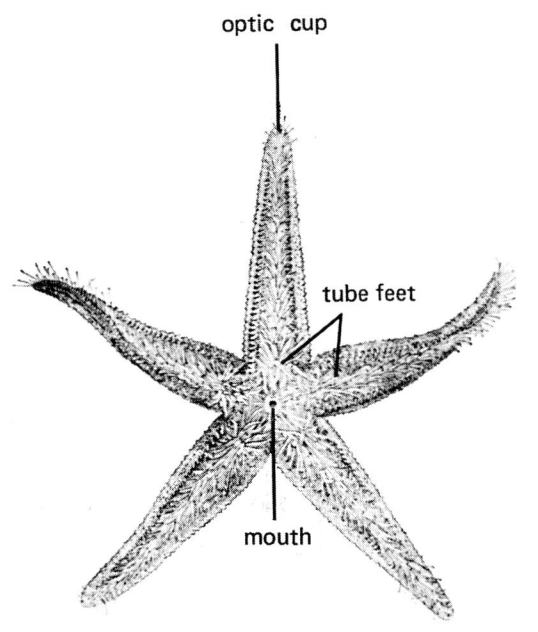

2.22. Asterias rubens. *Oral or lower surface; from Chadwick (1923, pl. 2, fig. 13).*

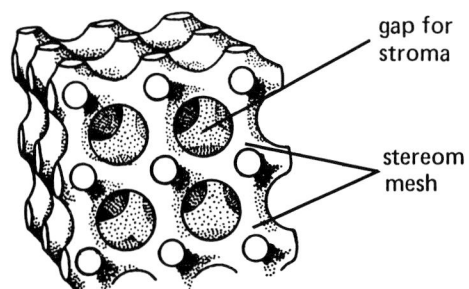

2.23. *Diagrammatic microstructure of the calcitic stereom mesh of an echinoderm; from Ubaghs (1967b, fig. 3.3, p. 513).*

three-dimensional network known as stereom mesh (fig. 2.23). The holes in this network are filled with soft tissue known as stroma which secretes the calcite of the plate. The stroma is of mesenchymal origin and the plates are therefore mesodermal. The skeleton is anatomically complicated (fig. 2.24). Thus a paired series of ambulacral plates roof each ambulacral groove, the tube feet emerge between neighbouring ambulacral plates and muscles between the plates act to narrow or widen the groove. Other muscles exist generally in the body wall. The

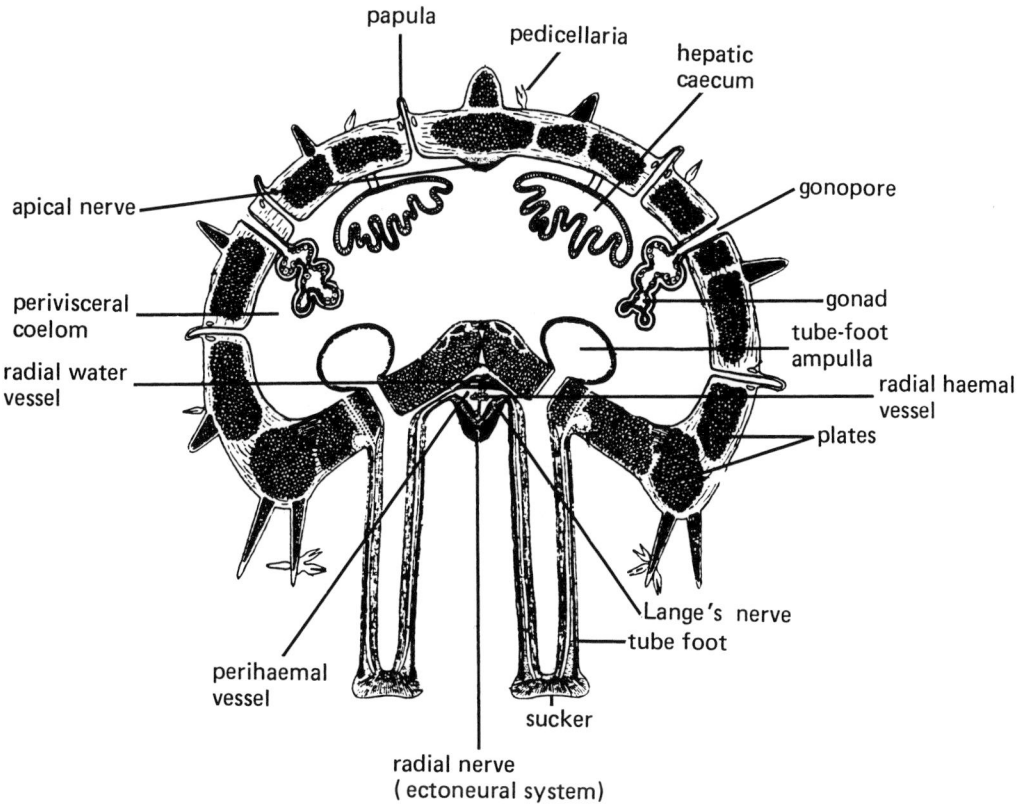

2.24. Asterias rubens. *Transverse section through an arm; from Chadwick (1923, fig. 1).*

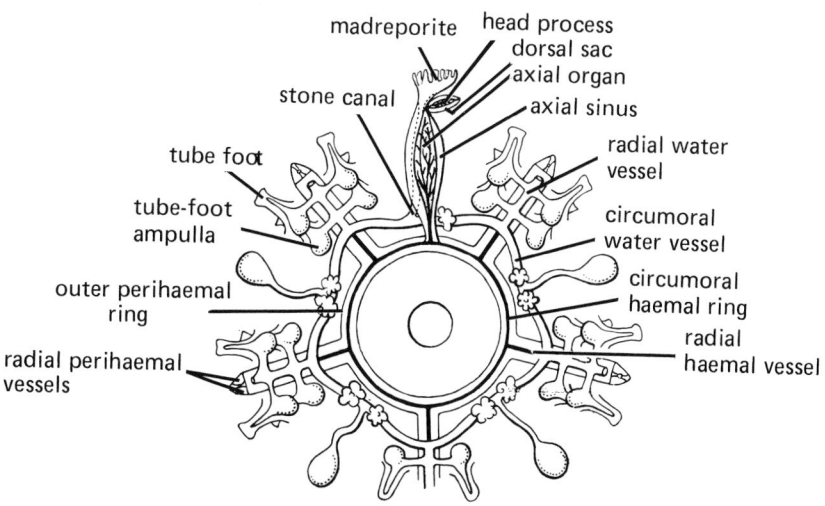

2.25. *Diagram of the organs round the mouth of an asteroid, seen from above; from Ubaghs (1967b, fig. 6).*

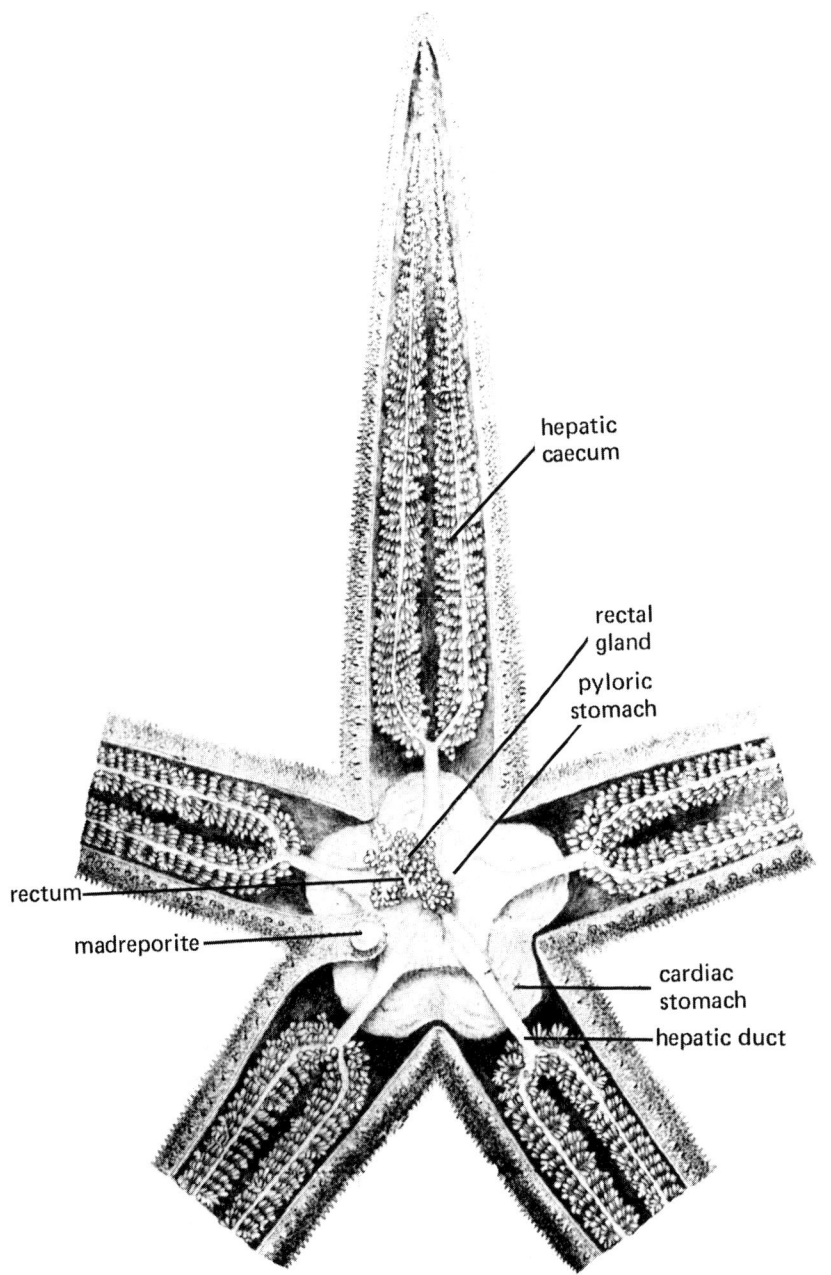

hepatic
caecum

rectal
gland

pyloric
stomach

rectum

madreporite

cardiac
stomach

hepatic duct

2.26. Asterias rubens. Alimentary canal from above, i.e. aboral aspect; from Chadwick (1923, pl. 4, fig. 31).

detailed anatomy of the skeleton and musculature, however, is irrelevant here. Attached to the plates of the body
wall are other plates forming spines and some spines are
modified in pairs as little tweezers or shears known as
pedicellariae which protect the animal from predators and
the settlement of epizoans. Histologically the spines and
hard parts of the pedicellariae are like all the other plates
of *Asterias* – each is at molecular scale a single crystal and
is made of stereom mesh and therefore mesodermal in
origin. The same characteristics are general for the plates

of echinoderms and, as I shall show later, also held for the calcite skeletons of primitive chordates. The skin between the plates give rise to tiny finger-like gills called papulae.

The tube feet are the effector organs of the water vascular system which as such is characteristic of echinoderms, though homologous with the left mesocoel of hemichordates. This system consists of hollow tubes of coelomic origin (fig. 2.25). Round the mouth is a circumoral water vessel which gives off a radial water vessel along the mid-line of each arm. The radial water vessels in turn give off lateral water vessels, each of which passes to a tube foot. Each tube foot is a hollow structure, rather like the finger of a rubber glove. It is thick-walled and muscular and very extensible and contractile. When extended, it ends in a circular sucker. Inside the body the tube foot is joined to a thin-walled bag or ampulla. When the tube foot expands, water passes from the ampulla into the tube foot, and when it contracts the water flows back into the ampulla. The circum-oral water vessel is connected to the so-called stone canal, hardened with spicules of calcite, which passes upwards inside the coelom and is attached to the lower surface of the madreporite. Near its top end the stone canal has a lateral pouch called the ampulla of the stone canal and a connection with the cavity of the axial sinus. The madreporite is penetrated by pores and consequently the cavity of the water vascular system (and also the cavity of the axial sinus) is hydrostatically continuous with the sea outside. This continuity allows the pressure inside the water vascular system to be equated with the ambient pressure, without the system changing in shape or volume (Fechter, 1965). The tube feet of *Asterias* are used in walking and in pulling apart the shells of the bivalve molluscs that the starfish eats. Each radial vessel terminates in a single pointed ampulla-less tube foot at the end of the arm.

The perivisceral coelom of *Asterias* is capacious (fig. 2.24), occupying much of the disc and extending along each arm almost to their ends. Into it project the various parts of the gut, the ampullae of the tube feet, the gonads, and the stone canal and axial sinus.

The alimentary canal (fig. 2.26) consists of a central portion inside the disc and five pairs of hepatic caeca in the arms. The central portion consists of a stomach and a short intestine. The stomach is bipartite with a capacious muscular portion, called the cardiac stomach, just inside the mouth, and a thin-walled pyloric stomach above this. The term 'cardiac' is by pure analogy with the vertebrate stomach; it does not relate to the heart of *Asterias*. The cardiac stomach is very muscular and can be protruded through the mouth to envelop the starfish's prey. The pyloric stomach is thin-walled and pentagonal. Each angle receives a hepatic duct by which the respective pair of hepatic caeca is connected to the stomach. From the centre of the pyloric stomach a short straight intestine, provided with a rectal gland, runs to the anus.

The gonads number ten – a pair in each arm (fig. 2.24). Each gonad opens to the outside by a gonopore situated near the base of the arm. The sexes are separate but not visibly different externally. Sperm and ova are shed into the water, chiefly in early summer, and fertilize there. Gemmill (1914, p. 219) reports that shedding is stimulated by the presence of sperm and ova in the ambient water.

The haemal or blood system is very complicated in *Asterias* and everywhere surrounded by coeloms of the so-called perihaemal system (figs 2.25, 2.27). The central and most massive part of the haemal system is the axial organ which partly envelops the stone canal. It is itself enveloped by a cavity, called the axial sinus, of the perihaemal system. The axial organ consists of anastomosing tubes and is interpenetrated by tubular extensions of the

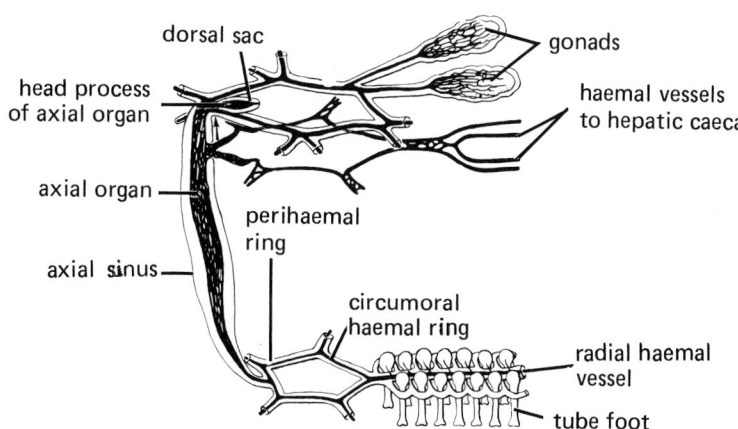

2.27. *Diagram of the haemal and perihaemal systems of an asteroid; from Ubaghs (1967b, fig. 5/2).*

axial sinus. Orally it is connected with a ring-shaped circum-oral haemal vessel from which radial haemal vessels go off into the arms. Aborally, just beneath the madreporite, it sends off a head process. This latter is surrounded, not by part of the perihaemal system, but by a separate and contractile capsule called the dorsal sac or madreporic vesicle. The beat of this dorsal sac has a cycle of about 30 seconds (Gemmill, 1914). It suggests that there is genuine, perhaps oscillatory, flow in the haemal system which would therefore be truly vascular. At its aboral end the axial organ is also connected, indirectly, to the gonads and to the stomach. The perihaemal system, as already mentioned, follows and encloses the haemal system in detail. Round the mouth there are an inner and an outer perihaemal ring which enclose the haemal ring between them, and paired extensions of the outer perihaemal ring pass down each arm, on either side of the radial haemal vessel. Similarly, perihaemal tubes accompany the aboral extensions of the haemal system to the stomach and gonads.

The radial haemal vessel and perihaemal canals are superficial in position forming a ridge in the mid-line of the ambulacral groove (fig. 2.24).

The physiology of the axial organ has been investigated by several authors, mainly in echinoids. It seems to have three main functions: (1) it is concerned in the degeneration of amoebocytes (Millott, 1966; Millott & Vevers, 1968) and presumably expels their degeneration products, by way of the axial sinus, through the madreporite; (2) it is certainly secretory (Millott & Vevers, 1968) and possibly acts as an endocrine gland (Millott, 1967, p. 63) – this endocrine function may explain its anatomical connection to the gonads; (3) it helps to circulate fluid in the haemal system by means of contractile vessels that it contains (Boolootian & Campbell, 1964). However, it is likely that the main pump that sends fluid through the haemal system is the dorsal sac, as already mentioned, and the enclosed head process of the axial organ. Pulsation has been observed in the dorsal sac in all the living echinoderm classes that have it, i.e. in ophiuroids (Narasimhamurti, 1933, p. 79; Gemmill (1919); in echinoids (Boolootian & Campbell, 1964; Narasimhamurti, 1931; Prouho, 1887, p. 331), and in asteroids (Narasimhamurti, 1931; Gemmill, 1914, 1919). The head process and dorsal sac of echinoderms are probably homologous with the heart vesicle and central sinus of hemichordates which have similar function and structure.

The nervous system of *Asterias rubens* (fig. 2.24) consists of three parts: (1) ectoneural, concentrated in the epithelium of the oral surface of the animal ; (2) hyponeural, concentrated in the wall of the perihaemal canals; and (3) apical or endoneural, concentrated in the gut and the coelomic epithelium in contact with certain aboral

muscles. Accounts can be found in Chadwick (1923) and in von Hehn (1970), and Cobb (1970), while Smith (1937) gave a detailed description of the nervous system of the related *Marthasterias glacialis* and Pentreath & Cobb (1972) have discussed the echinoderm nervous system in general.

The ectoneural system is entirely outside the basement membrane of the epidermis. It is mainly sensory and extends, at least in diffuse form, over the whole surface of the animal. However, it is concentrated on the oral surface where it develops into five radial nerves connected by a circum-oral ring. The radial nerves are thickenings of the epidermis in the mid-line of each ambulacral groove, external to the perihaemal canals (fig. 2.24). They are thus V-shaped in section and help to form the ridge in the mid-line of each ambulacral groove. The circum-oral ectoneural ring runs round the edge of the peristome, connecting the radial nerves together. Each radial nerve consists, as described for *Marthasterias glacialis* by Smith (1937), of three types of cell (fig. 2.28a,b): (1) supporting or glial cells; (2) sensory cells; and (3) ganglion cells (neurons). The supporting cells are basically like the epithelial cells of the rest of the surface. They extend from the basal membrane to the surface cuticle, which they secrete. They have expanded, nucleated outer ends which are in contact with each other and which bear a cilium. The expanded terminations taper inwards and send thin processes to the basement membrane which are strengthened by a fibre. Scattered among the expanded outer terminations of the supporting cells are the nucleated bodies of sensory cells. These cell bodies send a perceptor process outwards and a long central process inwards which synapses with the ganglion cells. The latter, finally, are of two types – multipolar and bipolar. The multipolar cells are concentrated just beneath the expanded ends of the supporting cells, receive the central processes of the sensory cells and presumably synapse with each other and with the bipolar cells. The latter form most of the deep part of the nerve and tend to run along the axis of the nerves, in bundles between the bases of the supporting cells. The circum-oral ring has much the same structure as the radial nerves, except that the bipolar ganglion cells are elongated concentric to the centre of the ring.

The ectoneural nerves are remarkably like normal epidermis (fig. 2.29). This also has epithelial cells with strengthening fibrillae, though the cells are cylindrical in shape rather than inwardly tapering. Also sensory cells are present in normal epidermis and send their central processes in among the epithelial cells and a few ganglion cells exist in the deeper parts, above the basement membrane. The radial nerves and circum-oral can therefore be seen as specializations of the epidermis, characterized mainly by the large number and relatively constant orientation of

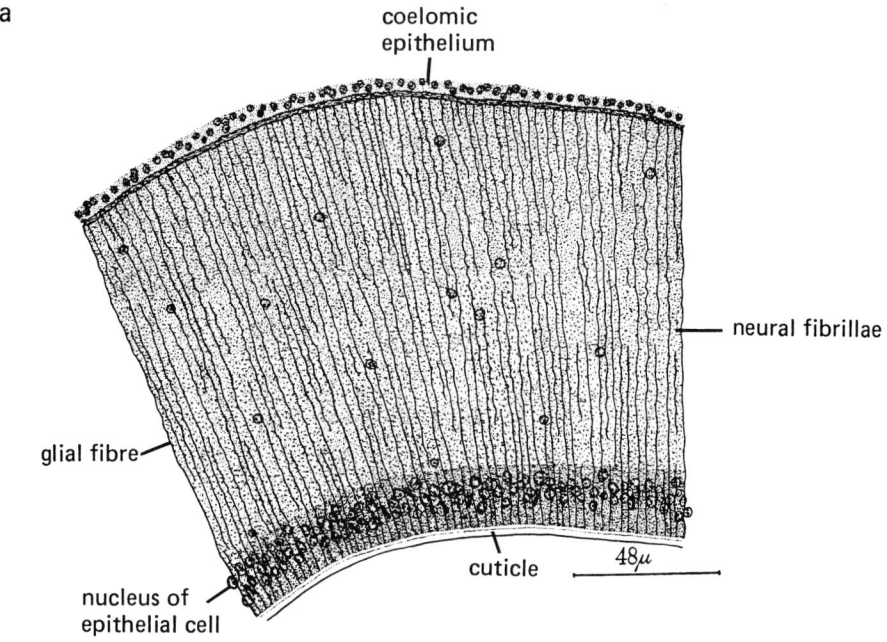

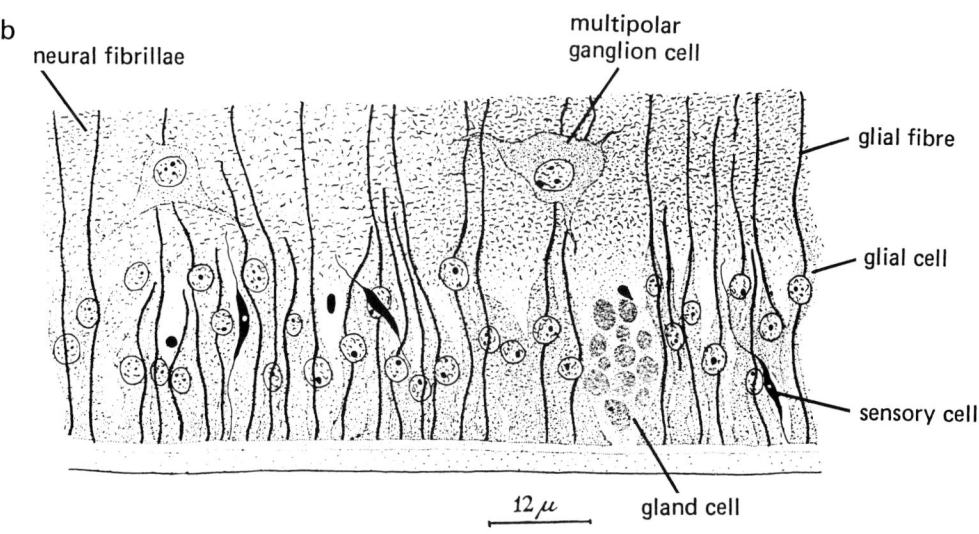

2.28. *Histology of transverse section through the radial nerve of an asteroid* (Marthasterias glacialis); *a) entire radial nerve; b) superficial parts of radial nerve; from Smith (1937, figs 1, 5).*

the ganglion cells. Indeed, the ectoneural nervous system of a starfish is histologically probably very like the primitive nervous system from which the vertebrate system evolved. Thus the supporting or glial cells, which supply the framework in which the ganglion cells run, are like the glial and ependymal cells of a vertebrate dorsal nerve cord (fig. 2.30). And the sensory cells with their peripherally situated cell bodies and long central processes, are like the

olfactory sense cells of a vertebrate. These starfish sensory cells, however, seem to be generalized in response, reacting to touch, light and smell (Pentreath & Cobb, 1972). The primitive nature of the ectoneural system justifies this rather detailed exposition.

Each radial nerve ends distally in an orange pad, called the optic cushion, just beneath the terminal tube foot (fig. 2.22). This optic cushion contains a number of optic

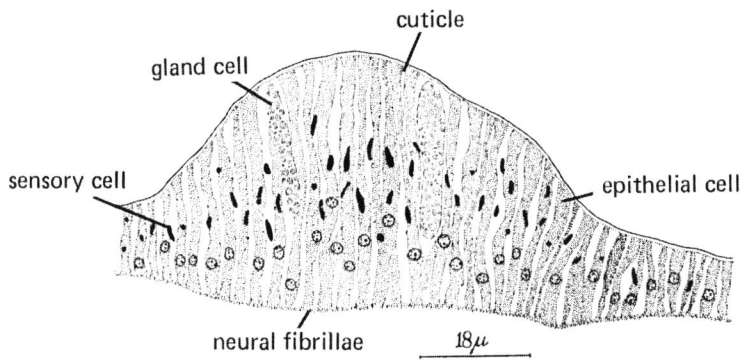

2.29. *Histology of transverse section through the entire aboral ectoderm of* Marthasterias glacialis; *from Smith (1937, fig. 8).*

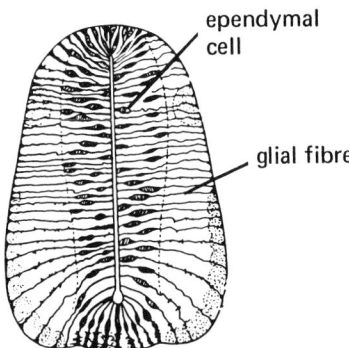

2.30. *Transverse section of dorsal nerve cord of three-day-old chick embryo to show fundamental resemblance to the radial nerve of a starfish (cf. fig. 2.28); from Nelsen (1953, fig. 353F).*

cups which seem to be directionally responsive to light (Hyman, 1959, p. 359). The orange colour of the pads comes from the pigment round each optic cup.

The hyponeural system of *Asterias rubens* (fig. 2.24) is developed in coelomic epithelium and is motor in function. Its most important parts are known as Lange's nerves and are situated just above (internal to) the radial ectoneural nerves in the oral wall of the radial perihaemal canals. They are always separated from the ectoneural nerves by the thickness of the epidermal basement membrane, but this may be only 0·01 µm (Cobb, 1970, p. 466). There are never any true synapses between the two systems. The fibres of Lange's nerve are mainly transverse to the arms. They innervate the ambulacral muscles (between the ambulacral plates) by sending nerves to them. And they innervate the ampullae of the tube feet by receiving processes of muscle tissue (the so-called muscle tails) from the ampullae. Such muscle tails are widespread in echinoderms, being found, for example in the pedicellariae of

echinoids (Cobb & Laverack, 1967). This is of interest, for muscle tails also occur in amphioxus connecting the somitic muscles with the dorsal nerve cord. There is no circum-oral ring of the hyponeural system but neighbouring Lange's nerves of adjacent ambulacra are connected by interradial nerves. The ontogenetic origin of the hyponeural nerves is unknown.

The apical nervous system (fig. 2.24) is developed in the epithelium of the perivisceral coelom, in the dorsal midline of each arm just internal to an apical muscle whose function is to flex the arm upwards. This apical system is probably continuous with the nervous system of the gut wall.

The embryology of *Asterias rubens* was described in detail by Gemmill (1914) and some of its earlier stages illustrated by Chadwick (1923). There is a good general account in MacBride (1914) and Gemmill (1915) described the strange and phylogenetically significant phenomenon of double hydrocoel in the species. The best fairly recent discussion of the phylogenetic causes of echinoderm ontogeny is found in von Ubisch (1957, 1958).

The ovum is slightly less than 0·2 mm in diameter, with uniformly distributed yolk. It is fertilized, free in the sea water, forms a fertilization membrane and begins to divide. The first two cleavages are meridional and the third equatorial with the four upper cells smaller than the four lower ones. Cleavage then continues, to form an almost spherical hollow blastula which becomes ciliated and is released from the fertilization membrane. Its wall is one cell thick. Shortly afterwards the blastula lengthens slightly, becoming ellipsoidal in shape, and then one end of the ellipsoid, the future posterior end of the larva, flattens and its constituent cells lengthen. Then the flattened end invaginates, the invaginated pouch becomes the archenteron and the blastula has become a gastrula (fig. 2.31). The blastopore is the anus of the larvae, so development is deuterostomatous.

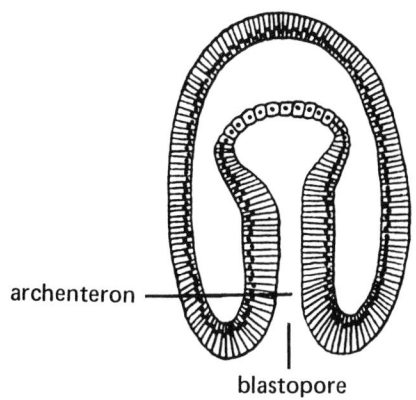

2.31. Asterias rubens. *Optical section of a gastrula; from Chadwick (1923, fig. III).*

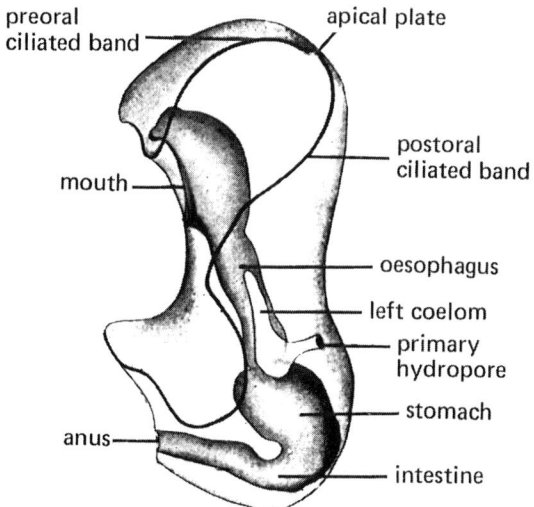

2.32. Asterias rubens. *Ten-day-old larva from left; from Gemmill (1914, pl. 18, fig. 2).*

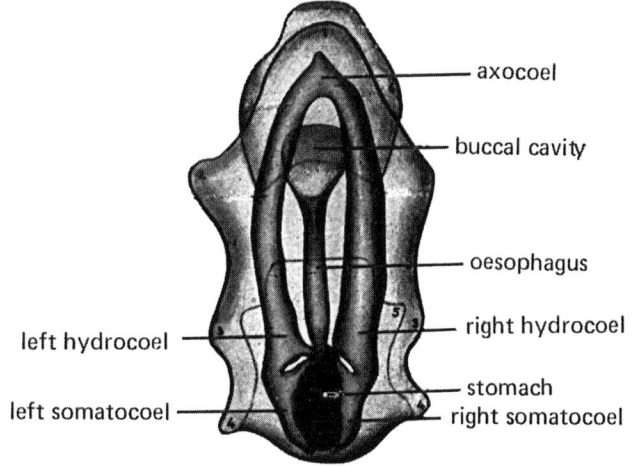

2.33. Asterias rubens. *21-day-old larva from dorsal surface. The right and left coeloms have elongated and fused anteriorly. The left and right somatocoels are symmetrical on either side of the stomach. A hydropore has appeared on the left but not on the right; from Gemmill (1914, pl. 18, fig. 5).*

The front end of the archenteron, farthest from the anus or blastopore, evaginates on left and right to form the left and right coeloms and amoeboid mesenchyme cells wander off into the blastocoel from this evagination. The remaining archenteron forms the larval gut, divided into oesophagus, stomach and intestine. The anus migrates towards the ventral side of the larva. On the ventral face near the oesophagus a field of ectoderm called the stomodaeum invaginates slightly, and the mouth arises by perforation where oesophagus and stomodaeum meet, in correct deuterostomatous fashion.

The right and left coeloms extend both forwards and rearwards and the left one only (in normal development) sends an extension to just left of the dorsal mid-line (fig. 2.32). This extension soon opens at the surface to become the primary hydropore. The right and left coeloms fuse anteriorly, in front of the oesophagus. Posteriorly both coeloms become divided into a posterior part (left and right somatocoels) and an anterior part (left and right axohydrocoels). Both axohydrocoels can be regarded as consisting of an anterior axocoel region and a posterior hydrocoel region, but there is no clear separation between axocoel and hydrocoel (fig. 2.33). The left and right axocoel regions communicate with each other across the median plane anteriorly. The left and right somatocoels are on either side of the stomach. A group of mesenchyme cells in the median line just right of the hydropore forms a little capsule that begins to pulsate. It becomes the definitive dorsal sac.

External changes in the larva are happening at the same time. Some of the cilia elongate along a tract which at first is topologically a ring surrounding the mouth and anterior to the anus. The ring is not circular, however, but consists of two transverse bands (pre-oral, post-oral) connected together by an almost straight band on right and left. Later the left and right halves of the ring meet at a point on the antero-dorsal surface of the larva, become interrupted there and fuse left to right so as to give an anterior pre-oral topological ring, and a posterior post-oral one. At the antero-dorsal point where the two rings still almost touch there is a thickened patch of ectoderm called the apical plate (fig. 2.32). The resulting larva, known as an auricularia, is remarkably like the tornaria of an enteropneust except that the apical plate of *Asterias rubens* has no tuft of cilia, nor eyes, there is no posterior ring of cilia (telotroch) round the anus and the latter opens ventrally rather than posteriorly (fig. 2.34). As the larva grows older the ciliated tracts become extensively scalloped, being borne on several sets of mobile, muscular arms (figs 2.35a,b; 2.36).

The larva has been symmetrical, up till now, except that the hydropore is developed on the left side only (fig. 2.33). At this time, however, the left somatocoel begins to encroach on the right side of the larva by sending a ventral horn round the ventral face of the stomach to fuse with the right hydrocoel region of the right axohydrocoel (fig. 2.35a). Slightly later it sends a dorsal horn forward over

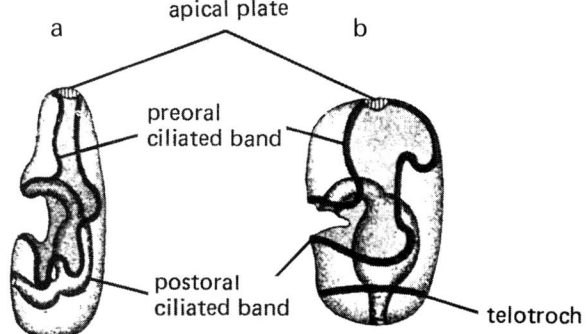

2.34. *Left aspects of the auricularia larva of a) a starfish and b) the tornaria larva of an enteropneust to show the close resemblance; from MacBride (1914, fig. 389, A–B).*

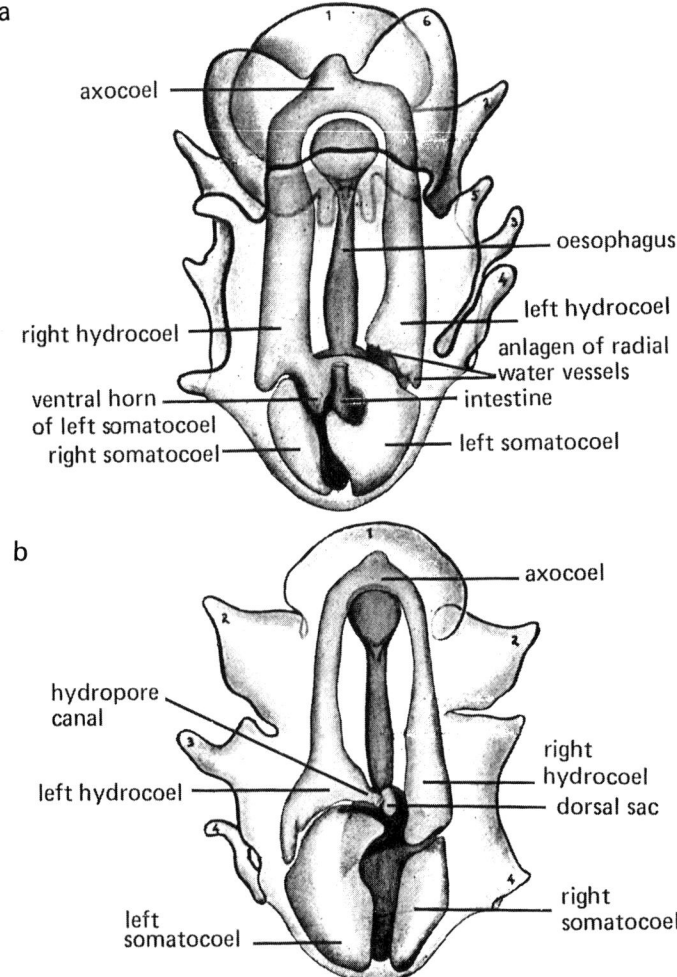

2.35. Asterias rubens. *a) 32-day-old larva in ventral aspect. The left somatocoel has sent out a ventral horn in front of the rectum and this has fused with the right hydrocoel preventing the further development of the latter. The left hydrocoel is beginning to develop into a water-vascular system, for it shows the anlagen of radial water vessels; from Gemmill (1914, pl. 18, fig. 7). b) Same 32-day-old larva as in fig. 2.35a but in dorsal aspect. Note the dorsal sac just right of the hydropore; from Gemmill (1914, pl. 18, fig. 7).*

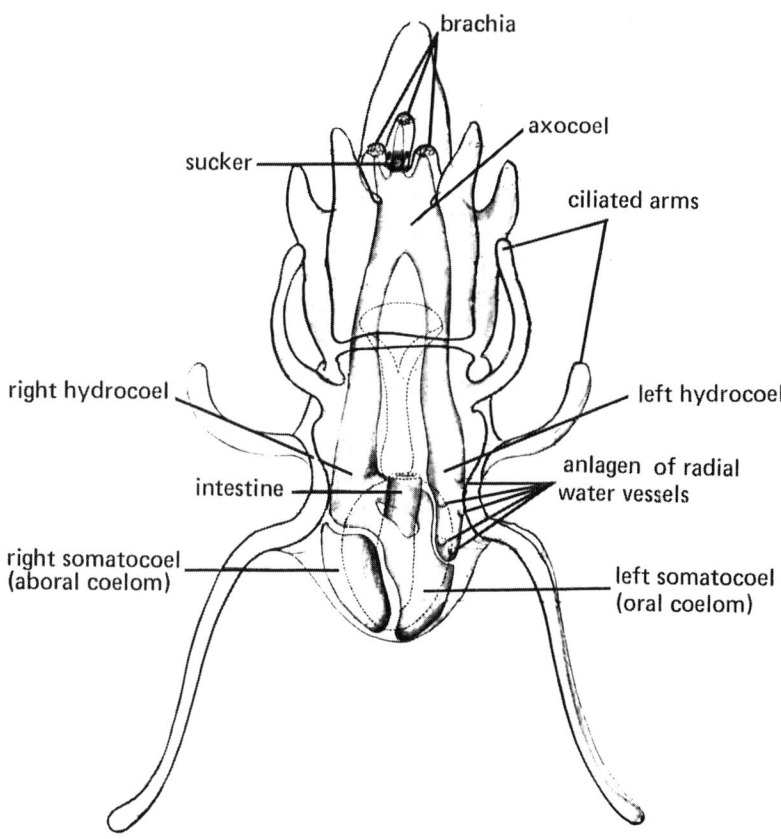

2.36. Asterias rubens. 38-day-old larva in ventral aspect. Three brachia and a sucker have developed at the anterior end, all in contact with extensions of the axocoel; from Gemmill (1914, pl. 19, fig. 9).

the dorsal face of the stomach and this horn comes to open into the left hydrocoel region (fig. 2.36). The left hydro-coel region develops five lobes at its posterior edge, being the anlagen of the radial water vessels (fig. 2.36). Soon the left somatocoel encircles the oesophagus completely. In sympathy with the five-lobed water vascular system the posterior outline of the larva also becomes five-lobed so that a 'starfish rudiment' begins to become evident at the hind end of the larva and the plates of the definitive skeleton begin to appear (fig. 2.37).

In preparation for attachment three brachia develop near the anterior end of the larva (figs 2.36, 2.37). They are left, right and median, contain extensions of the anterior median part of the axohydrocoels and have papillae at their ends for temporary attachment. There is a sucker, for longer lasting attachment, lodged in the valley between the three brachia.

At attachment the three brachia grip a hard surface, such as a frond of seaweed, by means of their terminal papillae and the sucker between them is pressed against the surface

and glues itself to it. The whole anterior part of the larva with the sucker, mouth and oesophagus rotates leftwards and rearwards relative to the median plane of the larva i.e. moves to the oral surface of the starfish rudiment. This movement closes the larval mouth; and the larval anus, which has come to lie on the oral face of the developing starfish, likewise closes (fig. 2.38a). The water-vascular system has by now acquired its basic definitive form with a circular circum-oral vessel from which five radial vessels run off, and these radial vessels have begun to develop tube feet. These latter are active and able to attach to the hard substrate. Finally the little starfish walks with its tube feet away from the point of attachment, the attachment stalk breaks under the tension, a new mouth is perforated inside the circum-oral ring, and a new anus develops on the aboral surface. Thus the meta-morphosis which converts the larva into a tiny starfish is complete.

The fate of the coeloms at metamorphosis is complex. The left somatocoel of the larva now partly underlies the

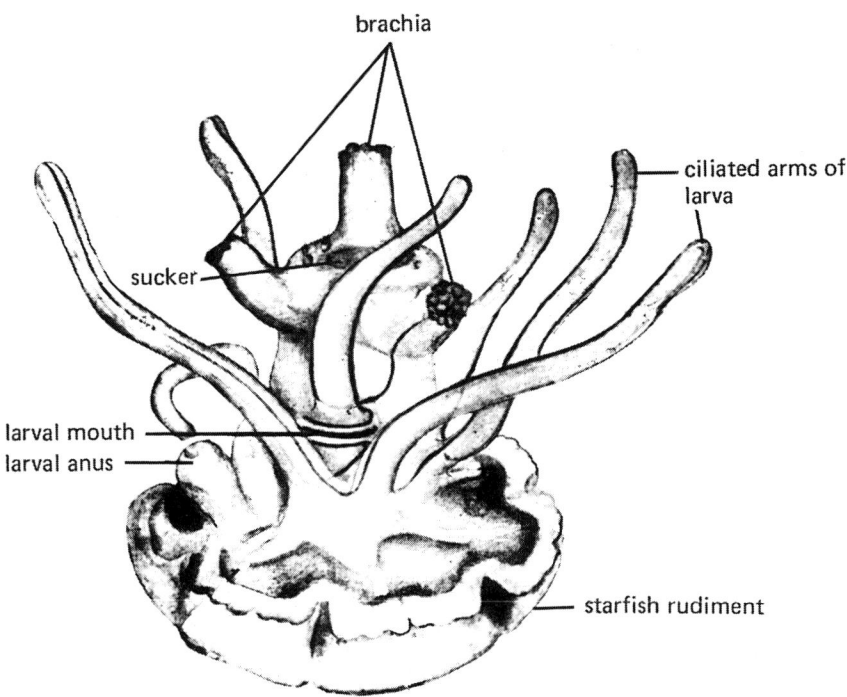

brachia

ciliated arms of
larva

sucker

larval mouth
larval anus

starfish rudiment

2.37. Asterias rubens. *Metamorphosing larva, ventral aspect. The brachia have rotated towards the left side of the larva, closing the mouth; from Gemmill (1914, pl. 21, fig. 13).*

stomach, extending onto the oral face of the latter, while the right somatocoel overlies the stomach on the aboral face (fig. 2.38b). In other words, the plane of extension of the starfish is achieved by rotating the larval sagittal plane through 90° on attachment, so that it lies parallel to the substrate with the right somatocoel upwards. The communication between the left somatocoel and the right hydrocoel region is severed and the right hydrocoel region of the coelom disappears without trace. The communication between the left somatocoel and the left hydrocoel region is also severed, but the left hydrocoel region does not disappear. On the contrary, in normal development it gives rise to the whole water-vascular system of the adult. It also produces the inner oral perihaemal ring of the adult, whereas the rest of the perihaemal system, including the axial sinus and stone canal, arises from the left somatocoel. As the starfish gets bigger the mesentery that originally separated the left and right somatocoels (oral and aboral coeloms) breaks down giving the extensive undivided perivisceral coelom of the adult.

To sum up, in the normal development of *Asterias rubens* an initial, bilaterally symmetrical, auricularia larva attaches to a substrate, converts only the left hydrocoel region into the definitive water-vascular system, develops a pentameral symmetry which is built round that system,

and undergoes a rotation, whereby the larval sagittal plane rotates through a right angle to lie parallel to the substrate, larval right side upwards.

Three abnormal conditions help in interpreting this embryology, being the states of double hydrocoel, double hydropore and inverse larvae. Double hydrocoel, according to Gemmill (1915) is sometimes frequent in larvae of *Asterias rubens* (fig. 2.39). It happens when the ventral horn of the left somatocoel grows less far to the right than usual, perhaps because the stomach is unusually large, and consequently does not fuse with the right hydrocoel region of the right axohydrocoel. As a result this region develops five protuberances at its posterior edge, just as the left hydrocoel does, and these are the anlagen of five right radial water vessels which normally never develop. Gemmill succeeded in rearing a few such specimens of *Asterias rubens* past metamorphosis (fig. 2.40). The resulting starfish has five right radial and five left radial vessels extending respectively from a pair of semicircular or arcuate water vessels near the mouth and with each radial water vessel borne on its own arm. The left semicircular or arcuate vessel is the homologue of the circum-oral water vessel of a normal starfish, whereas the right one has no normal equivalent. Usually the right water-vascular system is smaller and less well developed than the left (fig.

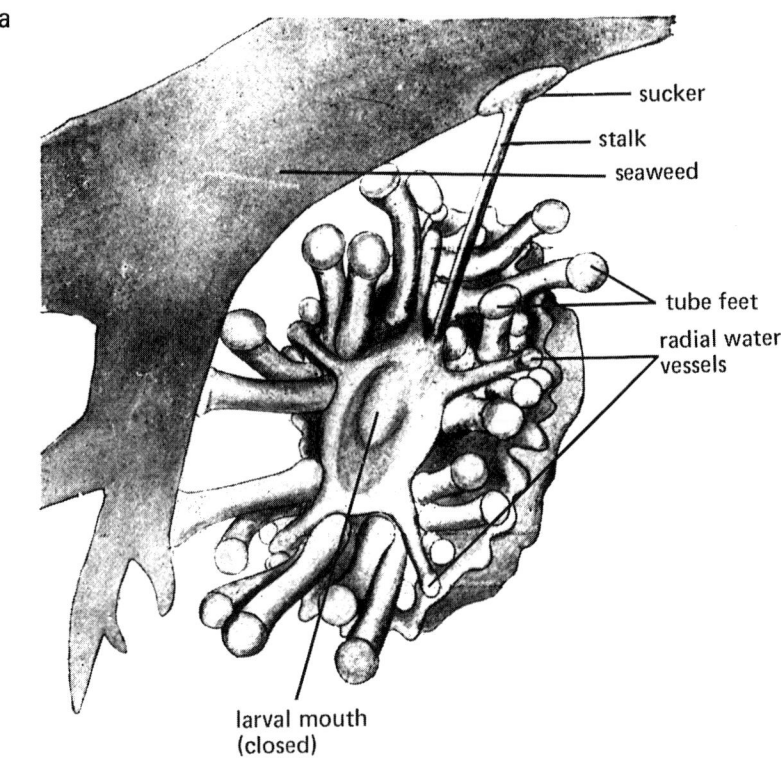

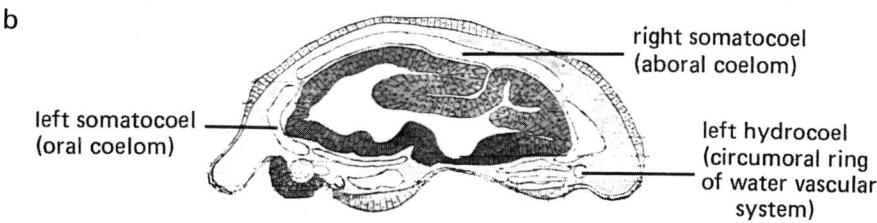

2.38. Asterias rubens. *Young starfish six or seven days older than in fig. 2.37. a) The starfish is pulling by means of its tube feet against the stalk and the latter will finally break; from Gemmill (1914, pl. 21, fig. 16). b) Transverse section through starfish of about the same stage as fig. 2.38a. Note that the right somatocoel is now situated aboral to the alimentary canal while the left somatocoel is more oral in position; from Gemmill (1914, pl. 24, fig. 31).*

2.40), but sometimes they are equal. Double hydropore, which can also be frequent in *Asterias rubens*, happens when the right coelom, as well as the left, sends an extension dorsalwards which opens to the surface just right of the larval mid-line (fig. 2.41). Both phenomena emphasize that an *Asterias rubens* larva has the potential to develop more symmetrically than it usually does. The two phenomena are not statistically associated, i.e. a double-hydropore individual is no more nor less likely to exhibit double hydrocoel than is a single-hydropore individual. Inverse larvae occur when the right hydrocoel gives rise to the definitive water-vascular system, but the left one does not (fig. 2.42). Such larvae, known in asteroids, ophiuroids and echinoids may give rise to adults indistinguishable from the normal ones (Oshima, 1922, Newman, 1921, 1925, von Ubisch, 1957, 1958). All these abnormalities indicate that the right side of the larva has the potential to produce structures which normally arise from the left side. Before discussing the phylogenetic significance of this embryology it is necessary to consider the crinoids.

The crinoids can be exemplified by the species *Antedon biflda* which is common in shallow British waters. It was

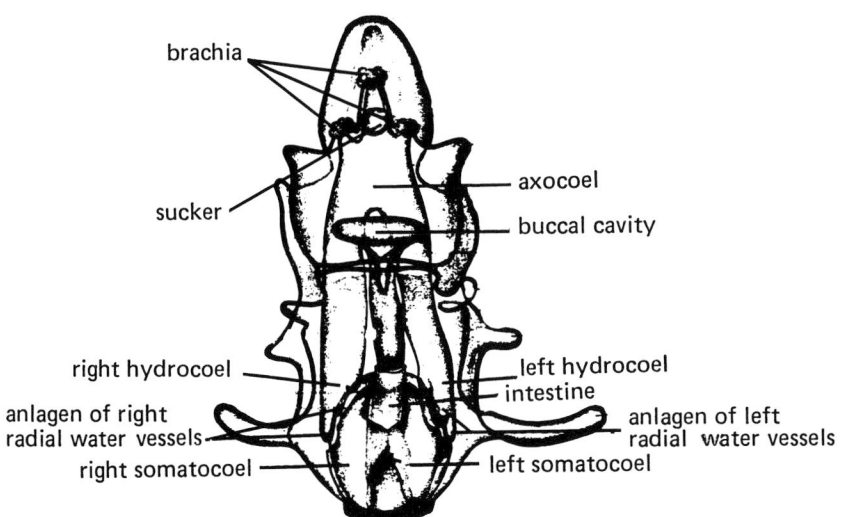

2.39. Asterias rubens. *Ventral view of a double-hydrocoel larva (compare normal larva in fig. 2.36). Right and left somatocoels have developed symmetrically and the left somatocoel has thus failed to fuse with the right hydrocoel. As a result the right hydrocoel, like the left one, has developed anlagen of radial water vessels; from Gemmill (1915, pl. 6, fig. 2).*

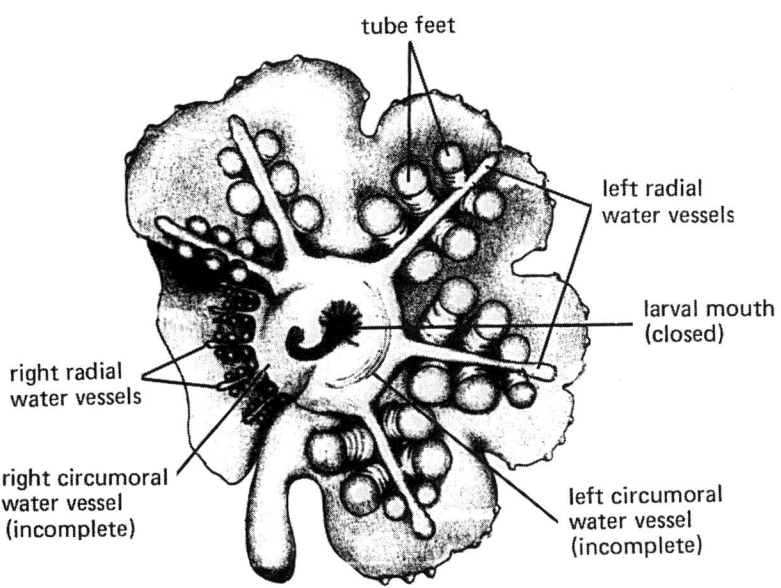

2.40. Asterias rubens. *A double-hydrocoel post-larva. The animal is derived from a double-hydrocoel larva and has five normal radial water vessels belonging to the left hydrocoel and five abnormal, stunted radial water vessels belonging to the right hydrocoel; from Gemmill (1915, pl. 7, fig. 9).*

well described at an elementary level by Chadwick (1907) and its embryology is also well known, though not recently investigated (Bury, 1888, Seeliger, 1892; MacBride, 1914, and Dawydoff, 1928 for general accounts).

Antedon consists of a small central disc or theca from which extend five pairs of feathery arms (figs 2.43, 2.44,

2.45). The whole animal is heavily skeletized, the skeleton consisting, as usual for echinoderms, of calcite plates of stereom mesh, each of which is a single crystal. The upper surface of the disc (fig. 2.44) is oral, by contrast with *Asterias*, and near its centre is the mouth. The lower surface of the disc (fig. 2.45) is aboral and carries many

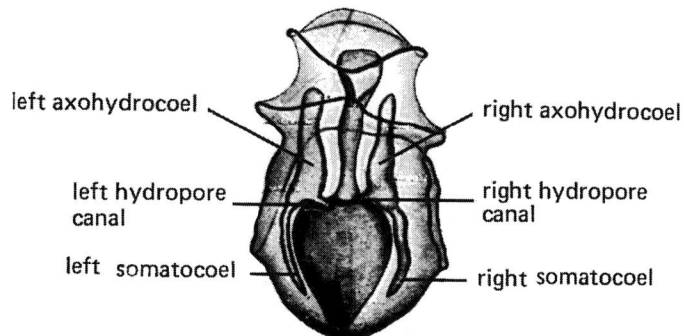

2.41. Asterias rubens. *An 18-day-old double-hydropore larva in dorsal aspect (compare slightly older larva in fig. 2.33); from Gemmill (1914, pl. 18, fig. 4).*

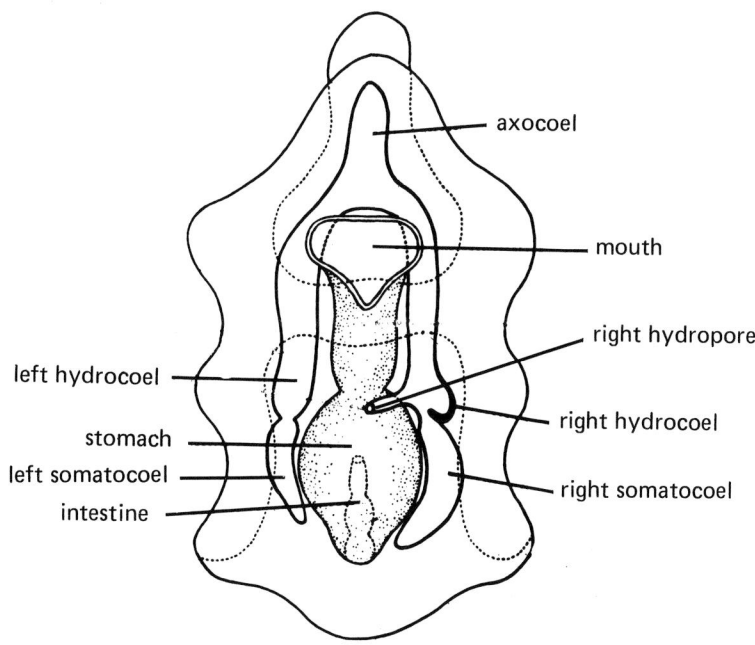

2.42. *Larva of the starfish* Patiria miniata *in dorsal aspect showing reversed asymmetry with the hydropore and water-vascular system developing on the right side rather than the left; from Newman (1925, fig. 11).*

jointed appendages, known as cirri, by which the *Antedon* clings to the substrate. The feathery look of the arms is caused by alternating terminal twigs known as pinnules. Each pinnule carries a food groove on the oral surface; the grooves are tributary to grooves on the oral surface of the central trunk of the arm; and the grooves of each pair of arms unite on the oral surface of the disc to form five grooves that run to the mouth. The anus is also on the oral surface of the disc, at the top of a protuberance called the

2.43. Antedon bifida. *General view from the side to show normal posture; from Chadwick (1907, pl. 1, fig. 1).*

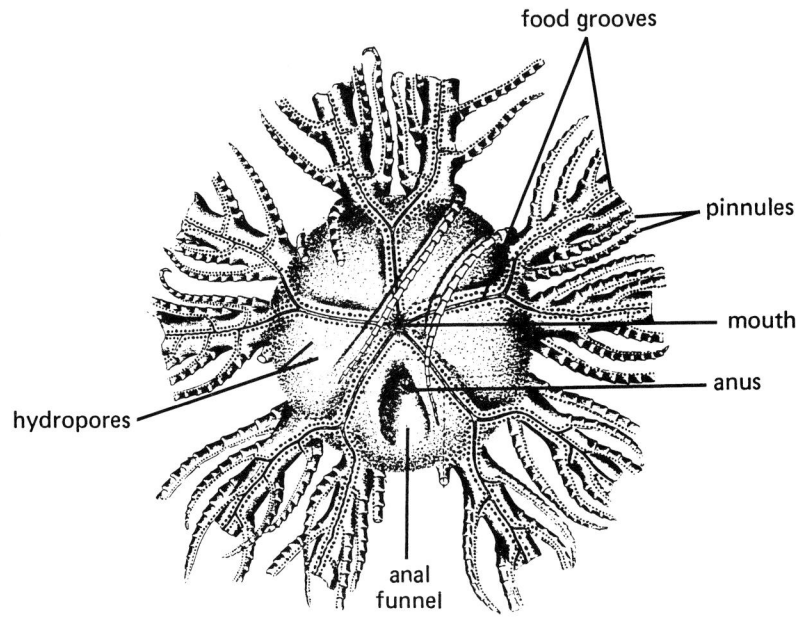

2.44. Antedon bifida. *Oral, upper aspect of disc and adjacent parts of arms; from Chadwick (1907, pl. 2, fig. 24).*

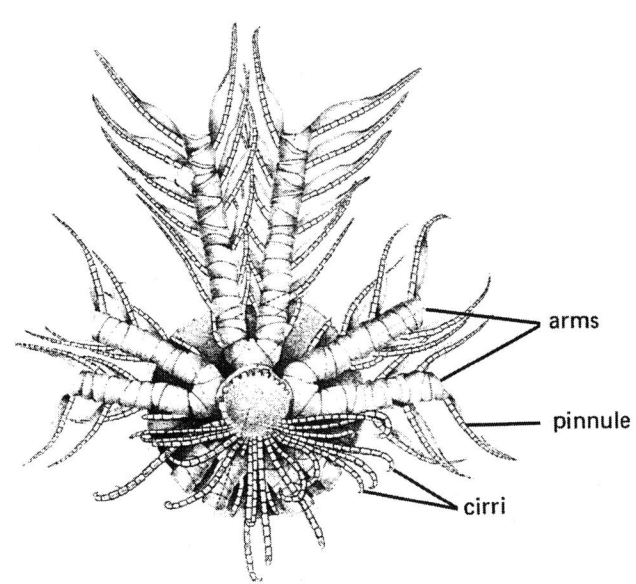

2.45. Antedon bifida. *Aboral, lower aspect of disc and adjacent parts of arms; from Chadwick (1907, pl. 1, fig. 3).*

anal funnel, and situated in an interradius between two of the five central food grooves. The soft oral wall of the disc is penetrated by a large number of ciliated hydropores – there is no madreporic plate.

The gut (figs 2.46, 2.47) is located inside the disc in a perivisceral coelom which is criss-crossed with slender bridges of soft tissue. The central mouth leads into an oesophagus and then into an intestine which runs through

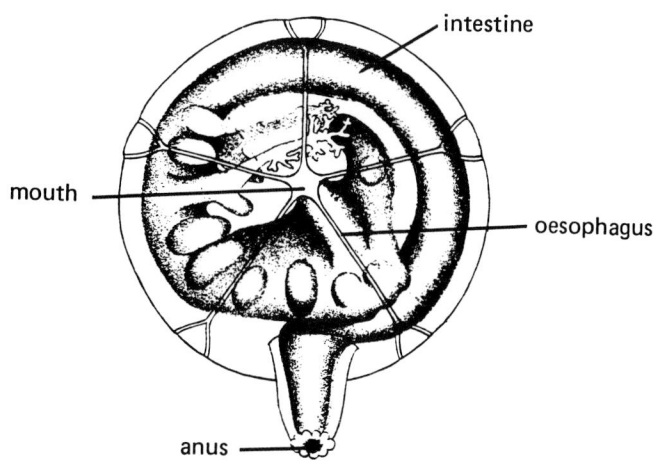

2.46. Antedon bifida. *Gut in oral aspect; from Chadwick (1907, pl. 2, fig. 23).*

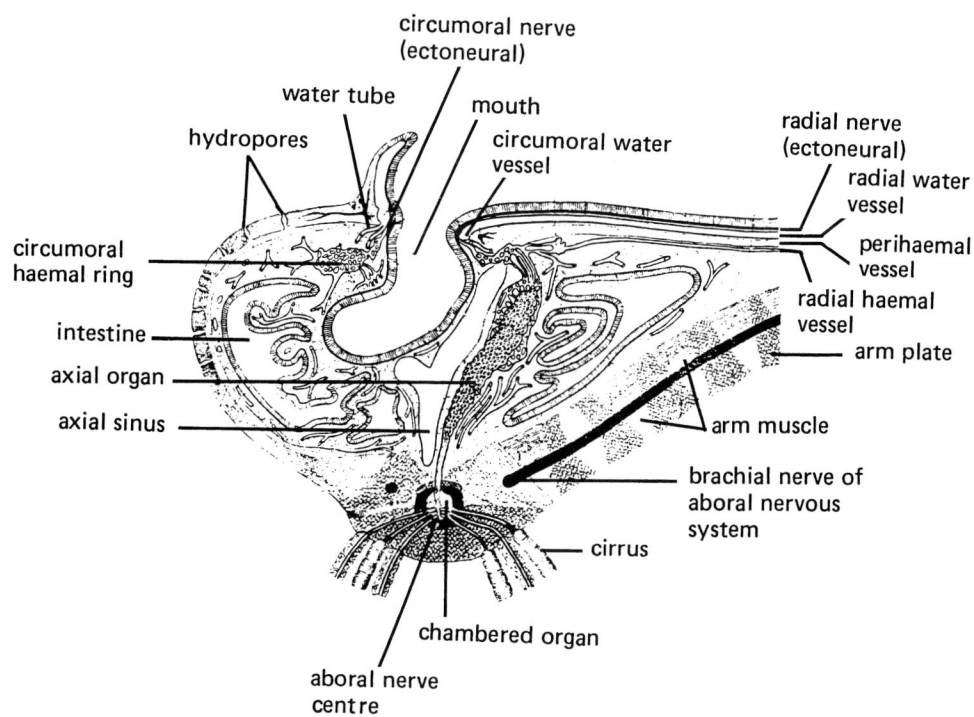

2.47. Antedon bifida. *Vertical section through the theca and the base of one arm; from Chadwick (1907, pl. 5, fig. 52).*

rather more than 360° clockwise round the disc (as seen from the oral surface) to end in the anus which is carried on the anal funnel. The gut encircles the central structure of the haemal system (the axial sinus). In *Antedon*, as in *Asterias*, there are three nervous systems – ectoneural, hyponeural and apical or aboral. The aboral nervous system, however, is much better developed than in *Asterias*. It consists of an aboral nerve centre and nerves supplied to the arms. The arms have muscles between the plates and *Antedon* can swim by waving the arms up and down. The large size of the aboral nervous system is presumably related to swimming and to the clasping cirri. The

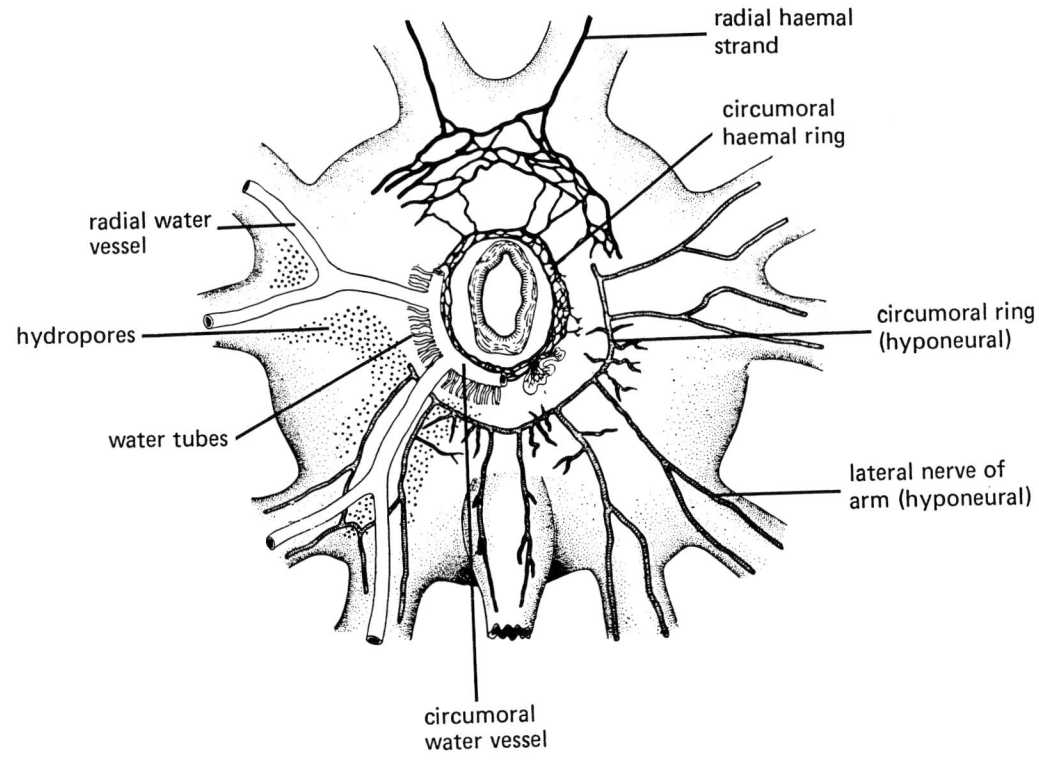

2.48. Antedon bifida. *Water-vascular system, haemal system and hyponeural system of the theca in oral aspect; from Cuénot (1948, fig. 52).*

cirri are built around small tubes which open into a hollow chambered organ situated inside the aboral mid-point of the disc. The chambered organ is so called because it is divided into five chambers. Each of these is a remnant of a closed tube that extends throughout the stem of the larva. Sexes are separate and the gonads are located in the pinnules of the arms.

The water-vascular system (figs 2.47, 2.48) consists, as usual in echinoderms, of a circum-oral ring, from which five radial water vessels go off situated beneath the five food grooves in the disc. As the food grooves divide, so the radial water vessels divide beneath them, to terminate like the food groove at the ends of the pinnules. The radial vessels give rise on either side to tube feet (sometimes called tentacles) in groups of three (Nichols, 1960). Unlike *Asterias*, however, the tube feet are used in filter feeding, not in walking. Small prey that collide with the tube feet are thrown by them into the food groove and carried, entrapped in mucus, by means of the cilia of the food grooves, to the mouth. From the circum-oral water vessel, which immediately encircles the oesophagus, a large number of open-ended water tubes extend into the peri-

visceral coelom (fig. 2.48). The ciliated hydropores that penetrate the oral surface of the disc likewise open into the perivisceral coelom. Consequently the cavity of the water-vascular system is connected to the ambient sea water as in other echinoderms, but the connection is more indirect, via water tubes, perivisceral coelom and ciliated hydropores. (It is also indirect in other echinoderms, for in them the water-vascular system is connected with the sea water by way of the axial sinus.)

The perihaemal system of the arms consists of a pair of tubes beneath the water-vascular system. These tubes pass inwards into the oral end of a vertical cavity known as the axial sinus in the central axis of the disc. The axial sinus is closely associated with the axial organ which sends haemal extensions into the arms just beneath the membrane that separates the paired perihaemal tubes. Extensions of the perivisceral coelom also pass into the arms. In correlation with the lack of a madreporic plate there is no madreporic vesicle (dorsal sac).

Antedon is a specialized crinoid for in the adult it is without a stem. It has a stemmed larva however, as explained later, and has relatives, known by the popular name of sea lilies, which are stemmed in the adult. Unfortunately the embryology of these stemmed forms is unknown.

The descriptive embryology of *Antedon*, as already mentioned, is based on work almost 100 years old, but there is no reason to doubt its correctness and a comparison with the embryology of *Asterias* is phylogenetically illuminating. As already mentioned, males and females carry their respective gonads on the terminal twigs (pinnules) of the arms. When the males emit sperm into the water from their testes, the ova migrate to the surface of the ovaries on the pinnules of the females and are fertilized there. Yolk is rather evenly distributed in the ovum and resulting embryo which are therefore opaque. Changes inside them can be followed only by means of serial sections.

The fertilized ovum divides in usual fashion – the first two cleavages are meridional, the third equatorial, and the third cleavage divides the embryo into four smaller cells at the animal pole of the egg and four larger ones at the vegetal pole. Cleavage continues, producing a hollow, spherical blastula with a wall one-cell thick.

At gastrulation one end of the blastula flattens and then invaginates to form the archenteron (fig. 2.49). The blastopore is a crescentic slit and, unlike *Asterias*, it soon

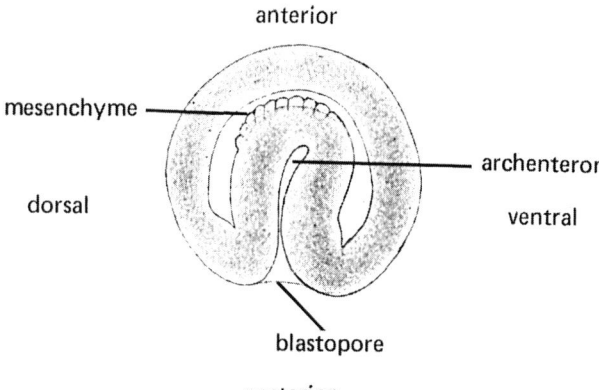

2.49. *Antedon bifida. Lateral aspect of gastrula in optical section; from Seeliger (1892, fig. 33).*

closes. It is situated at the posterior end of the embryo. Mesenchyme invades the blastocoel spreading out from the anterior end of the archenteron.

Then the archenteron divides. It constricts transversely to form a larger anterior and a smaller posterior vesicle. The posterior vesicle, which will form the somatocoels, stretches transverse to the embryo, becoming sausage-shaped (fig. 2.50a, d). The anterior vesicle, which will form

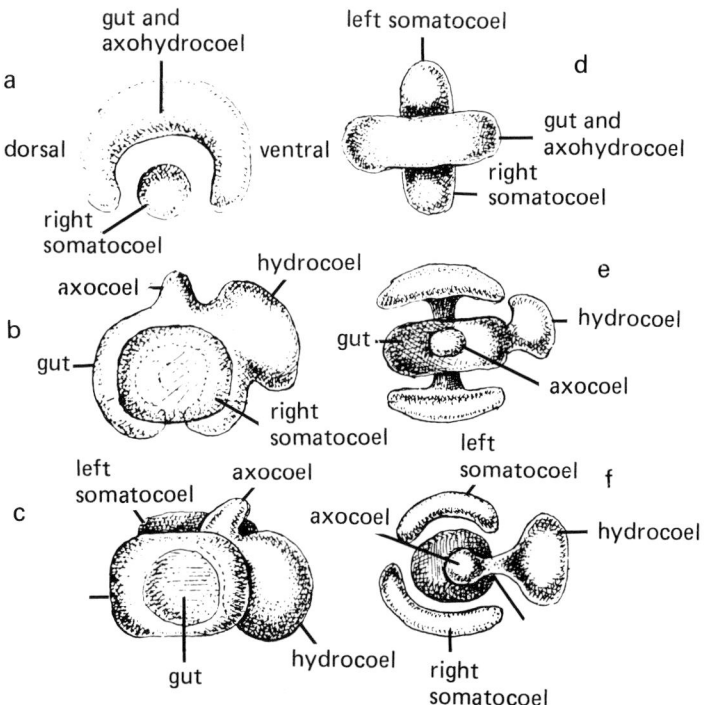

2.50. *Antedon bifida. Division of archenteron and production of coeloms. a), b) and c) Successive stages in left aspect, anterior end of larva upwards. d), e) and f) Corresponding stages seen from anterior end of larva. All from Cuénot.*

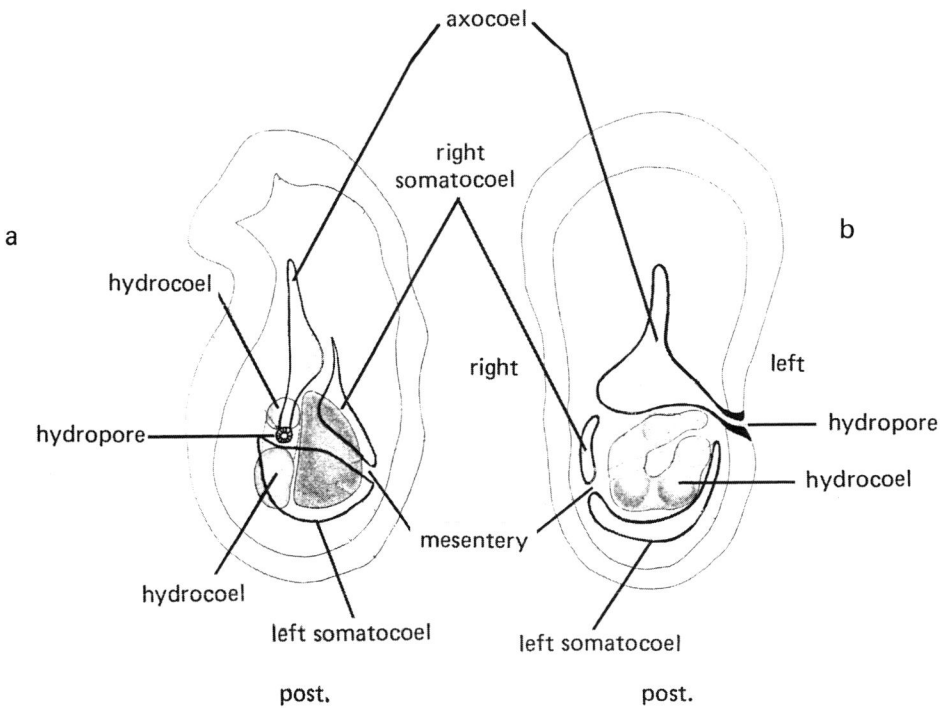

2.51. Antedon bifida. *a) Diagram of left aspect of 7-day-old larva; note the anterior extension of the axocoel towards the anterior end and attachment disc of the larva and the independence of the axocoel and hydropore from the hydrocoel. b) Diagram of ventral aspect of same larva; the gut is omitted and the edges of the right and left somatocoels in the dorsal part of the larva are shown by dotted lines; from Bury (1888, pl. 47, figs 60, 61).*

the axocoel, hydrocoel and gut, grows rearwards in the median plane of the embryo as a crescent round this sausage, with a dorsal and a ventral horn. These two horns meet each other behind the somatocoelic vesicle. The latter continues to stretch sideways, inflates at the ends to give an hour-glass shape (fig. 2.50b, e) and then separates into two symmetrically disposed bladders – the right and left somatocoels. Meanwhile the anterior and ventral part of the crescentic anterior vesicle is itself differentiating into an anterior median axocoel and a more posterior, more ventral and larger hydrocoel situated mainly left of the median plane. The axocoel develops a long extension anteriorly while another extension grows out of the axocoel leftwards and opens on the left face of the larva or embryo as the primary hydropore (fig. 2.51). The complex of coeloms and gut concentrated in the posterior part of the larva is known as the visceral mass.

Meanwhile, the embryo has become a ciliated larva and escaped from the ovary into the sea. The larva is approximately an ellipsoid of revolution with five rings of cilia transverse to its length and a tuft of cilia at its anterior end (fig. 2.52). At the base of this tuft there is a nervous apical plate. Somewhat ventral to the tuft is a fixation sucker. On

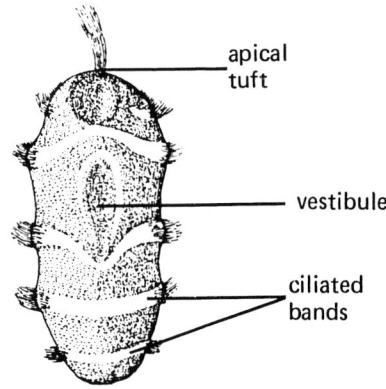

2.52. Antedon bifida. *External appearance of a ciliated larva; from Korschelt & Heider (1936, fig. 527, p. 494, after Reichensperger).*

the ventral face of the larva is a stomodaeal depression which, however, in larval life is never perforated to form a mouth. The primary hydropore is situated on the left side of the larva, rather posterior in position. There is a general resemblance to the larva of *Asterias* (apical plate; antero-ventral sucker; ventral stomodaeum; hydropore on left

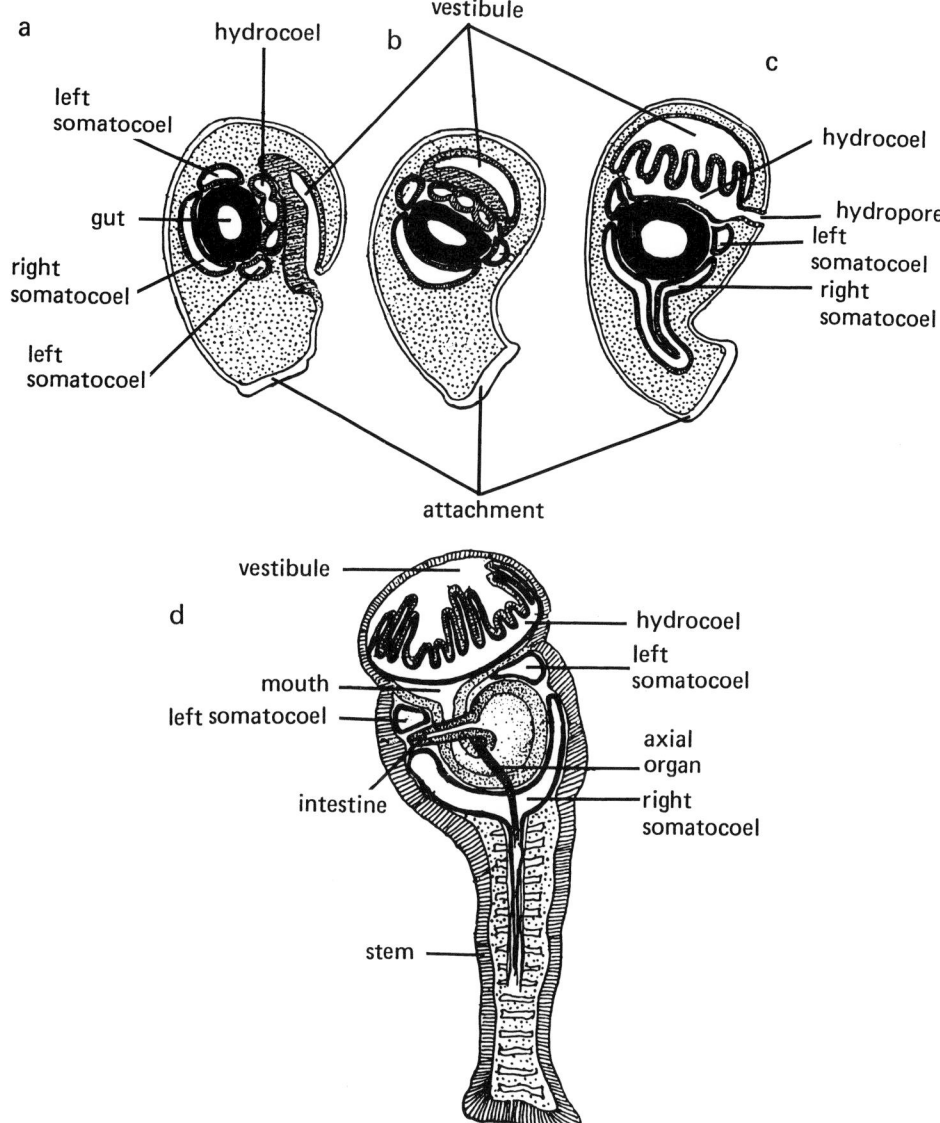

2.53. Antedon bifida. *Torsion in the larva after fixation; a), b), c) and d) represent successive stages. Note that the right somatocoel comes to be near the fixation point and the left somatocoel farther away; from Cuénot (1948, figs 394, 396, after Barrois).*

side) though the ciliary tracts are differently disposed and there is no mouth.

The visceral mass changes further. The gut becomes concavo-convex, with the concave face ventral. The hydrocoel comes to lie ventral to the gut, in the concavity of the latter, and becomes U-shaped in a frontal plane so that the concavity of the U is at first leftwards and posterior, but later leftwards and anterior (fig. 2.51b). Five little pouches develop on the ventral face of the hydrocoel, being the anlagen of the radial water vessels. The anlage of

the stone canal develops as an extension of one end of the U-shaped hydrocoel and grows towards the axocoel but does not yet open into it. The somatocoels shift, so that the right one embraces the gut dorsally and anteriorly while the left one embraces the gut and hydrocoel posteriorly (fig. 2.53a). This shift in the somatocoels later becomes more marked and has the same meaning as in *Asterias* and all other non-pathological echinoderms i.e. the left somatocoel becomes oral and the right one aboral.

Then the extension of the axocoel into the anterior part

of the larva disappears and the right somatocoel, now anterior to the visceral mass, sends out five parallel prolongations towards the same region. These are the anlage of the chambered organ of the fully developed larval stem and also produce the chambered organ of the adult. The plates of the skeleton appear at about this time as flat networks of calcite. The stomodaeum sinks inwards, towards the hydrocoel, and finally becomes completely separate from the surface as a so-called vestibule (fig. 2.53b).

The larva has meanwhile fixed to the substrate using the fixation pit or sucker which secretes a cement. At first the larva lies with the ventral face against the substrate but soon lifts itself up with the length at right angles to the substrate. The surface cilia are lost. The vestibule, and the visceral mass which is now associated with it, rotates, by a movement called torsion, from its primary ventral position to a posterior position in terms of the larva (figs. 2.53a–d). Thus the oral face of the future crinoid comes to be directed away from the substrate and is at the opposite extremity to the point of fixation. The larva now changes shape, differentiating into a jointed stem and a disc or theca. Meanwhile the hydrocoel, from being U-shaped, has formed a complete ring by fusing the two ends of the U; its five lobes have extended into the vestibule and each has developed two lobes on either side of them, these being the first paired tube feet. The stone canal has grown further out from one end of the U-shaped hydrocoel, comes to open into the axocoel, and thus, via the already formed hydropore, communicates indirectly with the sea water. An oesophagus grows out towards the vestibule at the centre of the hydrocoel ring and, where it meets the vestibule, the mouth arises as a perforation. The intestine grows out from the gut along the mesentery between the oral (left) and aboral (right) somatocoels and opens as the anus at the surface. The left or oral somatocoel insinuates itself between the hydrocoel and the gut in a crescent shape, so that the oesophagus, axocoel and stone canal are embraced by the horns of the crescent. Later the walls of the two somatocoels and of the axocoel break down so that something approaching the adult condition is reached with the stone canal opening into the general perivisceral coelom and the latter opening to the outside by the hydropore. Later still there will be a proliferation of hydropores to form the ciliated pits and of sand canals from the circum-oral ring of the water-vascular system. In *Antedon*, therefore, the primary hydropore at first has nothing to do with the water-vascular system. It develops from the axocoel with which the hydrocoel later communicates by producing the stone canal as an outgrowth.

The stemmed larva of *Antedon* (fig. 2.54) begins to feed when the roof of the vestibule opens. The larva gradually increases in size, develops arms and cirri and finally breaks

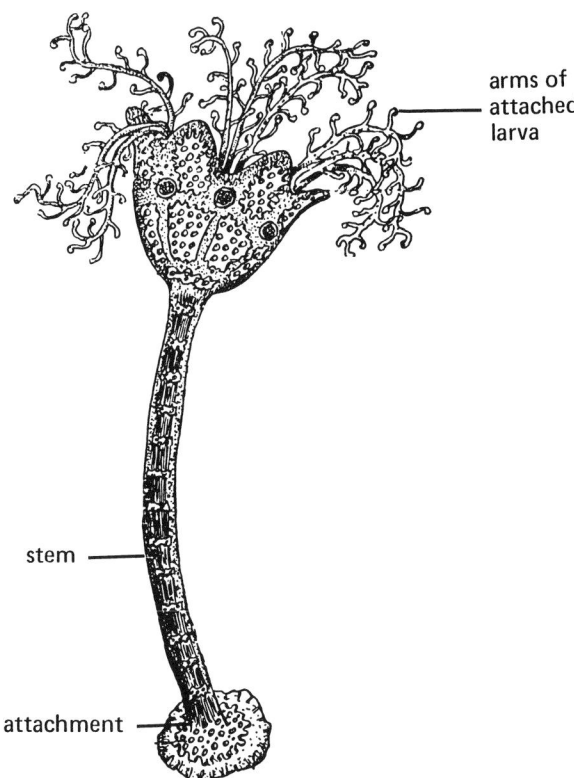

arms of attached larva

stem

attachment

2.54. Antedon bifida. *Stemmed larva after fixation; from Cuénot (1948, fig. 395, after Carpenter).*

away from the stem by a rupture at the base of the theca (disc), giving the adult, free crinoid.

The embryology of the echinoderms, exemplified here by *Asterias* and *Antedon*, allows a comparison with adult hemichordates which is by no means obvious starting from the adult echinoderms. Thus the axocoel of larval *Antedon* can be compared with the protocoel of adult *Cephalodiscus* because it is median and unpaired, has a hydrocoel pore comparable with the left protocoel pore of *Cephalodiscus*, and initially sends an extension towards the front end of the larva, just as the protocoel of *Cephalodiscus* extends towards the front end of the animal inside the protosome. Notably the larval axocoel of *Antedon* makes no connection with the hydrocoel (or left hydrocoel), just as the protocoel of *Cephalodiscus* does not communicate with the left mesocoel (equivalent to the left hydrocoel). The fixation sucker of *Antedon* and *Asterias*, towards which the axocoel extends, can be compared with the ventral surface of the head shield or protosome of *Cephalodiscus* which is likewise used for attachment (during crawling) and secretes a cement, though in *Cephalodiscus* this is used not for fixation but in making the horny tube. Moreover the madreporic vesicle of *Asterias*, which is pulsatile and contains the head process of the axial organ, can be compared with the pericardial or heart vesicle of hemichordates, which is pulsatile and contains the central blood sinus. This comparison is confirmed by the fact that the madreporic vesicle of *Asterias* is closely associated with the hydropore, just as the heart vesicle of *Cephalodiscus* is situated in the protocoel and therefore associated with the left (and right) protocoel pore. The comparisons already mentioned are strengthened by the presumably atavistic double-hydropore condition of *Asterias*, comparable with the paired protocoel pores of *Cephalodiscus*. Also the axial gland of *Asterias*, open by means of the hydropore to the outside, is perhaps homologous with the glomerulus of *Cephalodiscus* and other hemichordates, which voids to the outside through the protocoel pore or pores.

The water-vascular system of *Asterias* and *Antedon* can be compared with the tentaculate left mesocoel of *Cephalodiscus*. This is emphasized by the fact that, in the early embryology of *Asterias*, *Antedon* and all other non-pathological echinoderms, the circum-oral water-vascular ring does not begin as a circle round the oesophagus but as a crescent or U developed on the left side of the animal. Only later does the crescent close to form a ring. The comparison is confirmed by the double hydrocoel condition of *Asterias* and other echinoderms which mimics the paired mesocoels of *Cephalodiscus* and represents a most striking atavism. The water-vascular system functions as an organizer in echinoderm embryology much like the notochord in chordate embryology (von Ubisch, 1929). Doubling the hydrocoel may therefore double other organs in a non-atavistic manner, if these organs did not exist in the presumed symmetrical *Cephalodiscus*-like ancestor of echinoderms. The homology of the water-vascular system with the left mesocoel of *Cephalodiscus* makes good functional sense since the water-vascular system of *Antedon* and other crinoids is involved in filter feeding like the mesocoel tentacles of *Cephalodiscus*. The circum-oral ectoneural nerve ring of *Asterias* and *Antedon* can be homologized with the left half of the circumoral nerve ring of *Cephalodiscus*. For, because the hydrocoel induces the overlying radial ectoneural nerve, the closure of the originally U-shaped circum-oral water-vascular ring would cause the closure of the originally U-shaped ectoneural nerve above it. The hydrocoel lacks its own coelomic pore, as already mentioned, and in this respect differs from the left mesocoel of *Cephalodiscus*. Most probably this pore was originally present in phylogeny but has been lost and functionally replaced by the communication of the stone canal of the hydrocoel with the axocoel (protocoel). The suggestion that the hydropore originally had nothing to do with the hydrocoel is

confirmed by the fact that, in *Asterias*, double hydrocoel has no statistical association with double hydropore.

Finally the left and right somatocoels of *Asterias*, *Antedon* and other echinoderms, being situated in the larva left and right of the gut, can be homologized with the left and right metacoels of *Cephalodiscus* and other hemichordates. It is highly significant that, in all post-larval, non-pathological echinoderms, the larval left somatocoel has become oral in position, and the right somatocoel aboral. At the same time the left somatocoel has become crescent-shaped, with the mouth and axocoel embraced by the horns of the crescent, instead of the mouth being situated in the plane of junction of the right and left somatocoels. It is as if the mouth had migrated towards the left side, pushing against and distorting the left somatocoel as it did so. The situation is complicated, however, by the fact that the larval mouth of echinoderms, when it exists, does not transform into the post-larval mouth (except in some ophiuroids, von Ubisch, 1958). Rather the larval mouth closes and is functionally replaced by a newly perforated mouth in the post-larva. This probably represents a condensation of phylogenetic history, whereby a new mouth is produced in ontogeny, instead of the larval mouth migrating.

These resemblances, or indeed homologies, between juvenile echinoderms and adult hemichordates suggest relationship. This is confirmed, as already mentioned, by purely larval resemblances, in particular between the enteropneust tornaria and the asteroid auricularia, (apical disc and sometimes apical tuft of cilia anteriorly; morphology of ciliated tracts; division of gut into oesophagus, stomach and intestine; posterior position of larval anus, coinciding with the blastopore).

Hemichordates appear to be more primitive than echinoderms, for the larval plane of bilateral symmetry in hemichordates is vertical and, without change of orientation, becomes the vertical plane of symmetry of the adult. As against this, the larval plane of symmetry of echinoderms seems to be homologous with the plane of symmetry of hemichordates but rotates through 90° at metamorphosis becoming the horizontal plane of extension of the adult echinoderm. Thus the larval right and left somatocoels (metacoels) become the post-larval aboral and oral coeloms. As already mentioned, the aboral coelom is directed upwards in *Asterias* and other eleutherozoans, whereas the mouth and oral coelom are directed downwards. But in *Antedon* and other pelmatozoans the converse is true – the mouth and oral coelom are upwards and the aboral coelom downwards.

Which of these conditions is primitive for echinoderms? Probably the pelmatozoan mouth-upwards state is primitive and this for several reasons. Firstly, outgroup comparison suggests that the pelmatozoan mode of feeding is primitive since *Antedon,* like *Cephalodiscus,* uses the hydrocoel (left mesocoel) in filter-feeding (with the difference that *Cephalodiscus* uses the right mesocoel also). Secondly, a fixed filter-feeding stage is likely in the ancestry of all echinoderms to explain their most obvious characteristic – radial symmetry, which hemichordates show no sign of. For a fixed organism or organ, such as an oak tree, a sea anemone or a buttercup flower is likely to develop radial symmetry about the point of attachment so as to be equally exposed to influences, such as light, food or fertilizing insects, coming from all directions perpendicular to the axis of symmetry. With the mouth upwards such a fixed echinoderm could feed, whereas with the mouth downwards it could not. (Thus in *Asterias* fixation and torsion at metamorphosis go with the closure of the larval mouth.) Thirdly, an ancestral phase in which the larval right side was downwards against the substrate would explain the atrophy of the right hydrocoel which is characteristic of all normal modern echinoderms.

In the origin of echinoderms, therefore, it is likely that a *Cephalodiscus*-like animal fell over on the original right side, in what I call the dexiothetic orientation (fig. 2.55). This led to the disappearance of the right mesocoel (hydrocoel) which could not now function in filter-feeding since it was squashed beneath the body, probably into soft substrate. The change of orientation would block the right

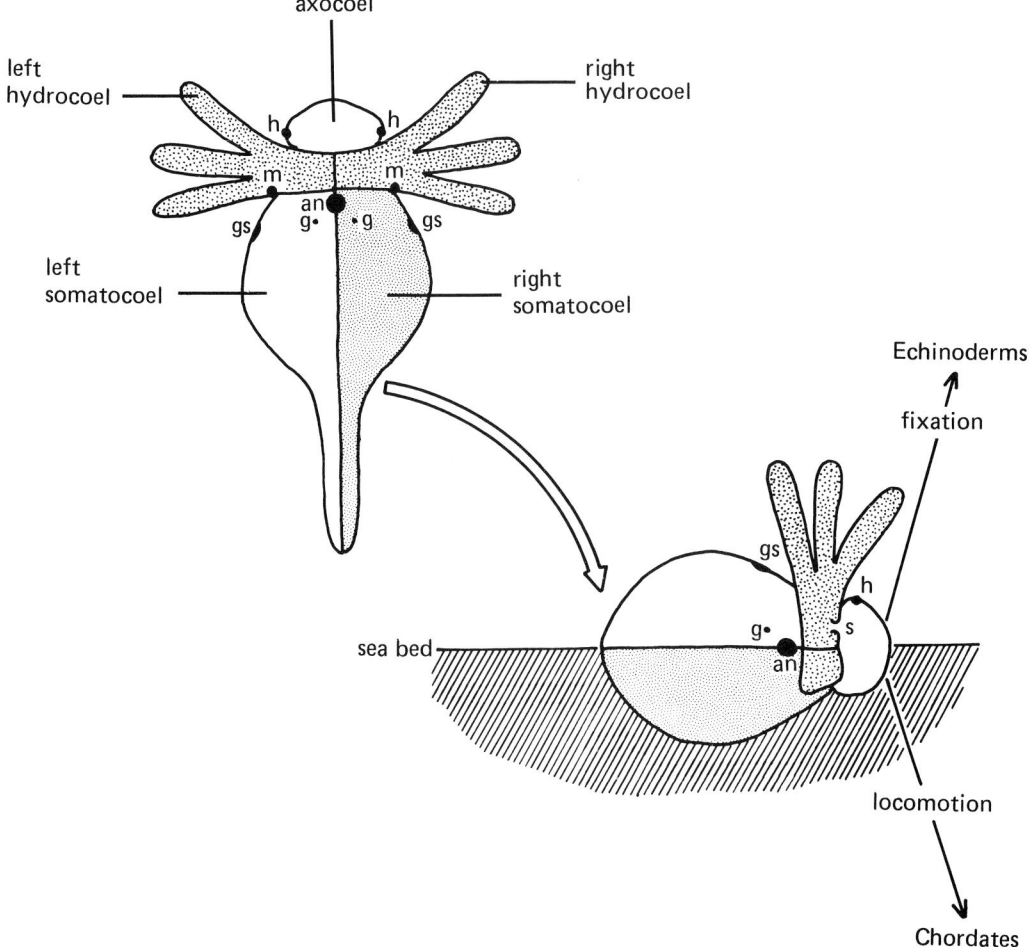

2.55. *The origin of dexiothetism. A Cephalodiscus-like ancestor loses its stem and falls over on the right side, so that the right somatocoel is downward and the left somatocoel upward as in a crinoid. As a result it loses the right hydrocoel and right tentacles and the openings of the right side of the body. A new canal (stone canal) appears between left hydrocoel and axocoel. The resulting animal, after acquiring a calcite skeleton, would represent the latest common ancestor of chordates and echinoderms; an = anus, g = gonopore, gs = gill slit, h = hydropore, m = mesocoel pore, s = stone canal. (New.)*

gill slit and right gonopore, so that these would also disappear, with the consequent loss of the right gonad. This dexiothetic form, later in the history of echinoderms, came to be fixed in position by the whole right side of the larval body (perhaps initially by the agency of the sticky head shield = larval attachment sucker of echinoderms), while the mouth and axocoel migrated upwards, onto the original left side of the animal, so as to take up a position in the centre of what was now the upper surface. The animal began to adapt for its fixed position by developing radial symmetry which began with a radially symmetrical filter-feeding system based on the persisting left hydrocoel. The anus also moved to the new upper side of the animal. Such a form, after acquiring a calcite skeleton and losing the original left gill slit, would be like the Middle Cambrian fossil *Stromatocystites* (fig. 2.56). Stephenson (1979) has

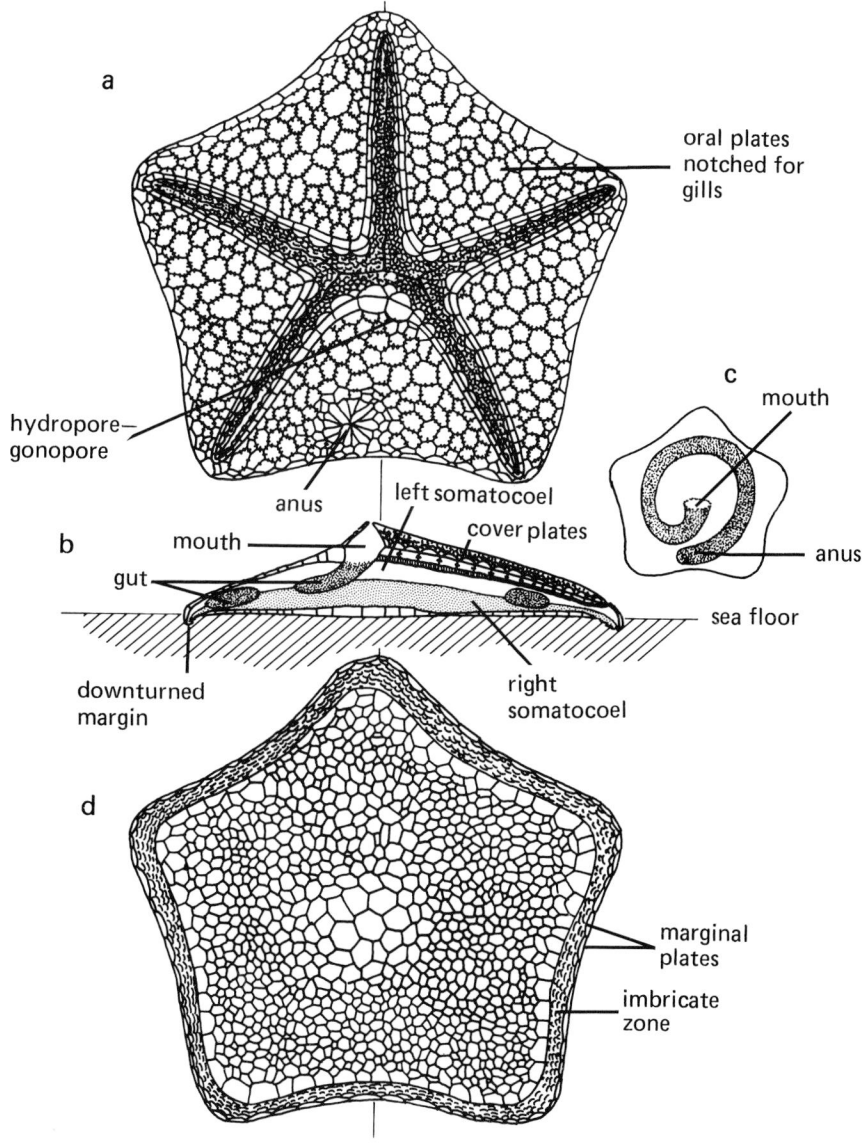

2.56. *Reconstruction of* Stromatocystites pentangularis Pompeckj *from the Middle Cambrian of Bohemia, based on work by Dr A. B. Smith. a) Oral or upper aspect; b) transverse section along line indicated in (a) and (d); c) possible course of gut in oral aspect (after Bather, 1915, fig. 4); d) aboral or lower aspect. The suggested positions of the right and left somatocoels in (b) are based on the embryology of living echinoderms such as* Asterias *and* Antedon. *By comparison with Recent forms, the two somatocoels were probably distinct in the post-larva but may have merged in the adult.*

discussed the reasons why the radial symmetry of such an animal should be based on the number five. (The condition of lying on the original right side is called dexiothetic, from *dexios* = right, *thetikos* = suitable for lying down; see Jefferies (1979).) Later the eleutherozoans arose, from this dexiothetic and pelmatozoan condition, by rolling over onto the original left side, mouth downwards, and using the tube feet for walking instead of filtration. The evolution of the echinoderms as a group is beyond the scope of this book, but great advances in its study have now been made by using cladistic methodology (Paul & Smith, 1984; Smith, 1984).

As concerns chordates, an early dexiothetic animal which had retained the original left gill slit, acquired a calcite skeleton, but had not acquired radial symmetry based on a pentameral hydrocoel (probably because it never became fixed), is a likely ancestor not for the echinoderms only, but also for the chordates as a phylum. It would be the latest common ancestor of echinoderms and chordates, as evidenced by the fact that the Middle Cambrian *Ceratocystis*, the most primitive known chordate, can be compared in its head openings with a *Cephalodiscus* lying on its right side (p. 221). But if this suggested origin for chordates is correct, then the echinoderms are more closely related to the chordates than the hemichordates are, with dexiothetism and the calcite skeleton as synapomorphies. Because of this, echinoderms and chordates together form a monophyletic group, for which I have proposed the name Dexiothetica, with the rank of sub-superphylum (Jefferies, 1979).

If the Dexiothetica are a monophyletic group, then the 'Ambulacralia' which Grobben (1908), following classical work by Metschnikoff, recognized to include hemichordates and echinoderms, is paraphyletic and should be abandoned. It is also possible that the hemichordates are paraphyletic for they appear as primitive members of the Deuterostomia, characterized by the lack of dexiothetism and the presumably primitive lack of a calcite skeleton. If the hemichordates are paraphyletic, then some of them will be closer related to the dexiothetes than others are. This may well be true, but the evidence is too weak to say.

The gonads of living echinoderms are mostly arranged on a five-fold plan. As already mentioned, this is true of crinoids and asteroids, and also holds for echinoids (five gonads inside the body) and ophiuroids (ten genital sacs with contained gonads). The sole exceptions among living echinoderms are the holothuroids, which always have one gonad only. The holothurians, however, probably represent the primitive condition in this respect, for many pentameral fossil groups have only one gonopore, and therefore probably only one gonad. This is true of the Lower Palaeozoic eocrinoids and rhombiferan cystoids, the Ordovician fossil *Aulechinus* which is regarded as a primitive echinoid, and other forms. It was also true of the calcichordates as I show later. The pentameral gonads of living crinoids, asteroids, ophiuroids and echinoids are therefore probably several parallel developments, reflecting merely the marked echinoderm tendency to multiply organs by five.

To sum up this chapter, the classical view that echinoderms are closely related to hemichordates is almost certainly correct, although most zoologists have forgotten the strong and subtle comparison on which it was based. More particularly the echinoderms can be derived from a form like *Cephalodiscus* which laid down on the right side and acquired a calcite skeleton. But such an animal was probably ancestral not to the echinoderms only, but also to the chordates.

Chapter 3   Living acraniates – amphioxus and its relatives

There are two living groups of non-vertebrate chordates, both given the rank of subphylum. They are the acraniates (Cephalochordata), represented by amphioxus, and the tunicates (Urochordata) or sea squirts. Amphioxus, which is the main subject of this chapter, is widely known and described in detail in all textbooks of vertebrate zoology. Tunicates, however, which are dealt with in the next chapter, have been neglected.

Amphioxus is used here as the informal name for the genus correctly called *Branchiostoma*. It, or rather the European species *Branchiostoma lanceolatum*, is much the best studied acraniate. The only other living genus in the subphylum is *Asymmetron* which is like amphioxus in most respects.

Some extraordinary claims have been made for amphioxus. Many seem to believe that it is an almost unmodified primitive chordate – an almost unchanged survivor of the latest common ancestor of itself and vertebrates. One elementary textbook quotes with approval the assertion that 'If amphioxus had not been discovered it would have to have been invented' (Grove & Newell, 1961, p. 324) – which echoes what Voltaire said about God. Again Drach wrote in 1948 that amphioxus was: 'l'animal le mieux connu après l'homme.' Subsequent work, in particular that of Flood (1966) on the supposed ventral spinal nerves, Welsch (1968) and Flood (1975) on the histology of the notochord, and Webb (1973, 1975) on the mode of swimming, shows that much of the older 'knowledge' about amphioxus was mistaken. It remains a very important animal, for the acraniates, as I believe, are probably the sister group of [tunicates + vertebrates] and some of its peculiarities, particularly its asymmetries, are primitive features which only acraniates among living chordates have retained.

In outside shape amphioxus is fish-like (fig. 3.1), but pointed at both ends as the name implies (*amphis* = both, *oxys* = sharp). A median fin starts at the anterior end just above the mouth, passes along the mid-line of the back, develops a wide blade at the posterior end and then runs forward, in the ventral mid-line, to stop at a point about one-third of the length from the posterior end. In front of this point there is a pair of protruding, ventro-lateral metapleural folds which separate the flattish sides of the animal from the flattish ventral surface. Where the metapleural folds exist, therefore, the animal is roughly an isosceles triangle in transverse section, with the dorsal median fin at the apex and the metapleural folds at the basal angles. The mouth is at the anterior end of the animal, just below and behind the terminal anterior lobe of the median fin. It is guarded by about 20 buccal tentacles (cirri). An opening called the atriopore exists in the ventral surface of the animal in the mid-line near the hinder ends of the metapleural folds. Near the posterior end of the animal, opening ventralwards but left of the mid-line, is the anus.

Amphioxus lives in the sea. As an adult and very young larva it inhabits the sea bottom, but during most of larval life it is planktonic and can drift thousands of kilometres carried by currents. As a planktonic larva it stays up in the water by alternately swimming upwards with its length vertical and sinking downwards with its length horizontal (Webb, 1975). Webb considers that such a larva sometimes catches large food (1969) but others regard it as entirely microphagous (Gosselck & Kuehner, 1973). When the planktonic larva swims it is said to rotate about its long axis, clockwise when viewed from behind (Wickstead, 1967). As an adult, amphioxus inhabits the sea bottom, preferring loose sand with rounded grains and abundant organic growth but no decay (Webb, 1975, p. 191). In such coarse substrates it buries itself entirely whereas in finer substrates the front end protrudes to some extent.

Some features of the internal anatomy are visible from outside. Most obvious of all are the muscle blocks which extend in an almost uniform series from near the anterior to near the posterior end of the animal. They number

a

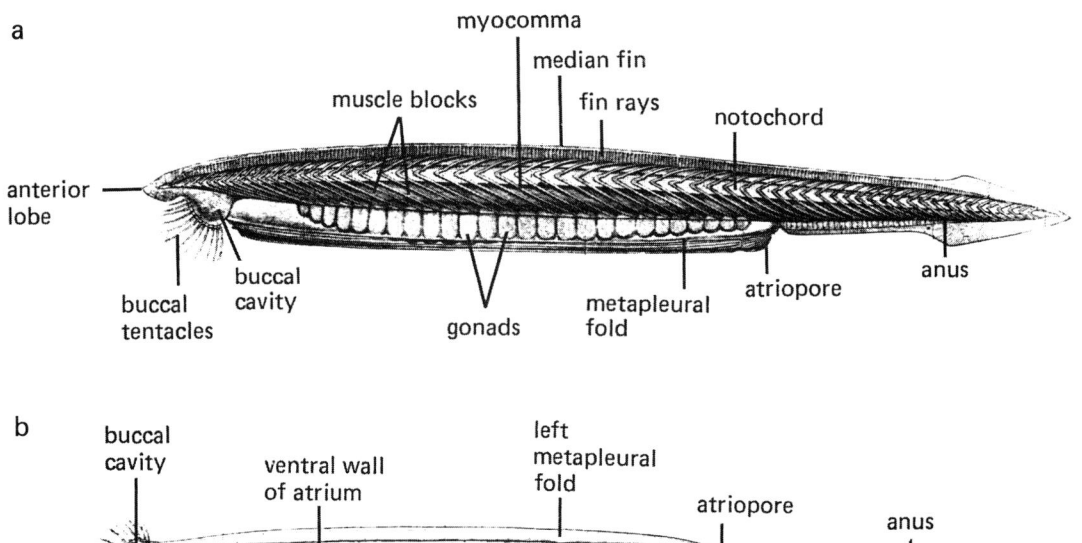

3.1. *External appearance of amphioxus* (Branchiostoma lanceolatum). *a) Left lateral aspect; b) ventral aspect; from Lankester (1889, pl. 1, figs 4, 5).*

about 60 on either side. Each muscle block is a lop-sided V in shape, with the point of the V anterior and its ventral arm longer than its dorsal arm. The walls of connective tissue separating the muscle blocks are called myocommata (or myosepta). Curiously the muscle-block series of right and left are not symmetrical, for each right muscle block is half a segment either in front of or behind the corresponding left muscle block. For reasons explained later, it is likely that the foremost pair of muscle blocks is homologous with the mandibular somites of vertebrates, conventionally numbered as the second somites. Also visible from outside are the transparent fin rays; these are a series of boxes which support the base of the median fin throughout its extent. At the ventral margin of the muscle blocks a series of white gonads is visible through the right and left body walls. Each one coincides in position with the ventral edge of a muscle block. (In *Asymmetron*, the only other acraniate genus, gonads exist on the right side only, whence the name.) The sexes are separate in amphioxus (and *Asymmetron*) but there is no obvious difference between male and female except in the sex of the gonads. Ventral to the series of muscle blocks it is possible to see the ventral half of a basket-like structure, which is the pharynx.

The alimentary canal runs fairly straight from mouth to anus (fig. 3.2). The mouth, guarded by the buccal tentacles or cirri, leads into the buccal cavity (fig. 3.3). This is a small

chamber limited behind by a vertical wall of flesh, the velum, which itself is penetrated centrally by a constrictable opening called the velar mouth, guarded by velar tentacles. The lateral walls and roof of the buccal cavity are convoluted to form a ciliated organ known as the wheel organ. This structure owes its name to early investigators who were reminded by the beating of its cilia of the 'wheels' on a rotifer or 'wheel-animalcule'. A shallow pit in this wheel organ, situated in the roof of the buccal cavity just right of the mid-line, is known as Hatschek's pit (fig. 3.4). The buccal tentacles are each supported by a rod of cartilage joined at their bases to a subdivided U-shaped cartilaginous hoop situated along right and left lips of the mouth.

Behind the velum and velar mouth is the pharynx (fig. 3.5), which has already been mentioned as visible from outside. The pharynx is an elongate chamber, taller than wide in transverse section. It extends rearwards to the transverse plane of the atriopore. The side walls of the pharynx are penetrated by about 80 or 90 parallel-sided gill slits which slope ventralwards posteriorly. Most of the gill slits extend from near the dorsal mid-line of the pharynx to near its ventral mid-line, except near the front end where the slits do not extend so far dorsally. In the dorsal mid-line of the pharynx, between the top ends of the right and left sets of gill slits, is a groove called the epipharyngeal groove. In the ventral mid-line, between

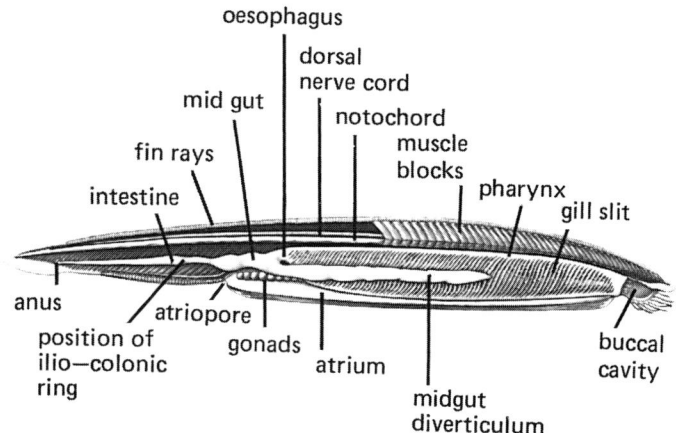

3.2. General internal anatomy of amphioxus – the right atrial wall and most of the right atrial wall and right muscle blocks have been removed; from Burn (1980 (Ed.), p. 188).

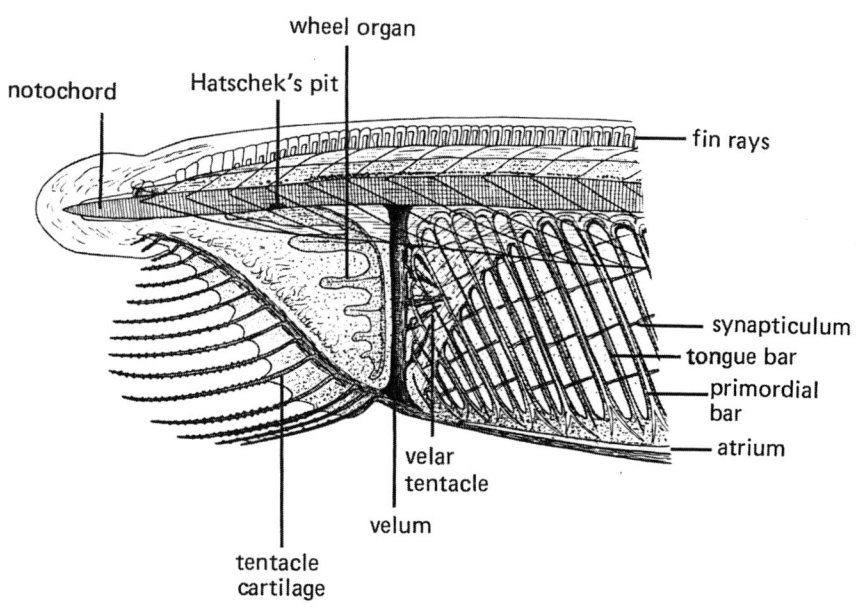

3.3. Anterior end of amphioxus to show features of buccal cavity, velum and the anterior part of the pharynx; from Franz (1927, fig. 44, p. 536).

the ventral ends of right and left gill slits, is another much shallower groove called the endostyle. Just anterior to the gill slits a pair of ciliated grooves, the peripharyngeal grooves, pass from the anterior end of the endostyle on right and left up to the anterior end of the epipharyngeal groove.

The gill slits do not open direct to the outside. Instead they debouch into a large chamber, named the atrium, which is approximately U-shaped in transverse section, surrounding the pharynx on right and left and ventrally. The atrium opens on the ventral surface of the animal through the atriopore but also extends behind the atriopore, on the right, to the transverse plane of the anus.

The gill bars between the gill slits are of two types – 'primary' or primordial, and 'secondary' or tongue bars.

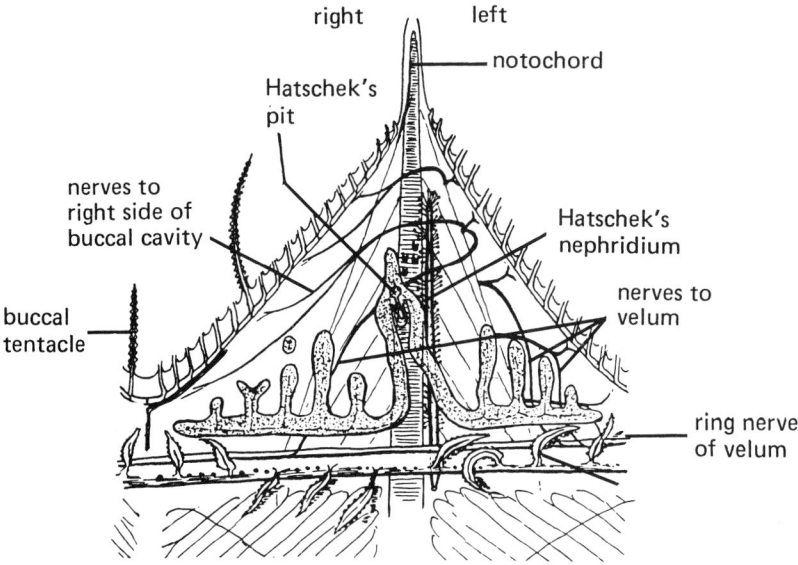

3.4. *Buccal cavity and velum of amphioxus, cut at the ventral mid-line and opened out. Note the asymmetrical nerve supply, from the animal's left only, to the buccal cavity and velum; from Franz (1927, fig. 44, p. 536).*

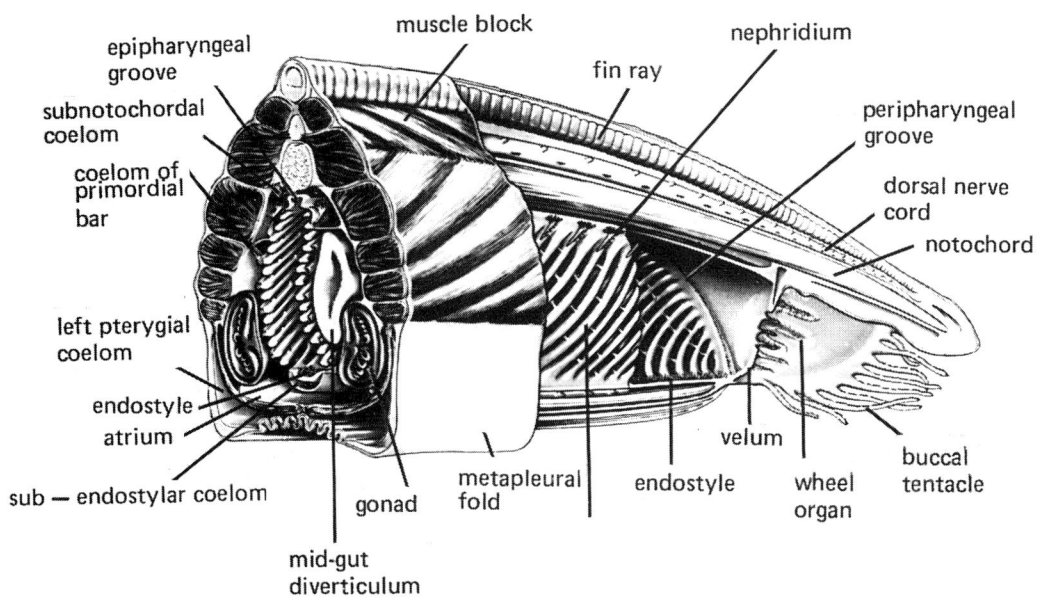

3.5. *Block diagram of anterior part of amphioxus; from Burn (1980, p. 189).*

These alternate with each other and differ embryologically. When the gill slits first appear in ontogeny they are circular and separated only by primordial bars. Soon, however, they become U-shaped (fig. 3.38), by the growth of a tongue bar down from the middle of the dorsal edge. Finally this tongue bar fuses with the ventral edge of the primordial gill slit, dividing the slit into two. Tongue bars and primordial bars become connected together by tiny bridges called synapticulae. The right and left series of gill bars are unsymmetrical in that each primordial bar on the

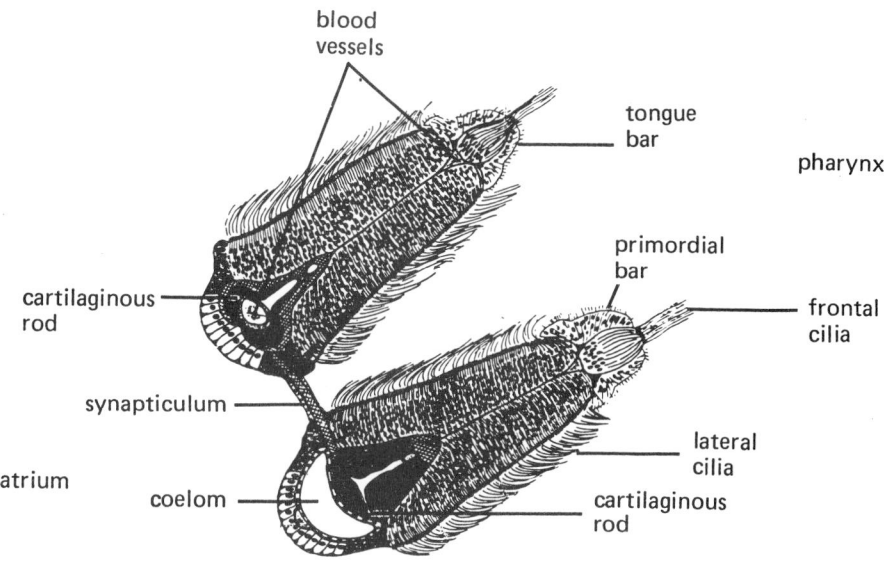

3.6.   Transverse section through a primordial bar and a tongue bar of amphioxus; from Franz (1927, fig. 47, p. 543). Compare fig. 2.16.

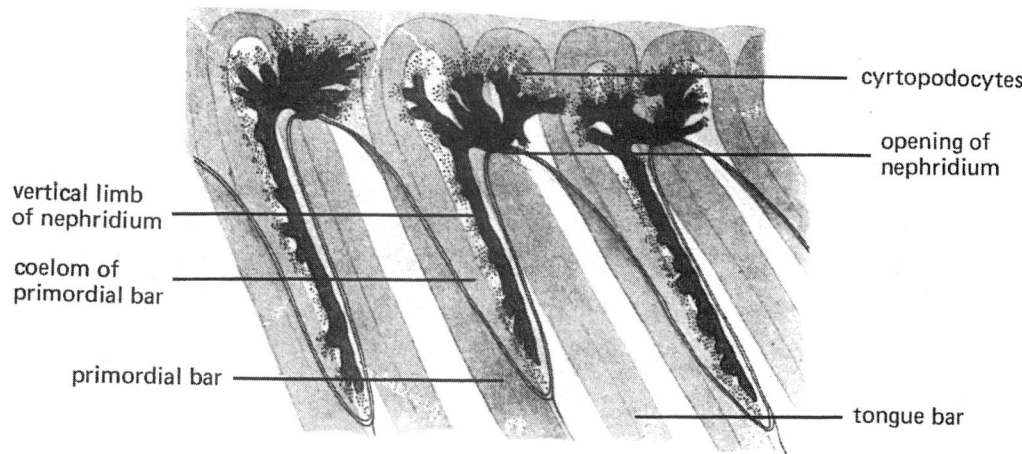

3.7.   The relations of the nephridia of amphioxus to the tongue bars, primordial bars and primordial-bar coeloms; from Goodrich (1933, text fig. 1, p. 725).

right is opposite a tongue bar on the left. I prefer to speak of 'primordial' and 'tongue' bars, rather than 'primary' and 'secondary' bars so as to avoid confusion with primary, secondary (and tertiary) gill slits (see below, p. 77).

The difference between primordial and tongue bars is recognizable in the adult by several structural details (fig. 3.6). Primordial bars contain a tubular coelom extending from the unpaired subendostylar coelom up to the paired subnotochordal coeloms, while tongue bars have no coelom. Primordial bars have blood vessels running up them from the endostylar vessel to the paired dorsal

aortae; tongue bars likewise have vessels running up them but these do not connect with the endostylar vessel except indirectly via vessels in the synapticulae. Primordial bars have a cartilaginous stiffener which extends along the whole length of the bar and ends beneath it in a fork; tongue bars likewise have a cartilaginous stiffener but it simply stops ventrally without a fork; each of the ⊤-shaped nephridia, which are situated dorsal to the gill slits (fig. 3.7), has a vertical limb extending down into the coelom of a primordial bar and a horizontal limb extending rearwards over the dorsal end of tongue bar. As already mentioned,

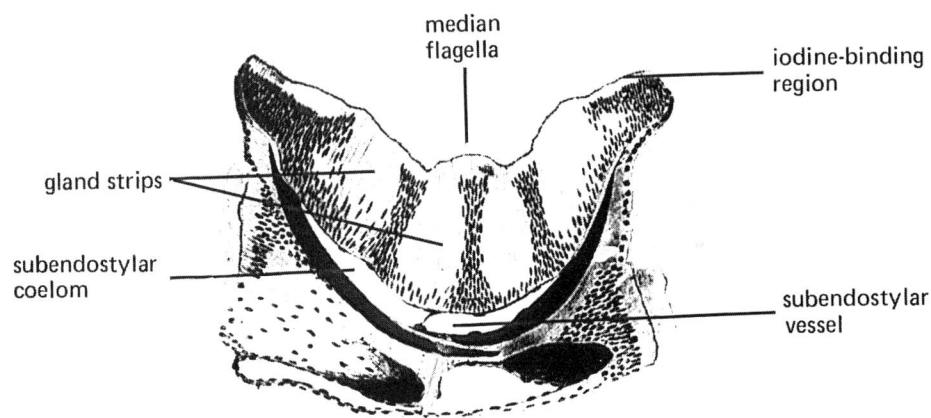

3.8. Transverse section of the endostyle of amphioxus; from Pietschmann (1929, fig. 70, p. 61).

neither the skeleton nor the coeloms of the gill bars agree with those of the U-shaped slits of enteropneusts. In particular, the coeloms of amphioxus are in the primordial bars, not in the tongue bars, whereas in enteropneusts the converse is true. It is likely that the gill slits of enteropneusts and acraniates are in a general way homologous but there is no reason to assume that the tongue bars of enteropneusts are homologous with those of amphioxus; both are probably parallelisms functioning to increase the number of pumping cilia round each gill slit.

Behind the pharynx in amphioxus there is a short oesophagus. Behind the oesophagus is a mid-gut diverticulum ('liver') (fig. 3.2) which extends forwards as a pouch right of the pharynx. Behind the origin of this diverticulum the alimentary canal passes back to the anus as a simple tube (the intestine), with a longitudinal fold (typhlosole).

The ciliary coating of the alimentary canal is functionally important. Cilia are abundant almost everywhere except in front of the peripharyngeal bands. Along the groove of the endostyle in the ventral mid-line of the pharynx there are two pairs of mucus-secretory gland strips separated, or flanked by, non-glandular ciliated strips (fig. 3.8). The cilia of the median ciliated strip of the endostyle are long enough to be called flagella. The pharyngeal face of the gill bars is coated with so-called frontal cilia (fig. 3.6). The sides of the gill bars (i.e. with respect to the animal their anterior and posterior surfaces) are covered with lateral cilia. The peripharyngeal grooves in front of the gill slits and the epipharyngeal groove along the dorsal mid-line of the pharynx are also strongly ciliated. Behind the pharynx the whole of the internal face of the intestine, mid-gut and mid-gut diverticulum is ciliated, but the cilia are especially long and numerous at the posterior end of the mid-gut in the so-called iliocolonic ring (fig. 3.2). Cells that bind iodine are located just dorsal to (external to) the more dorsal glandular strips of the endostyle in a position exactly corresponding to cells

of the same function in tunicates and lamprey larvae (Thorpe & Thorndyke, 1975).

The main function of the pharynx in amphioxus is to extract food particles from the water. How this happens (fig. 3.9) was first described by Orton (1913). The gland strips of the endostyle secrete mucus which flows dorsalwards as sheets on either side, driven by the frontal cilia on the pharyngeal surface of the gill bars. The anterior ends of the sheets of mucus are presumably held by the peripharyngeal grooves, though this has never been seen in amphioxus. When the mucus sheets reach the dorsal

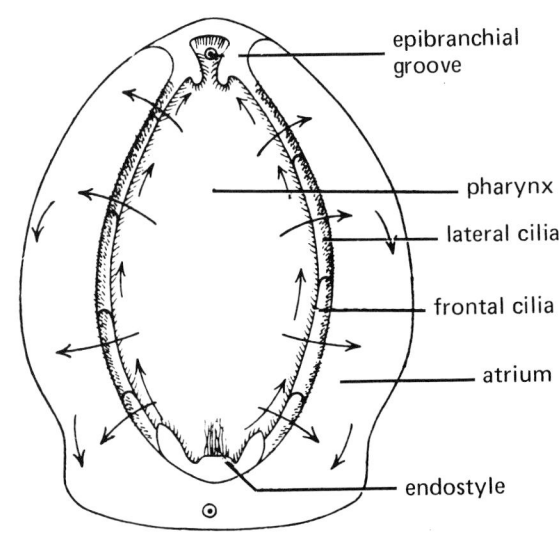

3.9. Diagrammatic transverse section of the pharynx and atrium of amphioxus to show mode of functioning. Thicker arrows show movement of water. Finer arrows show movement of food and mucus. Right is at right and left is at left; from Drach (1948, fig. 412, p. 955, after Orton, 1913).

mid-line they join each other and pass into the epipharyn-geal groove whose cilia beat the mucus rearwards and pass it to the oesophagus as a continuous mucous rope.

A stream of water is at the same time being pulled through the mouth, into the pharynx, through the gill slits into the atrium, and hence out through the atriopore, by the action of the lateral cilia of the gill bars (those lining the slits). This water current brings in particles which become trapped in the mucous sheets and thus incorporated in the mucous rope in the epipharyngeal groove and carried rear-wards to the oesophagus. This filter-feeding mechanism is very similar to, and no doubt homologous with, that of tunicates and the ammocoete larva of lampreys. The pharynx of amphioxus probably functions in respiration, as well as in catching food, for Azariah (1969) showed that the quantity of water that passed through it increased when oxygen tension was low.

Behind the pharynx the mucous rope is gripped by the cilia of the ilio-colonic ring which rotate it, anti-clockwise as seen from behind (Barrington, 1938). (This direction of rotation of the rope is the same as in tunicates.) The mid-gut diverticulum also secretes digestive enzymes (Barrington, 1965, p. 126). Behind the ilio-colonic ring the mucus is carried backwards by the cilia of the intestine to the anus, where its exit is controlled by a sphincter muscle.

The notochord of amphioxus extends from the anterior to the posterior end of the animal, except for the extreme anterior and posterior terminations of the median fin.

Its anterior and posterior ends are visible from outside, where they project beyond the series of muscle blocks. Its structure has recently been studied intensively by several authors (Flood, Guthrie & Banks, 1969) and has been reviewed by Flood (1975). It is circular, or laterally-compressed elliptical, in cross section and tapers towards the ends. It is surrounded by a sheet of connective tissue in which the collagenous fibres have a helical or circular arrangement.

The most obvious constituents of the notochord are the notochordal plates (fig. 3.10). These are discs stretched transversely across the cavity of the notochordal sheath and they have recently been shown to be muscles of the slow-acting catch-muscle type. The muscle fibres in the plates run transversely across the notochord. Each plate corresponds to a single cell in most cases, but sometimes two neighbouring plates belong to a single cell, being separated by a vacuole within the cell. Most of the spaces between plates are extracellular. The notochordal plates extend upwards as 'muscle-tails' through holes that perforate the notochordal sheath. These muscle-tails synapse with the overlying dorsal nerve cord. Innervation by muscle-tail is also known in echinoderms (Cobb & Laverack, 1967).

The function of the muscles of the notochordal plates is not known for certain. However, their contraction prob-ably increases the hydrostatic pressure within the noto-chord and hence makes it locally stiffer. Amphioxus moves

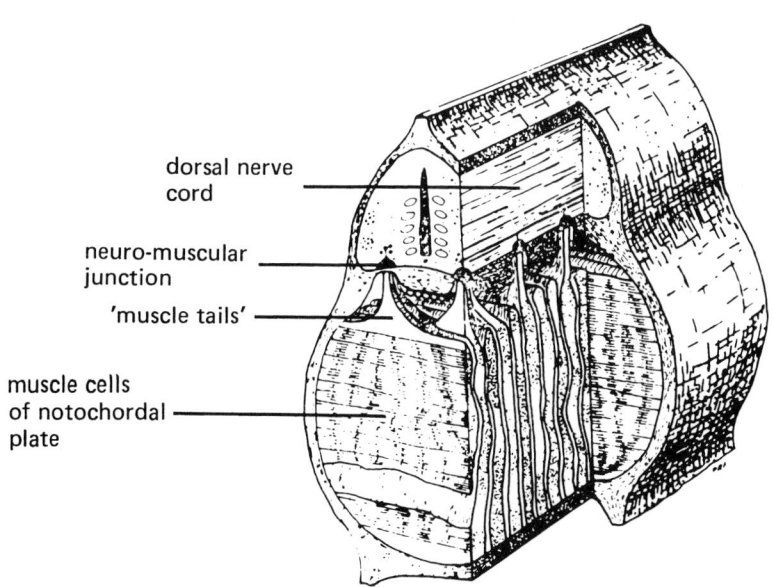

dorsal nerve cord

neuro-muscular junction

'muscle tails'

muscle cells of notochordal plate

3.10. *Histological structure of the notochord of amphioxus and its relations to the dorsal nerve cord; from Flood, Guthrie & Banks (1969, fig. 1).*

with equal facility forwards and rearwards. Webb (1973, 1975) has suggested that contraction of the notochordal muscles would stiffen the end of the notochord that happened to be leading. This would help amphioxus to penetrate the sediment in which it lives.

The muscular system of amphioxus (outside the notochord) can be divided into somatic and visceral parts. It will be described together with the coeloms, with which it is intimately connected. The somatic musculature consists of the externally obvious right and left series of muscle blocks. The two series run alongside the notochord from near the anterior to near the posterior end of the animal. Each muscle block is covered on its median face by a coelom called a sclerocoel by homology with vertebrates, though it is not skeletogenous. Each muscle block is covered laterally by a myocoel (Franz, 1927). The myocoels, however, are virtual (i.e. have no actual cavity) except near the anterior and posterior ends of the muscle block series. The muscle blocks function by pulling against the intervening myocommata. Since the notochord is incompressible, the contraction of the muscle blocks causes the body to bend from side to side as required in swimming and burrowing.

The muscles of each muscle block are made up of a number of single plates, each one a single cell. These muscle plates extend longitudinally from the anterior to the posterior myocomma of the block and in transverse section are arranged approximately radial to the notochord. The innervation of the plates of the muscle blocks is most remarkable (Flood, 1966, 1968). Until recently it was thought that each muscle block received a ventral motor nerve from the dorsal nerve cord as in a vertebrate. In fact, however, the supposed ventral nerves are made of muscle (fig. 3.11). Each muscle plate sends a cytoplasmic process (muscle-tail) to the lateral surface of the dorsal nerve cord and the actual synapse is situated at this surface. These muscle-tails are even more striking than those of the muscles of the notochord, and equally reminiscent of the ampullar muscle-tails of echinoderms.

It is historically interesting that Anton Schneider (1879) had already discovered that the 'ventral nerve roots' of amphioxus were made of muscle. Until Flood's remarkable research (1966) this result was disregarded by nearly all later workers, perhaps because it made amphioxus less vertebrate-like. The only exception was Rohde (1887) but this itself is not surprising, for Rohde was Privat-Dozent at Breslau, where Schneider was Professor. The nature of the muscle-tails ought to have been obvious, for they are actually striated like voluntary muscle. Fig. 3.11, taken from Schneider's work of 1879, is thus a belated recognition of his acuity.

The remaining coeloms of amphioxus comprise three major parts and several outliers. The major parts are the

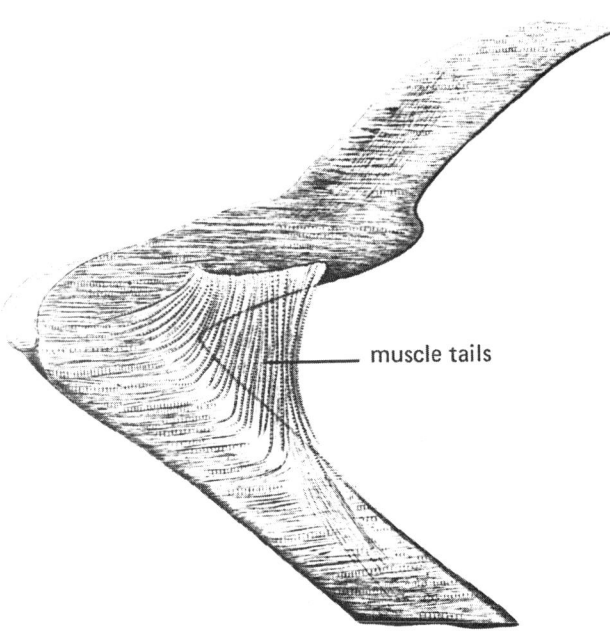

muscle tails

3.11. *A muscle block from the right side of amphioxus in left aspect, showing the muscle tails by which it receives innervation from the nerve cord. This striking illustration is taken from Schneider (1879, pl. 15, fig. 2). The anatomical relations which it shows were ignored until their rediscovery by Flood (1966).*

general coelom and the left and right pterygial coeloms. Posteriorly the general coelom consists of the postpharyngeal coelom which surrounds the intestine much like the abdominal coelom of a vertebrate. In the pharyngeal region (fig. 3.5), however, the general coelom is greatly reduced by the atrium and consists merely of a pair of subnotochordal coeloms dorsally, an unpaired subendostylar coelom ventrally, and branchial-bar coeloms which connect subnotochordal and subendostylar coeloms together, through the primordial gill bars. The pterygial pair of coeloms is developed in the ventral wall of the animal beneath the atrium. Left and right pterygial coeloms meet in the ventral mid-line where they are separated from each other by a median septum. The anterior ends of the pterygial coeloms are not symmetrical (fig. 3.12). For the left pterygial coelom passes forward and forms the outer lip coelom of the left and right sides of the buccal cavity. The forward continuation of the right pterygial coelom, on the other hand passes upwards in the inner face of the right side of the buccal cavity and contains the blood complex called the glomus. Right and left pterygial coeloms meet behind the atriopore around which they are modified to form an atrioporal sphincter muscle. The dorsal walls of the pterygial coeloms are muscular, being the pterygial muscles; and the left pterygial muscle, but not the right one, continues forwards in

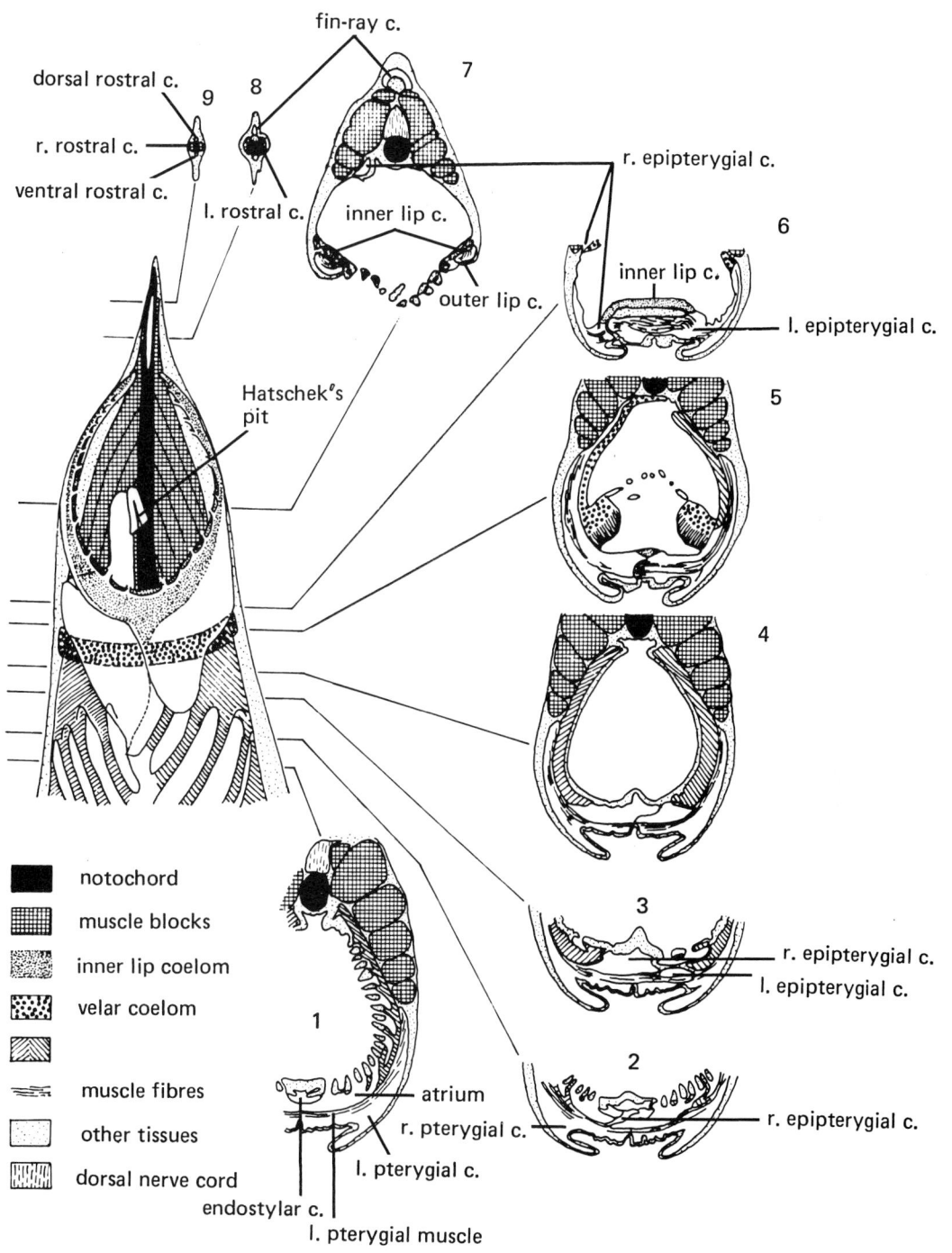

fin-ray c.

dorsal rostral c.

r. rostral c.

ventral rostral c.

l. rostral c.

inner lip c.

outer lip c.

r. epipterygial c.

inner lip c.

l. epipterygial c.

Hatschek's pit

notochord
muscle blocks
inner lip coelom
velar coelom
muscle fibres
other tissues
dorsal nerve cord

r. epipterygial c.
l. epipterygial c.

atrium
r. pterygial c.
l. pterygial c.
endostylar c.
l. pterygial muscle

r. epipterygial c.

3.12. Coeloms and coelom asymmetries at the anterior end of amphioxus (diagrammatic). Drawn on the basis of information and sections in Franz (1927). The three-dimensional reconstruction omits the pterygial coeloms, pterygial muscles and tongue bars.

the outer lip coelom to form the outer lip muscle on left and right. Near the front part of the pharynx the left and right pterygial coeloms develop portions dorsal to the pterygial muscle, these portions being distinguished as epipterygial coeloms from the metapterygial coeloms beneath the muscle (fig. 3.12). Epi- and metapterygial coeloms communicate with each other by way of holes through the pterygial muscle. Just in front of the foremost gill slit the epipterygial coeloms open into the subendostylar coelom, and right and left pterygial coeloms are thus continuous with the general coelom. The pterygial muscles can contract suddenly to force water out of the atrium through the mouth or the atriopore. The atrioporal sphincter contracts more gently to control outward flow through the atriopore (Bone, 1961).

Outlying coeloms, not continuous with the pterygial or general coeloms, are numerous. There is firstly the velar coelom, developed in the velum, with the velar muscle forming its anterior wall (figs 3.12, 3.14). Secondly there is the inner lip coelom, developed inside the left and right lips of the mouth, median to the lip cartilage, and with an inner lip muscle developed where the coelom makes contact with the cartilage. In the adult the velar coelom and the inner lip coelom are symmetrical, but in embryology both develop from the left mandibular somite (Legros, 1898, van Wijhe, 1914). It is therefore true that inner and outer lip coeloms and the associated muscles of right and left sides belong embryologically only to the left side of the animal.

Other coeloms in amphioxus are the cavities within the fin-ray boxes, the rostral coeloms at the anterior end of the animal and the gonocoels. The fin-ray coeloms are formed in ontogeny as dorsal extensions of the myocoels of the muscle blocks, being developed alternately from right and left sides (Drach, 1948, p. 1015). The rostral coeloms (fig. 3.13) are four in number – dorsal, ventral and left and right lateral. The lateral rostral coeloms, on either side of the notochord are anterior prolongations of the myocoel-sclerocoel of the left and right mandibular somites. The ventral rostral coelom, just ventral to the anterior portion of the notochord, is formed from the right first somite (the right portion of the premandibular somite = right anterior gut diverticulum) (Goodrich, 1917). The dorsal rostral coelom, just dorsal to the anterior end of the notochord, is not a serial homologue of the fin-ray coeloms, although exactly in line with them, for it appears earlier than they do in ontogeny (Hatschek, 1906, p. 6). Its first origin is unknown. The gonocoels are associated with the gonads in the body wall where it encloses the atrium. They are ventral extensions of the myocoels (Drach, 1948, p. 996).

The nervous system of amphioxus consists of a central nervous system (dorsal nerve cord and brain) and a peripheral nervous system (cranial nerves, segmental nerves and visceral plexi).

The dorsal nerve cord overlies the notochord but is separated from it by the notochordal sheath of connective tissue. This sheath, as already mentioned, is penetrated by holes through which the muscles of the notochord synapse with the dorsal nerve cord. In transverse section the dorsal nerve cord is approximately an isosceles triangle, higher than wide. It has a central canal which is cylindrical ventrally but passes dorsally into a narrow, vertical, median cleft, in transverse section being somewhat the shape of a key-hole upside-down. The dorsal nerve cord innervates the animal by immediate synaptic contact with the notochordal muscles and with the muscle-tails of the muscle blocks and also by means of the dorsal nerves.

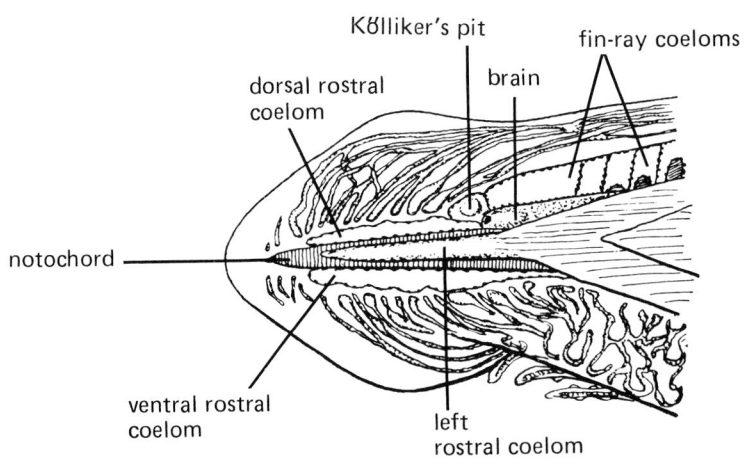

3.13. *Coeloms at the anterior end of amphioxus (left aspect); from Franz (1927, fig. 37a, p. 516).*

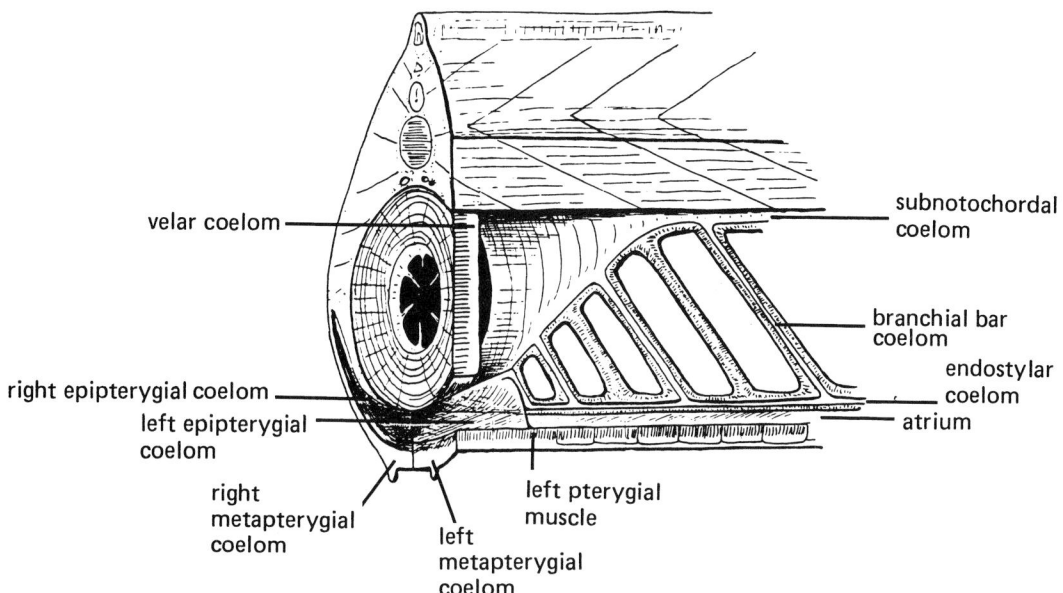

3.14. *Coeloms in the velar region of amphioxus; diagrammatic; from Franz (1927, fig. 53, p. 562).*

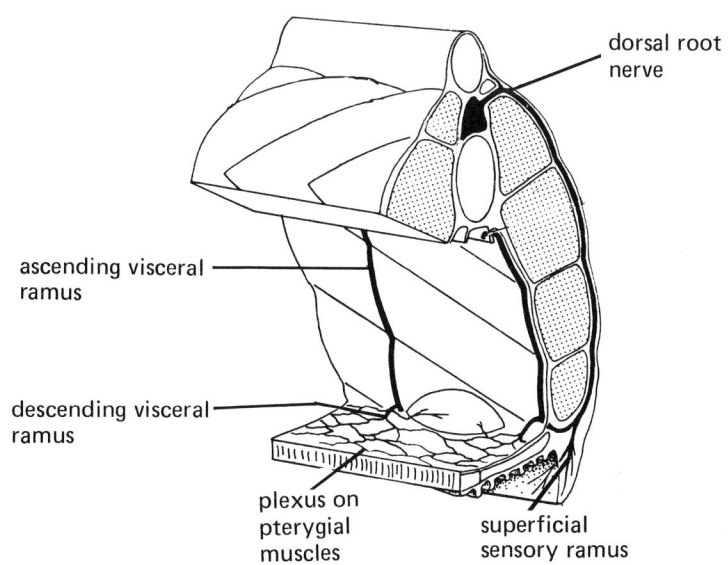

3.15. *Course of a dorsal root nerve in amphioxus; modified from Bone (1961, fig. 1, p. 244).*

A typical dorsal nerve (fig. 3.15) goes off from the dorso-lateral face of the dorsal nerve cord into each myocomma. It runs transversely in this between adjacent muscle blocks until it reaches the outer surface of the blocks. Here it divides into a dorsal and a ventral branch. The dorsal branch innervates the skin of the dorsal part of the animal. The ventral branch innervates the skin of the ventral part of the animal and also, in the region between the mouth and the atriopore, sends a visceral branch into the body round the ventral edge of the muscle block series. This visceral branch soon divides into an ascending and a descending branch. The descending branch connects with a nerve plexus on the dorsal surface of the pterygial muscle and controls the cough reaction by which this muscle blows unwanted particles away from the mouth or the atriopore. The ascending branch runs up the outer wall of the atrium where it breaks into a plexus from which fibres pass into the gill bars and down to the endostyle. According to Bone (1961) and earlier workers the cell bodies of all the sensory nerves of amphioxus are peripheral and send axons to the central nervous system. This result was denied by Bone & Best (1978) but confirmed on the basis of serial-section electron microscopy by Baatrup (1981). The dorsal-root nerves of the right side alternate with those of the left. This is a consequence of the corresponding alternation in the myocommata. According to Owsjannikow (1867) and Schneider (1879), the nth left myocomma is half a segment behind the nth right one, whereas according to Franz (1927) the converse is true. Probably both situations can exist.

The synaptic contacts made by the muscle-tails of the muscle blocks with the nerve cord (i.e. the contacts which most workers until recently supposed to be the bases of vertebrate-like ventral root nerves) are situated on the lateral faces of the dorsal nerve cord (fig. 3.21).

As regards function, the nerves to the dorsal nerve cord of amphioxus are partly sensory and partly visceromotor (in supplying the musculature and cilia of the gut and the pterygial and other non-somatic muscles). The direct synaptic contacts with the dorsal nerve cord, on the other hand are somatic motor in vertebrate terms since they supply the muscle blocks (as well as the notochordal muscles which vertebrates do not have).

The histology of the dorsal nerve cord and the nervous system in general is exceedingly complex. It was described in detail by Bone (1960, 1963) though this work, because of its date, now needs revision as concerns the innervation of the notochord and muscle blocks (see also Guthrie, 1975). Amphioxus differs from vertebrates most spectacularly in lacking spinal ganglia – or indeed any ganglia outside the central nervous system. The spinal ganglia of vertebrates carry bipolar cells of neural crest origin which are consequently also absent in amphioxus, whose sensory cell bodies are peripheral. (Fossil evidence indicates that the absence of ganglia in amphioxus is secondary, for trigeminal and spinal ganglia both existed in the stem-acraniate *Lagynocystis*.) Bone suggests that the absence of ganglia may be connected with the poor blood supply of amphioxus which requires all cell bodies to be in contact with the central canal of the dorsal nerve cord for nutritive reasons (1960, p. 35).

The giant-fibre system of the dorsal nerve cord of amphioxus has attracted much attention. The classical description of this system was by Rohde (1887) after whom the cells are named. There is an anterior group situated just behind the brain and a posterior group in the tail. The two groups are separated by a region without Rohde cells. The Rohde cell bodies are situated across the median plane and straddle the dorsal part of the central cleft of the dorsal nerve cord (fig. 3.16). The Rohde cells probably function in relaying sensory impulses to the muscle blocks. Guthrie (1975) suggested that the Rohde cells were homologous with a system of giant cells in the dorsal nerve cord of the lamprey, as described by

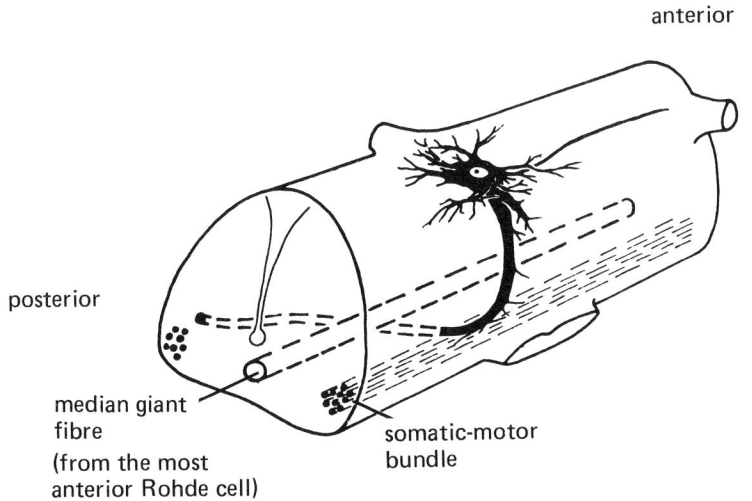

3.16. *An anterior Rohde cell in the dorsal nerve cord of amphioxus; from Barrington (1965, fig. 57, after Bone).*

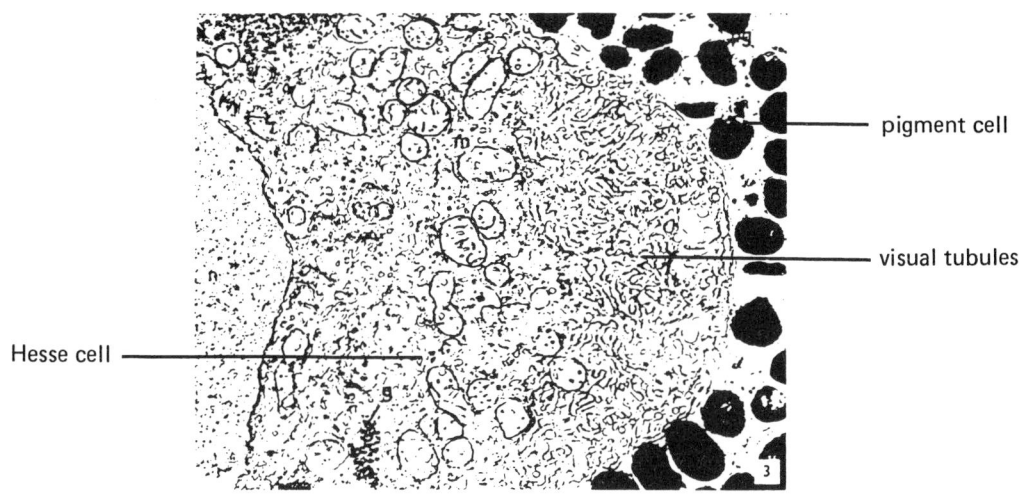

pigment cell

visual tubules

Hesse cell

3.17. A Hesse cell of amphioxus and part of the associated pigment cell; from Eakin & Westfall (1962, fig. 3).

Rovainen (1967) and of *Myxine*, as described by Bone (1963a, b).

Photoreception in the dorsal nerve cord is by means of the so-called Hesse cells. These are approximately spherical cells, each of which is half included in the concave surface of a cup-shaped pigment cell (melanocyte), like an egg in an egg cup (fig. 3.17). An axon leaves the Hesse cell from the surface opposite this melanocyte (i.e. from the uncovered half of the 'egg'). A Hesse cell plus its melanocyte can be referred to as a cup eye. These eyes are arranged in segmental groups lying right and left of the canal of the central nerve cord. In the middle part of the cord there is a region where cup eyes are rather sparse so that the eyes have roughly the same distribution as the Rohde cells. Electron microscopy of a Hesse cell shows that the surface nearest to the melanocyte is increased in area by many narrow tubules extending into the cell (Eakin & Westfall, 1962). This increased surface probably carries visual pigment, in the same manner as in the retinal cell of a vertebrate eye. Physiological experiment (Parker, 1908) shows that amphioxus is most sensitive to light in the anterior and posterior regions of the dorsal nerve cord where cup eyes are most abundant. This confirms that the eyes are optic in function.

The Hesse cells are not homologous with the photoreceptor cells of vertebrates (Eakin & Westfall, 1962, p. 534ff) because the photoreceptive organelles of vertebrates are the outer segments of the retinal cells and are modified cilia belonging to the epithelial cells that surround the lumen of the central nervous system (ependymal cells). As sign of their ciliary origin the outer segments of vertebrate retinal cells carry nine fibrillae arranged in a circle and passing down to a centriole in the cell body. The surface area of the outer segments is increased by plate-like invaginations perpendicular to the length of the cilium. By contrast, the Hesse cells are not ependymal in definitive position (though their origin is unknown) but are buried in the wall of the nerve cord. Moreover, the presumed photo-sensitive part – the region with tubulae next to the melanocyte – is not a modified cilium and the manner in which its surface area is increased is different from that of a vertebrate retinal cell. As discussed later, there are two further types of photoreceptive cells in the brain of amphioxus.

The brain of amphioxus (figs 3.18, 3.20, 3.21) is a poor thing and degenerate compared with that of the fossil *Lagynocystis*. Its posterior limit is defined as lying just in front of the most anterior Rohde cell. The most obvious feature of the brain is its internal cavity, or ventricle, which is an anterior expansion of the central canal of the nerve cord.

The histology of the brain of amphioxus has been described by Meves (1973). Some points deserve mention. Firstly, in the dorsal wall of the brain, just behind the ventricle, there are the two types of visual cells – the cells of Joseph and the lamellar cells. In each Joseph cell (fig. 3.19), according to Welsch (1968), there is a large vacuole parallel to the surface of the cell and the floor of this vacuole is thrown into innumerable projections called microvilli. These microvilli are presumably the light-sensitive portion of the cell. In the lamellar cells, on the other hand, the external surface of the cell is increased in

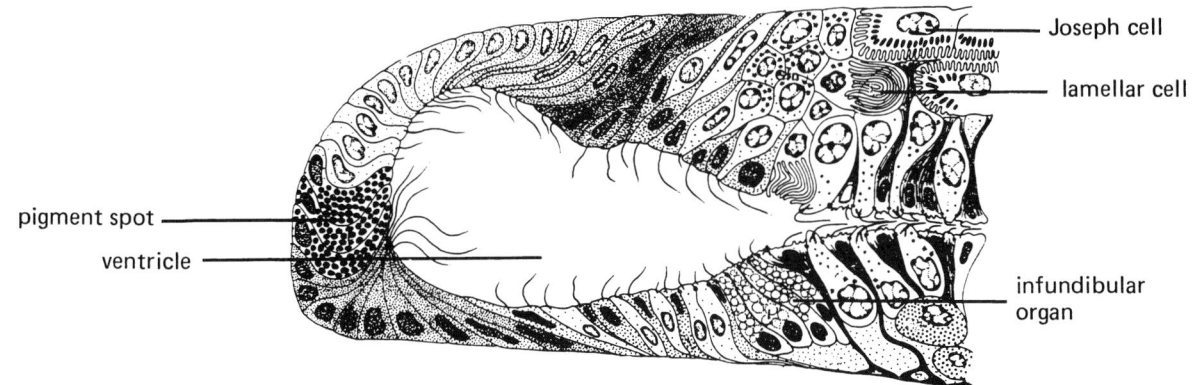

3.18. *Sagittal section through the 'brain' of amphioxus to show the histology (semi-diagrammatic); from Meves (1973, fig. 9, p. 527). The infundibular organ secretes Reissner's fibre.*

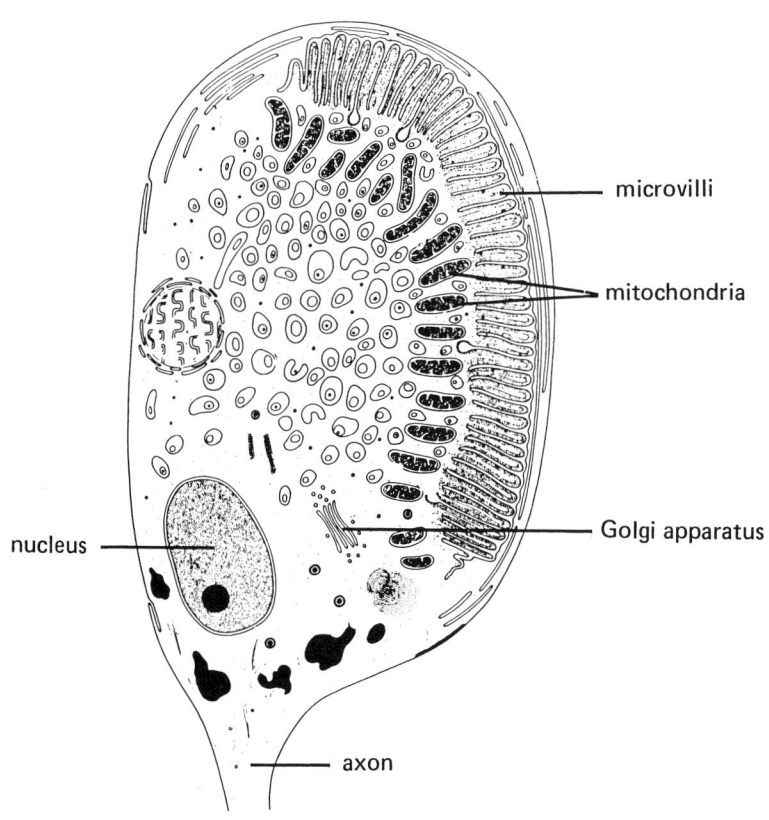

3.19. *A Joseph cell of amphioxus (diagrammatic); from Welsch (1968, fig. 1, p. 254).*

area by being thrown into lamellae. According to Eakin & Westfall (1962) and Eakin (1963, 1968) these lamellae arise in connection with a cilium, as do the membranes in the outer segments of vertebrates and tunicate retinal cells. If this is true the lamellar cells of amphioxus may be homologous with these retinal cells. Meves, however, found no evidence that the lamellae in amphioxus are connected with a cilium. A pigment spot is developed at the anterior

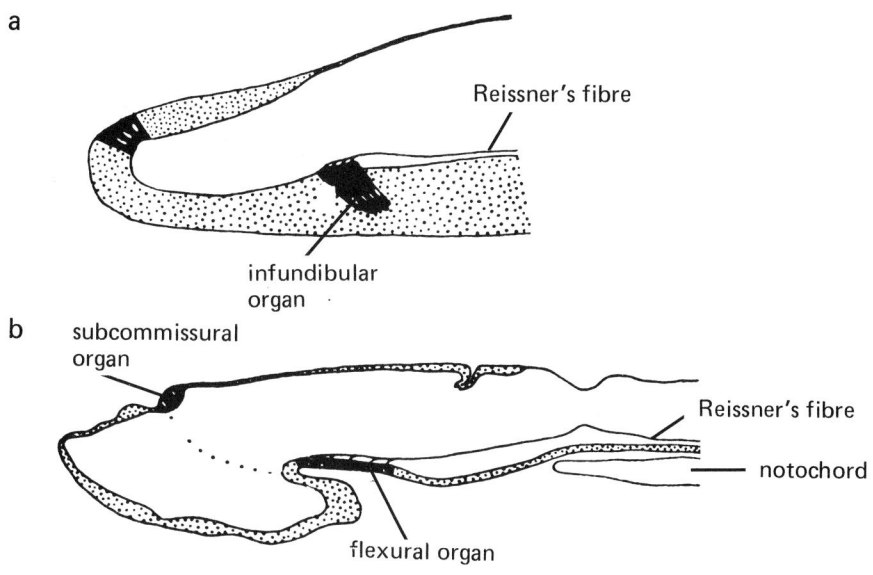

a

Reissner's fibre

infundibular organ

b   subcommissural organ

Reissner's fibre

notochord

flexural organ

3.20. Reissner's fibre in the brains of amphioxus a) and an embryo salmon b); from Barrington (1965, fig. 53) after Olsson & Wingstrand (1956).

end of the brain of amphioxus (fig. 3.18). Possibly this is connected with an optic function, by throwing a shadow on the visual cells of the posterodorsal portion of the brain.

Another peculiar feature of the brain is the so-called infundibular organ (fig. 3.20). This is a patch of specialized cells in the floor of the brain just behind the ventricle. Olsson & Wingstrand (1954) showed that these cells each secrete a non-living fibre known as a Reissner's fibre. Shortly behind the infundibular organ, the individual fibres unite to form a single fibre which passes posteriorly inside the central canal of the nerve cord to end as an agglomeration in a vesicle at the posterior end of the cord (Olsson, 1956). There is a Reissner's fibre in most vertebrates. In adult vertebrates this is secreted by the subcommissural organ which is a part of the *dorsal* ependyma between the diencephalon and the optic lobes. In an early salmon embryo, however, Reissner's fibre is secreted by a region of ventral ependyma (flexural organ) in the anterior floor of the mid-brain. This embryonic region of the salmon could be exactly homologous with the infundibular organ of amphioxus. Olsson has demonstrated (1972) that tunicate tadpoes also secrete a Reissner's fibre.

Nerves from the brain are numerous and complex. The first pair (1) comes off anteriorly from the ventral part of the brain and supplies the anterior lobe of the median fin. Its ganglion cells, according to Franz (1927, p. 528) are the corpuscles of Quatrefages which are situated peripherally in the skin of the rostral lobe. Just posterior and dorsal to the first pair of nerves, there is a tiny ciliated pit on the

left surface of the animal. This is Kölliker's pit which is connected with neurons whose axons pass to the brain (fig. 3.21). The function of Kölliker's pit is unknown. By ontogenetic origin the pit is a remnant of the neuropore.

The second pair of nerves (2) is also symmetrical. The nerves leave the dorsal surface of the brain and, like the first pair, help to innervate the anterior lobe. The third pair of nerves are the first to run out along myocommata. They follow the pair of myocommata between what are here regarded as the 2nd and 3rd somites, i.e. between the first and second muscle blocks. Like the myocomma that it runs in, the right 3rd nerve is either anterior or posterior to the left 3rd nerve. The 4th to 7th nerves emerge from the brain and run along myocommata out to the skin before branching into dorsal and ventral branches. Each pair is unsymmetrical in its origin from the brain, with the right nerve either anterior or posterior to the left one, corresponding to the myocommata.

An even more remarkable asymmetry affects the peripheral courses of nerves 3 to 7 (fig. 3.4). The left 3rd and 4th nerves, but not the right ones, send a 'visceral' branch round the ventral margin of the muscle blocks to the internal surface of the buccal cavity and, in the case of the fourth left nerve, to the right part of the velum. Moreover the 4th to 7th left nerves are connected by visceral branches to the ring nerve of the velum which has no other supply. The buccal cavity and velum are thus innervated entirely by the 3rd to 7th nerves of the left side. The corresponding right nerves have no visceral branches

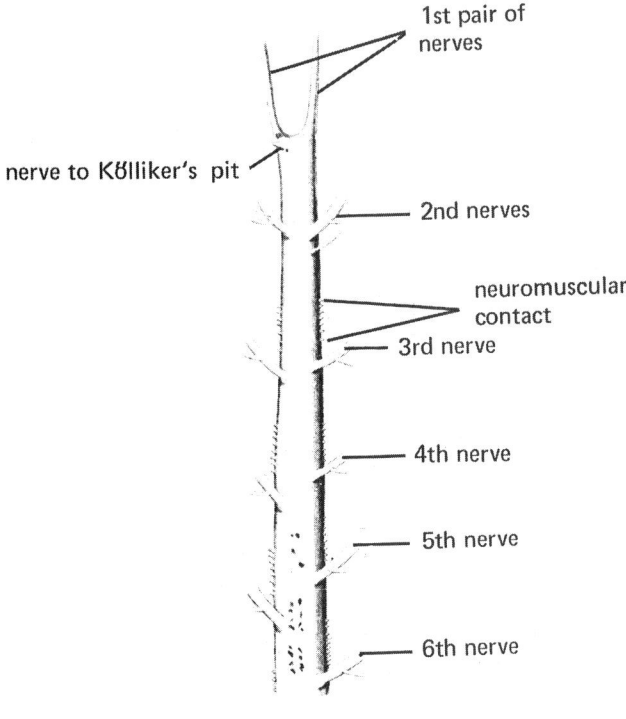

1st pair of
nerves

nerve to Kölliker's pit

2nd nerves

neuromuscular
contact

3rd nerve

4th nerve

5th nerve

6th nerve

of a single-layered epithelium, each cell of which has a cilium on the surface facing the lumen. Other cells constituting the nephridia are the solenocytes or cyrtopodocytes described later. These cyrtopodocytes are arranged in bundles each of which is attached to a diverticulum of the nephridial wall.

The nephridia are closely associated with coeloms – Hatschek's nephridium is just right of a special coelom derived from the myocoel of the left mandibular or second somite. The branchial nephridia are lateral to the branchial-bar and sub-notochordal coeloms except for the most anterior pair of branchial nephridia which are anterior to the foremost gill slits and which are therefore lateral to the epipterygial coeloms. The branchial nephridia are separated from the coeloms by the coelomic epithelium, except where this epithelium is penetrated by bundles of cyrtopodocytes. Hatschek's nephridium is likewise separated from its special coelom by a rather irregular coelomic epithelium (Goodrich, 1910, p. 197).

The nephridia are also closely associated with blood vessels, as might be expected of excretory organs. Hatschek's nephridium runs parallel to the anterior part of the left aorta, separated from it only by the special coelomic cavity. The branchial nephridia are each median to a small plexus of blood vessels. The cyrtopodocyte-carrying diverticula of the nephridia project laterally into the interstices of the blood plexi.

The cells known as solenocytes (Goodrich, 1902, solen = pipe) or better as cyrtopodocytes (Kümmel, 1967) are the most remarkable feature of the nephridia of amphioxus (fig. 3.22). Each has a cell body with a nucleus. From each cell body two structures extend. The first of these is a 'foot process' which spreads over the surface of the neighbouring blood vessel. The foot process has lobate edges, and the lobes of adjacent foot processes interdigitate, leaving only a narrow zig-zag strip of exposed blood-vessel wall between. The second structure extending from the cell body is a slender cylindrical cage of ten bars, with a flagellum in the central axis of the cage. The bars of the cage extend through the coelom from the cell body to the wall of the nephridium and penetrate the epithelium of this wall so that the central space of the cage opens into the lumen of the nephridium. Each bar is triangular in section, with one point of the triangle directed centrally and is formed of an electron-dense material which is not ordinary cytoplasm (Brandenburg & Kümmel, 1961; Kümmel, 1967; Nakao, 1965). The bars are separated by spaces somewhat less than 1 μm wide, so that the liquid in the coelom is actually continuous, via these spaces, with the liquid inside the cylindrical cage, and hence with the lumen of the nephridium and indirectly with the atrium and the outside world. The flagellum inside the cage has its proximal end fixed to the cell body

(van Wijhe, 1893, 1902: Franz, 1923, 1927). As already mentioned the coeloms and muscles of the buccal cavity are likewise peculiar since they arise embryologically from the left side and the outer lip coelom and muscle on left and right sides retain a connection with the left pterygial coelom, but not with the right pterygial coelom, even in the adult.

The excretory organs of amphioxus are small tubes known as nephridia. Two types of these exist, according to position: they are Hatschek's nephridium (fig. 3.4) and the paired or branchial nephridia (fig. 3.7). Hatschek's nephridium is a single straight tube left of the anterior part of the notochord; it ends blind anteriorly but opens posteriorly into the roof of the pharynx, just behind the velum. The number of branchial nephridia, by contrast, is equal to the number of primordial gill bars. Each branchial nephridium is ⌐-shaped, with a vertical and a horizontal limb. The vertical limb runs down a primary gill bar, posterior to the coelom of that bar, and ends blind. The horizontal limb runs over the top of the gill slit next behind the bar that carries the vertical limb, and likewise ends blind. The lumen of the nephridium opens into the atrium by a hole just external to a tongue bar in the ventral part of the horizontal limb of the nephridium. Both Hatschek's nephridium and the branchial nephridia are made up mainly

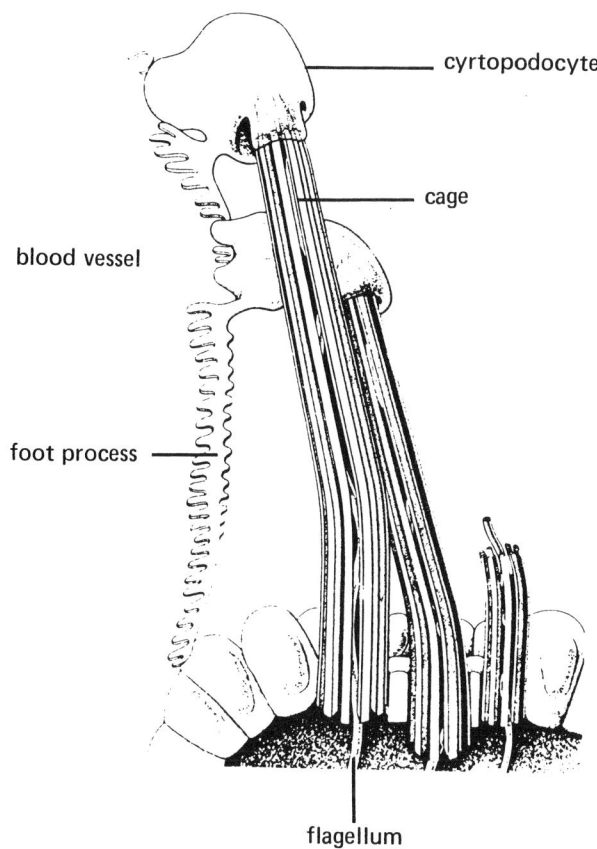

3.22. Diagrammatic reconstruction of some cyrtopodocytes of amphioxus; from Brandenburg & Kümmel (1961, fig. 19, p. 447).

and beats towards the nephridium. The flagellum would thus reduce the pressure in the coelom and cause excretory products to be filtered through the wall of the blood vessels, by way of the narrow zig-zag gaps between neighbouring foot processes (Brandenburg & Kümmel, 1961, p. 450). Nakao (1965, p. 10, fig. 4) has demonstrated that Hatschek's nephridium differs from the branchial nephridia in having an incomplete membrane between the bars of the cages of cyrtopodocytes.

The history of research on the nephridia is complicated. They are not easy to see and were first described, independently and almost simultaneously, by the Englishman Weiss and the German Boveri (both 1890). The first extended description, however, was given by Boveri (1892) who considered that the nephridia were homologous with the pronephric ducts of vertebrates. In 1902 Goodrich discovered the flagellum inside the cage of the cyrtopodocyte, but, having only an optical microscope, he described the cage as a tube with continuous walls. He homologized the cyrtopodocytes with the soleno-

cytes of annelids. Information on the ontogeny of the nephridia was supplied by Legros (1909, 1910) and by Goodrich (1910, 1934) and will be discussed below. In 1933 Goodrich compared the nephridia of amphioxus (*Branchiostoma*) with the rather similar ones of *Asymmetron*. The electron microscope brought new insight into the nature of the cyrtopodocytes (Brandenburg & Kümmel, 1967; Kümmel, 1967; Nakao, 1965) by showing that the flagellum is suspended inside a cage, rather than in a slender cylindrical tube. This cage is different from that of annelids in that the bars are separate from each other (for even in Hatschek's nephridium the membrane connecting the bars is incomplete) whereas in annelids the bars are connected by basement membrane. Goodrich's suggested homology between the cyrtopodocytes of amphioxus and the solenocytes (or cyrtocytes) of annelids thus became less appealing. The resemblances could well be convergent.

The reproductive system of amphioxus consists of a left and right row of gonads projecting into the atrium from its lateral wall. The sexes are separate but there seem to be no anatomical differences between male and female except within the gonads.

The medial face of the gonads is constricted by the cardinal vein of the left or the right side: a plexus of vessels goes off from each vein into the interior of each gonad.

The vascular system of amphioxus is difficult to investigate because the blood is colourless and blood cells are few, and because the endothelial lining of the vessels is discontinuous. Rähr (1979) has recently redescribed the system using a perfected injection technique and confirming his statements by photographs of whole mounts. His account is more complete and evidently more accurate than those of previous authors (e.g. Franz, 1933) and is followed here.

The vascular system (figs 3.23, 3.24) is divided into a venous and an arterial part connected to each other by plexis. The major vessels can mostly be homologized with those of vertebrates and on this basis can be called veins or arteries. However, there is no heart and also no certainty that the blood always flows in the same direction as in a vertebrate. Rähr writes as though it does, on the basis of his own observations (1979, p. 5), but Skramlik (1938), Wolf (1940, 1941) and Azariah (1965) described the flow as reversible and irregular.

The main vessels of the arterial system are as follows: (1) a 'sinus venosus' is located behind the pharynx, anterior to the meeting place of the paired Cuvierian ducts with the efferent 'liver' vessel – its position thus corresponds to that of a vertebrate or tunicate heart; (2) an endostylar vessel extends forward from the sinus venosus beneath the endostyle; (3) a pair of dorsal aortae is present above the gill bars, immediately beneath the notochord, on either

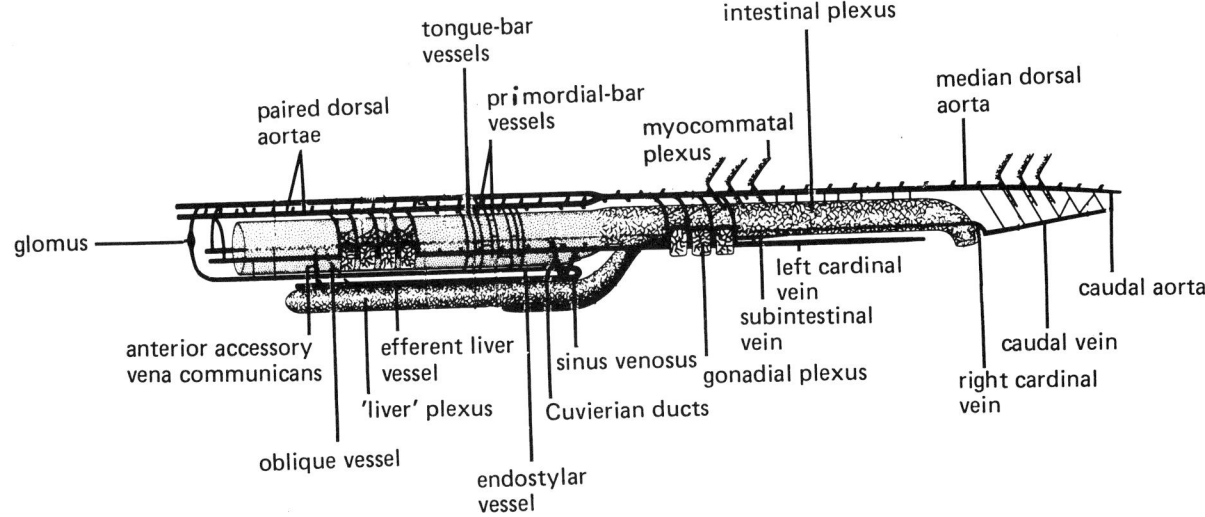

3.23. *Blood-vascular system of amphioxus; from Rähr (1979, fig. 1).*

side of the epipharyngeal groove; and (4) a median dorsal aorta lies in the mid-line immediately beneath the notochord behind the pharynx – at its anterior end it divides into the paired dorsal aortae, while posteriorly it is connected to a short vertical caudal aorta which opens into the caudal vein.

The branchial vessels within the arterial system include three vessels in each primordial gill bar, two of which are directly connected with the endostylar vessel. These vessels in the primordial bars are connected by vessels in the synapticulae to a pair of vessels in each tongue bar. Dorsal to the gills some of these branchial vessels break up into plexi associated with the nephridia. Other branchial vessels lead directly into the paired dorsal aortae, which also receive vessels from the nephridial plexi. A series of transverse connections above the epipharyngeal groove link the left and right aortae together. The endostylar vessel is connected at its anterior end to the right dorsal aorta by way of an expanded vessel called the glomus which is located just right of the velum in the right epipterygial coelom. There is no corresponding vessel on the left.

The main vessels of the venous system are as follows: (1) the right and left cardinal veins – these run longitudinally through most of the length of the animal. In the pharyngeal region they are situated just right and left of the atrium medial to the gonads. Behind the pharynx they run right and left of the gonadial coeloms. Each is divided into an anterior portion, in front of the junction with the left or right Cuvierian duct, and a posterior portion, behind this junction. (2) A caudal vein is situated in the median plane behind the anus – at its anterior end it joins the right

cardinal vein and at its posterior end joins a vertical caudal aorta, which leads into the median dorsal aorta. (3) A subintestinal vein lies just beneath the intestine in the mid-line and runs forward into the so-called 'afferent liver vein' which is on the ventral surface of the mid-gut diverticulum. (4) An 'efferent liver vein' runs along the dorsal surface of the mid gut diverticulum – at its posterior end this vein connects with the sinus venosus and the paired Cuvierian ducts. (5) A pair of Cuvierian ducts, or sometimes more than one pair, connect the right and left cardinal veins to the sinus venosus. (6) The vessels, the anterior accessory vena communicans and the obique vessel, connect the 'efferent liver vein' with the right cardinal vein.

The plexi of the vascular system are numerous: (1) there is a plexus associated with each nephridium as already mentioned; (2) in each myocomma there is a plexus connected with the dorsal aortae by way of a short dorsal segmental artery – Franz (1933) thought that the dorsal segmental arteries communicated, not with myocommatal plexi, but directly into the sclerocoels, which would thus have been filled with blood. Rähr's disproof of this view is his most striking result. Behind the anus (fig. 3.24) the myocommatal plexi are connected to segmental connecting vessels which run in the median plane from the dorsal aorta down to the caudal vein. Vertebrate analogies suggest that blood would flow in the post-anal region out of the median dorsal aorta, into the dorsal segmental arteries, though the myocommatal plexi, into the median segmental connecting vessels, and from these into the caudal vein. In front of the anus, the myocommatal plexi may be similarly connected with the cardinal veins, but

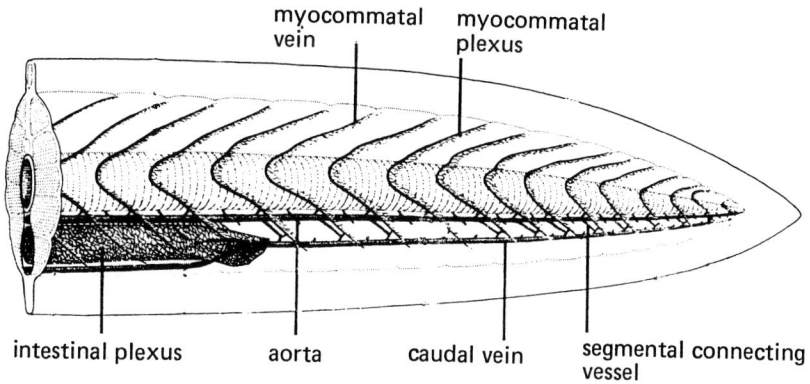

3.24. Blood-vascular system of the posterior part of amphioxus; from Rähr (1979, fig. 16).

Rähr was unable to find such a connection. (3) There is an atrial plexus, in the dorsal wall of the atrium; which is connected with the paired dorsal aortae by way of ventral segmental vessels. Ventrally this plexus is connected with the gonadial plexi and with the cardinal veins. (4) Gonadial plexi are connected with the atrial plexi and also with the cardinal veins. (5) An intestinal plexus covers the wall of the intestine; it is connected by vessels to the dorsal aorta and also to the subintestinal vein. (6) A 'liver' plexus in the mid-gut diverticulum is continuous posteriorly with the intestinal plexus and is connected with the 'afferent liver vein' beneath and the 'efferent liver vein' above (I put these terms in quotation marks because it is doubtful whether the mid-gut diverticulum is homologous with the vertebrate liver and also doubtful whether the flow of blood is regularly from the 'afferent' to the 'efferent' vessel).

In summary, the vascular system of amphioxus has no sinuses and most of its chief vessels can be homologized with those of vertebrates. In lacking sinuses it resembles gnathostomes but differs from tunicates, myxinoids and lampreys (branchial sinuses). Several asymmetries exist in the vascular system in the form of structures which are present on the right but not the left. Such are the glomus, the anterior accessory vena communicans, the oblique vessel and the junction of the right cardinal vein with the caudal artery. Unfortunately little can be said about the vascular system of calcichordates so the phylogenetic significance of these asymmetries is unclear.

The embryology of amphioxus is most remarkable. In particular a late larval amphioxus is a grotesquely unsymmetrical animal and most of its asymmetries, as I shall try to show later, are straightforward phylogenetic recapitulations. The first paper to give an accurate description of an amphioxus larva was that of Müller (1851). He noticed the asymmetry of the gill slits. Other early descriptions of the larva were those of Schultze (1852) and Leuckardt &

Pagenstecher (1858). The first extensive description was written by the brilliant Russian embryologist Kowalevsky (1867) and he corrected and expanded this first work in 1876. The Austrian Hatschek wrote a still more extensive study of the stages up to the appearance of the first gill slit (1881), and this is in many ways still the best account of the subject. Conklin (1932), however, covered very much the same ground, confirmed that Hatschek's account was almost completely correct and gave new illustrations. Lankester & Willey (1890) gave an account, more accurate than previously, of the development of the atrium and of the first gills (which belong morphologically to the left side). They also recognized the larval endostyle. Willey (1891) described in better detail than Kowalevsky the truly extraordinary development of the first right gill slits and the metamorphosis of the larva to the symmetrical adult condition. The best account of the development of the nephridia is that of Goodrich (1934), but see also Legros (1909). The best account of the development of the so-called club-shaped gland was given by Goodrich (1930). The origin of the coeloms of the mouth region was worked out by Legros (1898) and van Wijhe (1914).

The egg of amphioxus is small with little yolk and its fate map resembles that of ascidians in its distribution of endoderm, mesoderm, chordomesoderm, neurectoderm and ectoderm (Conklin, 1932). The first division splits the egg into two halves, corresponding to right and left halves of the adult. The second division separates two yolky presumptive endoderm cells from the less yolky presumptive mesoderm and ectoderm cells. Divisions continue synchronously until the 128 cell stage (7th division). By this time the embryo is a hollow spherical blastula (fig. 3.25), made of relatively large yolky endoderm cells at one pole that grade into smaller, non-yolky ectodermal cells at the other pole.

Gastrulation begins when one surface of the blastula,

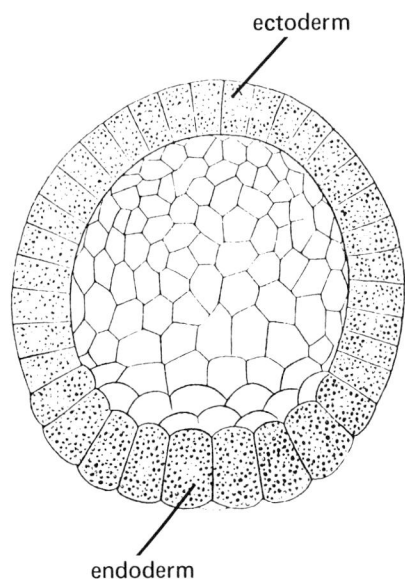

3.25. A blastula of amphioxus in optical section; from Hatschek (1881, pl. 2, fig. 18).

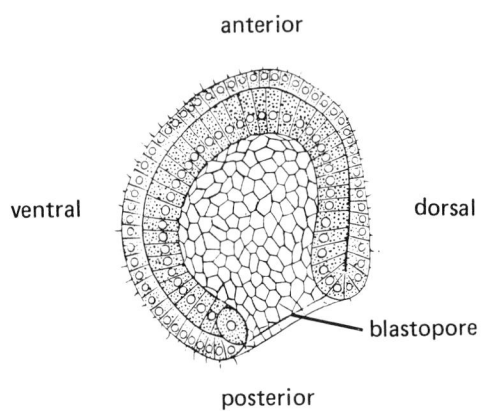

3.26. A gastrula of amphioxus from the left side; from Hatschek (1881, pl. 2, fig. 18).

made of the relatively yolky endoderm cells, becomes slightly flattened. The flattening deepens to form a depression, which pushes into and almost eliminates the internal cavity of the blastula (blastocoel). Into the depression the sheet of presumptive mesoderm and chordomesoderm flows. At the end of the gastrulation (fig. 3.26) the embryo is almost an ellipsoid of revolution, two cell layers thick, with a small blastopore at the posterior end, i.e. at what will become the posterior end of the animal. The postero-dorsal half of the surface layer of cells is neurectoderm, while the antero-ventral half is ectoderm. The inner cell layer of the gastrula consists of chordo-mesoderm dorsally, normal mesoderm dorso-laterally and posteriorly, and endoderm antero-ventrally. Each cell of the external surface of the gastrula, whether ectoderm or neurectoderm, has developed a single cilium. By means of the cilia the larva swims inside the egg membrane with a rotation anti-clockwise as seen from behind (Hatschek, 1881, p. 37). Shortly afterwards the egg membrane ruptures and the larva escapes into the water, still rotating, in the same sense, as it swims.

The next stage (figs 3.27a, b, c) shows on the outside of the embryo as neurulation, which is the sinking in of the neurectoderm to form the dorsal nerve cord. This involves at first a slight dorsal flattening of the gastrula. This flattening becomes a broad shallow groove ending posteriorly at the blastopore. The ventral lip of the blastopore is continuous with the ridges that form the side of the broad groove – these ridges form an approximate V with

the ventral lip at its point. Gradually the ridges coalesce from behind forwards and grow forwards over the floor of the original shallow groove. The V thus becomes progressively more obtuse (when seen in dorsal aspect) until finally only a small opening, the neuropore, remains near the front end of the animal. The floor of the original groove, which will be the future dorsal nerve cord, thus comes to be covered by a roof of ectoderm. The lateral edges of this floor curl upwards, meet together and fuse in the mid-line giving the tube which the nerve cord basically is. This tube for a long while remains connected posteriorly with the archenteron by the so-called neurenteric canal. The larva which is undergoing, or has just completed, neurulation is called the neurula.

In the inner cell layer of the neurula many changes are happening meanwhile. Along the mid-dorsal line of the archenteron the epithelium pouches upwards to form the notochord. Just to the left and right of the notochord the archenteric epithelium pouches upwards dorso-laterally to provide the anlagen of the dorso-lateral series of muscle blocks (i.e. the anlagen of the mandibular and more posterior somites). The remaining, more ventral parts of the inner epithelium represent the anlage of the gut and of the first pair of somites (premandibular) (figs 3.28a, b). The dorso-lateral somitic ridges soon develop a pair of transverse constrictions – the posterior limits of the future second somites. Afterwards other transverse constrictions develop behind these, being the posterior limits of the third somites, fourth somites and so on. The more anterior and first-formed few somites contain a coelomic cavity from the start, this being by origin a portion of the archenteron. (This mode of producing a coelom is known as enterocoely.) More posterior somites, however, are at first solid. Coeloms arise in them later, by a splitting process known as schizocoely. After a time the intersomitic

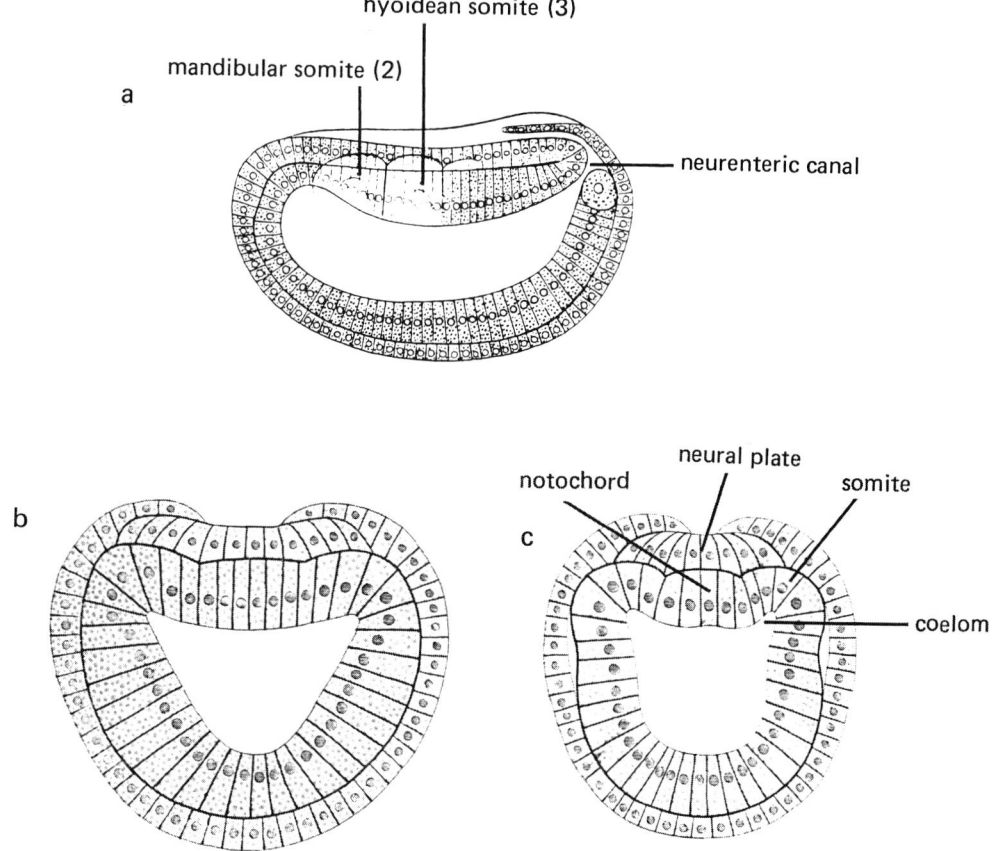

3.27. *Neurulation of amphioxus and the beginning of differentiation of somites 2 and 3. a) Left aspect – the special circular cell ('polar cell') drawn by Hatschek just ventral to the neurenteric canal does not in fact exist; b) transverse section at the transverse level of somite 2; c) transverse section in the posterior quarter of the body; from Hatschek (1881, pl. 3, fig. 37; pl. 7, figs 77, 78).*

boundaries of the right side are found to have shifted half a somite rearwards compared with the left side, or perhaps sometimes those on the left shift half a somite rearwards compared with those on the right. The asymmetry which thus arises persists in the adult.

The notochord at its first origin does not extend to the front end of the inner cell layer of the neurula (fig. 3.28), for initially there is a small portion of the archenteron in front of it. This portion dilates upwards at the same time as the notochord grows forwards. As a consequence two upward pockets of the archenteron arise at right and left of the extended front end of the notochord. These are the anterior gut diverticula (Hatschek's 'vordere Darmdivertikeln'). Their complicated fate will be described later. In vertebrates they can best be homologized with left and right premandibular (or first) somite *plus* the pre-oral gut (Koltzow, 1901, p. 412 ff). This is the reason why the

most anterior of the dorso-lateral somites are here called the second or mandibular somites. The anterior gut diverticula originate late. They do not start to develop until the 9th or 10th somites have formed at the posterior end of the dorso-lateral series.

The dorso-lateral somites begin to extend downwards on either side of the archenteron until finally they meet their antimeres in the ventral mid-line, ventral to the gut. The somites of left and right sides fuse together here and the transverse walls between successive somites break down (fig. 3.29). Consequently signs of segmentation persist only dorsally.

At the same time the first pair of somites (anterior gut diverticula) separate entirely from the archenteron and become asymmetrical (fig. 3.29). The left one becomes thick-walled without much increase in size while the right one becomes thin-walled and expands. The left one

a

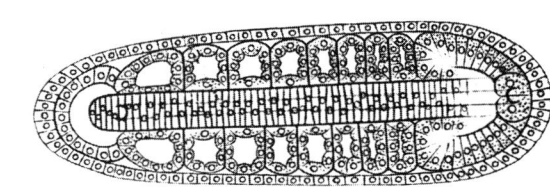

b

neuropore

mandibular somite = 2

anterior gut
diverticulum
(premandibular somite = 1)

somites 3,4

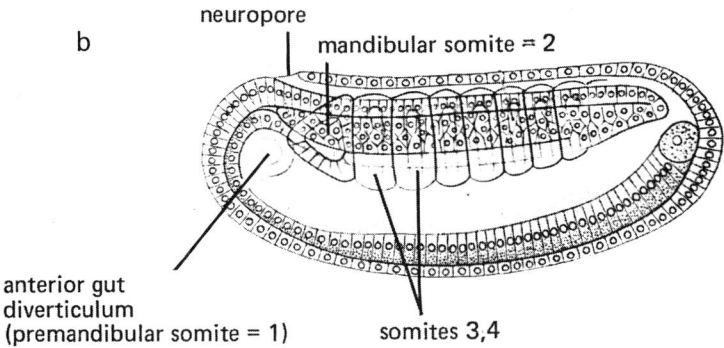

3.28. *Late neurula of amphioxus. The anlagen of the anterior gut diverticula have just formed; a) left lateral aspect; b) dorsal aspect. The 'polar cell' near the posterior end of the body is imaginary; from Hatschek (1881, figs 48, 49).*

a

right anterior gut
diverticulum
= ventral rostral
coelom

left anterior gut
diverticulum
(Hatschek's pit)

b

left ant. gut
diverticulum

right anterior gut
diverticulum

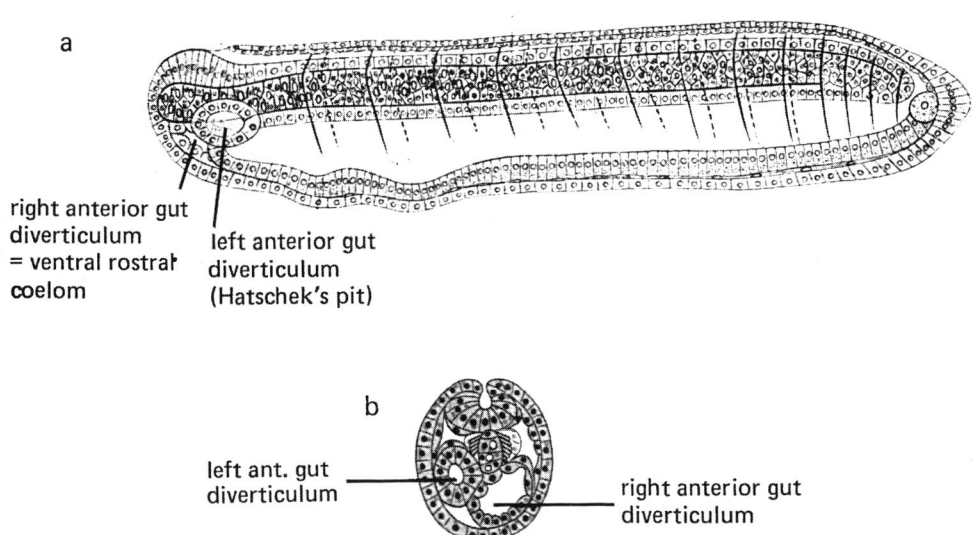

3.29. *Still later neurula of amphioxus. In the ventral part of the animal the division between the somites has disappeared; the right and left anterior gut diverticula are differentiating; the right somites have moved posteriorly with respect to the left ones. a) Left lateral aspect; b) transverse section through the gut diverticula; from Hatschek (1881, figs 54, 129).*

a

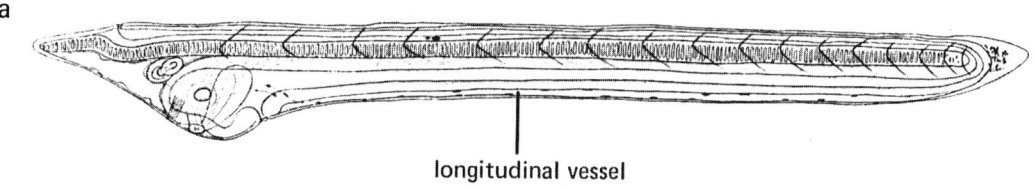

longitudinal vessel

b

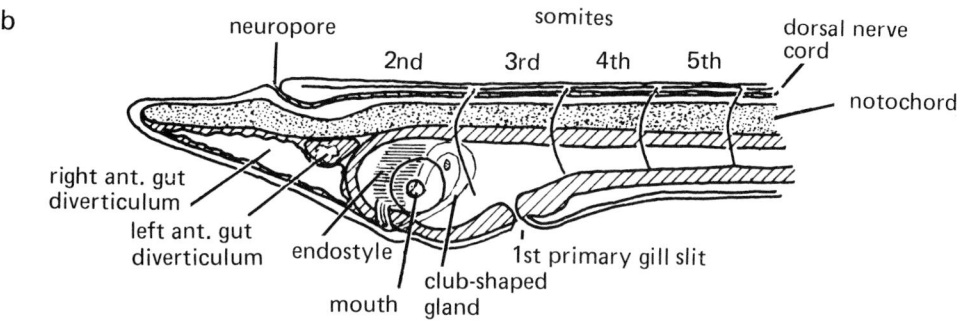

3.30. Larva of amphioxus in which the mouth has just originated in the ventral part of the left mandibular somite (2nd somite). a) General view of left aspect, from Hatschek (1881, fig. 63). b) Redrawing of anterior end of Hatschek's figure with endostyle inserted after Conklin (1932, fig. 110) and internal opening of club-shaped gland after Goodrich (1930); from Jefferies & Lewis (1978, fig. 22e).

later in larval life will develop an opening to the exterior just anterior to the mouth, becoming the pre-oral pit (Hatschek, 1881). The ectoderm ventral to this pit becomes ciliated and this ciliated ectoderm and the pit ultimately form the wheel organ inside the buccal head of the adult (Goodrich, 1917). The right first somite, on the other hand, in expanding forms the most anterior coelom of the larva, mainly ventral to the notochord, and this eventually becomes the ventral rostral coelom of the adult. In the larva this coelom is a totally distinct cavity but in the adult it has become continuous with the general coelom or splanchnocoel. Gradually the larva becomes more elongate and acquires a pointed front end and the beginnings of a caudal fin.

At the anterior end of the pharynx a number of complicated events now happen simultaneously. Firstly, the larval mouth originates. The first anlage of this is a circular thickening of the endoderm in the left wall of the archenteron near the anterior end of the animal in the ventral part of the second or mandibular somite (figs 3.30a, b). This thickening fuses with the ectoderm in contact with it and the mouth appears as a circular hole in the left side of the animal. This mouth is in fact homologous with the velar mouth of the adult, for the larva does not yet possess a buccal cavity. The fact that the mouth first appears in the left mandibular somite is very important by comparison with the mitrates, as will be discussed later (Chapter 8).

The first larval gill slit arises somewhat like the mouth as a thickening of the endoderm. This thickening is situated, however, in the ventral mid-line, ventral to the third somite. The centre of the thickening, as with the mouth, fuses with the ectoderm and develops a through-going hole, which is the gill slit itself. By the time that this first gill slit has originated, a longitudinal vessel is clearly visible running along the mid-ventral line of the larva but bending rightwards anteriorly, so that the gill slit is left of it (fig. 3.30b). This vessel will become the endostylar and sub-intestinal vessels of the adult. The fact that it is right of the first gill slit is the first indication that the slit, although mid-ventral in position, belongs morphologically to the left side of the animal. The left sidedness of this slit and of those which later emerge in a line behind it in the larva, is firmly corroborated by later evidence.

The embryological terminology of the gill slits must now be clarified. As will later emerge, amphioxus has three groups of gill slits, differing in their time of origin. The first (primary) group to originate arises in the larva and belongs morphologically to the left side of the animal. The second (secondary) group arises at the metamorphosis of the larva, thus later than the first group, and belongs morphologically to the right side. The third (tertiary) group, finally, is formed in the post-larva and adult and belongs to both left and right sides. The terms 'primary', 'secondary' and 'tertiary' were proposed by Willey (1891). They

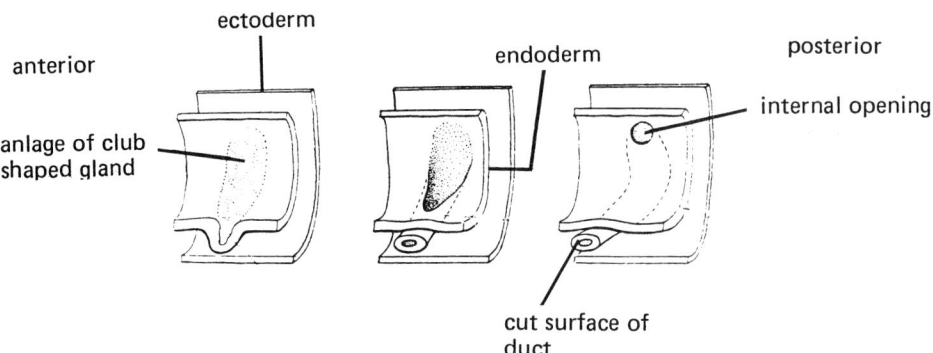

anterior

ectoderm

endoderm

posterior

internal opening

anlage of club
shaped gland

cut surface of
duct

3.31. The origin of the club-shaped gland of amphioxus; from Goodrich (1930, text fig 10). The gland begins as an outpouching of endoderm, like a visceral pouch. Goodrich therefore regarded it as the first but abnormal gill slit of the primary series.

are firmly anchored in usage and cannot be changed but are somewhat unfortunate. For the words 'primary' and 'secondary' are also widely used in the literature to distinguish what I call primordial gill bars from tongue bars.

The so-called club-shaped gland (figs 3.30, 3.31, 3.32) also forms in the pharynx at about this time. The best description of this process was given by Goodrich (1930). The first anlage of the gland (fig. 3.31) is a thickening of the endoderm in the right wall of the archenteron, almost opposite but slightly posterior to the nascent mouth. An evagination of the thickening appears ventrally, being the first anlage of the external duct of the gland. The thickening pouches out from the archenteron, so that the cavity inside the thickening finally becomes separated from the archenteron everywhere except dorsally. At the same time the end of the anlage of the external duct fuses with the ectoderm and acquires an opening on the left, just anterior to the mouth. The main part of the wall of the club-shaped gland becomes glandular. In its development the club-shaped gland resembles a gill slit. Later, it will disappear completely as the endostyle grows backwards dorsal to it between the primary and secondary groups of gill slits. In the adult, no trace of it remains.

The morphological significance of the club-shaped gland has been disputed. Willey (1894, p. 60) thought it was the earliest of the secondary gill slits, i.e. that it belonged morphologically to the right side of the animal. Van Wijhe (1914) (who thought, remarkably enough, that the velar mouth of amphioxus was by origin the most anterior gill slit of the primary group) regarded the club-shaped gland as the right antimere of this gill-slit-mouth. Like Willey, therefore, he regarded the club-shaped gland as an anomalously early member of the secondary group of gill slits. As against this, Goodrich (1930, p. 163) regarded it as the most anterior member, and second to

appear, of the primary group. The fact that its external opening is on the left side supports this view, as does its early time of origin. While these structures are appearing near the anterior end of the alimentary canal, the anus arises posteriorly, just anterior to the caudal fin. At its first origin it is in the ventral mid-line, but soon it comes to lie left of the caudal fin.

The first anlage of the endostyle (fig. 3.30a) arises as a thickening of the right wall of the pharynx just anterior to the club-shaped gland. At first it is an irregular pentagon in outline, but it soon becomes bean-shaped, with the concavity of the 'bean' anteriorly. The peripharyngeal bands of cilia were said by Willey (1891, p. 189) to grow out from the lower and upper (left and right) ends of this endostyle. Both were said to grow rearwards and join in the dorsal mid-line of the pharynx to form the epipharyngeal band. However, Willey also said that the left one became visible before the right one (1891, p. 189). Indeed the left band probably arises in fact before the other, since in the one-gill-slit larva described by Bone (1958), a left peripharyngeal band is recorded but no right one. It would be interesting to know just how the right peripharyngeal band arises but this would need new observations. Its origin seems never to have been properly described in the literature. It is difficult to see how, in the absence of right gill slits, the right and left larval peripharyngeal bands could be distinguished from each other.

The feeding of the one-gill-slit larva was described by Bone (1958a). The larva (fig. 3.32) is about 2 cm long at this stage and swims forward by means of its surface cilia, rotating, usually anti-clockwise as seen from behind but sometimes in the opposite sense. A feeding current is produced mainly by the big cilia round the edge of the gill slit. Particles are drawn in through the mouth by this current and become trapped by a string of mucus that is attached at its anterior end to the peripharyngeal band. Particles

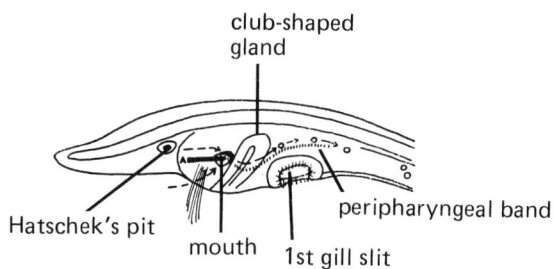

club-shaped
gland

peripharyngeal band

Hatschek's pit

mouth   1st gill slit

3.32. Feeding in the one-gill-slit larva of amphioxus, showing the course followed by the food particles (small circles); from Bone (1958, fig. 2, p. 462).

and mucus are carried upwards and rearwards by this band and thus enter the oesophagus. The mucus seems to be produced by Hatschek's pit, the club-shaped gland and probably by the endostyle. Bone supposes that the endostyle is the main source of mucus in the late larva.

At the one-gill-slit stage, therefore, a larval amphioxus has acquired somewhat the adult shape (fig. 3.30). It is very elongate, with a caudal fin, but protrudes ventrally in the anterior part of the pharynx. This protrusion carries the mouth as a hole in its left wall. Anterior to the mouth is the opening of Hatschek's pit, while ventral to the mouth is the opening of the club-shaped gland. Inside the pharynx the endostyle is developed in the right wall, and the left peripharyngeal band also exists, though the right peripharyngeal band has perhaps not yet appeared. There are up to 20 pairs of somites in the dorso-lateral series, the right intersomitic boundaries being usually posterior to the corresponding left ones, except at the posterior end of the animal. The so-called eye spot of black pigment exists at the anterior end of the dorsal nerve cord and a large mid-ventral blood vessel also exists.

After the one-gill-slit stage several changes occur concomitantly (fig. 3.33). The mouth increases in size by expanding rearwards so that it becomes first elliptical and then lens-shaped with a point anteriorly and posteriorly. It thus comes to occupy much of the left body wall beneath somites 2 to 8. Because the mouth expands, the single gill slit is forced out of the mid-line onto the right side of the animal. New slits are then added one behind the other in line, behind the first gill slits, up to the number of about 14, these being the primary group of slits. The more anterior of these slits, in the mouth region, are in the right wall of the body, but more posterior ones are approximately in the ventral mid-line. A superficial fold, being the anlage of the definitive right metapleural fold, begins to grow downwards over the gill slits while the left metapleural fold begins to grow upward over them.

The nerve supply to the mouth at this stage is not well known in amphioxus itself. Bone (1958, p. 463) stated that

in older larvae of *Asymmetron bassanum* (Günther) the upper lip was innervated from the left dorsal roots, which is what would be expected. He was also able to show that the muscle posterior to the mouth was innervated in *Branchiostoma* from dorsal roots of the left side (Bone, 1959, p. 519). This larval innervation is important in explaining the asymmetry of the nerves to the buccal cavity and velum of the adult, which as already mentioned, is entirely from the nerves 3 to 7 of the left side. As explained below, the larval mouth becomes the velar mouth of the adult; the immediately surrounding body wall becomes the velum of the adult; and the left and right walls of the adult buccal cavity develop as folds adjacent to the prospective velum on the left side of the animal. If these organs are innervated in the larva entirely from the left dorsal roots above them, then the adult asymmetry of innervation would be a direct consequence.

The muscles and nerve supply of these primary slits (fig. 3.34) were described in amphioxus larvae by Bone (1959, p. 518). The gill bars contain unstriated muscle fibres whose function is probably to decrease the size of the slits, thereby restricting the flow of water through them. Bone has shown that, although the primary slits belong morphologically to the left side, their motor nerve supply comes entirely from the dorsal root nerves of the *right* side. This asymmetry is probably not a phylogenetic recapitulation. It is most likely a larval specialization, dependent on the fact that the early gill slits are nearer to the right dorsal roots than to the left ones.

A longitudinal fold of tissue, as already mentioned, has developed in the right wall of the body, dorsal to the slits, and grown downwards so as partly to cover them. This fold is the future right metapleural fold. Then a left fold arises, opposite the posterior part of the right fold (fig. 3.35). On the median face of both folds a subsidiary, subatrial fold arises and these two subatrial folds grow towards each other, meet and coalesce in their posterior part. Gradually the coalescence spreads forward (figs 3.35a, b, c), so forming the floor of the definitive atrium. At first this floor is incomplete, for the more anterior gill slits remain unenclosed, but these also finally become covered by the forward growth of the atrial floor. Posteriorly a permanent opening remains, being the future atriopore. The atrial cavity when first enclosed by the atrial folds is a long narrow canal of ectoderm. Later this canal expands by outpouching laterally, and indeed most of the definitive atrium is formed by these evaginations.

Soon after the atrial folds have begun to coalesce, a thickening appears in the right wall of the pharynx dorsal to (i.e. morphologically right of) the existing primary gill slits and dorsal to the longitudinal blood vessel (fig. 3.36). The anterior end of this endodermal pharyngeal thickening does not extend to the front of the pre-existing gill-slit

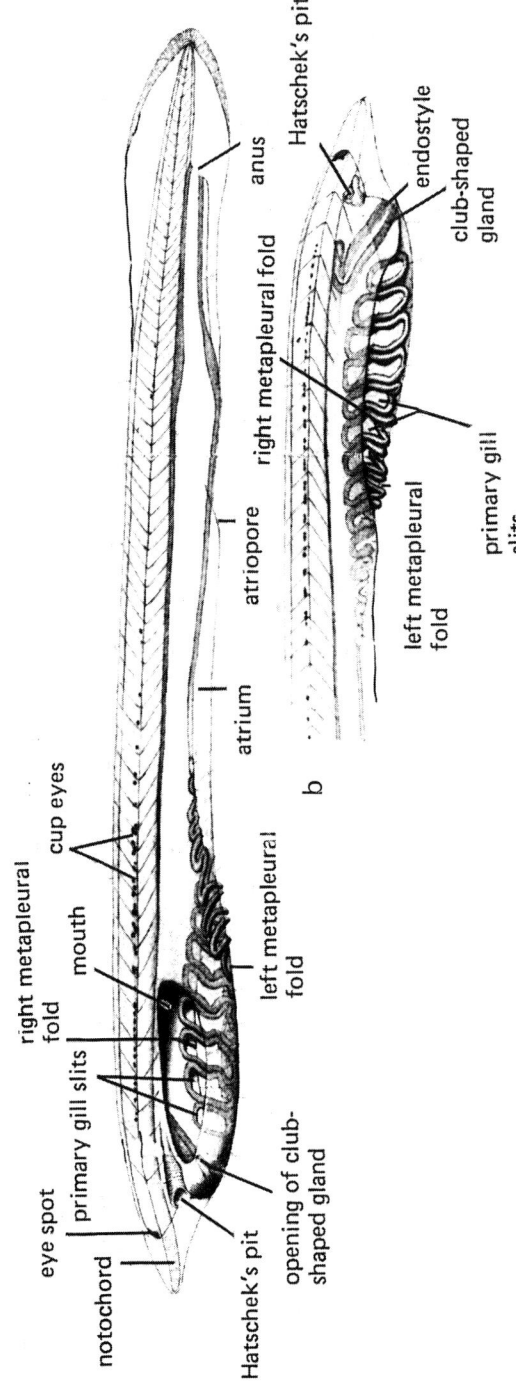

3.33. *The 14-gill-slit larva of amphioxus; a) left aspect; b) right aspect of the same individual. Note the left-sided mouth and the single series of left, primary gill slits; from Lankester & Willey (1890, pl. 19, figs 4, 6).*

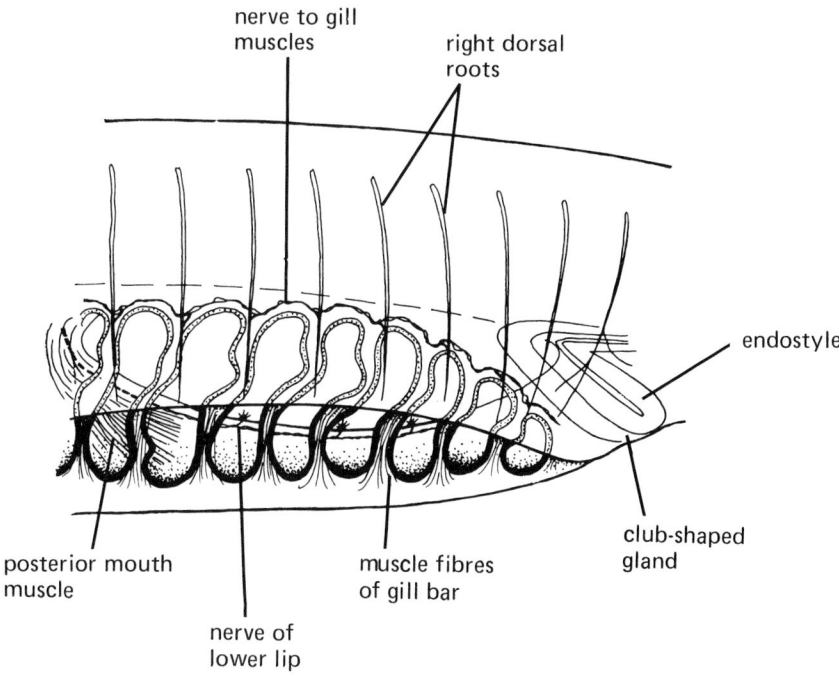

nerve to gill
muscles

right dorsal
roots

endostyle

club-shaped
gland

posterior mouth
muscle

muscle fibres
of gill bar

nerve of
lower lip

3.34. *Innervation of the mouth (probably from the left) and the primary gill slits (solely from the right) in a late larva of amphioxus; right aspect; from Bone (1959, fig. 6).*

series, but stops two or three slits short. Soon a series of dilatations appears in it, each one situated above a pre-existing gill bar. Then holes appear at the centre of four or five of the dilatations simultaneously – through-going holes which connect the pharynx with the atrium which by evagination has grown upwards lateral to them. These holes are gill slits of the secondary series. In the course of time, eight such secondary slits arise. In the light of the fossils it is phylogenetically highly significant that these secondary right gill slits appear more or less simultaneously and later than the primary left ones since right gill slits seem to have appeared suddenly in the mitrates. For most investigators, however, their asymmetrical origin in amphioxus has been an anomaly.

The appearance of the secondary slits marks the beginning of metamorphosis by which the asymmetrical larva gives rise to the almost symmetrical adult. This metamorphosis involves complex changes in several organs, all happening simultaneously, and which are best described organ by organ.

The primary slits are reduced in number during metamorphosis from 14 to 8, becoming equal in number to the secondary slits. This involves the closing and disappearance of the most anterior primary slit (which had been the first one to appear) and also the five most posterior ones

(fig. 3.37). At the same time the secondary slits change in shape, becoming elongate in the transverse plane of the animal. This lengthening drives the surviving primary slits into their definitive position in the left wall of the pharynx. The club-shaped gland likewise disappears.

A further change in shape affects all the slits except the most anterior pair (i.e. all except the most anterior secondary slit on the right and the most anterior surviving primary slit on the left, this being by origin and position the secondary primary slit). This change in shape is a downgrowth of the middle of the dorsal border (fig. 3.38), accompanied by a much shorter upgrowth of the ventral border. Finally, the downgrowth and upgrowth meet and coalesce, giving a tongue bar. Thus arises the distinction between the primordial and tongue bars of the adult.

The endostyle has meanwhile shifted into a mid-ventral position, concomitant with the symmetrization of the gill slits. From being bean-shaped it has now become V-shaped, with the V pointing backwards between the primary and secondary slits (fig. 3.38a). The V gradually becomes more acute, growing rearwards mid-ventrally between the primary and secondary slits to give the symmetrical bilateral definitive endostyle. It is likely that the flagellate median strip of the definitive endostyle develops

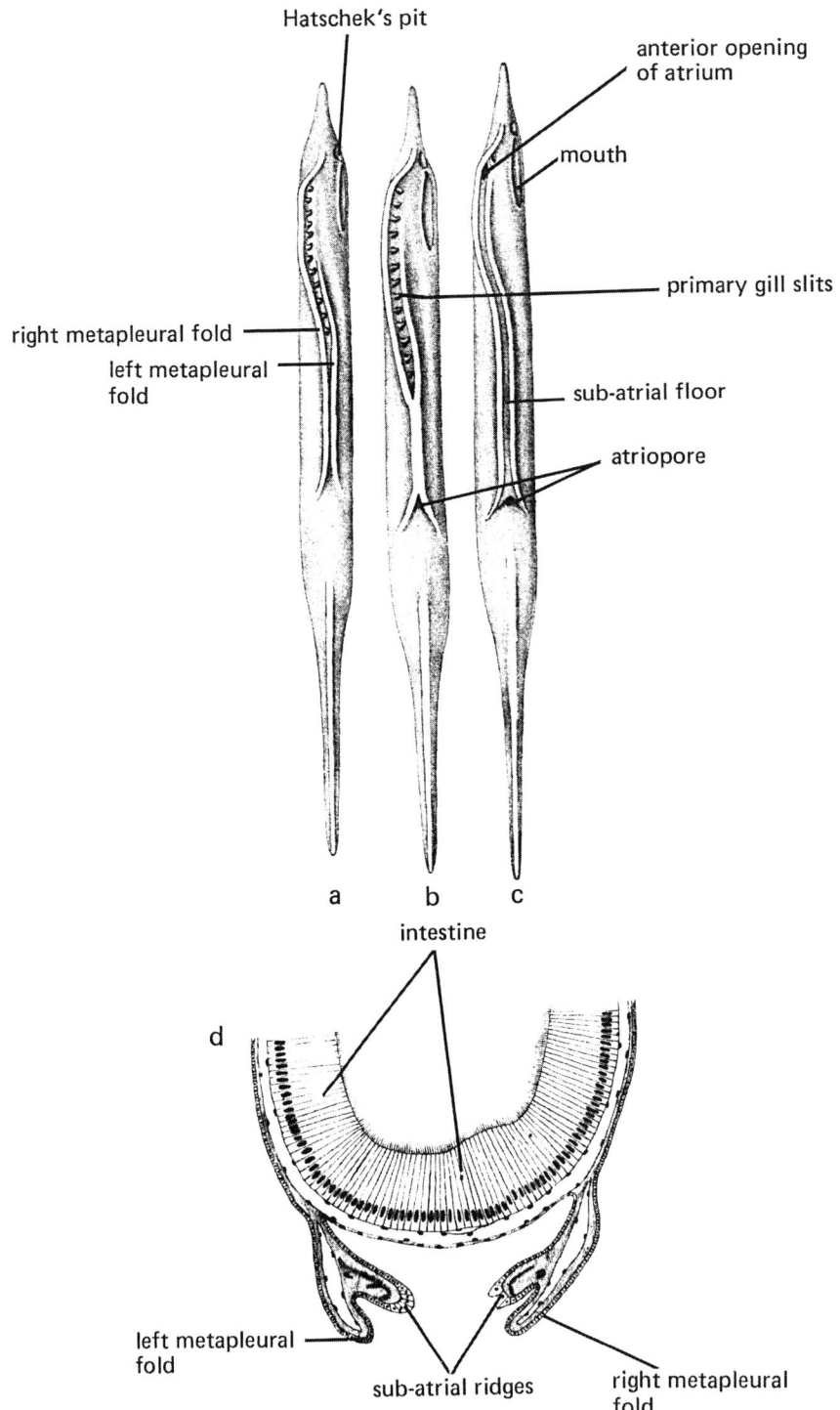

3.35. *Development of the atrial floor of amphioxus; a), b) and c) ventral views of three successive stages showing the coalescence of the right and left subatrial ridges; d) transverse section of the ventral part of a late larva to show the right and left subatrial ridges developing on the median faces of the metapleural folds; from Lankester & Willey (1890, pl. 30a, b, c and pl. 32, fig. 17).*

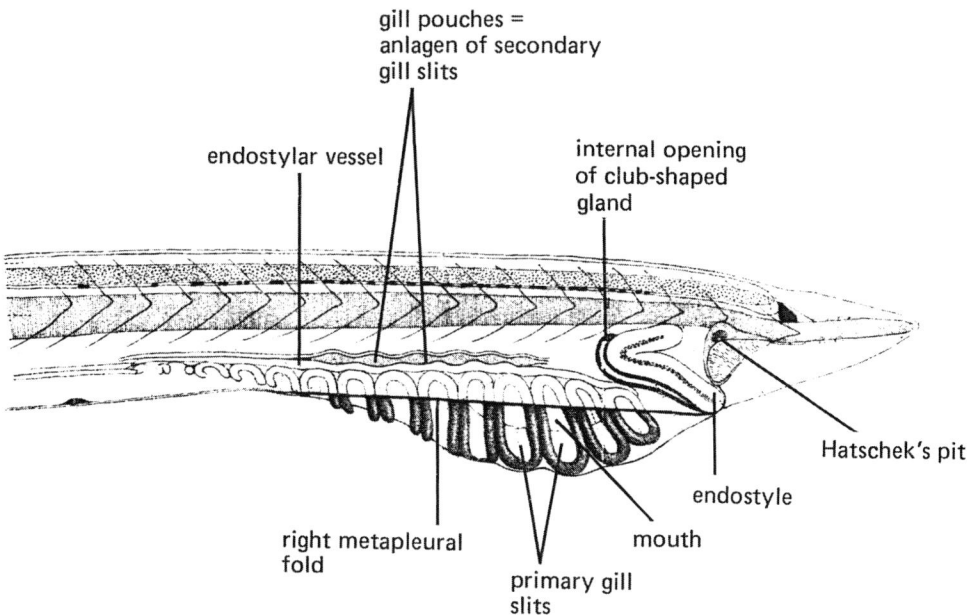

3.36. The first appearance of anlagen of the right, secondary gill slits in a late larva of amphioxus; right aspect; from Willey (1891, pl. 13, fig. 1).

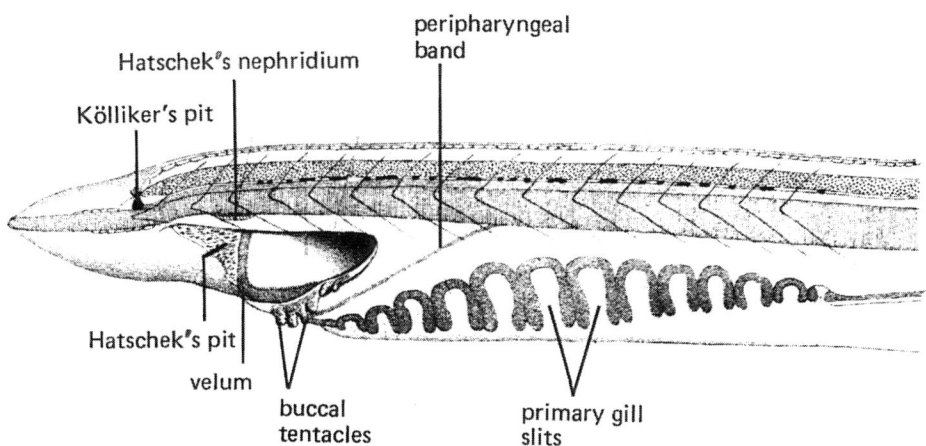

3.37. Metamorphosing larva of amphioxus, in left aspect. The left wall of the buccal cavity, and the anlagen of the buccal cirri (tentacles) have appeared and the velum is rotating into a transverse plane; from Willey (1891, pl. 13, fig. 6).

from a patch of pharyngeal wall which was initially situated between the two branches of the V anterior to the larval endostyle (Garstang, 1928, p. 90). During all these changes the right peripharyngeal band has appeared by an unknown process. The rearward growth of the endostyle between the primary and secondary series of gill slits gives final proof that the primary slits belong morphologically to the left side and the secondary slits to the right side.

Meanwhile the mouth is becoming symmetrical. The first change in this region is that the lens-shaped mouth loses its anterior and posterior points, becoming more elliptical. At the same time the buccal cavity begins to arise. Two superficial folds appear in this connection (fig. 3.37). One of them is longitudinal and dorsal to the mouth, and will be the future left side of the buccal cavity; the other, ventro-posterior to the mouth, is the future right side of the cavity. These two folds join behind the mouth,

a

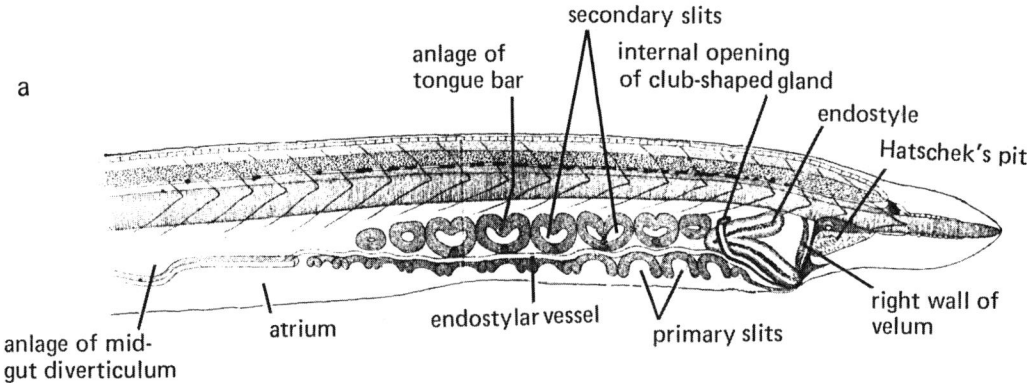

secondary slits
anlage of
tongue bar
internal opening
of club-shaped gland
endostyle
Hatschek's pit
right wall of
velum
primary slits
endostylar vessel
atrium
anlage of mid-
gut diverticulum

b

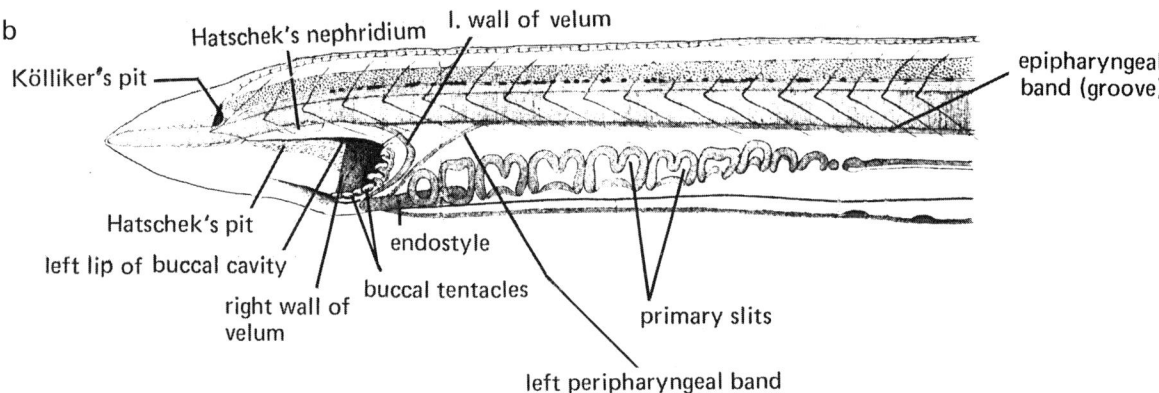

Kölliker's pit
Hatschek's nephridium
l. wall of velum
epipharyngeal
band (groove)
Hatschek's pit
left lip of buccal cavity
endostyle
buccal tentacles
right wall of
velum
primary slits
left peripharyngeal band

3.38. *Metamorphosing larva of amphioxus; a later stage than fig. 3.37 in a) right and b) left aspects. The velum is still rotating into a transverse plane and the endostyle has begun to grow backwards, as a* **V**, *between the primary and secondary gill slits; from Willey (1891, pl. 13, figs 7, 8).*

forming a V. The lower, or future right, limb of this V is continuous with the rostral fin that extends round the anterior lobe to join the dorsal median fin. In the lower, right fold a small number of pieces of cartilage appear near its edge in a line, and each cartilage grows outwards, producing a corresponding protuberance in the lip (figs 3.37, 3.38b). These protuberances are the definitive buccal tentacles. Gradually the two buccal folds grow forwards. At the same time the wall immediately surrounding the larval mouth (which wall will become the velum) rotates about a vertical axis so as to become transverse to the animal. The series of buccal cirri and cartilages extends at either end, invading the left (originally upper) lip and so finally gives the symmetrical array of buccal tentacles seen in the adult. At the same time the meeting place of the right and left buccal folds moves into a mid-ventral position. When these changes are finished the mouth region is as nearly symmetrical as in the adult, except that the left part of the ciliated organ has not yet developed.

The mid-gut diverticulum begins to develop during metamorphosis, being an outgrowth of the post-pharyngeal part of the gut on the right side. It gradually grows forward right of the pharynx. One result of its origin is the division of the mid-ventral blood vessel into a more posterior portion (sub-intestinal vessel and 'afferent liver' vessel) and a more anterior one ('efferent liver vessel' sinus venosus and endostylar vessel).

The ontogeny of the coeloms of amphioxus is difficult to follow from the literature. Anatomists have been hampered by fixation effects and influenced, as is inevitable, by their own preconceptions. Perhaps the simplest account is that of Smith & Newth (1917) who maintained that, in the larva, all the somitic cavities were in communication ventrally with the splanchnocoel except for the products of the left and right anterior gut diverticula (1st somites), i.e. except for Hatschek's pit (which opens externally) and the ventral rostral coelom. In many places, however, the communication of the somitic coeloms with the splanchnocoel

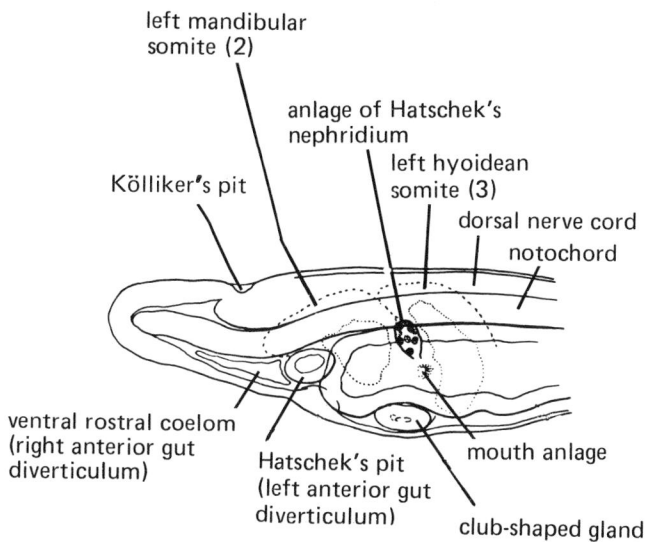

3.39. The origin of Hatschek's nephridium in amphioxus. The anlage of the nephridium is dorsal to the anlage of the mouth and may be endodermal in origin; from Goodrich (1934a, p. 504, fig. 1).

was purely virtual – two separate epithelia could be followed into the splanchnocoel but there was no actual space between them.

Van Wijhe (1914) (summarized by Franz, 1927, p. 597) recognized three coelomic cavities in the larva, i.e. the rostral coelom, the general splanchnocoel and the cavity of the left mandibular somite. According to van Wijhe, these three are probably all geometrically separate from each other in the larva. The left mandibular coelom was horse-shoe-shaped at first, with one limb of the horse-shoe above the larval mouth, one limb beneath it and the arc of the shoe anterior to the mouth. The muscle anterior to the larval mouth arose in this anterior arcuate portion (in confirmation of Legros, 1898, p. 533). Franz considered that the coeloms of the ventral body wall (meta- and epipterygial coeloms) arose as extensions of the general splanchnocoel into the atrial folds (1927, p. 601).

In the course of metamorphosis, according to van Wijhe, the ventral rostral coelom ceases to be separate from the splanchnocoel, when the wall between them breaks down. The left mandibular coelom, from being horse-shoe-shaped, becomes ring-shaped, by the growth together of the two limbs of the horse-shoe behind the mouth (i.e. morphologically left of it). This coelom forms the velar coelom of the adult and also sends off an extension which follows the inside of the cirral skeleton to give rise to the definitive inner lip coelom The outer lip coelom, on the other hand, is formed by the extension of the left epipterygial coelom outside the cirral skeleton, onto both lips of the mouth.

The ontogeny of the nephridia was described by Goodrich (1934a, b) who took account of, and partly refuted, the work of Legros (1909, 1910). Hatschek's nephridium and the paired nephridia have similar histories. The first recognizable anlage of each nephridium is a small group of cells situated between the coelomic epithelium and the endoderm. Each such group is located where the corresponding nephridium will be situated in the adult. Thus the anlage of Hatschek's nephridium (fig. 3.39) lies just dorsal to the endodermal thickening which later becomes the larval or velar mouth; and the anlagen of the paired nephridia are postero-dorsal to the endodermal thickenings which later become gill slits (figs 3.40a, b, c). The first origin of these groups of cells is unknown. When first observable they are far distant from any ectoderm. For instance the first visible anlagen of the nephridia of the secondary gill slits can be seen to lie dorsal to the endodermal gill-slit anlagen when the ectoderm of the atrium has not yet evaginated dorsalwards (Goodrich, 1934b, fig. 3.40 herein). Goodrich supposed that these anlagen must arise from cells given off from the ectoderm of the posterior end of the animal which by forward migration, later located themselves next to the gill slits (Goodrich, 1934b, p. 670). He pointed out, however, that this was assumption, based on the axiom that all nephridia must arise from ectoderm.

These anlagen, when first seen, are separated from the endoderm by a well-marked basement membrane, according to Goodrich (1934b, p. 663). Certainly, however, there is a later stage when this basement membrane is

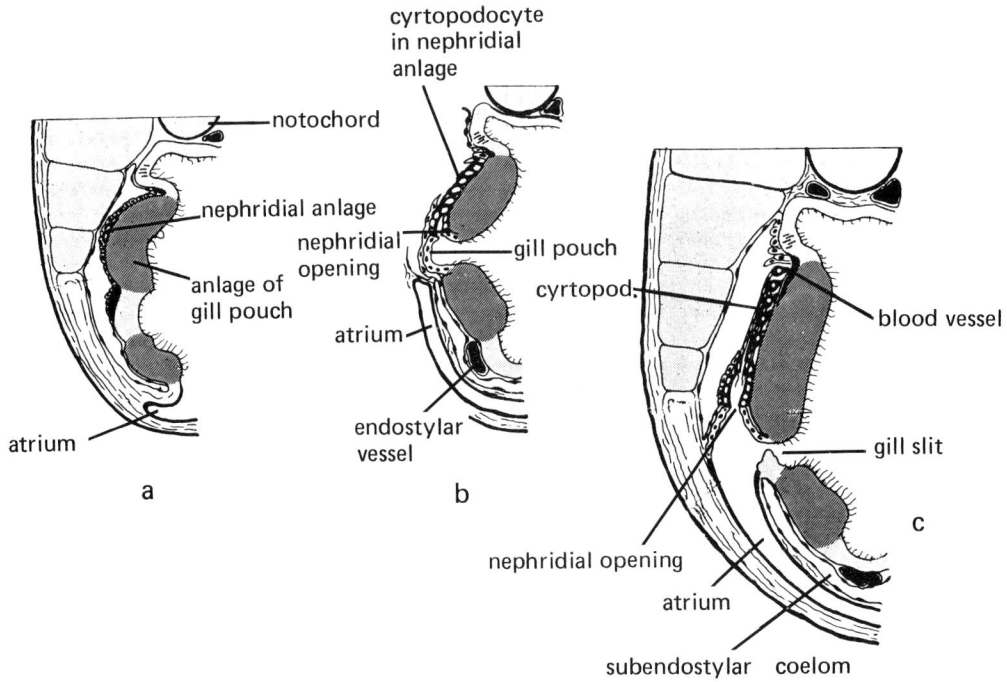

3.40. *The origin of the suprabranchial nephridia of amphioxus. a), b) and c) are transverse sections through the right, primary slits of successively older metamorphosing larvae. The nephridial anlage arises far distant from any atrial ectoderm and is probably endodermal in origin; from Goodrich (1934b, pl. 651, figs 1, 3, 4).*

interrupted and the nephridial anlagen intimately join the endoderm of the gill slit (Goodrich, 1934b, pl. 37, figs 10a, b). Since I consider the nephridia to have arisen in phylogeny by the fragmentation of an epicardium, and since epicardia arise as evaginations from the pharynx, I cannot help suspecting that this stage when the nephridial anlagen are intimately connected with the endoderm, indicates in fact that they arise in ontogeny from the walls of the pharynx. The first origin of the nephridia would repay investigation.

The nephridial anlagen later form tubes opening into the endodermal part of the gill anlagen (or into the pharynx in the case of Hatschek's nephridium). Subsequently, when the endodermal gill-slit anlagen break through into the atrium, the nephridia come to open, in effect, into the atrium, but always into that small part of the atrial wall which is of endodermal origin. The cyrtopodocytes break out through the coelomic epithelium that covers the nephridial anlagen initially. By this movement, the cell bodies of the cyrtopodocytes and the cage of rods enclosing the flagellum, come to lie in the cavity of the coelom.

When metamorphosis is complete amphioxus has become a relatively symmetrical animal. It differs from the adult by being smaller, by having only eight pairs of gill slits (being the primary and secondary series), by lacking

the left half of the wheel organ, and by having no gonads. The animal has become benthic, living buried or partly buried in sand. Willey (1894, p. 149, 174) referred to this stage as the critical stage, for he thought that it had great phylogenetic importance.

The development of these still absent structures happens during the post-larval period. Particularly striking are the tertiary gill slits. These penetrate the wall of the pharynx, behind the slits already present, in sequence from front to rear. The total number of slits increases as the body length rises (table in van Wijhe, 1914, p. 13). New tertiary slits are probably added throughout the animal's life. Their origin was described by Lönnberg (1902–5).

In summary, amphioxus is less vertebrate-like than elementary text-books would suggest. Also, despite a huge literature, there are still gaps in our knowledge – nobody, for instance, has yet described from observation the flow of blood in the dorsal vessels nor the origin of the blood system in ontogeny, and the ultimate source of the nephridia and the origin of the right peripharyngeal band are uncertain. Moreover, the embryology of the related genus *Asymmetron* has never been described, except for isolated stages, nor has even its adult anatomy been worked out in detail. From the present viewpoint one fact must be stressed; amphioxus, despite first appearances, is thoroughly unsymmetrical.

Chapter 4   The tunicates

The tunicates (or sea squirts) have received less attention in the textbooks than amphioxus. This is probably because they are usually thought to be less closely related to the vertebrates than amphioxus is. An additional discouragement is that tunicates, unlike the acraniates, are a large and diverse group. It takes years to understand them properly and tunicate systematists are in consequence very scarce.

The tunicates are marine animals mostly living attached to hard surfaces on the sea bottom though some live free in the sediment while others are pelagic. They owe their name to possessing a tough tunic, or test, consisting largely of a cellulose-like substance (tunicin) and water. They were first recognized as a group by Lamarck (1816) and during the first two-thirds of the nineteenth century were regarded either as 'worms' or as shell-less relatives of the bivalve molluscs. One of their strangest features, however, was the motile larva which was seen to be shaped like a tadpole, having a head and a tail. It was the brilliant Russian embryologist Kowalevsky who first showed, in 1867, that the resemblance of the larvae to tadpoles was not entirely accidental since the larval tail contained a notochord, a dorsal nerve cord and muscles. These facts, together with the presence of gill slits, peripharyngeal bands and an endostyle in the anterior part of the gut in both larva and adult, proved that the real affinities of tunicates were with the acraniates and vertebrates.

This relationship of the tunicates soon came to be formally recognized in zoological classification. When Kowalevsky made his discovery it was usual to class the acraniates as vertebrates, following the work of Johannes Müller in the 1840s (they were 'acraniate vertebrates' as opposed to 'craniate vertebrates'). The first step toward including tunicates in a larger group together with 'vertebrates' in this old-fashioned wide sense, was taken by Ernst Haeckel, father of the law of recapitulation, in 1874. He invented the word Chordonia to signify the exclusive common ancestors of tunicates and 'vertebrates'.

The Englishman Balfour (1880, p. 4, 41, 42) seems to have been the first to use the word Chordata. He applied it exactly in the modern sense to comprise tunicates, acraniates and vertebrates (s.s.), whether living or extinct. He also invented the word Urochordata (as Urochorda). This was a group coextensive with the tunicates, but having the rank of subphylum rather than class. The word implies that the notochord is restricted, in the larva, to the tail (oura is Greek for tail).

I shall illustrate the tunicates by describing a well-studied form in detail. I choose *Ciona intestinalis* (Linnaeus) for this purpose, since it is large, widespread, and primitive. Also, and above all, its anatomy has been very clearly described by Millar (1953) and was the subject of a beautifully illustrated monograph by Roule (1884). In physiological matters I largely follow Goodbody (1974). As it happens, *Ciona intestinalis* is one of the two species on which Kowalevsky made his original revolutionary embryological observations of 1867 (as *Ascidia canina*) and it is also the species on which these observations were first confirmed by von Kupffer (1870).

Ciona intestinalis is numerous world-wide in the sea, down to about 500 m, when there are suitable objects for it to fix to. Millar considers that it is probably the only species of its genus. However, it is very variable in size, shape and the numbers of repeated parts, so earlier authors had tried to recognize several species within it. For brevity I shall henceforth call it *Ciona*.

Ciona is approximately sausage-shaped (fig. 4.1), a few to several centimetres in length, translucent in appearance and usually greenish-grey in colour. One end of the sausage is attached to a suitable hard surface with the aid of little spreading processes called villi. The other end of the sausage is divided into two prongs (siphons), each prong ending in a large hole. These holes are roughly circular, but their edges are lobed and there are red and yellow pigment spots (the so-called but sightless 'ocelli')

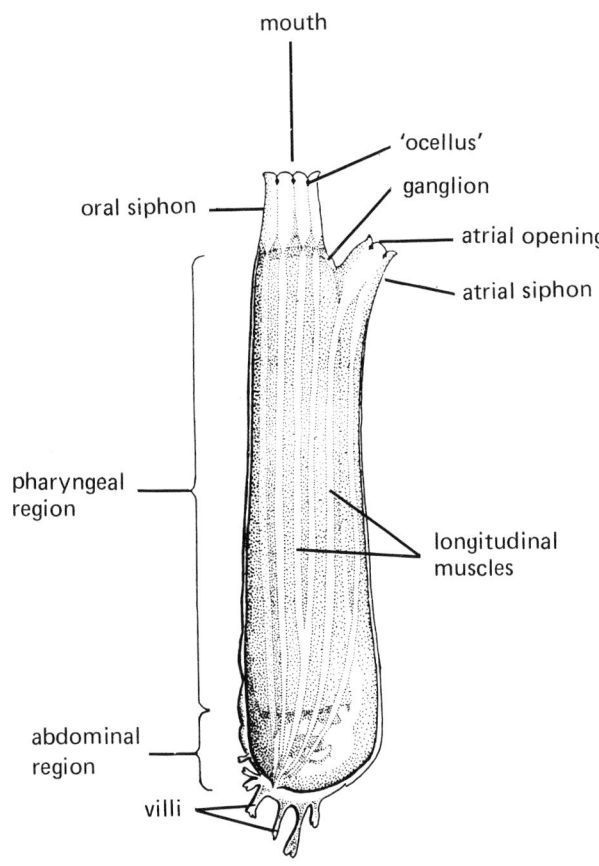

mouth

'ocellus'

ganglion

oral siphon

atrial opening

atrial siphon

pharyngeal region

longitudinal muscles

abdominal region

villi

4.1. Ciona intestinalis. *External appearance in left aspect; from Millar (1943, pl. 1, fig. 1).*

the whole, and a small posterior abdominal region near to the surface of fixation. The pharyngeal region contains the pharynx, as the name implies, and this is surrounded, right, left and dorsally, but not along the ventral mid-line, by a chamber called the atrium. The atrium is thus lateral and dorsal to the pharynx in tunicates, not lateral and ventral to it as in amphioxus. It opens to the outside by the atrial aperature. The wall of the pharynx is penetrated by a large number of gill slits (stigmata). The feeding current mentioned above passes through the mouth, into the pharynx, through the stigmata, into the atrium, and thence out through the atrial opening. The outer wall of the atrium is known, by analogy with bivalve molluscs, as the mantle. The abdominal region, posterior to pharynx and atrium, contains the non-pharyngeal gut, gonads, heart and some other organs.

The test, or tunic, which forms the external surface of *Ciona*, is secreted by the epidermis. It covers the pharyngeal and abdominal regions of the body. It also extends some distance inside the atrium and covers the whole internal surface of the oral siphon (i.e. the buccal cavity); this distribution is linked with the fact that the atrium and buccal cavity are both lined with ectodermal epithelium. The outer layers of the test tend to be tougher than the inner layers and to have a high protein content. The test consists mainly of a cellulose-like carbohydrate (tunicin) and water and is entirely dead except for vanadium-rich 'morula' cells from the blood which wander into it, passing through the epidermis that secretes it. The morula cells disintegrate in the test and may have some function in polymerizing carbohydrates to form tunicin (Goodbody, 1974, p. 81).

The muscular system, which allows *Ciona* to change its shape and to open and close its apertures, is situated within the mantle wall in the gap between the outer test-secreting epidermis and the epidermis which lines the atrium. Longitudinal muscles are arranged in a small number of bands, radiating out from the animal's area of attachment (fig. 4.1). Circular muscles, lying just internal to the longitudinal muscles, are best developed in the siphons, concentric to the mouth and to the atrial opening (fig. 4.4).

The alimentary canal of *Ciona* consists of buccal cavity, pharynx and non-pharyngeal gut (fig. 4.2a). The buccal cavity is simply the space inside the oral siphon. Consistent with its ectodermal origin, it is coated, as already mentioned, by a thin layer of test. The passage between it and the pharynx is guarded by a number of tentacles. These are in the same position as the velar tentacles of amphioxus and perhaps homologous with them. The passage itself corresponds to the velar mouth of amphioxus.

The pharynx of *Ciona* is a large organ extending from the buccal tentacles to the abdominal region. It is elliptical

between the lobes. One hole is at the extreme free end of the animal, and is the mouth, being the culmination of the oral siphon. The other hole is called the atrial opening; it is the culmination of the atrial siphon. When the animal is feeding a continuous current of water flows into the mouth and out of the atrial opening. A whitish organ visible just below the surface of the body between the two siphons is the ganglion or 'brain'.

As to morphological orientation, the mouth is anterior; the atrial opening is dorsal to the mouth; and a plane bisecting the mouth, the atrial opening and the area of fixation divides the animal into left and right halves.

Ciona can lengthen, shorten, bend, close or open the two apertures, either each one separately or both together, and cough out water, through one or both apertures, by suddenly shortening. It shows no external sign of its chordate relationships.

The body of *Ciona* (fig. 4.2) can be divided into a major pharyngeal region, constituting the anterior four-fifths of

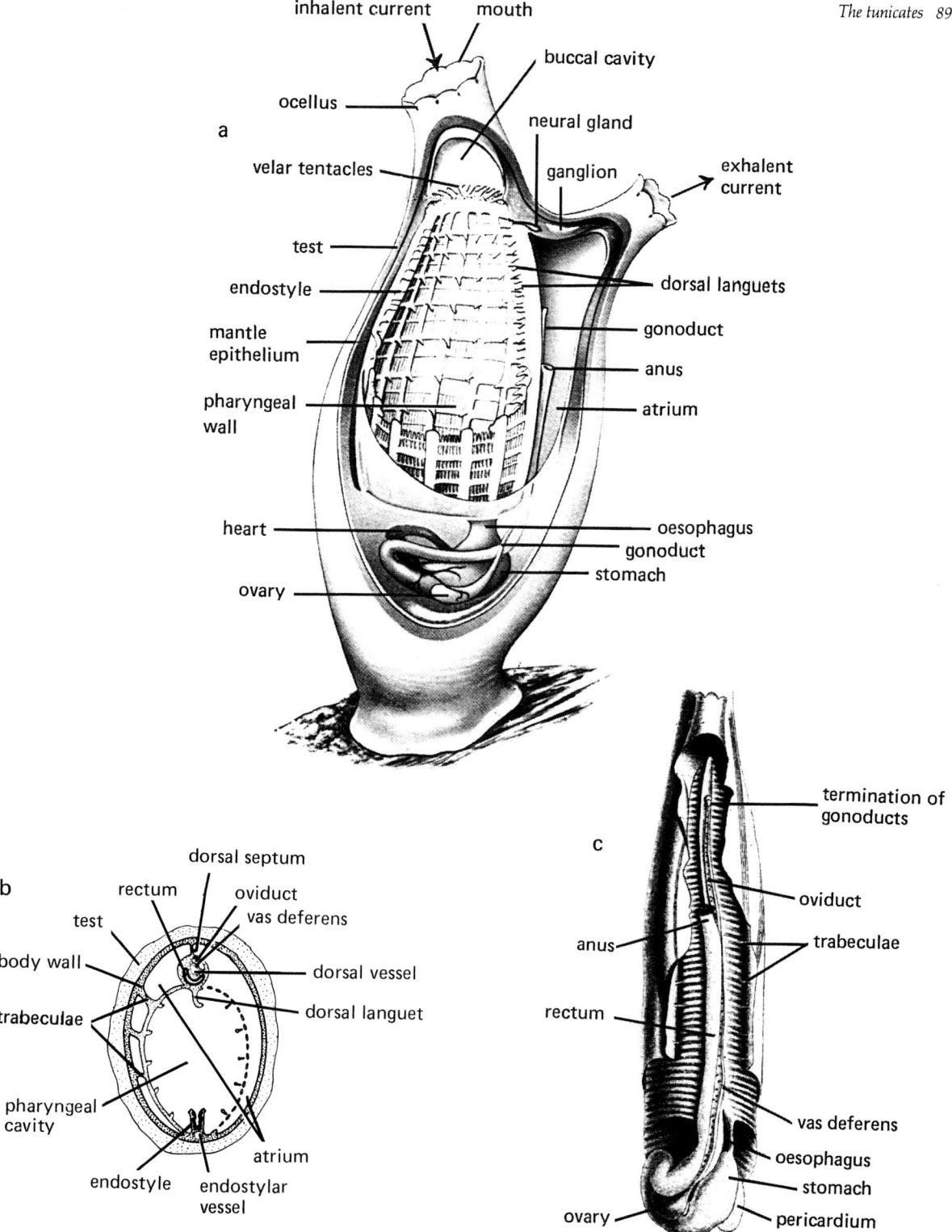

4.2. Ciona. *Main internal features. a) In left aspect from Burn (1980, p. 184); b) diagrammatic transverse section, passing along a transverse intrabranchial bar on the left and between two such bars on the right (left of figure is also animal's left) – note the rightward ventralward slope of the dorsal languet; from Millar (1953, pl. 11, fig. 51); c) dorsal aspect with the dorsal body wall removed to show the contents of the atrium and abdominal region – note that the rectum is left of the oesophagus; from Roule (1884, pl. 1, fig. 3).*

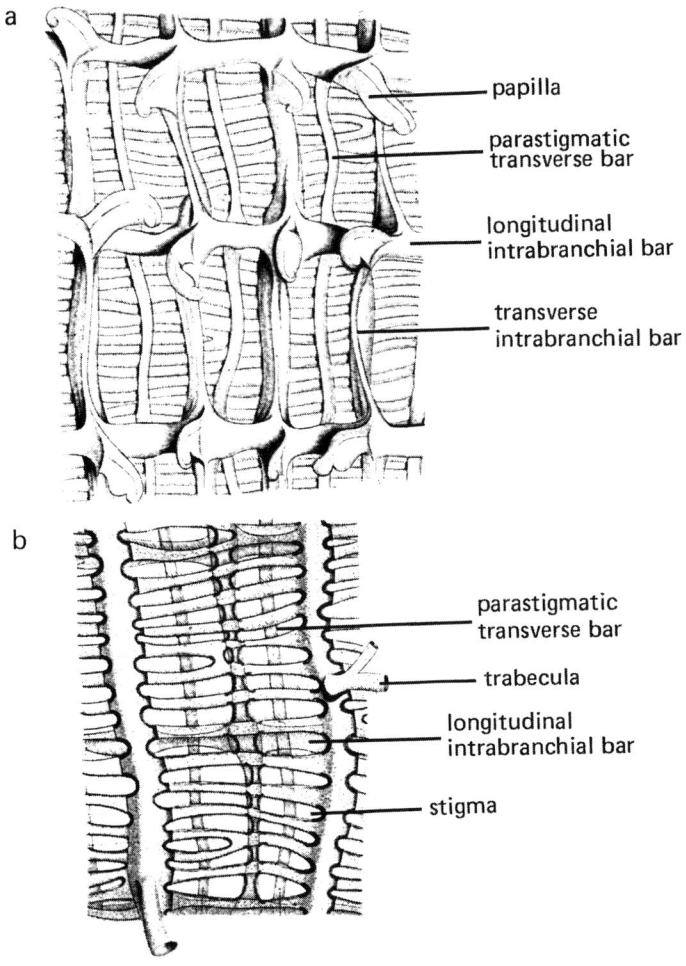

a

- papilla
- parastigmatic transverse bar
- longitudinal intrabranchial bar
- transverse intrabranchial bar

b

- parastigmatic transverse bar
- trabecula
- longitudinal intrabranchial bar
- stigma

4.3. Ciona. *Pharyngeal wall seen from: a) inside; b) outside. In both figures dorsal is at the top; from Roule (1884, pl. 2, figs 17, 18).*

in section and entirely surrounded by atrium except along the ventral mid-line. It is held in place in the atrium by strands of tissue called trabeculae (fig. 4.2b) which function somewhat like the guy ropes of a tent and pass to the pharynx from the outer wall of the atrium. Also the atrium is partly divided into right and left halves by the dorsal septum in the posterior dorsal mid-line. This is a sheet of tissue which has the rectum and gonoducts running along its ventral, pharyngeal edge. The dorsal septum is a relic of the fact that the atrium arises embryologically by the fusion of left and right atrial chambers. In the adult this fusion is complete only anteriorly, in front of the dorsal septum.

The wall of the pharynx is penetrated, as already mentioned, by a very large number of gill slits called stigmata (figs 4.2a, 4.3a, 4.3b). Each stigma is elongate parallel to the

length of the animal. Neighbouring stigmata are arranged in transverse rows, with up to about 200 pairs of rows in large individuals. The stigmatal rows are separated by transverse interstigmatic spaces. Each stigma is lined by cilia round the edge, and it is the beating of these cilia which pumps the water current into the mouth and out of the atrial opening.

The endostyle (fig. 4.5) is situated inside the pharynx, in the ventral mid-line, being developed in the valley that separates a pair of upstanding endostylar folds. It is formed from the epithelium of this valley and comprises three pairs of longitudinal gland strips with associated ciliated strips. The floor of the groove is formed of ciliated epithelium whose cilia are long enough to be called flagella. The two strips of non-glandular epithelium between the three gland strips on each side are also ciliated. Finally,

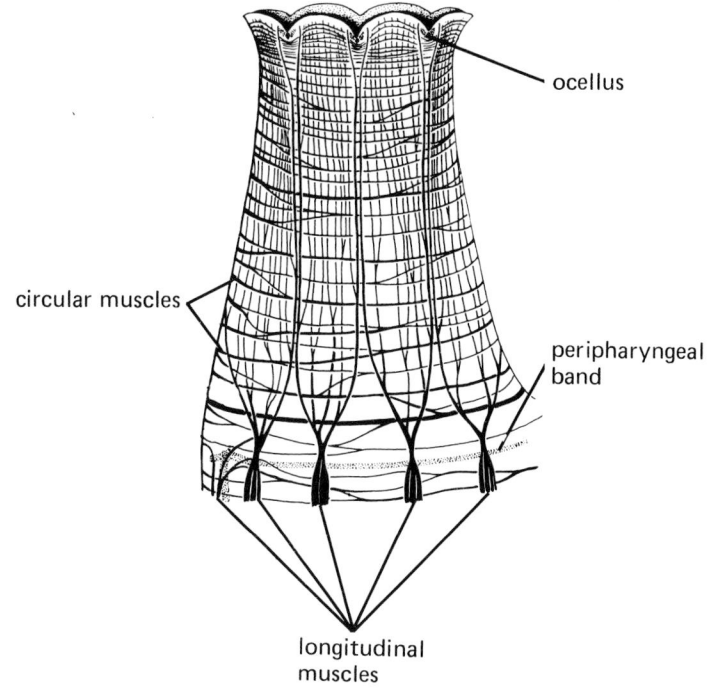

4.4. Ciona. *Muscular system around the mouth (in the oral siphon); from Millar (1953, pl. 4, fig. 13).*

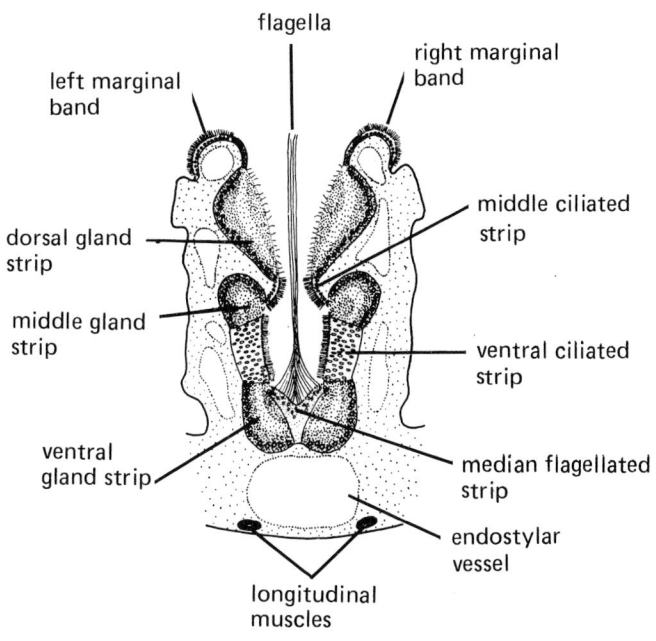

4.5. Ciona. *Transverse section of the endostyle; from Millar (1953, pl. 9, fig. 31).*

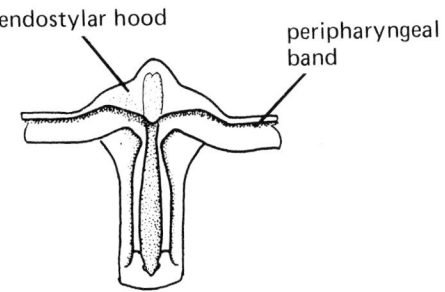

4.6. Ciona. *Anterior end of the endostyle in dorsal aspect to show the relation between the endostylar groove and the peripharyngeal bands; from Millar (1953, pl. 9, fig. 34).*

along the crests of the endostylar folds, run the most dorsal ciliated bands – called the marginal bands of the endostyle. These are separated from the most dorsal gland strips by non-ciliated, non-glandular strips of epithelium.

At its anterior end the groove of the endostyle projects forward beneath the pharyngeal surface as the so-called endostylar hood (fig. 4.6). The marginal bands of endostylar cilia turn to right or left behind the hood and run, as the ciliated peripharyngeal bands, up the sides of the pharynx, anterior to the most anterior stigmata, to meet each other in the dorsal mid-line (fig. 4.7). Each peripharyngeal band forms the posterior wall of a peripharyngeal groove. Each of the latter, like the endostyle, is the valley between two upstanding folds – to wit, in this case, the anterior and posterior peripharyngeal lips. The right and

left anterior lips and the right and left posterior lips also join each other in the dorsal mid-line. Here the united posterior lips project backwards to enclose a short ciliated epibranchial groove (fig. 4.7). Just anterior to where the peripharyngeal bands join dorsally is the opening of the duct of the neural gland into the roof of the pharynx; this opening is known as the ciliated funnel.

A complicated intrabranchial system of bars forms a grid on the internal surface of the pharyngeal wall of *Ciona* (fig. 4.3a, b). Between every two neighbouring rows of stigmata the wall is produced into a ridge known as an intrabranchial transverse bar. These bars are connected together by ciliated longitudinal bars continuous from the posterior to the anterior end of the pharynx. Finally, the longitudinal bars are also connected together by parastigmatic transverse bars, each one internal to the middle of a row of stigmata. Wherever the longitudinal bars meet interstigmatic transverse bars a ciliated process (papilla) projects into the cavity of the pharynx. The function of this intrabranchial grid, as discussed later, is to hold mucus away from the pharyngeal wall during feeding, and propel it dorsalwards by means of the cilia on the papillae and the longitudinal intrabranchial bars. Transverse and longitudinal bars carry blood vessels. The trabeculae are attached to the external surface of the interstigmatic bars.

In the dorsal mid-line of the pharynx there is a row of tentacles known as the dorsal languets (figs 4.2a, b, 4.7, 4.8). These are ciliated on their right and left side and curve characteristically ventralwards and rightwards. Each languet is situated between neighbouring rows of

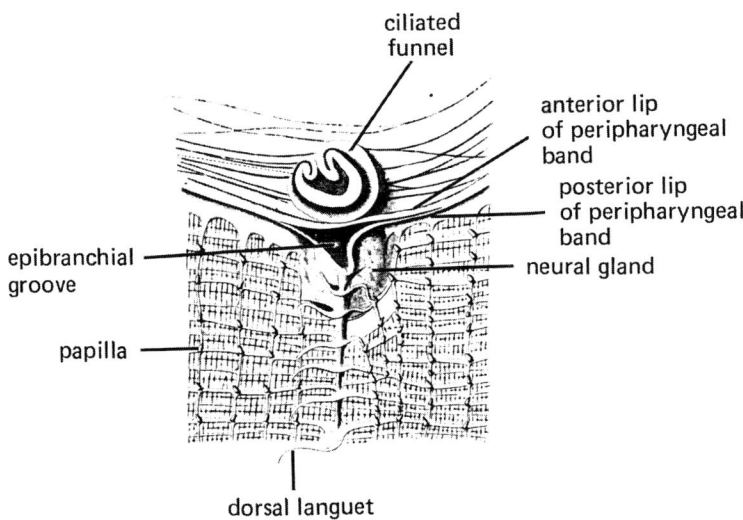

4.7. Ciona. *Anterior dorsal wall of pharynx in ventral aspect, to show relations of ciliated funnel, right and left peripharyngeal bands, epibranchial groove and dorsal languets. Note that the dorsal languets slope ventralwards towards the animal's right (left of the figure); from Roule (1884, pl. 4, fig. 32).*

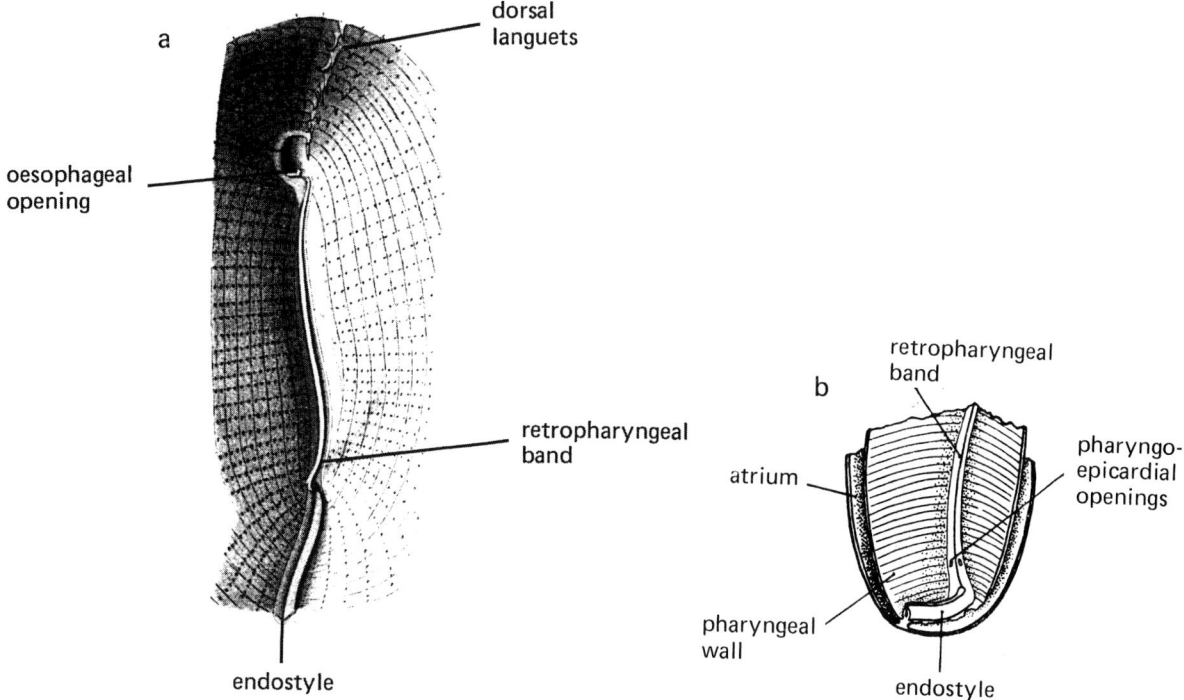

4.8. Ciona. *Features of posterior part of pharynx. a) Anterior aspect of posterior wall of pharynx — note that the dorsal languets slope downwards and to the animal's right (left of the figure), the oesophageal opening is right of the mid line; the retropharyngeal band enters the ventral left corner of the oesophageal opening and comes off the right posterior portion of the endostyle; from Roule (1884, pl. 3, fig. 29b). b) Diagram of same region viewed slightly from animal's left to show the pharyngo-epicardial openings on either side of the retropharyngeal band; from Millar (1953, pl. 8, fig. 29).*

stigmata, at the dorsal meeting place of a right and a left interstigmatic transverse bar. Small, immature dorsal languets are found in the dorsal mid-line coinciding with the parastigmatic bars. (In some tunicates the dorsal languets are replaced by a fold of tissue, with the same curvature and function, called the dorsal lamina.)

The posterior part of the pharynx is anatomically complex and important. The oesophagus opens into the pharynx at the postero-dorsal angle of the latter. The opening is not exactly median, however, but located just right of the line of dorsal languets (fig. 4.8b) at the posterior end of the partial channel embraced by the concavity of the languets. In the the ventral mid-line of the pharynx the endostylar groove is produced backwards into an endostylar appendix (fig. 4.9) and this is connected to the opening of the oesophagus by a retro-pharyngeal groove that runs up the pharyngeal wall in the median plane of the pharynx (figs 4.8a, b). The right wall of the retro-pharyngeal groove is ciliated (retro-pharyngeal band). With two exceptions the strips of ciliated and glandular epithelium which make up the endostyle continue back into the appendix. The exceptions are the ventral median

flagellated band, which wedges out, and the left marginal band, which stops short. Because the left marginal band has stopped, the roof of the endostylar appendix is asymmetrical in transverse section (fig. 4.9b), being ciliated on the right but not on the left. This right ciliated band of the endostylar appendix is continuous with the right marginal band of the endostyle, in one direction, and with the retropharyngeal ciliated band, in the other. Thus the retropharyngeal band is indirectly a continuation of the right marginal band of the endostyle. (In being ciliated on one side but not the other, the retropharyngeal groove is like the peripharyngeal grooves.) The retropharyngeal groove (and band) runs into the oesophageal opening in the median plane of the pharynx, at the ventralmost angle of the opening. Since the oesophageal opening is offset to the right of the mid-line, the retropharyngeal band enters its leftward, ventral portion (fig. 4.8b).

The asymmetries of the pharynx may seem trifling but they are almost universal in tunicates and very important for my subsequent argument (Chapter 8), so I shall repeat them here: the dorsal languets (or lamina) bend ventralwards and rightwards; the oesophageal opening is just

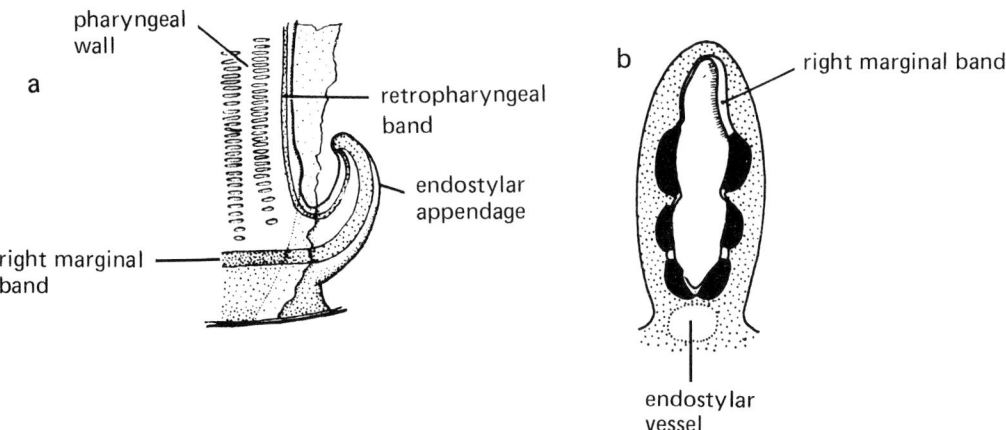

4.9. Ciona. *Anatomy of the posterior endostylar region. a) General aspect of endostylar appendage from left to show the continuity of right marginal band with the retropharyngeal band. b) Transverse section of endostylar appendage to show asymmetry of right marginal band (no counterpart on left); from Millar (1953, pl. 9, figs 35, 36).*

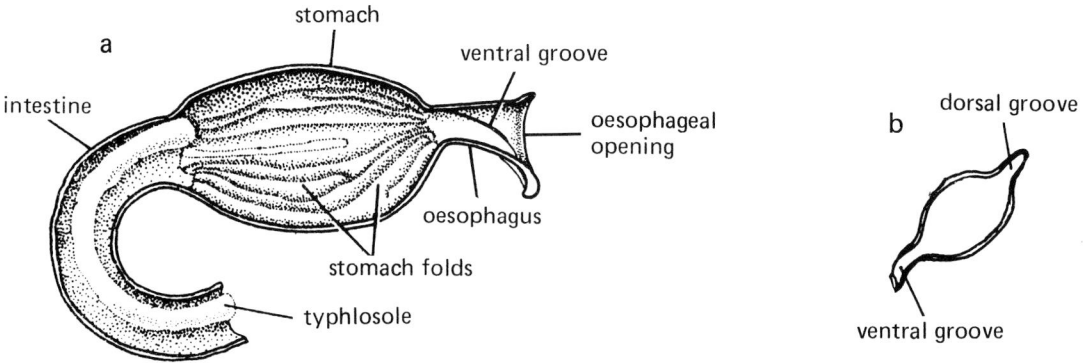

4.10. Ciona. *Internal features of the non-pharyngeal gut. a) Interior of oesophagus stomach and intestine in right aspect; b) transverse section of oesophagus near its opening into pharynx (animal's left at left of figure) – the ventral groove is continuous with the retropharyngeal ciliated band; from Millar (1953, pl. 10, figs 44, 37).*

right of the dorsal mid-line (as defined by the line of proximal attachments of the dorsal languets); the right marginal band, but not the left one, continues into the retropharyngeal band which enters the oesophageal opening on the left of the latter.

Two openings in the pharyngeal wall exist on either side of the retropharyngeal band (fig. 4.8a). These are the pharyngo-epicardial openings which connect the pharynx with the left and right epicardial chambers of the abdominal region (see below).

The non-pharyngeal gut of *Ciona* consists of oesophagus, stomach, intestine and rectum These are looped in such a way that the intestine is left of the stomach and oesophagus and leads into the rectum which is dorsal to the pharynx (figs 4.2a, c).

The oesophagus has two internal grooves, called dorsal and ventral by Millar in accordance with their position at the anterior end of the oesophagus (figs 4.10a, b). The ventral groove is the more marked and is a continuation of the retropharyngeal groove. It twists spirally round the inside left wall of the oesophagus so that, at the posterior end of the latter, it has become dorsal. The dorsal groove also twists round the oesophagus, but not so completely, so that at the posterior end of the oesophagus it is at the right side. The two grooves are mainly glandular and not ciliated, except that the ventral groove has a ciliated non-glandular strip forming part of its right side. This ciliated band is a direct continuation of the retropharyngeal band. The lateral walls of the oesophagus, outside the ventral and dorsal grooves, are ciliated.

The stomach is an ovoid sac, (fig. 4.10a) posterior to the oesophagus. Its internal surface is thrown into longitudinal folds. The crests of the folds are ciliated, whereas the troughs are glandular.

The intestine and rectum cannot be distinguished externally, but are histologically distinct. Also the rectum is suspended in the atrium, dorsal to the pharynx, whereas the intestine is in the abdominal region of the body, suspended in the epicardia. Both intestine and rectum have an internal fold or typhlosole, which makes them crescentic in section (figs 4.2b, 4.10a). Just posterior to the stomach the inside wall of the intestine is ciliated (ciliated ring). The rest of the intestine is not ciliated, however, consisting of gland cells and absorbtive cells. The rectum, on the other hand is ciliated, and the cilia presumably function in removing faeces.

There is a pyloric gland (fig. 4.11) which wraps itself round the outside of the rectum and opens by a duct into the lumen of the intestine just behind the stomach on the left side of the ciliated ring of the intestine. The function of this gland is unknown, but it may control the pH inside the lumen of the intestine (Goodbody, 1974, p. 43). Its position suggests homology with the pancreas of vertebrates.

The mode of feeding in *Ciona* can be observed in young transparent individuals and explains many of the anatomical details of the pharynx. The cilia associated with the pharynx are of two types, functionally speaking. Those lining the stigmata by their beating produce the water current that enters by the mouth and leaves by the atrial opening. The other cilia, on the internal surface of the pharynx, are concerned with moving the mucus.

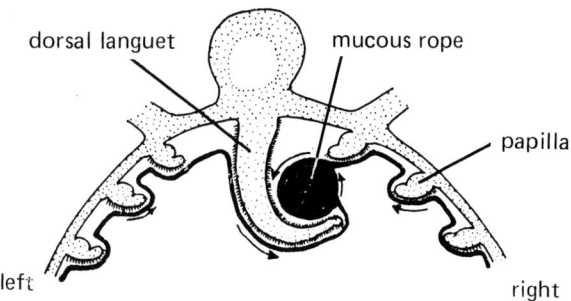

4.12. Ciona. *Transverse section of dorsal part of pharynx to show the formation of the mucous rope by the rolling together of the left and right mucous sheets. Note that the dorsal languets slope downwards and to the animal's right; from Millar (1953, text fig. 5).*

During feeding, mucus is produced continuously along the whole length of the endostyle. It flows dorsalwards as a sheet on right and left and is forced against the ciliated intrabranchial structures (papillae, transverse bars, longitudinal bars) by the outward pressure of the water current (fig. 4.12). The anterior ends of the sheets are gripped and moved dorsalwards by the peripharyngeal grooves (shown for the tunicate *Clavelina* in fig. 4.13a, b, c). The flagella of the median ventral strip of the endostyle probably serve to keep right and left sheets separate from each other. Chemically speaking, the mucus consists mainly of iodinated proteins (Goodbody, 1974, p. 30).

At the dorsal mid-line, right and left sheets behave somewhat differently, in accordance with the rightward and ventralward curve of the dorsal languets (fig. 4.12). The right mucous sheet passes directly into the partial channel inside the concavity of the languets. The left sheet passes over the left ventral surface of the dorsal languets and then also enters their partial channel. Inside this channel both sheets are rolled up together (like two carpets being rolled together) to form the so-called mucous rope (figs 4.13c. 4.14). The rolling-up happens by the action of the cilia of the languets which beat towards the tip of the latter. Because of the curvature of the languets the mucous rope is rotated anti-clockwise as seen from behind. In tunicates that have a dorsal lamina, the mucous rope is rolled up inside the cavity of the lamina, just as with the languets.

The mucous sheets act as a filter. Water can pass through them because of the pressure set up by the beating stigmatic cilia. Small particles in the water are filtered out by the mucus and are carried dorsalwards to be incorporated in the mucous rope. According to Jørgensen (1952) all particles larger than 1 μm in diameter are filtered out (Goodbody, 1974, p. 18). Flood & Fiala-Medioni (1979) have published electron micrographs of tunicate mucous filters and shown that they are delicate lattices with interstices about 1 μm across.

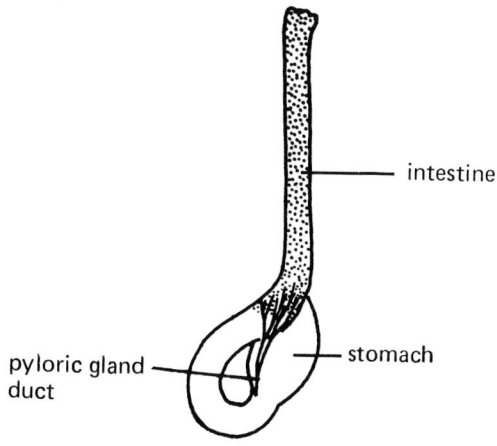

4.11. Ciona. *Distribution of pyloric gland in the wall of the intestine. The anterior end of the animal is upwards. Note that the intestine is left of the stomach; from Millar (1953, pl. 10, fig. 45).*

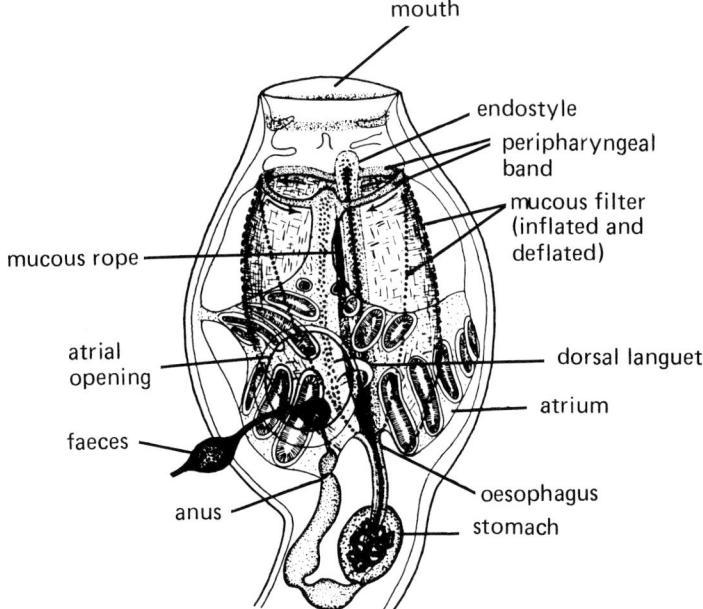

Fig. 4.13a.

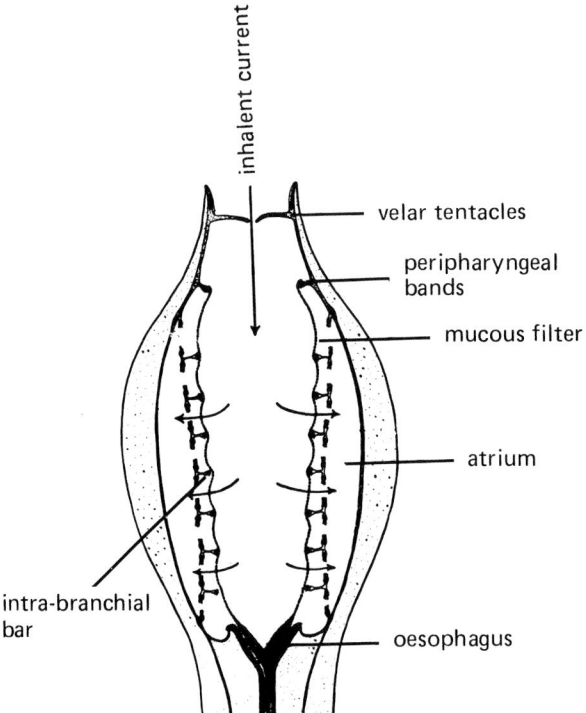

Fig. 4.13b.

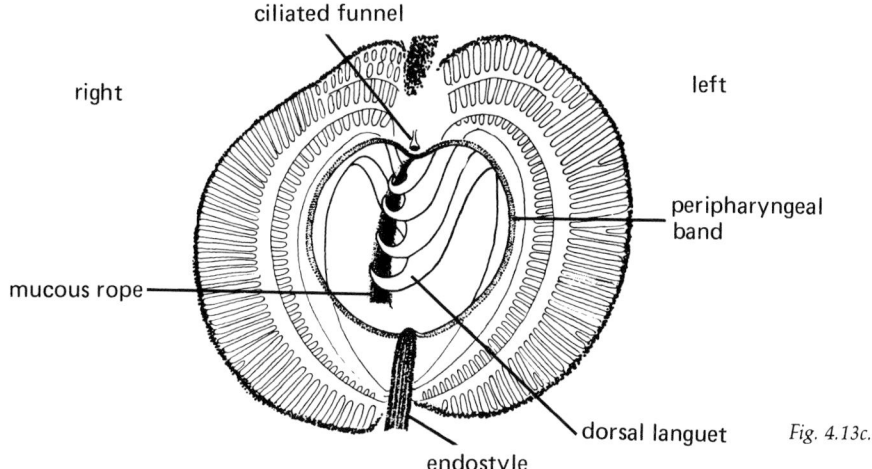

right

ciliated funnel

left

mucous rope

peripharyngeal
band

dorsal languet   Fig. 4.13c.

endostyle

4.13. *Feeding in the colonial tunicate* Clavelina lepadiformis. *a) Dorsal aspect of an individual zooid feeding; b) diagrammatic frontal section of a feeding zooid; c) sketch of pharynx of a feeding zooid, seen as though looking through the mouth – note that the dorsal languets slope downwards and to the animal's right; from Werner & Werner (1954, figs 3, 6, 11).*

The oesophageal opening, as already mentioned, is slightly right of the mid-line and therefore immediately posterior to the partial channel formed by the concavity of the dorsal languets. The mucous rope therefore passes into the opening of the oesophagus (fig. 4.14), and the cilia of the oesophageal wall are also actively involved in rotating it. The rope therefore rotates faster at its posterior than at its anterior end. The mucus with its contained food particles passes rearwards into the stomach and the anterior ciliated ring of the intestine probably helps to pass it back into the intestine (Goodbody, 1974, p. 38).

The role of the retropharyngeal groove in this process is not totally clear from the literature. It presumably helps to pass mucus up to the oesophageal opening at the start

of feeding. Also it probably holds the posterior ends of the right and left sheets together during feeding.

The concept of the alimentary ciliated loop (fig. 4.15a, b) should be mentioned here (Jefferies & Lewis, 1978, p. 242). This is a way of regarding some of the ciliated bands of *Ciona* and other tunicates as forming segments of a longer ciliated band. One end of the alimentary ciliated loop (arbitrarily chosen as distal) is the posterior end of the left marginal band of the endostyle; from this point the loop runs forward as the left marginal band, passes into the left peripharyngeal band, into the epipharyngeal band, the right peripharyngeal band, the right marginal band, the retropharyngeal band, and thus into the ciliated band of the ventral oesophageal groove,

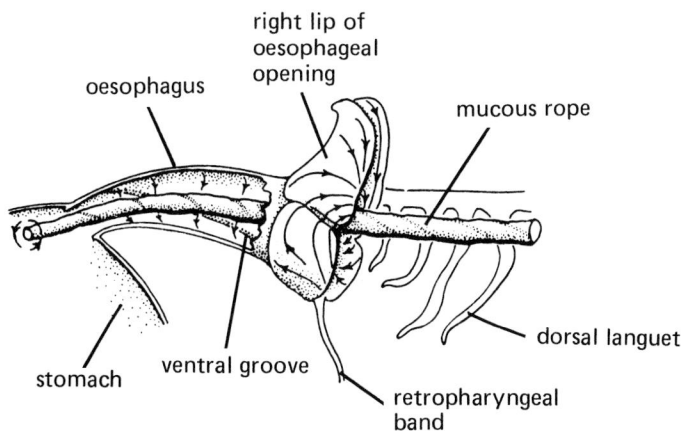

oesophagus

right lip of
oesophageal
opening

mucous rope

dorsal languet

stomach

ventral groove

retropharyngeal
band

4.14. Ciona. *The passage of the mucous rope from the pharynx, through the oesophagus to the stomach in right aspect. Note that the rope is right of the dorsal languets; from Millar (1953, text fig. 3).*

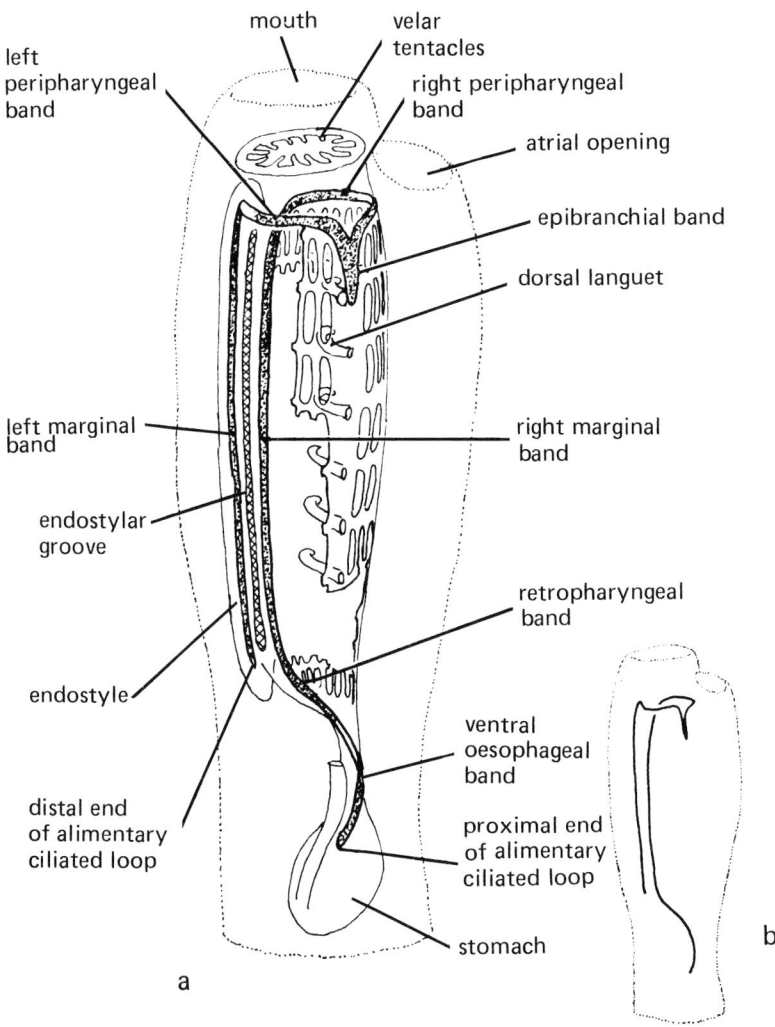

4.15.   *The alimentary ciliated loop of tunicates; a) in relation to other parts; b) on its own; from Jefferies & Lewis (1978, fig. 13, p. 243).*

stopping proximally at the posterior end of the latter. The left and right peripharyngeal bands may look like antimeres; indeed functionally they are such. Equally they can be regarded as segments, connected together by the epibranchial band, of the alimentary ciliated loop.

The epicardia (fig. 4.16a, b, c) are a pair of cavities behind the pharynx. The epithelial walls of the epicardia form mesenteries in which are suspended the viscera of the abdominal region, i.e. the oesophagus, stomach, intestine, ovary and heart. Anteriorly the two epicardia are roughly symmetrical, the mesentery between them being immediately behind the retropharyngeal groove. Posteriorly the left epicardium is bigger than the right one, because the inter-epicardial mesentery runs right-wards posteriorly.

As already mentioned, the right epicardium is connected with the right side of the pharynx by the right pharyngo-epicardial opening, and the left epicardium is connected to the left half of the pharynx by the left pharyngo-epicardial opening (fig. 4.8a). These openings are indications of the ontogenetic origin of the epicardia, as explained later.

Ciona is hermaphroditic, like almost every other tunicate. The testis (fig. 4.17a) is a white gland that branches over the outer surface of the stomach and intestine. The ovary is a relatively massive gland (fig. 4.17a) situated in the gut loop, left of the stomach. The courses of the vas deferens and oviduct are contiguous – the narrow vas deferens runs inside the wide oviduct, and both lie dorsal to the pharynx and just right of the rectum (fig. 4.2). Both

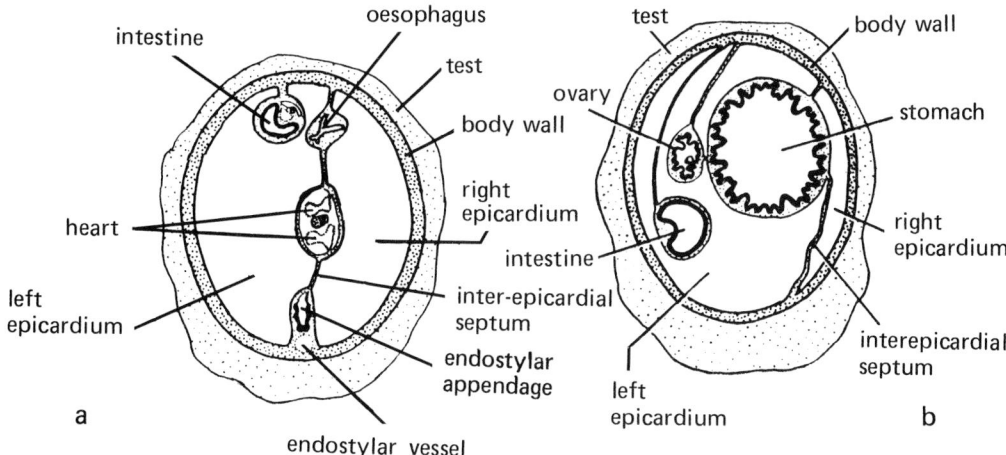

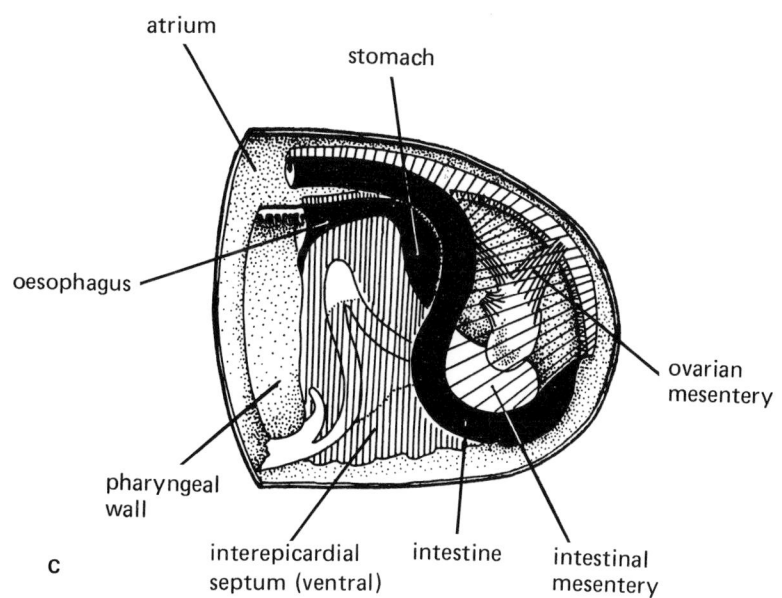

4.16. Ciona. *The epicardia and interepicardial mesenteries. a) Transverse section just behind the pharynx; b) more posterior transverse section, through the stomach; (c) diagrammatic relations in left aspect – note that the intestine is left of the stomach.*

ducts terminate a short distance anterior to the anus, just inside the atrial opening. At the end of the vas deferens there are about a dozen small apertures by which sperm can be released into the atrium (fig. 4.17b). The oviduct, on the other hand, has a single hole in its dorsal surface near the termination. Before the sperm are released the vas deferens is swollen near its end, so as to block the hole in the oviduct (fig. 4.17b). In this fashion, eggs cannot escape until the sperm have been released. The animal is thus

protandrous and this mechanism is a physical barrier to self-fertilization.

The vascular system of *Ciona* is complicated and its anatomy was worked out in detail by Millar (1953, p. 50 ff). It consists of a heart (inside a pericardium) (figs. 4.2a, 4.16c, 4.18, 4.19) and blood vessels (fig. 4.18). The latter have no endothelium except near the heart, being merely gaps in the mesoderm. Also, because of the remarkable fact that the heart periodically reverses its direction of beat, the

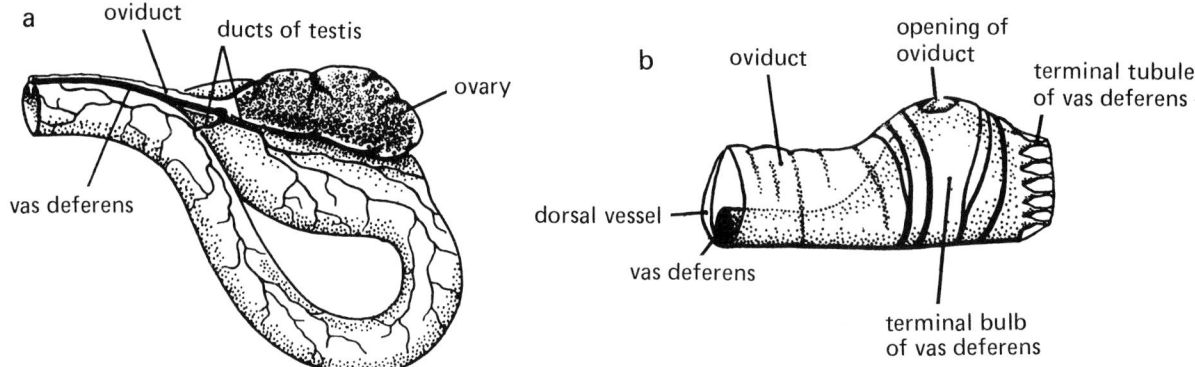

4.17. Ciona. *Reproductive system. a) Relations of gonads to gut; b) end of gonoducts to show how the swollen terminal bulb of the vas deferens blocks the opening of the oviduct, ensuring that sperm are released before eggs; from Millar (1953, pl. 19, figs 97, 96).*

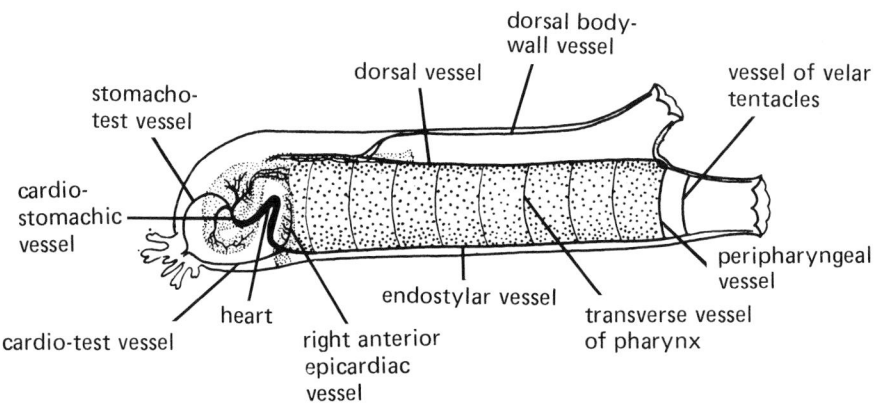

4.18. Ciona. *Blood system in right aspect; from Millar (1953, pl. 15, fig. 74).*

vessels are not physiologically separable into arteries, capillaries and veins. The blood of *Ciona* is a plasma containing eight types of blood cells. There is no respiratory pigment. The most remarkable blood cells are the 'morula' cells containing large quantities of vanadium and sulphate. These cells are destined to wander into the test where they disintegrate and probably play a part in depositing tunicin, as already mentioned.

The pericardium is a triangular sac situated in the mesentery dividing left from right epicardia, just ventral to the oesophagus. The base of the triangle is ventral and the apex dorsal, and the heart is a wide V-shaped tube that runs within the pericardium along the two apical sides of the triangle. One end of the heart leads to an endostylar vessel (fig. 4.18), the other to a vessel that spreads over the right wall of the stomach and supplies the viscera (fig. 4.19). There is an important vessel along the dorsal

mid-line of the pharynx and this communicates with the endostylar vessel by means of pharyngeal transverse vessels (in the interstigmatic transverse bars of the pharynx) and by vessels in the posterior lip of the peripharyngeal grooves (fig. 4.18). Vessels go to the mantle wall via the trabeculae and also by way of branches that leave the ventral endostylar vessel just anterior to the heart. Millar's account should be consulted for details. Periodic reversal of the heart beat is not peculiar to *Ciona* but universal in tunicates.

The physiology of the heart in tunicates has caused much discussion. There is no nerve supply to the heart and the origin and conduction of the waves of contraction seems to be exclusively through the muscle (myogenic conduction). The heart can only function properly, however, when it is inside the pericardium and under pressure from the pericardial fluid. This pressure probably is

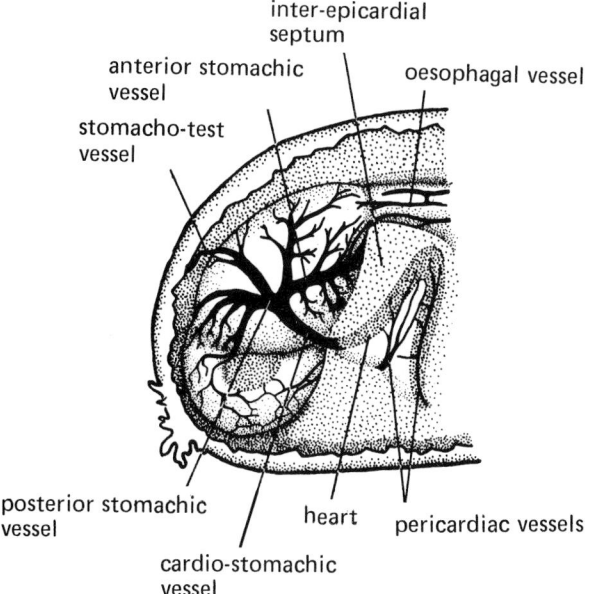

inter-epicardial
septum

anterior stomachic
vessel

stomacho-test
vessel

oesophagal vessel

posterior stomachic
vessel

heart

pericardiac vessels

cardio-stomachic
vessel

4.19. Ciona. *Heart, pericardium and associated vessels in right aspect. Note that the vessel to the heart comes off the right side of the stomach; from Millar (1953, pl. 16, fig. 77).*

under nervous control, for the pericardium is innervated. Pulsation of the heart seems to be controlled by a pace-making centre at each end of the heart – a visceral and a hypobranchial pacemaker. These act against each other and alternate in dominance. The visceral centre is more active when there is a high CO_2 tension in the blood, while the hypobranchial centre is not affected by this tension. Consequently, when CO_2 is abundant, the visceral pacemaker will dominate and blood will flow forward through the pharynx and become oxygenated. It is interesting that the contraction of the heart is also exclusively myogenic in myxinoids (in most respects the most primitive living vertebrates) and in the earliest ontogenetic stages of a chick (Goodbody, 1974, p. 57, p. 64).

The nervous system of *Ciona* consists of a ganglion (figs 4.1a, 4.2a, 4.20a, b, c), nerves radiating from the ganglion, a nerve plexus in the viscera and sense organs in the epidermis. The ganglion, as already mentioned, is situated between the siphons. The neural gland is in contact with its ventral surface. A pair of nerves goes off from the ganglion anteriorly and another pair posteriorly and each nerve breaks up to supply the body wall, especially the musculature of the siphons. In addition, the right posterior nerve from the ganglion sends a visceral nerve down to the viscera of the abdominal region (fig. 4.20b). This nerve runs in contact with the dorsal strand mentioned below and supplies a nerve plexus in the ovary, gut and peri-

cardium. In the ganglion the cell bodies of the neurons are concentrated just below the surface, in the cortex. The core or medulla of the ganglia is made of axons running to the nerves.

The ganglion functions in the so-called crossed reflex which is involved in coughing. When the outside of *Ciona* is touched its reaction is to shorten the body and close both siphons – for this reaction the ganglion is not necessary, as can be shown by cutting it out. If *Ciona* is tickled inside one of the siphons, however, its normal reaction is to close the other siphon and then contract the body. Under natural conditions this will remove the source of the tickle by ejecting water forcibly through the tickled siphon. This is the crossed reflex and it does not happen if the ganglion has been removed (Goodbody, 1975, p. 99 ff).

As to sensory receptors, *Ciona* responds to touch, light, temperature and chemical stimuli but, much as in echinoderms, no specialized receptors for these different stimuli have been definitely identified. There are sensory cells in the epidermis each of which has a process extending into, or possibly through, the test and an axon going off to the interior of the animal. These sensory cells are concentrated in the siphons and are probably mainly concerned with touch (Millar, 1953, p. 44, Goodbody, 1974, p. 94). The most obvious specialized sense organs are the cupular organs of the atrium (fig. 4.21a, b). As already mentioned, the internal surface of the part of the atrium nearest the siphon is covered with test, while more posterior parts are naked. In the generally naked region, however, small protuberances of epidermis exist, each with a 'flag' of test attached to it and many sensory cells within it. Cilia from the sensory cells are embedded in the flag and will move as the flag moves. Each protuberance, with its flag or cupula, is a cupular organ. These organs probably serve to detect vibration and may represent the vertebrate acustico-lateralis system in primitive form (Bone & Ryan, 1978). The 'ocelli' between the lobes of the mouth and atrial aperture are definitely not photoreceptors (Millar, 1953, p. 12; Goodbody, 1974, p. 98).

The neural gland of *Ciona* is immediately ventral to the ganglion (fig. 4.20a, b, c). Its lumen is connected to a duct which opens at the ciliated funnel anterior to the peripharyngeal bands in the dorsal mid-line of the pharynx. The main function of the gland is probably to provide a route for the escape of dead or moribund phagocytes into the pharynx.

From the posterior right-hand portion of the gland a strand of tissue (the dorsal strand) runs rearwards and ventralwards alongside the gonoducts to the surface of the ovary, where it ends in a small swelling (fig. 4.20b, d). Just anterior to the front end of the dorsal strand the right posterior portion of the neural gland is specialized to form

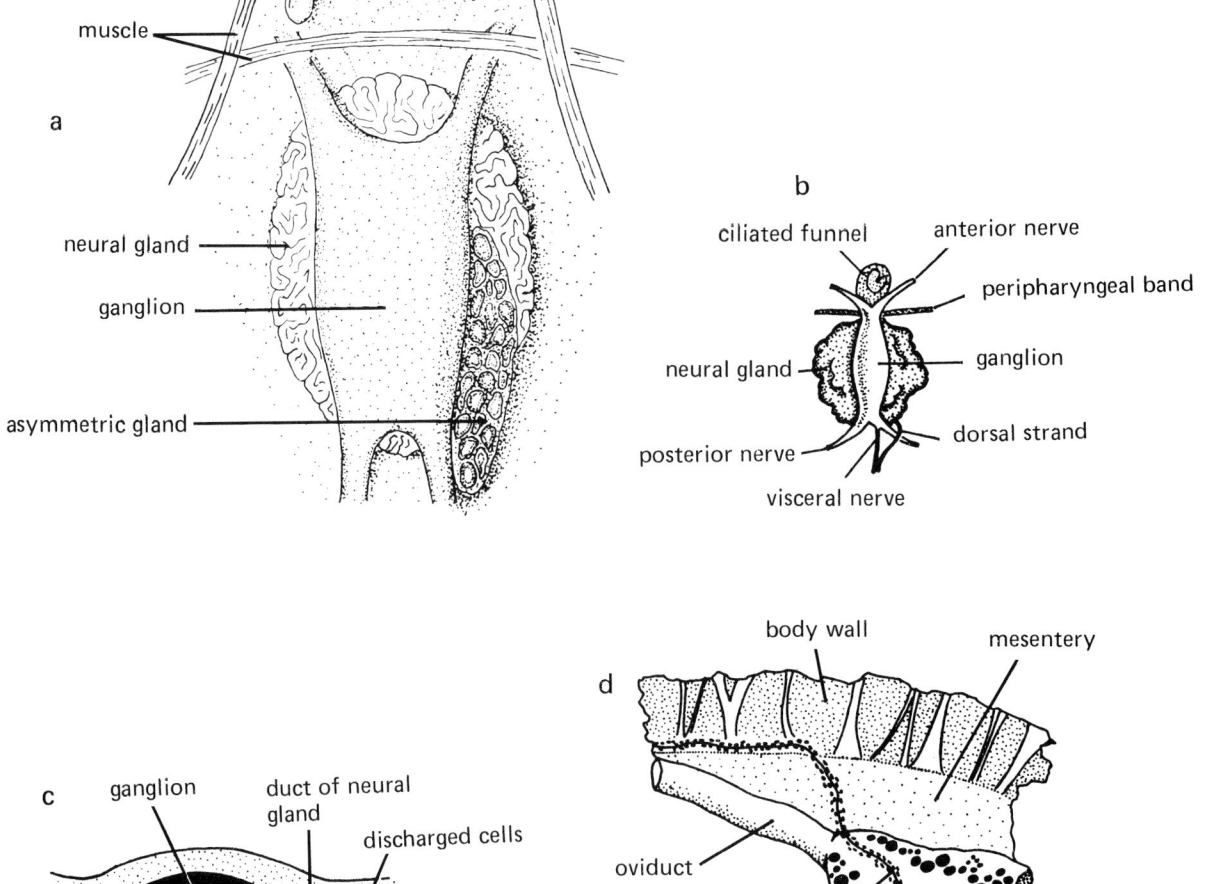

4.20. Ciona. *Nervous system, neural gland, asymmetric gland and dorsal strand. a) Ganglionar complex in dorsal aspect; from Pérès (1943, fig. 14, p. 263). b) The same, from Millar (1953, pl. 12, fig. 54), to show the dorsal strand emerging from the position of the asymmetric gland. c) Ganglionar complex in sagittal section; from Millar (1953, pl. 12, fig. 66). d) Posterior end of the dorsal strand on the left surface of the ovary (left aspect); from Millar (1953, pl. 12, fig. 63).*

the so-called asymmetric gland (fig. 4.20a). This was first observed by Pérès (1943) in *Ciona* who found it to occur only in young individuals and in old individuals outside the breeding season. It seems likely, therefore, that the asymmetric gland somehow controls and suppresses

breeding. This tends to confirm the belief that the neural gland in some way corresponds to the vertebrate pituitary. A study by Carlisle (1953 and earlier papers) provided physiological evidence for this view, since he found that extract of the neural gland had a gonadotrophic effect on

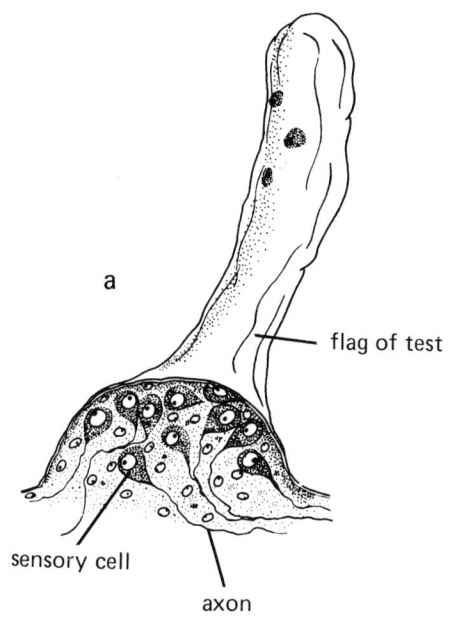

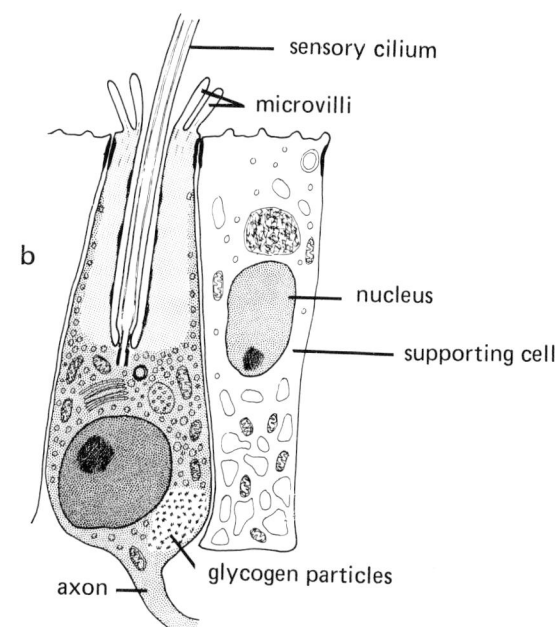

4.21. Ciona. *Structure of a cupular organ. a) General structure; from Millar (1953, pl. 12, fig. 60); b) histology of a sensory cell and a supporting cell; from Bone & Ryan (1978, fig. 1, p. 426).*

mice. Moreover he observed that the ciliated funnel sometimes contained spermatozoa of the same species. This led him to suggest an olfactory function for the neural gland, useful in making the animal cease to feed when conspecific sperm was in the water. Perhaps the dorsal strand conveys gonadotrophic hormone from the assymetric gland to the gonads. As against all this, Dodd & Dodd (1966) failed to find any gonadotrophic action from the neural gland of *Ciona*, despite extensive re-investigation.

The homology of the neural gland with the vertebrate pituitary is largely based on adult anatomy. Thus if the ganglion is somehow equivalent to the vertebrate brain, or to the diencephalon (Berrill, 1955, p. 245), then the position of the neural gland ventral to the ganglion and just dorsal to the alimentary canal can be compared with the adenohypophysis of an embryo vertebrate. Moreover, in myxinoids, which may well be the primitive sister group of all other living vertebrates, the adenohypophysis is endodermal in origin, arising from the dorsal wall of the pharynx (Chapter 6), just as the neural gland opens into the pharynx. However, there are embryological objections to homologizing the neural gland with the hypophysis. Firstly the adenohypophysis of non-myxinoid vertebrates (the monophyletic group of [lampreys + gnathostomes] arises from Rathke's pouch which is part of the buccal cavity and ectodermal in origin. Secondly the neural gland of tunicates arises in ontogeny,

along with the dorsal strand, by modification of the left part of the cerebral vesicle of the tadpole, being thus comparable with the vertebrate neurohypophysis rather than with the adenohypophysis (Willey, 1893; Elwyn, 1937). The mitrate situation, at least in the adult anatomy, was probably directly comparable with the tunicate adult condition, as I shall try to show later (Chapter 8). Concerning the embryological origin of the neural gland of mitrates there is, of course, no information.

The adult anatomy of *Ciona* is therefore more complex than the outside of the animal would suggest. In the structure of the pharynx, in particular, it is remarkably like amphioxus in a way that requires homology and confirms the chordate relationship that was first based on the tadpole larva.

Ciona is oviparous. The eggs and sperm are shot out into the sea water through the atrial opening and fertilize each other in the sea. Although hermaphroditic, *Ciona* is self sterile, and cross-fertilization is an almost absolute rule. Discharge of eggs and sperm usually happens about 90 minutes before sunrise (Castle, 1896, p. 210). Development is quick. The fully formed tadpole hatches about 25 hours after fertilization, has a free life of some 6 to 36 hours and then attaches to a hard object and metamorphoses into a small tunicate much like the adult (Berrill, 1935). Feeding does not begin until attachment, and indeed in the tadpole it cannot occur because the mouth and the paired atrial

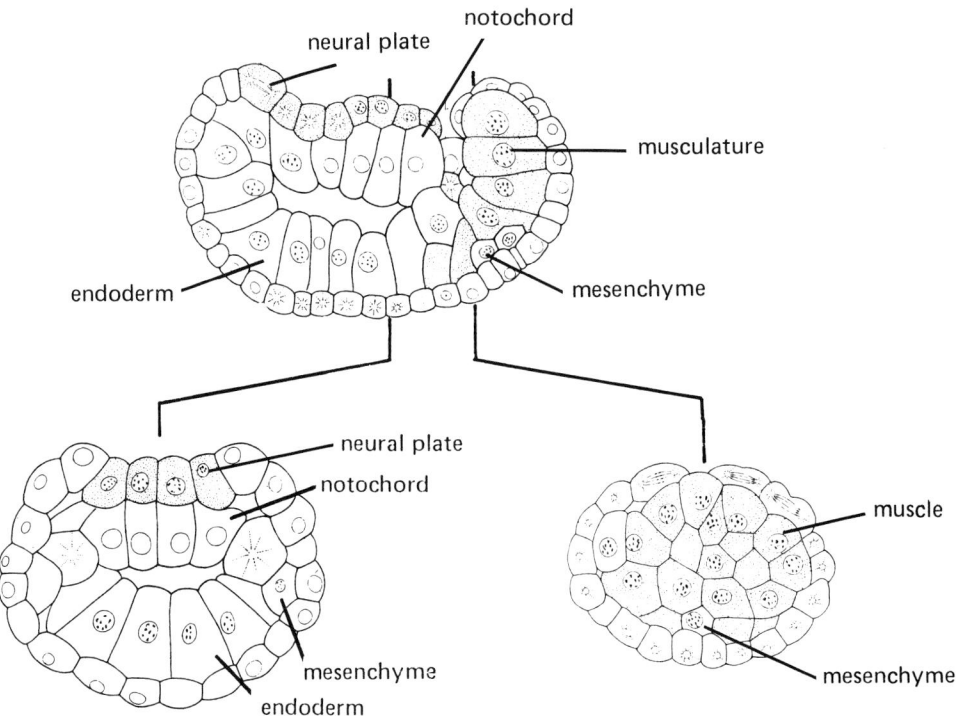

4.22.   Ciona. The neurula in sagittal section and with two transverse sections; from Castle (1896, pl. 12, figs 98, 95, 93).

apertures are covered with test. Moreover, except in the very latest stages of the tadpole, there are no gill slits and the rectum does not open into the left atrium, i.e. there is no anus.

The early development was described in great detail, from the unfertilized egg onwards, by Castle (1896) and Conklin (1905). The egg has an animal hemisphere with little yolk and a vegetal hemisphere with more yolk. The first cleavage plane is meridional, extending from the vegetal to the animal pole. It divides the egg into future left and right halves. The second cleavage is also vertical and divides the egg into anterior and posterior. The third cleavage plane is horizontal, separating four yolky cells in the vegetal hemisphere from four less yolky ones in the animal hemispheres. The fourth cleavage is vertical. In the animal hemisphere, where it divides four cells with little yolk, it often occurs earlier than in the vegetal hemisphere. In subsequent cleavages the less yolky cells always divide sooner than the yolkier ones, so that the animal hemisphere, having more cells, begins to overgrow the vegetal hemisphere. At the same time, anterior cells come to divide sooner than posterior ones so that the embryo elongates and bilateral symmetry begins to show. All this is very similar to the development of an echinoderm or of amphioxus.

Gastrulation involves the cells of the animal hemisphere overgrowing those of the vegetative hemisphere. The blastopore is at first an elongate depression, then becomes slit-like, then is reduced to a hole at the posterior end. Before gastrulation is finished, neurulation has already begun since the neural plate has started to form as a flattening on the dorsal surface in front of the blastopore. At the end of gastrulation the (fig. 4.22) embryo is slightly longer than wide and the external layer of cells is entirely ectodermal or neural. Inside, there is a small archenteric cavity surrounded by endoderm cells except dorsally where there is presumptive notochord. Immediately outside the endoderm, near the posterior end of the embryo, are right and left masses of presumptive mesenchyme. These are the cells which will later form the pericardium, heart and some other organs and the loose internal packing of the head of the tadpole. The right and left masses are connected by a small postero-ventral isthmus. Behind the point of closure of the blastopore the presumptive muscle cells and nerve cord of the tail are situated.

In neurulation, the dorsal flattening becomes first a shallow depression. Then the sides of the depression meet each other to form a tube under the ectoderm, beginning at the posterior end. The depression gradually closes up in this fashion. As seen from outside it becomes restricted

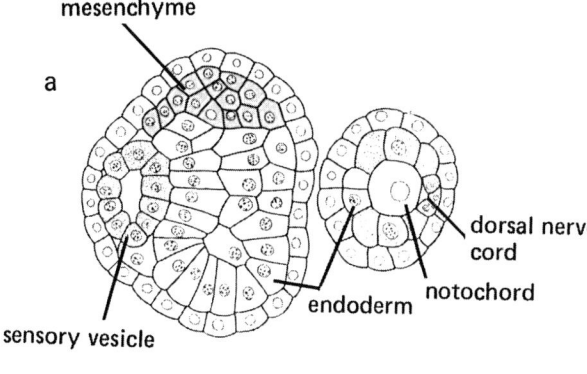

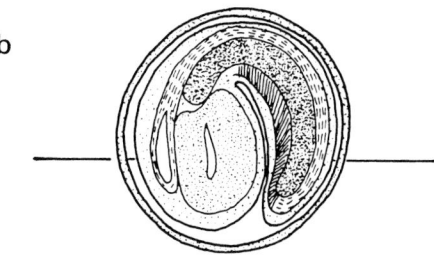

4.23. Ciona. *Late neurula just before release from egg. a) Transverse section through head and tail; from Castle (1896, pl. 13, fig. 106); b) left aspect of total larva to show disposition of parts; redrawn after Kupffer (1870, pl. 9, fig. 14).*

to a neuropore at the surface near the front end, and then closes completely. By the end of neurulation (fig. 4.23) the embryo is becoming distinctly tadpole-shaped with a massive head and a relatively slender, ventrally curved tail. The head contains: endoderm, a small archenteric cavity, paired masses of mesenchyme, and an expanded hollow anterior portion of the neural tube (cerebral vesicle). The tail contains: the notochord (as a single line of cells), presumptive muscle cells – three cells deep in section on either side, dorsal nerve cord, showing in section as a ring of four cells round a tiny cavity, and a ventral strip of endoderm two cells wide.

The fully formed tadpole of Ciona is about 1·2 mm long and hatches from the egg by means of a proteolytic enzyme. Its general anatomy (fig. 4.24a) was described by Willey (1893, 1894). Its tail is about four times as long as its head ('trunk'). The whole animal, except for the most anterior surface of the head, is covered with a thin layer of test, which in the tail is extended to form a vertical fin (Dilly, 1969a). The notochord is a turgid tube about 40 cells long that extends through the tail and projects a short distance forward from it (Berrill, 1947a, p. 251). As in the

preceding late neurula, it has muscle cells on either side, a dorsal nerve cord dorsal to it and an endodermal strand ventral to it (fig. 4.23).

The dorsal nerve cord of a Ciona tadpole (fig, 4.25) consists mainly of non-nervous epithelial cells corresponding to the ependymal cells around the central cavity of a vertebrate nerve cord. At left and right, however, there are axons which connect cell bodies in the brain to the muscles of the tail (Dilly, 1962, fig. 1; Olsson, 1969). Mackie & Bone (1976) have confirmed the existence of the same ventro-lateral axons for the tadpoles of the tunicate Dendrodoa, and in addition have recorded a bundle of axons dorsal to the nerve cord. Another noteworthy fact is that Olsson (1969) recorded the presence of a Reissner's fibre in the cavity of the dorsal nerve cord of a tadpole probably of Ciona and in 1972 found it in a tadpole of the tunicate Amaroucium. The presence of this fibre confirms, what can scarcely be doubted anyway, that the tunicate dorsal nerve cord is homologous with that of a vertebrate, apart from simplifications connected with small size.

The muscle cells of the tail of a Ciona tadpole are three deep in transverse section and are arranged in a left and a right muscle strip about six cells long i.e. about 18 cells in total on each side (Berrill, 1947b, p. 253). Each cell has a layer of muscle fibrils just beneath the lateral surface. The beating of the tail from side to side, combined with the fact that the muscle fibrils are oblique to the length of the tail, causes the tadpole to rotate as it swims forward (Pucci-Minafra, 1965, p. 304).

The head of a Ciona tadpole (fig. 4.24a) ends anteriorly in three adhesive papillae – two dorso-lateral and one mid-ventral. The papillae and the intervening surface are the only parts of the tadpole not covered by test and will later function in fixing the tadpole to a suitable substrate. The buccal cavity is a mid-dorsal invagination of the ectoderm beneath the test. And there is a pair of atrial invaginations, likewise ectodermal, beneath the test at the right and left sides of the head. Neither the buccal cavity, nor the atrial invaginations, communicate in the early tadpole stage with the lumen of the pharynx. The end of the intestine touches the left atrium but does not, in the tadpole of Ciona, yet open into it. In the late tadpole of Clavelina, however, such an opening already exists (fig. 4.26) (Julin, 1904).

The dorsal nerve cord expands anteriorly into a cerebral ganglion which as already mentioned receives the anterior ends of the tail axons. The cerebral ganglion itself passes anteriorly into a large, sensory vesicle that fills the whole width of the head and touches the buccal cavity (Willey, 1893). The sensory vesicle has a sensory function as its name implies; also its left dorsal wall evaginates to produce the neural gland, dorsal strand and definitive ganglion of the adult (Willey, 1894, p. 300ff).

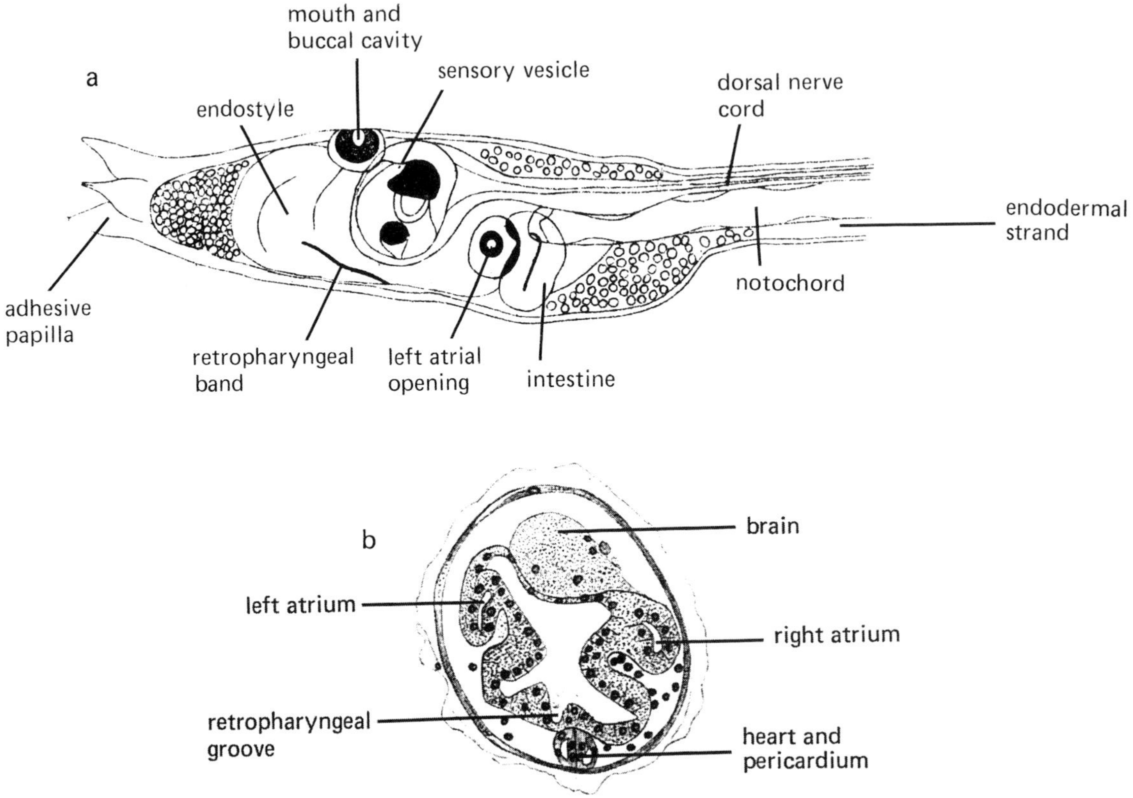

a
mouth and buccal cavity
sensory vesicle
endostyle
dorsal nerve cord
endodermal strand
notochord
adhesive papilla
retropharyngeal band
left atrial opening
intestine
b
brain
left atrium
right atrium
retropharyngeal groove
heart and pericardium

4.24. Ciona. *General anatomy of the head of a tadpole. a) In left aspect; from Willey (1893, pl. 30, fig. 1, retropharyngeal band identified after de Selys Longchamps 1900, pl. 17, figs 2–4); b) transverse section through atrial region; from de Selys Longchamps (1900, pl. 17, fig. 6).*

sheath
central cavity of dorsal nerve cord
ependymal cell
Reissner's fibre
neurons
ependymal cell

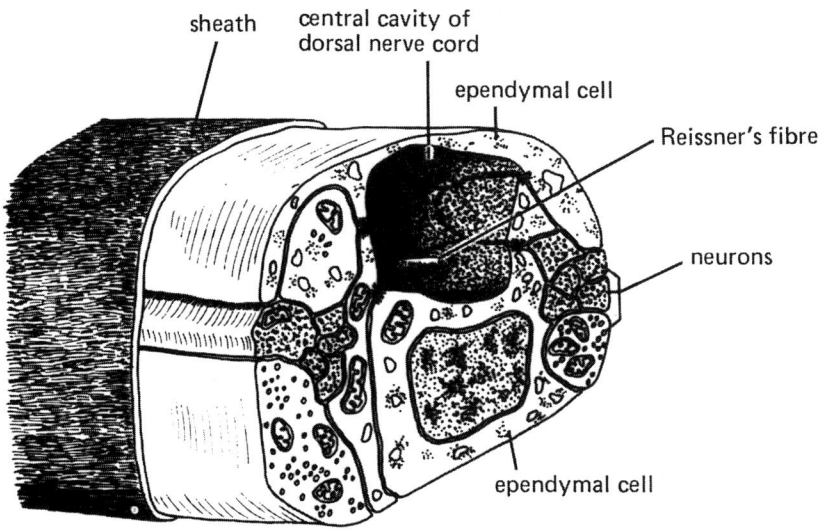

4.25. *Dorsal nerve cord of an ascidian tadpole; from Olsson in Sterba (1969, fig. 2, p. 295).*

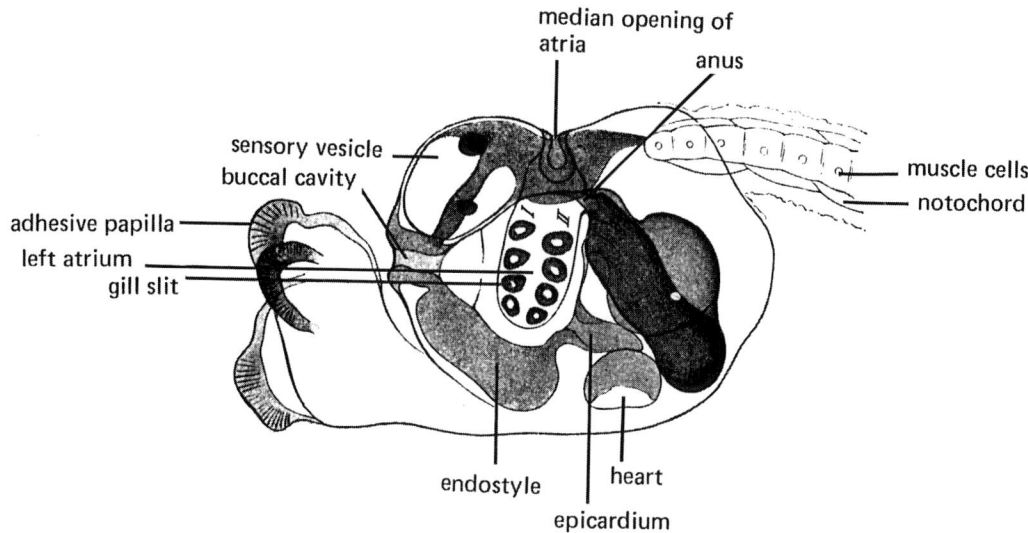

median opening of
atria

anus

sensory vesicle

buccal cavity

muscle cells

notochord

adhesive papilla

left atrium

gill slit

endostyle

heart

epicardium

4.26. A late tadpole of the colonial tunicate Clavelina *to show especially the rectum opening into the left atrium; from Julin (1904, fig. 12, p. 562).*

There are three types of sense receptor in the sensory vesicle – an ocellus, an otolith and probable pressure receptors. These receptors have been the subject of electron-microscopic studies by Dilly (1962, 1964, 1969, and in Young, 1981, p. 64) and by Eakin & Kuda (1971, 1972). The ocellus and otolith, have been known for a long time and in essentials were correctly interpreted by Kowalevsky (1867; see also Berrill, 1947, p. 617). The probable pressure receptors were discovered by Dilly (1969) who, however, thought them to be photosensitive.

The ocellus (illustrated for *Ascidia* in fig. 4.27) is situated in the postero-dorsal wall of the sensory vesicle. It consists of three lens cells, a pigment cell and about ten or twenty retinal cells. The three lens cells are arranged in line and each contains an intracellular lens formed of glycogen. The pigment cell is approximately conical in shape and is wrapped around the posterior part of the line of lens cells. The retinal cells are the actual transducers of the system. Each one consists of two parts: a cell body containing a nucleus, and an outer segment, which by origin is a non-motile cilium. Like other non-motile cilia it contains a ring of nine micro-tubules in its proximal part. The outer segments project through the cup-shaped pigment cell into the cavity of the cup, but posterior to the lenses, so that the beam of light focused by the lenses will fall on them. The surfaces of the outer segments are deeply invaginated to form a great number of closely packed lamellae which lie perpendicular both to the length of the outer segment and to the beam of light (Dilly, in Young, 1981, fig. 3.20). The light-sensitive pigment is probably concentrated on the surface of the outer segment, so the lamellae would serve to increase the quantity of the pigment. The tadpole is at first attracted by light and later repelled by it so that it tends to become attached in a suitable dark place, such as the underside of a rock. The ocellus is presumably important in these reactions.

The retinal cells of tunicate tadpoles could well be homologous with those of vertebrates. For vertebrate retinal cells are modified ependymal cells and their outer segments, which are the transducer element, are likewise non-motile modified cilia. Moreover, the outer segments of vertebrates are like those of tunicates in having the surface invaginated perpendicular to the length of the segment to form discs. It seems possible that the glycogen lenses of the tunicate ocellus are homologous with the certain intracellular glycogen lenses (ellipsoids) of vertebrate retinal cells, though this possible homology does not seem to have been discussed in the literature.

A paper by Froriep (1906) seems to have been entirely forgotten. In late tadpoles of the tunicate *Distaplia* Froriep found that the definitive ganglion and the neural gland and its duct were already well developed and exactly median in position (figs 4.28a, b). The cerebral or sensory vesicle with its ocellus was situated right of the definitive ganglion and neural gland. Left of the gland was an upward process of the brain which Froriep saw as a reduced antimere of the sensory vesicle. He therefore interpreted the sensory vesicle as homologous with the right eye of a vertebrate and the supposed left antimere as homologous with the left eye of a vertebrate. This result

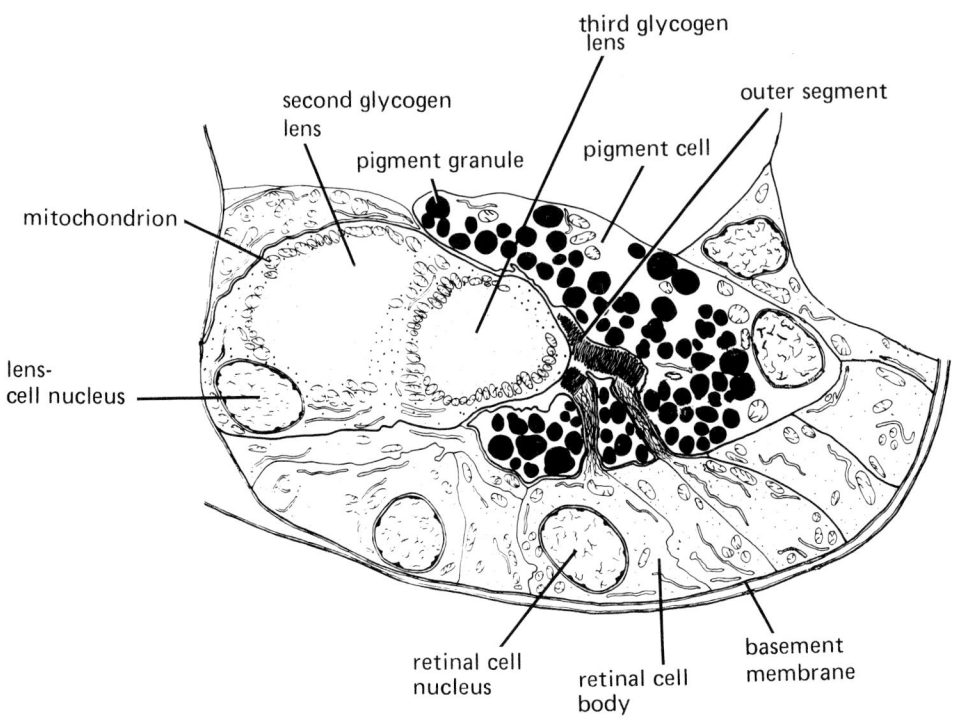

4.27. The ocellus of the tadpole of the tunicate Ascidia nigra; *from Young (1981, fig. 3.20, p. 64) after an electron-micrograph by P. N. Dilly.*

was mentioned by MacBride (1914) but without giving a reference. Grave (1922) could find no trace of any left eye in the tadpole of *Amaroucium* but, though aware of MacBride's statement, he was working in ignorance of Froriep's paper. Froriep was a brilliant anatomist and a careful observer, and the question of whether this vestigial left eye exists ought to be restudied.

The otolith is a dark grain contained in a cell in the floor of the cerebral vesicle (fig. 4.24a). The containing cell can be called a statocyte (Eakin & Kuda, 1971). The grain contains melanin, but it is not known whether it is denser or less dense than water. Presumably the otolith functions in accelerometry and detection of gravity. The otolith cannot be homologous with the functionally equivalent acustico-lateralis systems of vertebrates. For the transducers of the latter do not project into the cerebro-spinal fluid, but into enclosed portions of the external world, i.e. the cavity of the inner ear or of the lateral line.

The ocellus and otolith are present with almost constant structure in the great majority of tunicate tadpoles and doubtless represent a primitive feature of the group.

The alimentary canal of the fully formed tadpole of *Ciona* can be divided into the pharynx, anterior to the cerebral vesicle, and the non-pharyngeal gut beneath and behind the vesicle. The non-pharyngeal gut is already differentiated in the tadpole into oesophagus, stomach and intestine-rectum. The distal end of the rectum, as already mentioned, bends round to the left to touch the left atrial chamber. The pharynx contains a deep anterior vertical groove which is the future endostyle. The retropharyngeal groove (fig. 4.24a, b) runs across the floor of the pharynx from the ventral end of the endostyle to the opening of the oesophagus (Selys-Longchamps, 1900). The pyloric gland begins to arise in the tadpole as an outpouching from the pyloric end of the stomach.

The heart of *Ciona* has already begun to arise in the tadpole (Selys-Longchamps, 1900). Its first definitely recorded anlage is a pair of vesicles postero-ventral to the phaynx, symmetrically beneath the retropharyngeal groove (fig. 4.24b). These two vesicles come to make contact with each other and the walls thicken at the contact. These medial walls ultimately become muscular, to form the walls of the heart, and the cardiac cavity develops between them. The thin lateral walls of the vesicles form the pericardium. The cavity of the heart, however, does not develop until after metamorphosis. The paired cardiac

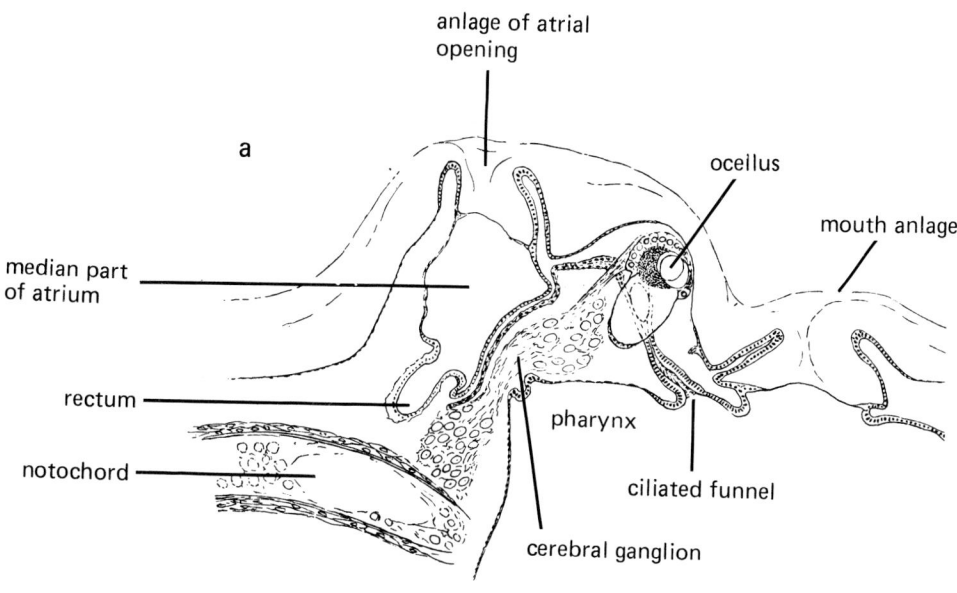

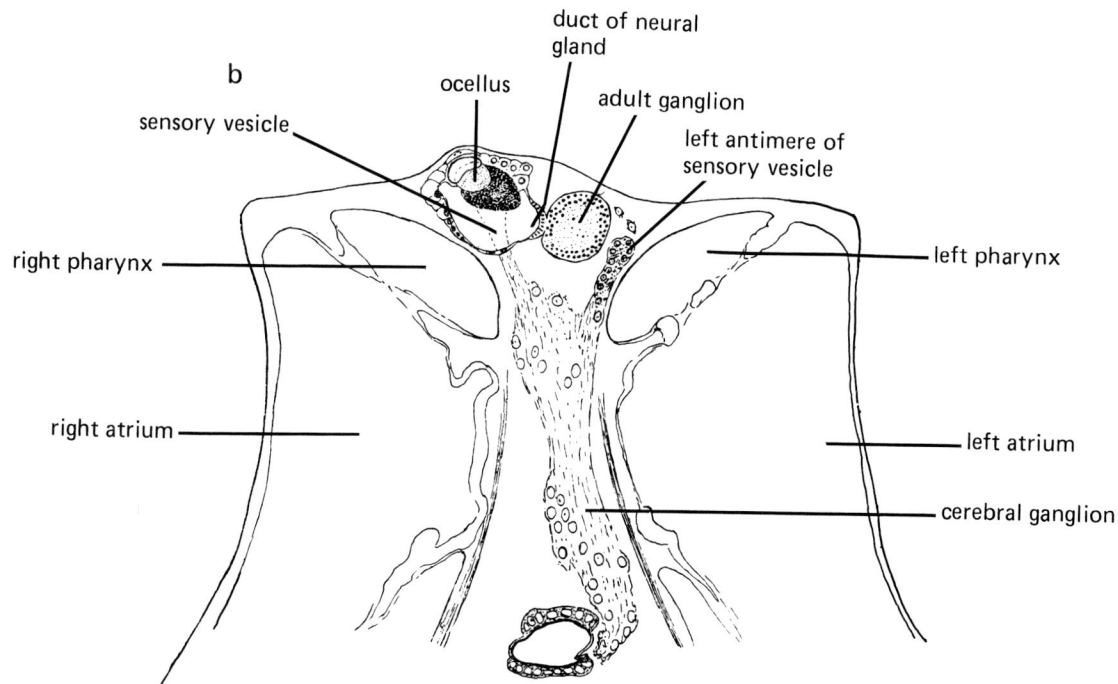

4.28. *The ocellus in the tadpole of the colonial tunicate* Distaplia, *to illustrate Froriep's theory that it is homologous to the right eye of a vertebrate. a) Sagittal section – a reconstruction based on eight adjacent sections; from Froriep (1906, fig. 1); b) transverse section to show the ocellus and its presumed vestigial left antimere – a reconstruction based on five adjacent sections; from Froriep (1906, fig. 2).*

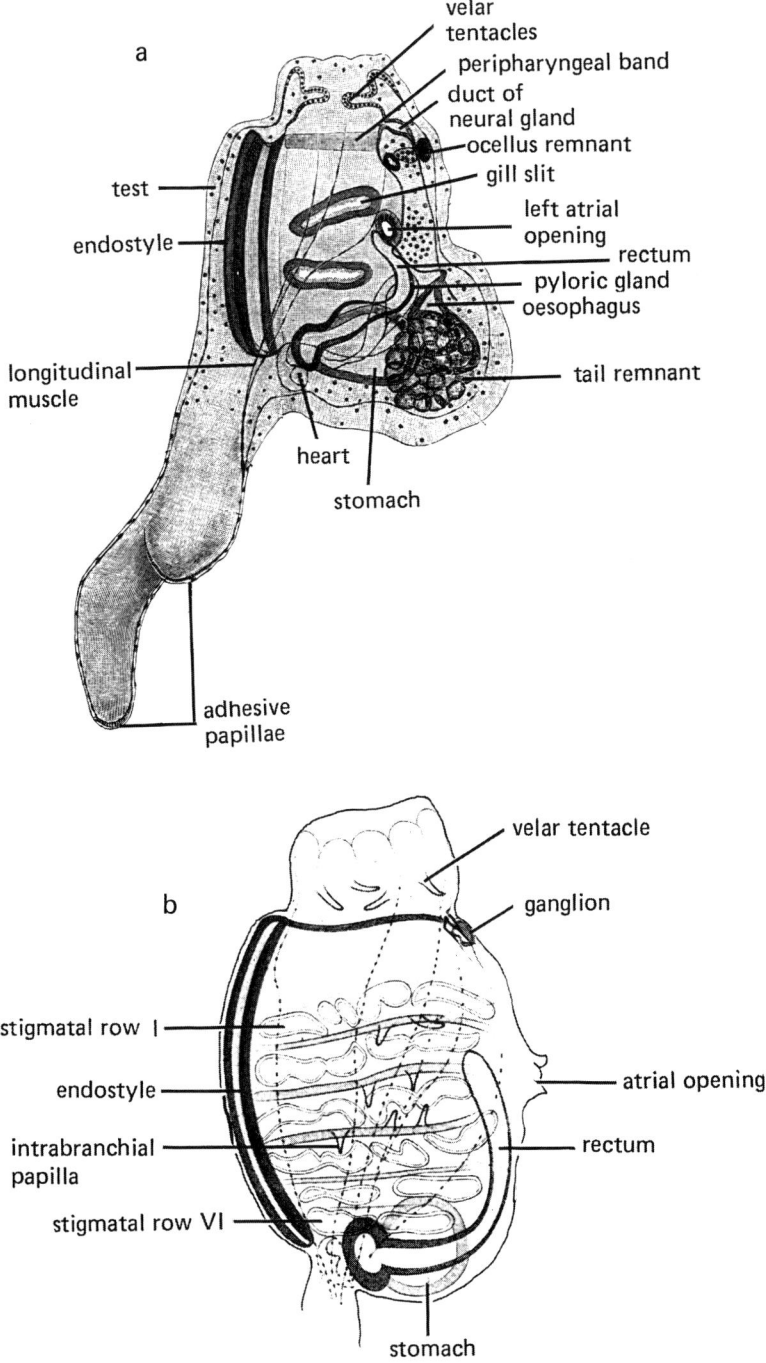

a

velar
tentacles

peripharyngeal band

duct of
neural gland

ocellus remnant

gill slit

test

left atrial
opening

endostyle

rectum

pyloric gland

oesophagus

tail remnant

longitudinal
muscle

heart

stomach

adhesive
papillae

b

velar tentacle

ganglion

stigmatal row I

endostyle

atrial opening

intrabranchial
papilla

rectum

stigmatal row VI

stomach

4.29. Ciona, *post-larval development. a) Post-larva just after attachment; the animal has undergone torsion so that the mouth is at the opposite end to the attachment; two pairs of primary protostigmata (I & IV) exist; the atrial openings are still separate; from Seeliger & Hartmeyer (1893, fig. 84, p. 375). b) A later post-larva in which the right and left atrial openings have joined to form a single, dorsal atrial siphon. There are six rows of stigmata, formed by subdividing six secondary protostigmata (I to VI); the intrabranchial vessels have begun to form; from Seeliger (1893, pl. 20, fig. 37).*

vesicles shift rightwards during development, even during the tadpole stage, away from their position immediately under the retropharyngeal groove. The anlage of the heart probably arises from a postero-ventral isthmus of mesenchyme which Castle recorded in the late gastrula between the right and left mesenchymal masses (fig. 4.22b). This is closely similar to the origin of the vertebrate heart at the mid-ventral meeting place of right and left extensions of the lateral plate (fig. 4.29).

At the end of the free-swimming stage the tadpole of *Ciona* attaches itself to a hard object by means of the three anterior papillae. Fixation is the signal for a number of changes that happen concomitantly: the tail disappears; the animal undergoes a process known as torsion which brings the mouth to the end of the body farthest from the point of fixation; the stigmata (gill slits) arise, apart from the two pairs that exist already, and the pharynx and alimentary canal become functional; and the other adult systems also are elaborated and become functional.

The disappearance of the tail after fixation takes place by contraction of the tail epithelium, which has used up all its food reserves (Berrill, 1947, b). The contraction of the epithelium begins at the distal end of the tail and spreads forward. It pushes the notochord, muscle cells and dorsal nerve cord into the head of the now attached tadpole. This leaves the test of the tail empty of contents, and finally it falls off. There is at first, no phagocytosis of the cells that have been ejected from the tail into the head.

Torsion in *Ciona* is an abrupt change in orientation which the organs of the animal undergo relative to the point of fixation. There is a contraction of the ventral side of the attached tadpole, presumably by active contraction of the muscle fibres of the head as described in *Amaroucium* by Scott (1952). Simultaneously there is an expansion of the dorsal side of the head. The combined effect of these changes is that the mouth, originally dorsal but near the point of fixation, shifts to the extreme free end of the animal and the paired atrial openings move through the same angle, to finish in a subterminal position. At the same time parts of the pharynx rotate through 90° in the opposite direction so that the endostyle becomes ventral rather than anterior, while the retropharyngeal groove becomes posterior rather than ventral (fig. 4.29a).

The origin of the gill slits (stigmata) is complicated. In all non-colonial benthic tunicates there is an ontogenetic stage with six pairs of slits in the pharyngeal wall, these being elongated dorso-ventrally (fig. 4.29b). These six pairs of slits are known as secondary protostigmata and are numbered I to VI from front to rear. In *Ciona* these six pairs of secondary protostigmata arise from three preceding pairs of primary protostigmata, these being numbers I, IV and V (fig. 4.30). Primary protostigmata I and IV appear earliest, as separate perforations of the pharyngeal wall,

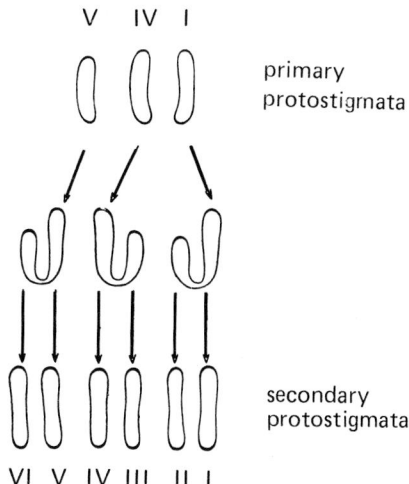

4.30. Ciona. *Origin of the secondary protostigmata – diagram of the successive stages in right aspect; from Brien (1948, fig. 169, p. 696).*

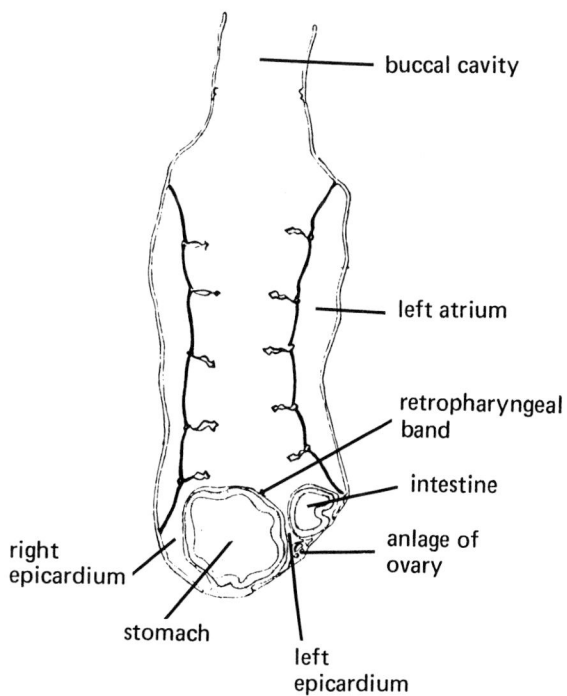

4.31. Ciona. *Frontal section of a young animal shortly after attachment to show the condition of the epicardia widely open to the pharynx and left and right of the retropharyngeal band; from Damas (1900, pl. 2, fig. 8a).*

and develop hooks at their ventral ends, J-fashion (fig. 4.31). These hooks become separate from the parent protostigma, and give rise to the secondary protostigmata II (developed from, and behind, I) and III (developed from, and in front of, IV). At the same time primary protostigma V arises behind IV as a separate perforation and, by means of a similar hook at its ventral end, produces secondary protostigmata VI behind it.

The six pairs of secondary protostigmata begin to divide across their length by the growth of processes from their front or rear walls (fig. 4.29b). The resulting six pairs of stigmatal rows subsequently divide along their length. Thus, by repetition of such subdivisions, the hundreds of stigmata in an adult *Ciona* have arisen from only three pairs of initial perforations.

The intrabranchial vessels develop at the same time as the slits. They start as papillae on the internal surface of the pharynx (fig. 4.29b). These send out subordinate papillae in a dorso-ventral or an antero-posterior direction, and these secondary papillae meet each other, and fuse where they meet to produce the intrabranchial network of transverse and longitudinal vessels.

Changes in the atria are taking place alongside the changes in the stigmata. The atria expand to accommodate the six pairs of secondary protostigmata; the test blocking the mouth and atrial apertures of the tadpole is lost; the rectum opens into the left atrium at the anus; and pharyngeal cilia develop – both the water-pumping cilia round the stigmata and the mucus-shifting cilia of the endostyle and alimentary ciliated loop. Finally the atria expand onto the dorsal surface and the two apertures fuse to form a single median anterior atrial aperture (fig. 4.29b).

The epicardia of *Ciona* develop as rearward expansions of the pharynx on either side of the pre-existing retro-pharyngeal groove (Selys-Longchamps, 1900; Damas, 1899). Each epicardium in its expansion wraps itself round the visceral organs with which it comes into contact (fig. 4.31). However, the left epicardium expands more than the right one, so that it alone forms the mesentery round the intestine and ovary. The pharyngo-epicardial openings are wide at first, but later constrict to produce the pair of small foramina found in the adult. The heart lengthens and, since the front end is attached to the endostyle and the rear end to the right side of the stomach, it acquires a V-bend in consequence. The pyloric gland develops as an evagination from the pyloric end of the stomach.

The common anlage of the definitive ganglion, neural gland and dorsal strand arises in *Ciona* as an evagination of the dorsal left wall of the cerebral vesicle of the tadpole (Willey, 1894). By the time of fixation this evagination has become a totally separate vesicle but this soon acquires an opening into the dorsal wall of the pharynx, anterior to the peripharyngeal bands of cilia (Elwyn, 1937, for *Ecteinascidia*). A dorsal thickening of the vesicle gives rise to the definitive ganglion. A ventral thickening gives rise to the neural gland. And a posterior extension, gradually growing down to the site of the future ovary, presumably becomes the dorsal strand, as it does in the ontogeny of *Corella* worked out by Hūus (1924) (fig. 4.32). During these events the larval nervous system has degenerated.

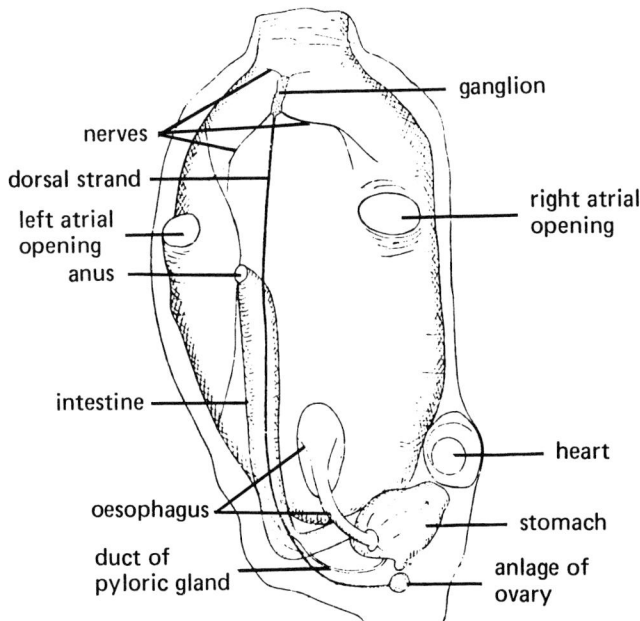

4.32. *Corella.* Young individual of this solitary ascidian in dorsal aspect to show the dorsal strand and its relations to the ganglion and to the anlage of the ovary. At this stage the right and left atria are still separate; from Hūus (1924, fig. 1, p. 17).

The gonads arise from a small concentration of mesenchyme cells situated in the gut loop. This concentration becomes a hollow vesicle which then becomes bilobed. The medial lobe gives rise to the testis and the lateral, left lobe to the ovary. At the same time the vesicle develops an anterior extension which grows forward right of the rectum and finally splits along its length to produce the vas deferens and oviduct (Hůus, 1937, p. 638).

Many Haeckelian recapitulations exist in the ontogeny of *Ciona*, as shown by a comparison with the fossils known as mitrates. Such include among others: the loss of the tail with its notochord, dorsal nerve cord and muscle blocks; the loss of the larval brain with its eye; and the fact that the atrium of the adult arises from a pair of larval atria.

The classification of tunicates is complicated, for the group is very diverse, with solitary and colonial, benthic and pelagic types. Only one aspect of the classification will be discussed here. This concerns the phylogenetic position of the appendicularians (Larvacea), which are small, solitary, pelagic tunicates whose adult is tadpole-like in possessing a tail. This adult lives in a complicated transparent 'house' which the animal repeatedly builds and abandons (fig. 4.33). The house is homologous with the test of other tunicates. Delicate sieves in the house pre-

filter nannoplankton from the sea water. The filtrate enters the mouth of the adult as food and is taken up, inside the pharynx, by an endostylar mucous filter of standard tunicate type. The feeding current of sea water is pumped through the house by the undulation of the tail.

Two views compete concerning the phylogenetic position of the appendicularians. The first supposes that the tadpole-like adult is primitive, that appendicularians have never had an attached phase in their ancestry, and that they represent, in modern terminology, the sister group of all other tunicates. This view was proposed by Haeckel (1874, p. 714) and adopted, for example, by Lohmann (1933) who was the greatest of all experts on appendicularians. The second view was proposed by Willey (1893, p. 350) who wrote that: '*Appendicularia* is probably not a primitive form, but represents a larval ascidian which has become secondarily pelagic and progenetic'. This means that appendicularians would once have had a tailless, benthic attached adult stage as in *Ciona* (for example), but have lost it by sexual precocity. Willey's view was elaborated in an extremely influential paper by Garstang (1928).

Garstang's argument, when re-cast in modern terminology, is that the closest living relatives of appendicularians among other tunicates are a small pelagic group known as Doliolids (fig. 4.34). These have a tadpole in the life cycle (fig. 4.35) which is followed by a tailless adult. This adult is not attached to a substrate, for the Doliolids are a pelagic group. But the ontogenetic transition from a tailed to a tailless condition is probably homologous with what happens in *Ciona* and other benthic tunicates. Moreover the tailless condition of adult tunicates in general probably arose phylogenetically as a result of attachment to a substrate in a benthic ancestor. It follows that, if appendicularians are the sister group of Doliolids, then the latest common ancestor of both groups had a tadpole in its ontogeny which was followed by a tailless adult. And that appendicularians have therefore lost the tailless adult in phylogeny.

The individual Doliolid animal (fig. 4.34) is barrel-shaped with an anterior mouth and a posterior atrial opening. The body wall contains ring-shaped muscles whose contraction forces water out of the atrial opening and moves the body forward by jet propulsion. The posterior wall of the pharynx is a diaphragm penetrated by a right and a left series of ciliated gill slits. The life cycle is much more complicated than already indicated. It includes individuals that bud but do not reproduce sexually, others that reproduce sexually but do not bud, and others that only feed (see Berrill, 1950, p. 276 ff).

Probable synapomorphies which connect the appendicularians and Doliolids together and distinguish both groups from all other tunicates are as follows:

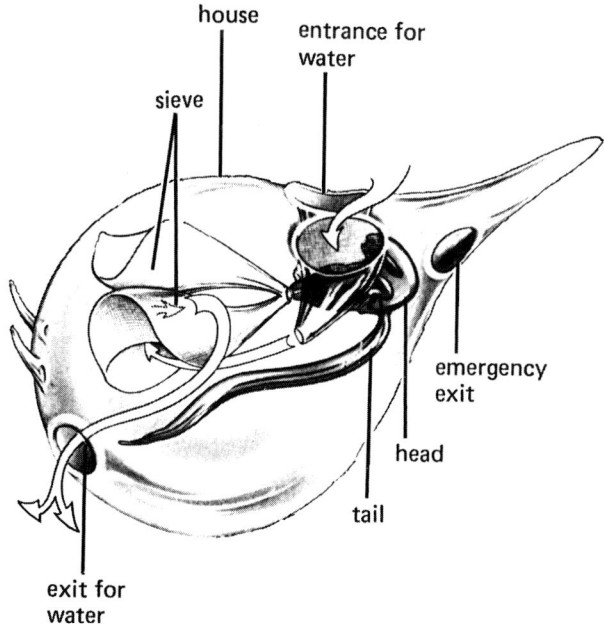

house
entrance for water
sieve
emergency exit
head
tail
exit for water

4.33. The appendicularian Oikopleura *inside its house. The animal draws a current of water through the house by wriggling its tail. Food particles are filtered from the water by means of sieves and then pass into the mouth. The animal can leave the house through the emergency exit; from Burn (1980, p. 186).*

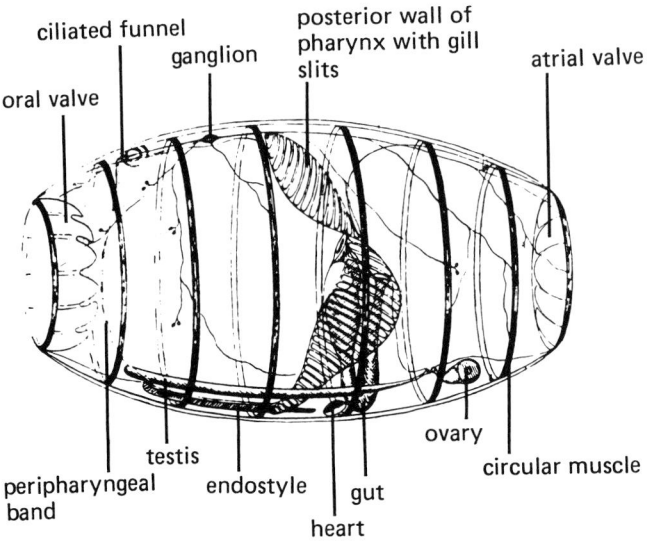

4.34. *A sexually reproducing individual (gonozooid) of a Doliolid (*Doliolum resistibile*), seen from the left. Only the left half of the branchial diaphragm is shown; from Neumann in Kükenthal & Krumbach (1933–1940, fig. 166, p. 204).*

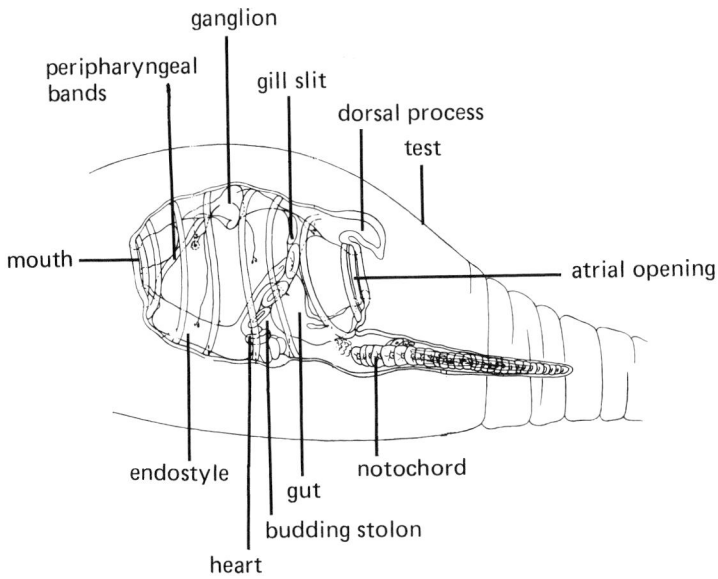

4.35. *Tadpole of a Doliolid (*Doliolum denticulatum*) to show the posterior position of the atrial opening (as in the adult) and the ventral position of the tail (as in appendicularians, cf. fig. 4.33); from Neumann in Kukenthal & Krumback (1933–40, fig. 174, p. 212).*

1. There is no cellulose (tunicin) in the test (house) (for data see: Hŭus *et al.*, 1933–40), pp. 141, 233, 332, 409, 557).
2. The Doliolids have the extraordinary habit, recorded by Uljanin (1884, p. 14), of periodically moulting the test, just as appendicularians periodically abandon the house.

3. The endostyle is simple in both groups, having only two pairs of gland strips, whereas all other tunicates have three such pairs.
4. Appendicularians, like some Doliolids, have the anus right of the oesophagus instead of left – this is unlike all other tunicates and also unlike the most primitive crown chordates (mitrates).

5. The tail is attached to the ventral face of the head in the tadpoles of Doliolids and in appendicularians, whereas in all other tunicates (and also in mitrates) it is posterior to the head.

These synapomorphies strongly confirm Garstang's theory that appendicularians arose by paedogenesis from a Doliolid-like ancestor. Garstang's views are further confirmed by the fact, pointed out by Berrill (1955, p. 137), that appendicularians have about half as many cells as a normal tunicate tadpole. Thus the appendicularian *Oikopleura*, for example has 20 cells in the notochord and 18 muscle cells in the tail, whereas the tadpoles of *Ciona* and other benthic tunicates have respectively 40 and 36. This is despite the fact that appendicularians are sometimes much larger than the tadpoles of benthic tunicates. Thus *Oikopleura* may be 36 mm long and the longest known appendicularian is about 80 mm long (*Bathochordaeus*, see Lohmann, 1922; Berrill, 1950, pp. 311, 312) whereas the largest known tadpole of a benthic tunicate is only 3·5 mm long (*Stolonica socialis*, Berrill, 1947b, p. 156; 1950, fig. 72). All this suggests that *Ciona*, in having a tailed tadpole followed by a fixed adult, is more primitive than the appendicularians are.

The phylogenetic position and phylogenetic significance of the tunicates will be dealt with in Chapters 8, 9 and 10. At this point, however, I shall summarize the main views. Sometimes the tunicate tadpole is seen as a recapitulation of an ancestral adult condition. This is the view advocated in this book and it goes back to Haeckel (1874).

It is also associated with the names of Macbride (1926) and Willey (1894). Sometimes the attachment of tunicates is regarded as primitive for chordates and the tunicate tadpole is seen as a phylogenetic interpolation into the life history, which subsequently by paedogenesis gave rise to adult vertebrates and acraniates. The tadpole, on this view, would be an adaptation for larval dispersal or for choosing suitable points of settlement on the sea bottom. This theory was propounded by Garstang (1928) and developed by a number of eminent comparative anatomists including Berrill (1955), Romer (1972) and Starck (1963). Finally, in an ingenious if somewhat perverse book, Løvtrup (1977, p. 138) has proposed that both amphioxus and tunicates are more distantly related to vertebrates than molluscs are, i.e. that the phylum Chordata is paraphyletic or possibly polyphyletic. This proposal has to be taken seriously because of the numerous facts listed in support of it and because Løvtrup's methodology has given valuable results when applied within the vertebrates. It is contrary to what most zoologists since Kowalevsky have believed, but this does not in itself show that it is mistaken (see Chapter 10, however).

To sum up, the tunicates have a complex anatomy and an interesting life history. They are highly relevant to the origin of the vertebrates but have not recently attracted much attention and many of the most striking facts about them have been forgotten. They have a number of asymmetries which I have tried to emphasize in this chapter.

Chapter 5   Adult anatomy and basic phylogeny of living vertebrates

The literature on living vertebrates is enormous, but the part that bears on the origin of the group is relatively small. In this chapter I shall survey the adult anatomy of primitive living vertebrates and discuss their likely phylogenetic relations with each other. I shall deal with embryology in the next chapter.

The vertebrates are classically divided into two groups: the agnathans and the gnathostomes. As the names imply, the gnathostomes have jaws whereas the agnathans lack them. The agnathans comprise two groups of primitive 'fishes': the hagfishes or myxinoids and the lampreys or petromyzonids. The gnathostomes, on the other hand, include the jawed fishes and also the tetrapods including man. The gnathostomes are almost certainly a monophyletic group since the presence of jaws is advanced compared to their absence. The agnathans, on the other hand, are defined on a primitive feature – the lack of jaws – and are therefore suspect of being paraphyletic.

Perhaps the most widely known myxinoid is *Myxine glutinosa* – the Atlantic Hagfish or Glutinous Hag (fig. 5.1). A compilative account of this species was edited by Brodal & Fänge (1963) and updated by Fänge (1973); there is a detailed description of its anatomy by Marinelli & Strenger (1956). A recent general account of all agnathans is Hardisty (1979), and there is an essay in Russian with the same scope by Balabai (1956). In what follows I shall refer to *Myxine glutinosa* as *Myxine*.

Myxine is an ugly, worm-like, almost unpigmented animal, about 30 cm long and 2 cm in diameter when adult. It lives in mud on the sea bottom, being particularly well known in certain Norwegian fjords. Like all other myxinoids it can flourish only in sea water, since its kidney cannot control the osmotic pressure of the blood. It burrows in the mud of the sea bottom and the burrow is U-shaped with the mouth and nostril showing at one end of the U and an exhalent current emerging from the other (Hardisty, 1979, p. 57).

A respiratory current enters at the front end through the single nostril and leaves the body by way of a pair of branchial openings situated on the ventral surface about one third of the length rearwards from the front end. *Myxine* can swim either rearwards or forwards, by left-right undulation. Like other myxinoids, but unlike any other vertebrates, it can also tie its body into simple knots (half-hitches). The knots can slide along the body and are used to give purchase when biting off lumps of food. *Myxine* will eat moribund fishes but is not correctly described as parasitic, for it will also ingest a variety of other animal food, including mud-dwelling worms (Strahan in Brodal & Fänge, 1963, p. 22ff).

In external shape *Myxine* is almost exactly cylindrical, except posteriorly where it is laterally compressed. The single nostril is at the anterior end. The mouth is a vertical slit ventral to the nostril and separated from the latter by a platform of tissue (fig. 5.2). The mouth is armed with horny 'teeth' that point rearwards into it. One tooth, the ethmoid tooth, is situated medially in the roof of the mouth. The other teeth are in a pair of groups and can be pulled in and out of the mouth. In fact they overlie the so-called dental cartilage (fig. 5.3) which is shaped like the covers of a book with the teeth on the inner surface. When the dental cartilage moves forward out of the mouth, the 'book' opens and the teeth are exposed (fig. 5.4). When the dental cartilage moves rearwards into the mouth, the 'book' closes, and the teeth of right and left sides gnash together. As already mentioned, there is a pair of branchial openings on the ventral surface. The left opening is always larger than the right one. About one sixth of the length forwards from the hind end of the animal is the cloacal opening. A median fin begins on the dorsal surface above the cloaca, runs backwards round the rear end of the animal and forwards on the ventral surface as far as the branchial opening, except for being interrupted at the cloaca.

The skin of *Myxine* is separated from underlying

caudal fin
nostril
mouth
left gill
opening
pores of mucous
sacs
cloaca
preanal fin

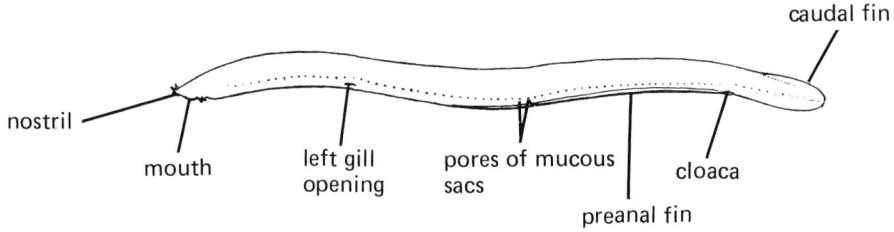

5.1. Myxine glutinosa, *external features; from Goodrich (1909b, fig. 15c, p. 30).*

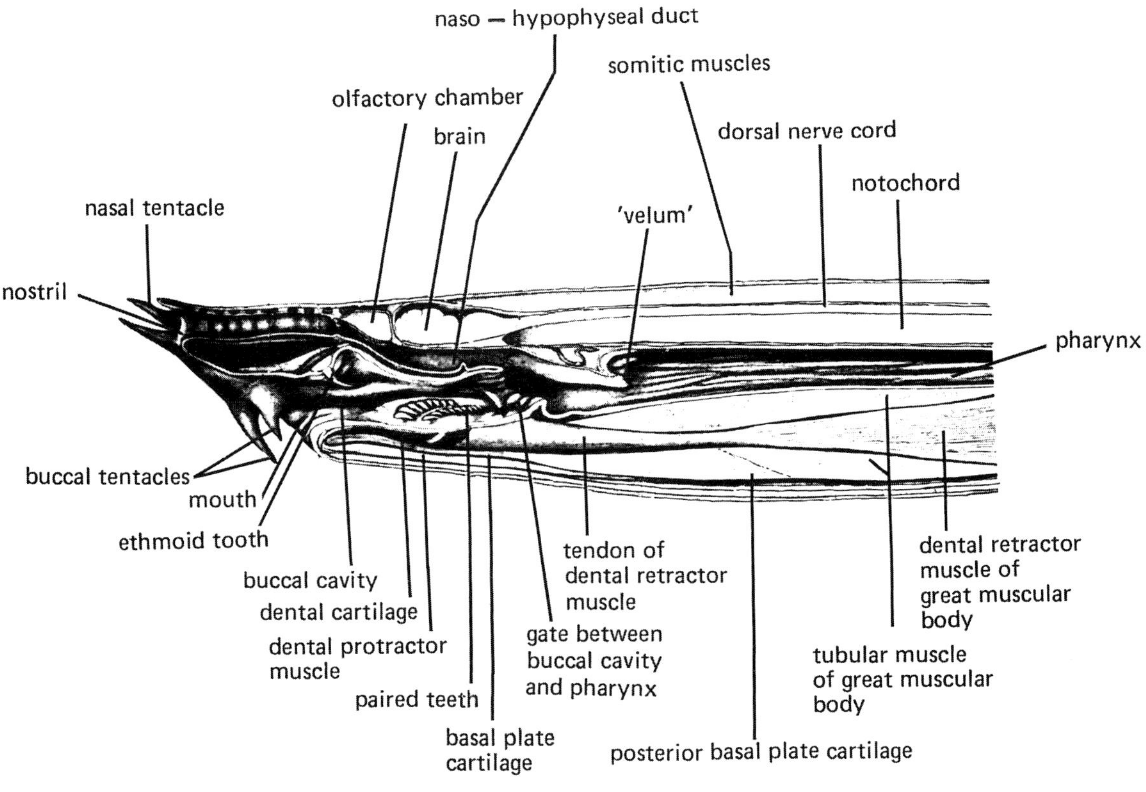

naso — hypophyseal duct
somitic muscles
olfactory chamber
brain
dorsal nerve cord
notochord
nasal tentacle
'velum'
nostril
pharynx
buccal tentacles
mouth
ethmoid tooth
buccal cavity
dental cartilage
dental protractor
muscle
paired teeth
basal plate
cartilage
tendon of
dental retractor
muscle
gate between
buccal cavity
and pharynx
posterior basal plate cartilage
tubular muscle
of great muscular
body
dental retractor
muscle of
great muscular
body

5.2. Myxine, *anterior part of animal in sagittal section; the teeth are in the resting position, inside the buccal cavity; from Marinelli & Strenger (1956, fig. 98, p. 122).*

tissues by an extensive subcutaneous blood sinus (fig. 5.11), filled with red blood. The segmented body muscles are directly visible through the skin. An obvious row of mucous glands is situated ventro-laterally on left and right (fig. 5.8). There is one gland per segment on each side and each opens by a pore (fig. 5.1). The glands can expel a huge quantity of mucus when the animal is irritated, whence the name *glutinosa*.

The nostril of *Myxine* leads into the naso-hypophyseal duct (fig. 5.2) which runs rearwards to open through the roof of the pharynx. A dorsal cul-de-sac of this duct contains the unpaired olfactory organ, which is in contact with the olfactory part of the brain. Just behind the place where the duct opens into the roof of the pharynx is a pump known as the 'velum'. By the action of the 'velum', aided by the muscles of the branchial pouches, water is drawn

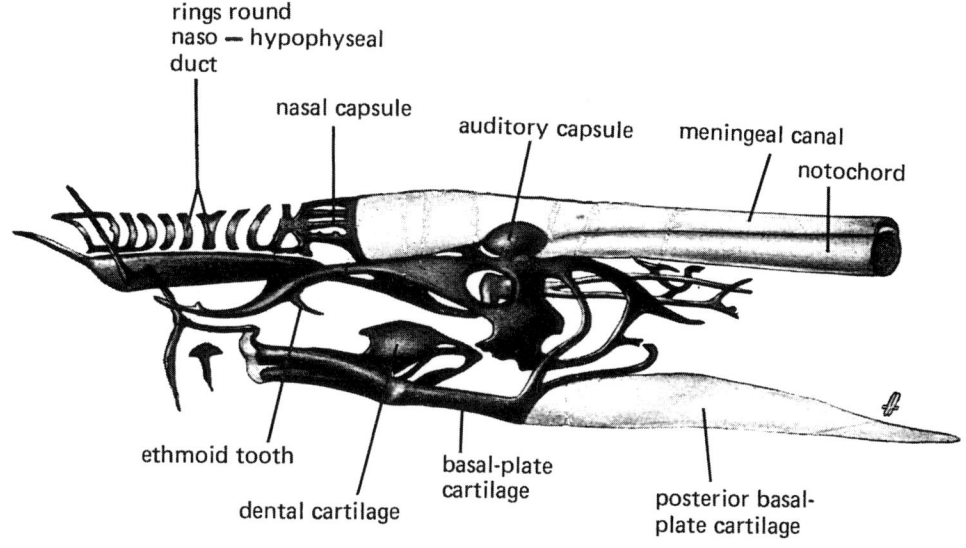

5.3. Myxine, *cartilages of the head; from Marinelli & Strenger (1956, fig. 130, p. 163).*

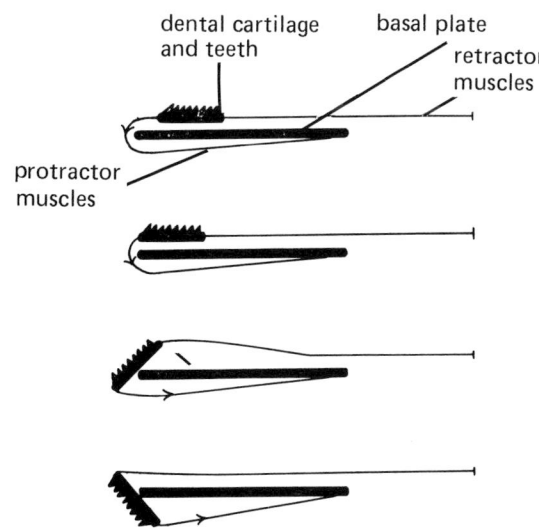

5.4. *Diagram of the mode of action of the 'lingual' muscles that actuate the teeth of a myxinoid; from Dawson in Brodal & Fänge (1963, p. 234, fig. 5).*

in through the nostril, passed into the pharynx to the branchial pouches and from these through ducts to the gill openings. Consequently the respiratory current flows when the nostril is open but the mouth is shut.

The skeleton of *Myxine* consists of a notochord and a number of cartilages (figs 5.3, 5.6). The notochord extends almost the whole length of the body but stops anteriorly just behind the otic capsules. It is a turgescent flexible rod, and is totally unconstricted, for, unlike lampreys and gnathostomes, myxinoids have no vertebrae (neuralia). The notochord is overlain by the dorsal nerve cord which is enclosed in a tough-walled meningeal canal. This canal continues anteriorly beyond the front end of the notochord where it expands to enclose the brain. The cartilages consist of fin rays, the 'skull', a number of cartilages associated with the teeth, suprapharyngeal cartilages and a pair of rings surrounding the branchial openings.

The neurocranium of *Myxine* (fig. 5.3) consists of a pair of otic capsules, a single unpaired nasal capsule, a series of rings round the naso-hypophyseal duct serving to hold the duct open, and a plate of cartilage just ventral to the brain and therefore corresponding in position to the parachordals and trabeculae of other vertebrates. The 'lingual' skeleton associated with the teeth consists of the book-shaped cartilaginous dental plate already mentioned, and also of a series of rods (the basal plate) in the ventral mid-line. A tendon passes from below round the front of the basal plate, like a rope round a pulley block, and runs rearwards to the front edge of the dental plate, while another tendon passes rearwards, from the rear end of the dental plate, just above the basal plate (fig. 5.4). Muscles pull these tendons (fig. 5.2): a weak ventral set pulls the anterior tendon and moves the dental plate forward while a strong dorsal set pulls the posterior tendon and thus moves the dental plate rearwards in the biting action. The muscles pulling the dental plate rearwards are surrounded

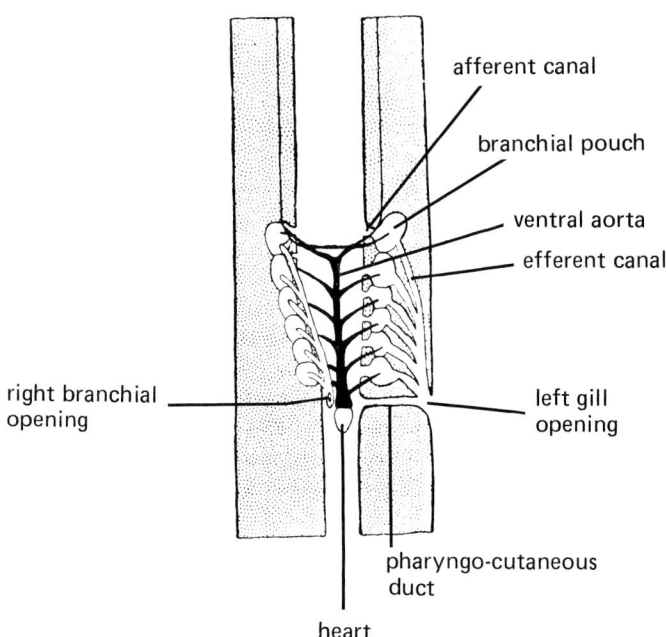

afferent canal

branchial pouch

ventral aorta

efferent canal

right branchial
opening

left gill
opening

pharyngo-cutaneous
duct

heart

5.5. Myxine. *The branchial apparatus and its afferent blood supply, ventral aspect, diagrammatic. The branchial structures on the animal's right are in their natural position, while those on the left have been rotated outwards; from Goodrich (1909, fig. 19, p. 33).*

by a tubular muscle which, by its own contraction, relaxes them. Together these muscles form a club-shaped mass – the great muscular body (fig. 5.2).

The alimentary canal of *Myxine* consists of buccal cavity, naso-hypophyseal duct, pharynx, intestine and cloaca. There is no obvious stomach. The buccal cavity leads rearwards from the mouth, beneath the floor that separates it from the naso-hypophyseal duct (fig. 5.2). It opens posteriorly by a 'pharyngeal mouth' into the floor of the pharynx and the curtain or gate of tissue round this pharyngeal mouth probably corresponds more closely to the velum of other chordates than does the muscular pump traditionally called the 'velum'.

The branchial apparatus (fig. 5.5) is connected to the pharynx. It consists of six pairs of branchial pouches. Water flows into each pouch through an afferent duct and leaves it through an efferent duct that passes to the branchial opening. The efferent ducts on each side join together before reaching the surface of the body, and in addition, the left group joins a so-called pharyngo-cutaneous duct that passes direct from the pharynx to the outside, without containing a branchial pouch (fig. 5.7). Because of this arrangement there is only one pair of externally visible branchial openings, of which the left one is bigger than the right. The pharyngo-cutaneous duct is said to provide passage for surplus water. The internal

surfaces of the gill pouches are folded parallel to the flow of water. Their walls contain capillaries in which the blood flows in a direction opposite to the flow of water. This gives a counter-current system advantageous for gas exchange.

The coeloms will be briefly discussed before continuing with the alimentary canal. There are two of them – the abdominal coelom and the pericardium. The abdominal coelom extends behind the pericardium to somewhat behind the cloaca and is connected to the pericardium by a large foramen on the right side. The gonad, intestine and liver are suspended in the abdominal coelom by mesenteries and the kidneys are situated in its dorsal wall. The abdominal coelom opens to the outside by the coelomic pore in the dorsal wall of the cloaca, and this pore is presumably the exit for gametes (fig. 5.6). The pericardium contains the systemic heart and also the hepatic portal heart that pumps blood into the liver (fig. 5.7).

The alimentary canal just behind the pharynx narrows somewhat and then widens again to run into the intestine (fig. 5.7). It fills most of the abdominal coelom. Where the gut widens again, the bile duct enters it, which suggests that the narrow portion behind the pharynx corresponds to the oesophagus and stomach of gnathostomes. The lack of a stomach is probably not primitive for chordates, since tunicates have stomachs as also do echinoderm larvae and

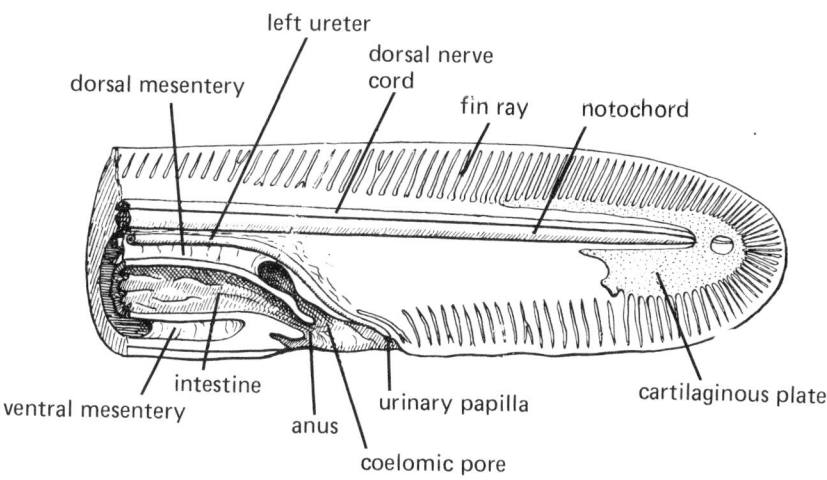

5.6. Myxine. *The cloaca and organs entering it and the cartilages of the tail. The single coelomic pore is the presumed exit for gametes; from Goodrich (1909, fig. 19, p. 33).*

pterobranchs. It is more likely to be an advanced feature caused by the snake-shape of *Myxine*, for pythons likewise often lack stomachs (Marinelli & Strenger, 1956, p. 137).

The liver of *Myxine* has an anterior and a posterior lobe (fig. 5.7). Ducts from these lobes lead into a large gall bladder which debouches through the bile duct into the intestine, as already mentioned. The endocrine pancreas, containing insulin-secretory cells, is a small lobe attached to the bile duct. The exocrine pancreas is diffuse and represented by cells in the walls of the posterior hepatic vein, in the mesentery of the gut and liver and in the wall of the intestine itself. It is not concentrated near the bile duct and in this respect differs from the exocrine pancreas of lampreys and of gnathostomes (cf. Barrington, 1945, 1968; Adam *in* Brodal & Fänge, 1963, p. 278–9).

The intestine opens posteriorly into the cloaca (fig. 5.6). It has no spiral valve or typhlosole and this again is a feature where myxinoids differ both from lampreys and from gnathostomes (but also from tunicates). The wall of the cloaca is also penetrated, as already mentioned, by a coelomic pore opening from the abdominal coelom and by the openings of a pair of urinary ducts from the kidneys.

The muscles of *Myxine* can be divided into those special to the head region and those of the rest of the body. The biggest group of muscles is the somitic muscles (fig. 5.8). These form most of the mass of the animal and extend as an uninterrupted series of myomeres from almost the front end to the rear end. Each myomere extends from the dorsal mid-line downwards to a ventro-lateral position level with the mucous glands. The muscular floor of the body consists of roughly transverse oblique muscles and

of longitudinal rectus muscles internal to these. The myomeres are built of fibres which extend parallel to the length of the animal from myocomma to myocomma like the muscle plates of amphioxus (Jansen & Andersen, p. 166ff *in* Brodal & Fänge, 1963). The fibres are innervated by spinal nerves running to them along myocommata (unlike amphioxus).

In the head region the largest muscles are the 'lingual' muscles which move the teeth into and out of the mouth, as already mentioned. They, together with the muscles of the velum, tentacles and naso-hypophyseal duct are supplied by the trigeminal nerve. There are no oculomotor muscles and *Myxine* lacks the oculomotor, trochlear and abducens nerves, which supply the oculomotor muscles in gnathostomes and lampreys. Muscles moving the basal cartilaginous plate, over which the dental plate slides, are supplied by the facial nerve. And the muscles contracting the branchial pharynx and heart region are supplied by the vagus.

The renal system of *Myxine* consists of a pair of kidneys. On left and right sides these kidneys are divided into two portions that are not connected in the adult – a pronephros (fig. 5.9) and a mesonephros (fig. 5.10). The pronephros is a small body dorsal to the pericardium on either side and suspended in the right or left anterior cardinal vein. Three tubules enter it from the pericardial cavity on each side and seem to provide a clear pathway connecting the pericardium, and therefore the abdominal coelom also, with the cavity of the vein. Posterior to the rest of the pronephros is a knot of capillaries called a glomus. This receives an arterial blood supply (Holmgren, 1950) made up of three glomeruli and is suspended in a

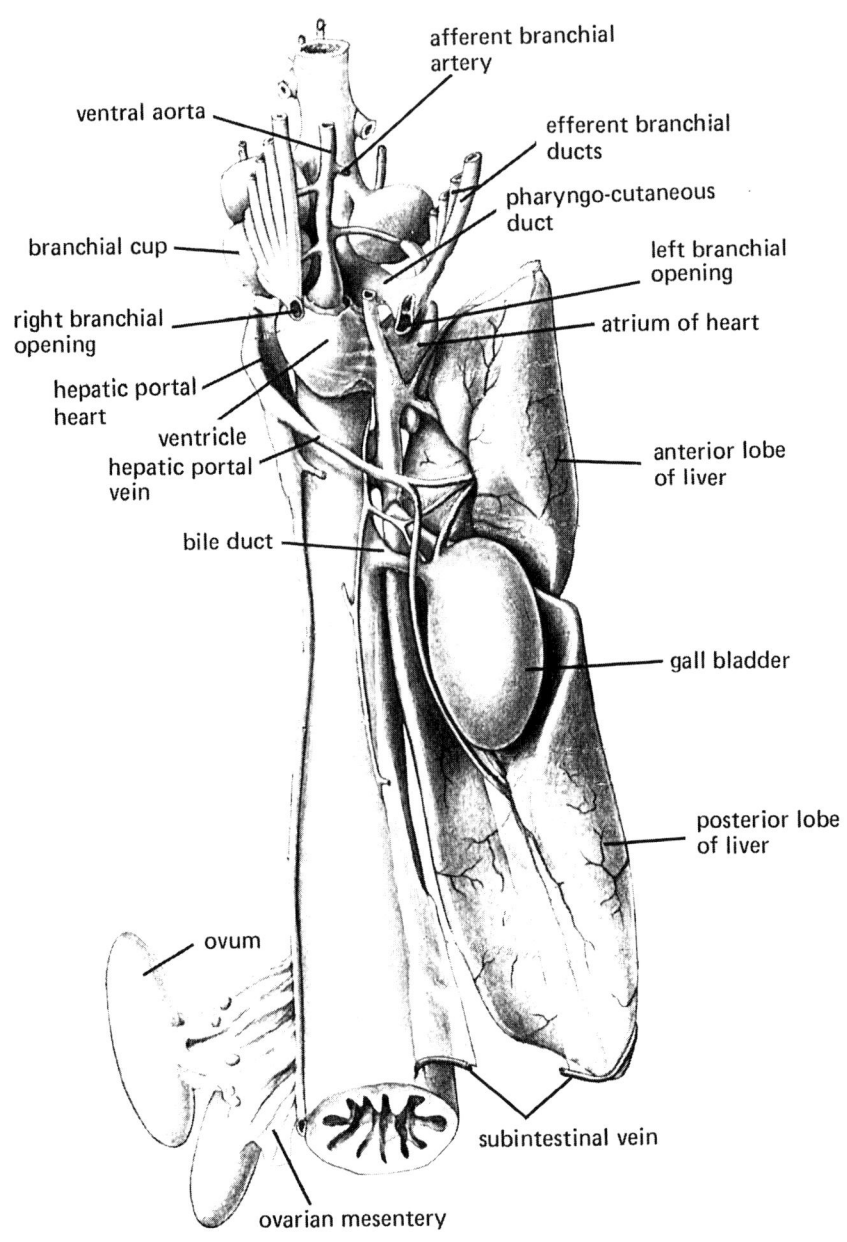

afferent branchial
artery

ventral aorta

efferent branchial
ducts

pharyngo-cutaneous
duct

branchial cup

left branchial
opening

right branchial
opening

atrium of heart

hepatic portal
heart

ventricle

hepatic portal
vein

anterior lobe
of liver

bile duct

gall bladder

posterior lobe
of liver

ovum

subintestinal vein

ovarian mesentery

5.7. Myxine. *Gut and associated organs in the region of the liver, ventral aspect; from Marinelli & Strenger (1956, fig. 107, p. 136).*

capsule formed from the wall of the pericardial cavity. Liquid excreted from the vessels of the glomus would enter the pericardium and could pass, through the opening between pericardium and abdominal coelom, into the abdominal coelom and thence to the outside through the abdominal pore of the cloaca. The pronephros has no duct but seems to be functional. The fact that it is so well developed in the adult (unlike lampreys and gnathostomes) is probably a primitive feature. The mesonephros is much more extensive. It consists of a series of glomeruli, one pair per segment. These are supplied with blood from the dorsal aorta and each one is suspended inside a Bowman's capsule. Each capsule opens directly, without any duct, into a pair of ureters. These run rearwards

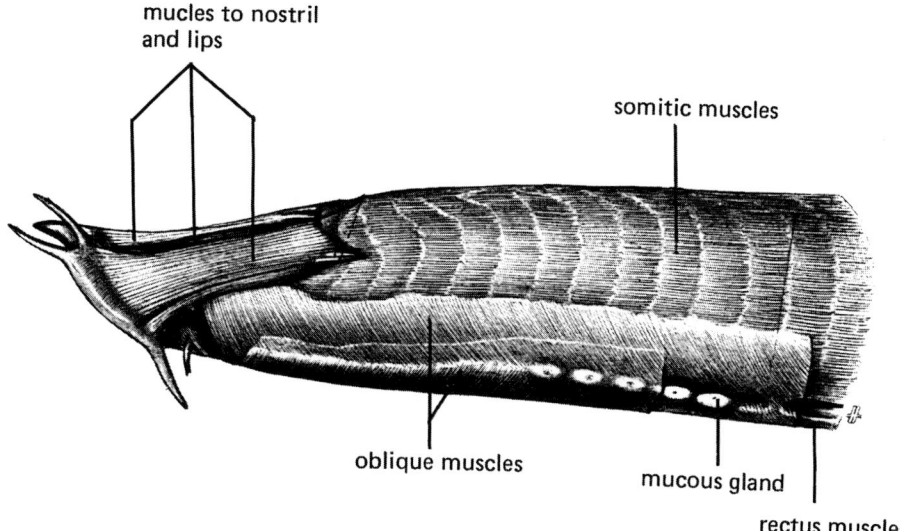

5.8. Myxine. *Muscles of the anterior part of the head. The eyes are covered by the large longitudinal muscle to the lips; from Marinelli & Strenger (1956, fig. 71, p. 91).*

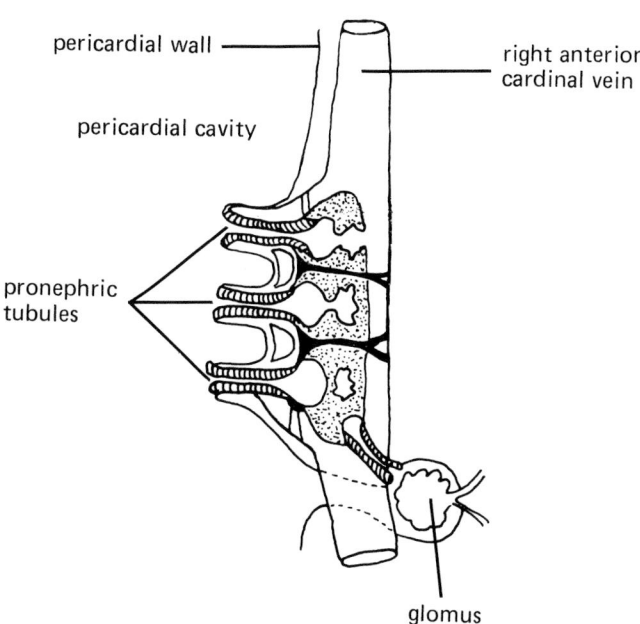

5.9. Myxine. *Diagram of the right pronephros in dorsal aspect. The three pronephric tubules open medially into the pericardial cavity while the central mass of the pronephros is suspended in the anterior cardinal vein; from Fänge in Brodal & Fänge (1963, fig. 1, p. 518, after Holmgren, 1950).*

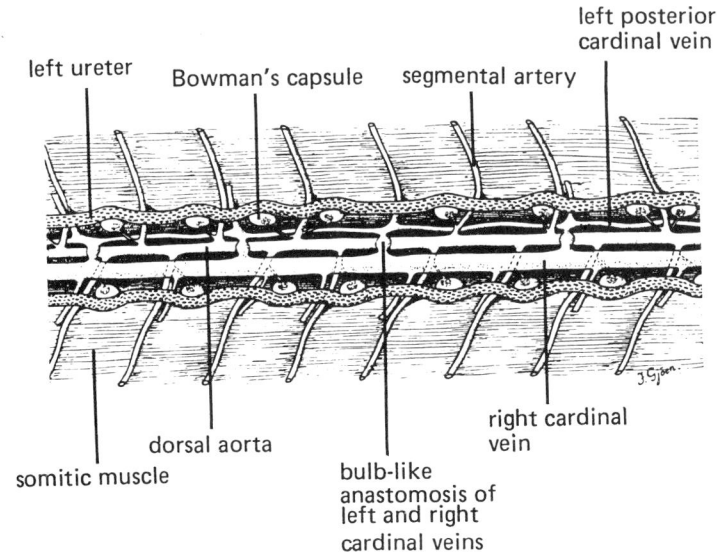

left ureter   Bowman's capsule   segmental artery   left posterior cardinal vein

somitic muscle   dorsal aorta   bulb-like anastomosis of left and right cardinal veins   right cardinal vein

5.10. Myxine. *The mesonephros and associated structures in ventral aspect as revealed by dissection. Note that the Bowman's capsules are directly in contact with the ureters; from Fänge in Brodal & Fänge (1963, fig. 2, p. 521).*

and open separately into the cloaca (fig. 5.6). The kidney of *Myxine* therefore differs from that of lampreys and gnathostomes where there is a long tubule between each Bowman's capsule and the ureter. This tubule puts back into the blood much of the salt secreted by the Bowman's capsule. It can be regarded as an adaptation to fresh water shared by gnathostome and lampreys. Its absence in myxinoids is probably a primitive feature related to the fact that myxinoids live in the sea and, most probably, have never had freshwater ancestors (Robertson, 1956). The kidneys of *Myxine* produce urea.

The blood salts of *Myxine* and other myxinoids are equal in concentration to those of sea water (Robertson in Brodal & Fänge, 1963). In this respect myxinoids are unlike all other vertebrates and, by outgroup comparison with tunicates and amphioxus, likely to be primitive. It is true that elasmobranchs also have blood isosmotic with sea water. Here, however, the osmotic pressure is in large part due to urea dissolved in the blood, so the situation is not truly comparable with myxinoids, and could well reflect freshwater ancestry.

The reproductive system of *Myxine* consists of an ovary or a testis, for the sexes are separate as in other vertebrates (contrary to the early opinion of Nansen, 1887; see Marinelli & Strenger, 1956, p. 141). Females are much commoner than males. The ovary or testis is developed in a fold of mesentery (fig. 5.7) that projects from the right side of the intestine and is continuous through the length of the abdominal coelom, from the pericardium back to the cloaca. The eggs are large and hard-shelled

and produced throughout the length of the ovary. They are released into the abdominal coelom and presumably escape to the outside through the coelomic pore in the dorsal wall of the cloaca. Sperm are produced mainly in the posterior portion of the testis and presumably escape to the outside by the same route as the eggs. The actual reproduction of *Myxine* has never been observed and even fertilized eggs have very seldom been found (Holgren, 1946).

The blood system of *Myxine* is unusual for a vertebrate in having very few capillaries. Instead, the arteries and veins are connected together by large blood-filled sinuses that Cole referred to as the 'red lymphatic' system (1926, 1912). These sinuses surround most of the important organs of the body and provide their blood supply. In correlation with this, the blood pressure produced by the main or systemic heart in the venous part of the system downstream of the sinuses must be very low and is raised locally by a number of accessory hearts which other vertebrates do not have. These are situated where the blood from the sinuses flows into the veins and into the hepatic portal system. Their beating is not synchronized together, nor with that of the systemic heart. *Myxine* also has a normal or 'white lymphatic' system filled with transparent lymph as in other vertebrates.

The systemic heart of *Myxine* consists of sinus venosus, atrium and ventricle. The latter is approximately in the mid-line and opens forward into the ventral aorta. Blood flows forwards in the ventral aorta and is given off into six pairs of afferent branchial arteries to the gill pouches. The

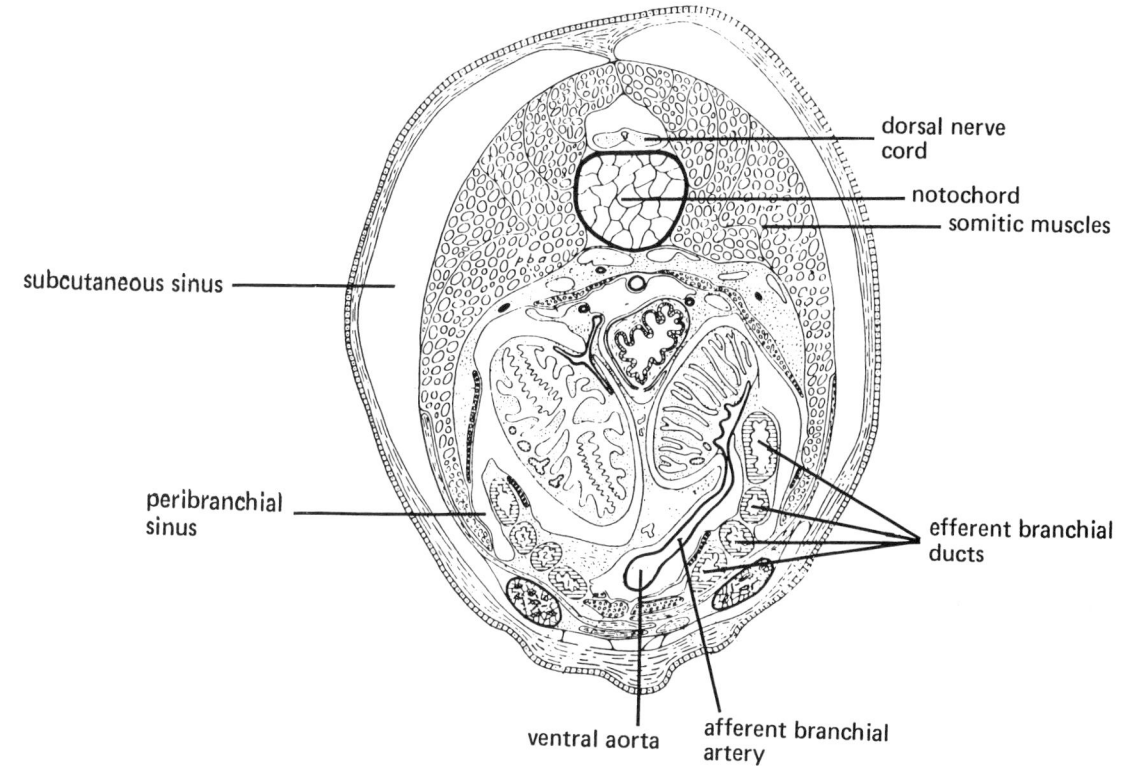

dorsal nerve cord

notochord

somitic muscles

subcutaneous sinus

peribranchial sinus

efferent branchial ducts

ventral aorta

afferent branchial artery

5.11. Myxine. *Transverse section through branchial region to show, in particular, the subcutaneous and peribranchial portions of the blood-sinus system ('red lymphatics' of Cole); from Cole (1926, pl. 4, fig. 8).*

muscles of the gill pouches and ducts help to pump blood through the branchial vessels.

The blood-sinus or 'red lymphatic' system is divided into two parts almost separate from each other (fig. 5.11). These can be called the subcutaneous complex and the peribranchial complex. The subcutaneous complex comprises the huge subcutaneous sinus and a number of cavities that penetrate between the organs of the head. The most posterior head sinus is connected to the peribranchial complex of sinuses. The blood sinuses open into the venous system at several points, guarded by valves. There are also openings between the 'red' and 'white' lymphatic systems.

The systemic and accessory hearts of *Myxine* have no nerve supply, except for a pair of caudal hearts near the tail which are controlled by the spinal nerves of the region. Contraction in these aneural hearts is myogenic like the heart of a tunicate or of an early chick embryo (Johansen, 1960). Outgroup comparison and Haeckel's law both suggest, therefore, that the aneural condition of the systemic heart is primitive, by contrast with all non-myxinoid vertebrates, including lampreys.

The nervous system of *Myxine* is much more complicated than that of a tunicate or amphioxus (see Bone and Peters *in* Brodal & Fänge, 1963). The dorsal nerve cord is flattened in shape and occupies only a small part of the hemicylindrical meningeal canal (fig. 5.11), the rest of which (as in lampreys) is filled with fatty connective tissue and the blood vessels that supply the nerve cord. Dorsal and ventral roots come off it alternately. The cell bodies most closely associated with the dorsal roots are, unlike amphioxus, situated in dorsal spinal ganglia.

Giant cells occur along the whole length of the dorsal nerve cord (fig. 5.12). The course of their axons and the position of their cell bodies resemble those of the Rohde cells of amphioxus, with which Bone believes them to be homologous. The myxinoids are the only vertebrates which have such a system of giant cells with cell bodies throughout the length of the dorsal nerve cord. By outgroup comparison with amphioxus this condition is probably primitive for all vertebrates. Lampreys and gnathostomes have a somewhat similar system in the so-called Müller cells. With these, however, the cell bodies are situated in the medulla oblongata of the brain and the giant

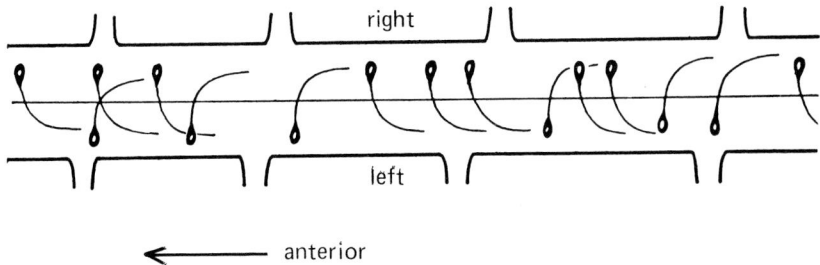

5.12. Myxine. *Giant cells in the dorsal nerve cord, in dorsal aspect; from Bone in Brodal & Fänge (1963, p. 63, fig. 6).*

fibres from them pass rearwards along the dorsal nerve cord. It is not certain whether myxinoids have Müller cells; Bone found Müller-like fibres in them but could not trace them to cell bodies. Perhaps the Müller cells of petromyzonts and gnathostomes are homologous with the most anterior Rohde cells of amphioxus, and with the most anterior giant cells of myxinoids, while equivalents of more posterior Rohde cells have been lost in petromyzonts and gnathostomes. The central canal of the dorsal nerve cord of *Myxine* contains a Reissner's fibre which is secreted by the subcommissural organ of the brain.

Before discussing the brain of *Myxine*, a question of terminology must be dealt with (fig. 5.13). The vertebrate brain is conventionally divided into three parts known as prosencephalon, mesencephalon and rhombencephalon (fore-brain, mid-brain, hind-brain). However, Starck (e.g. 1975, p. 356, 369ff) argued that the basic division should be into two parts which he called 'prosencephalon' and 'rhombencephalon'. (Henceforth I shall indicate his usage by inverted commas.) He denied that the conventional mesencephalon was a useful concept, for he saw its ventral part (tegmentum) as a forward continuation of the 'rhombencephalon' and its dorsal part (tectum – optic lobes) as a rearward continuation of the 'prosencephalon'. His views were based on ontogeny since, in man for example, the separation between 'prosencephalon' and 'rhombencephalon' roughly coincides in the early embryo with a strong ventral flexure of the brain (primary cephalic flexure) over the front end of the notochord. It also conforms to the external shape of the brain in an adult lamprey (fig. 5.13) or myxinoid. In a previous paper I accepted Starck's analysis and terminology (Jefferies & Lewis, 1978, p. 273).

However, although Starck's arguments seem reasonable when the adult myxinoid or lamprey brain is considered, the embryology suggests a different picture. The development of the outside of the brain of the myxinoid *Eptatretus* has been studied by von Kupffer (1900) and

Conel (1929). The early embryos have the anterior part of the brain bent ventrally over the front end of the notochord to form the primary cephalic flexure (fig. 5.14a). This flexure divides the brain into two parts, at first sight conforming to Starck's views. But in later development the tectum of the mesencephalon develops in the dorsal part of the brain *posterior* to the flexure (fig. 5.14b). The primary cephalic flexure therefore divides the brain, not into Starck's 'prosencephalon' and 'rhombencephalon', but into: (1) the conventional prosencephalon anteriorly and (2) the mesencephalon *plus* conventional rhombencephalon posteriorly. This complex of mesencephalon *plus* conventional rhombencephalon is the part of the brain that overlies the notochord and has been called the deuterencephalon. In the embryology of *Lampetra* there is a sulcus (fig. 5.38) on the internal surface of the brain which coincides with the posterior boundary of the mesencephalon and confirms that the mesencephalon exists as a unit (posterior intra-encephalic sulcus of von Kupffer (1895, p.19)). Starck's terminology can also be criticized on grounds of linguistic stability. The conventional definitions of prosencephalon, mesencephalon and rhombencephalon were established by von Baer early in the nineteenth century. It is probably impracticable to try to alter them now, even if it were desirable. Starck's intuition that the vertebrate brain is bipartite is important, but he seems, as I now think, to have misplaced the boundary between the two portions. The term 'archencephalon', sometimes used in the embryological literature, is a redundant synonym of the conventional prosencephalon.

The brain of mitrate calcichordates was divided into two parts which in Jefferies & Lewis (1978) I identified as 'prosencephalon' and 'rhombencephalon' in Starck's sense. It now seems more likely that they correspond respectively to the conventional prosencephalon and the deuterencephalon.

The walls of the brain of *Myxine* are very thick and the ventricles reduced to a few crevices. The paired olfactory bulbs are the most anterior part of the brain, as normal in

traditional divisions:

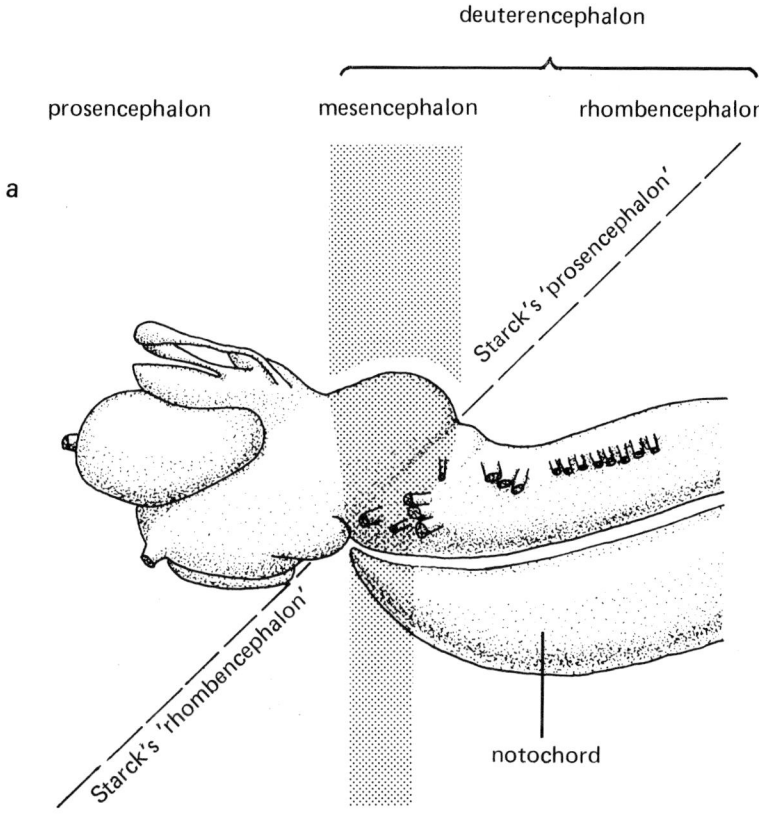

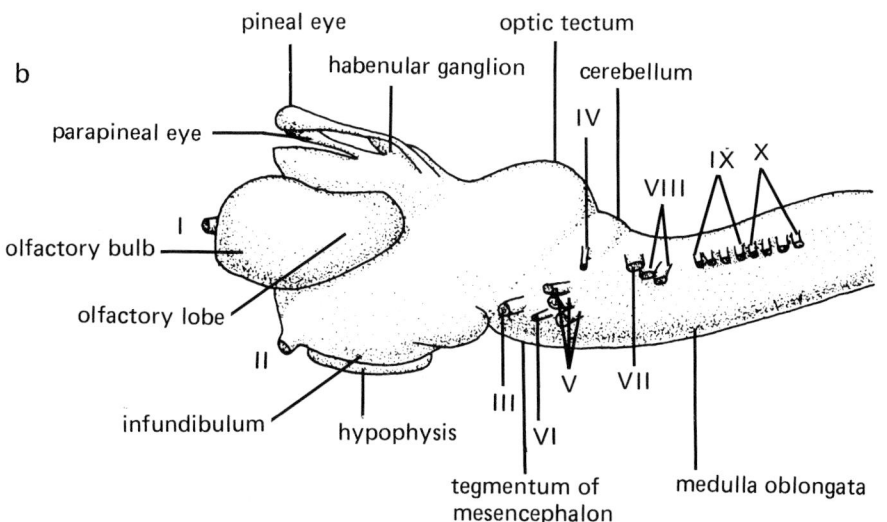

5.13. *Brain of the lamprey* Petromyzon *in left lateral aspect; a) to show the contrast between Starck's bipartition of the brain and the traditional subdivisions; b) to show the parts of the brain and the origins of cranial nerves. (New).*

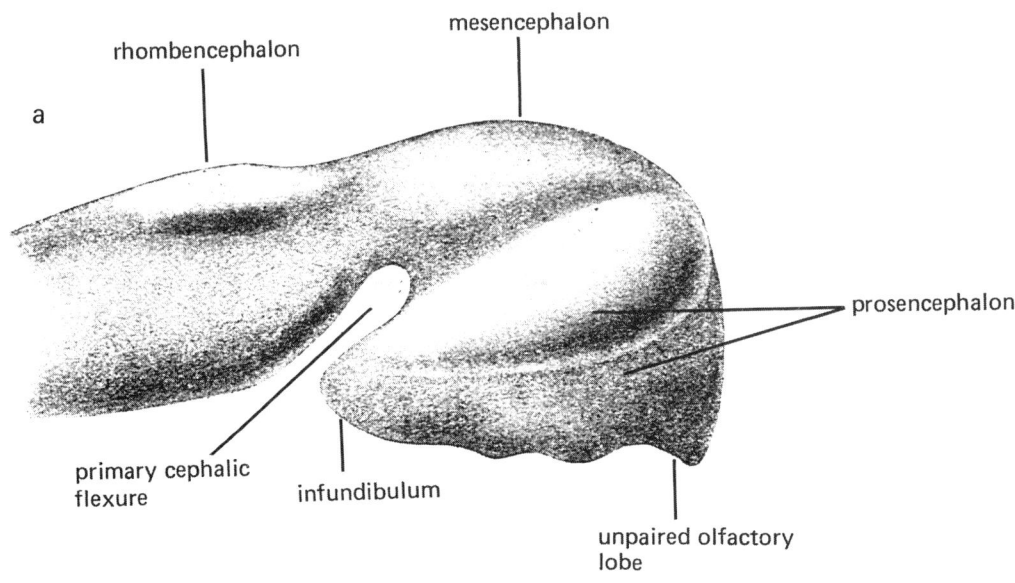

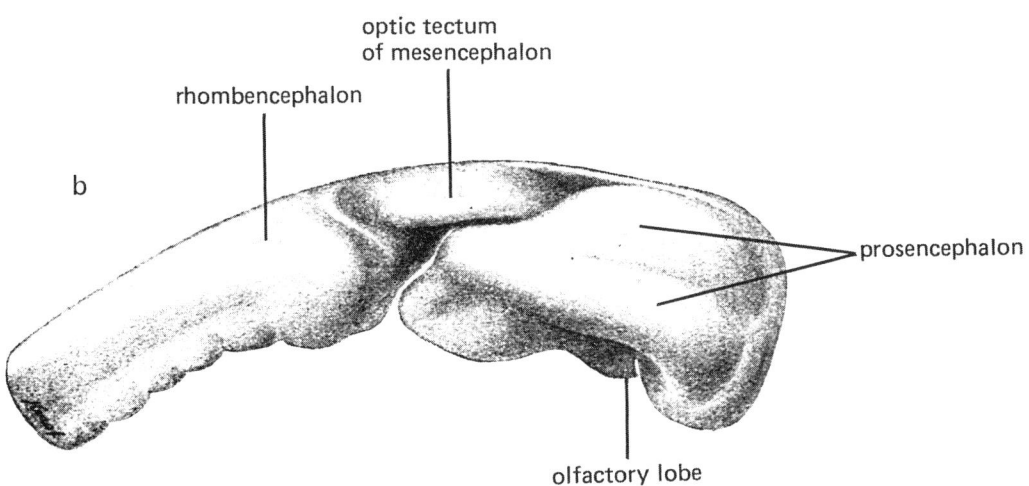

5.14. *The brain of a) an earlier and b) a later embryo of* Eptatretus *to show that the optic tectum (dorsal part of the mesencephalon) is situated behind the primary cephalic flexure, contrary to Starck's bipartition of the brain. The figures show the right aspect of wax models based on serial sections; from Kupffer (1900, fig. 43, p. 45 and fig. 47, p. 52).*

vertebrates, and receive fibres from the olfactory cup (fig. 5.15). Postero-dorsal to them are the hemispheres of the telencephalon and of the diencephalon. There are no pineal or parapineal eyes. Ventrally the diencephalon gives rise to slender optic nerves and, more posteriorly, is produced downwards to form the infundibulum. This is dorsal to the adenohypophysis, which is endodermal in origin.

The optic nerves end distally in small lensless eyes (figs 5.15, 5.16). These have the form of optic cups whose cavity is virtual (fig. 5.16) and the cups are surrounded by chorioid and scleral layers as in lampreys and gnathostomes. These cups are situated latero-dorsally in the head but are overlain by muscle. They may not be entirely non-functional as eyes, for Holmberg (1970) detected visual cells with irregularly lamellated outer segments and plausible histological signs of activity while Dücker (1924) reported a fairly normal histological stratification of the retina. Nerve fibres in the optic nerves are

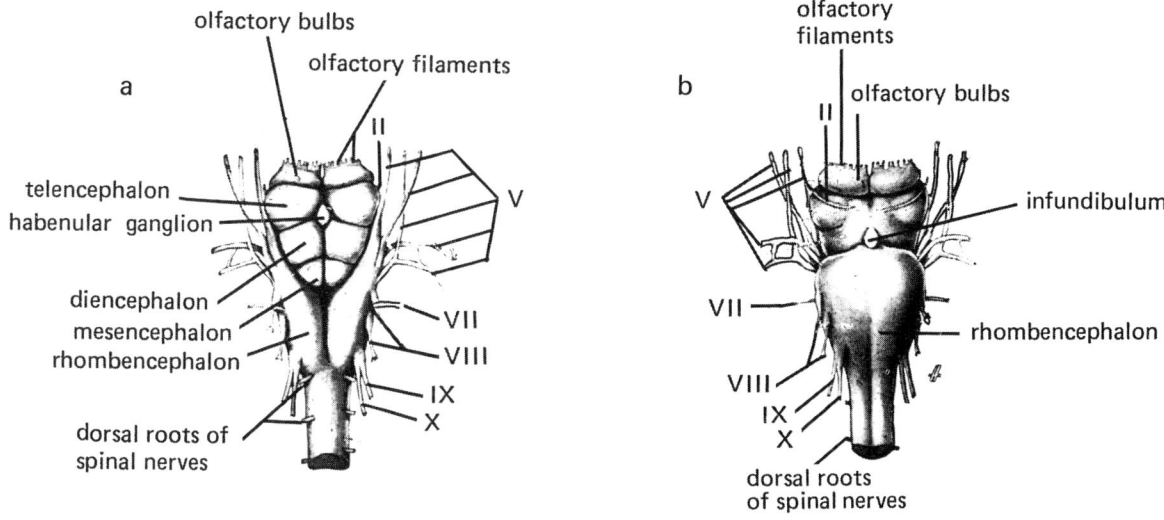

5.15. Myxine. *The brain in a) dorsal and b) ventral aspect; from Marinelli & Strenger (1956, fig. 118, p. 146).*

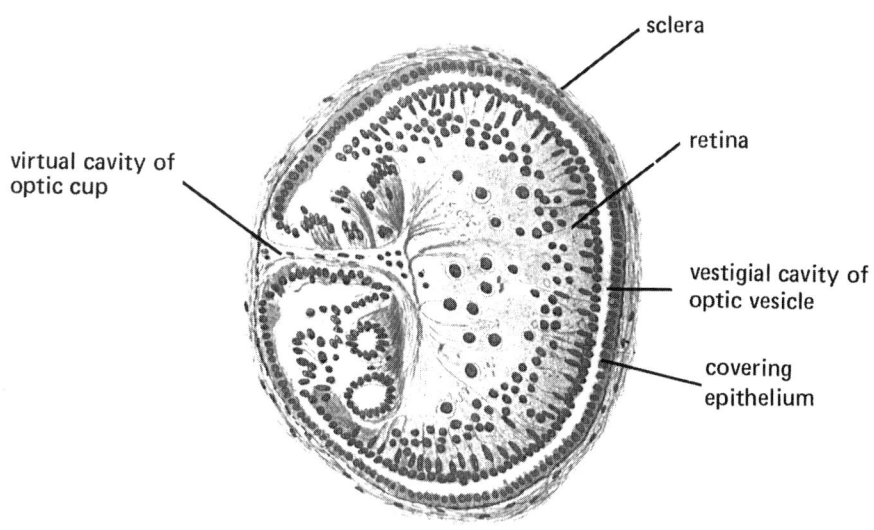

5.16. Myxine. *Horizontal section through the eye to show the virtual cavity of the optic cup; from Dücker (1924, pl. 29, fig. 31).*

few and the optic chiasma, just behind the bases of the optic nerves, is also poorly represented (Dücker, 1924, p. 520). The eyes of the myxinoid *Eptatretus* are better developed, as explained later. Perhaps to compensate for its poor eyes, *Myxine* has photo-sensitive skin (Newth & Ross, 1955). The subcommissural organ, producing Reissner's fibre, is situated in the mesencephalon (Afzelius & Olsson, 1957).

The medulla oblongata (rhombencephalon) is situated medial to the otic capsules. The trigeminal nerves have separate dorsal and ventral roots; the dorsal root is connected with a ganglion which is partly but not completely split into two parts. Behind this ganglion is the buccal ganglion and nerve; it is uncertain whether this should be assigned to the facial or the trigeminal complex. There are two pairs of ganglia associated with the auditory

a) Myxinoids

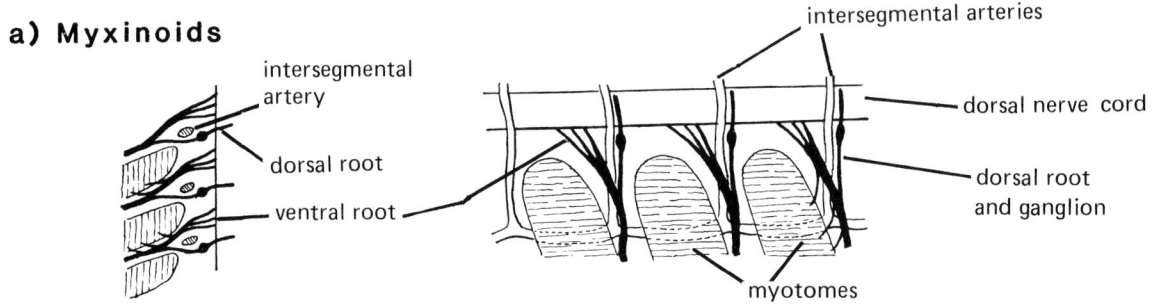

b) Lampreys

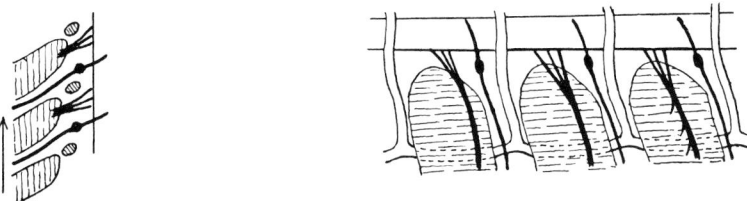

c) Gnathostomes

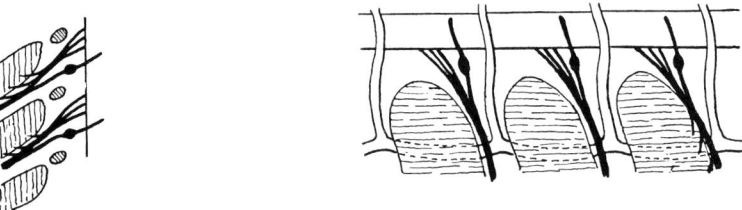

5.17. Dorsal and ventral roots of the spinal nerves and their relationship to the segmental arteries in myxinoids, lampreys and gnathostomes. On left, dorsal aspect of left side; on right, left aspect.

capsule. The glossopharyngeal and vagus nerves arise as a single trunk and the glossopharyngeal does not supply a gill pouch but innervates the muscles that constrict the pharynx. The vagus supplies the muscles that constrict the gill pouches and heart region. Near the heart region it breaks up into a plexus in the walls of the gut and then reforms as a single unpaired intestinal branch which follows the dorsal wall of the intestine.

The spinal nerves of *Myxine* have separate dorsal and ventral roots which join each other lateral to the noto-chord before running into the myocommata (fig. 5.17a). In thus possessing mixed spinal nerves, *Myxine* resembles

the gnathostomes and differs from the lampreys whose dorsal and ventral roots do not join. Moreover, the lamprey condition is probably primitive. For it resembles that of amphioxus, if we assume that the neuro-muscular contacts between the somites and dorsal nerve cord of amphioxus are homologous with the bases of ventral nerve roots in vertebrates. At first sight, therefore, it seems that the mixed spinal nerves of the myxinoids and the gnathostomes could represent a synapomorphy of the two groups which lampreys do not share.

However, Goodrich (1937) showed that the mixed nerves of myxinoids cannot be homologous with those

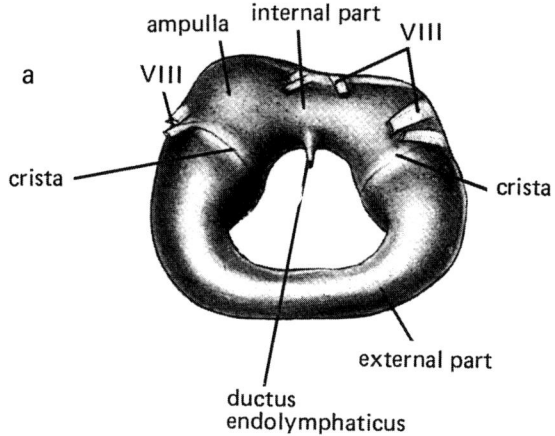

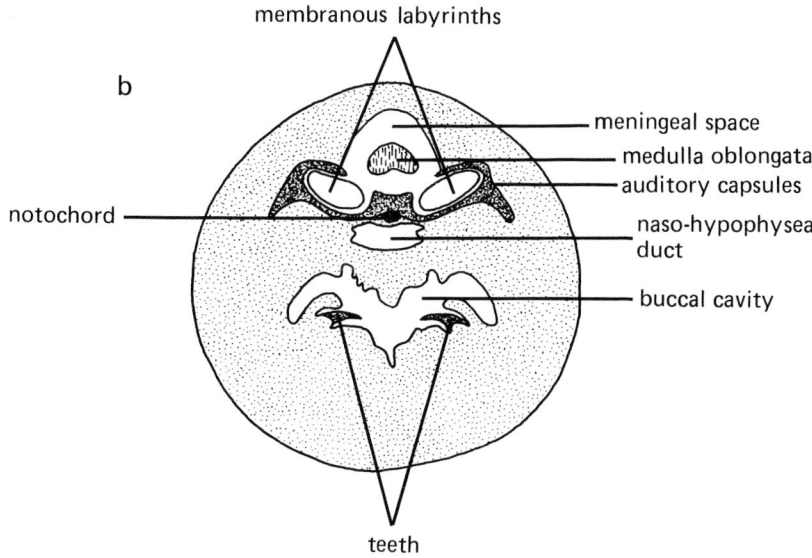

5.18. Myxine. *The ear (membranous labyrinth). a) Dorsal aspect of right membranous labyrinth (from Marinelli & Strenger, 1956, fig. 121, p. 153); b) transverse section to show position of labyrinths in the head. New drawing based on Marinelli & Strenger (1956, fig. 103, p. 129).*

of gnathostomes, for they are differently related to the segmental arteries which run dorsalward from the dorsal aorta (figs 5.17a, b, c). In *Myxine* and the myxinoid *Eptatretus* (= *Bdellostoma*), the dorsal and ventral roots which join to form a mixed nerve embrace the segmental artery. In gnathostomes, on the other hand they are anterior to this artery. Mixed spinal nerves are therefore not a synapomorphy of myxinoids with gnathostomes.

The auditory capsules are a pair of cartilaginous boxes situated right and left of the medulla oblongata (figs 5.3, 5.18b). The cavity of each capsule carries the membranous

labyrinth or inner ear which is bathed in a liquid known as perilymph. The membranous labyrinth (fig. 5.18a) is a circular hollow torus filled with endolymph and the plane of the torus slopes down medianwards at about 30° to the horizontal. The torus is thicker ventrally (internal part) than dorsally (external part) and inside it, at the places where thicker and thinner portion join, are two rings of sensory acoustic epithelium called cristae. A blind tube, the ductus endolymphaticus, ascends from the thick part of the torus. There is a patch of sensory epithelium, the macula communis, which lines the most median part of the

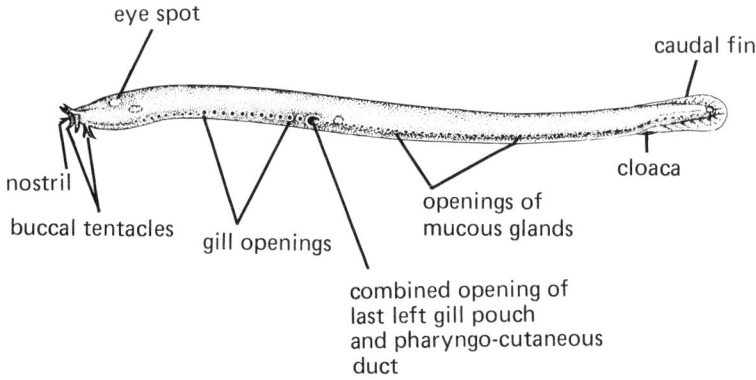

5.19. Eptatretus stouti, *external features; from Hardisty (1979, p. 2, fig. 1–2).*

torus, between the two cristae. The cristae and macula communis contain sensory 'hair cells' of the type standard for the vertebrate acustico-lateralis system, each cell having a tuft of cilia embedded, along with those of neighbouring sensory cells, in a 'flag' or cupula. I discuss such cells more fully later, in describing lampreys. Movements of the cupula are transferred to the contained cilia and cause a response in the sensory cells which are synaptically connected to the nervous system. The cupulae of the macula communis contain hard dense otoliths. The thin portion of the torus, ending ventro-medianwards at the cristae, corresponds to the anterior and posterior vertical semicircular canals of lampreys and gnathostomes, which likewise have cristae at their ventral ends. The thick ventral portion of the torus would correspond to the utriculus and sacculus of lampreys and gnathostomes. Lowenstein & Thornhill (1970) showed that the ear of *Myxine* can respond by means of the cristae to angular acceleration in all three planes of space (i.e. whether rolling, pitching or yawing) and also respond to tilting by means of the macula communis. It is not sensitive to vibration. There is no trace of a lateral line system in *Myxine*.

The closest relatives of *Myxine* do not share all its features. The best studied is *Eptatretus* (= *Bdellostoma*) *stouti* from the Pacific coast of North America (fig. 5.19). *E. stouti* is about 40 cm long and in general shape is like *Myxine glutinosa*. It is dark brown in colour, however, except for a white line along the edge of the fin and two unpigmented patches on the head which let light through onto the non-motile eyes inside. There are about 12 pairs of gills and, unlike *Myxine*, each gill opens directly to the outside. There is a large pharyngo-cutaneous opening on the left side, just behind the gill openings. A series of pores ventro-laterally are the openings of mucous glands.

The eyes of *E. stouti* are situated beneath the skin at the posterior edge of the paired unpigmented patches

already mentioned (fig. 5.20b). Unlike *Myxine* they are not covered with muscle; they are embedded in fat and sometimes touch the internal surfaces of the outer epithelium. As in *Myxine*, the eye has the structure of an optic cup, surrounded by vascular chorioid and tough scleral layers. The cavity of the cup is not virtual in *E. stouti*, however, but large and filled with vitreous (fig. 5.20a). The optic nerve is better provided with nerve fibres than that of *Myxine*. Dücker (1924) recognized well-defined layers in the retina which he homologized with those of the gnathostome retina. The existence of these layers was confirmed by Holmberg (1971) who also demonstrated, by electron-microscopical studies, that typical vertebrate cross-lamellated outer segments were present in the visual cells. All this strongly suggests that the eye of *E. stouti* is functional as a light detector. Indeed, in a Japanese myxinoid Kobayashi (1964) recorded electrical activity in the optic nerve when the eye was suddenly illuminated. Kobayashi called his fish *Myxine garmani* but Fernholm & Holmberg (1975) have re-identified it as *Eptatretus burgeri*. The eye of *Eptatretus* is therefore better developed than that of *Myxine*; indeed light is probably much more important for *Eptatretus*, for the skin of the latter is pigmented. A noteworthy fact is the entire absence of oculo-motor muscles in *Eptatretus*, both in the adult and in early ontogeny (von Kupffer, 1900). The absence of such muscles distinguishes myxinoids from both lampreys and gnathostomes and has been regarded as degenerate (e.g. Goodrich, 1909b, p. 50). Seeing that they do not appear at all in ontogeny, however, and are lacking even when the eyes are clearly functional, their absence is more likely to be primitive than due to loss, as Dücker pointed out. The eyes of myxinoids can be compared with the non-motile eye of tunicate tadpoles and the early larvae of lampreys and with the non-motile cispharyngeal eyes of mitrates.

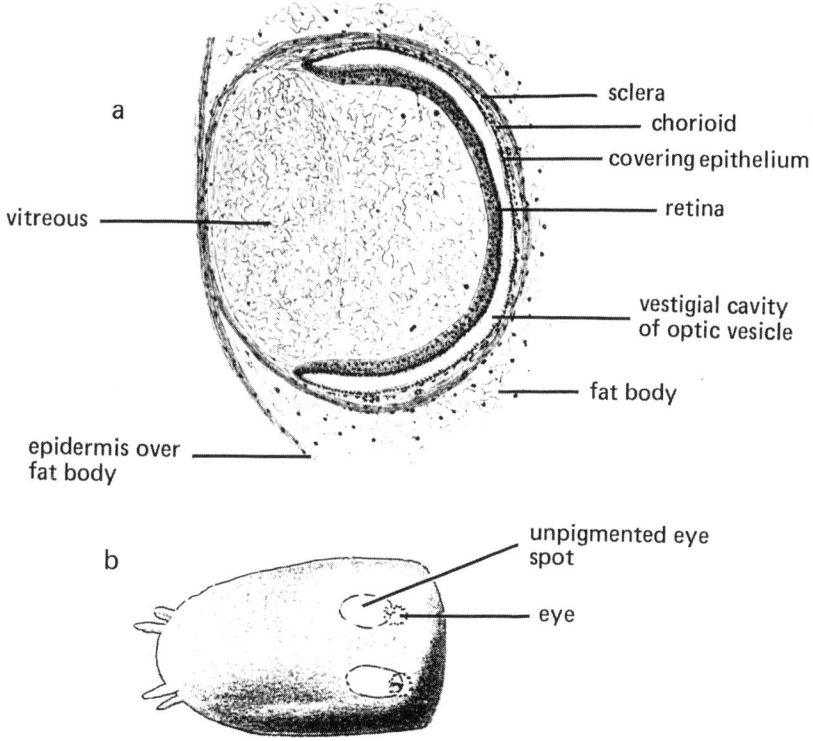

5.20. Eptatretus. *Anatomy of eyes. a) Transverse section; note that the cavity of the optic cup is patent and filled with vitreous, unlike* Myxine *(cf. fig. 5.16); from Dücker (1924, pl. 28, fig. 19). b) Dorsal aspect of head to show the eyes (dotted outline) in relation to the unpigmented eye spots; from Dücker (1924, pl. 27, fig. 17).*

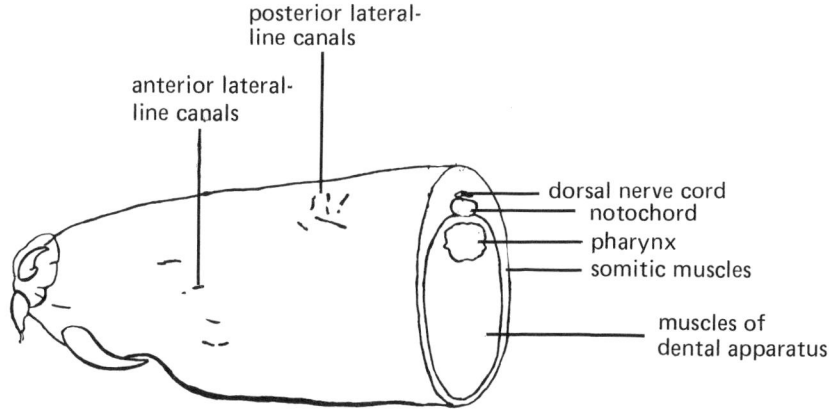

5.21. Eptatretus. *Left aspect of head to show possible lateral-line organs; from Ayers & Worthington (1907, fig. 5, p. 331).*

A lateral-line system in the head of *E. stouti* (fig. 5.21) was recorded by Ayers & Worthington (1907). Von Kupffer (1900, p. 83) recorded it as being present transiently in the embryo of the same species; he figured it but unfortunately never described it in detail.

The lampreys are the other great group of agnathans. They are also called Petromyzonida and are more generally known than myxinoids. This is for several reasons; they occur, for at least part of their life cycle, in fresh water; they are good to eat; one species has had a catastrophic

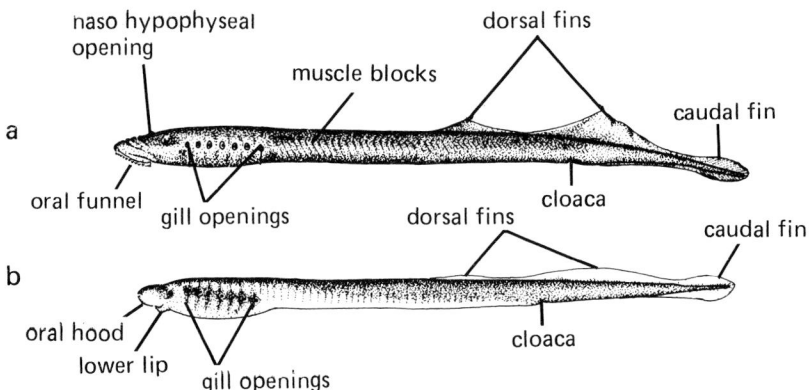

5.22. Lampetra fluviatilis. *External features of: a) adult; b) ammocoete larva; from Hardisty (1979, fig. 1.1, p. 1).*

effect, in recent years, on the fisheries of the North American Great Lakes; and finally, it was a 'surfeit of lampreys' which killed King Henry II of England. I shall concentrate mainly on the European species *Lampetra fluviatilis*.

Recent literature on the lampreys includes: a detailed anatomy of *L. fluviatilis* in Marinelli & Strenger (1954); five volumes of papers edited by Hardisty & Potter (1971, 1982); and general accounts in Sterba (1962) and Hardisty (1979). The lamprey central nervous system was reviewed in great detail by Niewenhuys (1977). The Russian text by Balabai (1956) includes some important observations which have been generally ignored.

The life cycle of *L. fluviatilis* (fig. 5.22) is complex and remarkable. The eggs are laid and fertilized in depressions ('nests') in the floors of the upper reaches of streams. The eggs become slightly buried in the nests and hatch into centimetre-long eel-shaped larvae (the 'pro-ammocoetes') which wriggle out of the sediment and are carried downstream to bury themselves, with the mouth exposed, in burrows in the sand of the river bottom. At this burrowing stage the animals are known as ammocoete larvae (fig. 5.22b) and are fascinating from the comparative viewpoint since they feed on microscopic food particles using a pharyngeal mucous filter homologous with that of a tunicate or amphioxus. The ammocoete stage lasts several years. It has small, non-functional eyes, hidden by pigmented skin, but the skin of the tail is light-sensitive.

At the end of the ammocoete stage the larva transforms into a large-eyed stage called a macrophthalmia. During this transformation the pharyngeal mucous filter is lost, the mouth acquires a circular sucker-like oral funnel lined with horny teeth, and a formidable toothed tongue develops. The macrophthalmia moves down to the sea where it becomes an active predator, hunting by sight and

smell. This predatory period is miscalled the 'parasitic' phase. During it the animal attaches itself to fishes and sucks their blood until they often die. The oral sucker fixes to the prey, the toothed tongue makes an incision, and coagulation of the blood is prevented by an anti-coagulant secretion from a pair of buccal glands.

After more than a year the animal ceases to feed and enters fresh water again, swimming upstream to the upper reaches of brooks. Here 'nests' are made, males and females copulate, fertilized eggs are shed into the nests, and the adults die.

The name 'ammocoete' recalls that the larval lamprey was at first thought to be a distinct genus (*Ammocoetus* Duméril, 1812; *Ammoecoetes* Cuvier, 1817). Not until 1856 did August Müller show that the eggs of lampreys become ammocoetes, and ammocoetes metamorphose into lampreys. The important fact that the endostyle of the ammocoete transforms into the thyroid gland of the adult lamprey was discovered by Wilhelm Müller in 1873.

Many species of lampreys, known as brook lampreys, have a simpler life cycle than *L. fluviatilis* in that the predatory 'parasitic' marine phase is absent and only the ammocoete feeds. This is probably always a secondary condition, several times acquired, since the anti-coagulatory buccal glands are always present in the non-feeding suctorial adult mouth.

Adult *L. fluviatilis* (fig. 5.22a) which I shall henceforth call *Lampetra*, are eel-shaped, about 50 cm long and 2·5 cm in diameter. At the anterior end is the mouth with the circular oral funnel. There is a pair of prominent motile eyes. Dorsal to the eyes, in the mid-line, is a single naso-hypophyseal opening. This leads, like the single nostril of myxinoids, both to the olfactory capsule and to the hypophysis (fig. 5.23). Unlike myxinoids, however, the naso-hypophyseal duct does not open into the pharynx.

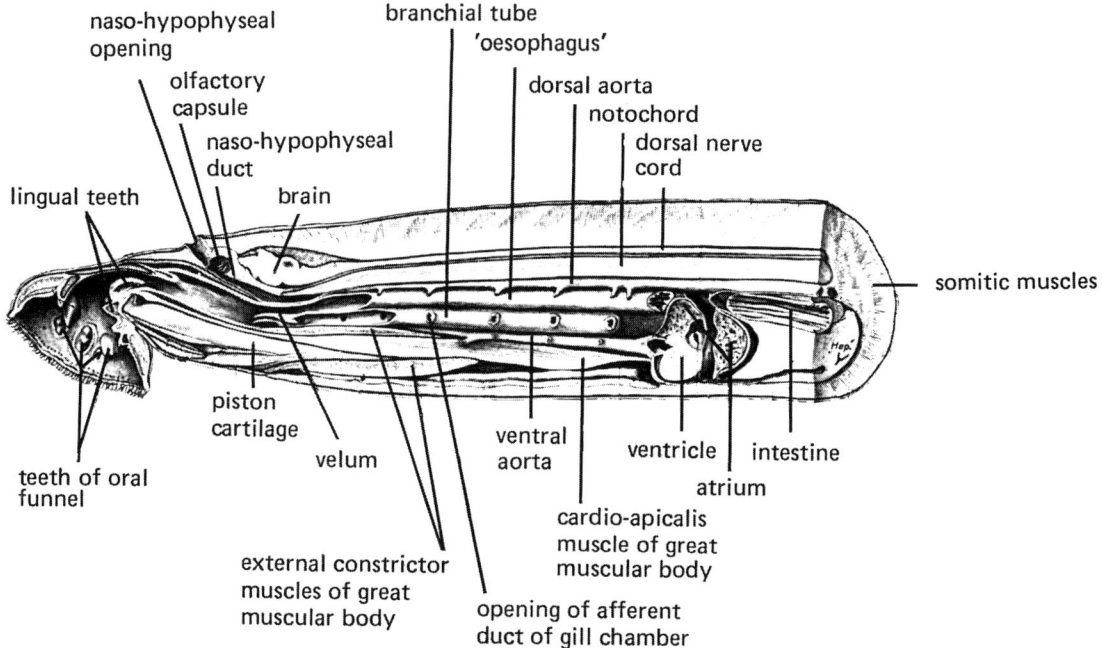

5.23. Lampetra. *Sagittal section of anterior part of adult. Note the lingual muscles that operate the lingual teeth; and the nasohypophyseal duct which does not open into the alimentary canal; from Marinelli & Strenger (1954, fig. 36).*

Behind the eyes, on each side, is a line of seven gill openings. There are numerous lateral-line pits on the head, trunk and tail. In the posterior third of the animal there is a median fin divided into two dorsal lobes and a caudal portion. About one fifth of the length forward from the posterior end, in the mid-ventral line, is the slit-shaped cloaca which contains a urogenital-papilla ('penis') in the male. The skin of *Lampetra*, unlike myxinoids, is not underlain by a huge blood sinus though there are small subcutaneous sinuses formed in ontogeny by the breakdown of a subcutaneous capillary network. Also there are no paired rows of mucous glands.

The branchial currents of the ammocoete and the adult lamprey are strikingly different. In the ammocoete the current enters the mouth, passes into the pharynx and then out through the gill slits. This is like the branchial current of amphioxus, tunicates, myxinoids and jawed fishes, and no doubt represents the primitive situation. The branchial current of the adult, on the other hand, is tidal, for it both enters and leaves the branchial pouches through the gill slits. This is necessary because the mouth is habitually blocked by attachment, sucker-fashion, to a prey fish or a stone. The change from unidirectional to oscillatory flow requires reorganization of the gill system of metamorphosis, as discussed later.

The main pump for the larval branchial current is a velum (fig. 5.38) consisting of two paddle-like muscular flaps situated between the buccal cavity and pharynx, though the velum is assisted by contraction and dilation of the pharyngeal walls. This arrangement recalls the pumping 'velum' of myxinoids, but the two structures need not be homologous, since the lamprey velum, like that of amphioxus and tunicates arises from the buccopharyngeal membrane of the embryo, whereas the myxinoid 'velum' is endodermal in origin.

The skeleton of adult *Lampetra* consists of a notochord and many cartilages (figs 5.24, 5.25, 5.26). The notochord tapers to a point anteriorly at the transverse level of the eyes but extends posteriorly almost to the rear end of the animal. Just dorsal to it, along its entire length, is a fatty meningeal body, in which the brain and dorsal nerve cord are buried. The cartilaginous skeleton consists of: the skull, the tongue and mouth skeleton, the branchial skeleton round the gills, a series of arcualia ('vertebrae') on either side of the meningeal body, and the fin rays.

The tongue skeleton (fig. 5.25) consists of a long piston cartilage to the anterior end of which an apical cartilage is flexibly attached. The front end of the apical cartilage is crowned with horny lingual teeth. When the lamprey has attached itself to its prey by the oral funnel, the piston cartilage shifts forward, bringing the lingual teeth against the prey's skin. Then muscles nod the apical cartilage up and down and the lingual teeth rasp a hole in the prey's side, through which blood is drawn out. The

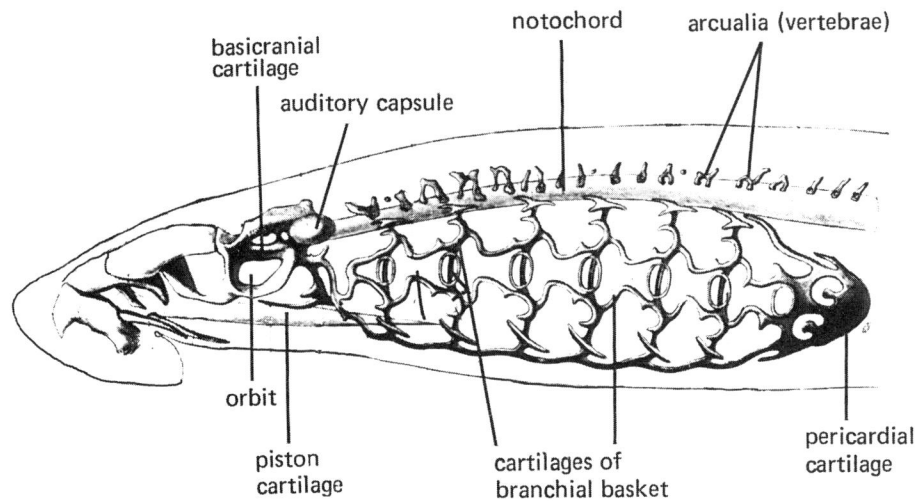

5.24. Lampetra. *Cartilages of head, anterior neuralia (vertebrae) and branchial basket; from Marinelli & Strenger (1954, fig. 64, p. 69).*

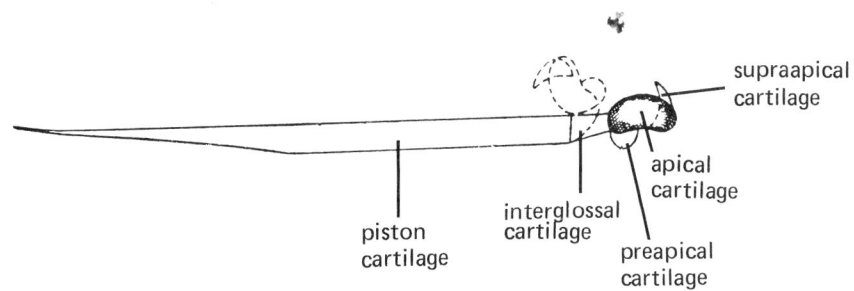

5.25. Lampetra. *The skeleton of the tongue and the movement of the apical cartilage and lingual teeth; from Balabai (1956, fig. 10, p. 20).*

apical cartilage is activated by a great muscular body somewhat like that of myxinoids (fig. 5.23). Indeed the apical cartilage with its teeth and muscular apparatus may be homologous with the dental cartilage and associated muscles of myxinoids.*

The branchial skeleton (fig. 5.24) is a springy cartilaginous basket without articulations. The bars of this basket are external to the gills, just beneath the skin, and posteriorly connect with a cartilaginous cup that contains the pericardium. As already mentioned, the scanty branchial skeleton of myxinoids is likewise outside the gills. The gills of all agnathans are thus endodermal in origin, and

internal to the branchial skeleton, and are described as endobranchiate. The gills of gnathostomes, by contrast, are ectodermal in origin, external to the branchial skeleton and ectobranchiate.

The arcualia, or vertebrae, of *Lampetra* are small irregular cartilages situated on either side of the meningeal body and thus on either side of the dorsal nerve cord (figs 5.24, 5.26). They correspond in position to the bases of the neuralia (neural arches) of gnathostomes and may well be homologous with them. However, they never connect together, arch-fashion, over the dorsal surface of the dorsal nerve cord, nor do they correspond on left and right. Anteriorly, they tend to be intersegmental, and the spinal nerves often perforate them. Posteriorly there tend to be two arcualia per segment on each side.

The fin rays are cartilages that stiffen the dorsal and caudal fins (fig. 5.26). Their arrangement is much as in myxinoids except that in lampreys pterygial muscles are

---

*Yalden (1985) has compared the tongues of *Myxine* and *Lampetra* and shown, in my view, that they are homologous. Contrary to his assertion, this does not prove that the agnathans are a monophyletic group. It is more likely that the tongue has been lost in the ancestry of gnathostomes (Chapter 9).

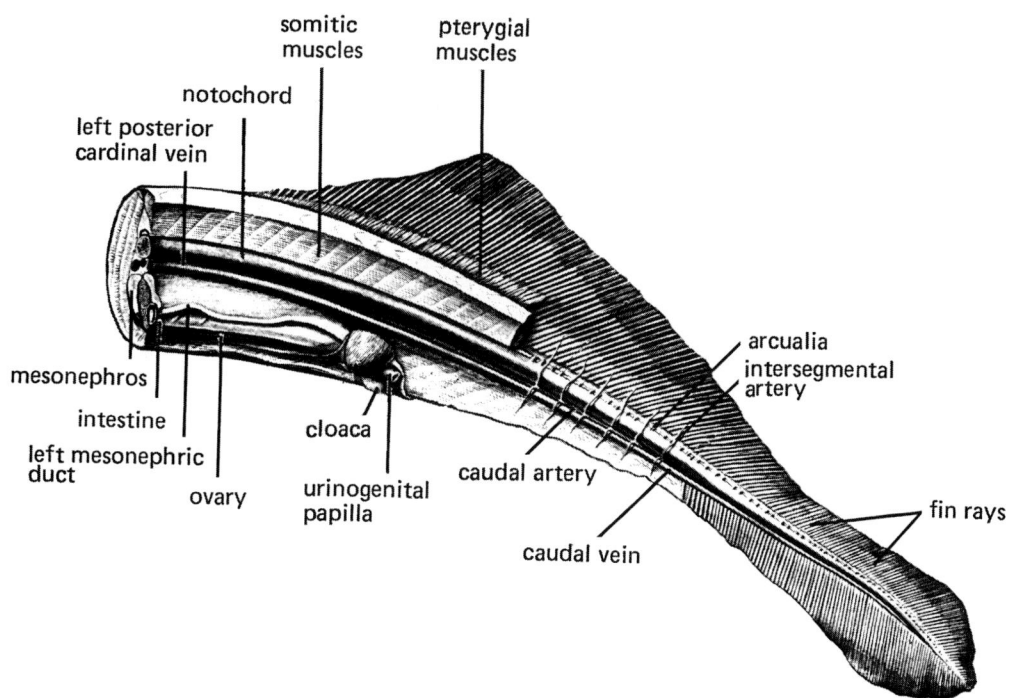

5.26. Lampetra. *Dissection of posterior part of body. Note especially the cartilaginous fin rays and their actuating muscles and the presence of arcualia (vertebrae); from Marinelli & Strenger (1954, fig. 46, p. 49).*

attached to their bases and extend into the proximal part of the fin. The fin rays share this feature with the cartilaginous fin rays of gnathostomes and this explains the name of myopterygians ('muscle-fins') which Janvier (1978) has proposed for the probably monophyletic group of [lampreys + gnathostomes].

The alimentary canal of the adult begins anteriorly with the oral funnel which is a flattish muscular cone lined inside with horny teeth (fig. 5.23). It represents the anterior part of the buccal cavity and passes back into the posterior part of the buccal cavity, which is sometimes, misleadingly, called the 'pharynx' (e.g. Reynolds, 1931). The central hole of the oral funnel is plugged by the 'tongue', armed as already said, with powerful cutting teeth. Behind the buccal cavity is the region which truly corresponds to the pharynx of other chordates and which, in post-larval *Lampetra*, but not in the ammocoete, is divided into two longitudinal tubes – a more dorsal 'oesophagus' and a more ventral branchial tube. The anterior openings of the oesophagus and branchial tube can be closed by the complicated velum which unlike that of an ammocoete, is a valve or gate rather than a pump. The branchial tube ends blind posteriorly, whereas the oesophagus passes straight back into the post-pharyngeal gut. The seven pairs of gill chambers open out of the branchial chamber and connect it

to the seven pairs of gill slits on the outside of the animal. During the predatory 'parasitic' phase the oesophagus allows blood from the prey to by-pass the branchial chamber and flow into the digestive part of the gut, so that *Lampetra* can breathe, by its tidal branchial current, even when feeding. After the predatory phase the 'oesophagus' becomes occluded, so that food physically cannot pass, but the branchial tube remains in action.

Behind the branchial region the 'oesophagus' passes into the intestine. At the point of junction there is, in the predatory phase, a bile duct which connects the gut to the gall bladder in the liver. An exocrine pancreas is developed in the gut wall in the same region, as in gnathostomes but unlike myxinoids – perhaps it is homologous with the pyloric gland of tunicates. The intestine has a typhlosole, or spiral valve, which again is like gnathostomes and unlike myxinoids (though it is also like tunicates and unlike acraniates). The intestine of *Lampetra* opens posteriorly into the cloaca. When the predatory phase is over, and the lamprey ceases to feed, the lumen of the oesophagus is blocked, as already mentioned, and the gall bladder and bile duct degenerate.

The branchial system of *Lampetra* contains seven pairs of ellipsoidal gill pouches. Each pouch opens admedian into the branchial tube and externally as a gill slit. The

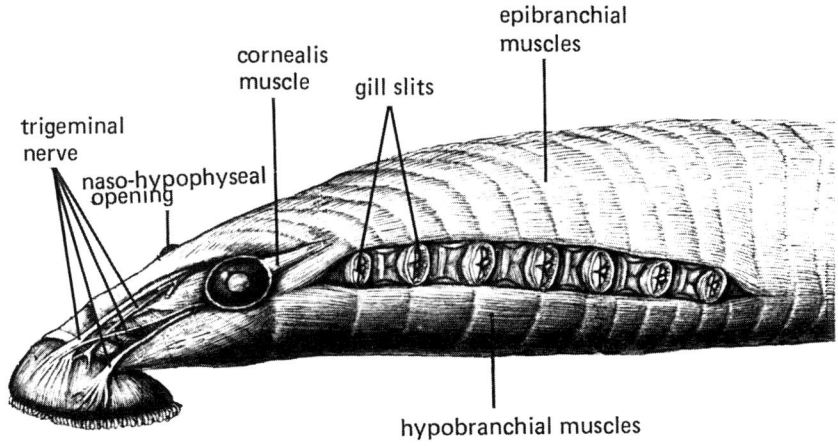

5.27.   Lampetra. *Muscles of head region.*

anterior end of the branchial tube, as already mentioned, can be closed by the velum. There are three external flaps over each gill slit. These act as adjustable valves. Each gill chamber is contained in its own peribranchial blood sinuses. The epithelium inside the gill chambers is folded into gill lamellae disposed parallel to the long axis of the chamber and the lamellae are crossed, in turn, by tiny plicae. The afferent branchial arteries run between successive gill chambers as in gnathostomes, not each to a gill chamber as in myxinoids.

As regards coeloms, the pericardium in *Lampetra* is completely separated from the visceral coelom by means of a septum (Marinelli & Strenger, 1954, p. 47). This differs from myxinoids where the two coeloms are connected by a foramen on the right side. The visceral coelom of *Lampetra* opens posteriorly into the cloaca by way of a pair of coelomic pores. Through these pores, sperm and ova pass out of the animal, as probably with the single unpaired pore of myxinoids. In gnathostomes, on the other hand, it is only the ova, and not the sperm, which pass through the coelom in leaving the animal. The most anterior organ in the main coelom of *Lampetra* is the liver, which underlies the intestine.

The most powerful group of muscles is the somitic muscles (figs 5.23, 5.26, 5.27). These consist of myomeres and extend from the rear end of the animal almost to the front end of the head. Behind the gill slits each myomere extends from the mid-dorsal to the mid-ventral line. Unlike myxinoids, and unlike amphioxus, but like gnathostomatous fishes, there is no ventral gap occupied by rectus and oblique muscles. Unlike gnathostomatous fishes, but like myxinoids and amphioxus, the myomeres are not divided into hypaxial and epaxial portions by means of a horizontal septum (Nursall, 1956). Above the

gill slits the myomeres continue anteriorly, as epibranchial muscles, to finish well in front of the eye (fig. 5.27). One of these epibranchial myomeres sends off a branch to the cornea of the eye on each side (cornealis muscle) and the contraction of this branch serves in focusing the eye. Beneath the gill slits the lateral muscles likewise extend anteriorly, as the hypobranchial musculature This musculature is segmented, like the rest of the lateral muscle, but the segments do not coincide in number, position or ontogenetic origin with those of the epibranchial muscles.

The nerve supply to the somitic muscles consists of spino-occipital nerves and of spinal nerves. The spino-occipital nerves supply the hypobranchial musculature and, except for the most anterior three pairs (which pass ventrally direct to their end organs), they follow a looped course, running rearwards round the gill slits and then forwards again. They thus resemble, and are no doubt homologous with, the hypoglossal nerve of the gnathostomes. This is unlike the myxinoid condition where the spinal nerves, even the most anterior ones, pass direct to their end muscles, and do not arch around the gill openings.

Other muscles in *Lampetra* include those operating the tongue and mouth, the branchial muscles, and the oculomotor muscles. The tongue and mouth muscles are innervated by the trigeminal nerve (V). The branchial muscles are supplied by the glossopharyngeal (IX) and vagus (X). The facial nerve (VII) has no branchial component, presumably because the first branchial pouch, which it supplies in gnathostomes, disappears during the ontogeny of *Lampetra*.

The oculomotor muscles of *Lampetra* (fig. 5.28a) are much like those of gnathostomes (fig. 5.28b), consisting in each orbit of four rectus and two oblique muscles. This

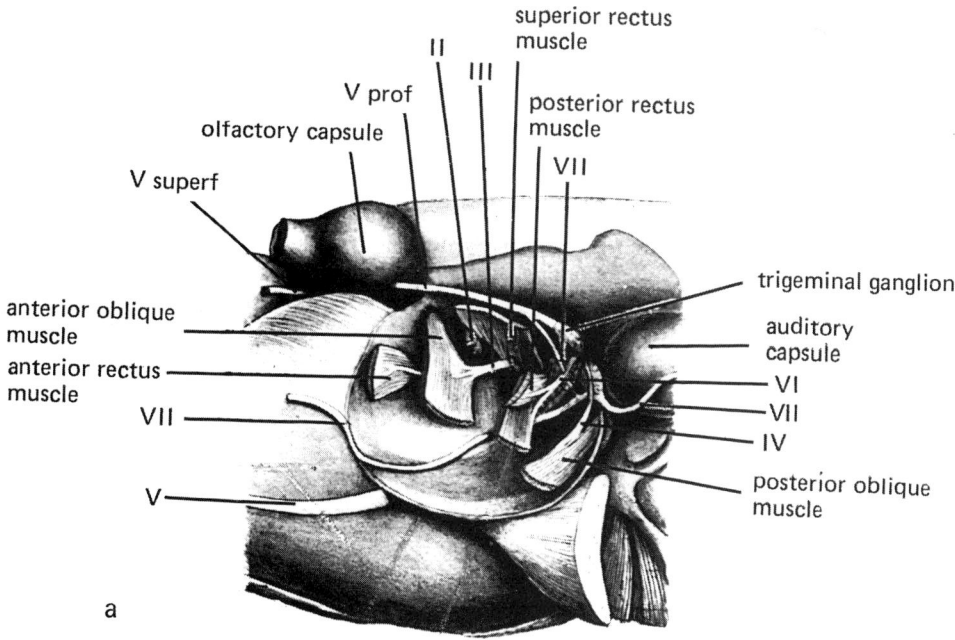

a

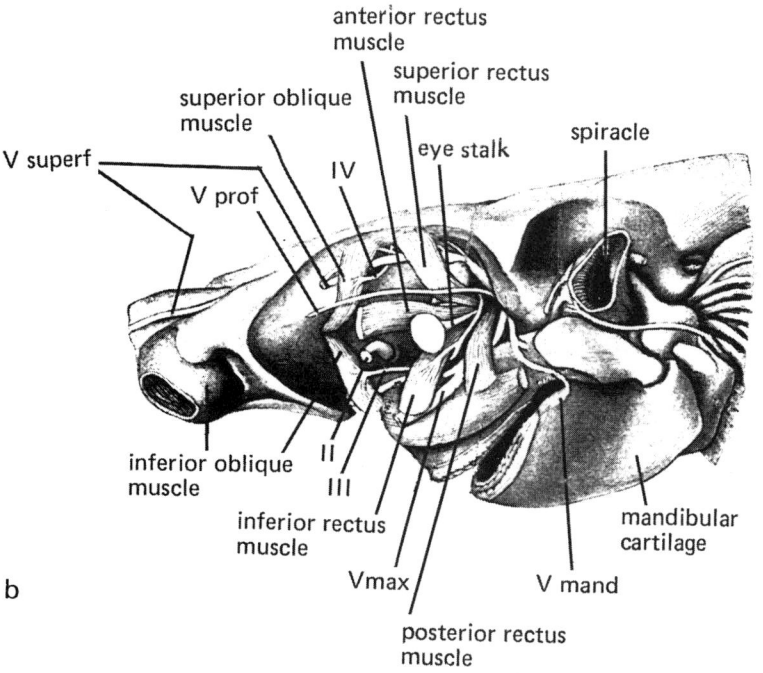

b

5.28. *Oculomotor muscles and nerves in a)* Lampetra, *b)* Squalus, *to show their essential correspondence as between lampreys and gnathostomes; from Marinelli & Strenger (1954, fig. 53, p. 58) and Marinelli & Strenger (1959, fig. 194, p. 263).*

a

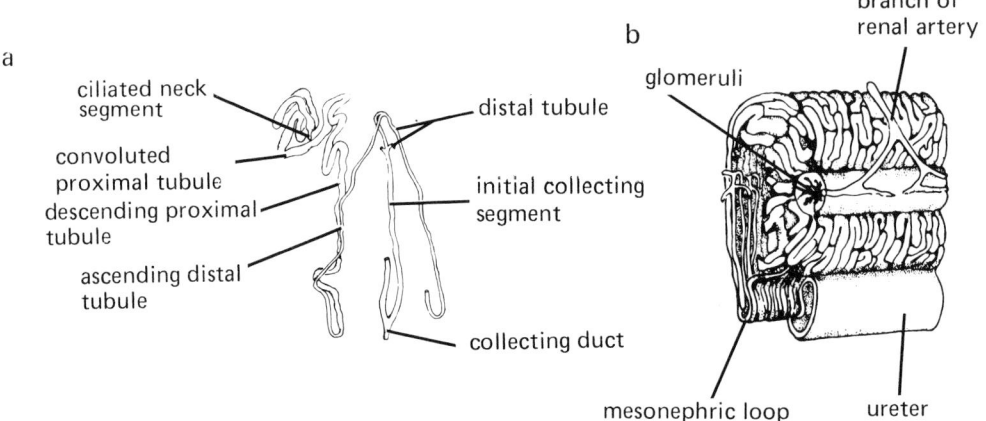

ciliated neck
segment

convoluted
proximal tubule

descending proximal
tubule

ascending distal
tubule

distal tubule

initial collecting
segment

collecting duct

b

branch of
renal artery

glomeruli

mesonephric loop   ureter

5.29.   Lampetra. *Structure of kidney. a) Isolated serial tubule, and b) portion of kidney seen, median aspect; from Morris in Hardisty & Potter (1972, fig. 8, p. 216).*

is a remarkable contrast with myxinoids which, so far as known, lack throughout their life any trace of such muscles. As already mentioned, this is true for myxinoids even when, as in *Eptatretus*, the eyes, though lensless and presumably not image-forming, seem to be fully functional as light detectors. The presence of oculomotor muscles in lampreys and gnathostomes is probably connected with the existence of a lens. For a lensed eye is image-forming, so there is benefit in being able to rotate it in its socket.

In gnathostomes the oculomotor muscles comprise superior and inferior oblique muscles anteriorly, and the four rectus muscles posteriorly, i.e. the anterior, posterior, superior and inferior rectus, arranged in the form of a cross (fig. 5.28b). The superior oblique of gnathostomes is supplied by the trochlear nerve (IV), the posterior rectus by the abducens nerve (VI), and all the other oculomotor muscles by the oculomotor nerve (III).

The oculomotor muscles of lampreys comprise anterior and posterior obliques (rather than inferior and superior) and anterior, posterior, superior and inferior recti. The posterior oblique (equivalent to the superior oblique of gnathostomes) is in fact rooted posterior to the rectus group (fig. 5.28a), though like its gnathostome equivalent it is supplied by the trochlear nerve. The rectus group is much as in gnathostomes, except that the anterior rectus is rooted further forwards in the orbit, and that the abducens nerve (VI) supplies, not only the posterior rectus, but also the inferior rectus (though this may also receive fibres from the oculomotor nerve (Lindström, 1949)). Muscles supplied by the oculomotor nerve (III) in lampreys therefore comprise the anterior oblique and the anterior and superior rectus and possibly the inferior rectus also.

Despite such differences, there seems little doubt that the lens, the oculomotor muscles and associated nerves are homologous in both groups – a complex of features that constitutes an important resemblance.

The osmoregulatory system of adult lampreys (fig. 5.29a, b) is also more gnathostome-like than that of myxinoids. Since lampreys live either most or all of their lives in fresh water, the ability to control the osmotic pressure of the blood is important. In fact the osmotic pressure is permanently higher than that of fresh water and about one third that of sea water (Morris, 1972). Osmotic regulation in fresh water takes place, as in gnathostomes, by passing salty urine out of the glomeruli, but then recovering most of the salt through the walls of the renal tubules. These tubules have descending and ascending portions as in gnathostomes. (In myxinoids renal tubules are short or absent). The pronephros is not functional in an adult lamprey whose functional kidney is a mesonephros. This again is like gnathostomes and unlike myxinoids and, if Haeckel's law holds, it constitutes an advanced resemblance. A functional pronephros, with funnels opening into the pericardium, does exist in the ammocoete larva (fig. 5.30).

The blood vascular septum of *Lampetra* consists of a heart, arteries, capillaries, venous sinuses and veins. The system of venous sinuses is less extensive than in myxinoids but still important.

The heart (fig. 5.23) lies in the pericardial coelom and has the standard four chambers, i.e. sinus venosus, atrium, ventricle and bulbus arteriosus. Unlike myxinoids, the heartbeat is neurogenic, not myogenic, and the heart receives a nerve supply from the vagus as in gnathostomes (Fänge, 1972, p. 246).

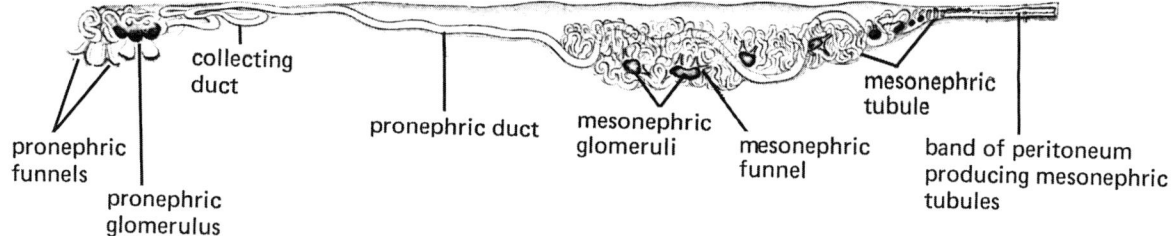

5.30. Lampetra. *The right kidney of an ammocoete larva in median aspect. The pronephric funnels open into the pericardial cavity; from Wheeler (1899, pl. 7, fig. 65).*

The ventral aorta passes forward in the ventral line (fig. 5.23) as a single trunk but, at the transverse level of the fourth branchial pouch, divides into right and left portions. This forking brings the ventral aorta on either side of the tongue musculature in the adult, and on either side of the endostyle in the larva. Between successive branchial pouches, and also behind the last pouch and anterior to the first one, the ventral aorta gives off afferent branchial vessels which supply the gill chambers with blood. Efferent branchial arteries convey blood from the gills to the single median dorsal aorta. The dorsal aorta runs rearwards in the mid-line immediately beneath the notochord, between the two cardinal veins (fig. 5.26). Posterior to the gills the dorsal aorta gives off arteries to the kidneys and also a mesenteric artery which runs along the spiral fold of the intestine and gives off a branch to the liver. The dorsal aorta also gives off paired segmental arteries.

The venous system contains anterior and posterior paired cardinal veins which open into the sinus venosus of the heart. In ontogeny, the sinus venosus arises in part from the right Cuvierian duct, while the left Cuvierian duct of the larva becomes occluded and then lost. Posteriorly the paired posterior cardinal veins unite to form the single caudal vein (fig. 5.26) which runs beneath the posterior part of the dorsal aorta. Ventrally the sinus venosus receives the unpaired jugular vein anteriorly, and the hepatic vein posteriorly.

In addition to the definite arteries and veins already mentioned, *Lampetra* has a system of venous sinuses filled with red blood. These are fed with blood through capillaries from the arteries and they drain into the above mentioned veins. They are located mainly in the head, where they include large spaces surrounding the tongue muscles, and the peribranchial sinuses. There are also subcutaneous sinuses, formed in ontogeny by the breakdown of the capillary network beneath the skin. The venous sinuses are comparable with those in myxinoids and are absent or much less well developed in gnathostomes. They could be a primitive vertebrate feature which gnathostomes have lost. As against this, the sinuses are

scarcely developed in the ammocoete larva but arise at metamorphosis, which would suggest, on the basis of Haeckel's law, that the sinuses of lampreys are a specialized condition, simulating the sinuses of myxinoids but not homologous.

The dorsal nerve cord of lampreys is band-shaped, being dorso-ventrally flattened. Its dorsal surface receives sensory dorsal-root nerves, while its ventral surface sends off motor, ventral root nerves. Unike myxinoids and gnathostomes, however, dorsal and ventral roots do not join to form mixed spinal nerves (fig. 5.17b). This lamprey situation, with separate dorsal and ventral spinal nerves, is probably primitive. For it resembles the situation in amphioxus, if we assume that the ventral neuromuscular junctions of the latter represent the primitive positions of ventral roots for purposes of comparison. As already mentioned, myxinoids and gnathostomes share mixed spinal nerves as a common advanced feature. But this is not a synapomorphy with respect to lampreys since the mixing of the spinal nerves is topologically different in myxinoids and gnathostomes and therefore cannot be homologous. There is a central canal throughout the length of the dorsal nerve cord. This is a posterior continuation of the ventricles of the brain and contains a Reissner's fibre (Olsson, 1955, pl. 74ff). Indeed, Reissner's fibre was first discovered in *Lampetra fluviatilis* (Reissner, 1860, p. 584).

The brain of *Lampetra* (figs 5.31, 5.32) is much thinner walled than that of myxinoids. The prosencephalon of the adult lamprey brain is divided, as in all other vertebrates, into the telencephalon anteriorly and the diencephalon posteriorly.

The telencephalon, which can be described as the olfactory brain, consists of a median and a pair of lateral portions. All three portions are hollow. The lateral portions of the telencephalon each consist of two conjoined spheres – the olfactory bulb anteriorly and the olfactory lobe posteriorly (figs 5.31, 5.32). The paired olfactory nerves enter the anterior ends of the olfactory bulbs, carrying the axons of the olfactory sense cells, whose cell bodies

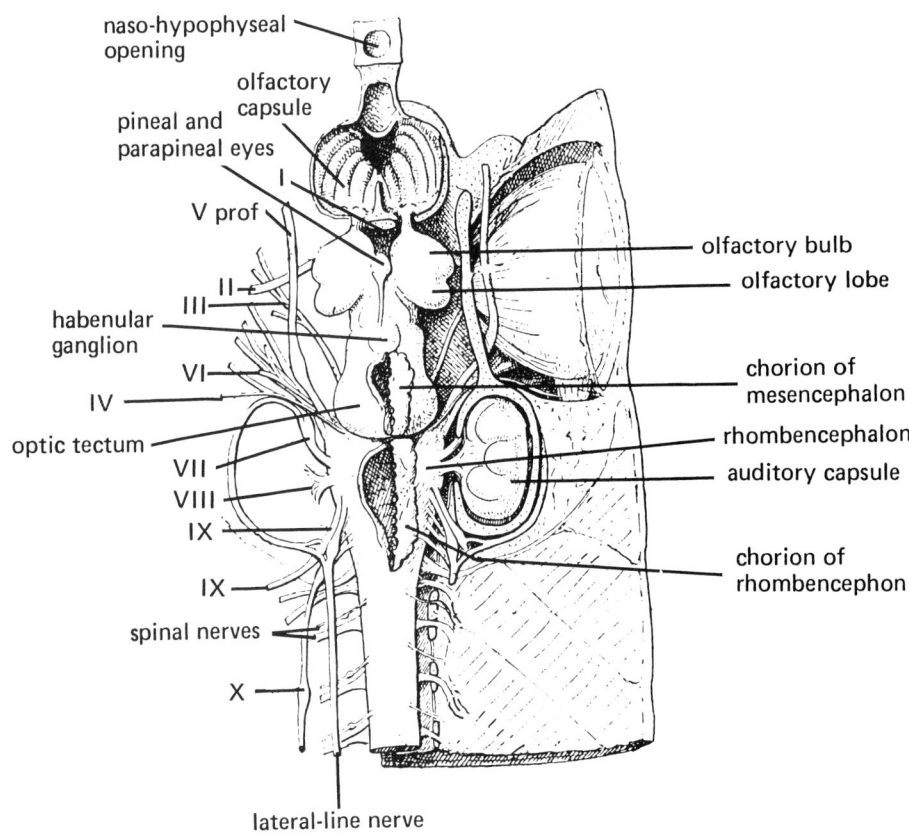

naso-hypophyseal opening

olfactory capsule

pineal and parapineal eyes

olfactory bulb

olfactory lobe

V prof

II

III

habenular ganglion

VI

IV

optic tectum

VII

VIII

IX

IX

chorion of mesencephalon

rhombencephalon

auditory capsule

chorion of rhombencephon

spinal nerves

X

lateral-line nerve

5.31. Lampetra. *Brain and cranial nerves in dorsal aspect; from Goodrich (1909b, fig. 25, p. 40).*

are situated in the lining of the olfactory capsule. Thus, though the olfactory organ itself is unpaired, the olfactory nerves, and the parts of the brain that receive them, are paired. The primary olfactory axons from the capsule form a synapse near the surface of the olfactory bulbs with the dendrites of the secondary olfactory neurons.

The diencephalon likewise has thick side walls and a thin roof and floor. Functionally it is the optic and hypophyseal brain. The optic tracts from the optic nerves (fig. 5.32) enter the floor of the diencephalon anteriorly, cross to the opposite side of the brain in the optic chiasma, in the floor of the diencephalon, and then pass upwards and rearwards to end in the optic tectum of the mesencephalon. The ventral part of the diencephalon forms a sac called the infundibulum. The most ventral part of this is the neurohypophysis which overlies and touches the adenohypophysis. Unlike the myxinoids, but like the gnathostomes, the adenohypophysis is ectodermal in origin, for the infundibulum is seen in early development to be anterior to the bucco-pharyngeal membrane. The

roof of the diencephalon extends upward to form the pineal and parapineal eyes. These are light sensitive; they are overlain by a patch of transparent skin and control diurnal colour change, for the lamprey is darker by day than by night.

The mesencephalon has a thin roof which is produced upwards as a folded vascularized chorioid lamina (fig. 5.31) – lampreys are the only vertebrates that have such a lamina in the mesencephalon. The ependyma ventral to the anterior part of the roof of the mesencephalon contains the subcommissural organ which, in the adult lamprey, secretes Reissner's fibres (Olsson, 1955, p. 177). Behind the chorioid lamina is the optic tectum which, as already mentioned, receives axons from the optic nerves (fig. 5.32). At the transition from ammocoete to adult the optic tectum expands greatly in size and becomes histologically much more complicated. This is related to the condition of the eyes. For the paired eyes of the ammocoete are lens-less, non-motile and functionless while those of the adult have lenses, are moved by oculomotor muscles and are

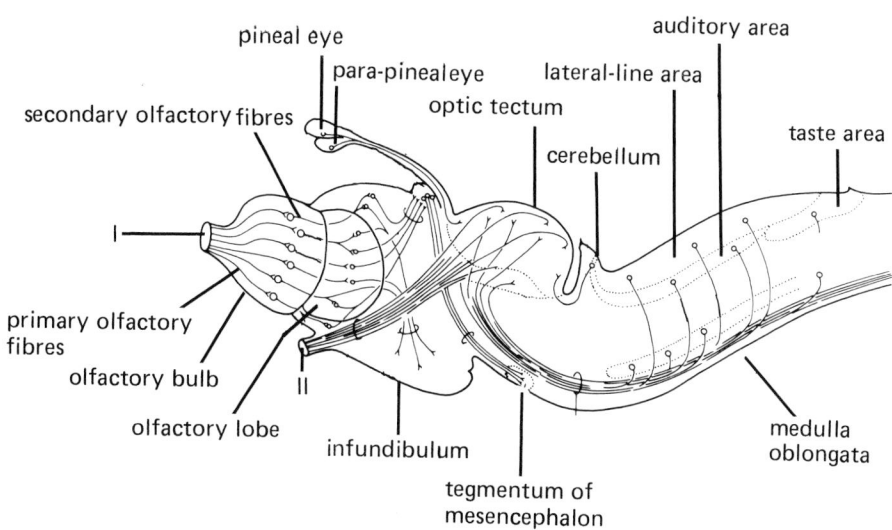

5.32. Lampetra. *Fibre connections inside the brain; from Nieuwenhuis (1977, fig. 12b, p. 124).*

used in hunting (Kennedy & Rubinson, 1977). The ventral portion of the mesencephalon is the tegmentum. It gives rise to the oculomotor nerves (III). Another pair of nerves to the eye muscles (i.e. the trochlear, IV) arises from the dorsal side of the brain immediately behind the optic tectum (fig. 5.31). According to Larsell (1947, p. 464) the dorsal position of the trochlear is secondary in ontogeny; in early ammocoetes the nerve and its nucleus are situated near the trigeminal roots and migrate dorsally into the region where cerebellum and mesencephalon join.

The rhombencephalon is divided into the cerebellum (metencephalon) and the medulla oblongata (myelencephalon). Its roof behind the cerebellum is thin-walled and elaborated into a chorioid lamina.

Many cranial nerves go off from the rhombencephalon (figs 5.13, 5.31). Most anteriorly are the roots of the trigeminal complex. These include two dorsal sensory roots (V_1 and V_2, i.e. the profundus and 'true' trigeminal) leading to the corresponding ganglia. Just ventral to this is the motor root of the trigeminal, closely associated with the root of the abducens nerve VI. In series behind come the roots of the facial nerve VII together with the acoustic-nerve VIII and with lateral-line components included. Behind this comes the common root of the glossopharyngeal and vagus (IX and X) which supply the gills and also have a lateralis component. And behind these come the roots of the spino-occipital nerves which supply the hypobranchial musculature.

The more dorsal parts of the walls of the medulla oblongata are largely sensory and include areas devoted to the lateral line, the ear, and to sensory impulses from the trigeminal.

The lateral eyes of *Lampetra* are well developed; they are much more complicated and gnathostome-like than those of the myxinoid *Eptatretus*. As already mentioned the eyeballs are large, moved by six oculomotor muscles homologous with those of gnathostomes, and are equipped with a lens. The lens is situated behind a non-motile iris, and anterior to the iris and lens is an aqueous body. The cornea of the eye is in two mechanically separate layers – the outer layer, which derives embryologically from the dermis and epidermis, forms a thin, convex window in the skin (fig. 5.33). This window can be bent inwards but not rotated or displaced. And the inner layer of the cornea, which derives from mesenchyme is an anterior transparent continuation of the scleral layer of the eyeballs, and can slide freely against this outer fixed transparent window as the eyeball rotates. To the posterior edge of the dermal window the so-called cornealis muscle is attached (figs 5.27, 5.33), whose contraction, by altering the convexity of the window, displaces and focuses the lens (Franz, 1932). The lens of *Lampetra*, unlike that of gnathostome, is not held in place by a ligament but simply jammed between the outer surface of the vitreous body and the inner surface of the scleral cornea. There are no muscles whatever inside the eye. The retina has the same histological stratification as in gnathostomes and myxinoids.

The embryology of the eye of *Lampetra* has been important in phylogenetic theory. Relevant papers include

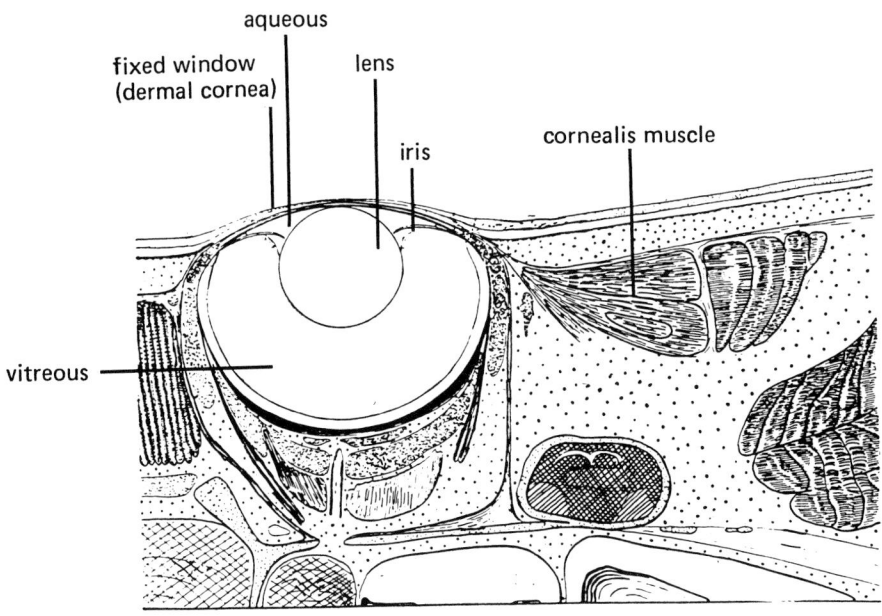

5.33. Lampetra. *Horizontal section through the eye and associated structures. Posterior is to the right of the figure. Note the cornealis muscle acting on the dermal cornea; from Franz (1932, fig. 3, p. 127).*

Dücker (1924), Keibel (1928) and above all Studnička (1912), who was led by his studies to recognize for the first time that the visual cells of vertebrate eyes are modified ependymal cells, i.e. they represent modifications of the internal lining of the brain. The first anlagen of the eyes are paired outgrowths from the ventral part of the diencephalon which grow sideways and upwards to terminate in slight expansions in contact with ectoderm. These expansions are the optic vesicles. They cause a lens to develop in the contiguous ectoderm (fig. 5.34a). The lateralmost epithelium of the vesicle, in contact with the lens, thickens and transforms into a retina with visual cells, bipolar cells and ganglion cells. And the ganglion cells send fibres to the ventral edge of this retina which grow down the optic nerve to form the optic chiasma and optic tracts. The outer epithelium of the optic vesicle acquires pigment and represents the anlage of the pigmented optic epithelium of the adult (fig. 5.34b). All these features are already developed in the centimetre-long pro-ammocoete, which has an unpigmented skin and does not live in burrows. The eye so formed is probably functional and probably serves as a directional eye. This means that it can tell which direction light comes from, but cannot form an image. It would function much like the eye of *Eptatretus*. As the larva of *Lampetra* increases in size, and changes from pro-ammocoete to ammocoete, it acquires pigmented skin totally hiding the lateral eyes which do not act as light receptors at this stage (Young, 1935, p. 257).

The optic vesicle becomes an optic cup during the ammocoete period by extending towards the skin round the dorsal, anterior and posterior margin of the vesicle (fig. 5.34c). The ventralmost point of the vesicle, from which the optic fibres pass into the optic nerve, does not extend outwards. A vitreous body arises between the retina and the lens. At the same time mesenchymatous tissue accumulates round the optic cup to produce the inner vascularized chorion and the tough outer scleral layer. The oculomotor muscles develop attached to the sclera. During the ammocoete stage the histologically differentiated retina with visual cells etc. remains very small. At transformation to the adult lamprey the skin pigment over the eye disappears and the skin forms the transparent corneal window; the size of the eye ball increases; and the histologically differentiated retina greatly increases in size to form the adult retina. The adult eye is not a directional, but an image-forming eye.

The eye of *Lampetra* thus shares with the gnathostome eye several features which are lacking from *Eptatretus*. These include the lens, pigment layer, (for absence in *Eptatretus* see Dücker, 1924, p. 510), iris, aqueous body, and oculomotor muscles. Features of the eye shared between *Eptatretus*, *Lampetra* and gnathostomes but absent from tunicate tadpoles include: the histological stratification of the retina, the optic cup, the vitreous body and the chorioid and scleral layers. Important differences between the eye of *Lampetra* and that of gnathostomes

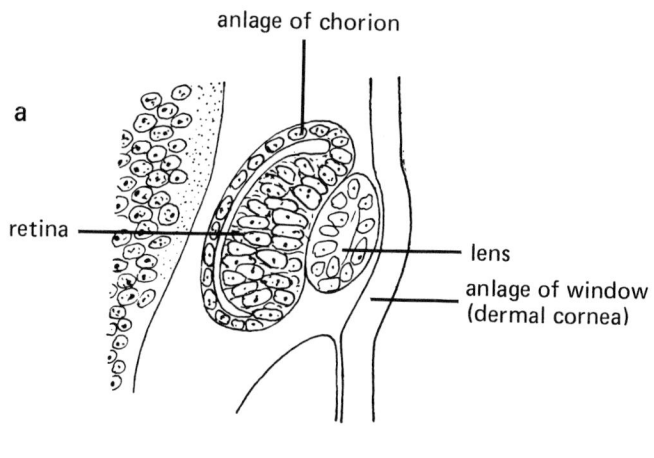

a

anlage of chorion

retina

lens

anlage of window
(dermal cornea)

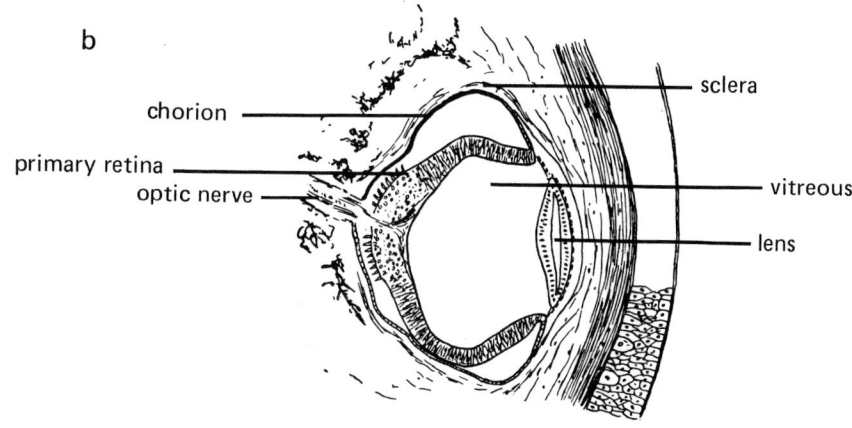

b

chorion

primary retina

optic nerve

sclera

vitreous

lens

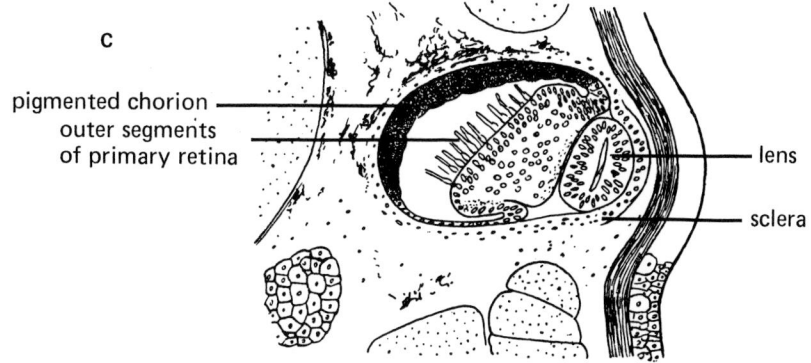

c

pigmented chorion
outer segments
of primary retina

lens

sclera

5.34. Lampetra. *Transverse sections of the eyes of the proammocoete and ammocoete. a) 8-mm-long proammocoete of* Lampetra fluviatilis *– the eye is an immobile directional eye; b) 18-mm-long ammocoete of* L. fluviatilis *– the optic cup is beginning to form; c) 50-mm-long ammocoete of* Lampetra planeri *– the optic cup has formed but the eye is functionless at this stage; from Studnička (1912, fig. 1, p. 563; fig. 3, p. 567; fig. 5, p. 570).*

include: the absence of muscles from the iris, of a ligamentary support for the lens, of ciliary muscles involved in accommodation, and of fusion between the dermal and scleral cornea; and accommodation by a cornealis muscle bending the dermal cornea.

The ear and lateral line of *Lampetra*, like those of other aquatic vertebrates, are traditionally held to be two parts of a wider acustico-lateralis system. This view goes back to Mayser (1882) (see Lowenstein, 1967) and is thoroughly confirmed by recent ultrastructural and physiological work. The main grounds for it are as follows. (1) Both systems are connected to nuclei in the antero-dorsal part of the medulla oblongata. (2) Both systems develop in embryology from a common dorso-lateral placode, of which the acoustic portion becomes deeply invaginated while the lateralis portion remains superficial or is only slightly invaginated. (3) Both are supplied with sensory organs of similar type whose primary function is to detect displacements of water. (4) In more detail, the sensory organs (maculae and cristae in the ear, neuromast organs in the lateral line) have the same structure – each of them contains ciliated sensory cells ('hair' cells) surmounted by an inert gelatinous structure known as a cupula; the cilia of each hair cell comprise a single kinocilium and a bundle of stereocilia; the kinocilium has a well-developed $9+2$ tubule system and is consistently situated at the edge of the bundle of stereocilia which are shorter than the kinocilium and have a much less well-developed tubule system (figs 5.35a, b); the cilia are embedded in the substance of the cupula, so that when the cupula bends, because of flow in the surrounding water or for any other reason, the cilia of the hair cells bend also; each hair cell receives several synapses, including afferent synapses with the bipolar cells of the supplying nerves. (5) The physiology of the sensory cells is much the same in the ear and the lateralis system – each hair cell in the resting condition emits a spontaneous signal which is transferred by synapse to the adjacent bipolar cells of the appropriate cranial nerves; when, by movement of the cupula, the ciliary bundle is bent towards the single kinocilium, the frequency of the signal increases and when the bundle is bent away from the kinocilium the frequency decreases (Lowenstein, 1967; Lowenstein & Osborne, 1964; Lowenstein & Wersäll, 1959; Flock, 1967). As mentioned in Chapter 4, the cupular organs in the atria of tunicates may be homologous with the groups of sensory cells in the vertebrate acustico-lateralis system.

Contrariwise, there are also differences between the ear and lateral-line system of vertebrates. Thus each hair cell in the maculae and cristae of the ear tends to have the same orientation as its neighbours, so that all adjacent cells will give the same signal at the same time. In the neuromasts of the lateral line, on the other hand, neighbouring cells tend

to have opposite orientations, i.e. if the kinocilium on one cell is 'east' of its bundle of stereocilia, then on the next cell the kinocilium will be 'west' of its bundle (fig. 5.35b). Corresponding to this is a double sensory innervation, with all 'east'-facing hair cells connected with one nerve, and all 'west'-facing hair cells connected with another. Also, the spontaneous signal of the ear varies very little in frequency being very near a pure tone, whereas that of the lateral line is highly variable.

Despite these differences the traditional view that acoustic and lateralis system are basically one and the same is clearly correct. Their common origin is recorded, as I shall show later, in the early fossil history (Chapters 7 and 8).

The lateral-line receptors of *Lampetra* are scattered circular neuromasts. There are several rows of them on the head and also two pairs of rows on the trunk. The more dorsal of these two pairs extends to the tail region and the supplying nerves, curiously enough, also carry impulses from the dermal light receptors of the tail (Young, 1935). The more ventral pair of rows of neuromasts, which corresponds roughly in position to the trunk lateral line of a gnathostomatous fish, does not extend so far rearwards. It is doubtful whether this line of neuromasts is homologous with the trunk lateral line of a jawed fish. For in such fishes the line is supplied by nerves running in the horizontal septum, which in lampreys does not exist.

The ear of *Lampetra* is very complicated. The paired membranous labyrinths (fig. 5.36), which are bathed in perilymph and enclosed in the cartilaginous otic vesicles, are situated just right and left of the anterior end of the medulla oblongata (fig. 5.31). Each membranous labyrinth has an anterior and posterior semicircular canal which embrace, and at either end open into, a space called the vestibule. The semicircular canals swell at their ventral ends to form ampullae. Inside each ampulla there is a ridge, or crista, of sensory hair cells. Another patch of sensory cells (macula communis) forms part of the horizontal floor of the vestibule – its cupulae contain hard, heavy otoliths, as in *Myxine* (Lowenstein, Osborne & Thornhill, 1968).

Lowenstein (1970) has shown, by electrophysiological study, that the canals can detect angular accelerations in all three planes of space, whether yawing, pitching, or rolling, although the horizontal canal of gnathostomes is lacking. The ear can also detect tilting and vibration. In ontogeny all the patches of sensory epithelium in the ear develop by division and elaboration of a single anlage (Thornhill, 1972). The anterior and posterior canals are homologous with the anterior and posterior vertical canals of gnathostomes and with the anterior and posterior halves of the thin part of the torus of *Myxine*. The ear of *Lampetra* is thus more complicated than that of myxinoids, and most of the extra complexity is gnathostome-like.

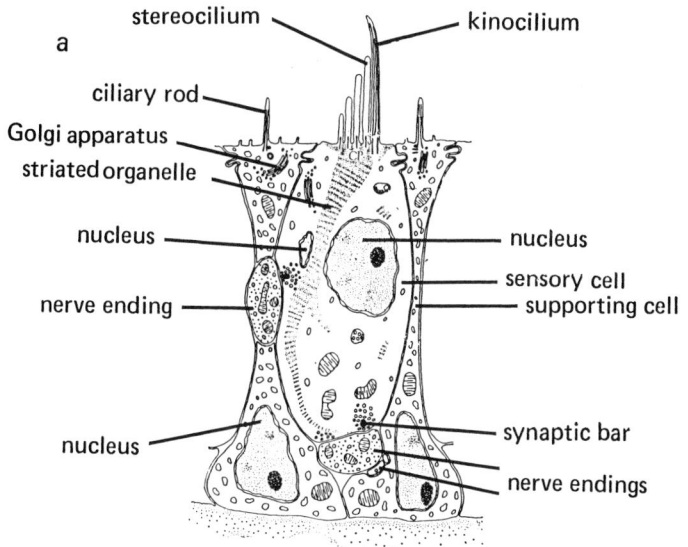

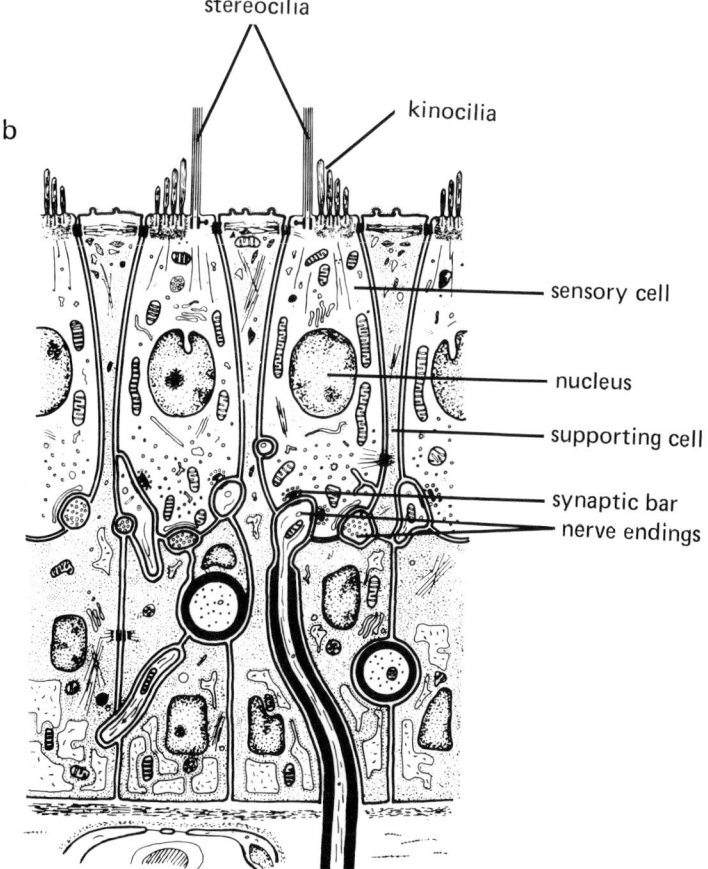

5.35.   *The histological resemblance between ear and lateral line. a) The ear of* Lampetra — *diagram of a sensory cell and two supporting cells from the sensory epithelium (crista) of the inner ear. b) Lateral line of a gnathostome (the burbot* Lota vulgaris); *from Lowenstein & Osborne (1964, fig. 1, p. 197) and Flock in Cahn (1967, fig. 6, p. 172).*

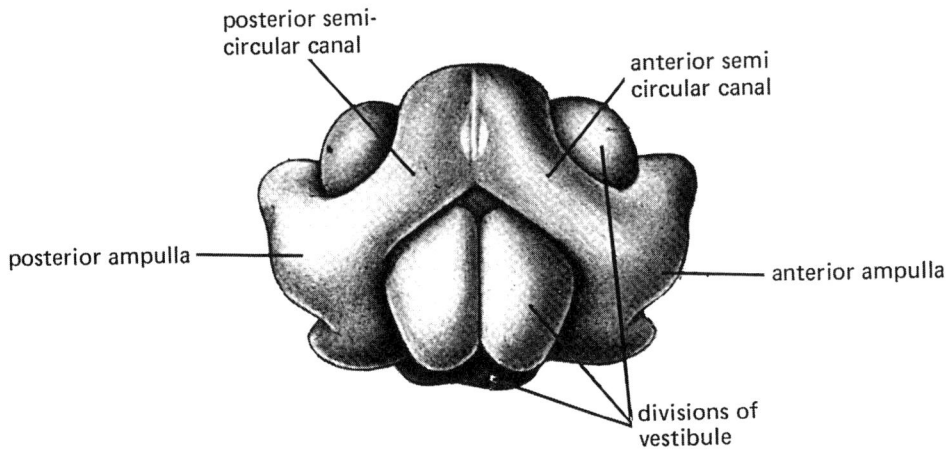

5.36. Lampetra. *The inner ear in right lateral aspect; from Marinelli & Strenger (1954, p. 59, fig. 54a).*

The olfactory organ of *Lampetra* is very well developed and presumably used in locating prey, as has been shown for *Petromyzon marinus* (Kleerekoper, 1972, p. 395). The olfactory capsule is unpaired and situated just in front of the brain (fig. 5.31). It opens anteriorly into the naso-hypophyseal duct (fig. 5.23). This duct runs downwards and rearwards from the nostril and the olfactory capsule straddles it dorsally. The duct ends blind in a pouch that overlies the posterior portion of the buccal cavity and underlies the brain. The contraction and dilation of the branchial tube pumps water in and out of the branchial pouches, as already mentioned. And the tidal branchial pumping action alternately squashes and dilates the naso-hypophyseal pouch, although there is no opening between the latter and the pharynx, so that water is pumped in and out of the nostril, keeping time with the branchial currents (Kleerekoper & Van Erkel, 1960).

The cavity of the olfactory capsule is divided almost into two by a vertical wall in the median plane (fig. 5.31), and each wall of each half capsule is also deeply folded. These folds bear olfactory epithelium on their surfaces and, by increasing the area of such epithelium, presumably improve the sense of smell. The olfactory epithelium contains sensory cells and supporting cells (fig. 5.37). Each sensory cell projects slightly into the cavity of the olfactory organ as an olfactory knob (Kleerekoper, 1969, p. 23). This knob is ciliated, as it always is in vertebrates (Kleerekoper, 1969, p. 27), and the cilia have been observed in other vertebrates to thrash randomly about (Kleerekoper, 1969, p. 32). The sensory cells make direct connection with the brain by sending off axons which join each other to form the very short paired olfactory nerves. These axons synapse just below the surface of the olfactory bulbs of the brain (fig. 5.32). The olfactory nerves of

vertebrates are unique among adult vertebrate nerves in that the cell bodies of axons are themselves sense cells, rather than synapsing with sense cells. They resemble in this respect the sensory nerves of tunicates, acraniates and echinoderms.

As with myxinoids, therefore, the olfactory capsule is incompletely divided into right and left halves, but the olfactory nerve is completely paired.

In embryology, the nasal placode from which the olfactory nerve of *Lampetra* develops is median and unpaired and closely associated with the future hypophysis, which is clearly ectodermal in origin, unlike in the myxinoids. Probably olfaction is relatively unimportant during the ammocoete stage. This is suggested by the fact that the olfactory epithelium remains unfolded in the ammocoete, except for a single sagittal fold, until near the end of larval life. Then, at metamorphosis, it suddenly becomes complexly folded with an abrupt increase in area. No doubt this sudden expansion of the olfactory system, like the simultaneous perfecting of the lateral eyes, is because the adult is a hunter, not a passive filter feeder.

The pharynx and buccal cavity of the ammocoete (fig. 5.38), as already mentioned, are strikingly different from the corresponding structures of the adult. The buccal cavity is not a funnel-shaped sucker but a horse-shoe-shaped oral hood, rather like the buccal cavity of amphioxus. It contains a ring of buccal tentacles and is separated from the pharynx by a pair of muscular flaps – the velum. The continuous rearwards and forwards beating of the velar flaps, in combination with expansion and contraction of the pharynx, draws water in through the mouth, pushes it into the pharynx and out through the gill slits. The pharynx, unlike that of the adult, is not divided into dorsal oesophagus and ventral respiratory tube with

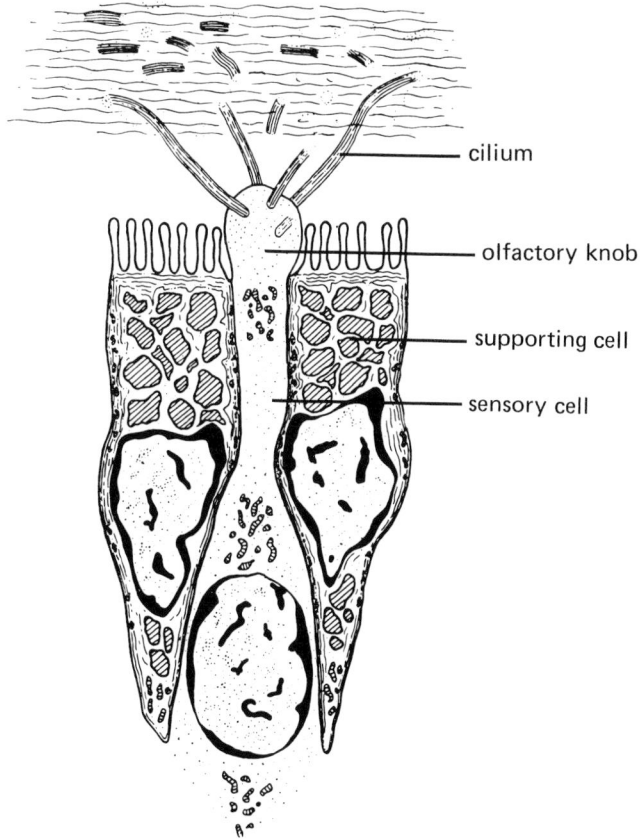

5.37. *Histology of the olfactory epithelium in a vertebrate (frog); from Kleerekoper (1969, fig. 15, p. 21).*

paired gill pouches. Instead it is a single cavity which extends on left and right by wide openings into seven pairs of tall gill chambers, each of which opens to the outside by a branchial slit. These chambers are separated from each other by partitions with gill lamellae on their anterior and posterior surfaces. The adult tongue, with its associated muscles, cartilages and horny teeth, is not yet developed in the ammocoete.

The endostyle, from the comparative viewpoint, is the most interesting structure associated with the ammocoete pharynx (figs 5.38, 5.39a, b, 5.40). It lies beneath the anterior part of the pharyngeal floor and when fully formed, is connected with the cavity of the pharynx by a duct in the mid-line. Its shape is complicated; in front of the duct it consists of a pair of tubes while behind the duct it has three of them – left, right and median, the median tube being coiled upwards and forwards in a plane spiral. The cavities of the two anterior tubes are ∩-shaped in cross section, because of a ventral invagination of their walls. These paired ventral invaginations continue behind the endostylar duct but transform there into the divisions

between the lateral tubes and the median tube. The total division between right and left tubes by a complete median septum in front of the duct is continued behind the duct by an incomplete median septum in the floor of the spiral median tube (fig. 5.39b). The endostyle is supplied by a pair of branches of the facial nerve (Alcock, 1898).

The duct between pharynx and endostyle is elongated in the median plane. Posteriorly it is continued in the floor of the pharynx by a ciliated groove which passes back to the opening of the oesophagus. Anteriorly the duct is ⊥-shaped in section. The side arms of the ⊥ are ciliated grooves which arise from the cavities of the paired anterior tubes of the endostyle. They pass upwards and forwards to the floor of the pharynx where they diverge from each other and run as ciliated grooves up the side walls of the pharynx in front of the most anterior pair of gill slits, to meet and fuse in the dorsal mid-line (fig. 5.38). These grooves are the peripharyngeal grooves which in position and function are remarkably like those of amphioxus or the tunicates and are almost certainly homologous with them.

The epithelium of the endostyle is differentiated into

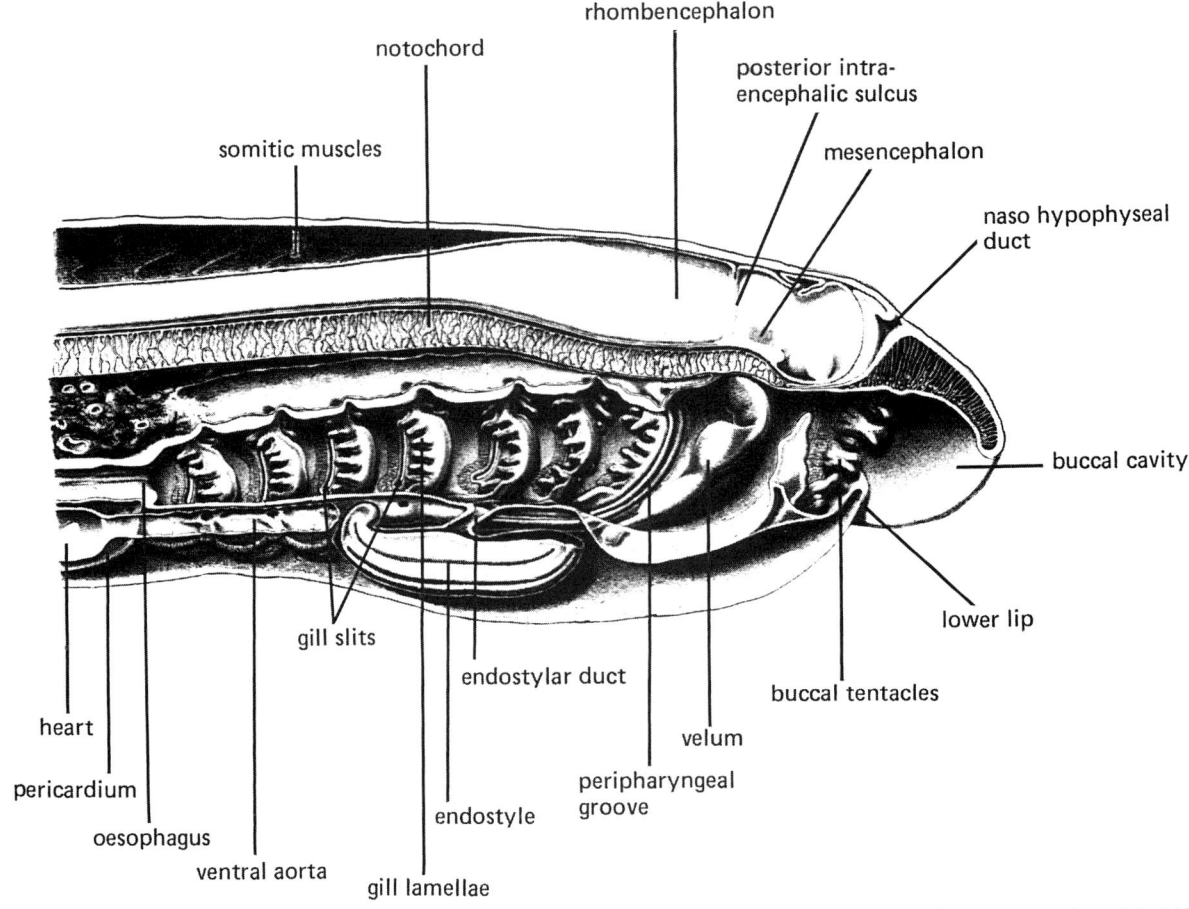

5.38. *Sagittal section of an ammocoete (species and genus not stated) to show the endostyle, peripharyngeal bands, velum, velar tentacles and the fold inside the brain that marks the posterior limit of the mesencephalon (posterior intra-encephalic sulcus); from Dohrn (1886, pl. 7, fig. 17).*

five different types of cell as shown in fig. 5.40 which is a transverse section through a right anterior endostylar tube (Barrington & Sage *in* Hardisty & Potter, 1972, p. 109). Type 1 cells secrete mucus into the endostylar cavity; type 3 cells are concerned with the fixation of iodine during larval life to form the hormone thyroxine; type 2 cells are immature and, depending on their position, continuously generate either type 1 or type 3 cells; type 4 cells form the crest of the ventral fold in each anterior tube – they are inactive in the larva but are the main source of the iodine-fixing cells which form the thyroid gland of the adult; finally, type 5 cells are an unciliated pavement epithelium which coats all that part of the endostylar cavity not covered by types 1 to 4.

The arrangement of the type 1 mucus-secreting cells which form the gland strips of the endostyle is at first sight very complicated. A transverse section through the endostyle anterior to the gut shows eight different patches of

them, for example; while a section behind the duct, passing through the planispiral and the lateral tubes, may show up to 16 such patches (12 in fig. 5.39a). However, as shown by Kieckebusch (1928), these apparently separate patches are connected together in space so that, in fact, there are only four gland strips in the endostyle, being left and right, dorsal and ventral (fig. 5.39b). This fact is important when comparing the ammocoete endostyle with that of tunicates and amphioxus.

The gland strips secrete mucus into the endostylar cavity from which it passes up the endostylar duct into the pharyngeal cavity, as has been shown experimentally by Sterba (1961, p. 108). In the cavity it forms a filter cone, the anterior ends of which are held by the peripharyngeal bands exactly as in tunicates (fig. 5.41). The apex of the filter cone passes back into the opening of the oesophagus by the action of cilia. These continuously rotate the cone, so producing a mucus rope which passes down the

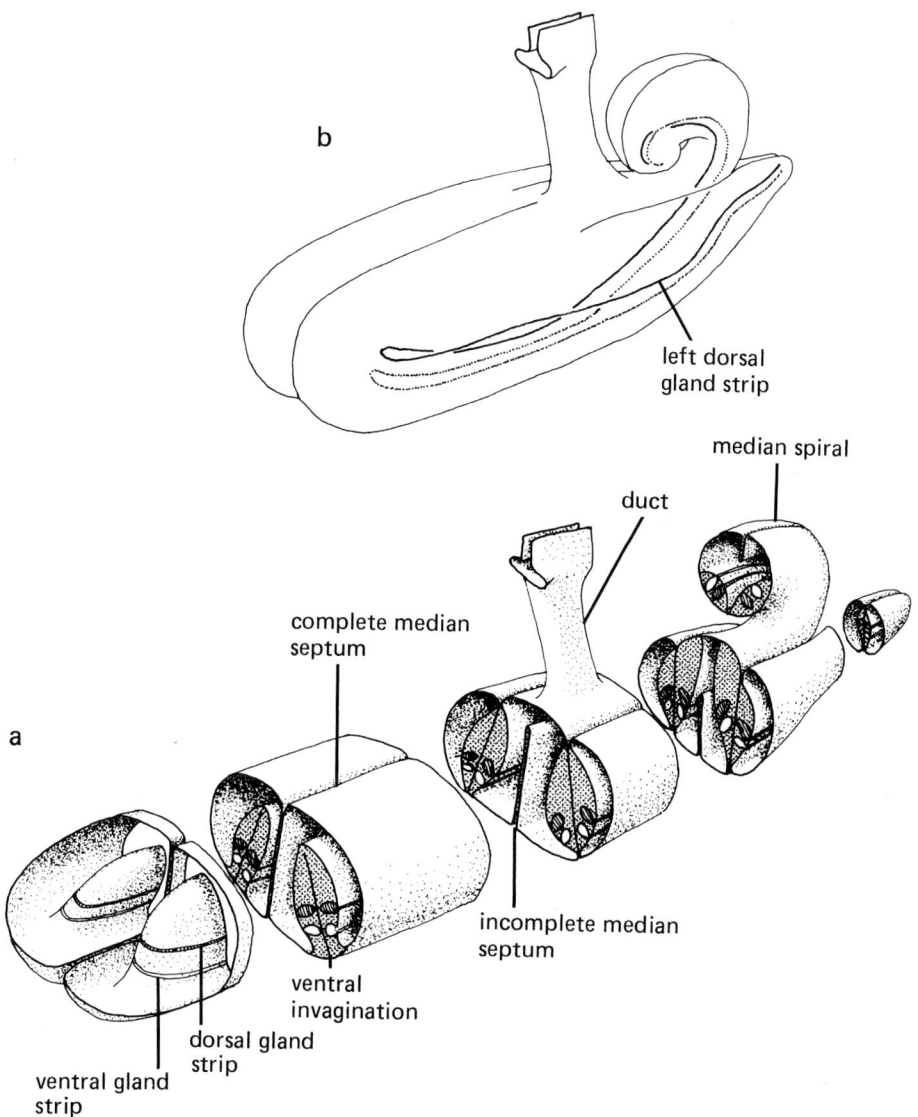

5.39 Lampetra. *Diagram of the endostyle of an ammocoete, based on Kieckebusch (1928). a) Left anterior dorsal aspect, exploded to show the relations of the two pairs of gland strips with the folds and septa. b) The same aspect, transparent, to show the positions in space of the left dorsal and ventral gland strips. (New.)*

oesophagus. The filter cone extracts particles from the inhalent water current and can be made visible by feeding the ammocoete or pro-ammocoete with particles of chalk, Indian ink or carmine (Balabai, 1951, 1956; Kieckebusch, 1928; Newth, 1930). Mallatt (1981) has given a different account of feeding in the ammocoete for he denies the existence of a filter cone. I cannot accept his results, however, since his experimental subjects were anaesthetized slowly, and his conclusions, unlike those of Newth (1930),

are not supported by photographs of the larvae actually feeding.

The early embryology of the endostyle was worked out by Dohrn (1886) and Reese (1902). It first appears, in the unhatched embryo, as a longitudinal median groove in the anterior part of the pharyngeal floor. The cavity of this groove becomes cut off from the pharynx by the rearward growth of an anterior lip and the forward growth of a posterior lip. These lips meet together and fuse. Where

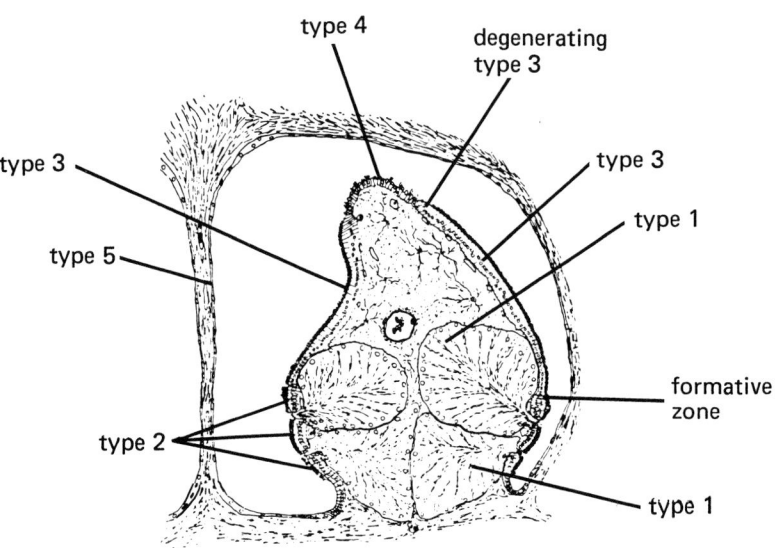

5.40. Lampetra planeri. *Cell types in the endostyle; transverse section through the right anterior chamber; from Barrington & Sage in Hardisty & Potter (1972, p. 109, fig. 4).*

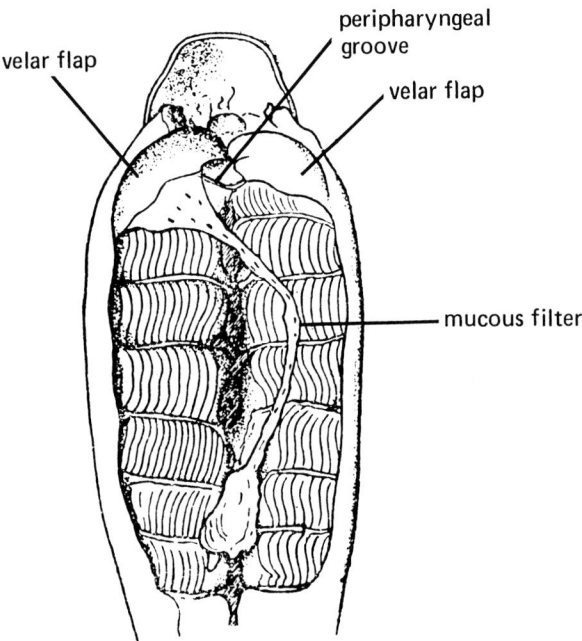

5.41. Ammocoete of Lampetra mariae. *Part of the mucous filter cone exposed by dissection of the pharyngeal cavity after feeding the larva with powdered chalk; from Balabai (1956, p. 73, fig. 46).*

they fuse they produce a solid vertical thread of cells which will later become the duct. Next, the endostylar cavity begins to be subdivided by longitudinal folds in its floor. The first such folds to appear are the paired ventral invaginations. These arise at the anterior end of the endostyle and grow rearwards; they will give rise in the fully formed ammocoete to the ventral invaginations of the paired endostylar tubes in front of the vertical duct, and to the divisions between right, median and left tubes behind the duct. The next fold to arise is a median fold which divides the endostylar cavity completely from floor to ceiling in front of the duct and forms the median septum of the median tube behind the duct. This fold likewise first appears at the front end of the endostyle and then grows rearwards. At about the same time the endostylar duct opens, by the excavation of a cavity in the pre-existing solid vertical thread of cells; the peripharyngeal bands arise; the velar membrane breaks down to form the velar flaps; and the gill slits open. Thus all conditions exist for the larval pharyngeal mucous trap to function.

Differentiation of the endostylar epithelium begins when the paired ventral invaginations have appeared, but the median fold is still absent (fig. 5.42). At this stage the endostyle is a distorted T-shape in transverse section and the dorsal and ventral gland strips first arise on the median faces of the ventral invaginations i.e. on either side of the upright of the T. This early epithelial differentiation can be compared directly with the endostyle of amphioxus (fig. 5.43). It suggests that the paired ventral gland strips of an ammocoete are homologous with the more median (and more ventral) gland strips of amphioxus and the dorsal gland strips of an ammocoete with the lateral (and more dorsal) gland strips of amphioxus. Olsson (1963) has confirmed this homology by showing that the dorsal gland

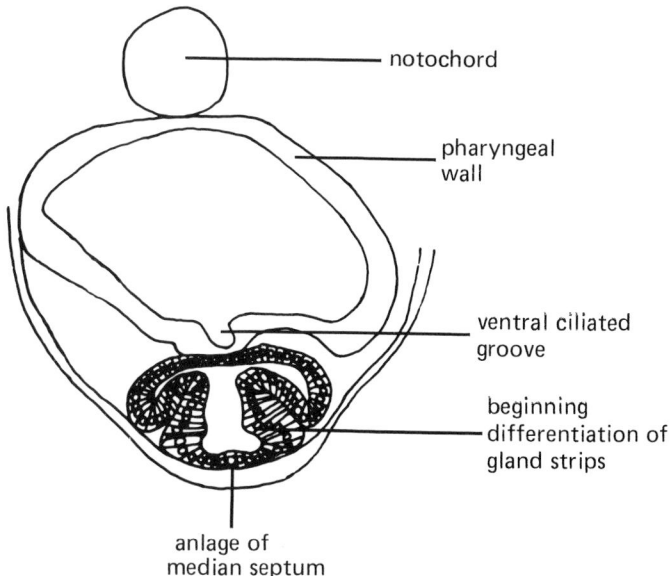

5.42. *Ammocoete of* Lampetra planeri. *When differentiation of the gland strips begins, the cavity is* **T**-shaped and the gland cells first appear on the upright of the **T**. This presumably represents the primitive condition and helps to identify the individual gland strips with those of amphioxus and tunicates; redrawn after Reese (1907, pl. 4, fig. 4b).*

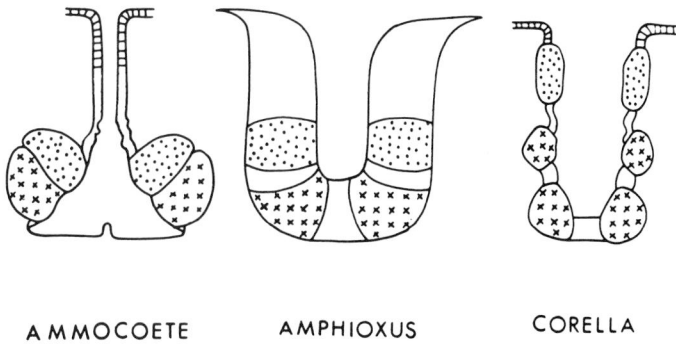

A MMOCOETE   AMPHIOXUS   CORELLA

5.43. *Diagrammatic cross sections of the endostyles of an ammocoete, amphioxus and the tunicate* Corella *to show the nature of the secretions of the various gland strips; crosses = mucus and proteins; dots = -SS or -SH-rich proteins and lipids; from Olsson (1963, fig. 11, p. 322).*

strips of ammocoetes, like the lateral gland strips of amphioxus, produce a secretion rich in cystine. Moreover, if the suggested homologies are correct, then iodine fixation is seen to occur in corresponding positions in the ammocoete and acraniate endostyle. Comparison with the tunicate endostyle is more difficult since it has three pairs of gland strips, not two. Here again, however, iodine fixation occurs just dorsal to the most dorsal gland strip which, as in the ammocoete and amphioxus, produces SS- and SH-rich proteins and lipids. This suggests that the dorsal gland strip of tunicates is homologous with its namesake in ammocoetes and amphioxus.

In summary, the pharynx of an ammocoete can be compared in many details with that of tunicates and of amphioxus. Corresponding features in all three groups include the endostyle, the dorsal and ventral gland strips, with their respective chemical products, the position of the iodine-fixating epithelium of the endostyle, the peri-pharyngeal bands, the mucous filter, the velum and the gill slits. These correspondences, contrary to Jollie (1973, 1977) prove homology. They show that lampreys retain in their larval life history a filter-feeding stage which corresponds to the adult of tunicates and acraniates, and which other vertebrates have lost.

5.44. Squalus acanthias. *External features; from Garman (1913, pl. 14, fig. 1).*

The endostyle and its duct degenerate completely at metamorphosis except that some cells, mainly of type 4, give rise to the thyroid gland of the adult. This consists of scattered cyst-like follicles situated near the tongue muscles, roughly where the endostyle used to be. The thyroid gland of myxinoids and gnathostomes arises as a ventral median invagination of the pharynx, and is active in producing thyroxine like the endostyle of ammocoetes, tunicates and amphioxus (for myxinoids see Gorbman, 1963). The traditional homology, originally proposed by Wilhelm Müller in 1873, of the thyroid gland with the endostyle, agrees with physiological evidence and with the ontogeny of lampreys. There seems little doubt that it is correct.

The gnathostomes are the third group of living vertebrates and can be illustrated by the Spiny Dogfish *Squalus acanthias* (=*Acanthias vulgaris*) (fig. 5.44). I shall describe this animal more briefly than the agnathans because the gnathostomes are more widely understood than the agnathans, and are slightly beyond the scope of a book on the ancestry of the vertebrates. Its present relevance lies in comparing it with myxinoids and lampreys and some phylogenetic conclusions that follow from this comparison. Also *Squalus acanthias* is well known embryologically, as discussed in the next chapter, and, as a shark it represents the group on which classical embryological ideas of the ancestral 'protovertebrate' were largely based. I choose it rather than the very similar Common Dogfish *Scylliorhinus canicula* because *Squalus acanthias* was anatomized in detail by Marinelli & Strenger (1959) and its embryology has been well described by Scammon (1911).

Squalus acanthias lives in the sea, mainly near the bottom, occurs in huge shoals and chiefly eats fishes which it hunts by sight and smell. The sexes are separate. Fertilization is internal and the young are borne alive. As to external appearance, the female *Squalus acanthias* is up to about 2 m long, but the males are smaller. The fish has an elongate streamlined shape with large mobile eyes.

There are two dorsal fins, each one with a strong fin spine anterior to it; the caudal fin is heterocercal, i.e. the upper lobe, containing the notochord, is much larger than the lower one; and there are two pairs of paired fins, pectoral and pelvic. In the male the medial edge of the pelvic fin is modified into a tubular clasper which functions as a penis in copulation, introducing sperm into the female cloaca. Body openings include: a pair of nostrils whose aperture is subdivided by a flap into an inlet and an outlet; a broad mouth with teeth; a pair of spiracles just behind the eyes; five pairs of gill openings guarded by outlet valves and arranged in line just in front of the pectoral fins; a median cloacal opening between the two pelvic fins flanked by abdominal pores which end blind in the female but lead into the abdominal cavity in the male; a pair of endolymphatic ducts, leading to the ears, on the dorsal surface of the head; and numerous minute holes connected with the lateral-line system. The surface of the body is set with numerous, small, sharp, rearward-pointing placoid scales which are much like the teeth both in shape and histology.

The branchial current enters through the mouth and leaves through the gill slits, each of which is guarded by a loose flap that serves as an outlet valve. This direction of flow differs from that of adult lampreys, but must be primitive for it resembles that of amphioxus, tunicates, ammocoetes and also myxinoids (except that in these the current enters through a nostril).

The mouth of *Squalus acanthias* is broad and lined with teeth. These are unlike those of lampreys and myxinoids, and like those of other gnathostomes in consisting of dentine covered with enameloid. Old teeth are constantly being replaced by new ones.

The skeleton of *Squalus acanthias* is cartilaginous except for: the teeth and placoid scales; remnants of the notochord; and the horny ceratotrichia, or fin rays, that support the fins. Cartilaginous elements include: well-developed vertebrae much more regular than the arcualia of *Lampetra* and supplied with centra and attached ribs (fig. 5.46); a chondrocranium which completely encloses

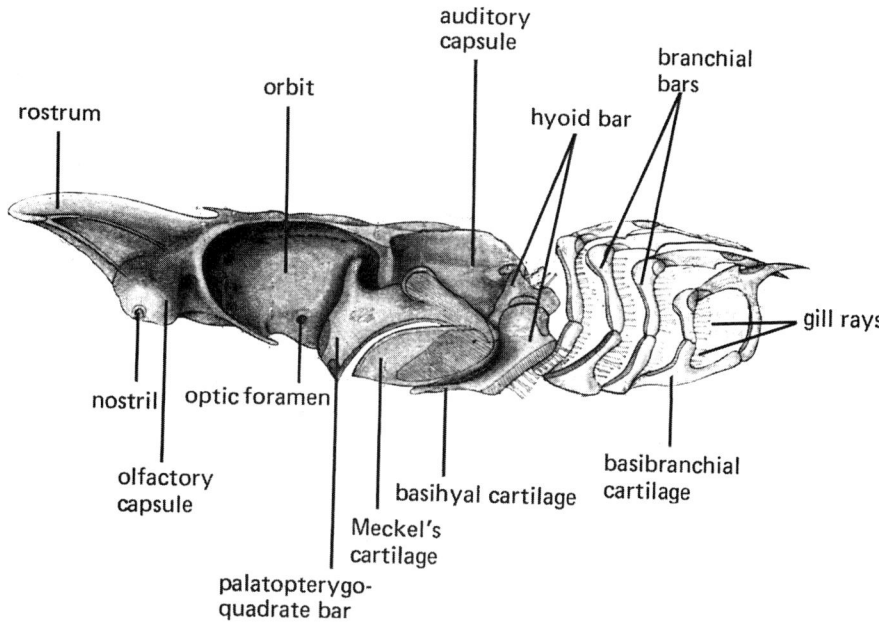

5.45. Squalus. *Neurocranium, mandibular and branchial skeleton; from Wells (1917, pl. 2, fig. 5).*

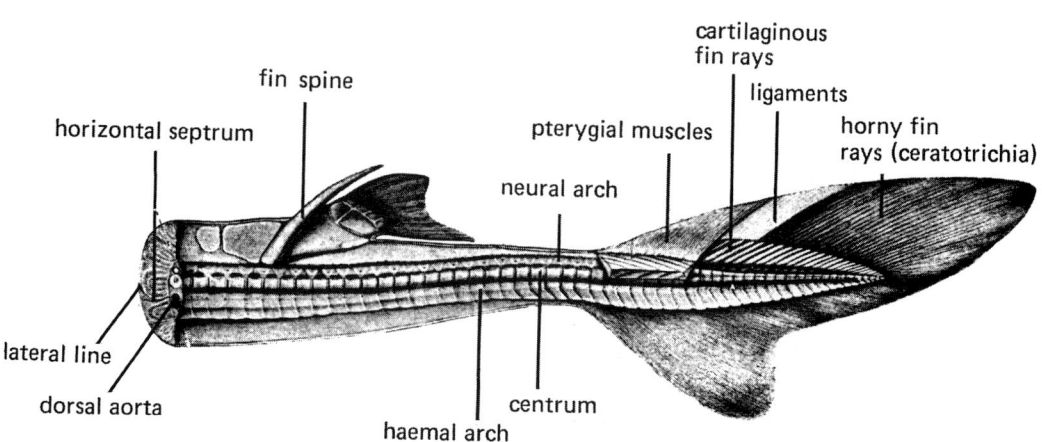

5.46. Squalus. *Skeleton and muscles of the tail in sagittal section. Note that* Squalus, *as a myopterygian, has pterygial muscles that alter the shape of the fin, like* Lampetra *but unlike* Myxine; *from Marinelli & Strenger (1959, fig. 206, p. 278).*

the brain except for a number of foramina (fig. 5.45); a visceral skeleton consisting of jaws, a hyoid arch, five pairs of branchial arches and a number of small cartilages (labials, extrabranchials, gill rays etc.), pectoral and pelvic girdles; and a complex cartilaginous skeleton in the paired and unpaired fins. The cartilages of the fin skeleton can be moved by means of pterygial muscles attached to them on either side (fig. 5.46). As already mentioned, fin muscles also exist in lampreys, but not in myxinoids.

The visceral skeleton is made up of cartilages articulated together to form seven pairs of visceral arches, separated by visceral slits (fig. 5.45). The most anterior, or mandibular, arch forms the upper and lower jaw. The second or hyoidean arch, is modified to support the mandibular arch. The more posterior, or branchial arches, are very like each other. The branchial arches are separated from each other by gill slits, and there is also a gill slit between the hyoidean and first branchial arch. The mandibular and hyoidean arches, on the other hand, are separated from each other by the spiracle. This is clearly a serial homologue of the gill slits, but has only very reduced lamellae, and does not function in gas exchange. The mandibular and hyoidean arches are in front of the ear and described as the pro-otic visceral skeleton. The branchial arches are metotic. The visceral skeleton, except for its unpaired most ventral elements (basihyal and basibranchial cartilages) arises from mesectoderm (neural crest) in ontogeny, along with the prechordal part of the skull, whereas all other cartilages are mesodermal in origin. The branchial arches are internal to the gill lamellae, not external to them as in lampreys. This suggests that gnathostome and lamprey visceral arches are not homologous with those of lampreys. Gill lamellae are present on the anterior and posterior border of all the gill slits, except for the posterior border of the last slit.

The alimentary canal of the dogfish begins with the buccal cavity and pharynx, which pass insensibly into each other, for there is no velum. The spiracles and gill slits open out of the pharynx. Next behind come: the oesophagus; the stomach with 'descending' and 'ascending' portions; the intestine which includes a spiral valve (as in lampreys, and tunicates, but unlike myxinoids); and the rectum which opens into the cloaca. The liver, with its gall bladder, opens by a duct (the bile duct) into the anterior part of the intestine. The pancreas, which embraces the anterior part of the intestine, likewise opens into the latter by a short pancreatic duct. A well-differentiated pancreas is a gnathostome feature. but ammocoetes have diverticula which are probably homologous with a pancreas whereas myxinoids do not (Barrington, 1972, p. 166). (The pyloric gland of tunicates, as already mentioned, may be homologous with the pancreas.)

There are two coeloms – the pericardium and the general body coelom as in myxinoids and lampreys. They are connected together by a Y-shaped pericardio-peritoneal duct, single in front and paired behind, which lies dorsal to the oesophagus. This duct is a remnant of the original ontogenetic continuity between the two coeloms (Goodrich, 1918a). As mentioned already, there is likewise a pericardio-peritoneal opening in myxinoids and ammocoetes, but not in adult lampreys.

The muscles of *Squalus acanthias* are in many ways similar to those of lampreys. There are powerful somitic muscles, used in swimming, that extend from the posterior face of the skull to the end of the tail. Unlike myxinoids or lampreys the somitic muscles are divided into a dorsal and ventral group (epaxial and hypaxial) by a horizontal septum which is a gnathostome feature (fig. 5.46). This horizontal septum is coplanar with the vertebral column, carries the lateral-line nerve to the trunk lateral line, and also carries the more proximal parts of the spinal nerves.

The visceral muscles, associated with the branchial arches, hyoidean arch and jaws, differ from those of lampreys and myxinoids by being external, not internal, to the somitic muscles. The fin or pterygial muscles have already been mentioned with the skeleton (fig. 5.46). The oculomotor muscles, as is well known, number six pairs homologous with those of lampreys (fig. 5.28a). The main difference from lampreys is that the oblique muscles are superior and inferior, rather than posterior and anterior, and are both sited in front of the rectus muscles.

The excretory kidney of the dogfish is a mesonephros, for the pronephros does not function in excretion at any stage of the life. This is unlike lampreys, where it functions in the larva, or myxinoids, where it functions throughout life. The blood of the dogfish is isotonic with sea water, like that of myxinoids but unlike lampreys. However, this is probably not a homologous resemblance with myxinoids. For the NaCl content of dogfish blood is only about half that of sea-water, the rest of the osmotic pressure being caused mainly by urea. Indeed, there is a region of the kidney tubules which returns urea to the blood after the glomerulus has extracted it. This suggests that *Squalus acanthias*, like all other gnathostomes and lampreys, had a freshwater stage in its ancestry and has re-adapted to sea water by increasing the urea in its blood. This means that, unlike echinoderms, acraniates, tunicates and myxinoids, its isosmoticity with sea water is probably advanced, not primitive (Young, 1981, p. 136).

The mesonephros is divided into two parts (fig. 5.47). A posterior portion, the opisthonephros, functions in excretion in both male and female, whereas an anterior portion is vestigial in the female but transports sperm in the male. There is a pair of ovaries in the female, hanging from the dorsal mid-line of the coelom, and paired oviducts. These are derived from the Wolffian ducts which arise embryologically in connection with the anterior, non-excretory mesonephros. The oviducts hang from the dorsal wall of the coelom, meet anteriorly to open by a common opening into the coelomic cavity and open separately into the cloaca. These posterior parts of the oviducts are enlarged to form uteri in which foetuses are nourished by placentas. Ova have to pass through part of the coelom to reach the median opening of the paired oviducts. The testes are paired and hang from the dorsal wall of the anterior part of the general coelom. Sperm directly enter the Wolffian duct of the non-excretory kidney and pass back to the cloaca. Unlike agnathans, the abdominal pores of dogfishes do not release the eggs or sperm.

The circulatory system is fundamentally like that of lampreys. However, there are no sinuses round the gills. As with lampreys, but unlike myxinoids, the afferent branchial arteries pass from the ventral aorta to the gill bars, not to the gill chambers.

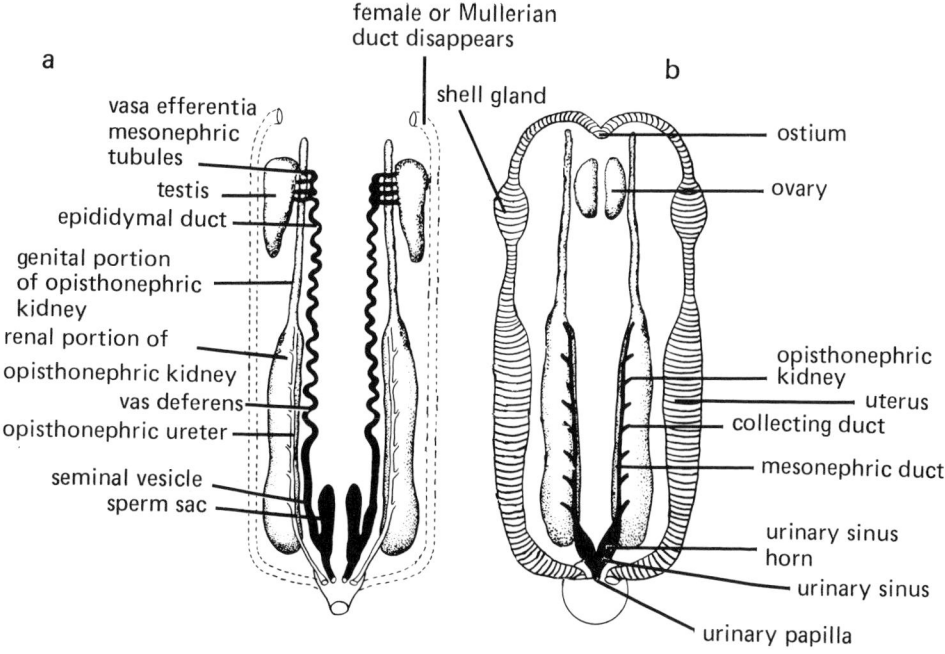

5.47. Squalus. *Diagram of the urinogenital system of a) male; b) female; from Nelsen (1953, fig. 351, a-3, a-4).*

As to the nervous system, once again the differences from the lamprey condition are rather slight. The dorsal and ventral roots of the spinal nerves join a short way from the dorsal nerve cord as in myxinoids, and unlike lampreys, but, as already noted, this is not a homologous resemblance to myxinoids (fig. 5.17c). A Reissner's fibre has been reported in *Mustelus canis*, which is a close relative of *Squalus acanthis* (Sargent, 1904).

The nerves of the hypoglossal complex run behind the rearmost gill slit (fig. 5.48) before turning forward again to supply the muscles of the ventral part of the gill bars. Unlike lampreys the most anterior roots supplying the hypoglossal nerve emerge from the skull (fig. 5.48). This suggests that several vertebrae have been incorporated in the gnathostome skull but not in that of lampreys. The vagus and glossopharyngeal have separate roots, which is unlike both myxinoids and lampreys. The cerebellum is much better developed than in a lamprey, perhaps in connection with the paired fins. There is no chorioid body developed from the roof of the mid brain (though lampreys and *Squalus* both have a chorioid body developed in the dorsal wall of the hind brain). There is only one pineal eye, for the parapineal is lacking. There is a thin touch-sensory nerve, the terminalis, running from the telencephalon to the olfactory epithelium (fig. 5.48). This nerve has not been found in agnathans.

As concerns the sense organs, the olfactory organ is paired (fig. 5.48), not median with partial pairing as in a lamprey. The eyes and oculo-motor muscles are like those of a lamprey in most respects, except for certain details, already mentioned. There is no cornealis muscle; instead accommodation is by contraction of muscles in the ciliary body that supports the lens (Walls, 1942, p. 260). The iris is muscular, and contractile according to the amount of light (Walls, 1942, p. 159). The ears have three semicircular canals (fig. 5.48), instead of two as in lampreys, for a horizontal canal is added. The lateral line consists of isolated neuromasts, as in lampreys, and also of extensive canals (Tester & Kendall in Cahn, 1967). In addition, certain lateral-line organs, the ampullae of Lorenzini are specialized to detect electricity (Dijkgraaf, 1967).

This account of myxinoids, lampreys and primitive gnathostomes will give a background for discussing how the three groups are related to each other.

The myxinoids, lampreys and gnathostomes represent a classic three-taxon problem in phylogeny, with three conceivable solutions, as shown in fig. 5.49. Which one of these cladograms is most likely to be correct? Løvtrup (1977), Janvier (1980, 1981b) and Hardisty (1979, 1982) have recently discussed this question and concluded that fig. 5.49a is more probable than the other two figures. In terms of the three forms that I have described above this

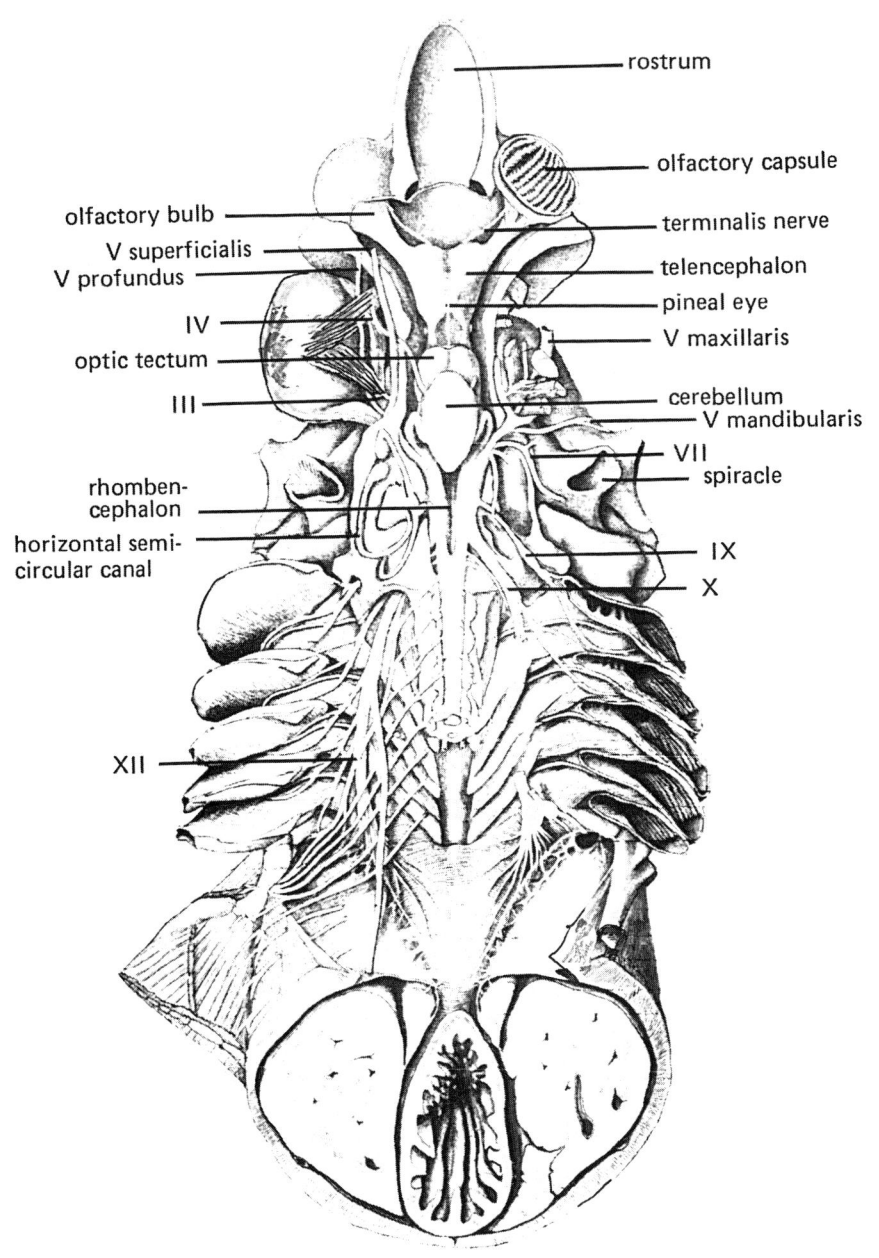

5.48. Squalus. Dorsal aspect of brain, cranial nerves and other organs of the head in dissection; from Marinelli & Strenger (1959, fig. 192a, p. 252).

would mean that *Lampetra* shared with *Acanthias* a number of advanced features which myxinoids do not have and which are also absent in tunicates, amphioxus and fossil mitrates suggesting that these features probably evolved in the segment **S** of the cladogram in 5.49a.

Such features include:

1. Motile eyes each with a lens and six pairs of eye muscles innervated, in almost the same manner, by cranial nerves III, IV and VI. All these features are absent in myxinoids and in tunicate tadpoles (which have a non-

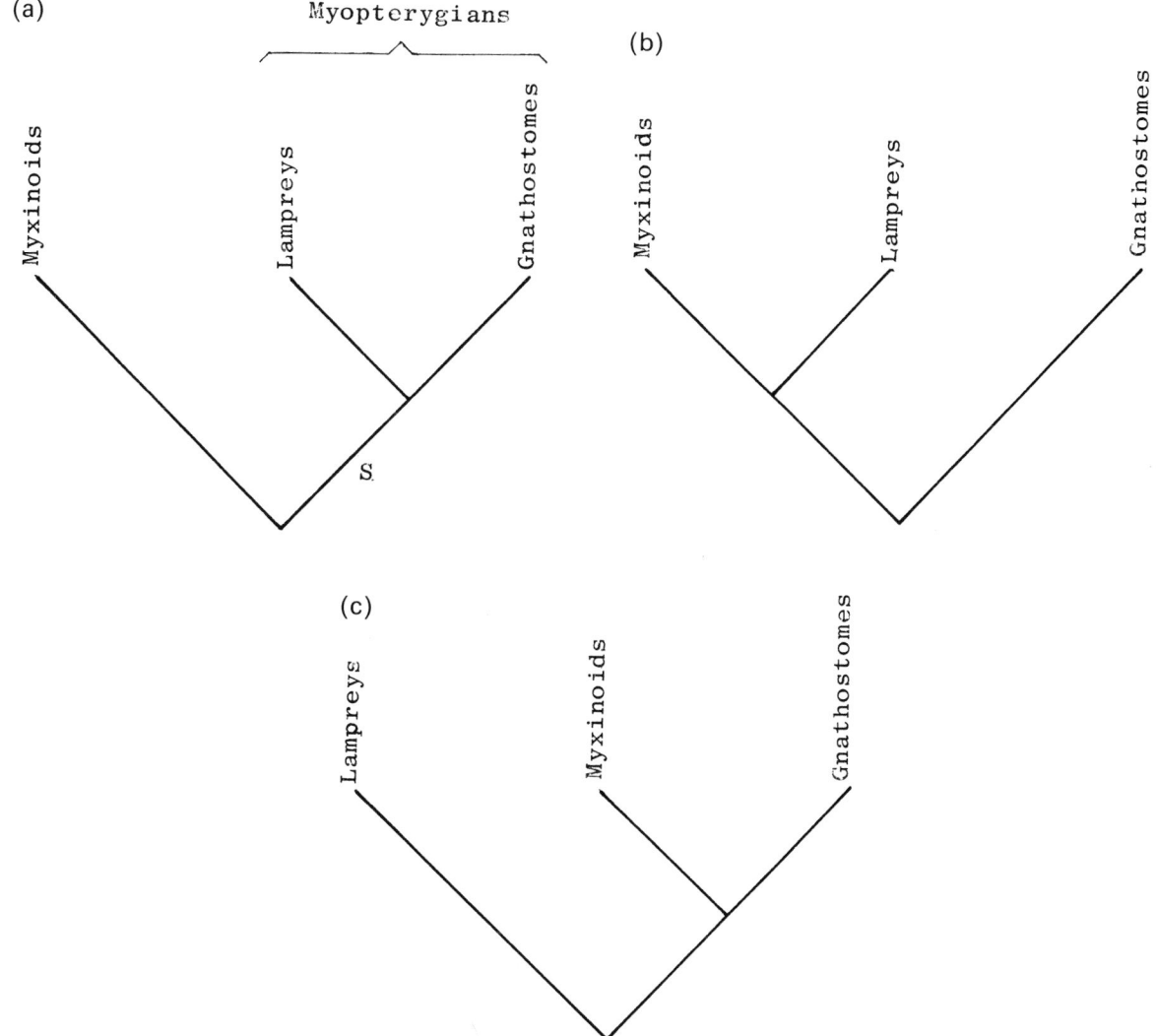

5.49. *Three alternative cladograms for the myxinoids, lampreys and gnathostomes.* S *indicates the segment in which the myopterygian synapomorphies arose.*

motile eye) and in mitrates, which also had non-motile eyes. Amphioxus has no eyes homologous with those of vertebrates, but this is likely to be a secondary condition.

2. A reduced salt-content in the blood, as compared with sea water. This is maintained by secreting from the glomerulus a solution isosmotic with, and chemically similar to, blood plasma and modifying this filtrate by extracting needed constituents as it flows down the kidney tubule – such as salt in fresh water and water in salt water. Lampreys and gnathostomes have probably all descended from an ancestor that was adapted to life in fresh-water,

whereas there is no sign that myxinoids have descended from such an ancestor. Comparison with tunicates and amphioxus suggests that the myxinoid condition of living isosmotically in sea water is primitive.

3. The pronephros has no excretory function in the adult of lampreys and gnathostomes. This is by contrast with larval lampreys and adult myxinoids, in both of which the pronephros is functional. Following Haeckel's law, the non-functional pronephros is likely to be an advanced condition.

4. The heart beat of gnathostomes and lampreys is

neurogenic, being controlled by a branch of the vagus. The heart beat of myxinoids, on the other hand, is myogenic as in tunicates and early chick embryos.

5. The adenohypophysis is ectodermal in origin in lampreys and gnathostomes, for it arises from Rathke's pouch in the buccal cavity. It is endodermal in origin in myxinoids, as it probably was in mitrates also to judge by its position in the fossils, and as its likely homologue – the complex of wheel organ, Hatschek's pit and head coelom – is in amphioxus. This complex arises from the anterior gut diverticula of Hatschek and is probably equivalent to the Seessel's pouch *plus* the premandibular somite of vertebrates. (An anomaly, however, is the origin of the neural gland of tunicates from the brain anlage.) The histological structure of the adenohypophysis is more complex in gnathostomes and lampreys than it is in myxinoids, and this may also be an advanced feature.

6. Vertebral arcualia are present in lampreys and gnathostomes, but not in myxinoids.

7. The fin skeleton has pterygial muscles attached to it proximally in lampreys and gnathostomes, but not at all in myxinoids.

8. The absence of Müller cells from the spinal cord (unlike myxinoids and the Rohde cells of amphioxus), such cells being retained only in the medulla oblongata.

These features, especially those of the eye and of the kidney, strongly support Løvtrup's and Janvier's contention that the first of the cladograms (fig. 5.49a) is correct, so that the myxinoids are probably the living sister group of the [lampreys + gnathostomes]. Janvier (1980) has given this latter group the name of myopterygians, in allusion to the muscles associated with the fins. Indeed it is almost inconceivable that the common special eye and kidney features of myopterygians had evolved in parallel in gnathostomes and lampreys On the other hand it *is* conceivable they once existed in the ancestors of all living vertebrates, but that myxinoids have lost them. This is unlikely, however, at least for the eye muscles, eye-muscle nerves and lens, for the eyes of the myxinoid *Eptatretus* seem to be fully functional although non-motile. Løvtrup and Janvier have brought many biochemical arguments to support this cladogram, but I do not feel competent to discuss them.

The cladogram just proposed would fail if it could be shown that advanced homologous features are shared between myxinoids and gnathostomes, or myxinoids and lampreys, which the remaining group primitively lacked. As concerns myxinoids and gnathostomes two possible such features must be considered. The first is the absence of the ammocoete larva i.e. the absence of a stage in the life history that feeds by means of an endostylar mucous filter in the pharynx. For outcrop comparison with tunicates and amphioxus shows that this feature of lampreys is primitive

for vertebrates or indeed for chordates, and its absence in myxinoids and gnathostomes must certainly be advanced. There is no good reason to think, however, that the lack of an ammocoete is homologous in myxinoid and gnathostomes. Its loss in myxinoids could well be due to the large-yolked egg of these animals; and its loss in gnathostomes could result from the efficiency of jaws as a food-collecting apparatus. The second conceivable synapomorphy between myxinoids and gnathostomes is the fact that the dorsal and ventral root of the spinal nerves fuse, which they do not do in lampreys. As already pointed out, however, Goodrich (1937) showed that the fusion is fundamentally different in the two groups and cannot be homologous.

As concerns myxinoids and lampreys, two features could be used to suggest that they constitute a monophyletic group: the tongue complex and the nature of the gills. In the tongue complex of lampreys, the apical cartilage, with its horny teeth is pulled downwards and forwards in a rasping action by contraction of muscles that form the core of a pear-shaped 'great muscular body'. Moreover, these muscles are relaxed by the contraction of the outer layer of this body (Balabai, 1956). In myxinoids a similarly constructed great muscular body pulls the dental cartilage forward, together with the horny teeth that overlie the cartilage, and it relaxes in the same way as in lampreys. This is a fairly complex resemblance which gnathostomes do not share and which is lacking in tunicates and amphioxus. Perhaps it once existed in the ancestry of gnathostomes but has been lost.

As to the gills, Schaeffer & Thomson (1980) have argued that the endobranchiate gills of myxinoids and lampreys can be regarded as a synapomorphy of agnathans, since, in their view, the ectobranchiate gill of gnathostomes cannot have evolved from endobranchiate gills, nor *vice versa*. I do not personally find this argument convincing. The endobranchiate condition, like the tongue system, could be a primitive state that gnathostomes have lost.

As regards classification within the vertebrates, therefore, the gnathostomes and lampreys probably form a monophyletic group of which the myxinoids are the living sister group. This monophyletic group, is characterized especially by: lensed eyes with oculomotor muscles and nerves, a kidney capable of controlling the osmotic pressure of the blood by means of descending and ascending tubules, a fin skeleton actuated by proximal pterygial muscle, vertebrae, neurogenic contraction of the heart, and an adenohypophysis arising from the buccal cavity and therefore ectodermal. It can conveniently be referred to as the Myopterygii.

As regards the history of the divergence of the vertebrates it is significant that a macrophagous adult exists in all living vertebrates, whether myxinoids, lampreys or gnathostomes. This is in striking contrast with acraniates or tunicates, where the adult is microphagous and feeds by means of a pharyngeal mucous filter. Most probably, the predatory adult phase is homologous in all three groups of vertebrates, or, in other words, the thyroid gland (as distinct from the endostyle) is homologous. Probably the latest common ancestor of all living vertebrates had an ammocoete larva in its ontogeny followed by a predatory adult. Moreover, this adult probably fed by means of rasping horny teeth actuated by a great muscular body, as in living myxinoids and lampreys. Later the ammocoete stage was lost independently in the stem groups of myxinoids and gnathostomes. And in the stem group of gnathostomes articulated cartilaginous branchial bars arose with gills external to them, and the most anterior of these bars, in front of the foremost pair of gill slits, became biting jaws with dentinal rather than horny teeth.

Chapter 6   The embryology of vertebrates and the head problem

In this chapter I shall describe the early embryology of the three great groups of vertebrates discussed in the previous chapter. I shall use *Squalus acanthias* to illustrate the gnathostomes, *Lampetra ffluviatilis* the lampreys, and *Eptatretus stouti* the myxinoids. I shall then discuss the various, and conflicting, phylogenetic conclusions that have been drawn from vertebrate embryology and try to decide between them. My viewpoint has been greatly influenced by fossils, especially mitrates, but in this chapter I shall not consider the fossil evidence.

As to the gnathostomes, the embryology of *Squalus acanthias* has been worked out in detail, except for the very earliest stages. Important works are: Scammon (1911), De Beer (1922), Neal (1898, 1914), Borcéa (1906), Holmgren (1940), Nelsen (1953) and Bjerring (1967).

The egg of *Squalus acanthias* is a huge yolky cell with a small circle of yolk-free cytoplasm, containing the fertilized nucleus, at one end. This clear cytoplasm divides producing a blastoderm. The latter contains a cavity, the segmentation cavity, which is roofed over by a skin of cells several layers thick and floored by a single layer of cells, which, however, are not entirely delimited, for their cytoplasm is continuous with the yolk of the egg. The blastoderm begins to thicken round the edge and to raise itself slightly above the general surface (the illustrations in fig. 6.1 and 6.2a refer to the Common Dogfish *Scyllium canicula*).

Gastrulation (fig. 6.2a–g) begins when one part of the thickened edge – known as the dorsal lip of the blastopore and corresponding to the future posterior end of the animal – becomes more marked, overhangs and begins to fold inwards. Areas which had been part of the external surface of the blastoderm pass rearwards in procession, roll downwards over the lip of the blastopore, and then extend forwards inside the embryo, obliterating the old segmentation cavity. In this way the archenteron is formed. Its roof consists mostly of endoderm. Posteriorly the noto-

chord overlies the endoderm, as a separate rod, and at first slopes downwards to join and form part of the roof of the archenteron anteriorly, but soon becomes separated completely by endoderm from the archenteric cavity. In front of the notochord the roof of the archenteron is formed by the mesodermal prechordal plate. On either side of the notochord, above the endoderm and below the ectoderm, a rod of mesoderm extends forwards from the blastopore lip to near the anterior end of the developing embryo, passing anteriorly into the roof of the archenteron.

During gastrulation the true embryo begins to be visible externally as a ridge in the blastoderm extending forwards from the blastopore lip (fig. 6.2a). At the same time the dorsal ectodermal covering of this ridge folds downwards, along the mid-line of the embryo, both above and in front of the notochord. This downfolded part is the neural plate from which the definitive dorsal nerve chord, brain and motor nerves will arise. The downfold becomes deeper and V-shaped in section and then its right and left dorsal edges curl towards each other, touch, and begin to fuse (fig. 6.3a–d). The fusion begins in the middle of the embryo's length and extends forwards and rearwards, to form the hollow tube of the dorsal nerve cord and brain. Anteriorly this tube for long remains open to the outside by the neuropore, and posteriorly it opens by a down-turned neurenteric canal into the archenteric cavity. The anterior and posterior openings will both finally close.

The dorsal parts of the columns of mesoderm on either side of the notochord come to be divided into somites. These first appear in the middle of the embryo's length and then individualize anteriorly and posteriorly. Unlike lampreys or amphioxus the divisions between the somites are from the first incomplete, for they never extend to the ventral edge of the mesoderm but only divide a dorsal strip of it (fig. 6.4). Below this, the mesoderm remains as undivided lateral plate. Thus the lateral plate does not arise by fusion of the ventral parts of somites, but is from the

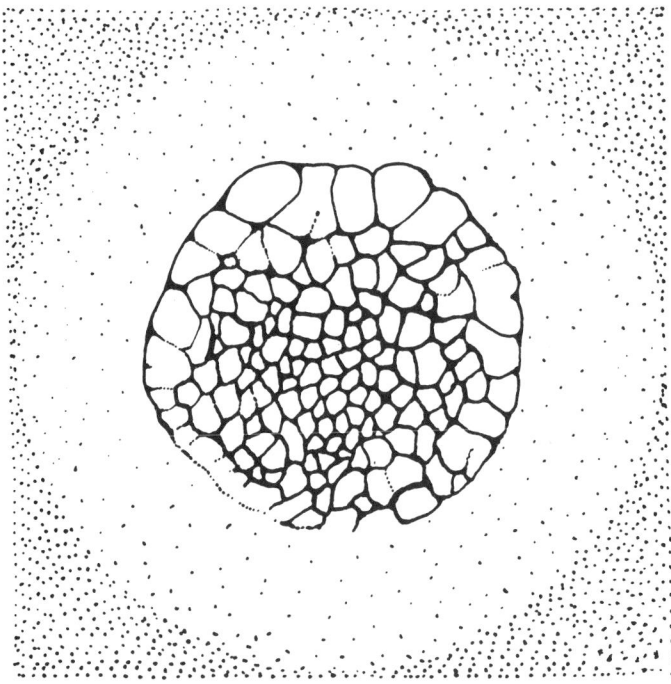

6.1. *The common dogfish* Scyllium canicula. *A blastoderm with 145 cells visible; from Nelsen (1953, fig. 158D, p. 318).*

beginning continuous. The delimitation of somites spreads forward along the most dorsal strip of the mesoderm until a complete series is present. The most anterior is the mandibular somite (2nd somite in the standard nomenclature), which is situated on either side of the front end of the notochord; behind this comes the 3rd or hyoidean somite; the 4th or 1st branchial somite; the 5th, 6th, 7th and 8th, also known as the vagal or 2nd to 4th branchial somites; and then the post-pharyngeal somites of the trunk and tail (figs 6.4, 6.5, 6.6). The names of the somites allude to the visceral bar that will arise beneath them in ontogeny. (The 1st somite of standard nomenclature is the premandibular somite which is at first unpaired and not continuous with the other conventional somites.)

The prechordal plate (figs 6.2d, 6.5b) is regarded as mesoderm like the paired somites and lateral plate, but forms the dorsal roof of the most anterior part of the archenteron (pre-oral gut or Seessel's pouch). It is in front of the front end of the notochord and gives rise in *Squalus* to two pairs of pouches – Platt's vesicles anteriorly and the premandibular or 1st somite more posteriorly (figs 6.6a, b, c, 6.7). Platt's vesicles are found only in sharks and rays and should probably be seen as specialized portions of the premandibular somite of other vertebrates. The right and

left halves of the premandibular somites long remain united by a bridge of tissue near the front end of the notochord. This bridge or commissure arises from prechordal plate anterior to the notochord, though at a later stage the front end of the notochord hooks over it (fig. 6.7).

As time goes on the embryo lifts up more and more from the yolk, the head extends forwards in front of the connection with the yolk (yolk canal) and bends downwards, the neuropore closes, and neural crest begins to form (fig. 6.5a). This arises from the dorsal mid-line of the neural canal and flows ventrally and sideways between the developing central nervous system and the ectoderm. It is first formed in the head region above the mandibular and hyoidean somites. Soon after, it starts to form over the branchial somites (fig. 6.8), behind the prospective ear region, and gradually its process of formation spreads rearwards along the future hind brain and dorsal nerve cord.

At the same time the optic vesicles are forming as lateral, upwardly directed, outgrowths near the front end of the brain (figs 6.5b, 6.6a, 6.8) and the auditory capsules arise as inpouchings of ectoderm (dorso-lateral placodes) which press downwards on the hyoidean and 1st branchial

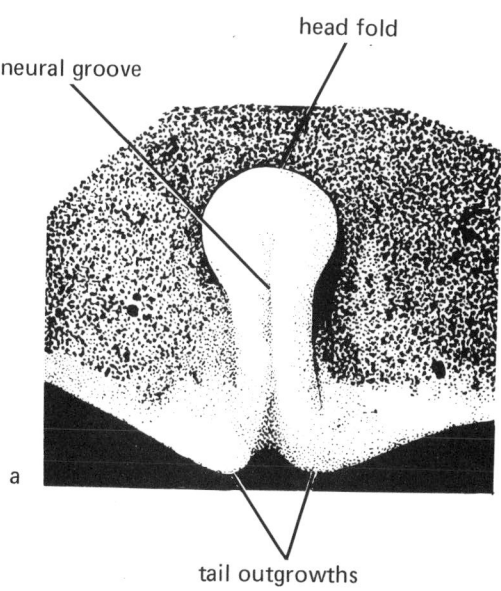

neural groove

head fold

tail outgrowths

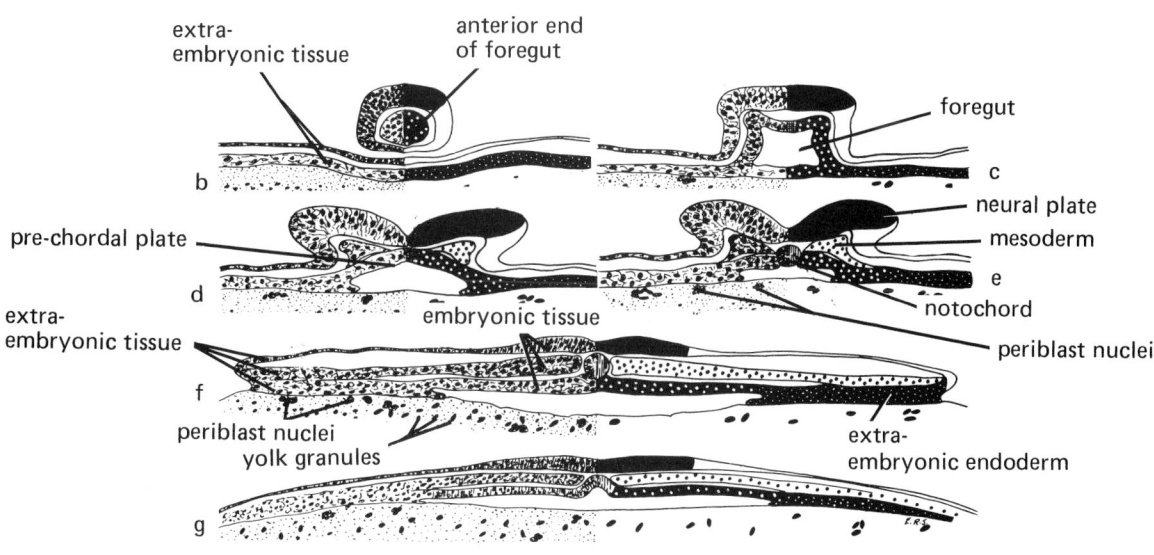

extra-embryonic tissue

anterior end of foregut

foregut

b

c

pre-chordal plate

neural plate

mesoderm

d

e

extra-embryonic tissue

embryonic tissue

notochord

periblast nuclei

f

periblast nuclei
yolk granules

extra-embryonic endoderm

g

6.2. a) Early embryo of Scyllium canicula *in dorsal aspect with (b–g) corresponding transverse sections through* Squalus acanthias; *from Nelsen (1953, fig. 212G and 213 H–M, pp. 442, 443).*

somites (fig. 6.6a). The auditory capsules thus arise in exactly the same position as in lampreys (figs 6.16, 6.17) and dorsal to the same somites and to the same gill bars.

The visceral pouches, which will later become the spiracle and gill slits, arise at about this time. The most anterior pair of visceral pouches, which is the first to differ-entiate and also the first to open, will give rise to the spiracles. It arises between the lateral-plate portions of the mandibular (2nd) and hyoidean (3rd) segments, ventral to the appropriate somites (figs 6.9a, 6.6 and 6.8). The second pair of visceral pouches from the front is also second to arise and will produce the 1st branchial slits. Its position,

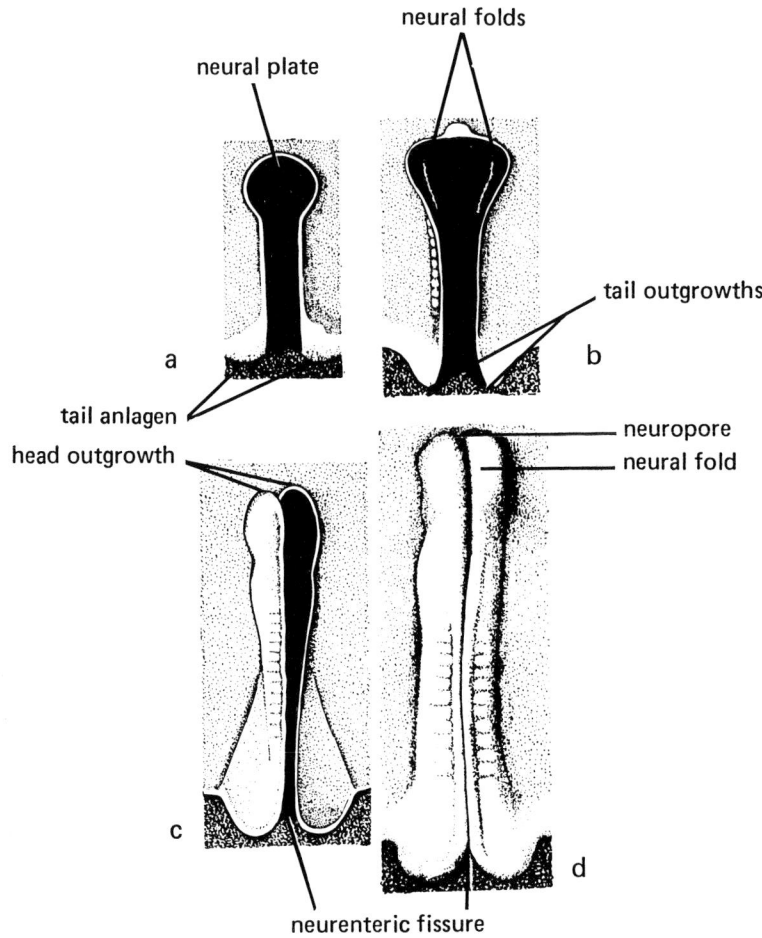

6.3. *a–d*) Squalus acanthias. *Successive stages of neurulation in dorsal aspect; from Nelsen (1953, fig. 229, p. 475).*

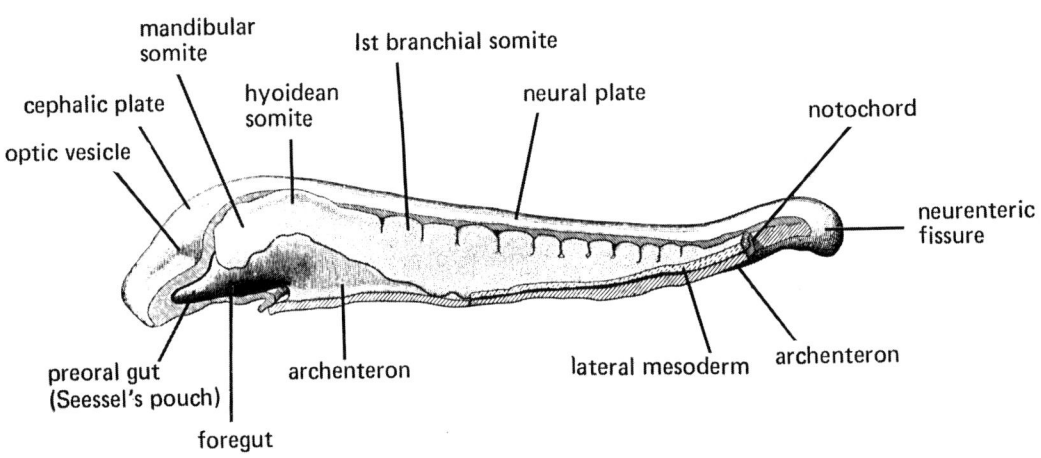

6.4. Squalus. *Early differentiation of somites; from Scammon (1911, fig. 3, p. 45).*

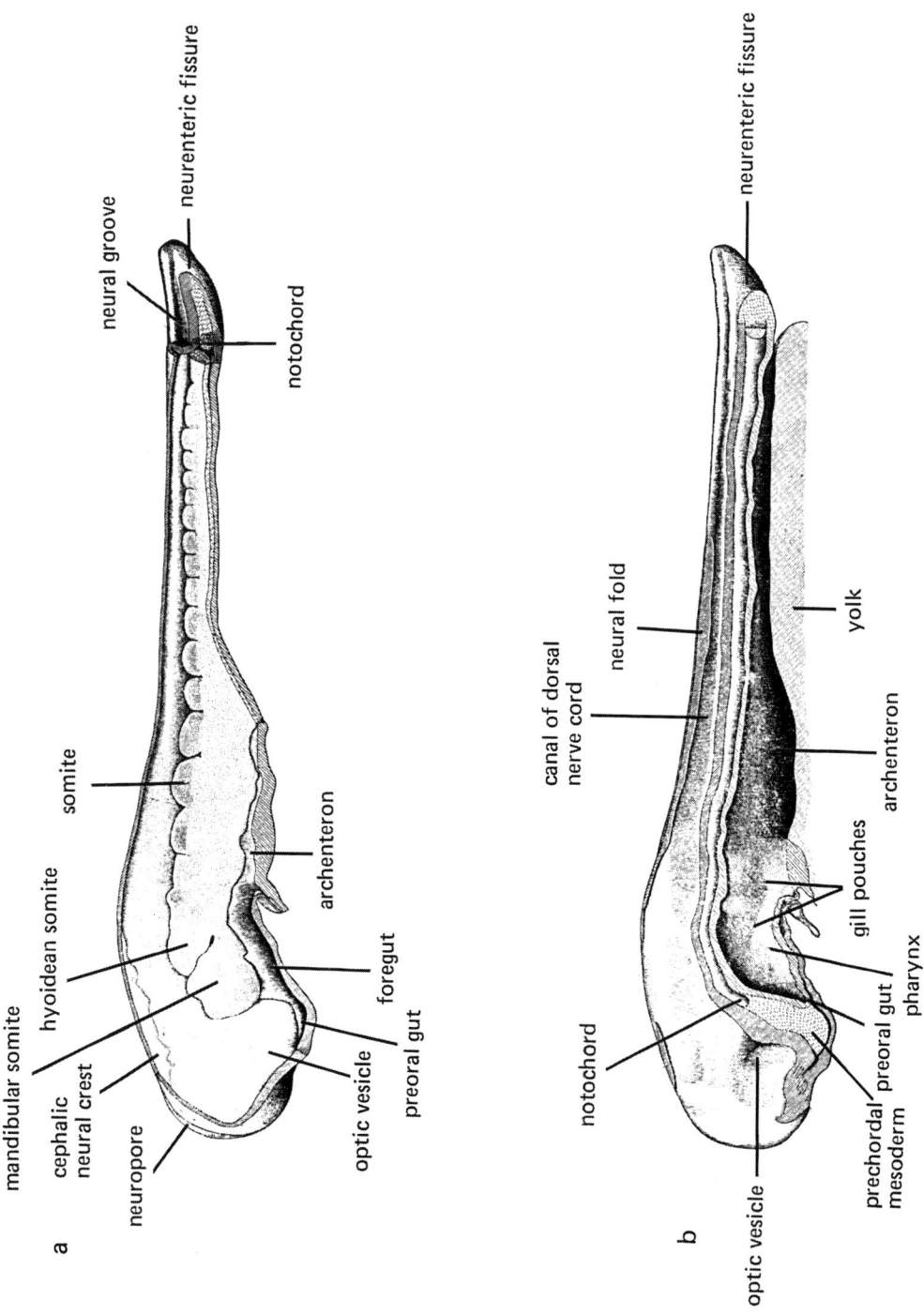

6.5. Squalus. *Separation of mandibular from hyoidean somites and respective positions of premandibular mesoderm, notochord and pre-oral gut (Seessel's pouch). a) Left aspect of embryo with ectoderm removed; b) sagittal section of same embryo; from Scammon (1911, figs 5 & 6, p. 47).*

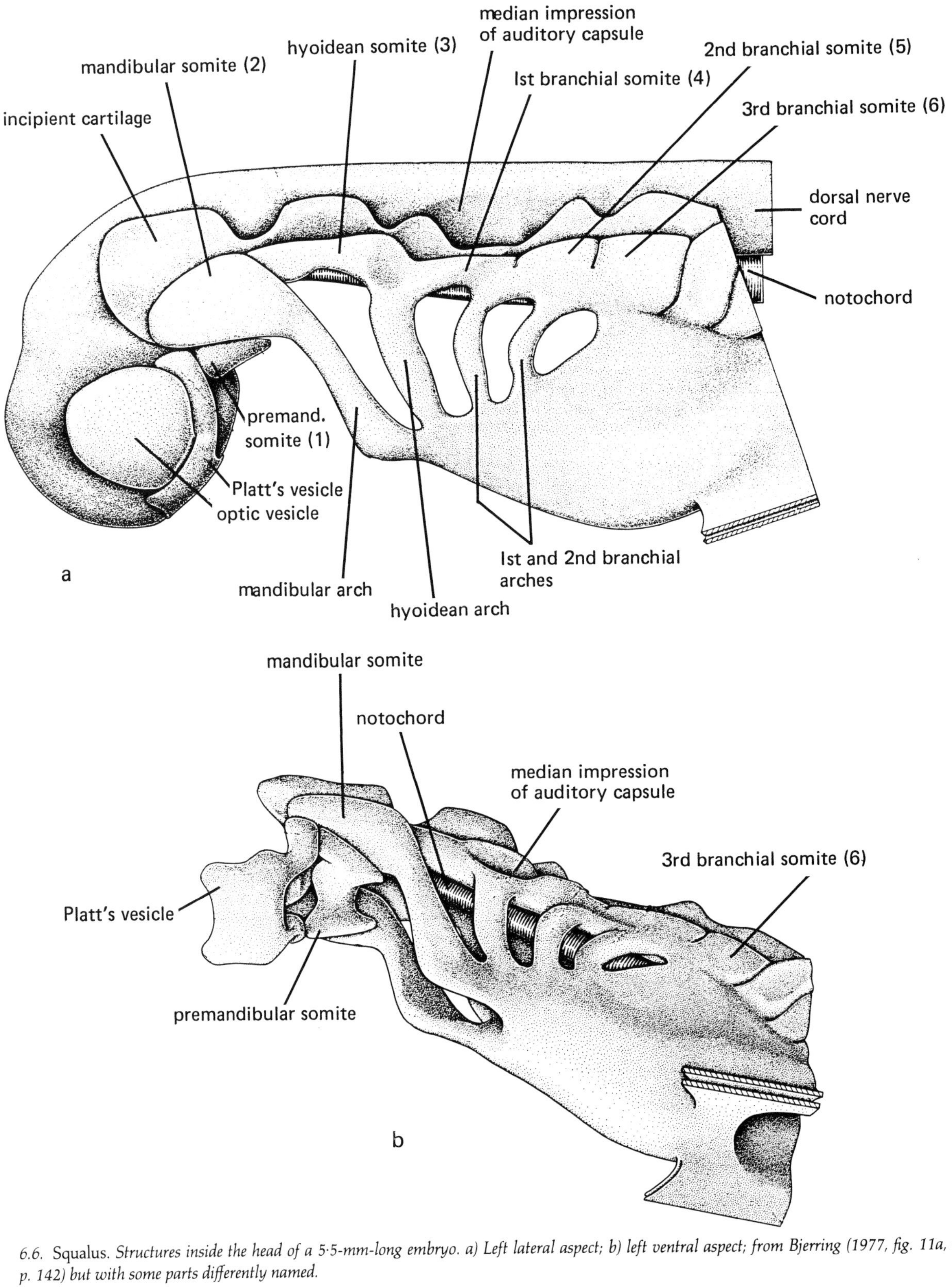

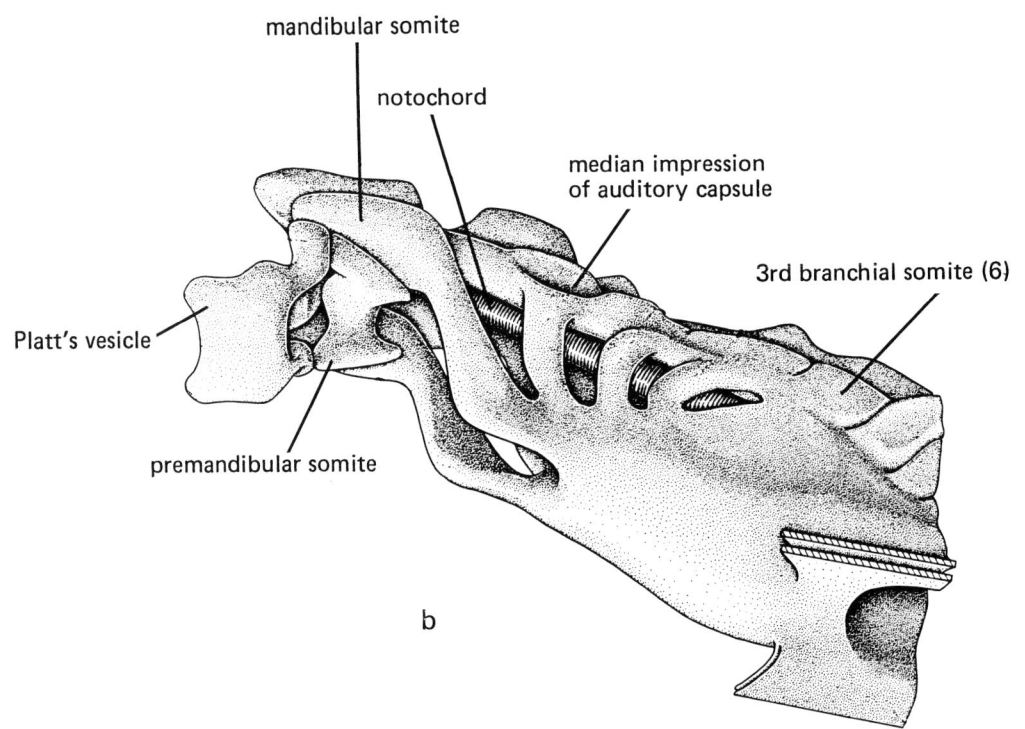

incipient cartilage

mandibular somite (2)

hyoidean somite (3)

median impression
of auditory capsule

Ist branchial somite (4)

2nd branchial somite (5)

3rd branchial somite (6)

dorsal nerve
cord

notochord

premand.
somite (1)

Platt's vesicle
optic vesicle

a

mandibular arch

hyoidean arch

Ist and 2nd branchial
arches

mandibular somite

notochord

median impression
of auditory capsule

3rd branchial somite (6)

Platt's vesicle

premandibular somite

b

6.6. Squalus. *Structures inside the head of a 5·5-mm-long embryo. a) Left lateral aspect; b) left ventral aspect; from Bjerring (1977, fig. 11a, c, p. 142) but with some parts differently named.*

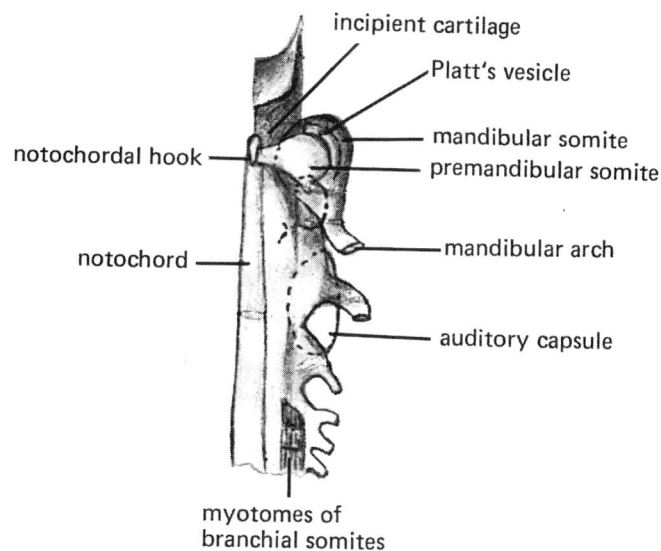

incipient cartilage

Platt's vesicle

notochordal hook

mandibular somite

premandibular somite

notochord

mandibular arch

auditory capsule

myotomes of
branchial somites

6.7. Squalus. *Structures inside the left dorsal part of the head of a 12-mm-long-embryo in ventral aspect; from Holmgren (1940, fig. 22).*

according to de Beer (1922), lies between the lateral plates of the hyoidean and the 1st branchial segment (the 4th somite of standard nomenclature). Four more pairs of visceral pouches arise behind these; they give rise to branchial slits 2 to 5 and lie between the lateral plate portion of segments 4 and 8 respectively (fig. 6.6a) (de Beer, 1922). The first post-branchial segment is therefore segment 8. There has been argument, which I discuss later, about whether particular visceral bars really correspond to particular somites.

The blood vascular system of *Squalus acanthias*, as first seen in an embryo about 4·8 mm in length, consists of: (1) a pair of vitelline veins draining the yolk and ventral to the gut, converging forward into (2) a median vessel, ventral to the gut, being the future heart and ventral aorta, which runs forward and splits into (3) a pair of vessels (the first aortic arch) which run up on either side of the gut, in front of the anlagen of the spiracles, to join dorsally into (4) a median vessel dorsal to the gut, the dorsal aorta. At a later stage (fig. 6.8) the heart arises in the posterior portion of the median ventral vessel where a lateral plate from either side embraces this vessel in the mid-line. The layer of lateral plate in contact with the median vessel becomes muscular and gives rise to the heart wall, whereas the outer layer of the lateral plate forms the pericardium.

The oculomotor muscles of *Squalus acanthias* arise from the premandibular, mandibular and hyoidean somites (fig. 6.9a, b). The premandibular somite gives rise to the inferior oblique and to the internal, superior and inferior

recti; the mandibular somite gives rise to the superior oblique; and the hyoidean somite gives the external rectus. These groups of muscles come to be innervated respectively by the oculomotor nerve (III), the trochlear nerve (IV) and the abducens nerve (VI). Neal (1914) showed in *Squalus acanthias* that these nerves arise by outgrowth of the neurons of the ventral motor column of the mid- and hind-brain, which is much like the origin of the ventral roots of spinal nerves.

The neural crest in *Squalus acanthias* flows ventralward and at the posterior boundary of the pharynx displays a remarkable difference in behaviour (fig. 6.10). In front of the middle of somite 6, that is to say, it flows downwards outside the somites, and where it comes in contact with them, breaks them down into mesenchyme. Behind that point, however, it flows downwards on the median surface of the somites without destroying them. This difference in behaviour corresponds to a remarkable difference in the fate of the neural crest, for the more anterior crest helps give rise to the sensory and visceromotor cranial nerves (V, VII, VIII, IX, X) and to the cartilages of the visceral bars and much of the skull. Posterior to the middle of somite 6, on the other hand, the neural crest gives rise only to the dorsal nerve roots and ganglia of the hypoglossal and spinal nerves. The anti-segmentationist Romer (1972) has plausibly argued that this distinction indicates a fundamental division between head and trunk, corresponding to the division between head and tail in a tunicate tadpole. According to Holmgren (1940), neural

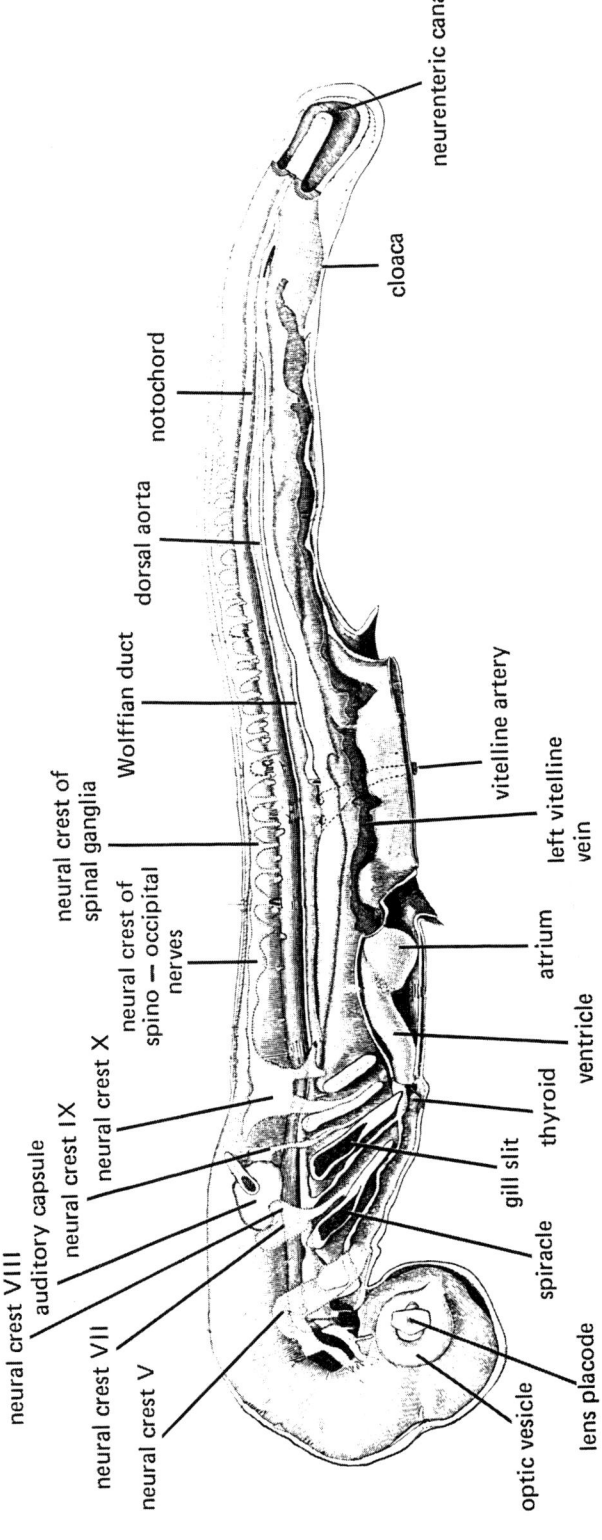

6.8. Squalus. Reconstruction of a 7·5-mm-long embryo; from Scammon (1911, fig. 10, p. 52).

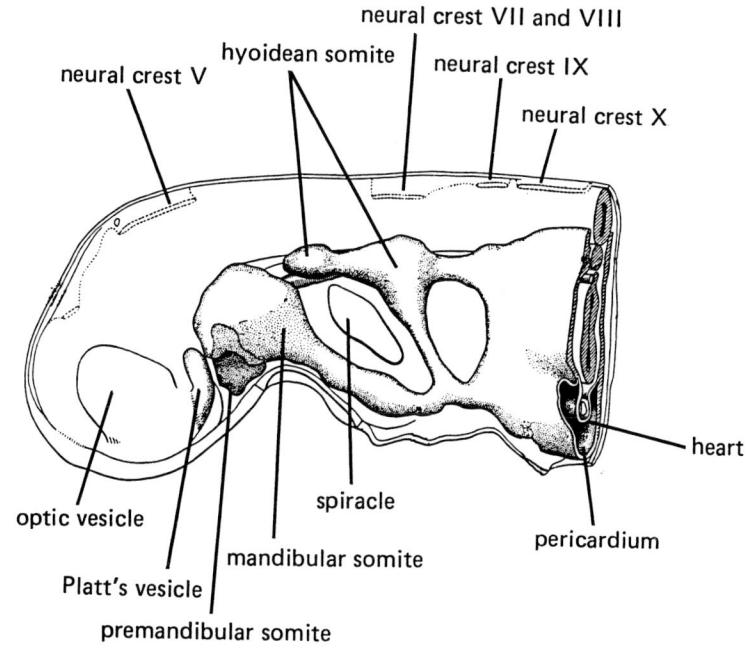

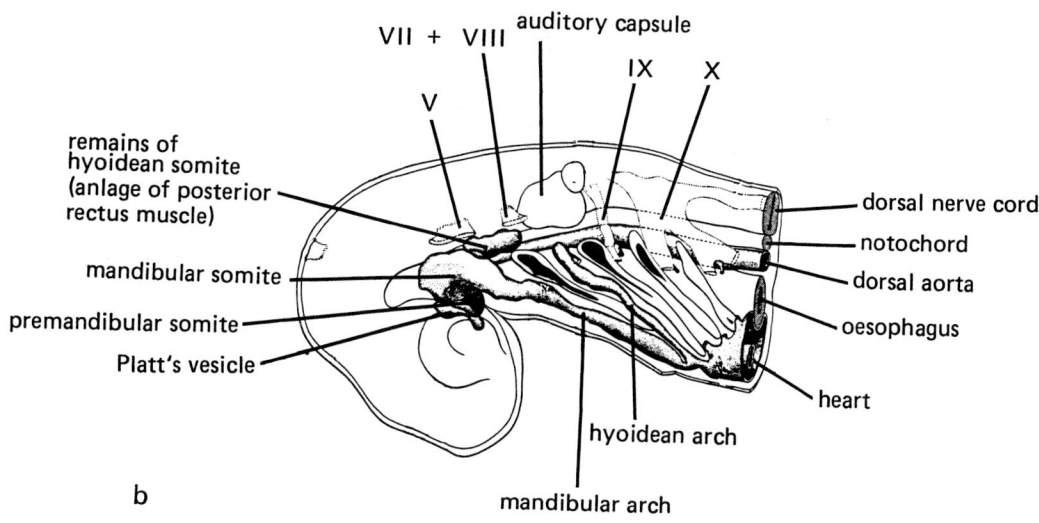

6.9. Squalus. *Successive stages in the origin of the oculomotor muscles. (a) 4·8-mm-long embryo; (b) 9-mm-long embryo; from Scammon (1911, figs 18, 20, pp. 65, 66).*

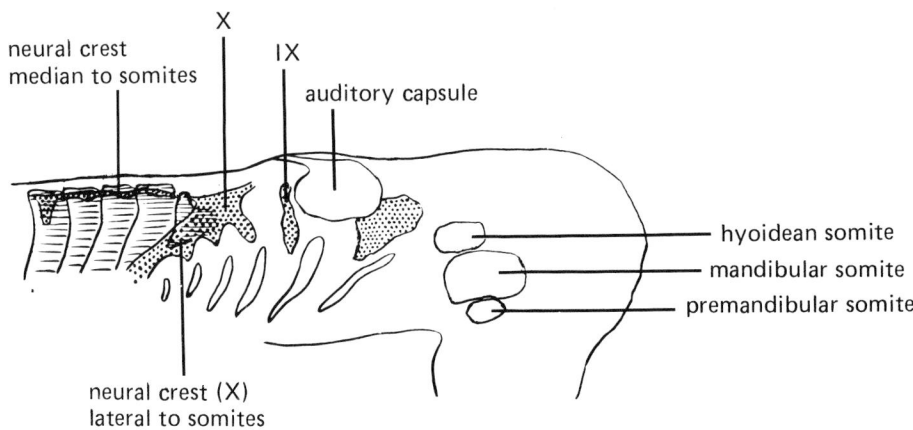

6.10. Squalus. *The difference in behaviour of neural crest in front of and behind somite 6; from De Beer (1922, text fig. 6, p. 463).*

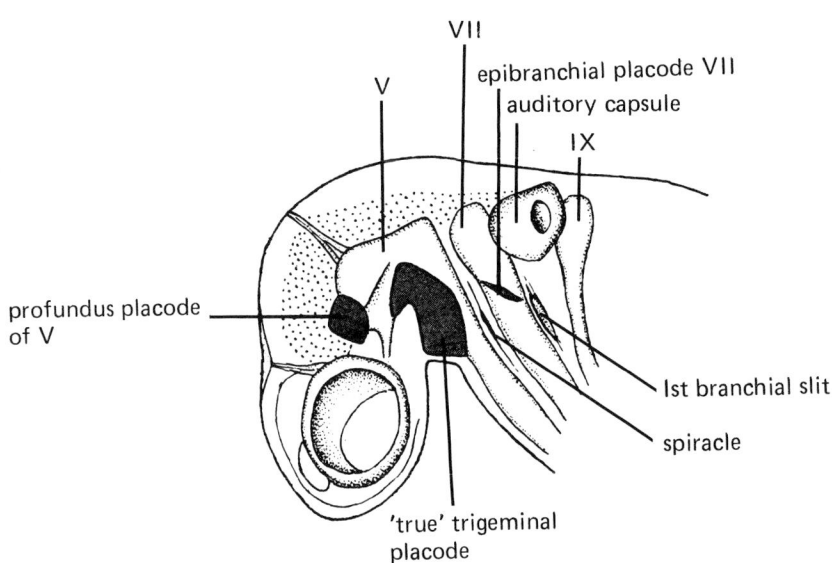

6.11. Squalus. *Reconstruction of a 7–8-mm-long embryo to show the placodes of the profundus branch of the trigeminal and of the non-profundus ('true') trigeminal; from Holmgren (1940, fig. 14, p. 64).*

crest in *Squalus acanthias* arises from placodes in the head region (fig. 6.11) as well as from the dorsal parts of the neural plate, and it gives rise to most of the skull anterior to the front end of the notochord (except for mesendodermal contributions from Platt's vesicle and the mandibular somites) and to the cartilages of the visceral bars. The rest of the skull, including the otic capsules and parachordal cartilages, arises from mesenchyme of somitic origin (of somites 2 and posteriorly). This part of the skull,

at least behind the auditory capsules, can therefore in the broad sense be seen as fused vertebrae, since the vertebrae likewise arise from mesenchyme of somitic origin.

The kidneys of *Squalus acanthias* arise as individual outpouchings (segmental tubules) from the lateral wall of the coelom which begin from the points where the somites pass ventralwards into lateral plate. These outpouchings grow laterally and rearwards; they fuse with each other; and the Wolffian ducts so produced extend rearwards to

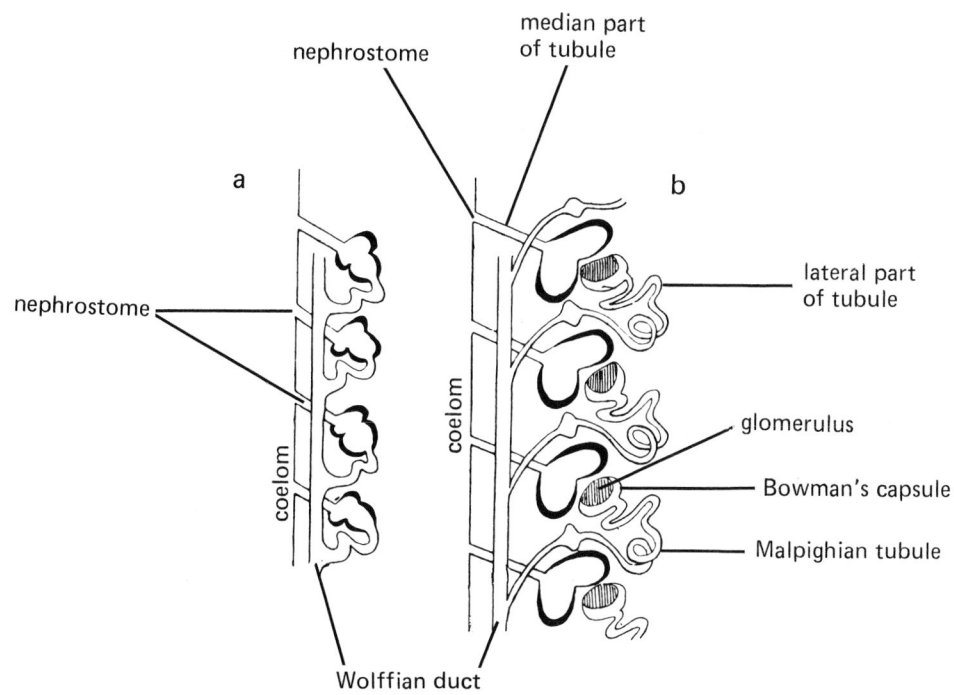

6.12. Squalus. Development of kidneys. a) Stage with continuous segmental tubules. b) Stage with tubules divided into a lateral portion ending in a Bowman's capsule and a median portion opening into the coelom; from Borcéa (1906, fig. 16, p. 272).

the future cloacal region (fig. 6.12a). Later each segmental tubule divides into two parts – a lateral part which opens into the Wolffian duct and has a Bowman's capsule and renal tubule at its closed end, and a more median part which opens by funnels (nephrostomes) as a series of segmental tubules, one per segment, into the coelom (fig. 6.12b).

As development proceeds the isthmus connecting the embryo with the yolk becomes more and more constricted. The posterior end of the gut pouches ventralwards to form the future cloaca, which is in contact with ectoderm but not yet open to the exterior. Later the intestine starts to twist, producing the dextral spiral of the spiral valve, and the liver and pancreas arise as separate evaginations (fig. 6.8).

A complete account of the development of *Squalus acanthias* can be got from the excellent work of Scammon (1911) supplemented in particular by Holmgren (1940). I have said enough here to provide a basis for my later discussion.

The embryology of *Lampetra fluviatilis* is well known. The earliest stages were described by Glaesner (1910), Weissenberg (1934) and Veit (1957). Later development,

with particular reference to head segmentation, was analysed by Koltzoff (1901) and his conclusions were confirmed and elaborated in a masterly paper by Damas (1944) on which the following account is largely based. Koltzoff's and Damas's conclusions are of fundamental importance for the whole of chordate comparative morphology.

The egg of *Lampetra*, unlike that of myxinoids, is small, soft-shelled and without much yolk, though a clear animal pole and a yolkier vegetal pole can readily be distinguished. The first cleavage is total, runs from animal to vegetal pole, and separates the egg into the equal and symmetrical halves. The second cleavage is also total, runs from pole to pole and is perpendicular to the first. The third cleavage is total and equatorial, separating four small clear micromeres from four large yolky macromeres (fig. 6.13a). Thereafter cleavage occurs more readily in the micromeres than the macromeres so that a blastula is formed (fig. 6.13b). This is a hollow ball of cells; an upper hemisphere of numerous micromeres is joined to a lower hemisphere of larger, less numerous, yolky macromeres and the cavity (blastocoel) of the blastula is located mostly in the upper hemisphere.

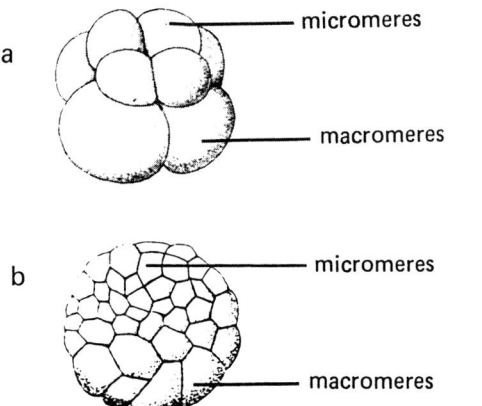

6.13. Lampetra fluviatilis. *a) 8-cell stage with four micromeres and four macromeres; b) lateral aspect of blastula with small non-yolky cells and larger yolky ones; from Glaesner (1910, pl. 11, figs 16, 28).*

Gastrulation begins when a group of cells situated in the vegetal hemisphere, near the junction with the animal hemisphere, begins to divide rapidly, forming an eminence. A fold appears just above this eminence and begins to lap downwards over it. The crevice beneath this fold is the blastopore. The fold extends sideways and also advances over the yolky hemisphere until finally the whole of the latter has disappeared inside the embryo. At the end of gastrulation the dorsal wall of the archenteron is formed of notochord, except anteriorly where it is formed of mesoderm of the prechordal plate (fig. 6.14). Left and right of the notochord and of the archenteron is the still unsegmented somitic mesoderm. Even before gastrulation

is complete the embryo is beginning to convert into a neurula, for the ectoderm above the notochord depresses to form the anlage of the dorsal nerve cord, and the ectoderm above the prechordal plate depresses to form the brain. The blastopore will become the definitive anus of the animal. The dorsal nerve cord and brain do not sink downwards as an invagination but as a solid rod of cells which later becomes hollow by splitting. The roofing-over of this rod by skin ectoderm proceeds, as in amphioxus or *Ciona*, from behind forwards so that a neuropore for a long time remains open anteriorly and a neurenteric canal, opening to the archenteron, remains posteriorly. Gastrulation and neurulation are very much as in amphioxus, except that there is somewhat more yolk. Likewise they are similar to the same processes in *Squalus acanthias* except that the latter has a very large quantity of yolk. At the end of gastrulation and neurulation the embryo of *Lampetra fluviatilis* is almost spherical except for the blastopore and a linear ridge which is the developing head. In the archenteron, endoderm soon grows inwards from left and right just beneath the notochord to form the roof of the archenteron, except anteriorly where the roof is still formed by prechordal plate. At the same time the somitic mesoderm begins to segment. The first somites to appear are the definitive somites 4, 5 and 6 but from these segmentation advances forwards and rearwards. Unlike *Squalus acanthias* and other gnathostomes, partitions between segments divide the mesoderm completely, extending from its dorsal to its ventral edge.

At the 15-somite stage (meaning 15 pairs of somites), all the anterior somites have become distinct (fig. 6.15). The most anterior pair of visceral ('branchial') pouches have

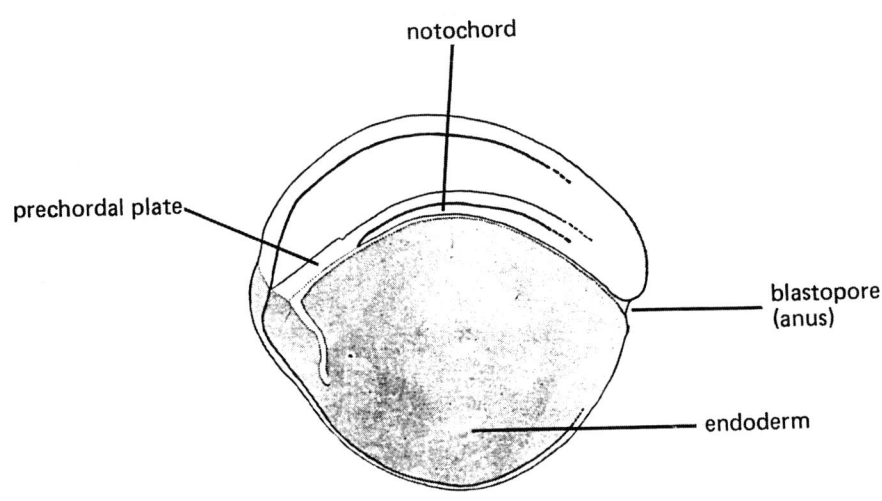

6.14. Lampetra. *Neurula reconstructed in left aspect; from Damas (1944, pl. 1, fig. 1).*

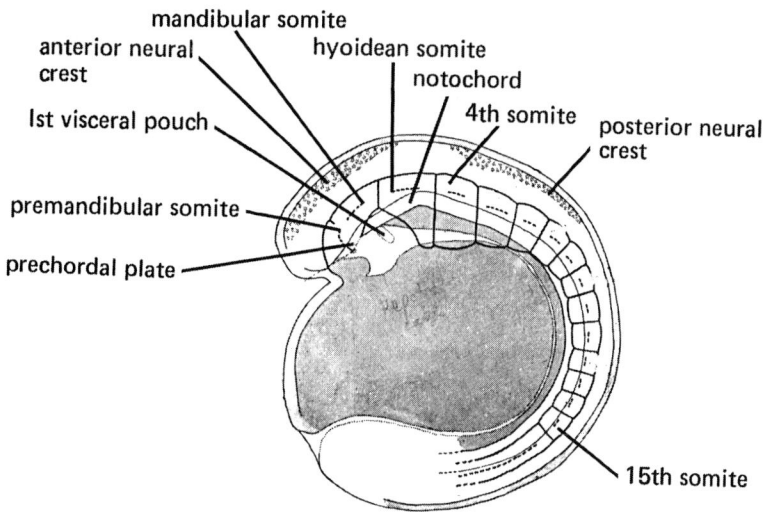

anterior neural crest
mandibular somite
hyoidean somite
notochord
Ist visceral pouch
4th somite
posterior neural crest
premandibular somite
prechordal plate
15th somite

6.15. Lampetra. 15-somite embryo (i.e. 15 pairs of somites). The first visceral pouch has arisen ventral to the division between somite 2 (mandibular) and somite 3 (hyoidean). The light stars in this and subsequent figures represent superficial mesectoderm of crestal or placodal origin (crestal in this figure); from Damas (1944, pl. 1, fig. 4).

arisen as dilations of the gut on either side of the foremost part of the notochord. Behind them the hyoidean or third somite has individualized and consists, like the somites 4, 5 and 6 at this stage, of a more dorsal part (the myotome) continuously connected with a more ventral part (lateral plate). The prechordal plate is beginning to divide, by means of a transverse constriction, into a more anterior premandibular segment, and a more posterior mandibular one. The premandibular somite is entirely in front of the notochord and left and right portions are connected together by a median bridge of unaltered prechordal plate which here forms the roof of what is known as the pre-oral gut. The mandibular somite is partly anterior to the notochord but also extends rearwards on either side of it. Important for what follows is the position of the first visceral pouch, in that it separates the mandibular or second somite from the hyoidean or third somite. Another important change seen at this stage is the migration, from the dorsal portions of the neural plate, of cells of the neural crest. There are two patches of these neural crest-cells (or, rather, two pairs of patches). The separation between them is at the posterior margin of the hyoidean somite, in the future ear region, and the posterior margin of the hinder patch is at somite 7.

At the 21-somite stage (fig. 6.16) several more changes have happened. There are now three visible pairs of visceral pouches, and the ventral, lateral-plate portion of the hyoidean somite is held between the first and second pouch. The position of the third pouch coincides exactly with intersomitic boundary 4/5. A more striking change

affects all somites posterior to somite 4 – the ventral or lateral-plate part of somites 5 and behind has been constricted off from the dorsal part. The lateral plate at this stage however, is still totally segmented, even at its ventral edges. In this it resembles the totally segmented lateral plate seen in the early stages of amphioxus and differs from the lateral plate of myxinoids and *Squalus acanthias* and other gnathostomes which is unsegmented from its beginning. The premandibular somite has largely been overridden by the mandibular somite. Indeed, the latter is now the most anterior somite and will remain so until the animal dies. The ectoderm of the future ear region, over the junction of somites 3 and 4, has thickened to form the ear placode. Another placode is that of the non-profundus part of the trigeminal nerve (V_2) which has appeared at the hind edge of the mandibular somite.

At the 26-somite stage (fig. 6.17) the divisions between the somites of the lateral plate have disappeared. The ear has become a cup and is beginning to crush the dorsal part of somites 3 and 4. The optic nerves pouch out from the anterior end of the prosencephalon. In segment 10 and behind a kidney anlage arises at the contact between myotome and lateral plate of each segment and begins to grow rearwards. The neural crest has spread farther ventrally and several new placodes have appeared – that of the profundus branch of the trigeminal (V_1), that of the 'true' trigeminal, the epibranchial ganglion on the second visceral pouch, and the vagus ganglion (X).

At the 50-somite stage (fig. 6.18) the mouth has invaginated, a median nasal placode has appeared and the

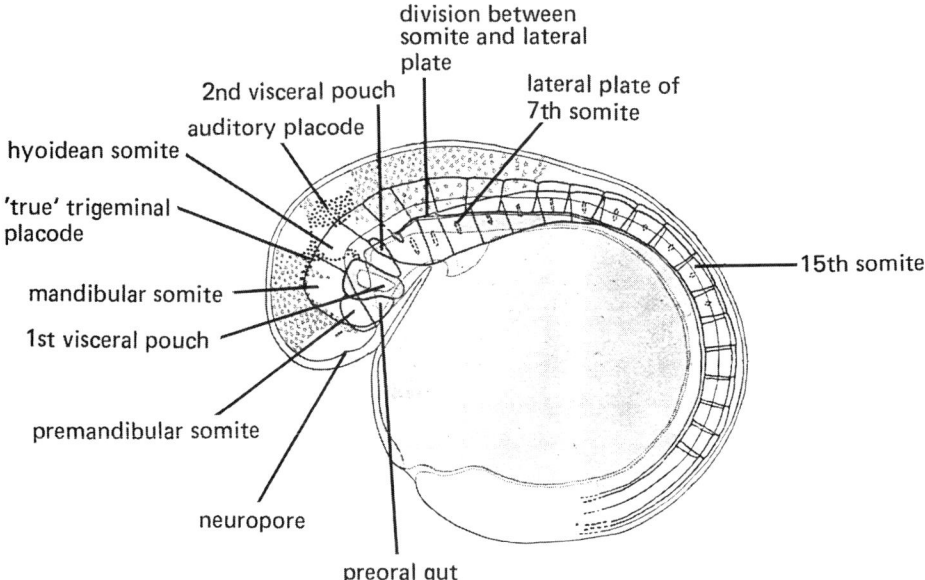

6.16. Lampetra. *21-somite embryo. The original somites have divided into a dorsal part (myotome) and a ventral part (lateral plate) and the lateral-plate portions have moved dorsally so as to lie left and right of the myotomal portions. Note that the ventral part of the hyoidean somite (3) is now held between the first and second visceral pouches while the third visceral pouch is situated just ventral to the junction between somites 3 and 4. Dark stars indicate deep-lying mesectoderm. The front of the mandibular somite is in the same transverse plane as, or very slightly anterior to, the premandibular somite; from Damas (1944, pl. 1, fig. 5).*

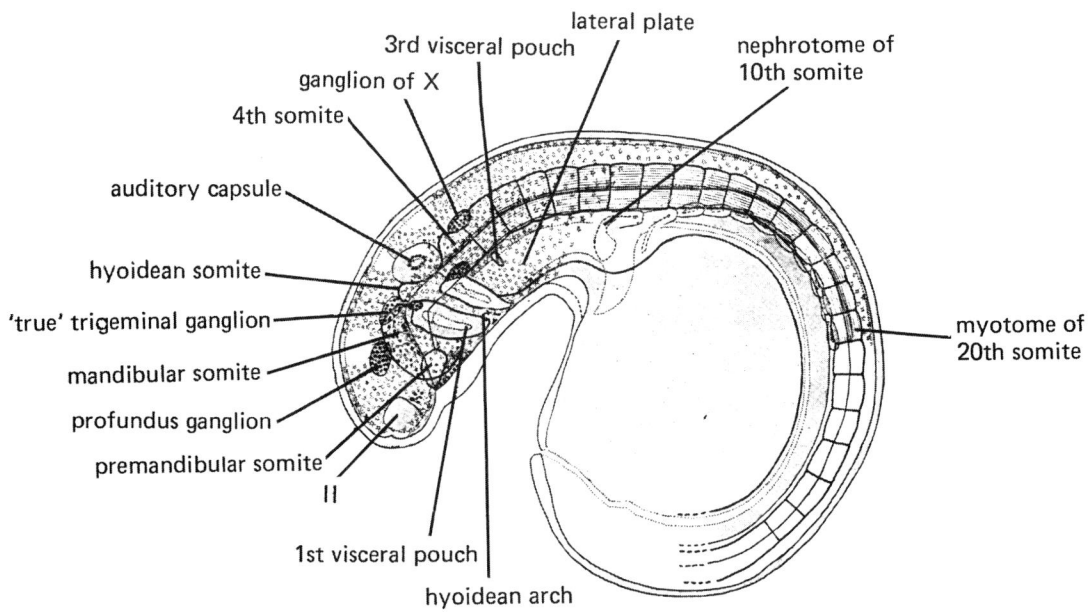

6.17. Lampetra. *Reconstruction of 26-somite stage. The somitic divisions have disappeared in the lateral plate. The auditory capsule has formed and is crushing the dorsal part of the hyoidean somite (third) and the antero-dorsal part of the fourth somite. Nephrotomes are growing posteriorly external to the myotomes and at the ventral ends of the latter; from Damas (1944, pl. 1, fig. 6).*

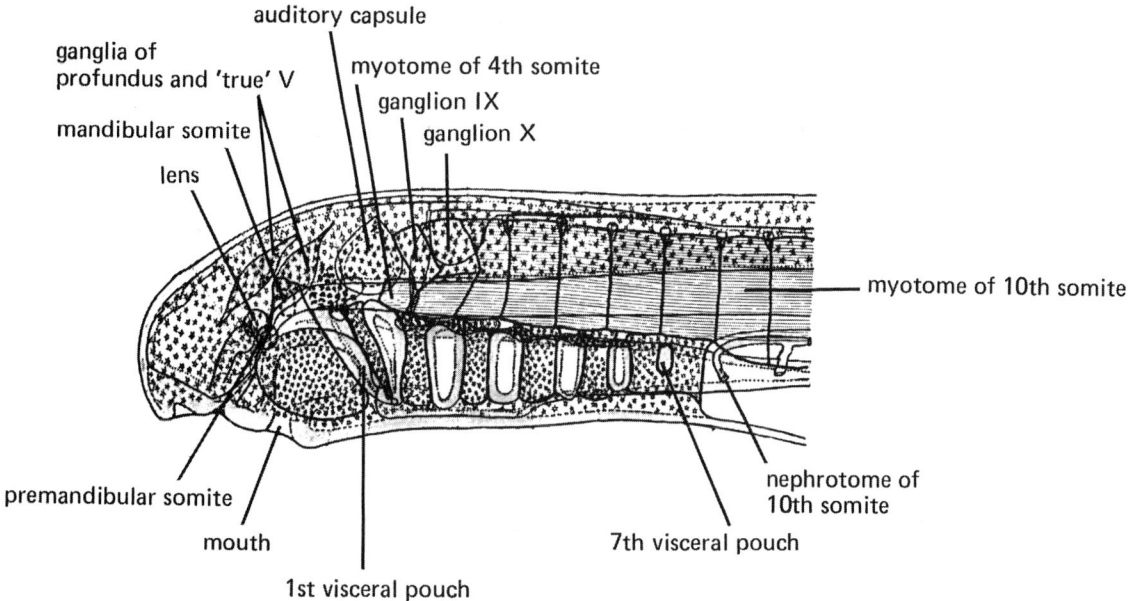

6.18. Lampetra. *Reconstruction of 50-somite stage. The hyoidean somite has disappeared, crushed by the auditory capsule, except for the portion destined to form the posterior rectus muscle. The lateral-plate of the mandibular somite now extends well in front of the premandibular somite; from Damas (1944, pl. 1, fig. 9).*

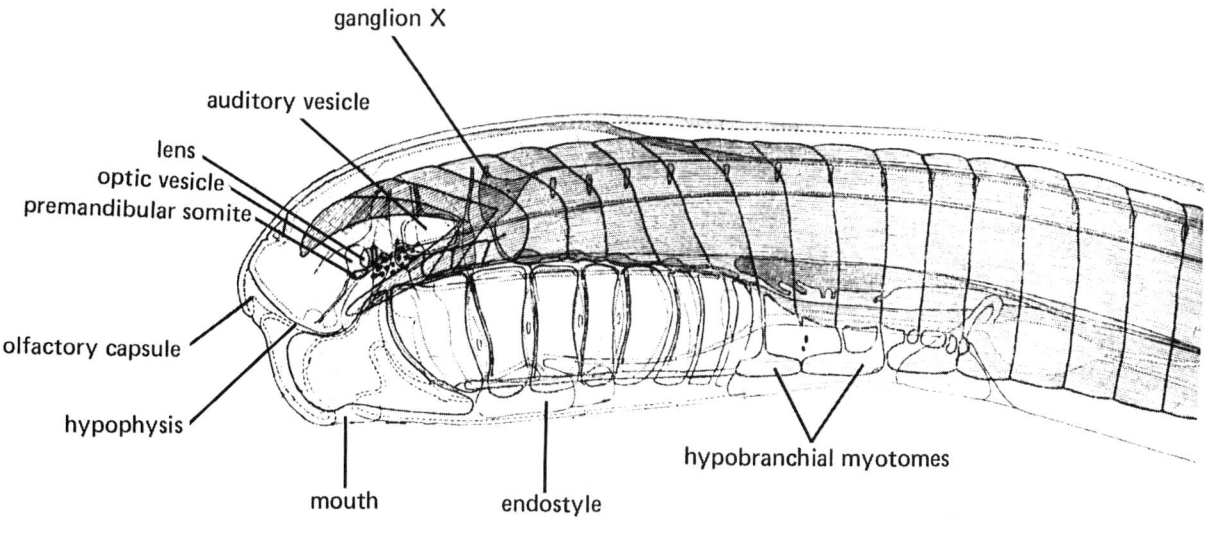

6.19. Lampetra. *Reconstruction of a 5-mm-long pro-ammocoete. The velar mouth has opened, as also have the second to fifth visceral pouches so that the animal can feed. (The first visceral pouch never does open.) The remains of the fourth somite have grown forward, above and below the auditory capsule. The hypobranchial musculature is beginning to develop from somites 12 to 17; from Damas (1944, pl. 3, fig. 13).*

optic nerve has grown upwards so as to induce the future lens and touch the premandibular somite, which will produce most of the oculomotor muscles. The dorsal part of the hyoid somite is reduced to a mass of mesenchyme and seven pairs of visceral pouches exist. At this stage the posterior boundary of the head can be recognized dorsally at about segment 6, behind the hind end of the vagus ganglion (X). Ventrally, however, the boundary of the head is more posterior; it lies at myotome 10 since here is the posterior limit of the ventral part of the neural crest, the anterior limit of the kidney, pericardial cavity and heart, and also the posterior limit of the pharynx since the eighth visceral pouch will later arise just beneath myotome 10. Also, myotome 10 will later be the most anterior myotome to produce hypobranchial musculature. At the 50-segment stage the lateral-line nerve of the trunk has begun to grow rearwards from the vagus ganglion.

At the 5 mm-stage (fig. 6.19) the embryo has hatched from the egg and become a pro-ammocoete. The bucco-pharyngeal membrane has been pierced, as also have four pairs of gill slits, and the endostyle is functional so that the larva can filter-feed. The first pair of visceral pouches has been suppressed but the seventh pair (of the original series) is now well developed, though not yet pierced. The developing eye muscles deserve notice, for the pre-mandibular somite, the dorsal portion of the mandibular somite and the dorsal, mesenchymatous portion of the hyoidean somite now stand near the eye-ball ready to give rise to their respective oculomotor muscles. Moreover, Damas and Koltzoff concluded that the segmental origins of the various oculomotor muscles were the same as in gnathostomes i.e. the hyoidean somite produces the pos-terior rectus, the mandibular somite produces the posterior oblique (superior oblique of gnathostomes), while the other muscles (anterior oblique and anterior, superior and

inferior recti) arise from the premandibular somite (fig. 6.20). At the 5 mm-stage of the proammocoete there is, as Damas points out, a numerical correspondence between the eight pairs of visceral pouches (including the anterior pair which are disappearing) and the overlying inter-somitic boundaries between segments 2 and 10, which likewise number eight (fig. 6.31).

A fully grown ammocoete shows several changes com-pared with the 5 mm-long larva. In particular, the corre-spondence between somites and visceral pouches has broken down, for the eighth pair of visceral pouches (i.e. the seventh pair of gill slits) have migrated rearwards to the transverse level of somite 17. Also an extensive hypo-branchial musculature has arisen largely from myotome 10. And the opening of the naso-hypophyseal duct has migrated rearwards and upwards onto the dorsal surface of the head.

In the metamorphosis of the larva into the adult other great changes happen, including: the loss of the endostyle and other extensive changes in the pharynx, the origin of the tongue, the sudden development of functional eyes (or redevelopment, for the eyes of the pro-ammocoete are functional, though those of the ammocoete are not), an increase in the complexity of the nose, the loss of the functional pronephros, and acquisition of a mesonephros. For present purposes, however, the early development of the ammocoete is more relevant than this metamorphosis.

As concerns the origin of the skeleton, Damas con-cluded that the trabeculae of the skull (fig. 6.21) and the branchial skeleton arose from mesectoderm i.e. were derived either from neural crest or from invading placodal material. The parachordals of the skull, on the other hand, were thought by Damas to be somitic in origin. As already noted, the same distinction seems to hold for *Squalus acanthias* and other gnathostomes (Holmgren, 1940, p.

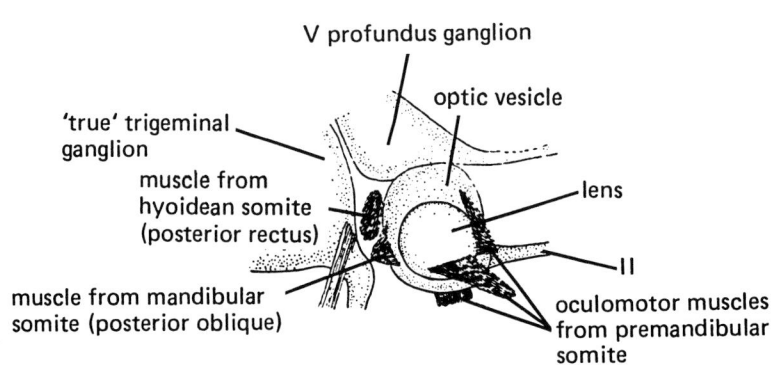

6.20. Lampetra. *Anlagen of the oculomotor muscles and their somitic origin – reconstruction of the eye of an 8-mm-long proammocoete in right lateral aspect; from Damas (1944, fig. 76, p. 168).*

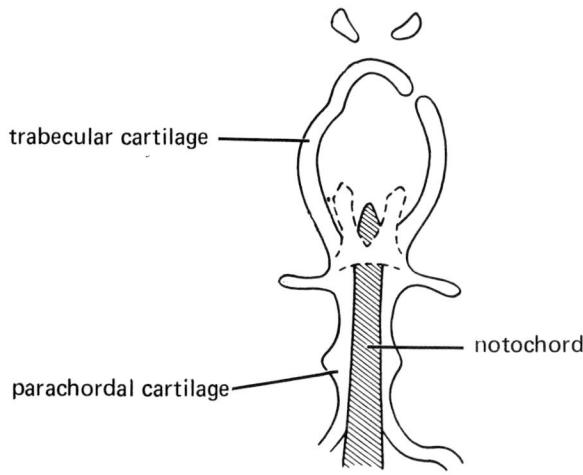

6.21. Lampetra. *Head skeleton of an 8-mm-long pro-ammocoete, in ventral aspect; from Fontaine (1958, fig. 83, p. 168).*

246). Newth at first suspected (1951) on experimental grounds that cartilages never arose from neural crest in *Lampetra*, but in 1956 he concluded, with Damas, that the branchial cartilages were crestal in origin, though he still doubted the crestal origin of the trabeculae.

Koltzoff and Damas also proposed a series of important homologies between *Lampetra* and amphioxus, as follows: (1) premandibular somite of *Lampetra* = anterior gut diverticula of amphioxus (fig. 3.28a, b), for both are the most anterior, initially symmetrical evaginations of the archenteron and both are initially anterior to the front end of the notochord, (2) mandibular somite of *Lampetra* = first dorso-lateral somite of amphioxus (fig. 3.28), for both, when they arise, are immediately behind the premandibular somite and both give rise to the velum and the muscles of the mouth and buccal cavity, although in amphioxus the velar muscles are in fact produced solely by the *left* first dorso-lateral somite, (3) the hyoidean somite of *Lampetra* = second dorso-lateral somite of amphioxus, the first branchial somite = third dorso-lateral somite of amphioxus, and so forth. In amphioxus there is never a convincing alternation of somites with visceral pouches. And in a tunicate tadpole such an alternation is impossible, for by comparison with mitrates (Chapter 8), the head would contain the mandibular pair of somites and carry all the gill slits, whereas the hyoidean and more posterior somites are located in the tail.

Among myxinoids, the embryology of *Eptatretus stouti* was studied by Dean (1899) using whole embryos. Von Kupffer (1900) examined the development of the head using serial sections and Conel (1929, 1931) reconstructed in detail the development of the brain. A new

study would be highly desirable since Dean's work is old and sketchy. So far as known, the development of *Myxine* is like that of *Eptatretus.*

The eggs of *Eptatretus* are about 22 mm long and hard-shelled. Sperm probably enters through a small hole in the shell (the micropyle). At fertilization the egg consists of a large yolky mass with a cap of tissue (germinal hillock) situated just beneath the micropyle. After fertilization the hillock begins to cleave, and to grow outwards forming a blastoderm. This outward growth is not uniform and finally, at the point where the edge of the blastoderm is farthest from the animal pole, gastrulation takes place and an embryo beings to form (fig. 6.22a). As the embryo forms, its head is situated towards the animal pole of the egg (i.e. beneath the micropyle) and its tail towards the vegetal pole. A blood vessel grows out from the head over the surface of the yolk (fig. 6.22b). The nerve cord and brain develop as an infolding of ectoderm along the mid-dorsal line of the embryo. At the stage shown in fig 6.23 the head has five pairs of visceral pouches, optic vesicles, and auditory capsules situated above the hyoidean visceral bar as in *Squalus* and *Lampetra.* About 77 pairs of mesodermal somites are present. The lateral plate is from the first undivided and identical with the mesoderm spread over the surface of the yolk lateral to the somites. It includes a coelom contained between two sheets i.e. a splanchnopleur in contact with the endoderm and a somatopleur just inside the ectoderm. The visceral pouches are ventro-lateral to the more anterior of these somites, but there is no sign of alternation between pouches, perhaps because the two types of structure are far distant from each other. The hepatic vein anterior to the head is beginning to produce the anlage of the heart. In adult *Eptatretus* the gill slits have migrated far posterior to their primary position.

The kidney develops from nephrotomes (fig. 6.24). When these first arise there is one of them just lateral to each somite from the head region rearwards to the future position of the cloaca. Initially therefore, there is no distinction between pro- and mesonephros. Each somite develops a tubule and a glomerulus and all become connected to a continuous nephric duct.

As concerns the head region, the main features of development are summarized in figs 6.25a, b, c, and 6.26. These represent sagittal sections taken from von Kupffer (1900). The head lies in a groove in the animal pole of the egg. At the stage represented by fig. 6.25a there is still a continuous wall separating the archenteron from the buccal cavity and the unpaired olfactory placode is developing as a specialized region of the skin in the most anterior part of the buccal cavity. The heart is anterior to the head. Note that the infundibulum, which in adult life will be dorsal to the adenohypophysis, is far posterior to

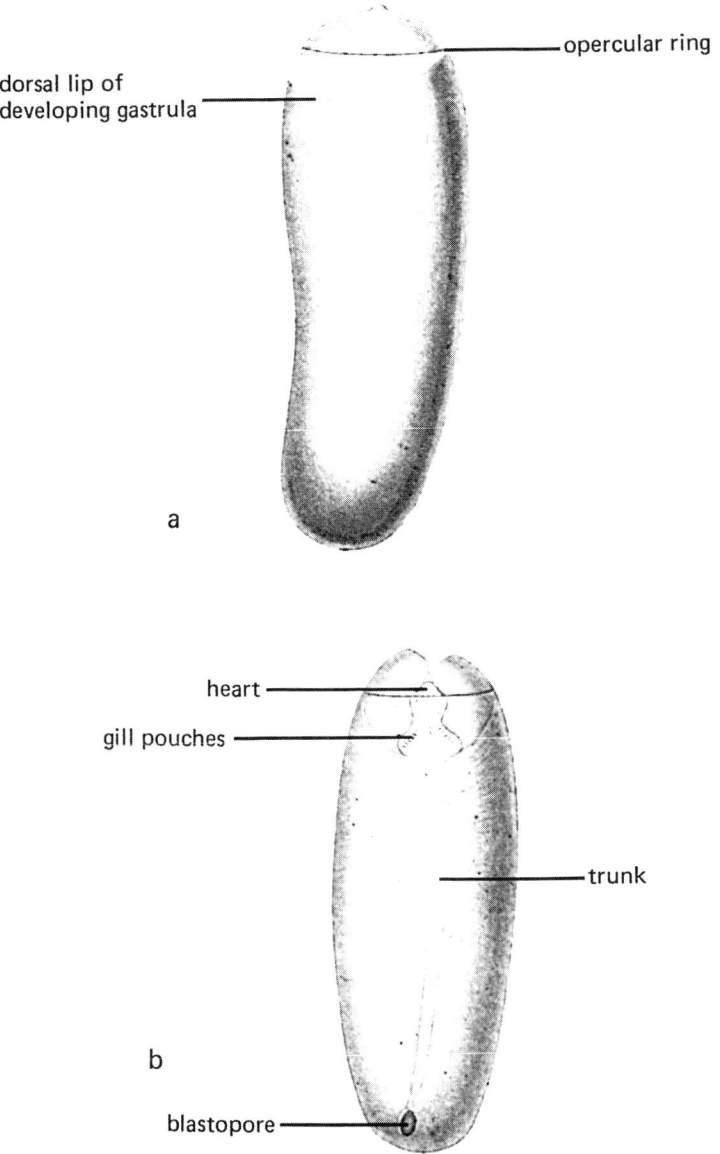

dorsal lip of
developing gastrula

opercular ring

a

heart

gill pouches

trunk

b

blastopore

6.22. Eptatretus stouti; *embryos developing inside the egg. a) Early gastrula. b) Embryo with gills and heart; from Dean (1899; pl. 17, fig. 19; pl. 18, fig. 43).*

the stomodaeal membrane. This indicates that the adeno-hypophysis of myxinoids, unlike that of lampreys and gnathostomes is endodermal in origin, as confirmed by Prof. Gorbman, of Seattle, on the basis of his own observations (personal communication, 1983). At a later stage the stomodaeal membrane has disappeared and the nasal placode has pouched inward (fig. 6.25b). At a later stage

still, the nasal pouch and hypophyseal pouch constrict off from the dorsal wall of the archenteron as a common cavity. At a still later stage a 'secondary stomodaeal membrane' has arisen anterior to this naso-hypophyseal cavity (fig. 6.25c) and the latter has become almost entirely separated from the gut beneath by a floor (palate) that has grown in from the posterior limit and from the sides. The

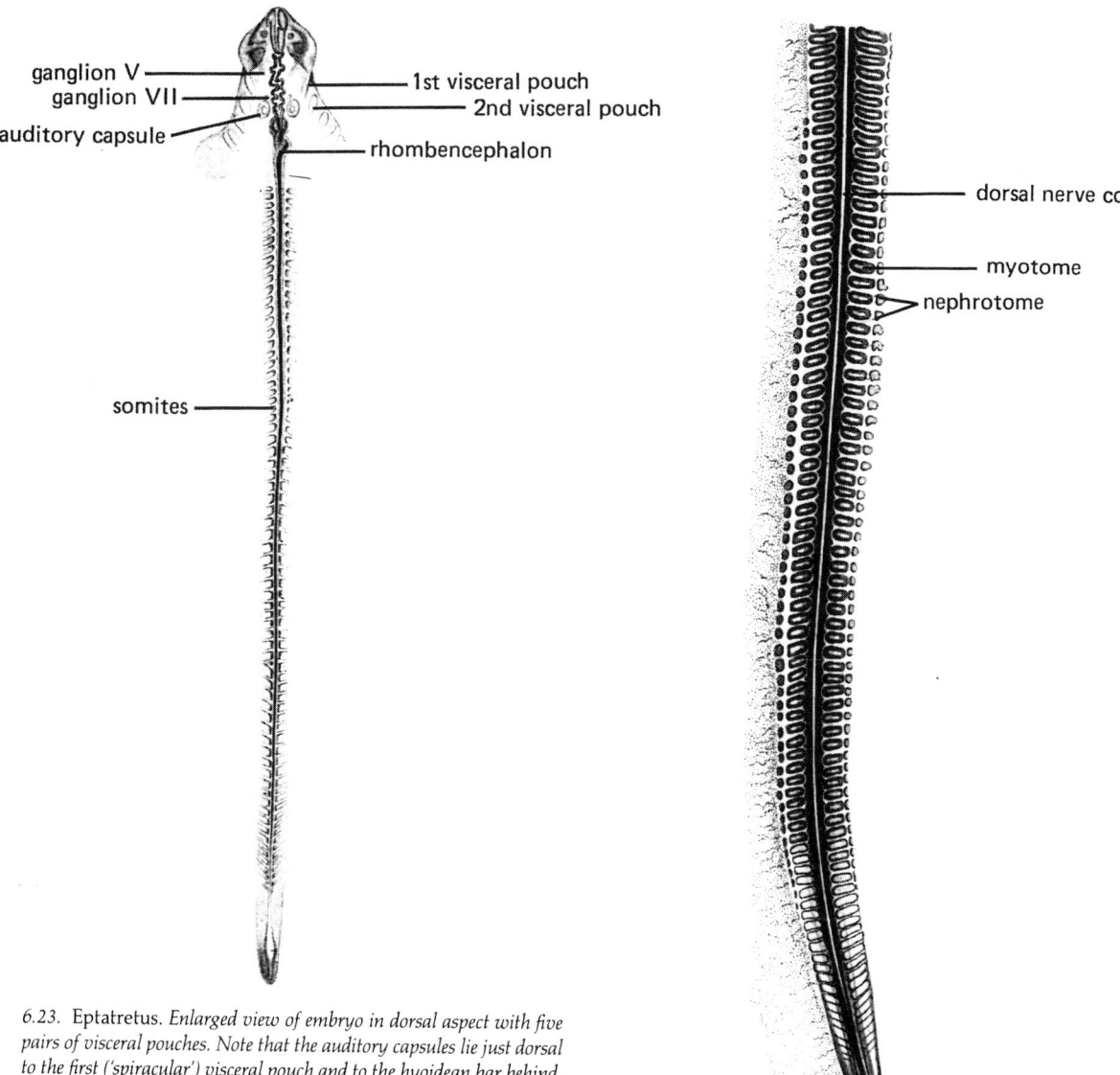

ganglion V
ganglion VII
auditory capsule
1st visceral pouch
2nd visceral pouch
rhombencephalon
somites

dorsal nerve cord
myotome
nephrotome

6.23. Eptatretus. *Enlarged view of embryo in dorsal aspect with five pairs of visceral pouches. Note that the auditory capsules lie just dorsal to the first ('spiracular') visceral pouch and to the hyoidean bar behind. The well-developed somites, numbering about 78 pairs, contrast with the undivided lateral plate; from Dean (1899, pl. 22, fig. 94).*

6.24. Eptatretus. *The first anlagen of the kidney (nephrotomes) in an embryo slightly older than that of fig. 6.23; from Dean (1899, pl. 26, fig. 130).*

posterior opening of the naso-hypophyseal canal into the pharynx is pierced through this floor later.

The 'velum' of the adult myxinoid does not arise from the stomodaeal membrane, but well behind this in the endodermal region of the gut. Perhaps it is not homologous with the velum of other sorts of chordate.

Kupffer showed the existence of placodal components in the trigeminal nerves and of epibranchial placodal components in the facial, glossopharyngeal and vagus nerves. He found no sign of extrinsic eye muscles nor of the oculomotor nerves (i.e. III, IV and VI). From the pre-oral

gut in an early embryonic stage, however, a pair of pouches ascend on either side of the infundibulum (fig. 6.26). It seems possible, in my view, that this pair of pouches is homologous with the premandibular somite of lampreys and gnathostomes.

The thyroid gland in adult myxinoids is a series of scattered alveoli in the ventral wall of the pharynx. It arises

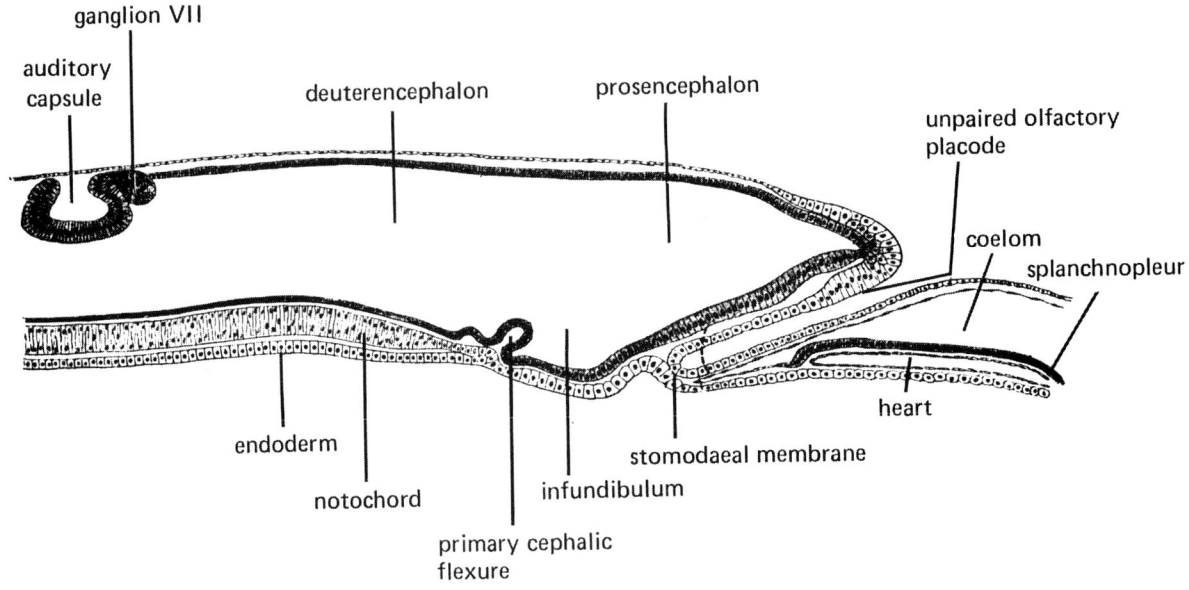

auditory capsule

ganglion VII

deuterencephalon

prosencephalon

unpaired olfactory placode

coelom

splanchnopleur

endoderm

notochord

primary cephalic flexure

infundibulum

stomodaeal membrane

heart

Fig. 6.25a.

deuterencephalon

infundibulum

primary cephalic flexure

prosencephalon

notochord

olfactory capsule

somatopleur

coelom

ventral aorta

prospective adenohypophysis

boundary of ectoderm with endoderm

splanchnopleur

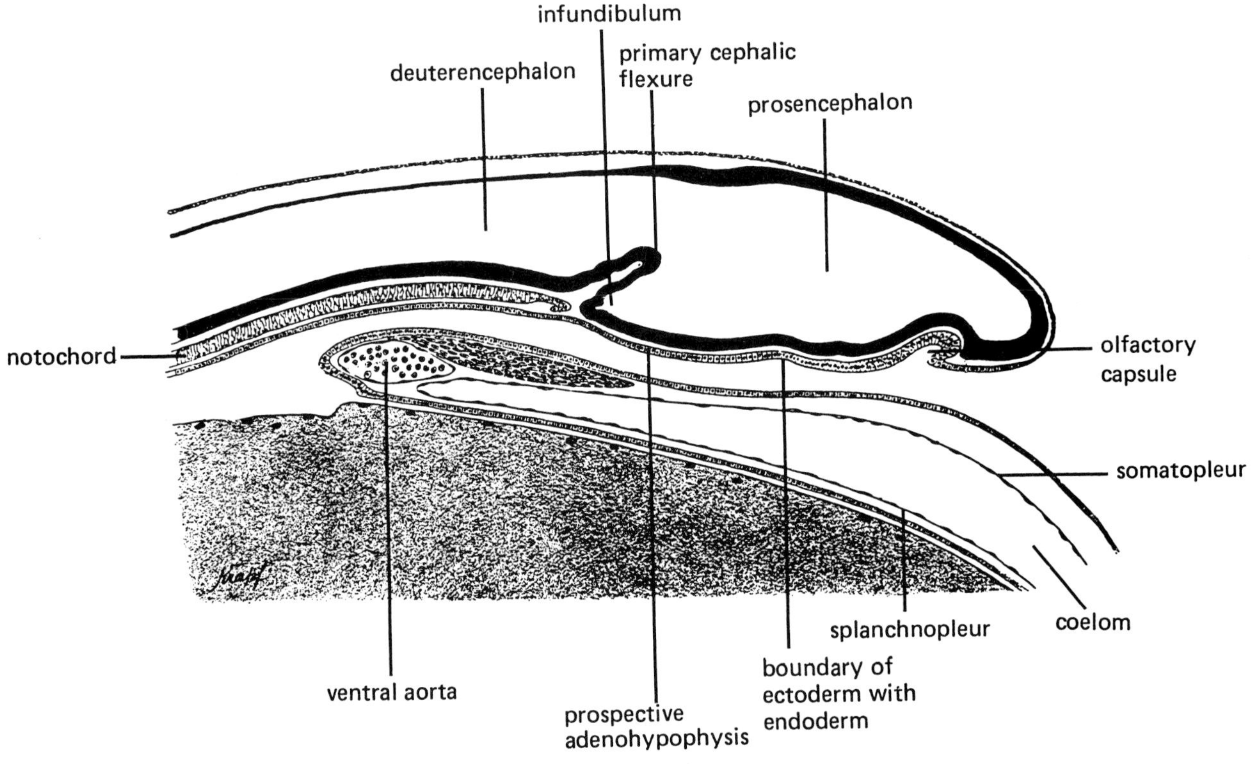

Fig. 6.25b.

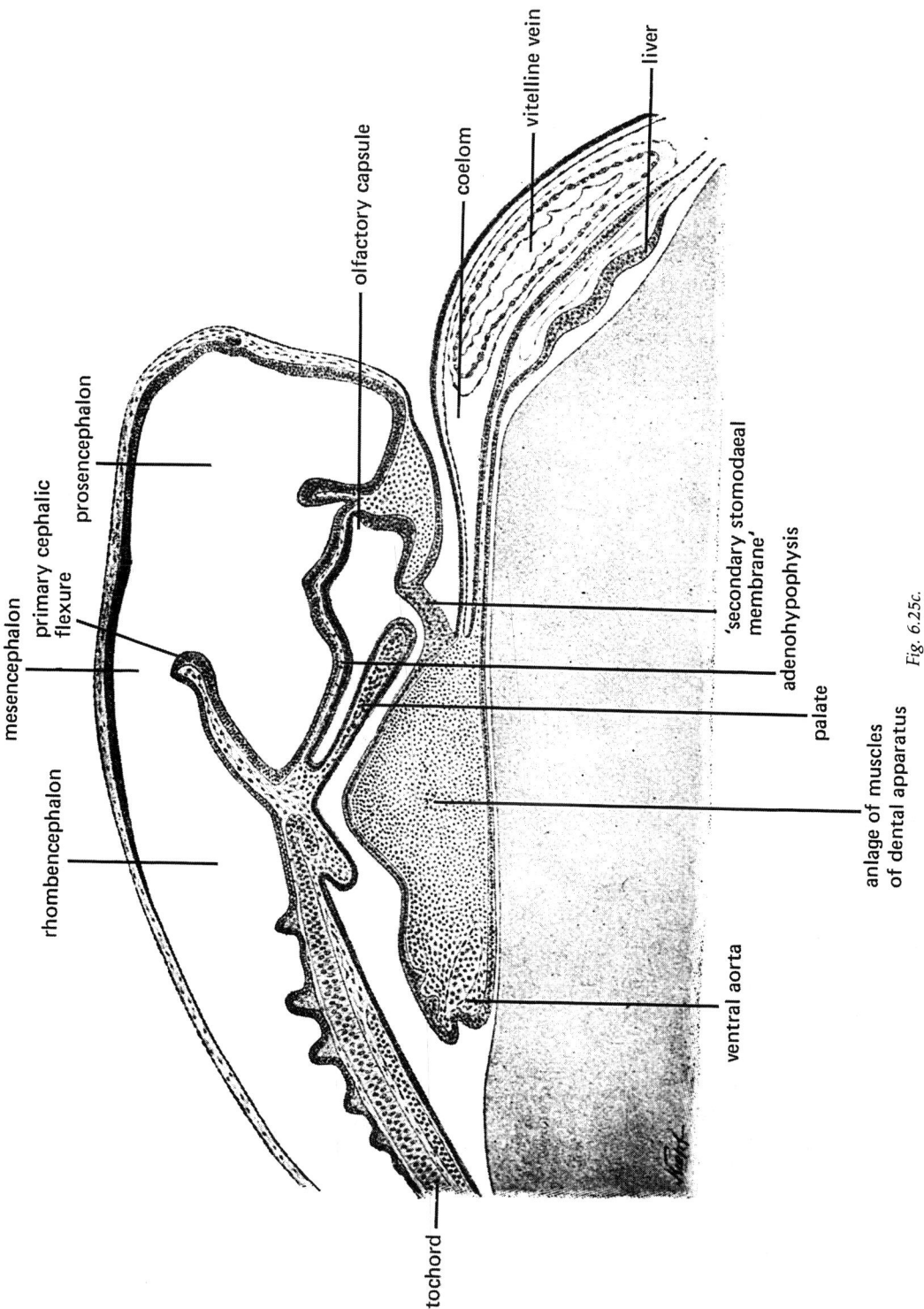

Fig. 6.25c.

6.25. Eptatretus. *Sagittal sections of successive embryos. a) Before the breakdown of the stomodaeal membrane. Note that the infundibulum is behind the stomodaeal membrane, and makes contact with the endoderm which will later become the adenohypophysis. b) After the breakdown of the stomodaeal membrane. c) After the formation of the 'secondary stomodaeal membrane' – this later breaks down in turn and a perforation in the palate allows the naso-hypophyseal canal to open into the pharynx; from Kupffer (1900, fig. 17, p. 20; fig. 30, p. 37; fig. 45, p. 48).*

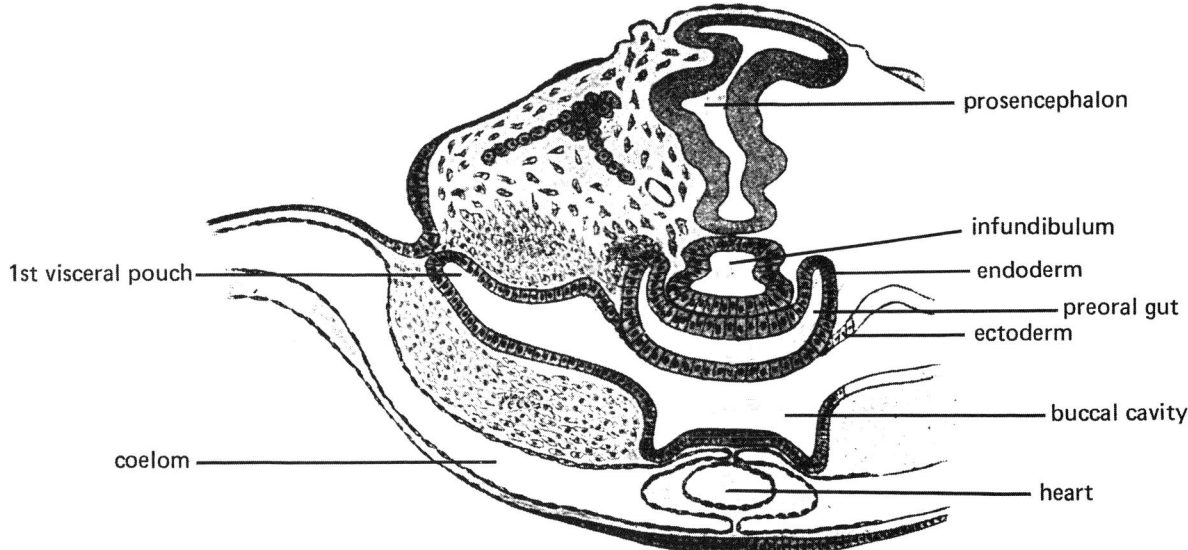

6.26. Eptatretus. *Transverse section through head at a stage slightly older than 6.25b. The pre-oral gut sends extensions dorsalward left and right of the infundibulum. These extensions may be homologous with the left and right portions of the pre-mandibular somite; from Kupffer (1900, fig. 19, p. 24).*

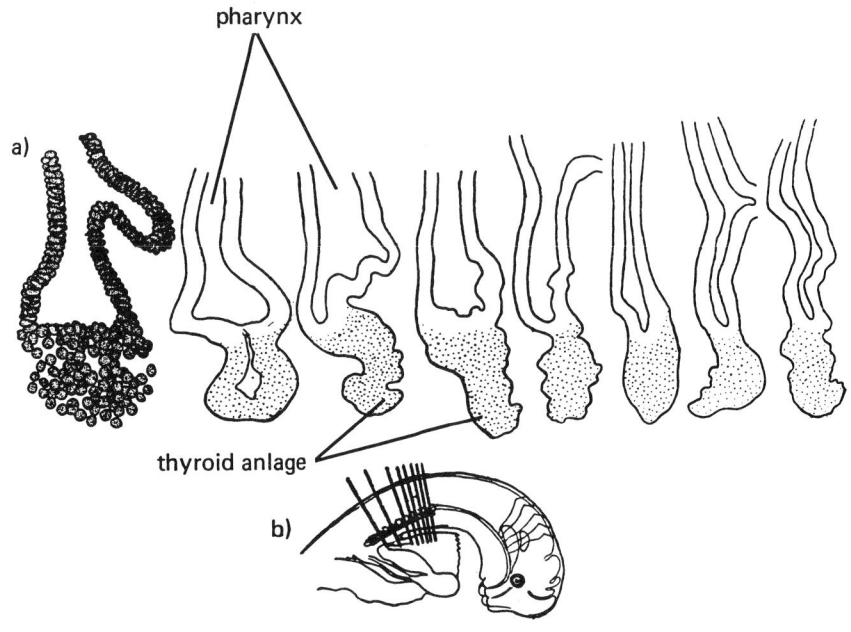

6.27. Eptatretus. *The anlage of the thyroid gland (stippled). a) From left to right are shown successively more posterior transverse sections as indicated in b); from Stockard (1906, p. 94, fig. 3).*

in *Eptatretus* as a trough, hollow anteriorly, formed in the ventral mid-line of the pharynx (fig. 6.27; Stockard, 1906). Its embryonic disposition is therefore like the endostyle of amphioxus, tunicates or an ammocoete. It may well be primitive in its great longitudinal extent since the thyroid (or endostyle) of ammocoetes and gnathostomes arises, by contrast, from an anlage restricted to the fore part of the pharynx.

The most important results from these embryological researches on *Eptatretus* are as follows. Firstly, the gills migrate rearwards in ontogeny, relative to the head; this suggests that the posterior position of the adult gills in myxinoids is an advanced condition. Secondly, the adenohypophysis of myxinoids is endodermal in origin. Thirdly, the separation of the nasohypophyseal duct from the mouth happens by horizontal subdivision of the buccal cavity *plus* anterior pharynx. Fourthly there is no trace of anlagen of any oculomotor muscles, despite the functional eyes. Fifthly the auditory capsule at its first origin is dorsal to the hyoidean gill bar, as in lampreys and gnathostomes.

The vertebrate head problem is one of the most complex in classical zoology, and also one of the vaguest. It is closely linked with attempts to reconstruct the ancestral vertebrate, generally but mistakenly conceived as being a single animal rather than a lineage of successive animals. Its history has been well reviewed by Goodrich (1930), Veit (1947), Romer (1972) and Starck (1963, 1975), though these authors thoroughly disagree with each other in the answers they prefer. Here I shall not write yet another history of the head problem but shall try to distinguish and solve the several lesser problems included in it. The use of embryology in deciding the nature of the vertebrate head has greatly depended on intuitive feelings of what is important in ontogeny. But criteria of importance have been lacking.

The head problem is basically phylogenetic and has been given two opposing (though in my view partly compatible) answers – anti-segmentationist and segmentationist. The anti-segmentationists conceive the ancestral vertebrate as being much like a tunicate tadpole with clearly distinct head and tail. The segmentationists, on

the other hand, see the ancestral vertebrate as being like amphioxus – fairly uniform from front to rear – from which uniformity the vertebrate condition has developed by cephalization. The anti-segmentationist viewpoint goes back to Froriep (1882) and has recently been advocated by Starck (1963, 1975) and Romer (1972). The segmentationist viewpoint begins with Gegenbaur (1871) but was elaborated by Balfour (1878) and van Wijhe (1882) and later by Koltzoff (1901), Goodrich (1930) and Damas (1944). In the work of Bjerring (1977) and Northcutt & Gans (1983) it is still very much alive.

The anti-segmentationists look into the kaleidoscope of vertebrate embryology and see a fundamental distinction between front and rear (head and tail) which they suppose was at one time clearer. To the front of the animal would belong: the pharynx with visceral (gill) slits and the endostyle or thyroid gland; a cartilaginous skeleton derived from neural crest and including the trabeculae of the skull and the cartilage of the jaws and visceral bars (except the median ventral elements of these bars); the part of the brain anterior to the notochord i.e. the prosencephalon or fore-brain; a prominent placodal element in the origin of the sensory nerves; the sense organs including nose, eyes, taste buds and lateral line (whose trunk branch always grows rearwards out of the head); a tendency for neural crest to flow out dorsal to any somites that it meets and to destroy them – though this holds only for gnathostomes (fig. 6.10) and not for agnathans. To the rear of the animal would belong: notochord, dorsal nerve cord and muscle blocks; kidney; the brain overlying the notochord (i.e. the mid + hind brain = deuterencephalon); segmental ganglia derived entirely from neural crest (i.e. with no placodal component) and situated between muscle blocks; a tendency for neural crest to flow out medial and ventral to any muscle blocks that it encountered and not destroy them; formation of cartilages not from neural crest but from mesenchyme derived from somites where these touched the notochord. The tunicate-tadpole-like ancestor favoured by the anti-segmentationists has been called by Romer the 'somatico-visceral animal' (fig. 6.28). Experimental evidence can be held to support the anti-

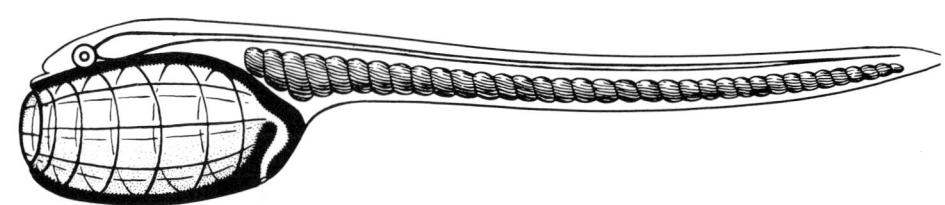

6.28. *Romer's 'somatico-visceral animal'. This postulated imaginary ancestor of the vertebrates is anatomically like a mitrate; from Romer (1972, fig. 8a, p. 148).*

segmentationist viewpoint (Starck, 1975, p. 144ff), for the development of the embryo seems to be controlled by the antagonism and interplay of two basic organizers. There is a heat-stable neuralizing factor associated with the front of the animal and a heat-labile mesodermalizing factor associated with the rear (fig. 6.29). Both factors are probably proteins. Pure neuralizing factor produces fore-brain; pure mesodermalizing factor produces somites, notochord and kidney; mixtures, according to the predominance of one factor or the other, produce the organs of intermediate position in the body (fig. 6.29).

The same basic distinction between front and rear is supported by the intuition, seen 'through a glass darkly' by many authors, that the front end of the notochord, which is also the position of the adenohypophysis or Rathke's pouch, is somehow an important landmark. Thus Goodrich (1930, p. 214) wrote: 'To this place reaches the anterior end of the notochord, above it arises the meso-

cephalic flexure of the brain, through it pass the internal carotids, and into it grows the hypophysis from below . . . Behind it extends the basal plate of the cranium, and it is embraced by the divergent posterior ends of the trabeculae cranii.' Anti-segmentationists, of whom Goodrich was not one, would hold that this place was once, in the earliest stages of vertebrate phylogeny, even more obvious than now.

The segmentationist view on the other hand depends on the recognition of embryological muscle-block segmentation forward of the region where it is obvious in the adult. It also asserts that these anterior, vestigial somites are simply related in position to the visceral slits. The strength of this view can be illustrated both from *Lampetra* (fig. 6.16) and *Squalus* (fig. 6.5a) where, in early embryos, there is a continuous ridge of mesoderm on either side of the notochord. The dorsal part of this ridge (in *Squalus*), or its whole depth (in *Lampetra*), comes to

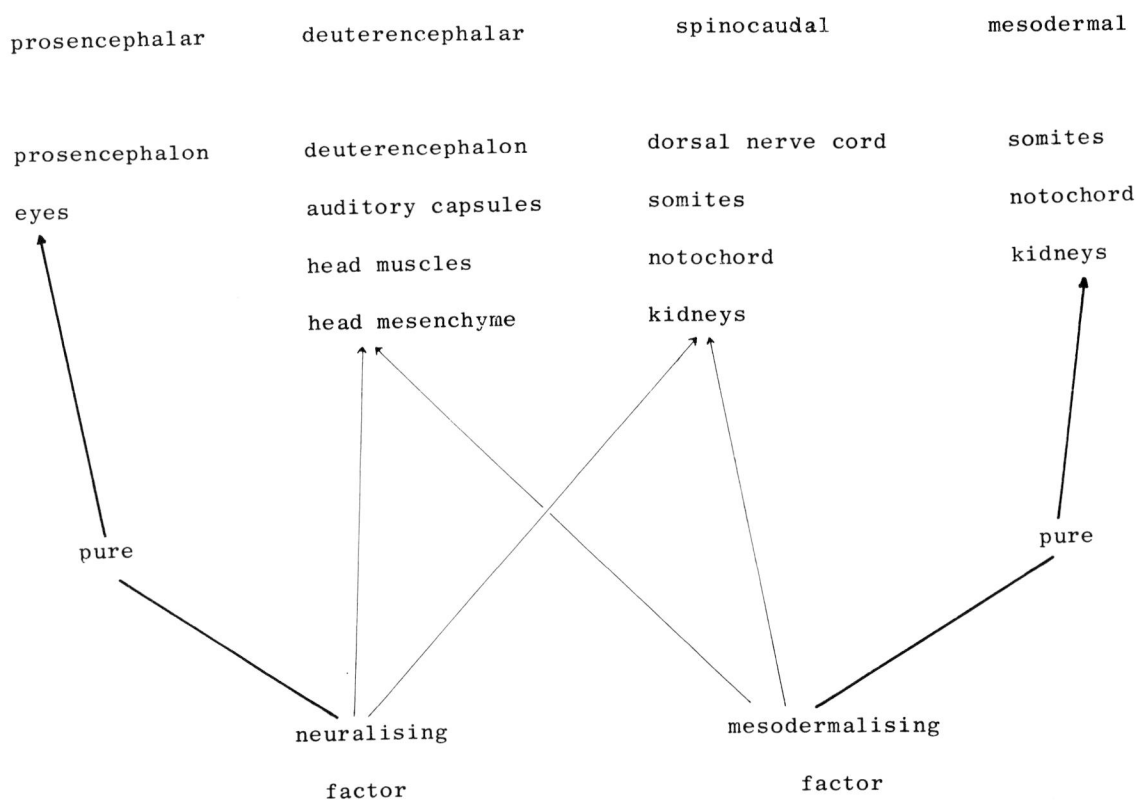

Inductions:

prosencephalar	deuterencephalar	spinocaudal	mesodermal
prosencephalon	deuterencephalon	dorsal nerve cord	somites
eyes	auditory capsules	somites	notochord
	head muscles	notochord	kidneys
	head mesenchyme	kidneys	

pure   pure

neuralising   mesodermalising

factor   factor

6.29. *Interplay of anterior neuralizing and posterior mesodermalizing factors in the induction of the vertebrate body. Modified and translated from Starck (1975, p. 145).*

be divided into somites and the series continues forward into the hyoidean somites (conventionally no. 3), and the mandibular somites (no. 2). Conventionally, but without naturalness, the premandibular somites are regarded as the beginning of this series (no. 1). The hyoidean somite is right and left of the notochord like the somites behind it. The mandibular somite, on the other hand flanks the front end of the notochord and arises from prechordal plate. And the premandibular somite is anterior to the noto-chord, and its right and left portions are connected together by a bridge or commissure across the median line. The observed position in series of the various somites in the mesoderm of early embryos has suggested that they are all serially homologous and that the adult differences between them have arisen by differentiation from primi-tive sameness, which is the essential segmentationist position.

Segmentationists also believe, as already stated, that there is, or once was, a simple relationship between visceral-slit segmentation (branchiomery) and somitic segmentation (myomery). However, there have been two segmentationist views on what this simple relationship is. Thus Balfour (1878) and his followers such as Ziegler (1908) and Goodrich (1918, 1930), believed in simple alternation. They held that the somites in the branchial region are fundamentally Γ-shaped (inverted-L-shaped), with the vertical limb of the Γ passing down a visceral bar and the horizontal limb extending forwards over the visceral slit next in front (fig. 6.30b). Van Wijhe (1882) and his followers such as Neal (1898), held on the other hand, that there were two somites dorsal to the hyoidean arch

and that somites behind that arch had indeed the form of an Γ with the vertical limb of the Γ projecting down a visceral bar, but that the horizontal limb extended over the visceral slit next *behind* (fig. 6.30a). The existence of the two discrepant views shows that the delimitation of somites in the branchial region of early embryos is not clear-cut, which has led some authors to reject any simple relationship between branchiomery and myomery and to throw out the whole segmentationist interpretation (Starck, 1963; Romer, 1972). This rejection gains strength from the fact that enteropneusts have branchiomery in the form of serial repetition of gill slits, but show no sign of myomery. The most economical explanation of this would be that branchiomery is older than myomery, and in that case why should a simple topological relationship between the two types of segmentation be expected in vertebrates? (On the other hand serial repetition of gill slits in chordates and enteropneusts could be mere parallelism.) The two rival segmentationist schemes of relationship for muscle blocks and visceral slits were based at first on gnathostomes, and in particular on sharks such as *Squalus.*

In lampreys, however, the situation seems to be clearer than in sharks, as pointed out by Koltzoff (1901) and Damas (1944). For *Lampetra* differs from *Squalus* or other gnathostomes in two ways which make its embryology informative. Firstly, there is no initial distinction between myotome and lateral plate. Instead the mesoderm is at first completely divided dorso-ventrally by the myocommata (fig. 6.15). Only later does the future lateral-plate region be-come separate from the muscle blocks by the development

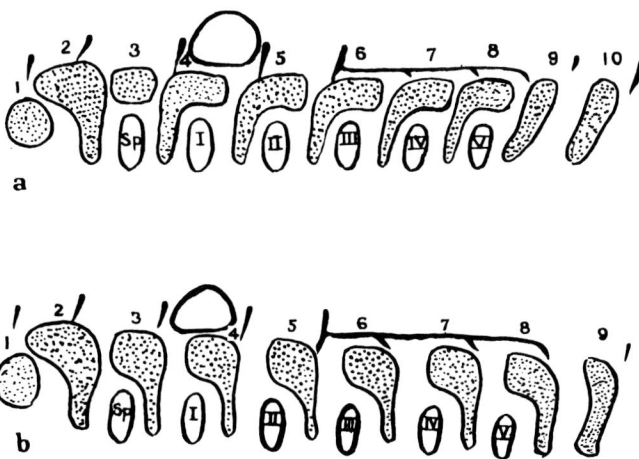

6.30. *Relationship of somites to gill bars in a vertebrate embryo according to: a) van Wijhe; and b) Balfour. 1–10 = numbers of the somites; Sp = spiracular visceral pouch; I–V = successive branchial pouches; the unstippled oval outline is the auditory capsule; from De Beer (1922, p. 469, text fig. 11).*

of a horizontal septum between the two regions (not homologous with the horizontal septum between the epaxial and hypaxial muscles of gnathostomes) followed by the disappearance of the somitic boundaries ventral to the septum. Secondly the somites just posterior to the ear do not disintegrate to form part of the skull as they do in gnathostomes. Rather they persist into the adult, remaining as muscle blocks and shifting forwards (fig. 6.18). These favourable circumstances show rather clearly that, in the matter of the relationship of visceral bars to muscle blocks, the Balfour or Goodrich school is right and the van Wijhe school mistaken. For in *Lampetra* there is only one hyoidean somite on each side, not two as van Wijhe required, and it grows downward between the first and second visceral pouches (fig. 6.16). More posterior visceral pouches appear after the lateral plate has formed by fusion, so that if the same basic relationship held for them it could not be established by the same observation. However, the second and third visceral pouches appear exactly ventral to the appropriate segmental boundaries, respectively between somites 3 and 4, and somites 4 and 5. And there is a numerical correspondence between the myocommata dorsal to the branchial region and the total number of

visceral bars when these have first all appeared (fig. 6.31). For there are eight pairs of visceral pouches, and eight overlying myocommata, or seven visceral bars and seven overlying myotomes. Ambiguities remain, even in *Lampetra* for: (1) the myotomes do not hold their initial positions, (2) the hyoidean or third myotome and part of the fourth myotome are suppressed (as in gnathostomes) by the auditory capsule, and (3) above all, the first visceral pouch, corresponding to the spiracular pouch of *Squalus*, never produces a slit but disappears in the early pro-ammocoete. But *Lampetra* strongly suggests that the segmentationists are right to assert a primitive simple relationship between branchiomery and myomery. Moreover, this relationship is probably a simple alternation as Balfour, Goodrich and Ziegler believed.

The most anterior somites – premandibular, mandibular and hyoidean – have created particular problems. They develop into the oculomotor muscles whereby the hyoidean somite gives rise to the external rectus (posterior rectus in lampreys); the mandibular somite gives the superior oblique (posterior oblique in lampreys); and the premandibular somite gives all the others (inferior oblique – or anterior oblique in lampreys – superior,

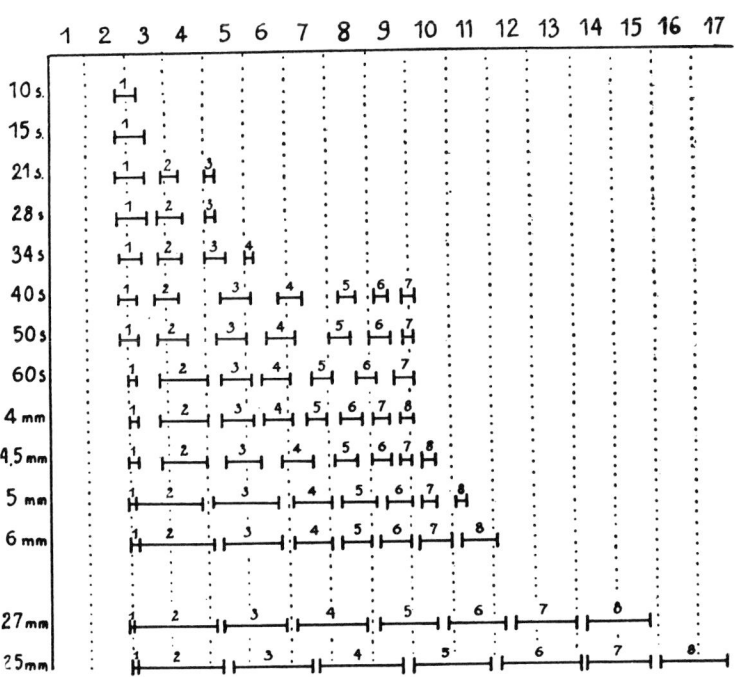

6.31. *The numerical correspondence between myomery and branchiomery in* Lampetra. *The horizontal axis shows the conventional numbers of the somites. The vertical axis shows the stage of development either as number of pairs of visible somites or as length in mm of the embryo. The short horizontal lines show the extent of the visceral pouches at successive stages in relation to the intersomitic boundaries dorsal to them. The diagram suggests that, up to and including the 6-mm stage, there is a fundamental alternation between somites and visceral pouches; from Damas (1944, p. 223).*

internal and inferior rectus). Moreover, seeing that the hyoidean and more posterior somites are associated with visceral pouches or slits, it has been supposed that the mandibular and premandibular somites were once likewise associated with slits which have since disappeared by fusion with the mouth. This view began with van Wijhe (1882) and was still maintained by Goodrich (1918b, 1930). It sees (fig. 6.32) the trigeminal nerve complex as consisting by phylogenetic origin of two pairs of branchial nerves – a profundus nerve (V_1) associated with a lost gill slit anterior to the premandibular somite; and a true trigeminal nerve represented by the maxillary and mandibular branches of the trigeminal (V_2 and V_3) and associated with a lost slit between the premandibular and mandibular somites. The corresponding motor nerves would be the oculomotor, being the ventral root related to the dorsal root of the profundus, and the trochlear being the ventral root related to the 'true' trigeminal. This scheme is plausible assuming that somites in the head region could only exist or arise in association with visceral slits. Without this assumption, however, the whole elaborate structure

collapses, for there is simply no solid evidence to support it. In *Lampetra* the notion of lost gill slits anterior to the spiracular pouch is particularly difficult to maintain. For visceral pouches arise from pharynx and never from buccal cavity, but in *Lampetra* there is no pharynx anterior to the spiracular pouch and the velum arises from the mandibular somite. Conceivably space once existed here and has been lost by compression but such is purest supposition.

A feature of all vertebrate embryos deserves emphasis at this point. The auditory capsule arises dorsal to the hyoidean and first branchial bars and by its development, at least in lampreys and gnathostomes, suppresses most of the hyoidean and first branchial somites (3 and 4) (figs 6.6a, 6.17). This situation is easy to understand on the basis of the fossil evidence from mitrates.

To summarize, there are valid arguments for the antisegmentationist view that vertebrates derive from a tunicate-tadpole-like ancestor with clearly distinct head and tail and no alternation between gill slits and muscle blocks. There are also valid arguments for the moderate segmentationist view that muscle blocks and gill slits at

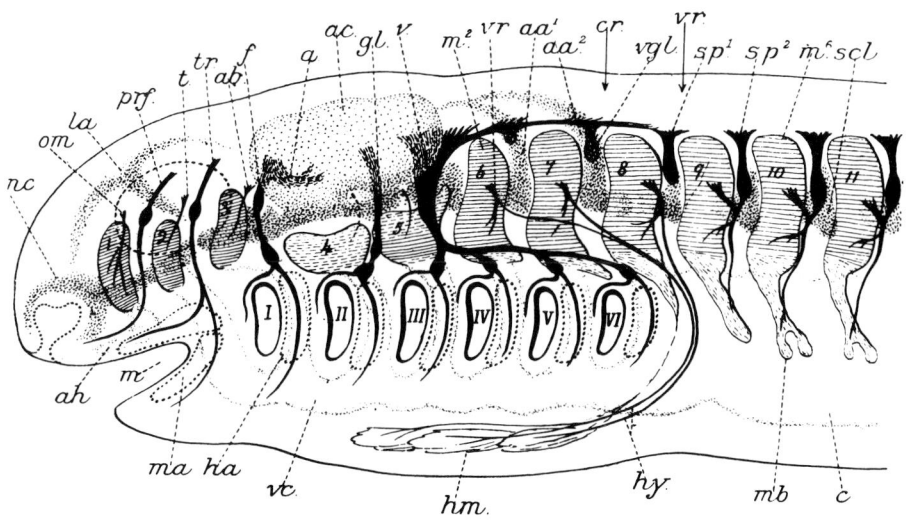

6.32. *Goodrich's diagram of the segmentation of the head in the Common Dogfish* Scyllium canicula. *This represents the text-book view of vertebrate head segmentation as accepted in much of the English-speaking world. Explanation:* 1–10 = *standard numeration of somites;* I–VI = *spiracular and five branchial pouches;* a = *auditory nerve;* aa^1, aa^2 = *first and second visceral arches of segments 6 and 7;* ab = *abducens nerve;* ac = *auditory capsule;* ah = *anterior head cavity (Platt's vesicle);* c = *coelom in lateral plate;* cr = *limit of cranial region;* f = *facial nerve;* gl = *glossopharyngeal nerve;* ha = *hyoid arch;* hm = *hypoglossal muscles from somites 6, 7, 8;* hy = *hypoglossal nerve;* la = *lamina antotica;* m = *mouth;* m^2, m^6 = *second and sixth metaotic myotomes;* ma = *mandibular cartilaginous arch;* mb = *muscle bud to pectoral fin;* nc = *nasal capsule;* om = *oculomotor nerve;* prf = *profundus branch of trigeminal nerve;* scl = *sclerotome of segment 10;* sp_1 = *vestigial dorsal root and ganglion of first spinal nerve;* sp_2 = *second spinal nerve;* t = *trochlear nerve;* tr = *'true' trigeminal nerve (i.e. non-profundus part of trigeminal complex);* v = *vagus nerve;* vc = *ventral coelom extending up each visceral bar;* vgl = *vestigial dorsal root and ganglion of segment 7;* vr = *ventral root of segment 6 supplying the second metaotic myotome and hypoglossal muscle. The myotomes are longitudinally striated, the nerves black and the scleromeres dotted. The cartilaginous visceral arches are represented by dotted outlines, as also are the optic capsule and the nasal sac; from Goodrich (1918, p. 23).*

one time alternated in the ancestry of vertebrates. There are no valid arguments for the extreme segmentationist view that two pairs of gill slits have disappeared anterior to the spiracle, despite the impressive support that this opinion has received. Later I shall try to show (Chapter 8), by means of fossil evidence, that the anti-segmentationist and moderate-segmentationist viewpoints are compatible but refer to different animals. The anti-segmentationist animal was probably ancestral to the segmentationist one.

Chapter 7　　Stem chordates – the cornutes

Fossil evidence on the relationship of chordates and echinoderms is the subject of this chapter and the next while Chapter 9 gives a phylogenetic overview. The fossils in question I have called calcichordates because, although they have many chordate characteristics, they show a calcite skeleton of echinoderm type. The calcichordates were strange-looking animals, being highly asymmetrical. Indeed the most primitive forms were shaped like boots. They were all marine and are found in rocks of Middle Cambrian to Middle Devonian age – about 400 to 600 million years old. The arguments in Chapter 7 are interlinked, in all possible directions, with those in 8 and 9.

Many different names have been given to the calcichordates as a group. Traditionally they were placed as echinoderms as part of Jaekel's class Carpoidea (1900). More recently they have been assigned to the echinoderms as part of the subphylum Homalozoa, and within it they form a class Stylophora, as for example in Ubaghs (1967b). But the class Stylophora, set up as an order by Gill & Caster (1960), is exactly co-extensive not only with the calcichordates but also with Jaekel's group Heterostelea (1900). I invented the name Calcichordata, originally with the rank of a subphylum of the phylum Chordata, to emphasize my view that the fossils in question, though with undoubted echinoderm affinities, should be placed in the chordates (Jefferies, 1967, 1968a). Informally I still use the word 'calcichordate', but under Hennig's influence it is no longer logical to regard the calcichordates as forming a subphylum, for subphyla ought to be monophyletic groups with some members still extant. I mention these different names because all are still in use. Also, in view of the different competing interpretations, an outside reader might easily suppose that the names referred to different animals.

Stem groups and crown groups have already been mentioned in Chapter 1 as a way of classifying fossils in relation to extant groups. In this and the next two chapters

I shall argue that the echinoderms are the living sister group of the chordates and that both together comprise a larger monophyletic group called the Dexiothetica (Jefferies, 1979). If so, the fossils traditionally regarded as primitive echinoderms without radial symmetry should include not only true stem echinoderms, but also stem chordates and stem dexiothetes. Allotting such fossils to their respective stem groups is a major task ahead for palaeontologists. At present it has scarcely begun and will involve careful reconstruction of the anatomy of many of the forms in question. Here I shall discuss only those fossils that have already been worked out in detail. It would be premature to consider the others.

The calcichordates are defined on a primitive feature – the retention of a calcite skeleton. This raises the suspicion, almost certainly correct, that as a group they are paraphyletic. They are divided into two traditional groups: the cornutes, set up by Jaekel (1900) as an order Cornuta of his echinoderm class Carpoidea; and the mitrates, set up by Jaekel (1918) as an order Mitrata of the Carpoidea. The names 'cornute' and 'mitrate' are useful vernacular terms but both groups are paraphyletic and there is no sense in arguing about their categorial rank. (Indeed, there is never any sense in arguing about categorial rank.) Cornutes are the particular subject of this chapter and mitrates of the next one.

There are three competing interpretations of the fossils that I call calcichordates. All three agree that these fossils consist of a massive part and an appendage and have a calcite skeleton of stereom mesh, but beyond that agreement almost stops, except as to factual details. The first interpretation is that of Ubaghs (1961–1983) who regards the appendage as a feeding arm, called an 'aulacophore' because a groove is prominent in its anatomy (*aulakos* = groove). Ubaghs' views have been accepted by several workers (Caster in Ubaghs, 1978; Chauvel & Nion, 1977; Kolata & Guensburg, 1979; Sprinkle, 1976; Bonik,

Gutmann & Haude, 1978). The second interpretation is mine (Jefferies, 1967-in press). It sees the appendage as a locomotory chordate tail and the massive part as a head, tail and head being accurately homolologous with those of a tunicate tadpole and broadly homologous with those of other chordates. In my early work on the subject (1967–1973) I saw the massive part as also homologous with a crinoid theca and the appendage with a crinoid stem but I abandoned these crinoid homologies in Jefferies (1975). And a third interpretation, proposed by Philip (1979, 1981) and accepted by Kolata & Jollie (1982) and Jollie (1982), regards the appendage as a locomotory stele, analogous but not homologous with a crinoid stem, and calls the massive part a theca. In some ways this is a return to the views of Bather (1913). These basic differences as to the nature of the appendage are compounded with other differences of intepretation, especially concerning the head openings and orientation, as I shall explain in dealing with individual animals. Ubaghs and Philip regard the calcichordates as echinoderms, not as chordates.

The cornutes are characterized by the presence of gill slits on the left side of the head only (if my interpretations are correct). In this they resemble larval amphioxus whose gill slits likewise belong morphologically to the left side (fig. 3.33). The mitrates, on the other hand, probably had right gill slits as well as left ones, like post-larval amphioxus with its primary left and secondary right gill slits, and also like living vertebrates and tunicates which have right gill slits as well as left ones. The presumed existence of right gill slits in mitrates can be regarded as a synapomorphy shared by them with all living adult chordates, but which cornutes lack. If so, all cornutes can be assigned to the stem group of chordates. All known mitrates on the other hand, can be assigned, on the basis of special features, to the three extant chordate subphyla, being either stem acraniates, or stem tunicates or stem vertebrates. All known mitrates are therefore primitive members of the chordate crown group. Theoretically, there should have been mitrates which were advanced stem chordates, and probably others which belong to the stem group of the group [tunicates + vertebrates]. These theoretically existent forms, however, have not been found.

The homologies and notation of head plates in the cornutes and mitrates will not be discussed in this book, except the marginal plates near the tail. These may straddle the median plane of the tail, or be left or right of it, and they may be dorsal or ventral. They are given the notation M (= marginal) with the subscripts: $_P$ = posterior (i.e. in the median plane); $_R$ = right; $_L$ = left; $_{1,2,3}$ = position in sequence away from median plane; $_D$ = dorsal; $_V$ = ventral. Thus M_{1LD} is the first left dorsal marginal plate. Marginal plates near the tail are homologous in

different species if given the same notation. This is not true for plates farther from the tail, where separate notations are needed for comparison and for description (see Jefferies & Lewis, 1978, p. 223) but these plates are not considered here.

In this chapter and the next I shall take individual cornutes and mitrates in the sequence required by the morphological argument. In Chapter 9, I shall place the various calcichordates on a cladogram, together with the living animals described in previous chapters, and shall try to present a phylogenetic history of the deuterostomes as a whole. In Chapter 10 I aim to refute a number of opinions which differ from my own.

Cothurnocystis elizae is one of the best studied cornutes. It is known from a few hundred specimens all of which came from a single bed at a single locality near Girvan in Scotland. Nearly all the specimens were found in the late nineteenth and early twentieth centuries by devoted amateur collectors. One of the most active of these was Eliza Gray of Edinburgh, after whom Bather named the species when he described it in 1913. It has since been studied by Ubaghs (1967b) and by me (1968, 1969; Jefferies & Lewis, 1978). The bed that yielded all the known specimens is known as the Starfish Bed, because of the abundance of complete starfishes in it, and belongs to the latest Ordovician, Ashgill Series. *C. elizae* is thus about 440 million years old. It and *Scotiaecystis curvata*, which comes from the same locality and horizon, are the youngest cornutes known. In morphology it is nevertheless rather a primitive cornute in most ways. The rock in which it occurs is a hard, fine-grained siltstone and preserves surprisingly good detail. Fossils associated with *Cothurnocystis elizae* include starfishes, ophiuroids, primitive echinoids, trilobites, bryozoa, brachiopods and molluscs and indicate a shallow marine environment. The preserved specimens of *C. elizae* are articulated and complete. They were probably killed by sudden burial on the sea floor and never afterwards uncovered. Goldring & Stephenson (1972) have discussed the conditions of deposition of the Girvan Starfish Bed.

As to preservation, most of the specimens of *Cothurnocystis elizae* do not retain the actual skeleton but are external moulds i.e. holes in the rock where the original plates have been dissolved away. A very good idea of the shapes and surface texture of the individual plates can be got by making latex casts of these holes. The species was reconstructed by drawing several different aspects simultaneously on a drawing board, using information from several specimens as need was. In the rest of this chapter I shall refer to *Cothurnocystis elizae* as *Cothurnocystis*, although several other species of the genus are known.

Cothurnocystis consists, like all other calcichordates, of a head and a tail (fig. 7.2a–e). The head was shaped

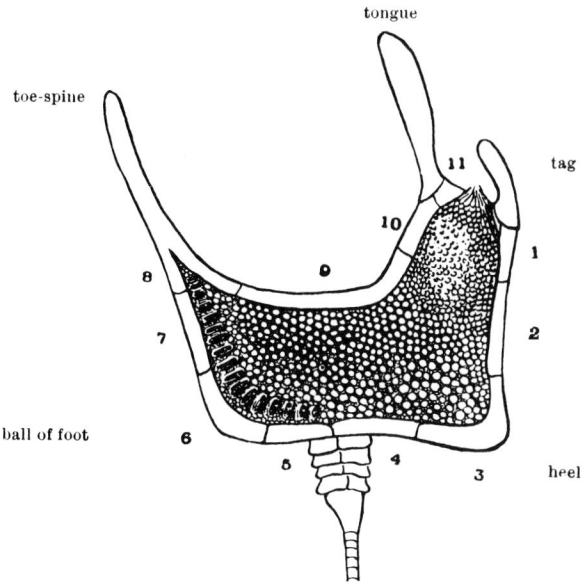

tongue

toe-spine

tag

11

10

1

8

9

7

2

ball of foot

6

5

4

3

heel

7.1. Bather's reconstruction of the dorsal aspect of Cothurnocystis elizae *to show his terms for the parts, based on a comparison with a boot; the numbers are his plate numbers; from Bather (1913, fig. 15, p. 399).*

like a long-toed medieval boot (*cothurnos* = boot). And, indeed, in his original description, Bather recognized parts that he called the toe-spine, tongue, tag, the heel and the ball of the foot (fig. 7.1). The tail is inserted in the middle of the 'sole' of the 'boot'.

As to orientation, the head is morphologically anterior, by comparison with a tunicate tadpole or a fish, and the tail morphologically posterior. One 'side' of the 'boot' probably lay on the sea floor and was morphologically ventral while the other 'side' faced upwards and was dorsal. And this also defines morphological right and left. However, there are many indications that *Cothurnocystis*, like all other calcichordates, habitually crawled rearwards, over the sea floor, pulled by the tail. Philip, who regards the tail as a locomotory stele, agrees that it is morphologically posterior. Ubaghs, for whom it is a feeding arm, sees it as anterior.

The skeleton of the head consists of a frame of marginal plates to which were attached a dorsal and a ventral flexible plated integument like the skins on a drum. The ventral integument was crossed by a strut connecting the anterior and posterior parts of the frame. Some of the most posterior marginal plates bear spikes on their ventral surfaces which would have served to lift the head above the sea floor and are evidence that this surface was indeed downwards and ventral. The anterior frame is produced into three appendages (right oral, left oral and left). These

correspond respectively to what Bather called tag, tongue and toe-spine. They are knife-like with sharp horizontal edges and, if the head was oriented in life with the ventral surface resting on the sea floor, they would slope anteriorly downwards. This slope of the appendages, together with the fact that the spikes of the posterior part of the ventral surface are sharpened anteriorly, would hinder the head from moving forwards and suggests that any locomotion was rearward and probably somewhat rightward, parallel to the two fixed appendages. The right oral appendage was articulated at its base so as to wave from side to side whereas the other appendages were rigid parts of the frame.

The tail consists of three parts – fore, mid and hind. The hind tail ends abruptly, as if a more distal part had been lost, and this, so far as known, is a universal rule for cornutes and mitrates. Four marginal plates form the tail insertion, being the left dorsal, left ventral, right dorsal and right ventral first marginals (M_{1LD}, M_{1LV}, M_{1RD}, M_{1RV}, fig. 7.10).

The openings of the head are anatomically crucial. Most striking of all is a series of about 16 openings situated in the left part of the dorsal integument. Each of them has a complicated structure (fig. 7.3). It is surrounded by a frame built of an anterior and a posterior U-shaped plate, and the two U's articulate together by their free ends. The anterior U is shorter than the posterior one and in life there was fixed to it a flexible, outwardly convex flap, stiffened with tiny platelets. The free end of the flap extended rearwards to close the opening of the posterior U. The fact that the flap was attached to the anterior U but not to the posterior one is shown by some specimens where, after death, the anterior U's have slipped slightly sideways compared with the posterior U's. The flaps have then moved with the anterior U's and are displaced relative to the posterior ones. The integument immediately round these openings is very fine-plated.

These openings can readily be interpreted as outlet valves. If pressure were high inside the head, the flexible dorsal integument where these openings were located would bulge upwards; the two U-plates around each opening would rotate about the horizontal articulation between them so that the articulation arched upwards relative to the anterior and posterior ends of the opening; the flap of the anterior U would lift away from the posterior U – partly because of the flap's flexibility and partly because of the mutual rotation of the two U plates; and water would escape. Pressure inside the head would therefore fall and the process would reverse; the integument where the openings were would deflate; the two U-plates of each opening would rotate about the articulation between them so that the articulation moved downwards relative to the anterior and posterior ends of the

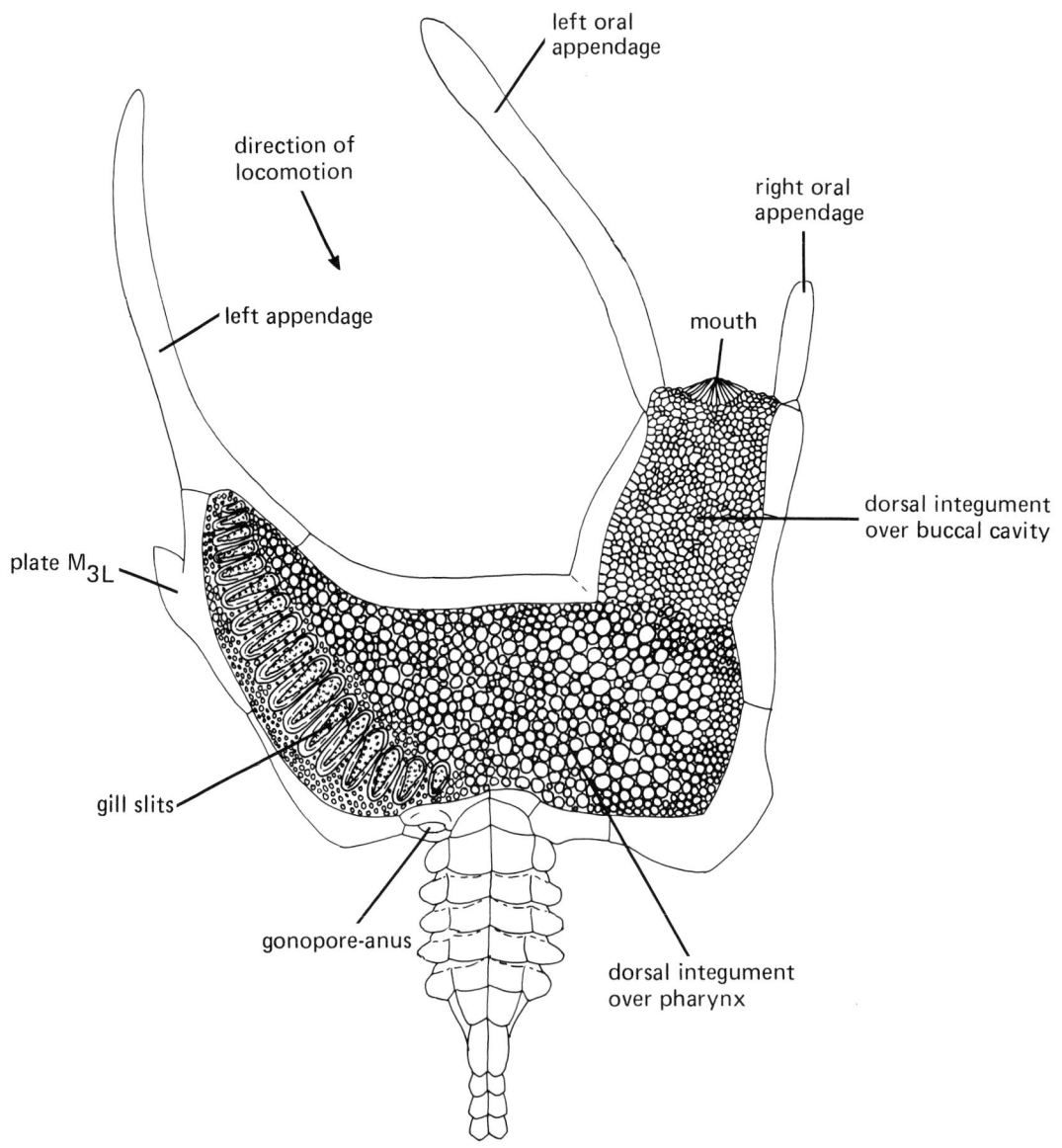

left oral
appendage

direction of
locomotion

right oral
appendage

left appendage

mouth

dorsal integument
over buccal cavity

plate M$_{3L}$

gill slits

gonopore-anus

dorsal integument
over pharynx

Fig. 7.2a.

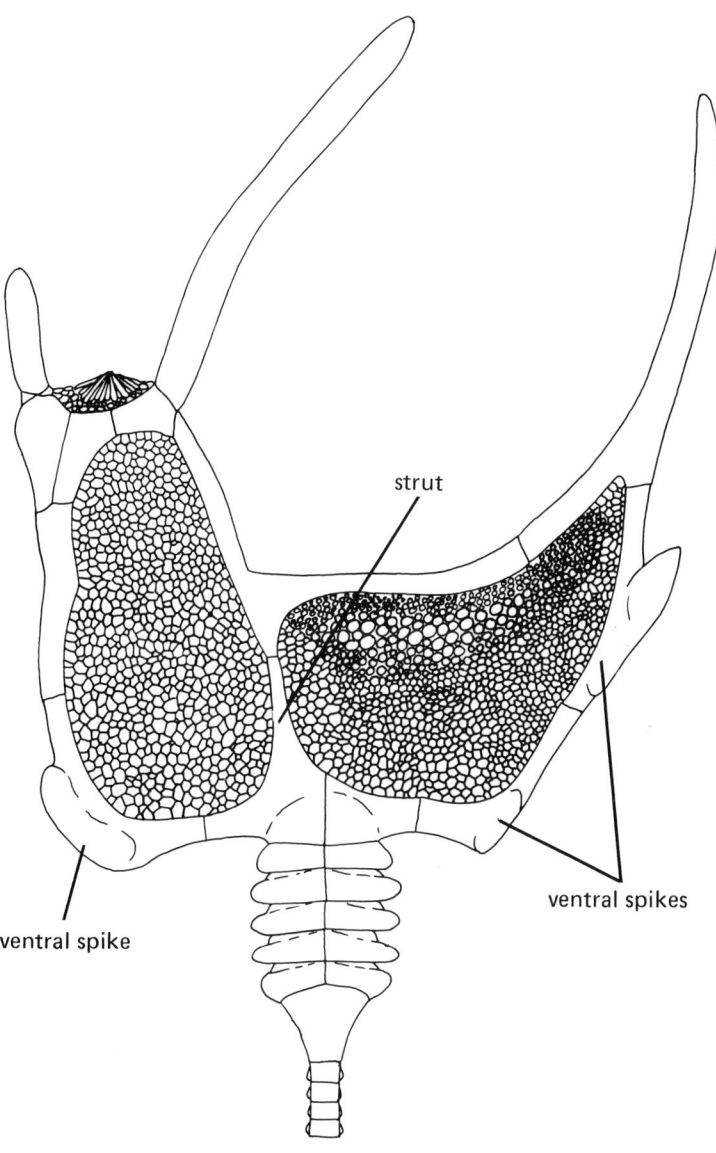

Fig. 7.2b.

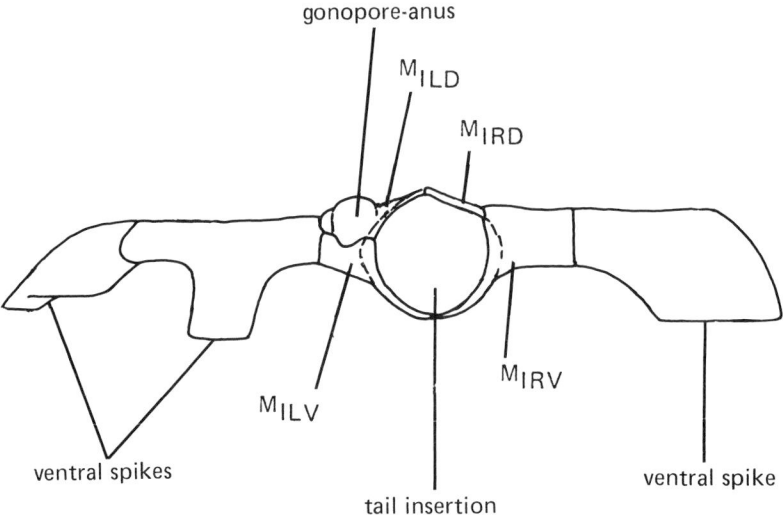

gonopore-anus

M$_{ILD}$

M$_{IRD}$

M$_{IRV}$

M$_{ILV}$

ventral spikes

ventral spike

tail insertion

Fig. 7.2c.

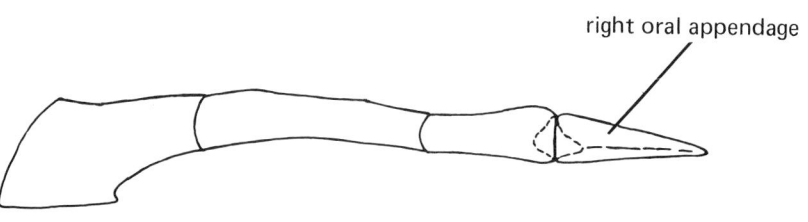

right oral appendage

Fig. 7.2d.

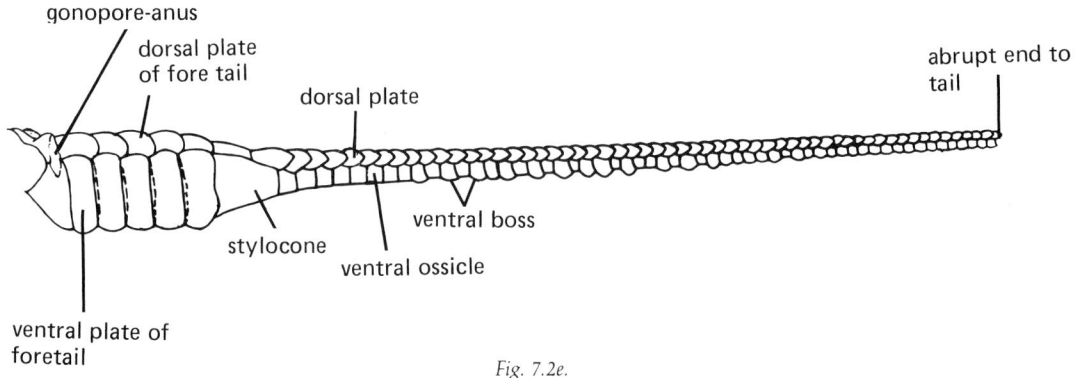

gonopore-anus

dorsal plate
of fore tail

dorsal plate

abrupt end to
tail

ventral boss

stylocone

ventral ossicle

ventral plate of
foretail

Fig. 7.2e.

7.2. Cothurnocystis elizae. *Reconstruction of external features in various aspects: a) dorsal, b) ventral, c) right lateral, d) posterior and e) left aspect of tail and adjacent parts of head; from Jefferies (1981a, figs 3a, 3b; 1968, figs 1a, d, g).*

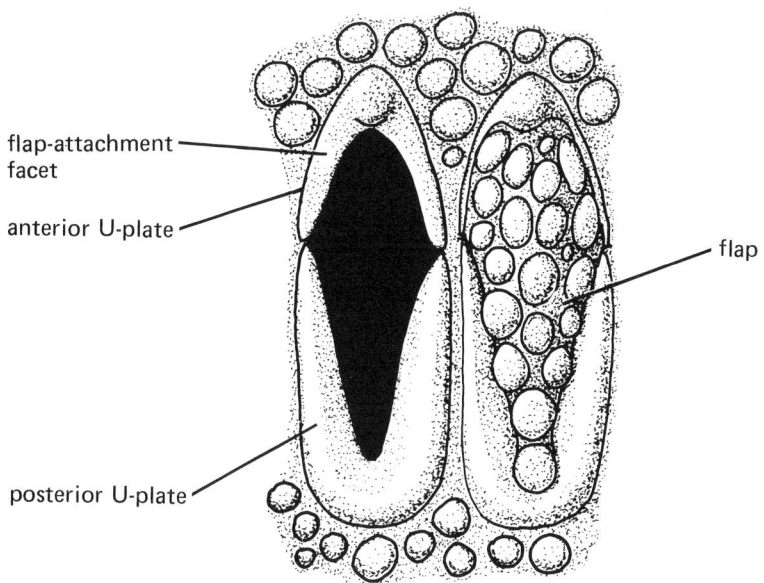

flap-attachment facet

anterior U-plate

flap

posterior U-plate

7.3. Cothurnocystis elizae. *Reconstruction of two gill slits in dorsal aspect, on right with flap and on left without it; from Jefferies (1981a, fig. 4, p. 494).*

opening; the flap of the anterior U would fall back in place on the posterior U and, aided by its outwardly convex shape, the flap would prevent water from entering.

Since the openings suggest outlet valves, they can plausibly be seen as gill slits. They can be compared in function with the passive outlet valves that guard the gill slits of a shark such as *Acanthias*. As already mentioned, being on the left side of the head, such gill slits compare with the primary gill slits of amphioxus which would thus represent a Haeckelian recapitulation. Gislén (1930) was the first to interpret these openings as gill slits, after Bather (1913) had specifically warned against it. If they are gill slits then the branchial current would issue from them almost horizontally rearwards, parallel to the length of the flaps.

Another opening (fig. 7.2a,c), clearly of different nature, was situated just left of the tail, between plates M_{1LD} and, M_{1LV}. It would thus be in the outwash from the rightmost gill slits which suggests that it was likewise an outlet, releasing something into the branchial current (fig. 7.19b,e). Inside the head there are two grooves (fig. 7.4), one large and one small, in the internal face of the marginal plates, running vertically down from this opening. These grooves pass downwards and rightwards into a larger groove which runs horizontally across the floor to a point just right of the tail and then turns forward to disappear into the general cavity of the head. This place just right of the tail is where the primitive genus *Cereatocystis* had its

gonopore and anus, as identified by comparison with primitive echinoderms (see below, fig. 7.17d). This suggests that near the opening in *Cothurnocystis* the smaller groove carried the gonoduct, the larger groove the rectum and that the opening itself was a gonopore–anus. The material washed away by the branchial current would be gametes and faeces. The genus *Scotiaecystis* provides additional evidence that the opening was an outlet (see below, fig. 7.14).

Yet another opening (fig. 7.2a,b) is at the anterior end of the head, between the marginal plates already referred to as the right and left oral appendages. This is the largest opening of the head and it is guarded by a pyramid of spike-shaped plates. Its anterior position, large size and the fact that all the other openings seem to have been outlets suggest that it was the mouth. The resemblance of the plates round it to the anal pyramid of some echinoderms led Bather (1913) and Ubaghs (1967b, p. S513) to see it as an anus. Against their interpretation, however, is the fact that only in some cornutes is its skeleton developed as a pyramid of spike-shaped plates. More often it is flexible below and rigid above as in the cornute *Ceratocystis* and the mitrate *Mitrocystella*. Sometimes, as in the mitrate *Placocystites*, two of the spike-shaped plates of this flexible boundary have developed into long curved sabre-like plates which is difficult to explain if the opening were indeed an anus. A pyramid of plates does not in itself suggest an outlet. It probably signifies only an opening

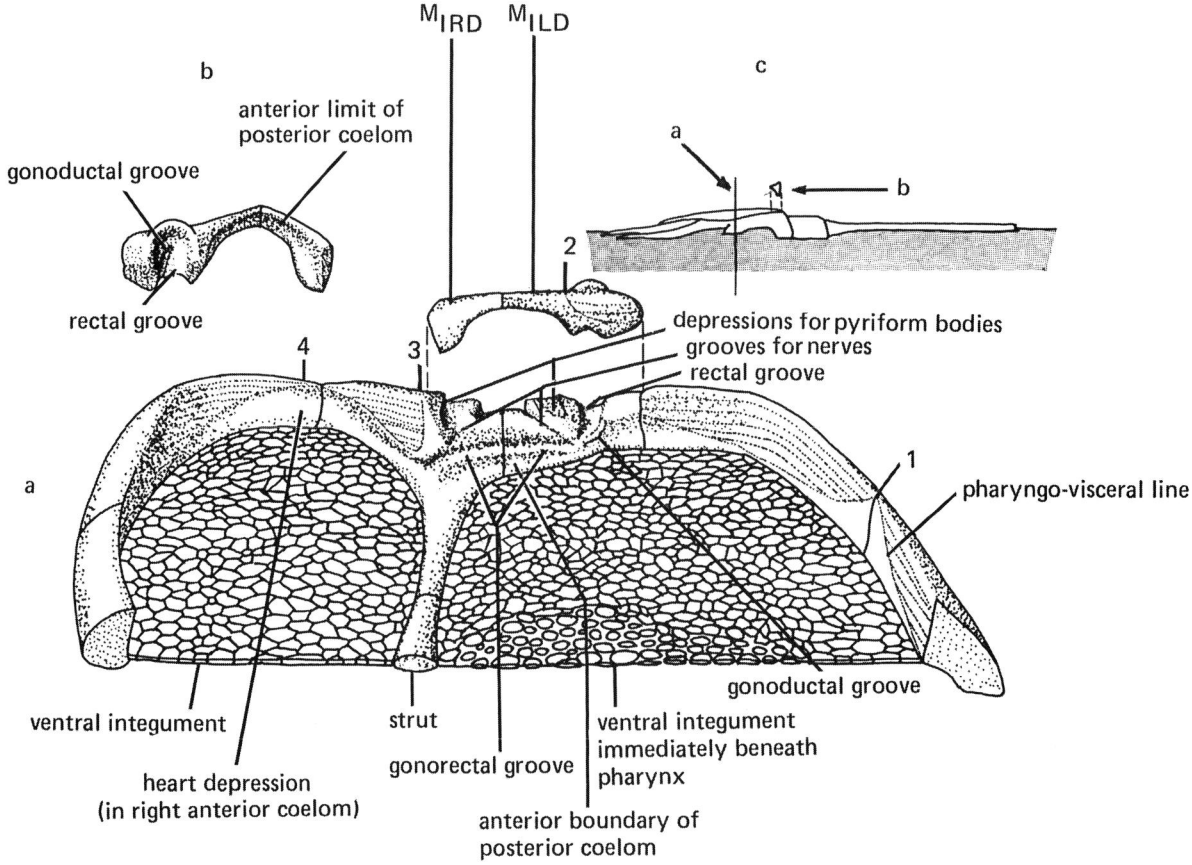

7.4. Cothurnocystis elizae. *a) Internal anatomy of the skeleton of the posterior part of the head in antero–dorsal aspect with the dorsal integument removed and plates M_{1LD}, M_{1RD} lifted up. b) Plates M_{1LD}, M_{1RD} seen from behind. c) Animal in left lateral aspect; X–X is the plane of section in (a) and A is the line of sight; from Jefferies (1969, text fig. 6a, b, c, p. 506).*

with a flexible integument and a sphincter. Thus in tunicates both the mouth and atrial opening are guarded by pyramids formed from spike-shaped lathes of stiffish tunicin, as, in particular, in the plated genus *Chelyosoma* (fig. 7.5a). Again, the holothurian *Psolus* has spike-shaped orals round the mouth (fig. 7.5b,c), associated in its case with a retractable introvert. The arguments for seeing this opening as an anus are therefore invalid. Everything suggests it was the mouth.

Many differing interpretations of the other head openings have been given. As to what I have called gill slits, Bather (1913) thought they were multiple mouths, but, working before latex casting had been invented, he had not properly understood their structure. Jaekel (1918, p. 122) thought they were reproductive. Ubaghs thinks they may have been respiratory (1967b, p. S541). As to the gonopore–anus, this was discovered almost simultaneously by Ubaghs and myself and Ubaghs sees it as

probably a hydropore (1967b, p. S518). (In my view only one known cornute had a hydropore, and this was *Ceratocystis*.)

The anatomy inside the head of *Cothurnocystis* can be reconstructed from the detailed sculpture of the internal surfaces of the marginal plates, variation in the plating of the integuments, and comparative and functional arguments. There is direct evidence for four chambers inside the head and comparative evidence for a fifth (fig. 7.6c,d).

The first chamber lies just behind the mouth, in the 'ankle' part of the 'boot'. On the right it is delimited behind by a low ridge present in all specimens on the inner face of the frame. On the left it is limited behind by the sharp bend in the marginal frame that separates the 'ankle' from the 'foot'. In addition there is in one specimen, just anterior to this sharp bend, a low ridge on the inner face of the frame which probably marks the posterior limit of the chamber. Being immediately behind the mouth the

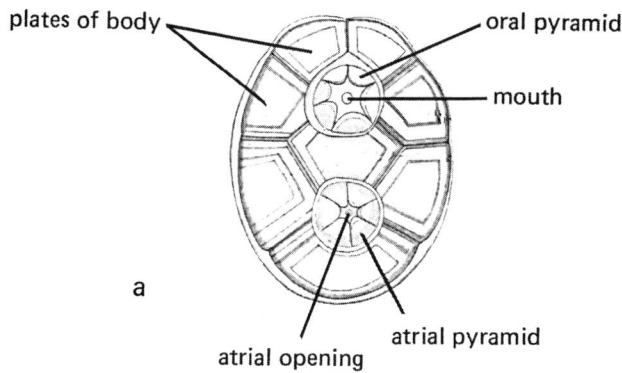

plates of body

oral pyramid

mouth

a

atrial opening

atrial pyramid

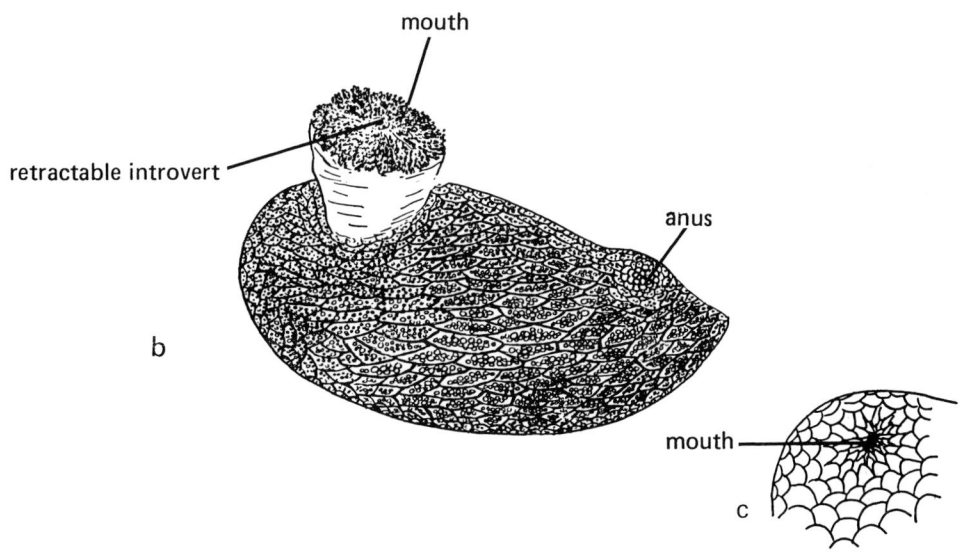

mouth

retractable introvert

anus

b

mouth

c

7.5. The functional significance of pyramids of spike-shaped plates. a) The benthic tunicate Chelyosoma *in which the tunic is developed as a plated skeleton and cones of stiff plates exist around both the mouth and the atrial opening; from Seeliger & Hartmeyer (1893–1911, pl. 34, fig. 1). b) The spike-shaped plates around the retracted introvert at the mouth of the Holothurian* Psolus. *c)* Psolus *with the introvert extended; Hyman (1955, fig. 50, p. 127).*

chamber can be called the buccal cavity. It was probably homologous with the buccal cavity of amphioxus, tunicates and vertebrates.

The second chamber is almost hemispherical and was situated just in front of the tail. It was in contact with the four first marginal plates (M_{1LD}, M_{1LV}, M_{1RD}, M_{1RV}). The ventral plates (M_{1LV}, M_{1RV}) are drawn out anteriorly into a shelf which forms part of the floor of the head and continues forward as the strut that crosses the ventral integument. The gonorectal groove runs across this shelf from right to left (fig. 7.4a) and passes at the left into the separate gonoductal and rectal grooves which climb verti-

cally upwards to the gonopore–anus. The ventral anterior boundary of the chamber is defined by a low, anteriorly convex, curved ridge on this shelf. In front of this ridge, but not behind it, some specimens show the characteristic pharyngeal sculpture (see below) and there are traces which suggest that the chamber opened forward by a funnel into the pharynx at right and left (Fig. 7.7c). The dorsal anterior surface of the chamber would have contacted the concave, ventro-posterior surface of the two dorsal first marginal plates (M_{1LD}, M_{1RD}). The chamber can be called the posterior coelom. It seems to have enveloped the dorsal surfaces of the gonoduct and rectum.

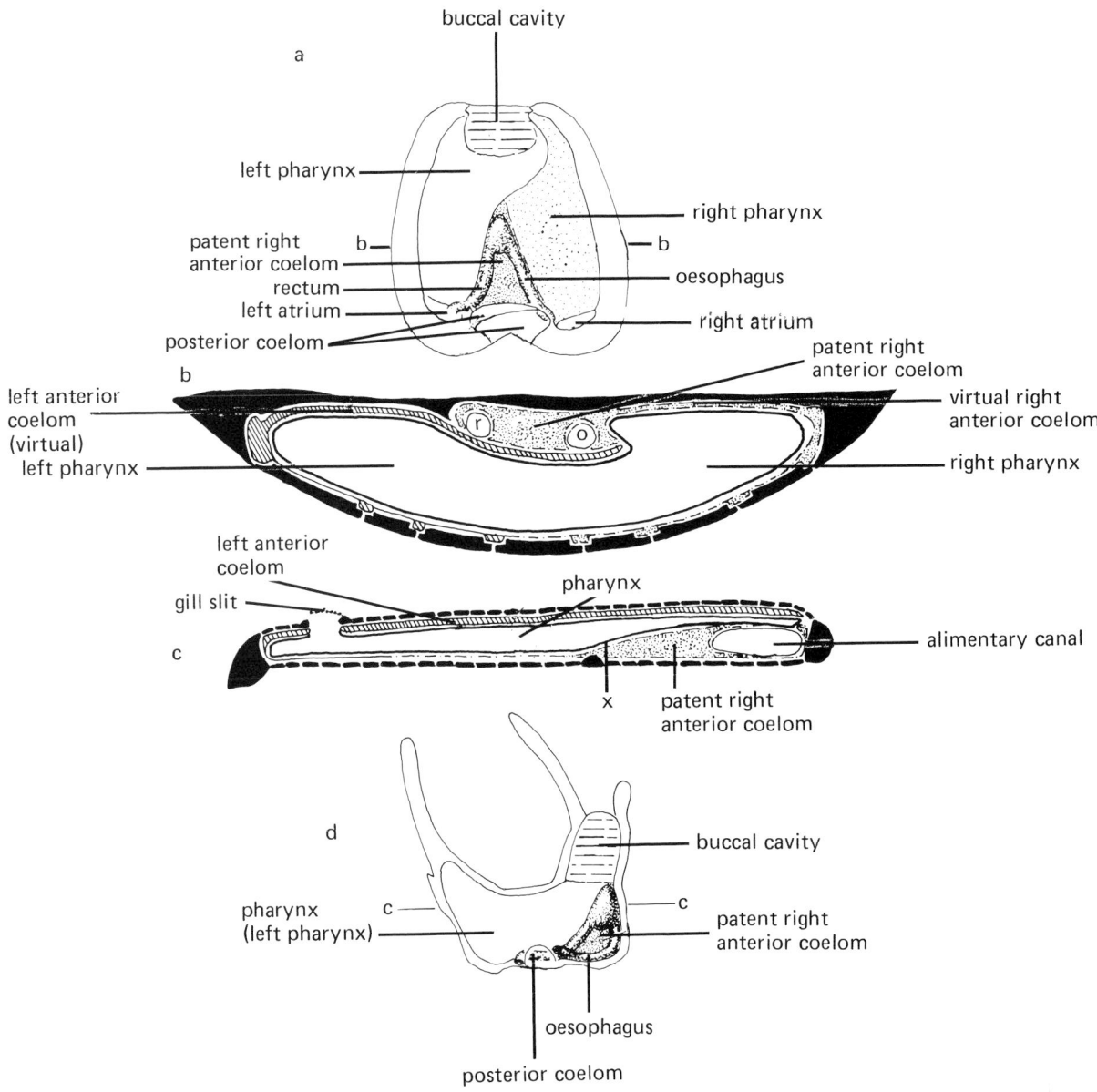

7.6. *The head chambers in a mitrate and a cornute. a) Dorsal aspect of the mitrate Mitrocystites, b) transverse section of the same through b–b in (a), c) transverse section of the cornute Cothurnocystis through c–c in (d), and d) dorsal aspect of the same; from Jefferies (1981a, fig. 5, p. 496).*

Comparison with mitrates suggests that the posterior coelom of cornutes is homologous with the left epicardium of tunicates.

The remaining two chambers based on direct evidence were situated, one overlying the other, between the buccal cavity and the posterior coelom, in the 'foot' part of the 'boot' (figs 7.4, 7.7). The upper of these chambers in very

large specimens sometimes shows a characteristic sculpture on the inner faces of the marginals consisting of distinct horizontal lines crossed by weaker vertical ones. This sculpture suggests the unity of the upper chamber and has a precise but undulating lower boundary. The sculpture is developed in particular on the front surface of the left dorsal marginal plate (M_{1LD}) (fig. 7.4). However, it

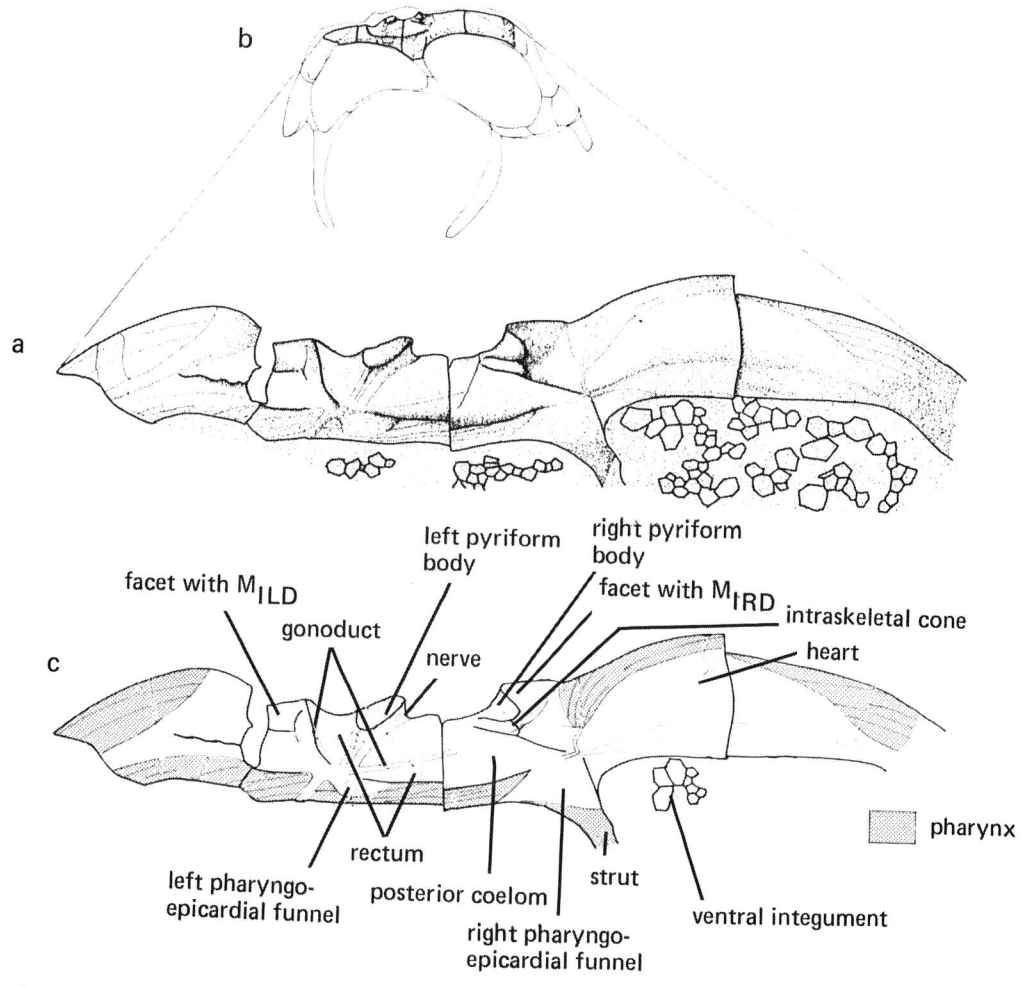

b

a

c

facet with M_{1LD}

gonoduct

left pyriform
body

nerve

right pyriform
body

facet with M_{1RD}

intraskeletal cone

heart

left pharyngo-
epicardial funnel

rectum

posterior coelom

right pharyngo-
epicardial funnel

strut

ventral integument

pharynx

7.7. Cothurnocystis elizae. *Superficial internal anatomy of the posterior part of the natural internal mould of the head, in postero–ventral aspect and simulating the soft parts. a) Camera-lucida drawing, b) position within animal, and c) interpretation; from Jefferies & Lewis (1978, fig. 26, p. 282).*

does not extend over the whole front surface of this plate but is delimited ventrally and to the right by a curved line. This reaches the dorsal edge of the plate, and therefore the dorsal integument, just right of the rightmost gill slit, at point 2 in fig. 7.4. The sculpture coincides here, therefore, with the presence of gill slits in the integument, which suggests that the sculptured chamber is the pharynx with gill slits opening through its roof. To judge by the distribution of the sculpture, the pharynx was most capacious in the posterior left part of the head where the gill slits were.

The lower of the two chambers for which there is direct evidence in the 'foot' part of the 'boot', always lacks the pharyngeal sculpture and seems to have been most capacious in the right part of the 'foot' near the 'heel' (figs 7.4, 7.7). The gonorectal groove emerges from it, which

suggests that it carried the gonad and most of the nonpharyngeal gut. It probably also contained the heart and the equivalent of the axial gland of echinoderms. The argument for the presence in it of these two organs, however, is complicated, and is based on *Ceratocystis*, on the one hand, and mitrates on the other. The likely position of the heart is a spindle-shaped groove in the inner surface of the posterior right marginals (M_{1RV}, M_{2RV}, figs 7.4, 7.7). The lower chamber in the 'foot' can be called the right anterior coelom. It is the fourth of the chambers based on direct skeletal evidence.

The plating of the integuments helps to confirm the existence of right anterior coelom and pharynx. In certain middle-sized specimens there are two types of plates in the integument – polygonal and circular (figs 7.2a,b). The polygonal plates are in close contact with their neighbours whereas the circular plates are separated from them by gaps which in life would have contained soft tissue, presumably including muscle since the integuments were flexible. The circular plates are best developed, in contradistinction to the polygonal ones, where the pharynx could reasonably be supposed to make contact with the integument – in the anterior left part of the ventral integument and over the whole dorsal integument except the region of the buccal cavity. Presumably the pharynx had particularly muscular walls so that it could contract, either intermittedly or continuously, like a bellows. In large specimens of *Cothurnocystis*, however, all the integument plates are circular with gaps, presumably muscular, between them; and in small specimens, although the integuments were already obviously flexible, all integument plates are polygonal and in close contact with their neighbours.

The fifth chamber, based on comparative evidence only, is the left anterior coelom (fig. 7.6c). It was probably purely virtual (i.e. without a cavity) and overlay the pharynx forming the roof of the head. I shall argue for its existence when dealing with *Ceratocystis*.

The tail of *Cothurnocystis* as already mentioned comprises three regions – the fore, mid and hind tail, as with every other known calcichordate (fig. 7.2a,b,e). The fore-tail skeleton consists of about five rings of plates, each ring of four plates each (left and right dorsal and left and right ventral). The rings are so constructed that successive rings touch each other in the dorsal and ventral mid lines but overlap laterally where they are separated from each other by gaps. This indicates that the fore tail could flex to left and right but not up and down (or scarcely so). The skeleton of the hind tail (fig. 7.2e) consists of about 56 ventral, approximately hemicylindrical ossicles and a roughly equal number of pairs of dorsal plates. The ventral ossicles are of two types: either they bulge ventrally and have curved posterior surfaces or else they are rather accurately hemicylindrical ventrally and have plane posterior surfaces. The curved interossicular surfaces are cylindrical about a transverse axis and, if the ossicles were connected by soft tissue in life, this would allow the tail to flex in a ventral plane (fig. 7.8b). The bulbous ventral surfaces presumably served to grip the sediment in downward flexion. The skeleton of the mid tail consists of a massive ventral element – the stylocone – with two pairs of dorsal plates above it. The stylocone resembles two hind-tail ossicles fused together and is deeply excavate anteriorly.

The dorsal surface of the ventral hind-tail ossicles has a complicated sculpture (fig. 7.8a). There is a very obvious median groove from which a pair of transverse grooves go off in each ossicle to finish in rather ill-defined lateral pits.

The dorsal plates are arcuate, attached ventrally to a facet of the ventral ossicle and meet together in the dorsal mid line, in my opinion at a suture.

As to the soft parts of the tail, the fore tail, to judge by its plates, seems to be adapted for flexion to left and right. For this purpose it would need muscles and the segmental nature of the skeleton suggests they might have been present as muscle blocks – a suggestion which is supported by the mitrate situation. The compressive forces involved in flexion would in *Cothurnocystis* be largely taken up by the skeleton. In most other cornutes and all mitrates, however, the skeleton was loose. In such forms some anti-compressional structure would be required in the soft parts of the fore tail and most probably it was a notochord in the central axis. This would end anteriorly at the point where it would no longer serve an anti-compressional function – where the tail joined the head. Probably the notochord extended rearward into the hind tail, filling the median groove. Evidence from the mitrates, where the ossicles of the hind tail are dorsal and mould the dorsal surface of the notochord, indicates the presence of a dorsal nerve cord and spinal nerve ganglia in these animals which presumably existed likewise in cornutes. The hind tail with its imbricating dorsal plates and its in-part cylindrical interossicular joints seems to be adapted for flexing up and down. For this it would need muscles which were presumably situated in the lumen between the plates and the ossicles, and these muscles would act against the column of ossicles, rendered elastic by its interossicular connective tissue. The segmented structure of the hind-tail skeleton, especially of the dorsal surface of the ventral ossicles, suggests, once again, that the soft parts were segmentally repeated and that the muscles were probably divided into muscle blocks.

The transverse groove coming off the median groove may well have carried blood vessels which in turn were connected with a longitudinal blood vessel inside the notochord. This is unlike the situation in any living chordate and the conclusion depends on a complicated argument as follows (fig. 7.9a,b). It is likely that the hind tail would need a blood supply and that this would be a direct continuation of what existed in the fore tail. The notochord of the cornute hind tail had ossicles immediately beneath it, so the blood supply ran either in, or dorsal to, the notochord. The notochord of the mitrate hind tail (the latter being equivalent to the hind part of the cornute fore tail) had ossicles immediately dorsal to it so here the blood supply ran either in, or ventral to, the notochord (disregarding forms like *Mitrocystella* which have a dorsal longitudinal canal as a specialization). It follows that: (1) either the blood supply moved about in phylogeny in a complicated manner which would involve passing through the notochord; or (2) it ran, both in cornutes and mitrates,

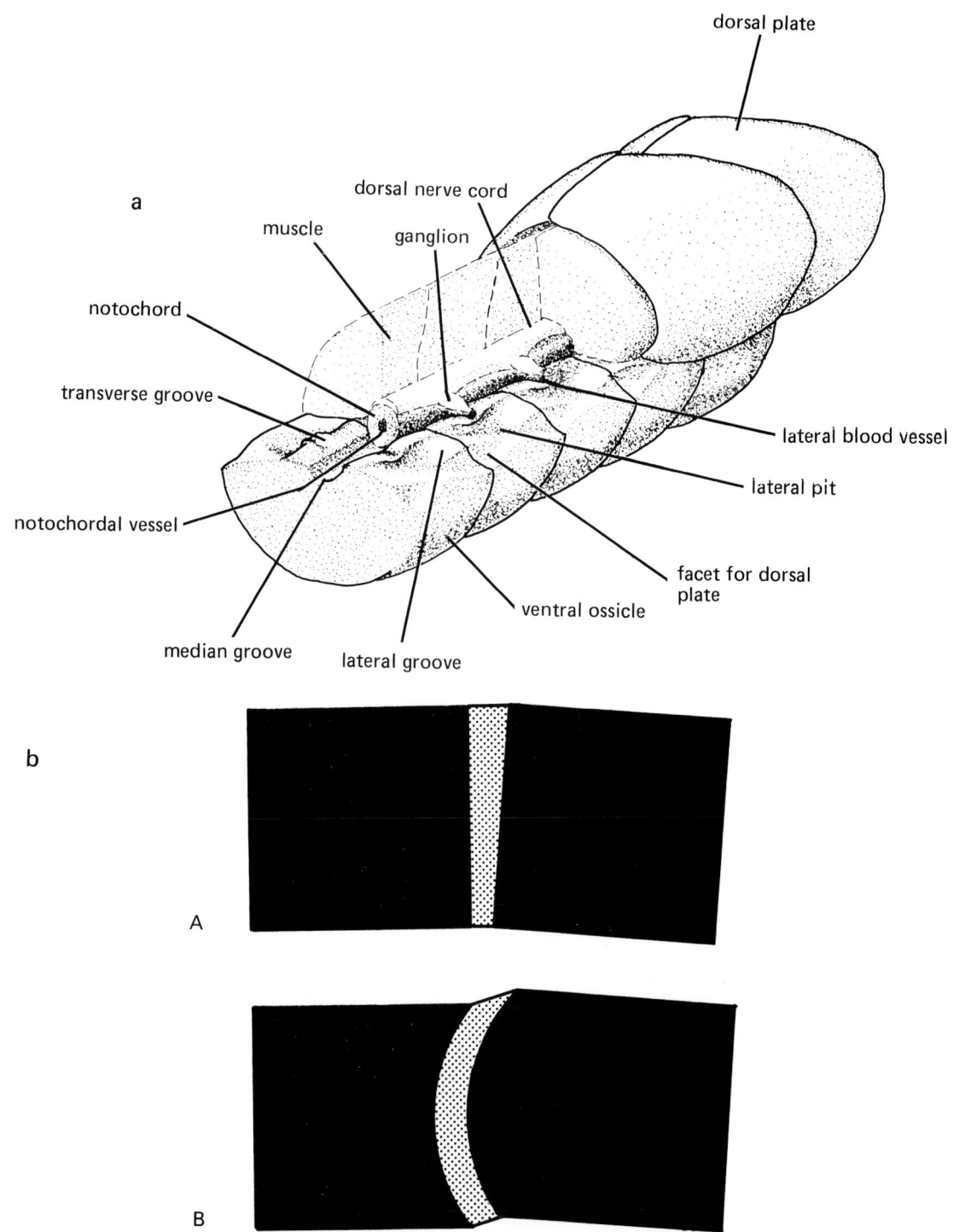

a

dorsal plate

dorsal nerve cord

muscle

ganglion

notochord

transverse groove

notochordal vessel

median groove

lateral groove

ventral ossicle

facet for dorsal plate

lateral pit

lateral blood vessel

b

A

B

7.8. Cothurnocystis elizae. *Structure of the hind tail, a) Block diagram. The position of the dorsal nerve cord and ganglia is based entirely on the situation in mitrates; the blood vessels in the transverse grooves are conjectural; from Jefferies (1981a, p. 498, fig. 6). b) Mechanism of downward flexion at a curved interossicular joint; in A, bending at the joint requires compressor or dilation, or both, of the intervening soft tissue; in B, bending takes place by distortion of the intervening tissue, but without compression or dilation.*

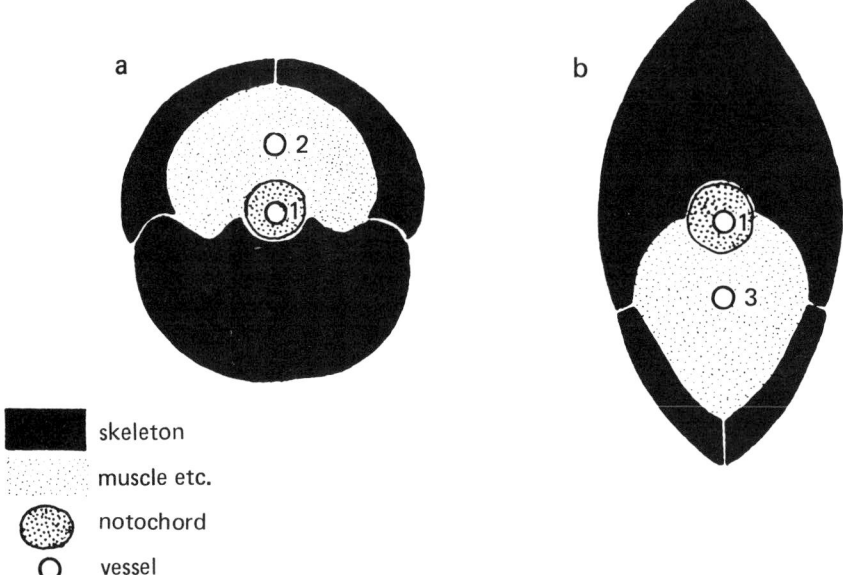

skeleton

muscle etc.

notochord

vessel

7.9. *Argument for the existence of a blood vessel inside the notochord of: a) a cornute, b) a mitrate. If the blood supply to the tail was in the median plane (as suggested by the interossicular vessels that come out of the notochord dorsally in* Placocystites, Mitrocystella *and* Mitrocystites*) then the longitudinal vessel was: inside the notochord (1) or above it (2) in cornutes; and inside the notochord (1) or beneath it (3) in mitrates. If the mitrate condition evolved from the cornute condition, then the simplest supposition is that the vessel was inside the notochord both in cornutes and in mitrates.*

along the long axis of the notochord. Assuming the latter, the transverse grooves could well have carried blood vessels to segmentally arranged muscle blocks. In mitrates such as *Placocystites* and *Mitrocystella* the existence of apparent interossicular vessels which leave the dorsal surface of the notochord in the mid line, probably confirms the presence, in the midline of the notochord, of a longitudinal vessel that gave rise to these interossicular vessels.

The stylocone of the mid tail of *Cothurnocystis*, with its deeply excavate anterior surface, probably received the posterior end of the muscles of the fore tail so that these could move the mid and hind tail as a unit. The stylocone would thus serve much like the cup at the top of an old-fashioned wooden leg.

As to how the tail functioned in locomotion, the fore tail was adapted for lateral flexion, the hind tail would behave as a stiffish rod or else, especially in its distal half where the curved interossicular surfaces are concentrated, would flex ventrally, by the elasticity of the interossicular connective tissue, or dorsally, by the contraction of muscles in the lumen, and the head seems to be suited for sliding rearwards over the sea floor. Most probably the power-source for this rearward movement was the muscles of the fore tail, flexing the tail from side to side. The hind tail could have gripped the sea floor by flexing downwards into it

distally, whereby the ventrally bulging surfaces of some of the ventral ossicles would help to grip the sediment. Thus *Cothurnocystis* would move across the sea floor by waving the tail from side to side, using the downcurved end, much like a punt pole or a mud anchor, to pull the head rearwards.

Ubagh's interpretation of the tail is thoroughly different. He looks for echinoderm resemblances and therefore compares the hind tail, in particular, with an ophiuroid arm. The median groove would house a radial water vessel. The lateral pits would receive the bases of tube feet and the transverse grooves would represent lateral vessels running to the tube feet. Furthermore Ubaghs sees the dorsal plates as corresponding to the cover plates of a crinoid and able to open outward, so that the postulated tube feet could function. Ubaghs' interpretation implies that the fore, mid and hind tails of cornutes are homologous with their namesakes in mitrates. Moreover, in the mid tail, the cornute stylocone would correspond to the mitrate styloid. And in the hind tail the cornute ventral ossicles and dorsal plates would correspond to the ossicles and plates of the mitrate hind tail. (For me the mitrate ossicles were dorsal and plates ventral.) Ubaghs' interpretation is plausible since the skeleton is echinoderm-like and all living echinoderms have a water-vascular system. The arguments against it are as follows:

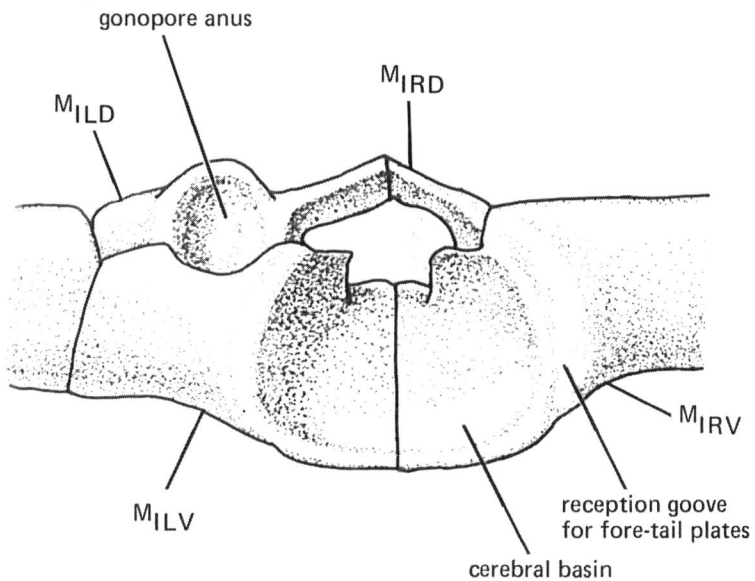

7.10. Cothurnocystis elizae. *The cerebral basin and adjacent structures in posterior aspect; from Jefferies (1968, fig. 3d, p. 254).*

(1) In some cornutes, such as *Scotiaecystis curvata*, the contact between the hind-tail plates and ossicle would prevent the plates from opening outwards like cover plates (fig. 7.16); the same is true in some mitrates where there is often a complicated overlap between ossicles and plates (*Mitrocystella barrandei*) and the median contact between the plates is sometimes (*Lagynocystis pyramidalis*) visibly a broad flat suture. (One of Ubaghs' chief American followers has recently admitted that the plates of mitrates could not open (Parsley, 1982).)

(2) There is evidence that the fore, mid and hind tails, and the hind-tail ossicles and plates, of cornutes are not homologous with those of mitrates. This conclusion depends on reconstruction of the soft parts inside the head and the basically identical asymmetry of the head chambers in cornutes and mitrates. The correctness of the reconstruction of the mitrate head chambers is shown by comparing the reconstructed mitrate pharynx with that of tunicates. For there is a point-by-point resemblance, with several asymmetries the same. And the correctness of my orientation of the mitrates is confirmed by functional morphological arguments, which indicate that mitrates crawled rearward by thrusting the tail forwards and ventralwards into the mud (Jefferies, 1984).

Cothurnocystis was probably a filter feeder, extracting microscopic food particles from suspension by means of a mucous filter, produced by an endostyle, inside the pharynx. The mouth of *Cothurnocystis* was nearly level with the sea floor or slightly raised above the sea floor by the ventral spikes of the head. This position of the mouth suggests that *Cothurnocystis* was a 'deposit feeder', grazing the topmost food-rich layer of sediment by drawing a current in through its mouth. The sediment would probably be sent into suspension from this topmost layer by the side-to-side wagging of the right oral appendage, whose base was articulated, and by the sideways movement, caused by yaw of the head during crawling, of the fixed left oral and left appendages.

The brain of *Cothurnocystis* probably lay in a smooth almost hemispherical cerebral basin, excavated in the first ventral margin plates (M_{1LV}, M_{1RV}) where the tail joined the head (fig. 7.10). It would thus be at the anterior end of the notochord, like the brain of a vertebrate. It was also in the same general position as the recognisably fish-like brain of a mitrate (though this was in contact with dorsal, not ventral, marginal plates). Just in front of the brain of *Cothurnocystis* a pair of grooves diverge from the median line, probably carrying nerves, and there is a pair of pits for the paired pyriform bodies (trigeminal ganglia) (figs 7.4, 7.7c).

I once thought that the stem of crinoids was homologous with the tail of calcichordates, or of chordates in general (Jefferies, 1967–73) though as from Jefferies (1975) I rejected this view. It is necessary to discuss this comparison because Philip (1979) has partly revived it, believing that the tail ('stele') of calcichordates is 'parallel' with the stem or crinoids. Also it has been expounded in a

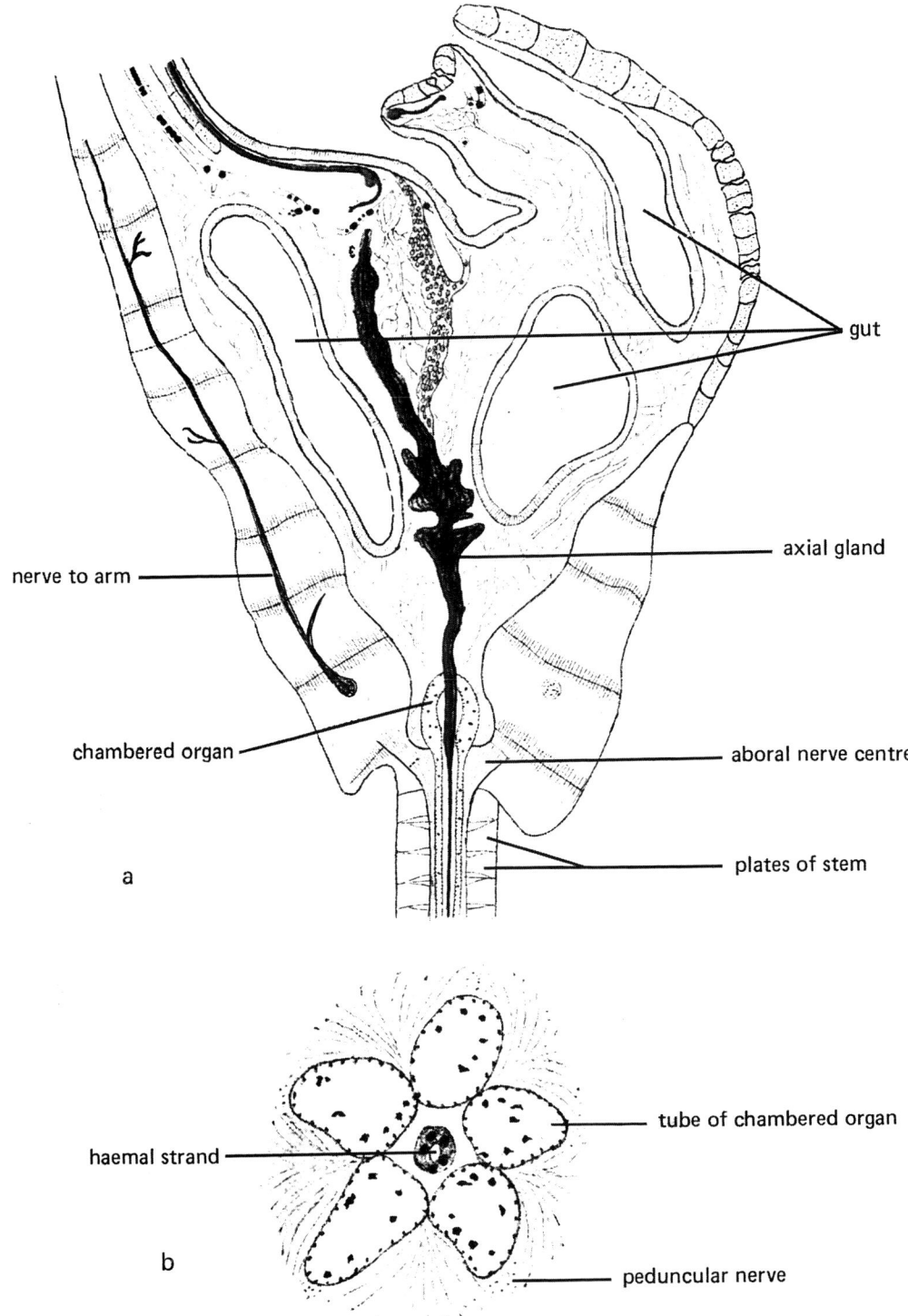

gut

axial gland

nerve to arm

chambered organ

aboral nerve centre

plates of stem

a

tube of chambered organ

haemal strand

peduncular nerve

b

7.11. Structures in the theca and stem of the Recent crinoid 'Pentacrinus' decorus. a) Sagittal section passing through an arm base and the anus; b) transverse section through the soft parts of the central canal of the stem; from Reichensperger (1905, pl. 3, fig. 1; pl. 4, fig. 7c).

stimulating, but now somewhat outdated, French text-book by Babin (1971) of which an English version was published in 1980. The tail of *Cothurnocystis* looks like a crinoid stem externally, for both are slender, somewhat flexible, segmented structures, containing calcitic ossicles and sharply delimited from the rest of the animal. Further-more, as regards soft parts, the crinoid stem, as described by Reichensperger (fig. 7.11a,b; 1905), contains a cylindri-cal central cavity in which are found, from outside inwards: (1) the peduncular nerve, (2) the chambered organ – this consists of five turgid tubes closed distally and proximally and in contact along their lengths so as to form the wall of a hollow cylinder, (3) a cylindrical cavity central to the chambered organ, and (4) a hollow haemal strand (blood vessel) running the length of the cylindrical cavity of the chambered organ. At the proximal end of the chambered organ, where the stem joins the theca, there is an aboral nerve centre which functions as the brain (Langeloh, 1937). And so the peduncular nerve can be compared with the dorsal nerve cord of chordates; the chambered organ with the notochord; the haemal vessel with the blood vessel that probably existed inside the calcichordate noto-chord; and the aboral nerve centre can be compared with the vertebrate brain.

These suggested homologies, proposed by me in 1967 and 1968, were questioned by Eton (1970) because the stem of crinoids arises in ontogeny at the anterior end of the larva. They were really made untenable by the palaeontological work of Sprinkle (1973) who showed that the stem, as a structure with a segmentally repeated skeleton, had arisen at least three times in early echino-derms among forms which, because of their radial sym-metry and lack of gill slits, were disqualified as ancestral to calcichordates. Also the chambered organ arises in ontogeny as five outgrowths from the right somatocoel (Seeliger, 1892), unlike the notochord which pouches directly from the archenteron. All this, as I now think, shows that any resemblance between crinoid stems and calcichordate tails is analogous, not homologous.

To sum up, *Cothurnocystis elizae* can be described, without chance of denial, as one of the strangest-looking animals that ever existed. Having gill slits (left gill slits only, like larval amphioxus), a post-anal tail and an echinoderm-like skeleton it is also zoologically one of the most significant.

Scotiaecystis curvata (fig. 7.12a,b,c), as already men-tioned, is the second cornute found along with *Cothurno-cystis elizae* in the Upper Ordovician Starfish Bed of Girvan. It, too, was originally described by Bather (1913) and has since been studied by me (1968) and by Ubaghs (1967). In many ways it is a more specialized form than *Cothurnocystis* but helps to confirm some of the conclu-sions drawn from that animal. In shape it is basically like

Cothurnocystis with a head and a tail. The head is boot-shaped, but less strikingly so than in *Cothurnocystis*, for there are only two anterior appendages – left and right. The right appendage, fixed to the frame, is homologous with the immobile left oral appendage of *Cothurnocystis*, the right oral appendage having been lost. The anterior part of the frame is curved dorsally in an arch, as is particu-larly obvious seen from in front (fig. 7.2b). This arch explains the name *curvata* given by Bather. Henceforth I refer to *Scotiaecystis curvata* as *Scotiaecystis*.

The skeleton of the head of *Scotiaecystis* consists of a frame of marginal plates on which dorsal and ventral flexible plated integuments are stretched, and the ventral integument is crossed by a strut, much as in *Cothurno-cystis*. The tail shows the same division into fore, mid and hind portions.

The openings of the head are again crucial. There is a series of about 40 slits in the left part of the dorsal integu-ment (figs 7.12a, 7.13) stretching from just left of the tail insertion to near the 'toe' part of the head, in much the same position as the gill slits of *Cothurnocystis*. The structure of the slits is different from that of *Cothurno-cystis* in that adjacent slits are separated by a chevron-shaped plate with the apex of the chevron pointing upwards. The line of chevron apices is curved and roughly bisects the area of dorsal integument in the region. Neigh-bouring chevrons fitted tightly against each other, but the whole chevron apparatus was carried flexibly in the dorsal integument.

This chevron apparatus can again be readily interpreted as an outlet valve. When water pressure was high inside the head the dorsal integument would inflate upwards. The chevron complex, bisecting the integument and therefore situated along the line of maximal stretching, would itself be stretched and gaps between the chev-rons would open, allowing water to escape. This outflow would release the internal pressure, deflate the dorsal integument, the chevrons would clap together and conse-quently water could not enter. Mechanically this is a different structure from the gill-slit series of *Cothurnocystis* but seems adapted to serve the same exhalent function. The slits between the chevrons can reasonably be inter-preted, therefore, as branchial slits and this, conversely, confirms the identification of the slits in *Cothurnocystis*.

An external gonopore–anus is lacking in *Scotiaecystis*. An obvious groove, presumably carrying the gonoduct and rectum, crosses the floor of the posterior coelom from right to left but does not open to the outside (fig. 7.14a,b). Instead it runs forward, left of the tail, as a canal within one of the dorsal marginal plates (M_{1LD}). The anterior end of this canal, where the gonopore–anus was presumably placed, is just behind the rightmost gill slits. This is easy to explain functionally (fig. 7.19c,f). The branchial current

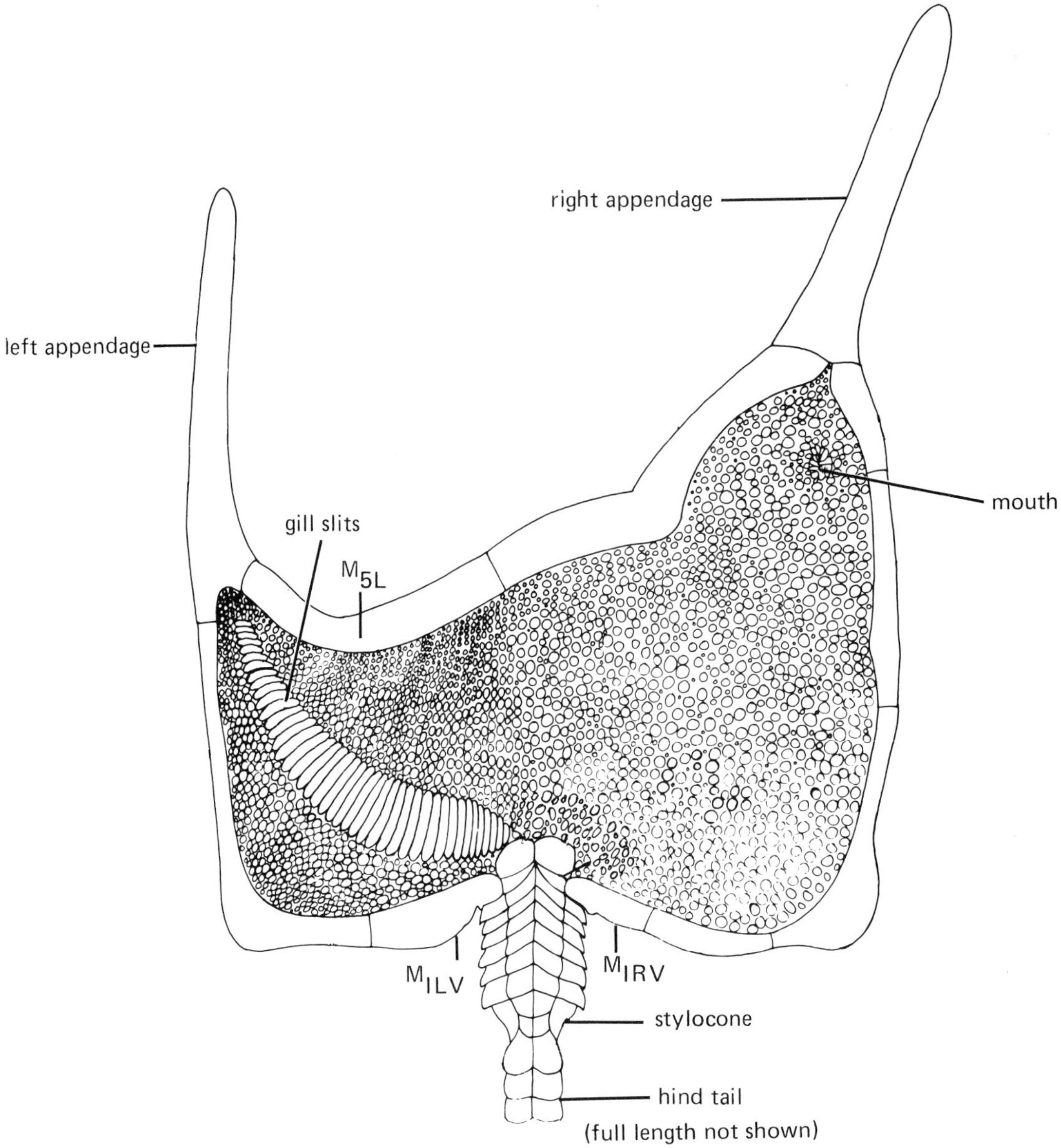

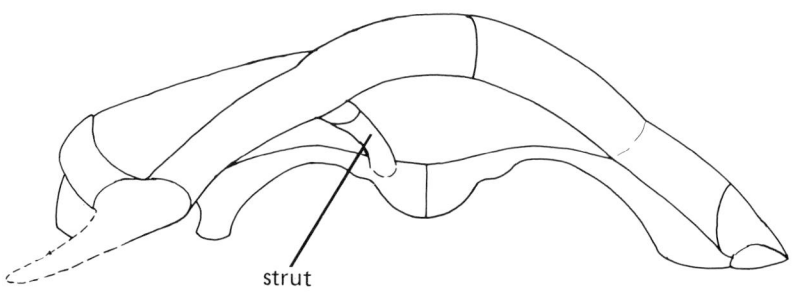

right appendage

left appendage

gill slits

M$_{5L}$

mouth

M$_{ILV}$

M$_{IRV}$

stylocone

hind tail
(full length not shown)

Fig. 7.12a.

strut

Fig. 7.12b.

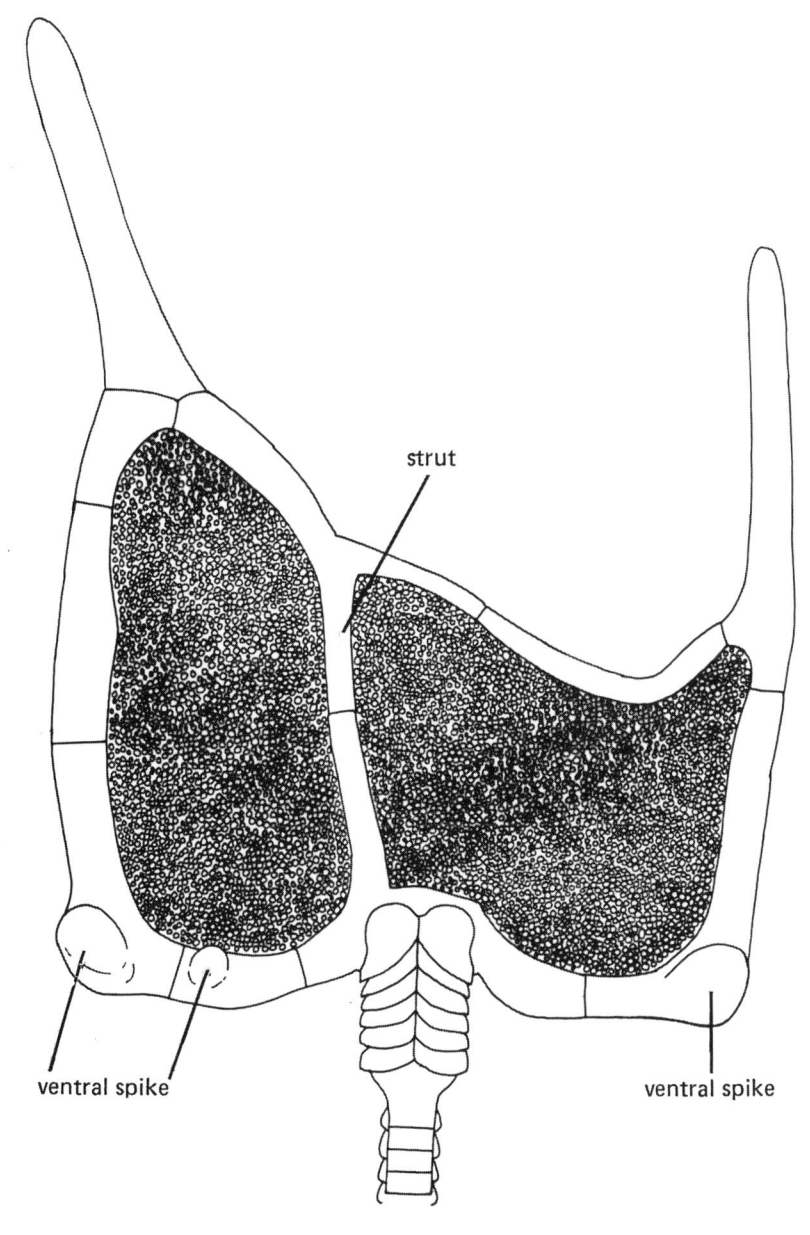

Fig. 7.12c.

7.12. Scotiaecystis curvata. *Reconstruction of external features in: a) dorsal, b) anterior, c) ventral aspects; from Jefferies (1968, fig. 8a, b, c; p. 267, 268, 269.*

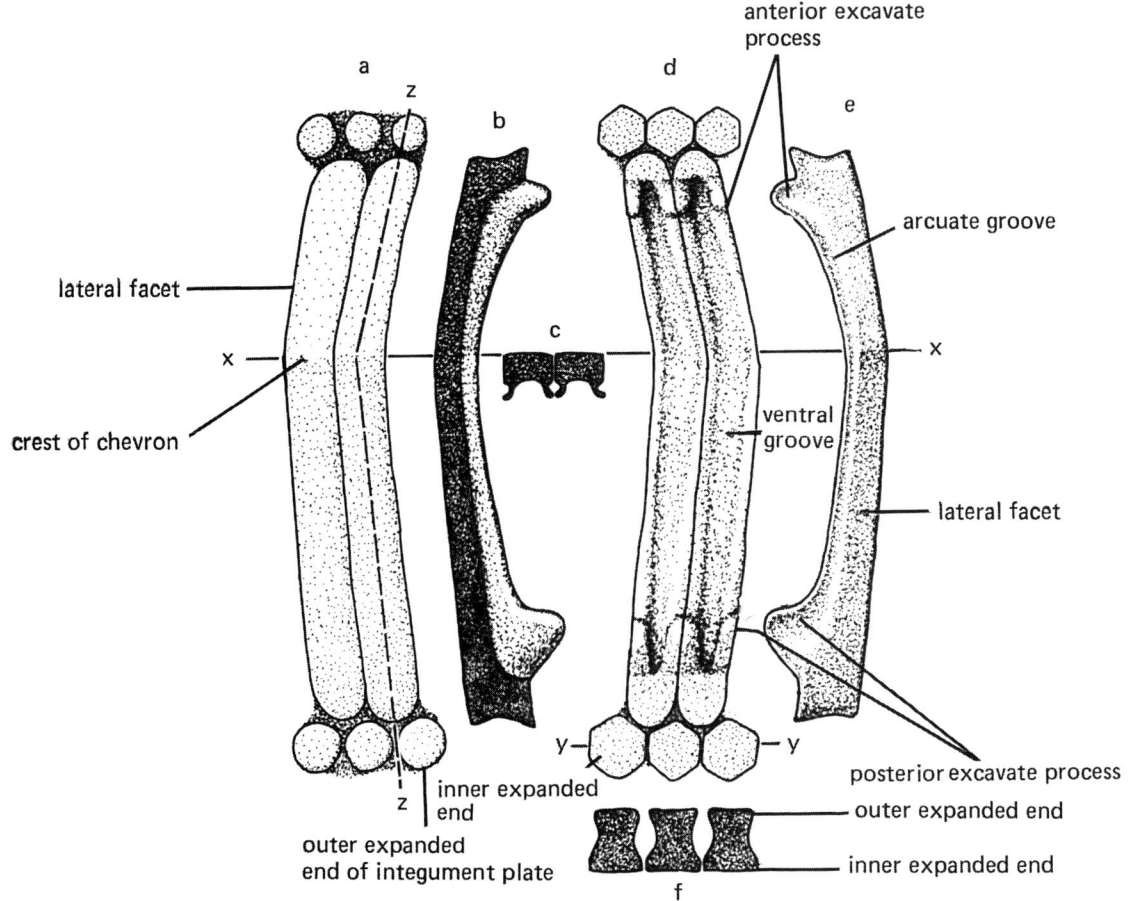

7.13. Scotiaecystis curvata. *Branchial chevrons and integument plates. a) Dorsal aspect of two chevrons, b) section through a chevron along z–z in (a), c) transverse section through two chevrons along x–x, d) ventral aspect of two chevrons and integument plates, e) gill-slit surface of a chevron, and f) transverse section of integument to show bobbin-shaped plates in section; from Jefferies (1968, figs 9a, d, e, p. 270).*

of *Scotiaecystis* would flow out roughly perpendicular to the chevron complex, i.e. approximately vertical and upwards, not horizontal and rearwards as in *Cothurnocystis*. The forward-running distal gonorectal canal would serve to introduce faeces and gametes directly into this vertical current. A cladistic analysis of the cornutes shows that in the structure of the gill slits, and in having the gonopore-anus debouching into them, *Scotiaecystis* is more advanced than *Cothurnocystis*.

The mouth of *Scotiaecystis* is represented by a small opening in the dorsal integument guarded by a pyramid of spike-shaped plates. This position of the mouth, opening almost vertically upwards, contrasts with that of *Cothurnocystis*. It suggests that *Scotiaecystis* was a suspension feeder, obtaining food particles from the water above it, rather than being a 'deposit' feeder like *Cothurnocystis*.

Like the other features already mentioned, the dorsally opening mouth is almost certainly an advanced condition compared with *Cothurnocystis*.

The integuments of *Scotiaecystis* were probably thicker than those of *Cothurnocystis*. This is suggested by the integument plates, each of which is bobbin-shaped with the axis of the bobbin perpendicular to the integument (fig. 7.13f). The inner expanded ends of each integument plate are polygonal and in contact with neighbouring polygons. The outer expanded ends are smaller, circular, and separated from neighbouring plates by gaps. Probably the spaces between neighbours external to the polygonal bases, were filled in life with soft tissue, probably muscular, and perhaps there were also sheets of muscle internal to the integument plates. On the internal surface of the head frame three zones are always

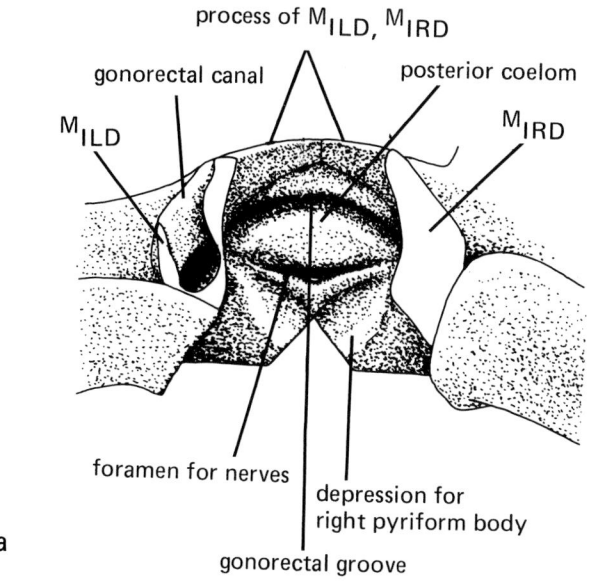

process of M_{ILD}, M_{IRD}

gonorectal canal

posterior coelom

M_{ILD}

M_{IRD}

foramen for nerves

depression for
right pyriform body

gonorectal groove

a

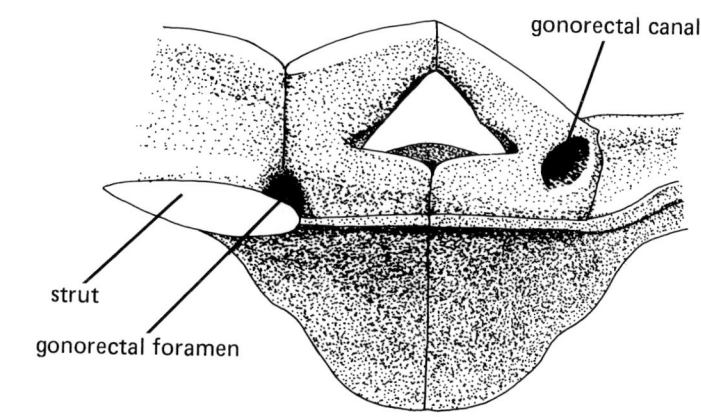

gonorectal canal

strut

gonorectal foramen

b

7.14. Scotiaecystis curvata. *Posterior coelom and gonorectal groove. a) Dorsal aspect of the posterior coelom with the 'roof' removed, b) anterior aspect of skeleton of posterior coelom; from Jefferies (1968, figs 11a, b, pp. 273, 274).*

recognizable – upper, middle and lower (fig. 7.15a,c). The upper and lower zones are relatively flat and seem to represent the attachment facets for the thick dorsal and ventral integuments. The middle zone varies considerably in depth at the expense of the other two zones and takes the form of a groove in the marginal plate. Expansions in the height of the middle zone (measured perpendicular to the dorsal and ventral edges of the marginal plates) indicate where chambers existed between the integuments.

The buccal cavity (fig. 7.15) shows clearly in *Scotiaecystis*, being defined posteriorly by a vertical ridge on the inside of the frame at right and left and by a lens-shaped

expansion of the middle zone of the internal surface of the marginal plate at right and left. The posterior coelom is also well defined as a distinct little chamber just in front of the tail (fig. 7.14) floored by the ventral first marginal plates (M_{1LV}, M_{1RV}) and roofed by the dorsal first marginals (M_{1LD}, M_{1RD}) and partly delimited anteriorly by processes of these plates. The gonorectal groove runs across its floor in a curved course from right to left. The right and left anterior coeloms and pharynx cannot be distinguished from direct evidence but probably lay much as in *Cothurnocystis*. This is suggested by the middle zone of the marginal plates being higher right than left of

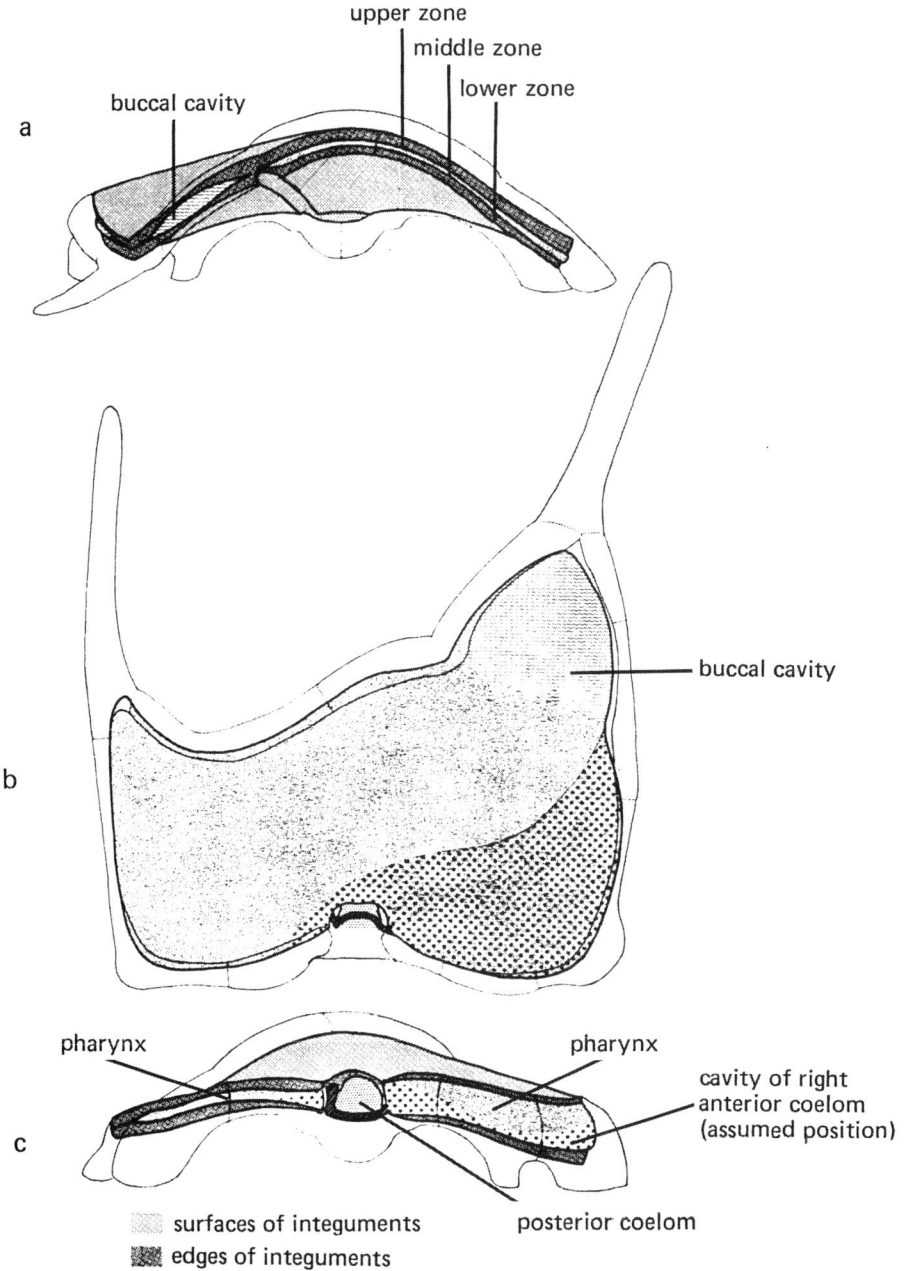

7.15. Scotiaecystis curvata. *Head chambers and zones for attachment of integuments. a) anterior aspect, b) dorsal aspect and c) posterior aspect; from Jefferies (1968, fig. 10a, c, e).*

the tail (fig. 7.15c) and by the fact that the gonorectal groove, as in *Cothurnocystis,* emerges from the region right of the tail where the gonad and non-pharyngeal gut would therefore presumably lie.

The tail of *Scotiaecystis* was much like that of *Cothurnocystis,* with fore, mid and hind portions. The skeleton of the fore tail (fig. 7.12a,c) consists of left and right, dorsal and ventral series of plates but, unlike *Cothurnocystis,*

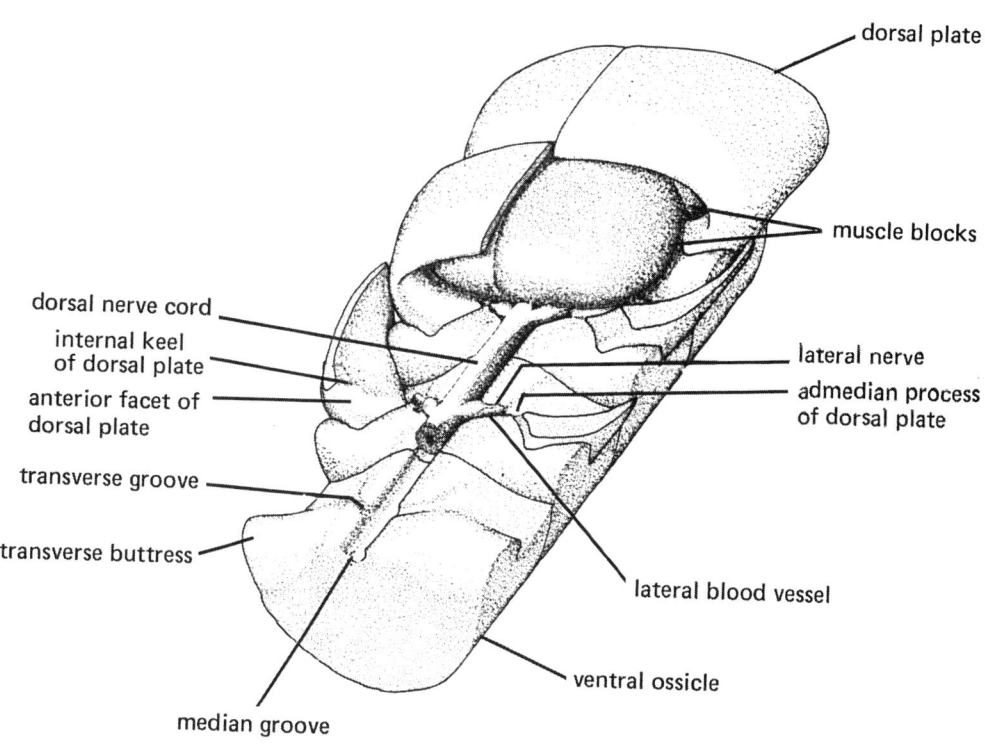

dorsal plate

muscle blocks

dorsal nerve cord

internal keel
of dorsal plate

anterior facet of
dorsal plate

transverse groove

transverse buttress

lateral nerve

admedian process
of dorsal plate

lateral blood vessel

ventral ossicle

median groove

7.16. Scotiaecystis curvata. Block diagram of hind tail in left antero-dorsal aspect. The dorsal nerve cord and ganglia are based on comparison with mitrates Note how the dorsal plates are locked onto the transverse buttresses in a way that would allow slight rocking about a transverse axis but would prevent the plates from rotating outwards from the mid line; from Jefferies, (1968b, fig. 12, p. 275).

does not seem to be adapted mainly for right-left flexion. The stylocone of the mid tail is relatively short, compared with *Cothurnocystis*. The hind tail contains up to about 16 segments and ends abruptly. It seems to have functioned as a rigid rod since all the interossicular surfaces are plane (fig. 7.16). The dorsal plates are joined to the ventral ossicles by a complicated surface of a form which would have prevented them opening outwards from the mid line. This fact argues strongly against Ubaghs' interpretation of the plates as cover plates. The left and right appendages of *Scotiaecystis* slope downwards anteriorly, as in *Cothurnocystis*, and would have hindered forward movement. The same effect would have been produced by the spikes of the posterior part of the head frame which are truncated by planes that slope downwards anteriorly. Like *Cothurnocystis*, *Scotiaecystis* seems to be adapted for moving backwards over the sea floor, pulled by the tail.

The brain of *Scotiaecystis* probably lay in a deeply excavate basin at the front end of the tail, much as in *Cothurnocystis*.

Scotiaecystis therefore confirms a number of deductions made from *Cothurnocystis*: the outlet-valve nature of the gill slits and the outlet character of the gonopore–

anus, adapted here to open into an upward-flowing branchial current; the existence of two of the head chambers (buccal cavity, posterior coelom); and the likelihood that most of the viscera probably lay in the posterior right part of the head; *Scotiaecystis* also makes it improbable that the tail functioned in the manner that Ubaghs has suggested. *Scotiaecystis* was in most ways more specialized than *Cothurnocystis* and was probably a suspension feeder, rather than a deposit feeder.

Ceratocystis perneri (fig. 7.17) is the next cornute to be considered. It is older than *Cothurnocystis elizae* or *Scotiaecystis curvata*, coming from Middle Cambrian rocks near the village of Skryje in Bohemia (Czechoslovakia) and is thus about 530 million years old. It was originally described in 1900 by Otto Jaekel, the friend and rival of Francis Bather, and was the first cornute to be made known. (Small scraps of it had already been mentioned and figured by Pompeckj (1896).) It has been further studied by Bather (1913), Ubaghs (1967) and me (1969). It occurs in a hard siltstone rather similar in appearance to the Starfish Bed that yielded *Cothurnocystis elizae*, and is accompanied by an abundant fauna, mainly of trilobites, which again indicates a shallow marine environment. Being

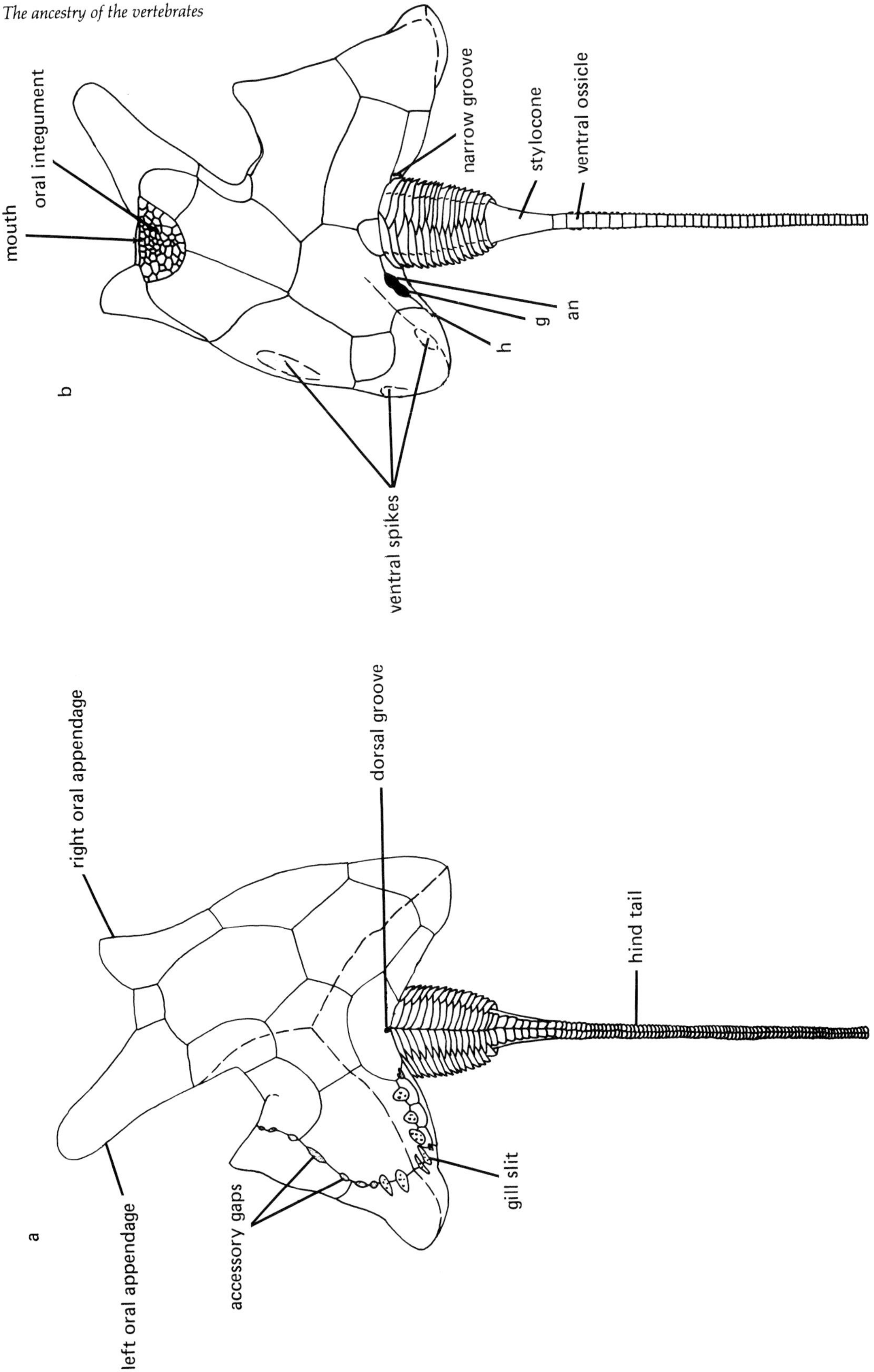
a
left oral appendage
accessory gaps
right oral appendage
gill slit
dorsal groove
hind tail
b
mouth
oral integument
ventral spikes
h
g
an
narrow groove
stylocone
ventral ossicle

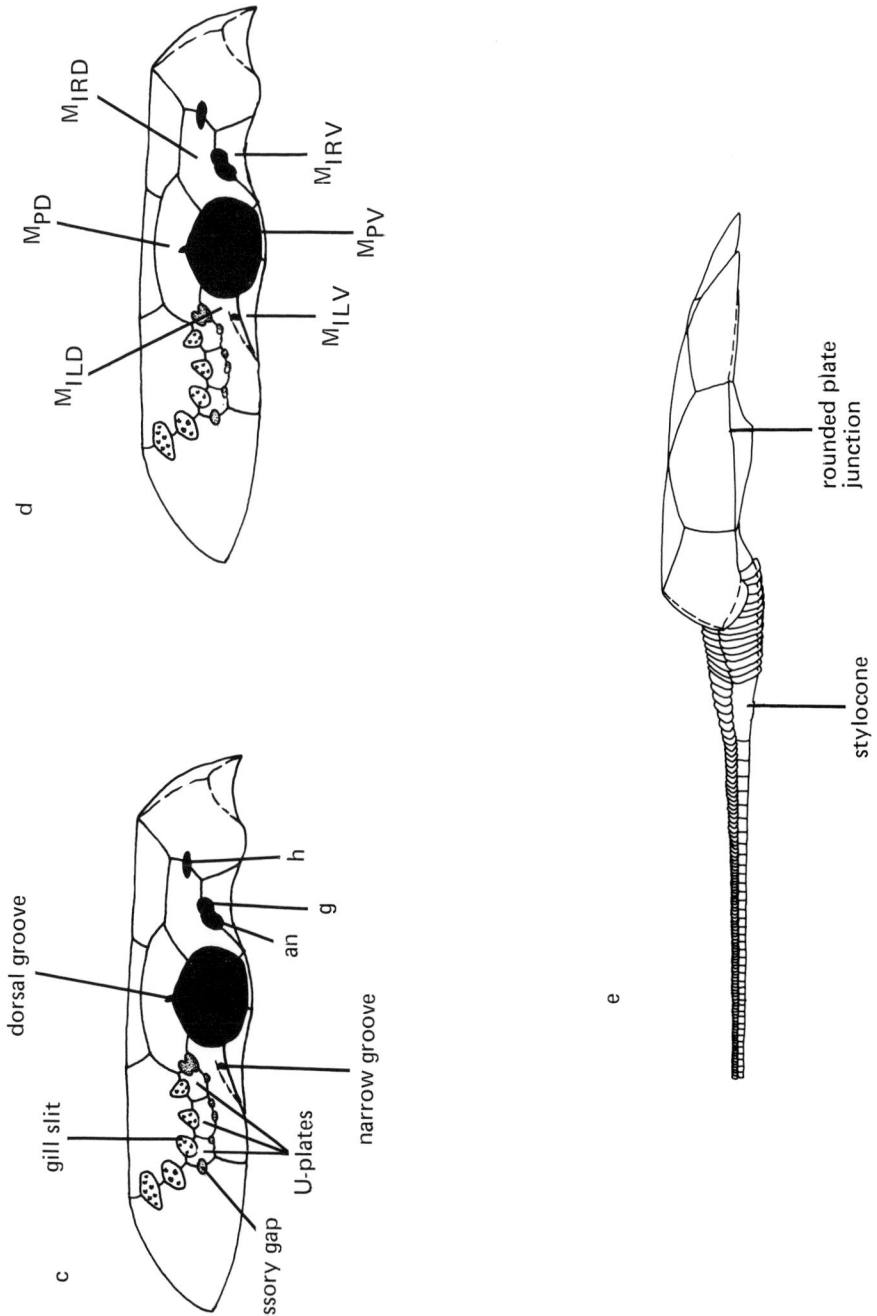

7.17. *Ceratocystis perneri. Reconstruction of external features. a) dorsal aspect, b) ventral, c) posterior, to show openings and gaps, d) posterior aspect to show plate notation, and e) right lateral aspect; from Jefferies (1969, text fig. 2a, b, c, d, pp. 498–499).*

stratigraphically early, *Ceratocystis perneri* happens also to be primitive. However, this agreement of morphology with stratigraphical position does not signify, for a much more advanced and *Cothurnocystis*-like cornute is known from Utah in rocks of much the same age (Sprinkle, 1976). These two cornutes are the only ones known from the Middle Cambrian, while *Nevadaecystis*, which in most ways is intermediate between *Ceratocystis* and *Cothurnocystis*, is known only from late Upper Cambrian rocks. The record is therefore clearly incomplete; no accurate agreement between primitiveness and stratigraphy occurs, nor would be expected. Henceforth I shall refer to *Ceratocystis perneri*, the only known species of its genus, as *Ceratocystis*.

The head of *Ceratocystis* is boot-shaped, like *Cothurnocystis*, though the right oral appendage, like the left one, was fixed not motile. The skeleton of the head consists mostly of large plates, unlike *Cothurnocystis* or *Scotiaecystis*, for there is no separation into a frame and dorsal and ventral integuments. Nevertheless it is possible to homologize many of the plates with particular marginal plates of *Cothurnocystis* and *Scotiaecystis*. The dorsal surface of the head is convex and traversed by three low keels which meet near its centre. The ventral surface is relatively flat, though low spikes are developed on some of the plates. Six plates surround the tail insertion – those at the first right and left dorsal and ventral marginals as in *Cothurnocystis elizae* (M_{1LD}, M_{1RD}, M_{1LV}, M_{1RV}) and also a median dorsal and ventral plate (M_{PD}, M_{PV}).

The openings of the head are numerous, though not all gaps between the plates represented openings in life. Firstly there are seven elliptical openings situated across plate contacts in the left dorsal part of the head. The three of these openings nearest the tail have U-shaped plates ventral and posterior to them (fig. 7.17a,c). This fact, and the general position of all seven openings, suggests that they are homologous with the gill slits of *Cothurnocystis* or *Scotiaecystis*. In life these openings were probably guarded by flaps of soft tissue. The evidence for this statement is two-fold. Firstly the observed dorsal edge of each opening is bluntly truncated, as if in life it carried a soft continuation, whereas the ventral edge is rounded. Secondly, in one specimen tiny platelets were observed in the gaps, comparable with, though sparser than, the platelets on the branchial flaps of *Cothurnocystis*.

Smaller and less regular gaps exist at certain plate contacts in the same general region as the gill slits. They probably represent incipient flexibility. For they can be shown, by plate homologies, to correspond in position to the edges of the dorsal integument of the left side of the head in *Cothurnocystis*, *Scotiaecystis* or *Nevadaecystis* i.e. they notch the dorsal edges of plates homologous with marginal plates of these other cornutes. The more pos-

terior accessory gaps, ventral to some of the gill slits, are surrounded by a shallow excavation on the internal surface of the plates (fig. 7.20b 'muscle associated with accessory gaps'). If the gaps, or the shallow excavation, or both, contained muscle, then *Ceratocystis* would have been able, given a minimal looseness between the head plates, to jerk the roof of the head downwards slightly, so as to eject a small quantity of water forcibly through the gill slits – an action which could evolve into vigorous flexion of the soft dorsal integument in other cornutes. Such a coughing action is common in filter feeders, as, for example, in tunicates (Hoyle, 1953). Ubaghs (1969) did not recognize the gill slits of *Ceratocystis* as differing in nature from what are here called accessory gaps. I think he was mistaken in this since the gill slits, numbering seven, are much more constant in position and occurrence than the accessory gaps. Moreover, the same distinction occurs in the cornute *Nevadaecystis*, as discussed below (fig. 7.27), where seven gill slits of *Cothurnocystis* type are clearly distinct from gaps which, so to speak, resemble an exaggeration of the accessory gaps of *Ceratocystis*. Another sign of incipient flexibility in the head of *Ceratocystis* is seen in the nature of the horizontal plate junctions near the right edge of the head. These junctions are rounded to form a hinge (fig. 7.17e) which again, given some looseness between the head plates, would allow the animal slightly to depress the right-hand part of the head roof. As with the accessory gaps, there is a shallow depression in the internal faces of the plates, internal to the hinge, and this depression could have contained muscle (fig. 7.20c). In confirmation of the view that these rounded junctions represent incipient flexibility, their position is partly homologous with the dorsal edge of the frame of *Cothurnocystis elizae*, though not entirely so.

Other openings exist, on the right side of the head (fig. 7.17). There is firstly the mouth, which is the largest opening of the head and situated much as in *Cothurnocystis*. It has a rigid upper lip whereas the lower lip is formed from a flexible plated semi-circular integument. Three other openings are present right of the tail near the 'heel' of the 'boot' and in the diagram these are labelled *h*, *g* and *an*. These do not resemble the accessory gaps of the left side of the head, being connected to the cavity inside the head by well marked grooves inside the plates (fig. 7.20a) – grooves which probably carried tubes in the soft parts. The openings *g* and *an* are close together right of the tail, and indeed often join to form a single bifid opening. The opening *h* is more to the right. All these openings straddle sutures.

These three openings can be identified by comparison with primitive fossil echinoderms. In the diploporitan cystoid *Glyptosphaerites* (fig. 7.18), from the Ordovician of Estonia, for example, four openings are present in the

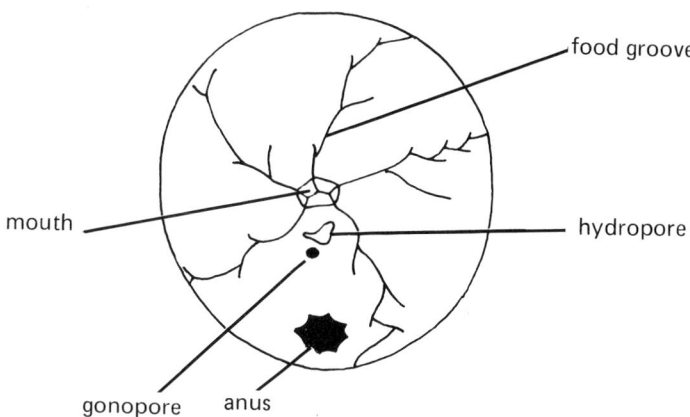

7.18. *The diploporitan cystoid* Glyptosphaerites leuchtenbergi *(Ordovician, Estonia) to show the four thecal openings which often occur in primitive echinoderms; Jefferies (1969, text fig. 5, p. 505).*

theca. The mouth is at the summit of the animal and towards it the ambulacral grooves converge. Some distance from the mouth, on the side of the theca, is a large circular gap in the rigid thecal plating. This probably represents the position of the anus, though most of the gap would have been covered in life by an anal pyramid or by a flexible membrane (the periproct). On a line between mouth and anus two further openings exist. The upper of these is probably a hydropore (the opening of the axial sinus and water-vascular system) for it is developed as a little sieve, like the madreporite of echinoids and asteroids except that it is developed from portions of more than one plate. One of the main functions of the hydropore in living echinoderms is to equalize the water pressure in the water-vascular system with that of the environment and this hydrostatic balance is not impeded by development as a sieve (Fechter, 1965). The remaining opening of *Glyptosphaerites*, between the hydrophore and the anus, is probably a gonopore since: (1) in some cystoids it is guarded by a tiny pyramid of plates, suggesting that it could be intermittently opened and closed (as in *Echinosphaerites*, see Jaeckel, 1899, pl. 8, figs 5, 19 and *Caryocystites*, Jaeckel, 1899, pl. 9, fig. 4a), (2) in holothurians, which are the sole living echinoderms with a single gonad only, the gonopore is closely associated with the hydropore when both open externally (as in *Benthodytes*, Hyman, 1955, fig. 53). Accordingly in the direction away from the mouth the sequence of openings in *Glyptosphaerites* and other cystoids was: hydropore, gonopore, anus. It is plausible to suppose that the same sequence held for *Ceratocystis*, in which case *h* was the hydropore, *g* the gonopore and *an* the anus.

These identifications are consistent with the situation seen in *Cothurnocystis* and *Scotiaecystis*. For the migration of gonopore–anus to left of the tail, either downstream of, or opening into, the gill slits respectively, would improve the 'plumbing' of the animal (fig. 7.19). On this supposition the gonorectal groove beneath the posterior coelom of these forms would connect the new position of the gonopore–anus with its original position. Moreover, the lack of a hydropore in all cornutes except *Ceratocystis* can be regarded as a synapomorphy of these forms with living chordates. In other words, the hydropore (which, as seen later, was probably the opening of the axial sinus) was lost in evolving toward the normal chordate condition.

Two further head openings were situated near the tail – 'the dorsal groove' just above the tail and the 'narrow groove' just left of it (fig. 7.17c). These will be discussed in connection with the brain.

As to head chambers, I have mentioned direct evidence of four such in *Cothurnocystis* and the same four are suggested by details of the natural internal moulds in *Ceratocystis* (figs 7.20, 7.21). (For internal moulds, being negatives of the skeleton, can be seen as positives of the soft parts.) Thus the posterior boundary of the buccal cavity is indicated by a number of grooves on the natural mould (fig. 7.20a,b). The anterior boundary of the posterior coelom is indicated by a shallow curved groove on the ventral surface of the natural mould (fig. 7.20a). The boundary between the pharynx and the right anterior coelom is indicated by a groove on the ventral surface (fig. 7.20a). A well-marked groove in the left part of the animal ventral to the gill slits perhaps indicates an extension of

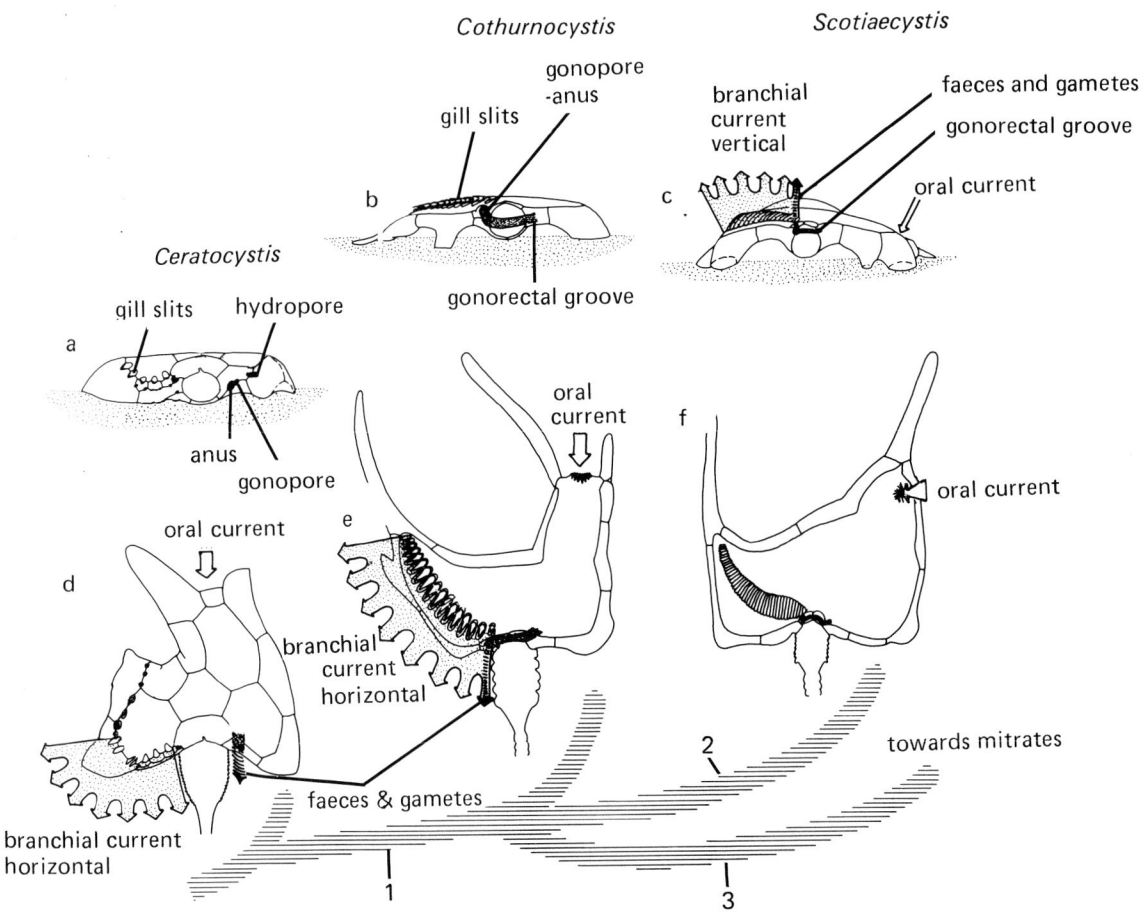

7.19. *Branchial currents and the voiding of faeces and gametes in three cornutes. a) and d)* Ceratocystis perneri *in posterior and dorsal aspect — the gonopore–anus is in the primitive position right of the tail, not connected with the branchial current; b) and e)* Cothurnocystis elizae *– the gonopore–anus has shifted to left of the tail (synapomorphy 1 in the cladogram) so as to lie in the almost horizontal branchial current from the rightmost gill slit; c) and f)* Scotiaecystis curvata *– the gonopore has migrated forwards (synapomorphy 2 in the cladogram) from the* Cothurnocystis *position, so as to open into the almost vertical branchial current. Synapomorphy 3 represents the mitrate acquisitions such as right gill slits; from Jefferies, 1981b, text fig. 5, p. 360.*

the right anterior coelom into the left side of the head (fig. 7.20a,c). Other grooves, delimiting raised areas on the natural moulds, suggest, as already mentioned, the position of muscles: (1) associated with accessory gaps ventral to the gill slits, and (2) connected with the rounded horizontal plate junctions near the right edge of the head. A further groove, on the right side of the natural mould, ventral to the hydropore, gonopore and anus, probably indicates the ventral boundary of some internal organ (fig. 7.20a,c). In any case, the main distribution of head chambers, so far as can be seen, was much the same in *Ceratocystis* as in *Cothurnocystis*. I shall deal later with the arguments for a purely virtual left anterior coelom.

The contents of the right anterior coelom are indicated

by the openings which emerge from it, i.e. the anus, the gonopore and the hydropore. The anus and gonopore imply the presence of the non-pharyngeal gut and a gonad. The hydropore would imply the presence of at least some of the organs associated with that opening in Recent echinoderms.

Such organs were discussed by Fedotov (1924) in a detailed comparative study (fig. 7.22). They consist of: (1) the axial sinus (axocoel) and the axial organ developed in its wall – the water-vascular system opens via the stone canal into the axial sinus near the hydropore so that the latter, though primarily the opening of the oxocoel, is often thought of chiefly as the opening of the water-vascular system; (2) the madreporic vesicle (dorsal sac),

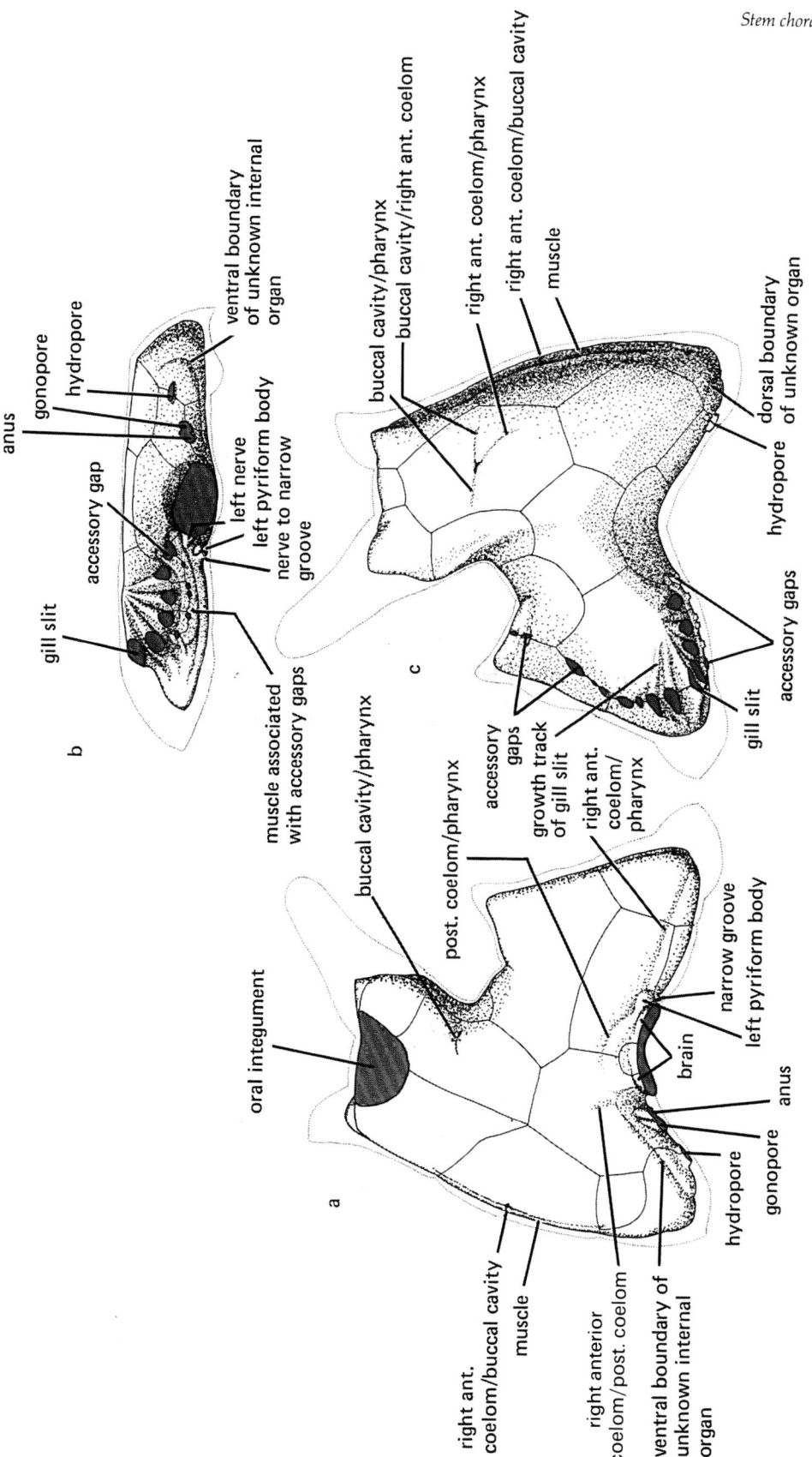

7.20. *Ceratocystis perneri. Reconstructed internal mould, representing the soft parts. The presumed muscle at the right margin of the head probably depressed the roof about the hinge-like horizontal suture of this region. a) Ventral aspect, b) dorsal, and c) posterior (cf. fig. 7.21): from Jefferies (1969, fig. 9, p. 511).*

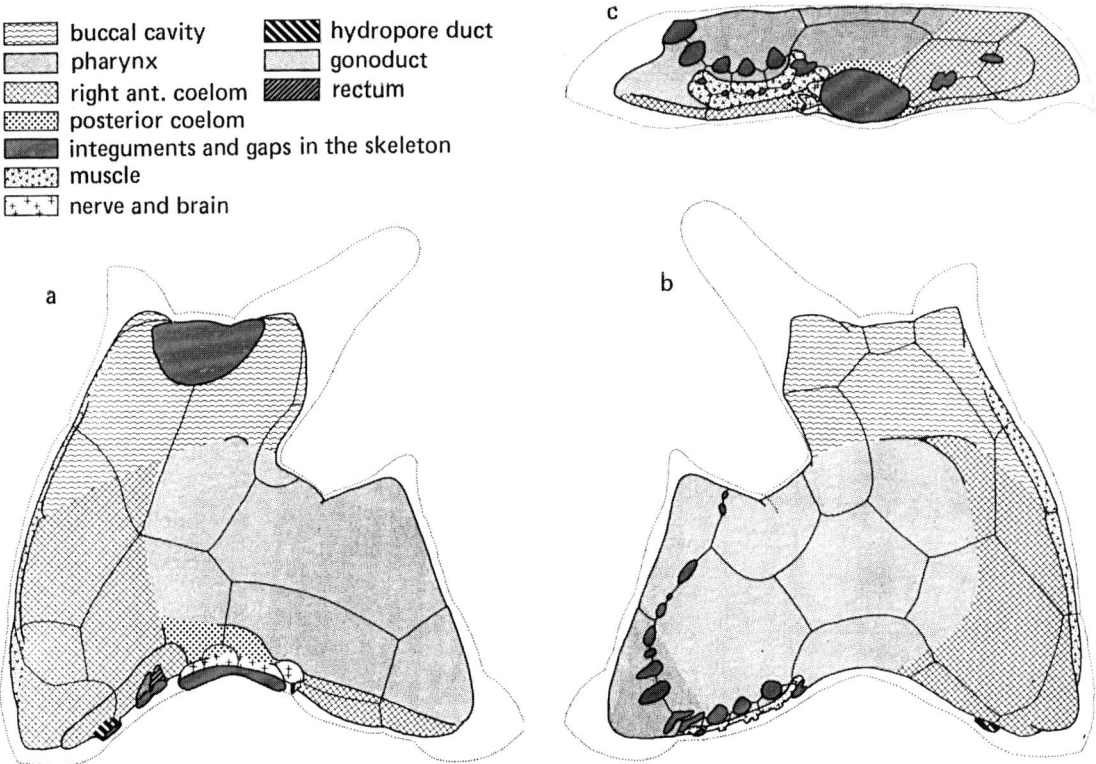

Legend:
- buccal cavity
- pharynx
- right ant. coelom
- posterior coelom
- integuments and gaps in the skeleton
- muscle
- nerve and brain
- hydropore duct
- gonoduct
- rectum

7.21. Ceratocystis perneri. *Reconstructed positions of head chambers (cf. fig. 7.22). a) Ventral aspect, b) dorsal, and c) posterior; from Jefferies (1969, fig. 10, p. 512).*

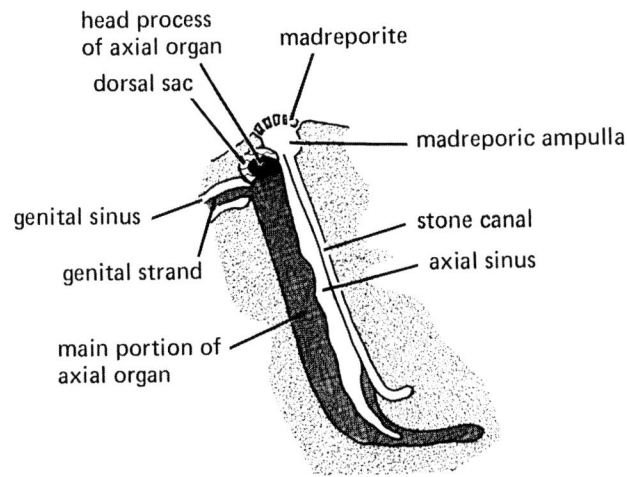

head process
of axial organ
madreporite
dorsal sac
madreporic ampulla
genital sinus
stone canal
genital strand
axial sinus
main portion of
axial organ

7.22. *The axial complex of an echinoid from Jefferies (1969, fig. 11, p. 514 after Fedotov, 1924).*

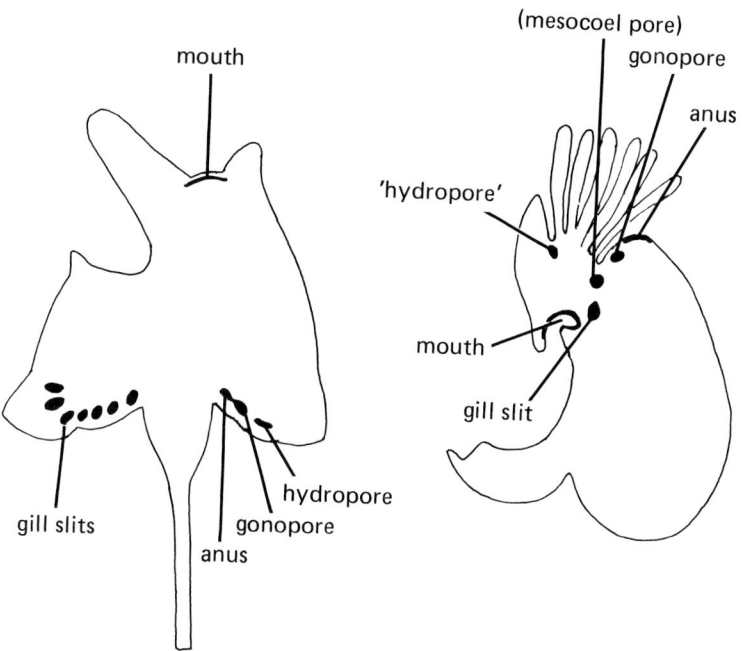

7.23. *Comparison between* Ceratocystis *in dorsal aspect and* Cephalodiscus *in left aspect to show a basic similarity in the disposition of the openings; from Jefferies (1981a, fig. 1, p. 506).*

which is just internal to the hydropore, surrounds an organ known as the head process of the axial organ and is often pulsatile, probably functioning as a heart. It follows that the axial sinus would probably open through the hydropore of *Ceratocystis* and that some equivalent of the madreporic vesicle and head process could well have existed near this opening. These organs, as already mentioned (Chapter 2) are probably homologous with the pericardium and heart in hemichordates. Their likely position in *Ceratocystis*, close to the presumed locations of the non-pharyngeal gut and gonad, suggests that they could likewise be homologous with the pericardium and heart of chordates. *Ceratocystis* shows no sign of a water-vascular system, presumably having lost it in phylogeny. There is therefore no reason to suppose that it had a stone canal.

· The axial gland of echinoderms is in some ways like the neural gland of tunicates. Both are probably endocrinal (Millott & Vevers, 1968, for the axial gland; Godeaux, 1964, for the neural gland). Both are concerned in expelling dead coelomocytes from the animal (Millott & Vevers, 1968, for the axial gland; Pérès, 1943, Millar, 1953, p. 47, for the neural gland). Both are connected by a special strand of tissue to the gonads (genital strand of echinoderms; dorsal strand of tunicates). These resemblances suggest that the axial gland and neural gland are hom-

ologous. The main difference, apart from differences in ontogeny, would be that the neural gland opens into the pharynx of tunicates whereas the axial gland opens to the outside through the hydropore. This change presumably happened in phylogeny in the segment of the cladogram between *Ceratocystis* and *Nevadaecystis* (fig. 9.2b) the latter being the most primitive known cornute without a hydropore.

A comparison between *Ceratocystis* and the hemichordate *Cephalodiscus* is illuminating. Figure 7.23 shows *Ceratocystis* in dorsal aspect, especially the openings, and *Cephalodiscus* seen from the left side. In moving clockwise around the diagram of *Ceratocystis* we pass: gill slits, mouth, hydropore, gonopore, anus. In moving clockwise around that of *Cephalodiscus* we pass: a pair of gill slits, mouth, a pair of hydropores (the left and right protocoel or axocoel pores), a pair of mesocoel pores (not represented in *Ceratocystis* or echinoderms), a pair of gonopores, anus. These two sequences resemble each other which suggests that the dorsal surface of *Ceratocystis* corresponds to the left side of *Cephalodiscus*. This is interesting because, as already argued in Chapter 2, the left side of *Cephalodiscus* corresponds to the oral, primitively upward, face of echinoderms. Stated otherwise, the extraordinary asymmetrical shape of *Ceratocystis* demands some explanation, and suggests descent from a

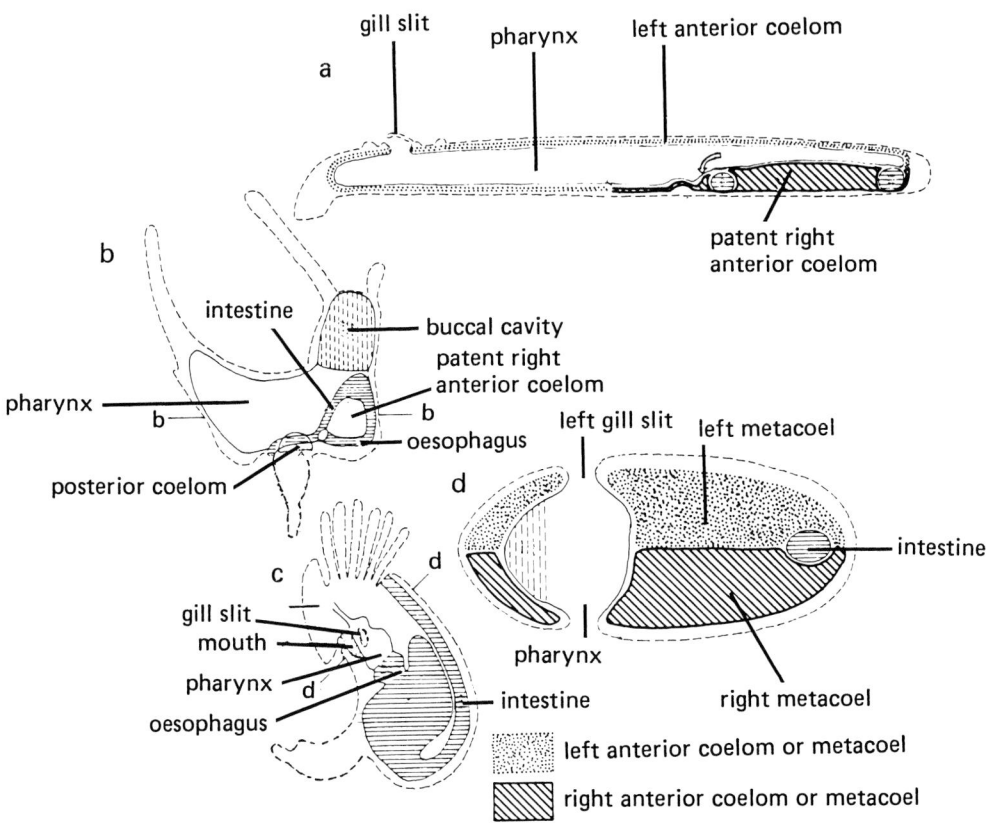

7.24. *Comparison of* Cephalodiscus *with a cornute (*Cothurnocystis*) to illustrate the argument for the existence of a left (dorsal) anterior coelom in the latter. This coelom corresponds to the left metacoel of* Cephalodiscus *and also to the left somatocoel of an echinoderm such as* Stromatocystites *(fig. 2.56b) or* Antedon *(fig. 2.53). a)* Cothurnocystis *in dorsal aspect, b) section along b–b in (a), c)* Cephalodiscus *in left aspect, and d) section along d–d in (c); from Jefferies & Lewis (1978, fig. 15b, c, p. 246).*

bilateral ancestor that laid down on one side. This ancestor was probably rather like a *Cephalodiscus* resting on its right side in what I have already called the dexiothetic orientation. And such a hypothetical ancestor could well have been ancestral to the echinoderms also. This would imply that echinoderms and chordates are closer related to each other than to *Cephalodiscus* or to other hemichordates and that echinoderms and chordates together constitute a monophyletic group – the Dexiothetica – which can be given the rank of subsuperphylum within the Deuterostomia.

As concerns the head chambers, the gut of *Cephalodiscus* is carried in a mesentery between the left and right metacoels (or left and right somatocoels in echinoderm terms). If *Ceratocystis* or *Cothurnocystis* correspond to *Cephalodiscus* lying on its right side (fig. 7.24), then the right anterior coelom of the cornutes, underlying the pharynx, corresponds to the right metacoel of *Cephalodiscus* and to the right, aboral primitively downward

somatocoel of echinoderms such as the early form *Stromatocystites* (Fig. 2.56). This would imply the existence in the cornutes, presumably as a virtual coelom overlying the pharynx, of a left anterior coelom corresponding to the left metacoel of *Cephalodiscus* and to the left, aboral, primitively upward, somatocoel of echinoderms such as *Stromatocystites*. The chambers of the head of *Ceratocystis* were therefore the buccal cavity, the pharynx, the posterior coelom, right anterior coelom (all based on direct evidence) and the left anterior coelom (based on comparative evidence).

The tail of *Ceratocystis* (figs 7.17, 7.25) consisted of fore, mid and hind portions and ended abruptly. The hind tail is like that of *Cothurnocystis* or *Scotiaecystis* in most respects except that each ventral ossicle corresponds to two or three pairs of dorsal plates and that right and left dorsal plates were not accurately opposite either. There is a median groove on the dorsal surface of the ventral ossicles of the hind tail and on either side of the groove are

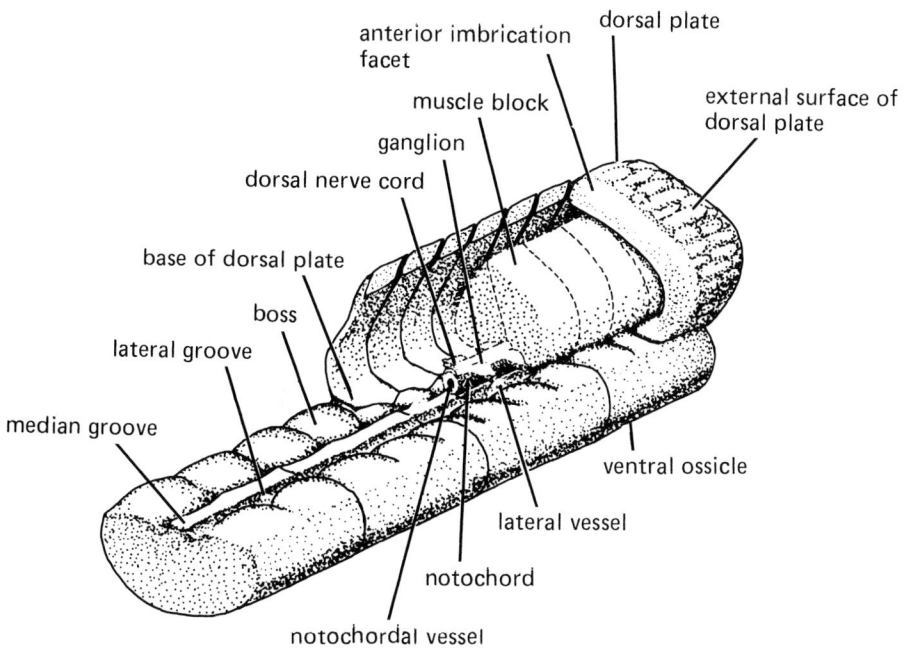

7.25. Ceratocystis perneri. *Block diagram of hind tail; the dorsal nerve cord and ganglia are based on comparison with mitrates; from Jefferies (1969, fig. 17, p. 517).*

two or three bosses per ossicle. The bosses of left and right sides do not coincide. The dorsal plates are attached to the ossicles by broad extensions that straddle the posterior and anterior faces of successive bosses. In the mid-dorsal line there is a lax suture where the plates of left and right sides meet. In the fossils the plates of left and right sides have often separated at this suture and Ubaghs (1967, p. 13) has taken this observed separation (fig. 10.5) as evidence that the plates could part from each other in life. In my view this does not follow. Most probably the separation was caused by explosive release of gases from the decaying soft parts after death and burial. The mid tail contains a massive, ventral, anteriorly excavate stylocone overlain by about eight dorsal plates on each side. The fore tail has a loose skeleton consisting of about 17 rings made of four plates each (right, dorsal and ventral; left, dorsal and ventral). The rings imbricate inside each other and the individual plates of a ring also imbricate together. The fore tail had a large lumen which presumably contained muscles and, in flexion, would have needed an anti-compressional structure, presumably the notochord, which would have extended rearwards into the hind tail along the median groove. The muscles of the fore tail were probably inserted into the deep anterior excavation of the stylocone and could thus move the mid and hind tail as a unit. The soft parts of the tail were presumably like those of *Cothurnocystis*.

The brain of *Ceratocystis* (fig. 7.26a,b) was situated where the tail joins the head and is more reconstructable than that of other cornutes, being in contact with skeleton both dorsally and ventrally. Its form is reflected by facets on the inner surface of the skeleton and can therefore be visualized directly by examining natural moulds of the internal surfaces of the marginal plates of the tail insertion. It can be anatomically interpreted largely by comparison with mitrates. To right and left of the brain were bodies which can be identified with the right and left pyriform bodies of mitrates, and therefore with the right and left trigeminal ganglia of living vertebrates. From the median dorsal region of the brain a dorsal process extends onto the posterior dorsal marginal plate (M_{PD}). It ends as a slight swelling inside the above-mentioned dorsal groove just above the tail on the dorsal surface of the head. The region of the brain from which this dorsal process emerges corresponds to the dorsal part of the prosencephalon of mitrates, and this region is optic in function in living vertebrates. This suggests that the dorsal process was a median eye, which is consistent with its dorsal and superficial position. It cannot be homologous with the pineal nor parapineal eyes of living vertebrates, for a median eye was lacking in closer relatives of living vertebrates than *Ceratocystis*, as, for example, in the mitrate-like cornute *Reticulocarpos* and the mitrates themselves. To left and right of the dorsal process were dorsal lobes which end

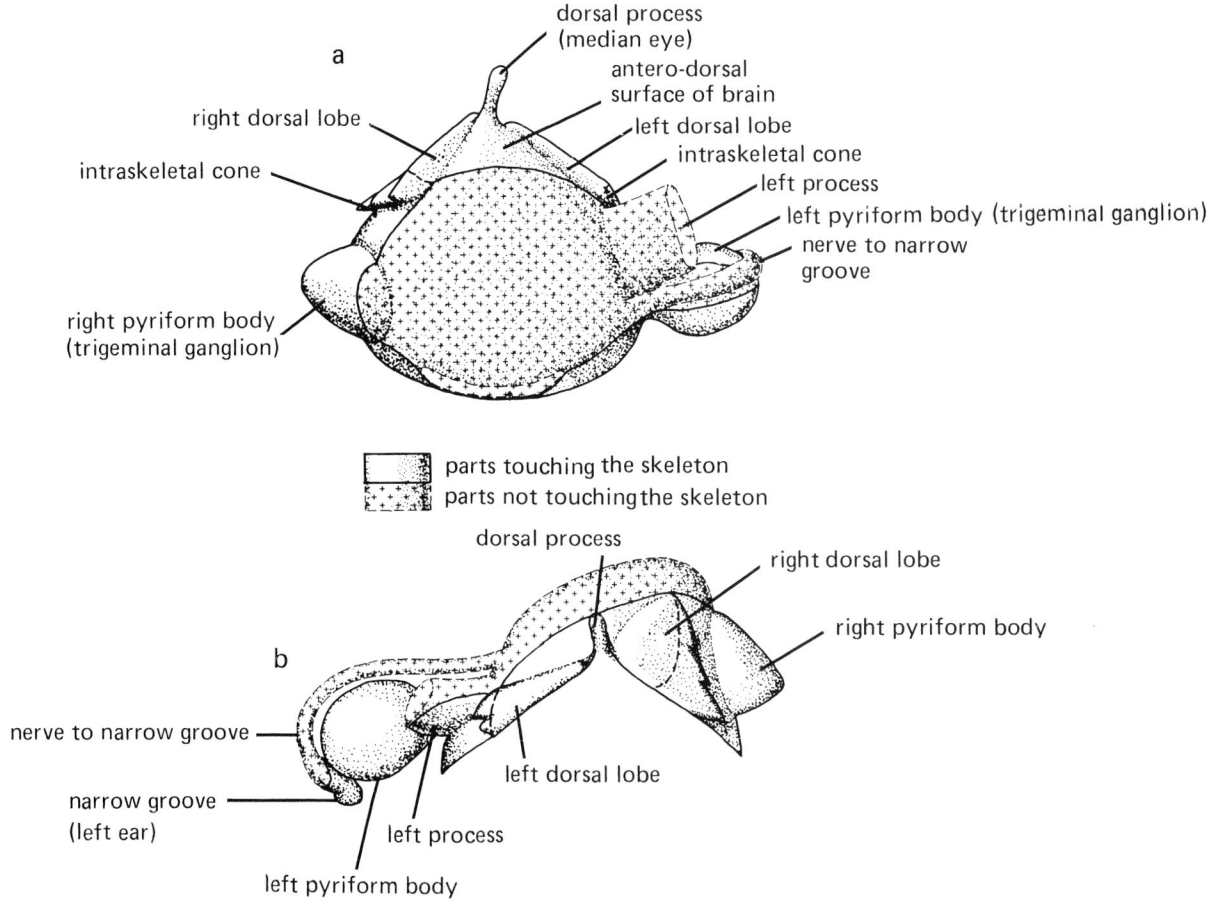

7.26. Ceratocystis perneri. *Reconstruction of the brain. a) anterior, and b) dorsal aspect. The front surface of the brain may well have been more complicated than shown, but there is no information; from Jefferies (1981a, fig. 9, p. 505).*

ventro-laterally in irregular conical projections that point into the skeleton. These intraskeletal cones resemble the precerebral knot of mitrates such as *Mitrocystella*, which probably represents the meeting place of terminalis and olfactory fibres before these entered the brain. The same probably helds for the intraskeletal cones and dorsal lobes of *Ceratocystis*.

The 'narrow groove' on the external surface of *Ceratocystis* (fig. 7.17c) is situated just left of the tail on the ventral surface of the animal. It is connected to the internal cavity of the head by a canal, presumably for a nerve, which passes into the skeleton just left of the left trigeminal ganglion. The narrow groove probably represents the beginnings of the acustico-lateralis system. Being superficially placed it presumably functioned as lateral line to detect movements in the surrounding water. It is probably homologous, however, not to the lateral line but to

the left ear of Recent vertebrates. This conclusion is based on a complicated argument, as follows. Firstly, the position of the narrow groove and its presumed nerve supply, just behind the left trigeminal ganglion (pyriform body) is comparable with the left acoustic ganglion of a fish. Secondly, in the mitrate *Mitrocystites mitra*, where the left gill slits, as in all mitrates, were enclosed in a left atrium, there is sometimes in the natural moulds a small lump of rock in the rightmost part of the atrium connected with a ridge that comes to it round the left pyriform body (left trigeminal ganglion) (fig. 8.10, p. 251). These structures can be compared in position with the narrow groove and supplying canal of *Ceratocystis* and probably represent the left acoustic ganglion and left auditory nerve. Not being superficial but invaginated into an atrium, these organs in mitrates would presumably have had an acoustic function in detecting sound, acceleration or tilting thus resembling the cupular organs in the atria of tunicates, which may be homologous with the acoustic system of vertebrates (Bone & Ryan, 1978) (fig. 4.21a,b). Again, the position of the supposed acoustic ganglion relative to the left trigeminal ganglion supports its acoustic nature. Thirdly, in the mitrate *Placocystites* there is clear evidence of a right antimere of the acoustic ganglion, located in the right atrium (fig. 8.25a, p. 288). Fourthly, in the mitrate *Mitrocystella* the position of this right acoustic ganglion was located just anterior to a probable lateralis ganglion which in turn was internal to a groove in the surface that could well represent lateral line, homologous with that of a fish. The lateral line of fishes and amphibians begins in ontogeny on the head near the ear and only later extends to the trunk and tail (Starck, 1975, p. 410). It is therefore not surprising to find that phylogeny followed the same course. Nor is it surprising, as already seen above (Chapter 5), that the acoustic and lateralis systems should be closely related to each other, and should first arise in the functional form of lateral line (Pumphrey, 1950). If an acustico-lateralis receptor existed in cornutes other than *Ceratocystis*, as its probable presence in the left atrium of *Mitrocystites* would imply, then it would be located near the gonopore–anus and not separately recognizable in the fossils.

A ridge on the natural mould of *Ceratocystis* goes off leftwards from the left face of the brain (left process in fig. 7.26a,b). It probably represents the base of some cranial nerve.

The habits of *Ceratocystis* were presumably much like those of *Cothurnocystis*. It probably pulled itself rearward across the sea floor by waving the tail from side to side and using the free end of the tail as a hook for gripping the sea floor. Since its mouth was level with the sea floor, *Ceratocystis* was probably a deposit feeder like *Cothurnocystis*. The fact that the specimens studied are complete and almost articulated suggests that they were killed by burial.

Ceratocystis is thus basically similar to *Cothurnocystis* but more primitive, especially in possessing a hydropore. It allows a fairly direct comparison with *Cephalodiscus*, and thus provides evidence for the existence of the Dexiothetica as a monophyletic group, and for the likely presence of a virtual, dorsally situated, left anterior coelom in the cornutes. Also it throws some light on the primitive condition of the chordate brain and sense organs.

The next cornute to be discussed is *Nevadaecystis americana* (Ubaghs) from the Upper Cambrian of Nevada (fig. 7.27). Only one specimen is known and it has been studied by Ubaghs (1963, 1967b) and by me (1969). Henceforth I shall call it *Nevadaecystis*. The specimen was found in a shelly limestone. It is imperfect and only visible from the dorsal aspect, though parts of the upper surface of the floor of the head can be seen. American palaeontologists have searched for additional specimens but without success. *Nevadaecystis* seems to be rare, at least in associated condition.

The head is boot-shaped, but more similar in outline to *Cothurnocystis elizae* than to *Ceratocystis*. The floor of the head is not entirely visible but is constructed, at least in large part, of big plates like the floor of *Ceratocystis*. The roof of the head, however, was formed of plated integument. The plates of this integument were larger than in *Cothurnocystis* and many of them can be homologized with those of the roof of *Ceratocystis*. In particular, the three keels of the dorsal surface of *Ceratocystis* can be recognized, and some of the plates that carried the keels. The integument plates are of two types – star-shaped ones situated mainly over the pharynx (in the 'foot' part of the 'boot') and polygonal ones situated over the buccal cavity (in the 'ankle'). This is reminiscent of the distinction seen in middle-sized *Cothurnocystis elizae* between the circular plates of the pharynx and the polygonal plates of the buccal cavity. It probably has the same functional meaning – that the pharyngeal wall was more muscular than the wall of the buccal cavity, because of the presence of muscles in or beneath the gaps between the plates. It is easy to see how the accessory gaps of *Ceratocystis*, by increasing in size and spreading rightwards over the dorsal surface, could result in the star-shaped plates of *Nevadaecystis*.

The occurrence of keels in the plates of the dorsal integument of *Nevadaecystis* makes no direct functional sense. They should probably be seen as vestiges of the keels of *Ceratocystis* which stiffened the roof in that genus. As vestiges, they indicate that the large-plated rigid roof of *Ceratocystis* is more primitive than the flexible dorsal integument of *Nevadaecystis*. (This argument, however, is based on the adaptational criterion of primitiveness and cannot be regarded as strong.)

As to head openings, seven gill slits existed in *Nevadaecystis*, as in *Ceratocystis*, and this helps to show that the gill slits of *Ceratocystis* are by nature different from the accessory gaps of that animal. The gill slits of *Nevadaecystis* are built on the *Cothurnocystis* plan with an anterior and a posterior U-plate and a flap stiffened with ossicles. The two U's are more equal in size than in *Cothurnocystis elizae*, however, and this is probably a primitive condition. For it is more like the gill slits of *Ceratocystis*, which notch the plates above and beneath them almost equally. It is hydrodynamically less efficient than the pattern seen in *Cothurnocystis elizae*, for the flap would obstruct outward flow to a greater extent

The mouth region of *Nevadaecystis* is not preserved in the only known specimen but the integument seems to have extended as far forward as the mouth. There was no

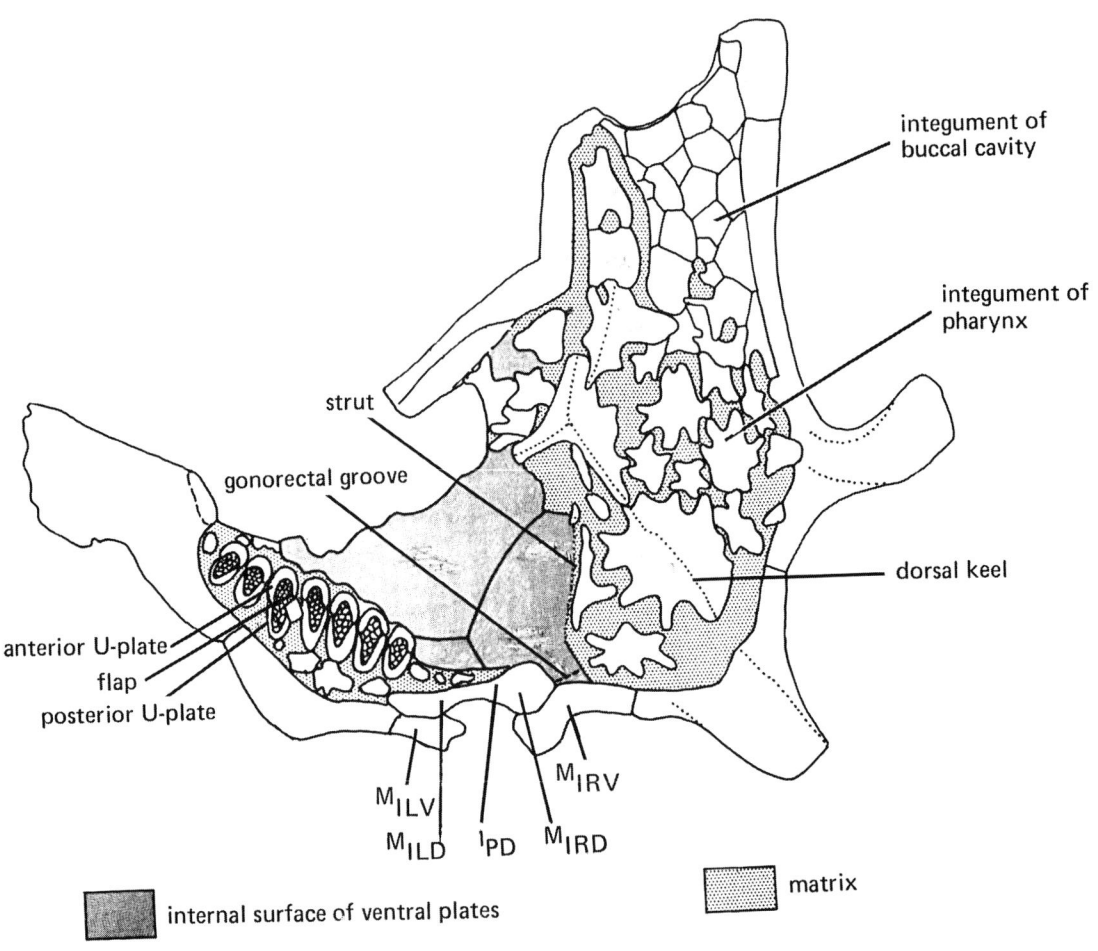

7.27. Nevadaecystis americana. *Camera-lucida drawing of the only known specimen in dorsal aspect; from Jefferies (1969, fig. 4, p. 502).*

hydropore. The position of the gonopore–anus is uncertain because the plates just in front of the tail are ill preserved. It was probably situated left of the tail as in *Cothurnocystis*, since the right end of the gonorectal groove is visible on the floor of the head and seems to pass leftwards beneath the probable location of the posterior coelom. Because of preservation it is also not possible to say whether *Nevadaecystis* had a median eye (dorsal groove) or an acustico-lateralis groove (narrow groove).

A strut, taking the form of a ridge on the internal surface of the floor plates, existed in *Nevadaecystis*, and ran forward from just right of the tail as in *Cothurnocystis* or *Scotiaecystis*. *Ceratocystis* lacks such a ridge. Presumably it was unnecessary there because the box-like skeleton of the head would be rigid whereas the head of *Nevadaecystis* was weakened by the flexible roof. It follows that the presence of a strut is a more advanced condition than its absence.

The tail of *Nevadaecystis* had fore, mid and hind parts much like those of *Cothurnocystis*. In the only known specimen it ends abruptly, but this could easily be caused by recent breakage of the fossil.

To sum up, *Nevadaecystis* confirms the distinction between the integument of the buccal cavity and of the pharynx as already seen in some specimens of *Cothurnocystis elizae*. Further, by having seven gill slits like *Ceratocystis*, it helps to show that the gill slits of that genus are fundamentally different from the accessory gaps. Moreover it shows how the accessory gaps of *Ceratocystis*, by increasing in size and spreading rightwards over the dorsal surface, could produce a fully flexible dorsal integument. And this in turn confirms that the accessory gaps of *Ceratocystis* probably contained, or were associated with, muscle. Lastly its morphology, intermediate in most ways between *Ceratocystis* and the only known Middle Cambrian *Cothurnocystis*, combined with the fact that it comes from the Upper Cambrian, shows that the cornute record for the Cambrian is incomplete and that stratigraphical sequence ought to be disregarded when deciding what features are primitive. When placed on a cladogram, as in Chapter 9, *Nevadaecystis** emerges as the most primitive known cornute apart from *Ceratocystis*. With luck further specimens will be found so that it can be studied in more detail.

Reticulocarpos hanusi is the next cornute to be discussed (fig. 7.28a,b,c). It was described by Rudolf Prokop of the National Museum, Prague, and myself (Jefferies & Prokop, 1972) on the basis of ten specimens collected in the early twentieth century by the outstanding Czech collector František Hanuš. All ten are preserved in siliceous nodules thrown up by the plough in fields near Prague and are preserved in the National Museum. They come from the Lower Ordovician Šárka Shales (Llanvirn Stage) and are thus about 470 million years old. *R. hanusi*, though a cornute, is closely related to the mitrates and is important for the light that it throws on their origin. It is younger than the oldest known mitrates which is another indication that stratigraphical age in calcichordates is no guide to primitiveness. The fact that *R. hanusi* occurs in shales indicates that it lived on a muddy sea bottom, unlike more primitive cornutes such as *Cothurnocystis elizae* or *Ceratocystis* which lived on fine sand. Many features of *R. hanusi* suggest that the bottom mud was extremely soft, and that the animal was always in danger of killing itself by burial.

The head of *R. hanusi* is almost bilaterally symmetrical in outline and small – only about 10 mm long instead of the 20–30 mm common in cornutes and characteristic of the forms discussed already. The head consists of a marginal frame with dorsal and ventral flexible, plated integuments. The ventral surface of the head was flat and was crossed by a partly hollow strut which, however, was incomplete anteriorly, not reaching forwards to the frame. The dorsal surface of the head was probably regularly convex. The dorsal surface of the head was crossed anteriorly by a dorsal bar between the left and right parts of the frame and this was more posterior on the left than on the right. Unlike *Cothurnocystis* (fig. 7.2b) there was no ventral bar behind the mouth and the dorsal bar was probably a mechanical substitute for the ventral bar. The plates of the frame are histologically very lightly built, except near the sutures between them, and this dense sutural calcite suggests that the animals were adult, despite their small size. The integument plates are thin, two-dimensional

*The species *Protocystites menevensis*, from the Middle Cambrian of South Wales, was originally described by Hicks in 1872. It has recently been re-studied and proves to be a cornute in most ways intermediate between *Ceratocystis* and *Nevadaecystis* (Jefferies, Lewis & Donovan, in press).

networks of calcite. The plates of the frame have a sharp edge ventro-laterally which extends, especially near the front part of the head, into a ventro-lateral peripheral flange sloping outwards and downwards. The tail seems to have been very short.

As to head openings, the gill slits are represented by a number of almost circular holes in the posterior left part of the dorsal integument (fig. 7.28a). In life these were perhaps partly covered with soft tissue. The gonopore – anus was situated just left of the tail, much as in *Cothurnocystis elizae* and between the same two plates (M_{1LV}, M_{1LD}). The mouth was at the anterior end of the head but the plates above and beneath it were unspecialized. The frame does not extend anterior to the mouth, i.e. there were no right and left oral appendages. At left and right posteriorly there is a pair of embayments in the dorsal edges of the marginal plates, corresponding to a gap in the dorsal integument on the left and perhaps on the right. These embayments correspond to depressions in the inner faces of the marginal plates just below the dorsal surface (fig. 7.29) and to protuberances of the natural moulds or of the soft parts. These dorsally situated protuberances were probably the vesicles of a pair of transpharyngeal eyes comparable in position to, and homologous with, those of mitrates.

The head chambers were probably much as in other cornutes (fig. 7.30). The posterior boundary of the buccal cavity is marked by the thick posterior edge of the dorsal bar. For this same bar occurs in the boot-shaped cornute *Galliaecystis*, as discussed below (fig. 7.33), and there it is visibly in the same position as the posterior boundary of the buccal cavity in *Cothurnocystis elizae* or *Scotiaecystis*, i.e. at the rear end of the 'ankle' part (buccal lobe) of the 'boot'. The posterior coelom, as in *Cothurnocystis elizae* or *Scotiaecystis*, was roofed over by the two dorsal marginal plates just in front of the tail (fig. 7.29, M_{1LD}, M_{1RD}). As in *Cothurnocystis* or *Scotiaecystis*, a gonorectal groove crosses the floor of the posterior coelom. On the left this leads up to the gonopore—anus. On the right it emerges from the right posterior part of the head where the cavity of the right anterior coeloms with its contents, would have lain. The pharynx and left anterior coelom were probably located as in *Cothurnocystis*.

The tail of *R. hanusi* (fig. 7.31a,b) consisted of fore, mid and hind portions, as usual in cornutes, but was in many ways peculiar. The skeleton of the fore tail consisted of major plates connected together by membranes carrying intercalary plates. The major plates were in the usual four series (dorsal, left and right; ventral, left and right). However, the plates are not joined to form rings. Instead the ventral plates of a pair are sutured together in the mid-ventral line while the dorsal plates are intercalated in the dorsal gaps between successive ventral plates. The ventral

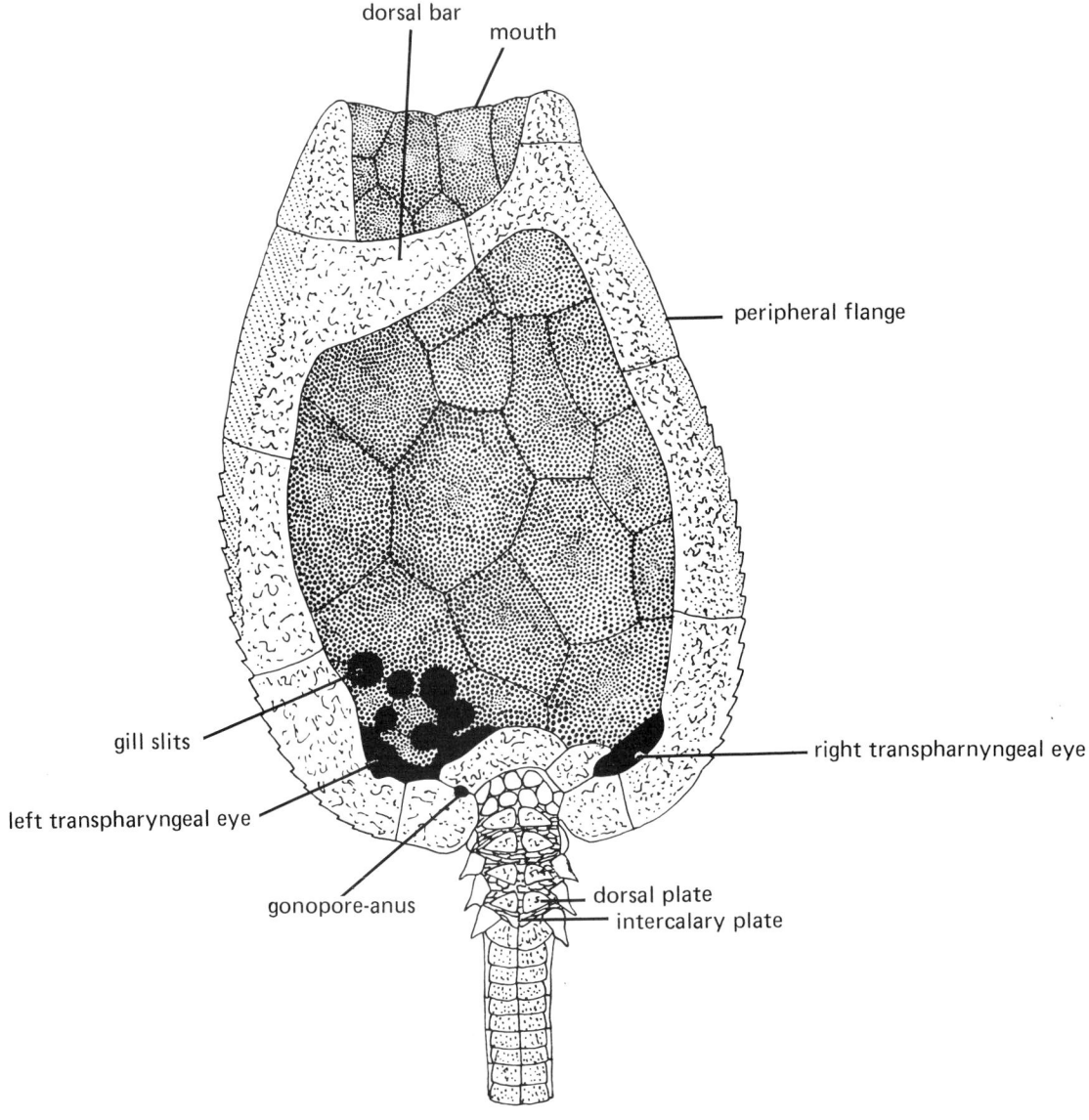

Fig. 7.28a.

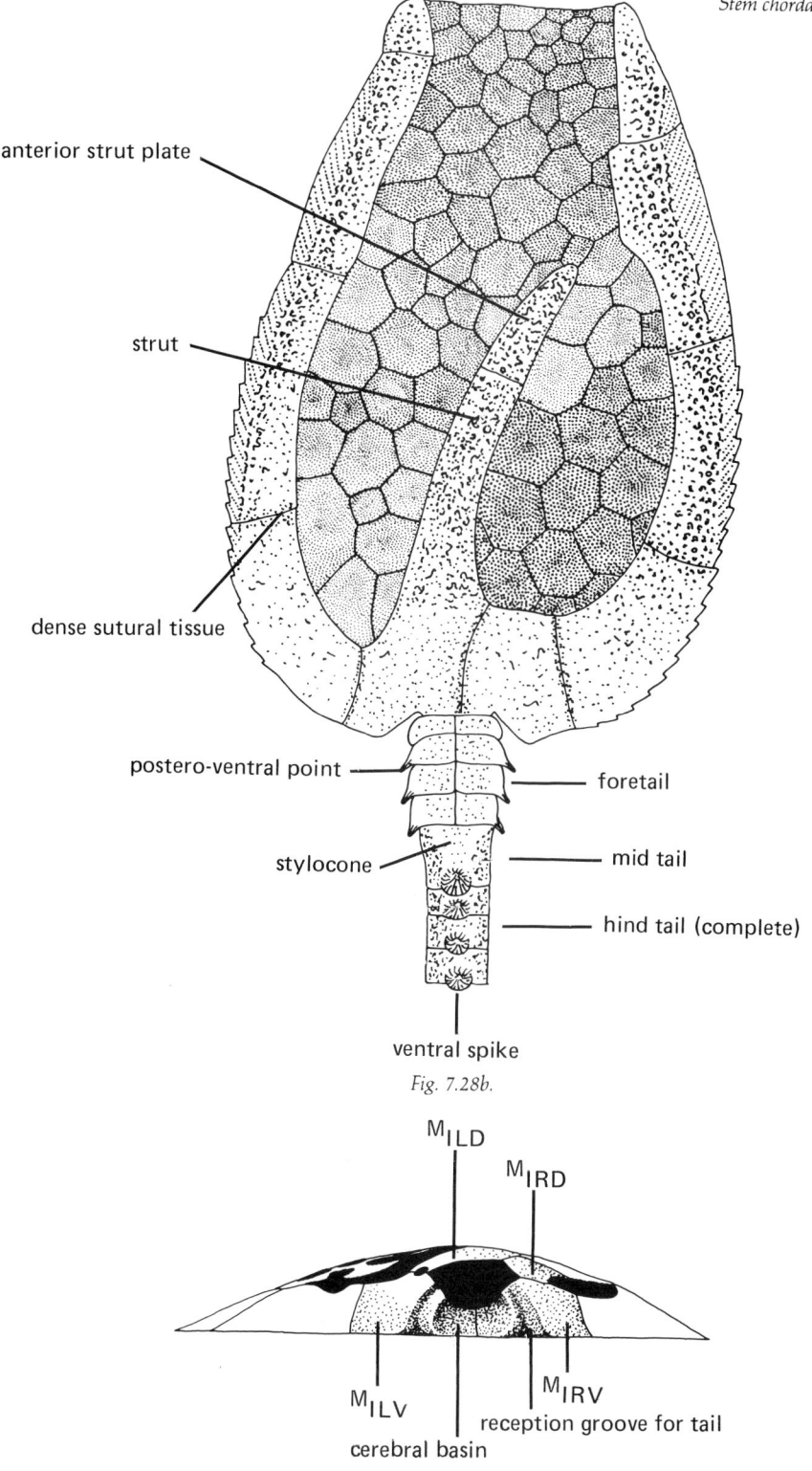

anterior strut plate

strut

dense sutural tissue

postero-ventral point

stylocone

foretail

mid tail

hind tail (complete)

ventral spike

Fig. 7.28b.

M_{ILD}

M_{IRD}

M_{ILV}

M_{IRV}

cerebral basin

reception groove for tail

Fig. 7.28c.

7.28. Reticulocarpos hanusi. *Reconstruction of external features. a) dorsal aspect, b) ventral, and c) posterior; from Jefferies & Prokop (1972, fig. 3a, b, fig. 4c, pp. 79–80).*

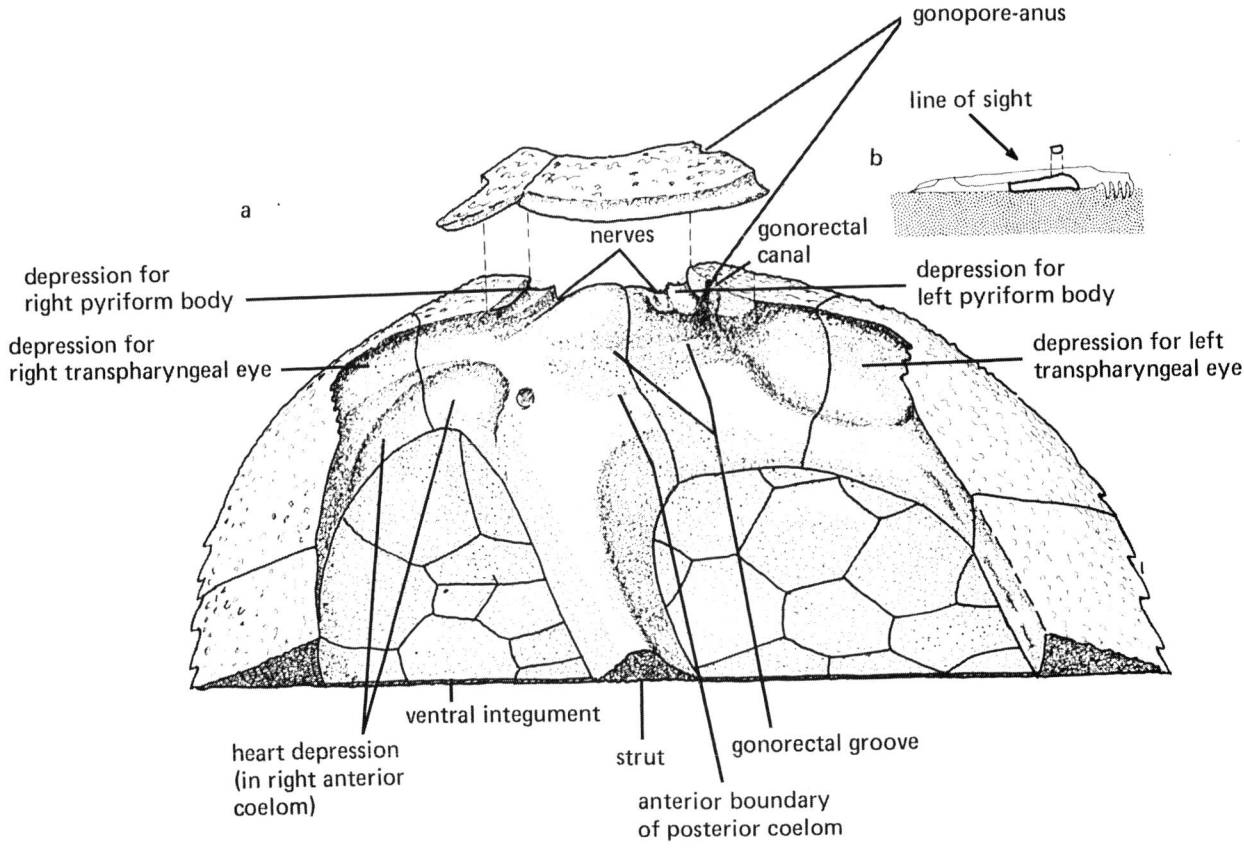

7.29. Reticulocarpos hanusi. *a) Internal anatomy of the skeleton in the posterior part of the head; the anterior part of the head has been cut away, the dorsal integument removed and plates M_{1LD} and M_{1RD} lifted upwards. b) Left lateral aspect to show line of sight in (a).*

plates, except the most anterior pair, extend at right and left into postero-ventral points. The ventral surface of these ventral plates is broad and flat. The most anterior tail plates are inserted in a reception groove excavated in the marginal plates of the head (fig. 7.28c).

The skeleton of the hind tail comprises approximately hemicylindrical ventral ossicles and paired dorsal plates numbering two pairs per ossicle. The hind tail ends abruptly and, so far as known, there are only two to four ossicles in it. This compares with about 60 in *Cothurnocystis elizae*, for example, and probably indicates that the autotomy of the tail, which seems to have been normal in cornute ontogeny, cut off the tail of *R. hanusi* nearer the head than in other cornutes. Each hind-tail ossicle carries posteriorly, in the mid-ventral line and pointing downwards, a stout, recurved, longitudinally striated, hollow spike. The dorsal surfaces of the ossicles are sculptured much as in *Ceratocystis perneri* with a deep median groove and about two pairs of bosses on each ossicle. The

bosses would have served to receive the corresponding dorsal plates. The skeleton of the mid tail comprised a stylocone and four pairs of superimposed dorsal plates. The stylocone had a posterior mid-ventral spike like the hind-tail ossicles. There are two types of surface ornament on the stylocone – a smooth median ventral patch contrasts with a more rugose surface elsewhere that resembles the ventral surface of the hind-tail ossicles. The smooth ventral patch would have been exposed when the tail was extended horizontally and is without counterpart on the hind-tail ossicles – it suggests that the stylocone slid inside the ventral plates of the fore tail in ventral flexion. And such a direction of flexion is consistent with the recurved, downward pointing spikes of the stylocone and hind tail. In flexing downwards (and forwards), the mid and hind tail would behave as a rigid rod, for the interossicular surfaces are plane, not curved. The ossicles are broader for their depth than in *Ceratocystis* or *Cothurnocystis*, which would adapt them for pushing against mud

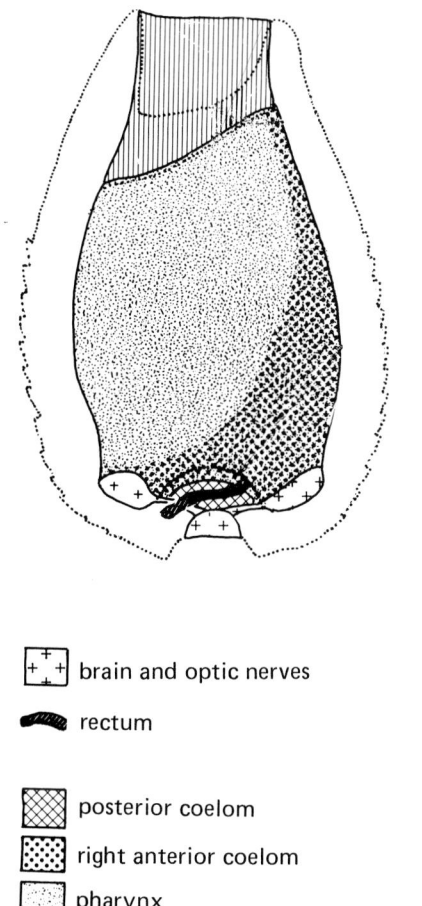

+ + brain and optic nerves

rectum

posterior coelom

right anterior coelom

pharynx

buccal cavity

7.30. Reticulocarpos hanusi. *Reconstructed head chambers in dorsal aspect; from Jefferies & Prokop (1972, fig. 6).*

in ventral flexion. The ventral spikes of *R. hanusi* are probably homologous with the ventral bulges of some of the hind-tail ossicles of *Cothurnocystis elizae*.

R. hanusi seems adapted to life on a sea bottom of soft mud, relying for support on the very weak strength of the mud (fig. 7.32). This is analogous to the use of snow shoes on snow or of mud shoes (plattens) on mud. It depends on reducing the load and spreading it as much as possible so as not to exceed the bearing capacity of the mud. (This is the pressure at which the mud collapses under superimposed load and can be measured in weight per unit area.) This method of support used by *R. hanusi* should be contrasted with the 'buoyancy' method employed by the mitrates.

Adaptations for load reduction in *R. hanusi* include small size (as compared with *Cothurnocystis elizae*, for example), the lightly built histology of the marginal plates and the reticulation of the integument plates, the hollowness of the strut and tail spikes and the shortness of the tail. Adaptations for load spreading include the broad flat ventral surface of the head and fore tail, the absence of ventral spikes on the head and the extension of the marginal plates into a ventro-lateral peripheral flange. As a consequence of these features, the load imposed on the sea bottom, assuming that gravity had its present value, would be about 10 mg cm^{-2} (Jefferies & Prokop, 1972, p. 102). It is possible to make a mud about as weak as this in the laboratory by precipitating clay particles in salt water.

Locomotion would have been by ventral flexing of the tail, pulling the head rearward. Adaptations for such ventral flexion have already been mentioned. The rather broad, flattish ossicles of the hind tail, as well as the broad flat ventral surface of the fore tail, would press downwards and forwards against the mud, pulling the head rearwards and upwards. The spikes would prevent the tail from slipping sideways, acting like the spikes of a running shoe. The bilaterally symmetrical outline of the head probably has to do with the fact that yaw-prevention devices, such as the spikes of a boot-shaped head, would be ineffectual on soft mud. It would be more efficient to avoid yaw by flexing the tail ventrally and acquiring bilateral symmetry. Also ventral flexion is easier in soft mud than in sand because the substrate is so weak.

The tail of *R. hanusi* can be compared with that of the primitive mitrate *Chinianocarpos thorali* (figs 8.26, 8.27, pp. 291, 293). The latter likewise has a fore, mid and hind portion, but, in my view, these are not homologous with their namesakes in cornutes. For the ossicles of the hind tail and the massive styloid of the mid tail are dorsal, not ventral, and the paired plates of the mid and hind tail, conversely, are ventral not dorsal. The tail of *Chinianocarpos thorali* ends abruptly, as if cut off by autotomy like the tail of *R. hanusi*. As in *R. hanusi* the major plates of the fore tail alternate with plated imbrication membranes. Very significant is the fact that the same plated imbrication membranes also exist in *Chinianocarpos thorali* between the ventral plates of the hind tail (fig. 8.27c). Moreover, it seems that the whole tail of *Chinianocarpos thorali* was able to flex in a vertical plane. In both these respects, the whole tail of *Chinianocarpos* is comparable with only the fore tail of *R. hanusi*, for the mid and hind tail of the latter was rigid and lacked plated imbrication membranes. The comparison suggests that the tail of mitrates is homologous only with the fore tail of *R. hanusi*; that the cornute mid and hind tail were lost by autotomy; and that the mitrate regionation into fore, mid and hind tail arose by modification of the remaining stump. Small spike-shaped

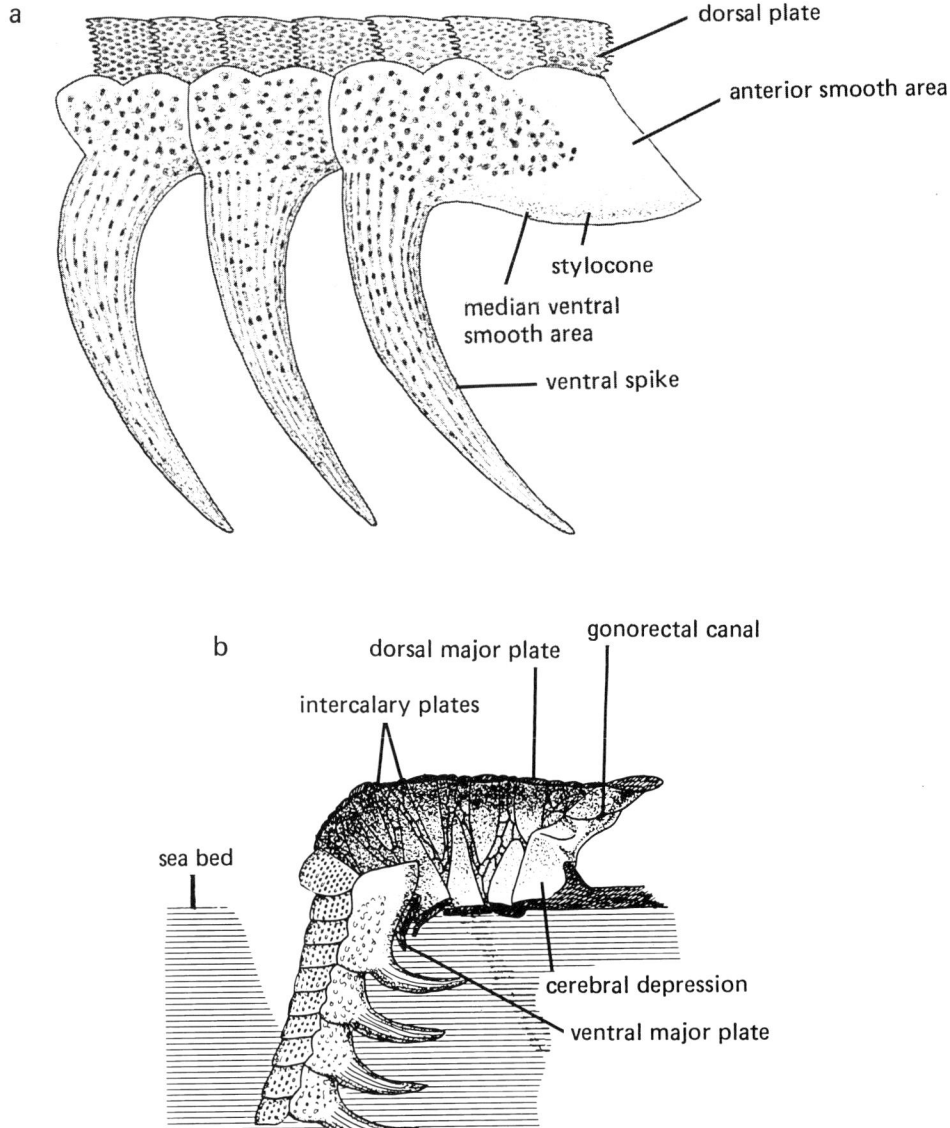

7.31. Reticulocarpos hanusi. *The tail and its mode of action. a) Right aspect of mid and hind tail; note the median smooth area on the stylocone which has no counterpart on the hind-tail ossicles. b) The tail at the end of the power stroke – the right half of the fore tail skeleton has been removed and also all of its soft parts, so showing the inside of the skeleton of the left half; note how the ventral smooth area of the stylocone would have slid upwards inside the ventral plates of the fore tail so that the hind tail was rotated forwards and downwards.*

plates in the dorsal surface of the fore tail of *Chiniano-carpos thorali* are probably serially homologous with the styloid and the dorsal ossicles of the hind tail.

Comparison with the tail of the mitrate *Lagynocystis pyramidalis* (figs 8.28c, 8.34, pp. 294, 305) confirms that

the whole mitrate tail is homologous with the cornute fore tail only. For in this species the plates of the hind tail have a flat or concave ventral surface and are drawn out into a postero-lateral spike, as are the ventral plates of the fore tail of *R. hanusi*. Moreover the plates of the mid tail and

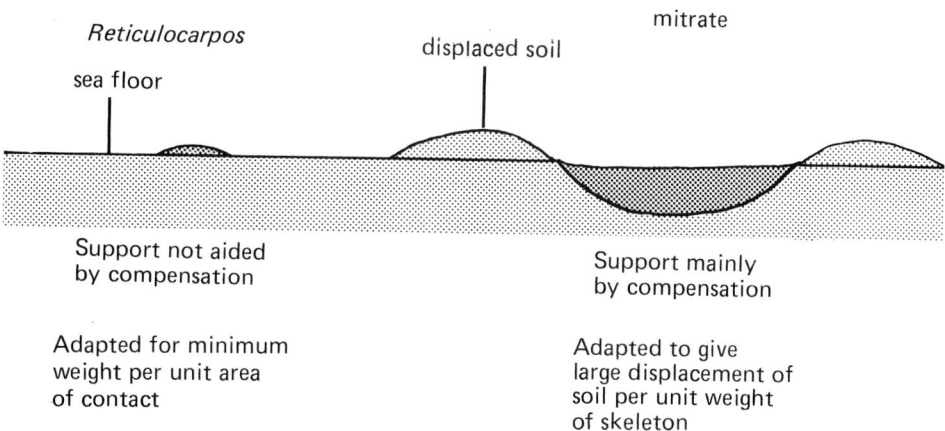

Reticulocarpos

sea floor

displaced soil

mitrate

Support not aided
by compensation

Support mainly
by compensation

Adapted for minimum
weight per unit area
of contact

Adapted to give
large displacement of
soil per unit weight
of skeleton

7.32. *Suggested mode of support on mud of* Reticulocarpos hanusi *contrasted with that of a mitrate (or of* Reticulocarpos pissotensis); *from Jefferies & Prokop (1972, fig. 13, p. 103).*

the most anterior plates of the hind tail alternate with plated imbrication membranes, like those of *Chinianocarpos*. Also, the fore tail of *Lagynocystis pyramidalis* has dorsal spikes much larger than those of *Chinianocarpos*. These large dorsal fore-tail spikes give the strong impression of being serially homologous with the spikes on the styloid (the dorsal ossicle of the mitrate mid tail) and the dorsal points of the dorsal ossicles. Probably the primitive mitrate tail resembled throughout its length the fore tail of *Lagynocystis*, and the dorsal ossicles and the stylocone of more advanced mitrates were formed by modification of dorsal spikes. Later, except in *Lagynocystis* and *Chinianocarpos*, the dorsal spikes disappeared from the mitrate fore tail.

Synapomorphies of *R. hanusi* which indicate that it is closer related to the mitrates than is, for example, *Cothurnocystis elizae*, include: the bilaterally symmetrical outline of the head; the dorsal bar which is comparable with the front edge of the head shield of the mitrate *Peltocystis* (fig. 8.38a, p. 310), for in both forms it involves the same two plates and is more posterior on the left than the right; the incompleteness of the strut, which in mitrates is absent; the peripheral flange, though in mitrates this is dorso-lateral in position, not ventro-lateral; the ventrally flexing tail; the shortness of the (cornute) hind tail, which in mitrates is absent; the presence of plated imbrication membranes between the ossicles of the fore tail; the presence of broad flat bearing surfaces on the ventral plates of the fore tail extending into paired postero-lateral points, comparable with the ventral bearing surfaces and postero-lateral points on the hind tail of *Lagynocystis*; the small absolute size, comparable with that of the primitive (and stratigraphically old) mitrates *Peltocystis cornuta* and *Chinianocarpos thorali*.

Thus *R. hanusi* suggests how the mitrates evolved from cornutes, though other forms, discussed below, help to fill the story in. It also suggests that the origin of mitrates had to do with a migration from fine sand to mud. *R. hanusi* is still a cornute, however, since it has only left gill slits and the cavity of the right anterior coelom, to judge by the origin of the gonorectal groove, probably lay in the cornute position on the right posterior floor of the head. Also the gill slits are external and the tail is cornute, having ventral ossicles in the mid and hind tail and dorsal plates. In all these respects mitrates show significant advances over *R. hanusi*.

The morphological gap between *R. hanusi* and *Cothurnocystis elizae* is partly filled by the genera *Galliaecystis* and *Amygdalotheca*. These forms were described by Ubaghs (1969) from the basal Ordovician of the Montagne Noire, southern France.

The genus *Galliaecystis* (fig. 7.33) is represented only by the species *Galliaecystis lignieresi* and this is only known from two specimens so that Ubaghs' description, although very thorough, is necessarily incomplete. The specimens are preserved in siliceous nodules and came out of shale which implies that the animals lived on a muddy sea bottom.

The head of *Galiaecystis* is boot-shaped and rather large for a cornute, being 25 mm long. The frame is of slender plates. It differs from that of *Cothurnocystis elizae* in having a bar just behind the buccal lobe and in lacking a ventral bar behind the mouth. Both of these peculiarities are advanced features compared with *C. elizae*. On the frame there are three ventral spikes which would have protruded into the sea bed as also would a short downward pointing left appendage. The spikes and appendage would have the usual function of helping the

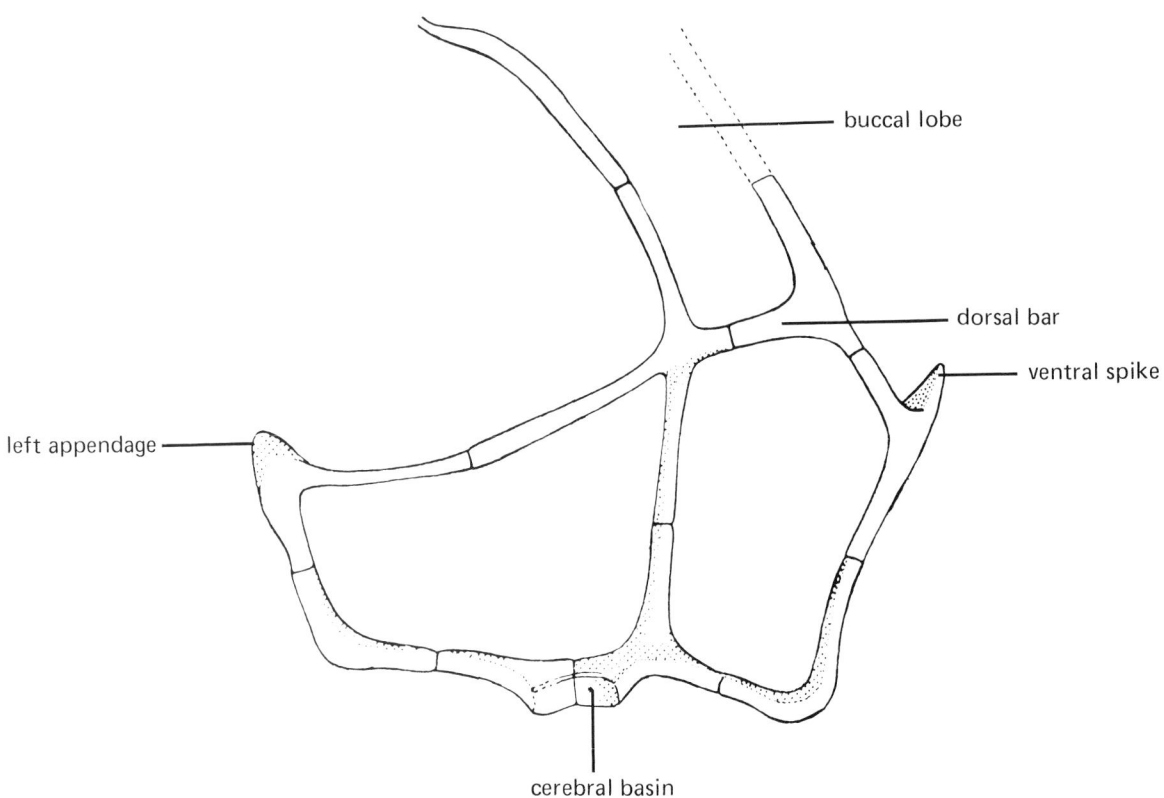

7.33. Galliaecystis lignieresi. *Reconstructed marginal frame of the skeleton in dorsal aspect. Note the dorsal bar, comparable with that of* Reticulocarpos hanusi; *from Ubaghs (1969, fig. 28, p. 69).*

head to slip rearward. A strut is present in the usual position and its anterior portion is probably an extension of a marginal plate as in *Cothurnocystis* or *Scotiaecystis*, rather than a separate plate as in *R. hanusi* or *Amygdolotheca.* The ventral surface of the head was by no means plane, unlike *R. hanusi.* The posterior part of the frame was arched upwards, so that the tail insertion was considerably higher than the left and right posterior angles of the head, though the tail insertion itself corresponded to a local downward bulge in the frame. In all these respects *Galliaecystis* is like *Cothurnocystis* and therefore primitive.

The integuments of *Galliaecystis* are covered with large polygonal plates which, although not two-dimensional networks of calcite, are histologically more like those of *R. hanusi* than like those of *C. elizae.*

As to gill slits, Ubaghs says they are absent in *Galliaecystis.* However, the dorsal integument seems to be loose just left of the tail, to judge by Ubaghs' photographs (Ubaghs, 1969, pl. 13, fig. 1b, 4) and in one of the two known specimens there are suggestions of part of a circular

gap in the dorsal integument like one of the gill slits of *R. hanusi* (see Ubaghs, 1969, pl. 12, fig. 1, just left, in my orientation, of the worm track that penetrates the dorsal integument). I therefore suspect that gill slits of *R. hanusi*-type were present. There was a transverse gonorectal groove and the gonopore–anus was presumably just left of the tail.

The tail is much as in *Cothurnocystis* (fig. 7.34). There are six rings of major plates in the fore tail. The longer of the two known hind tails shows at least 16 ossicles, though it is impossible to say whether this is a maximum. On each ossicle distally in the mid-ventral line there is usually a small boss or point, probably homologous with one of the ventral spikes of the hind tail and stylocone of *R. hanusi* and with one of the ventral convexities of the hind tail of *C. elizae.* In both the known specimens of *Galliaecystis* the tail is flexed strongly sideways – rightwards in one specimen and leftwards in the other. This suggests (though the posture of a corpse need not be the same as that of the living animal) that the tail pulled the head rearwards by sideways flexion, as probably in

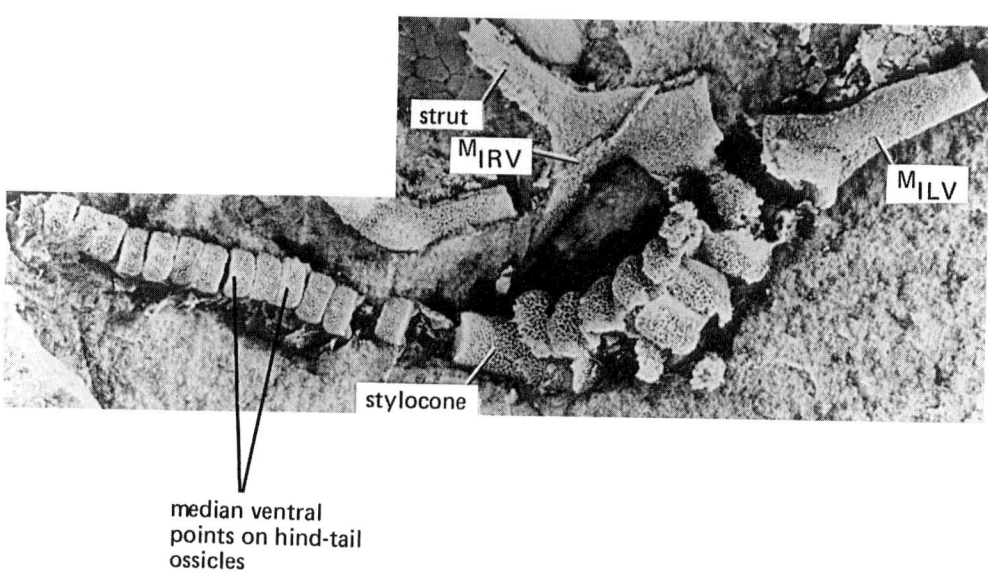

strut
M_{IRV}

M_{ILV}

stylocone

median ventral
points on hind-tail
ossicles

7.34. Galliaecystis lignieresi. *Ventral aspect of tail and adjacent parts of the frame to show the numerous hind-tail ossicles and their median ventral points; from Ubaghs (1969), pl. 13, fig. 2).*

Cothurnocystis, not by ventral flexion as in *Reticulocarpos*. Each hind-tail ossicle has a median groove on its dorsal surface and a single pair of transverse grooves. In the latter respect these ossicles are like those of *C. elizae* and differ from *R. hanusi* or *Ceratocystis perneri*. The ventral ossicles are roofed over by paired plates, as in other cornutes.

Galliaecystis, when compared with *R. hanusi*, thus has many primitive resemblances to *Cothurnocystis* including: the boot-shaped head with its non-planar ventral surface and ventral spikes; the long, perhaps sideways flexing tail; the small size of the points on the hind-tail ossicles; and the likelihood that the anterior part of the strut was a posterior extension of the marginal frame rather than a separate plate. On the other hand, when compared with *Cothurnocystis* it has advanced resemblances to *R. hanusi* including: the relatively large flat integument plates; the absence of any obvious gill-slit skeleton; the absence of a ventral mouth bar; the fact that the integuments extend to the anterior ends of the frame in the mouth region; the presence of points, though weak, on the hind-tail ossicles; and above all the presence of a dorsal bar behind the buccal lobe. These advanced features are synapomorphies with *R. hanusi* and show that *Galliaecystis* is closer related to it than to *C. elizae*.

As to internal structure, the main significance of *Galliaecystis lignieresi* lies in the position of the dorsal bar at the posterior limit of the buccal lobe and therefore of the buccal cavity. As already mentioned, this suggests that the dorsal bar of *R. hanusi* likewise represents the position of the posterior limit of the buccal cavity.

Amygdalotheca griffei (fig. 7.35a,b) is likewise known from only two specimens which are preserved in siliceous nodules from the same lowest Ordovician horizon as *Galliaecystis lignieresi* and the same general region in the south of France (Montagne Noire).

The head is bilaterally symmetrical in outline like *R. hanusi*, but longer, being 18 mm in length. There is no ventral bar behind the mouth, but neither is there a dorsal bar. The anterior part of the strut is a distinct plate, as in *Reticulocarpos*, but this makes contact with the frame anteriorly, by a suture, so that the strut is connected with the frame as in *Galliaecystis* or *Cothurnocystis*. The marginal plates of the frame are extended ventro-laterally into a peripheral flange as in *R. hanusi* and there are no ventral spikes. Indeed the ventral surface of the head is almost plane, as in *R. hanusi*, except for a ventralwards bulge at the tail insertion. The dorsal surface of the head was probably regularly convex upwards. The plates of the ventral integument are rather large and distinctly reticulate in histology, though not so thin as those of *R. hanusi*. Those of the dorsal integument are somewhat similar to the integument plates of *Scotiaecystis curvata*, each having a polygonal base the centre of which carries a process that expands upwards towards the dorsal surface (fig. 7.35b, cf. fig. 7.13f). It is likely that the dorsal integument was thick and muscular – just below the top edge of the inner surface of the frame there is a concave zone for its probable reception comparable with the upper zone on the inner face of the frame of *Scotiaecystis curvata*. The mouth was presumably situated anteriorly, where the

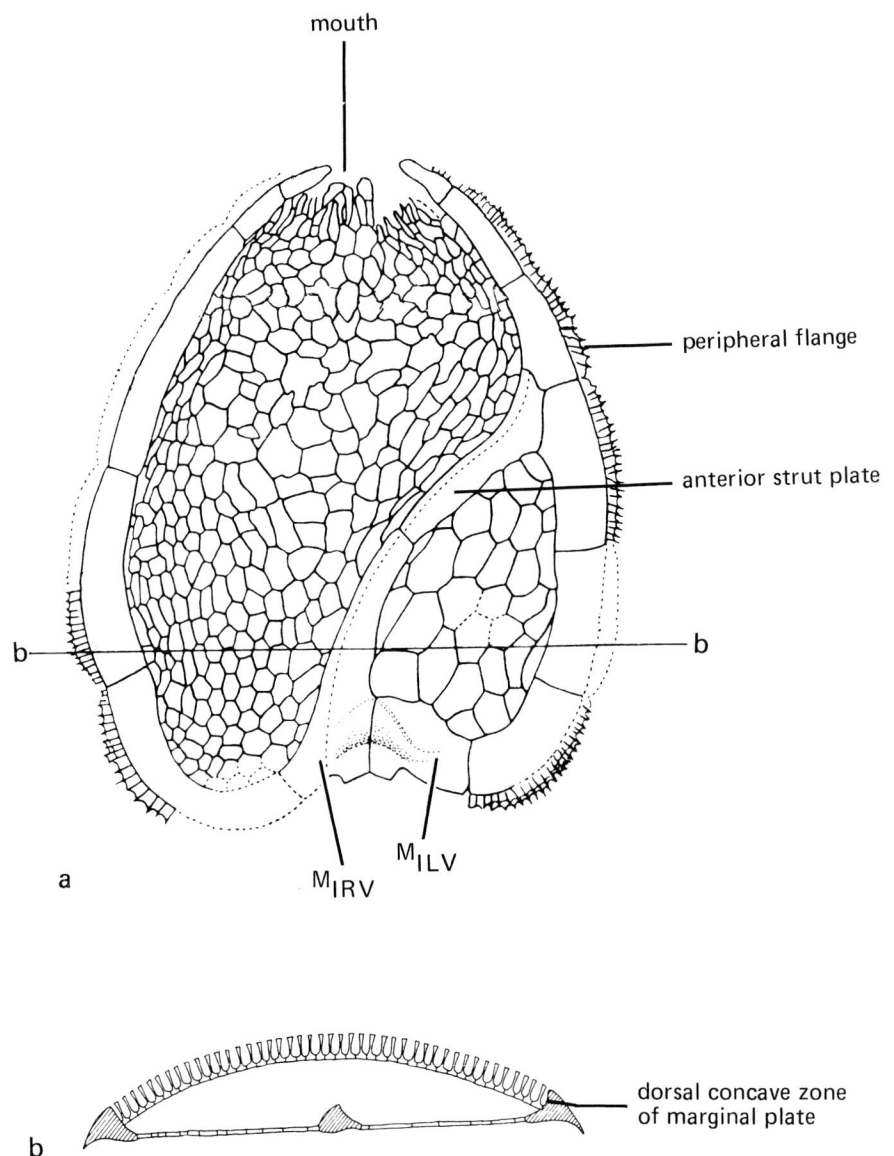

mouth

peripheral flange

anterior strut plate

b ———————————————————— b

M_{ILV}

M_{IRV}

a

b

dorsal concave zone
of marginal plate

7.35. Amygdalotheca griffei. *a) Camera-lucida drawing of ventral aspect; note that the anterior part of the strut is developed as a separate plate; from Ubaghs (1969, fig. 26/1, p. 64). b) Transverse section through b–b in a; based on Ubaghs (1969, fig. 26/2) but redrawn and modified.*

frame is interrupted. The ventral integument plates behind the mouth are not completely preserved but some of them are elongate, radiating from the position of the mouth, suggesting that the lower lip was especially flexible. No gill slits have been found in *Amygdolotheca*, and Ubaghs asserts that they were absent, but, to judge by his photographs, the relevant posterior left parts of the dorsal integument are not well enough preserved on either speci-

men to support his assertion. Gill slits of *Reticulocarpos*-type may well have been present but are easy to overlook.

The tail of *Amygdalotheca* is very badly known. Only the first three rings of the fore tail are visible and in only one specimen. The ventral surfaces of the ventral plates of the fore tail extend laterally into spikes like those of *R. hanusi*. Such spikes are lacking in *Galliaecystis*.

Amygdalotheca therefore shares with *R. hanusi* many

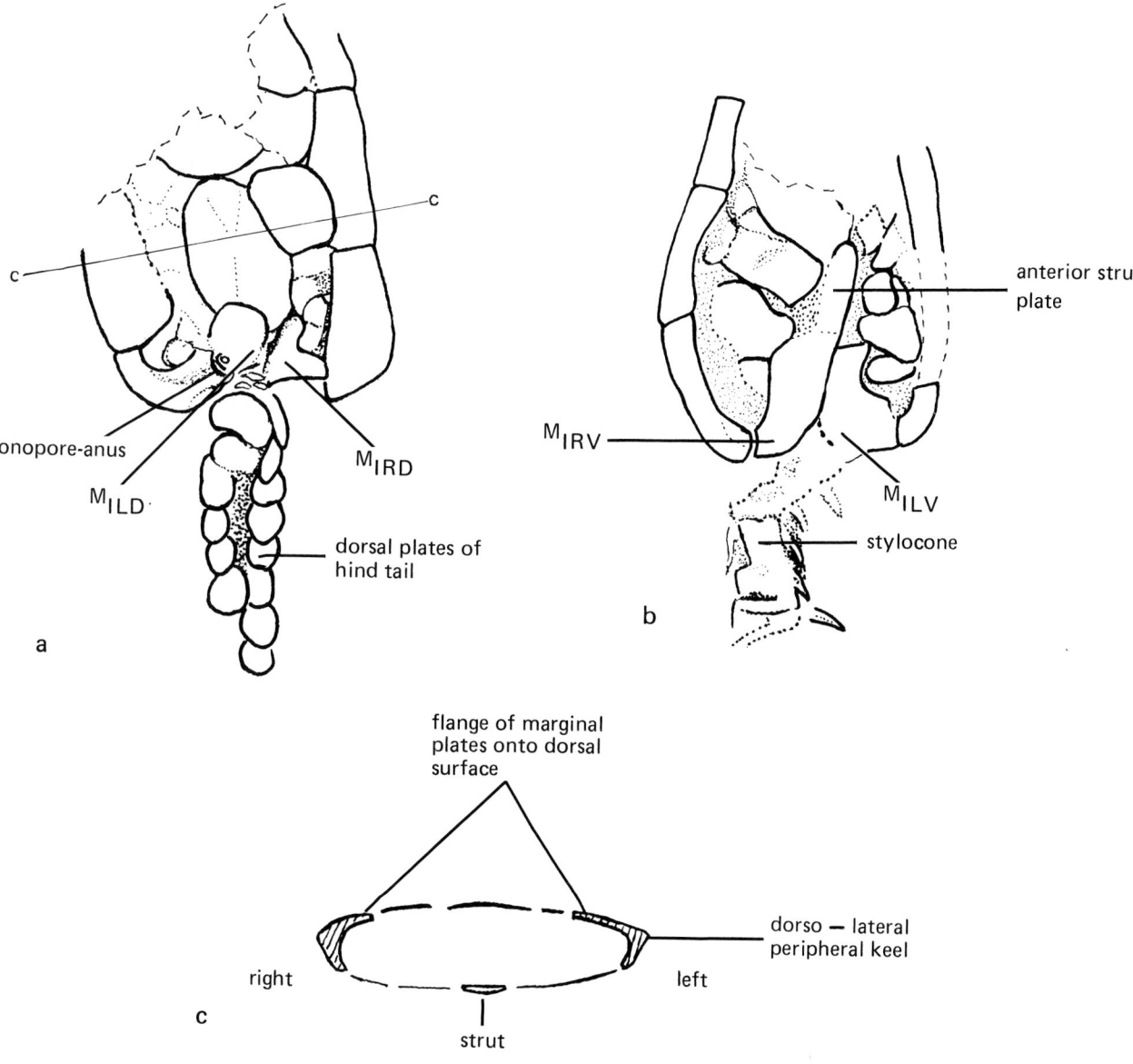

7.36. Reticulocarpos pissotensis. *Camera-lucida drawings of: a) dorsal surface, and b) ventral surface. c) Reconstruction of a transverse section through c–c in (a). This species is the only described cornute which is ventrally convex and dorsally almost plane, like a mitrate; from Chauvel & Nion (1977, fig. 2a, b, d, p. 38).*

advanced features which *Galliaecystis* lacks. These include: the bilaterally symmetrical head; the peripheral flange; the almost plane ventral surface and absence of ventral spikes on the head; paired spikes on the ventral plates of the fore tail; and the fact that the anterior part of the strut is developed as a separate plate. On the other hand, *Amygdalotheca* is more primitive than *R. hanusi* in its large size; the downward bulge of the tail insertion; the strut connected with the frame; and probably the presence

of elongate plates in the lower lip. And again there are features of *Amygdalotheca* which are peculiar to it and must be regarded as private specializations (autapomorphies). Such are: the absence of the dorsal bar; the *Scotiaecystis*-like dorsal integument plates (presumably this resemblance to *Scotiaecystis* is convergent, since such plates are not developed in more primitive members of the Scotiaecystidae); the dorsal concave zone of the inner surface of the frame plates, which received a thick muscular dorsal integument; the presence of two extra plates on either side of the frame as compared with *Galliaecystis* or *R. hanusi* (though in this book I shall not discuss plate homologies). The relatively flat ventral surface, and extensive peripheral flange can be seen as adaptations for staying up on mud using the same snow-shoe system that *R. hanusi* employed and bilateral symmetry can be seen as an adaptation for moving on such mud without yaw. However, the large size of the head, the fact that the ventral surface is not totally plane, and the relatively dense histology of the plates show that the snow-shoe system was less extremely developed than in *R. hanusi*.

The last cornute that I shall discuss at some length is the species *Reticulocarpos pissotensis* described by Chauvel & Nion from the Middle Ordovician shales (Llandeilo Stage) near the hamlet of Pissot near Domfront, Normandy, France. Only two incomplete specimens and some fragments are known.

In most ways *R. pissotensis* (fig. 7.36) is like *R. hanusi*. The head is small, being somewhat longer than 7 mm, and almost bilaterally symmetrical in outline. The tail is short with, so far as seen, only two ossicles in the hind tail, and there are spikes on the hind-tail ossicles and stylocone as in *R. hanusi*. However, there are some important differences from *R. hanusi*. In particular the ventral surface of the head, as judged by the shape of the marginal plates and of the anteriorly terminating strut, was convex downwards (fig. 7.35c), while the dorsal surface was flat or slightly convex upwards. Also the marginal plates extend much farther onto the dorsal than onto the ventral surface of the head. Unlike *R. hanusi* there is no extended peripheral flange, but there is a sharp keel where dorsal and ventral surfaces of the marginal plates meet. Because of the shape of the marginals this keel is dorso-lateral in position, not ventro-lateral. Some of the dorsal integument plates are very large compared with the size of the head, so resembling the centro-dorsal plates of mitrates. Indeed all these differences from *R. hanusi* are mitrate-like. They can be regarded as synapomorphies with the mitrates and suggest that *R. pissotensis* was closer related to the mitrates than *R. hanusi* was.

There are many uncertainties about *R. pissotensis*, for the known specimens, as mentioned, are incomplete. They are particularly badly preserved anteriorly so that it cannot be decided whether a dorsal bar was present or not. The nature and position of the gill slits, if indeed they opened externally, are also uncertain and it is therefore not entirely certain that *R. pissotensis* was a cornute rather than a mitrate. The gonopore–anus, however, was in the usual cornute position just left of the tail and was situated between the two usual plates (M_{1LD}, M_{1LV}). It is to be hoped that more specimens of this interesting species can be found so as to clarify its morphology.

From the functional point of view, *R. pissotensis*, with its convex ventral surface, had evidently begun, like the mitrates, to use the buoyancy or displacement method of staying up on mud (fig. 7.32), rather than the snow-shoe method used by *R. hanusi* or *Amygdalotheca*.

A number of other cornute genera should be mentioned. They include *Thoralicystis* described by Chauvel (1971) and Ubaghs (1969) and *Bohemiaecystis* described by Caster in Ubaghs (1968), both being primitive relatives of *Scotiaecystis*, as also is '*Cothurnocystis*' *melchiori* recently described by Ubaghs (1983) from the Lower Ordovician of the Montagne Noire. The genus *Phyllocystis* (fig. 9.5, p. 326) was established by Thoral (1935) and well described by Ubaghs (1968, 1969). It is like *Cothurnocystis elizae* in most ways but its head is bilaterally symmetrical in outline – this, however, is almost certainly a convergence with *Reticulocarpos* and *Amygdalotheca*, not a homology. *Chauvelicystis* (fig. 10.4) is like a *Cothurnocystis* but has a fringe of moveable spines; it was described by Ubaghs (1969) and Chauvel (1966). All these various genera come from the Lower and Middle Ordovician of Bohemia, Morocco and southern France.

To sum up, the cornutes were extraordinarily significant animals. They have a skeleton histologically like that of a starfish combined with a chordate tail and the solely left gill slits of larval amphioxus. The most primitive of them, *Ceratocystis*, suggests a flatfish-like dexiothetic stage in the exclusive common ancestry of echinoderms and chordates, while the most advanced of them, despite having left gill slits only, are closely related to the crown chordates. In Chapter 9 I shall return to the cornutes, allocate them to their various positions in the chordate stem group and write a history of how this stem group evolved.

Chapter 8   Primitive crown chordates – the mitrates

The mitrates are the more advanced subgroup of the calcichordates and, since all known mitrates can be referred to the stem groups of the extant chordate subphyla, all known mitrates are primitive crown chordates. The soft parts of mitrates can be reconstructed in more detail than in cornutes, having left abundant traces in and on the skeleton.

Mitrocystites mitra is the first mitrate to be discussed (fig. 8.1). It is found in the Lower Ordovician Šárka shales of Bohemia, Czechoslovakia, in the same beds as *Reticulocarpos hanusi*, and is thus about 470 million years old. Several hundred specimens have been collected and the best of them, as with *R. hanusi*, are in siliceous nodules which have been picked up from ploughed fields. *Mitrocystites mitra* was originally described by the great French emigré palaeontologist Joachim Barrande, though the description was only published in 1887, four years after his death. (Barrande is greatly honoured in his adoptive country; there is even a suburb of Prague called Barrandov.) The species has been restudied by Chauvel (1941), by Ubaghs (1967b) and by me (1968b, 1969, 1973; Jefferies & Prokop, 1972; Jefferies & Lewis, 1978). Henceforth I refer to *Mitrocystites mitra* as *Mitrocystites*, though other species of the genus are known.

Mitrocystites consists of a head and a tail. The head is up to 30 mm long with a flat dorsal surface and a convex ventral surface. The tail is in three portions which I call fore, mid and hind tail, though in my view the whole tail is homologous with the cornute fore tail only. The dorsal face of the head is made up of 13 marginal plates and five centro-dorsal plates. The ventral surface is mainly formed of plated integument which was probably fairly flexible anteriorly, where the plates overlap, but stiffer posteriorly where they juxtapose without overlap. Posteriorly there are two ventral first marginal plates (M_{1RV}, M_{1LV}) right and left of the tail insertion. Dorsal and ventral surfaces of the head meet at a dorso–lateral peripheral keel or flange on the marginal plates.

The openings of the head are numerous. In ventral aspect (figs 8.1b, 8.2) a large mouth, comparable in position with that of a cornute, is visible anteriorly. It had a stiff upper lip formed of marginal plates and a flexible lower lip guarded by spike-shaped oral plates. There are no external gill slits, but near the right and left posterior angles, at the right and left ends of the pair of ventral marginal plates, are the right and left atrial openings (the skeletal evidence for the left atrial opening in *Mitrocystites* in shown in fig. 8.2). Through these water would have issued from chambers known as the right and left atria, which would have had gill slits penetrating their front walls. On the right ventral marginal plate (M_{1RV}), just right of the tail insertion, is a groove which, for the moment, I shall refer to as the 'narrow groove'. In dorsal aspect two pairs of openings are visible, in the posterior part of the head (fig. 8.1a). The anterior pair were the openings onto the dorsal surface of the nerves n_3 of the palmar complex (which is discussed below), while the posterior pair were the openings of the combined nerves n_4 and n_5 of the palmar complex. Those latter openings are located in the deepest parts of arcuate peripheral grooves on the dorsal surface of the head. I shall return below to the likely nature of these nerves, or presumed nerves.

In the rival interpretations put forward by Ubaghs (1968) and Philip (1979) many of the statements just made would not hold. Both authors regard what I call the dorsal surface as ventral, and ventral as dorsal. Both see the fore, mid and hind portions of the tail as homologous with their namesakes in mitrates. Neither would accept the designations just given of the openings of the dorsal surface, using instead the name of paripores for the openings of n_3 and lateripores for the openings of n_4 and n_5. Philip categorically denies the existence of atrial openings (1979, p. 457). What I have called the narrow groove is for Ubaghs probably a hydropore (1968, p. S518), homologous with the gonopore–anus of cornutes which

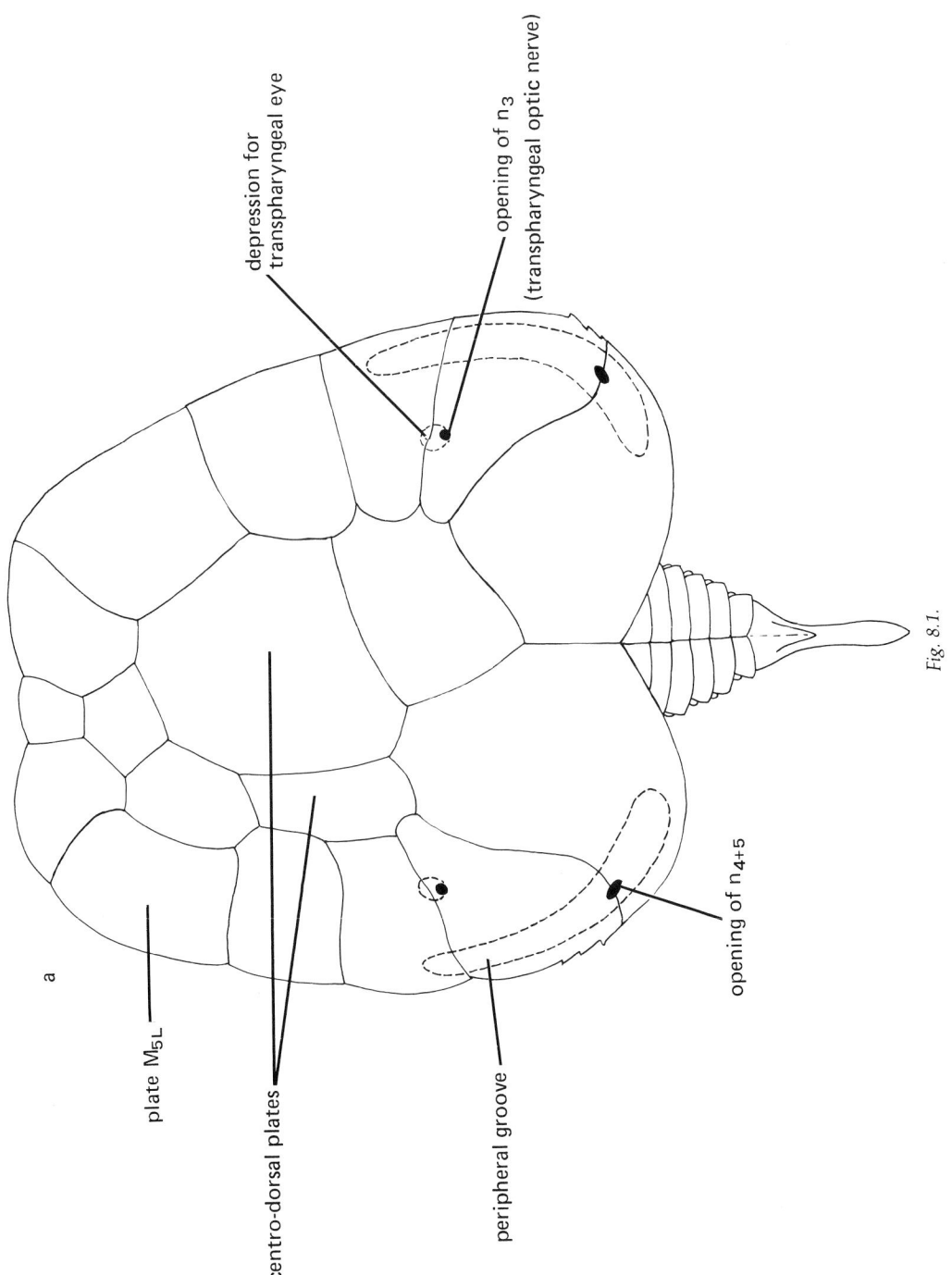

depression for
transpharyngeal eye

opening of n₃
(transpharyngeal optic nerve)

plate M₅ₗ

centro-dorsal plates

peripheral groove

opening of n₄₊₅

a

Fig. 8.1.

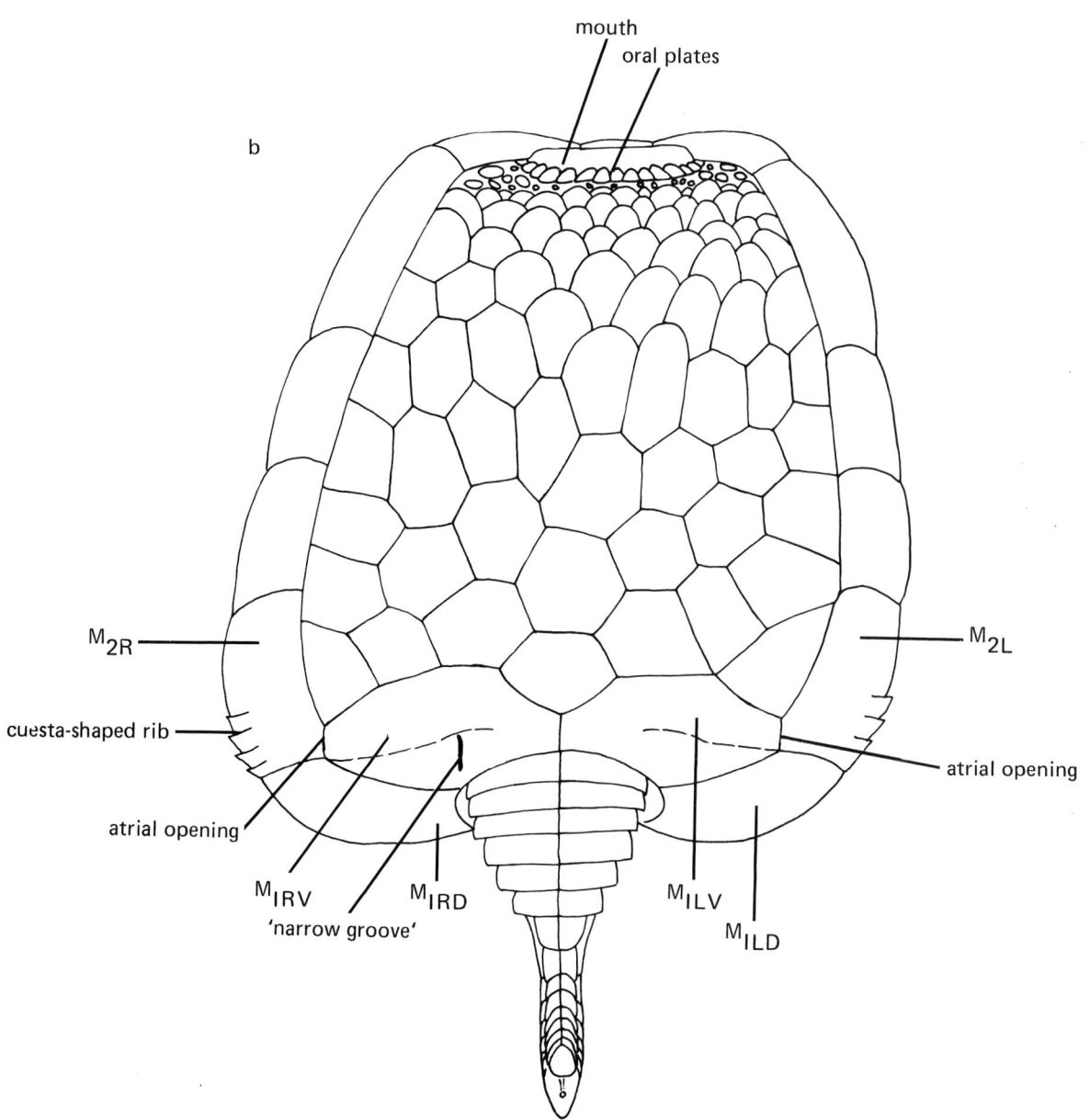

Fig. 8.1.

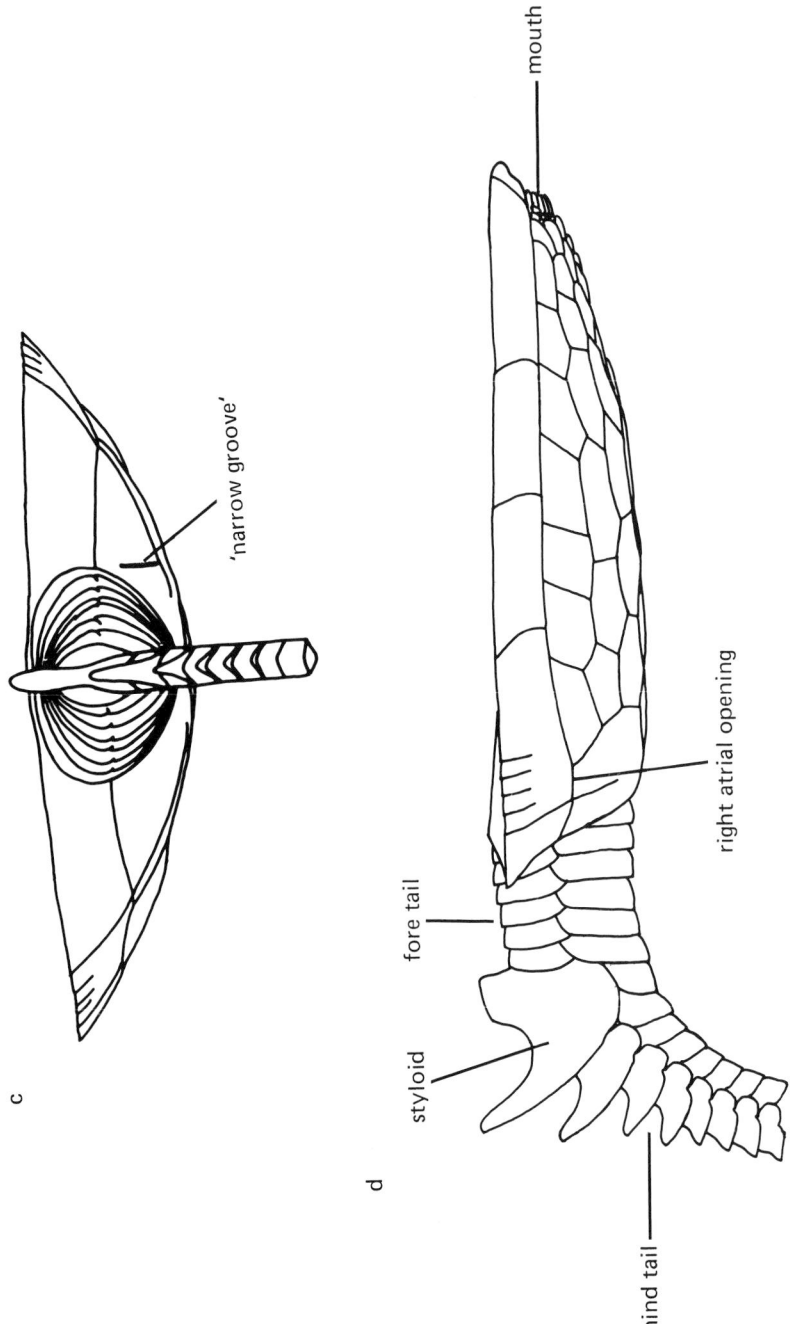

'narrow groove'

c

mouth

right atrial opening

fore tail

styloid

hind tail

d

8.1. Mitrocystites mitra. Reconstruction of external features. a) dorsal aspect, b) ventral, c) posterior, and d) right lateral; from Jefferies (1968, fig. 23a, b, c, e, p. 311 ff).

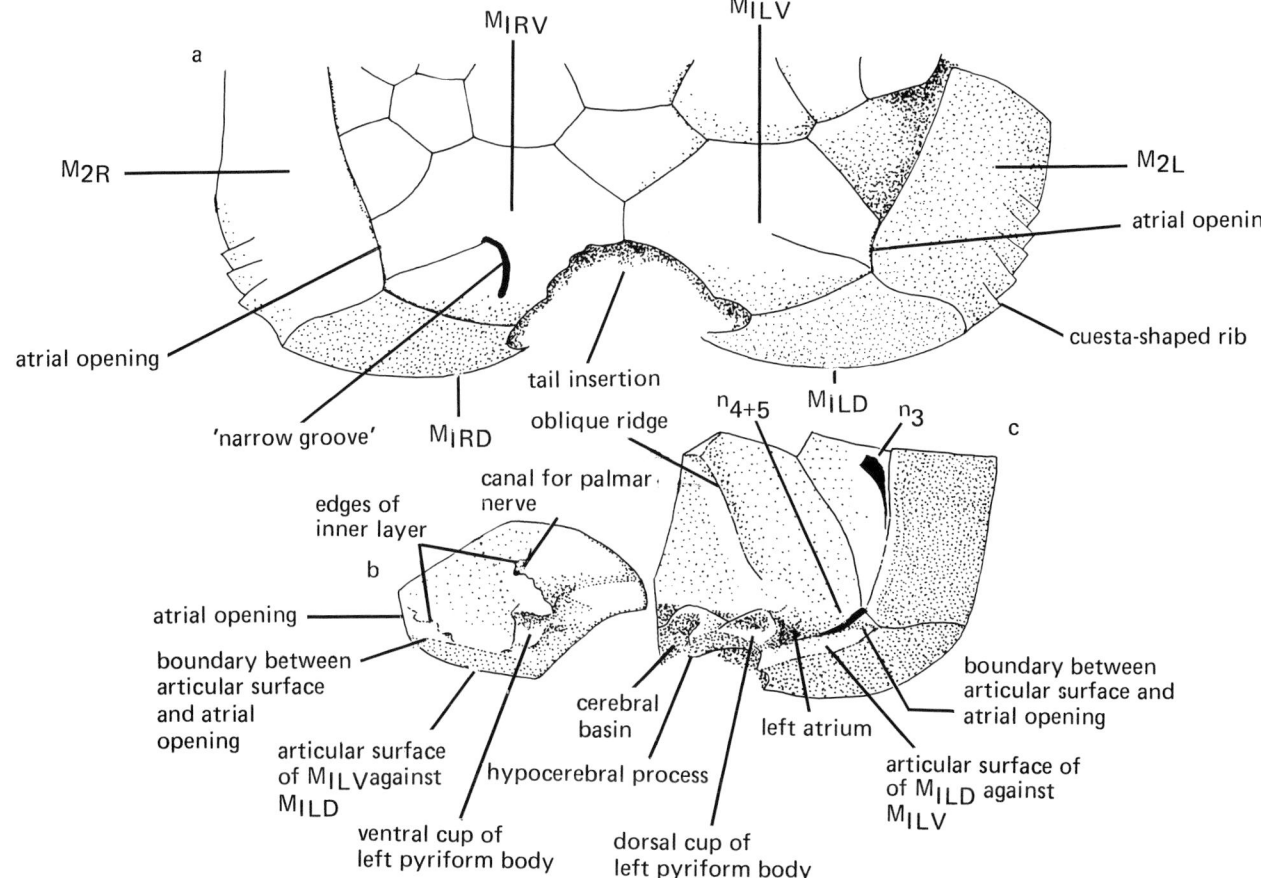

8.2. *Mitrocystites mitra. The atrial openings. a) Camera-lucida drawing of ventral aspect of posterior part of head. b) Dorsal (internal surface) of left postero-ventral plate M_{1LV}. c) Ventral aspect of dorsal skeleton of same individual as b – the articular surface is limited anteriorly against the atrial opening by a break in slope, which shows that the contact of M_{1LV} with M_{2L} is not a normal suture; from Jefferies (1981b, fig. 13, p. 375).*

he likewise regards as a hydropore (fig. 10.2). For Philip the tail is a 'stele', probably locomotory and serving to push the head forwards (Philip, 1979) or to pull it rearwards (Philip, 1981). For Ubaghs, above all, the tail is an 'aulacophore', to be interpreted as a feeding organ with an included water-vascular system. Ubaghs' interpretation has been incorporated in the *Treatise of Invertebrate Paleontology* (1967a).

Inside the head, several chambers can be recognized (fig. 8.3), mainly by examining natural internal moulds which, being negatives of the inside surface of the skeleton, can be seen as positive replicas, in rock, of the soft parts. When a natural internal mould is examined in dorsal aspect the most striking feature is an oblique groove that runs, sinuously and with varying strength, from near the right end of the mouth obliquely rearwards to just left of the tail insertion. It divides the head into an anterior left field and a posterior right one. When compared with a cornute such as *Cothurnocystis* (fig. 7.6) this groove is grossly parallel to the line separating: (1) the pharynx and buccal cavity and the presumed, virtual left anterior coelom overlying both, from (2) the right anterior coelom. The main difference is that the line in *Cothurnocystis* follows the right and posterior walls of the head whereas the oblique groove of *Mitrocystites* runs across the ceiling. This comparison suggests that the field left of the oblique groove would contain organs equivalent to the pharynx and buccal cavity of cornutes, together with the virtual left anterior coelom presumed to overlie them; and the field right of the oblique groove would contain the equivalent of the cornute right anterior coelom. Henceforth I therefore refer to the regions separated by the oblique groove as the fields of the left and right anterior coeloms respectively, or left and right fields for short.

The oblique groove varies in strength and direction. Its anterior portion is in places weak, as if its trough had been

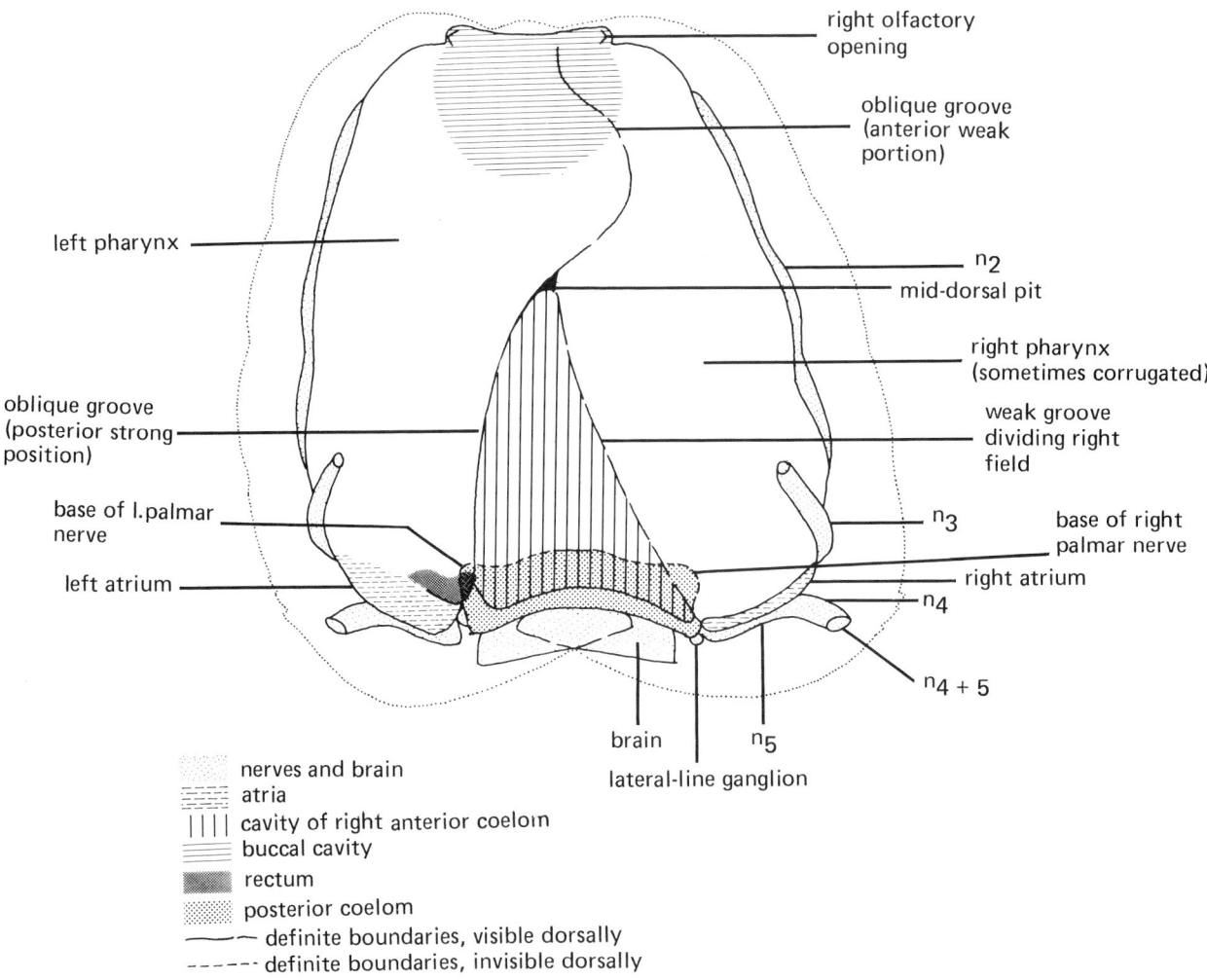

8.3. Mitrocystites mitra. *Reconstruction of the head chambers in dorsal aspect. The posterior limits of the buccal cavity are not known in* Mitrocystites. *(New.)*

partly smudged out, and its course forms a curve convex rightwards. Where the groove crosses the mid-line of the head, which is about one-third of the longitudinal distance from the front to the rear of the head, the groove sends off a short mid-dorsal pit rightwards and posteriorly. And followed rearwards and slightly left-wards from this pit the groove is very clearly marked, finishes sharply at depth, and is asymmetrical with a steeper slope on the right than on the left. Moreover the right field is divided into two parts directly related to the variation in the oblique groove. For across the right field a weak groove can sometimes be seen in the internal mould extending rearwards and somewhat rightwards from the mid-dorsal pit back to just right of the tail insertion. Right of this weak groove the right field is in one specimen longitudinally

corrugated (fig. 8.14b). Left of the weak groove it is always smooth. The variations in the oblique groove must be causally related to this subdividing of the right field.

The appearances can be explained if, in early ontogeny, the mitrates had the same arrangement of chambers as in cornutes. On this assumption a larval mitrate would have had a primary left pharynx best developed in the left part of the head (and clothed dorsally with a virtual left anterior coelom) and the primary pharynx would overlie a right anterior coelom best developed in the right posterior part of the head. Then a right pharynx pouched out of the anterior part of the left pharynx extending to the right posterior angle of the head (fig. 7.6). This right pharynx would pass beneath the cavity and contents of the right

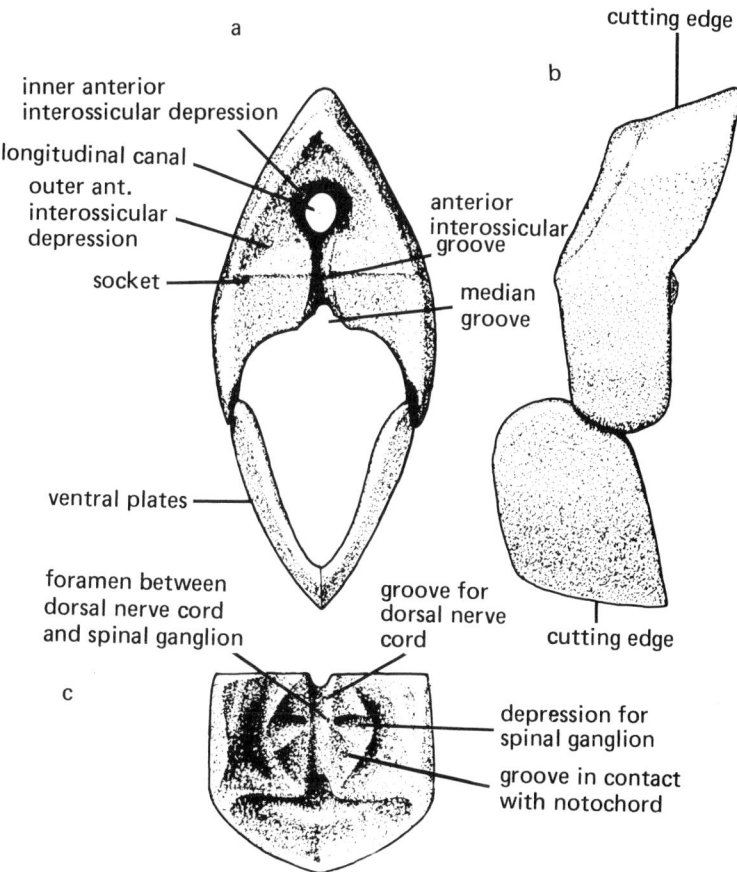

8.4. Mitrocystites mitra. *Reconstruction of a hind-tail ossicle and associated ventral plates. a) anterior aspect, b) left lateral, c) posterior, d) ventral aspect of the dorsal ossicle; from Jefferies (1968, fig. 25a–d, pp. 316–17).*

anterior coelom and would lift them up, squashing them against the ceiling of the head and pushing them towards the mid-line. Where the new right pharynx crossed the oblique groove it would weaken it and bend it rightwards, as is observed in fact. The right pharynx would inflate rearwards, towards the right posterior angle of the head, with the weak groove across the right field marking its left boundary. At the posterior end of the right pharynx gill slits would develop and, behind them, a right atrium. The longitudinal corrugations of the presumed right pharynx of *Mitrocystites* can be compared with similar corrugations in the posterior part of the left pharynx of *Mitrocystella* (see fig. 8.14). These corrugations have an important functional meaning as discussed later under *Mitrocystella*. At the present stage in the argument they can merely be taken to suggest that the right and left pharynxes are of the same nature. The observed relations thus imply that the primary left pharynx appeared before the secondary or right pharynx in ontogeny. This recalls the embryology of

amphioxus where the primary left gill slits appear before the secondary right ones (figs. 3.33, 3.36, pp. 80, 83). Thus most of the head of *Mitrocystites* would be filled with the same chambers as in cornutes, but modified by the development of a right pharynx. The position of a buccal cavity in the anterior part of the field of the left anterior coelom of *Mitrocystites* (fig. 8.3) is not based on direct evidence in that form but inferred from the situation in *Placocystites*, as shown later. The asymmetry of the posterior part of the oblique groove, deeper sided on the right than the left, suggests that the left field passed rightwards beneath the right field.

The posterior coelom of *Mitrocystites* was situated much as in a cornute. Its anterior limit on the dorsal aspect of the internal mould is defined by a transverse groove just in front of the tail insertion. On the ventral surface of the internal mould its anterior limit is a deep, narrow, anteriorly convex, arcuate groove. The distinctiveness of the anterior limit of the posterior coelom

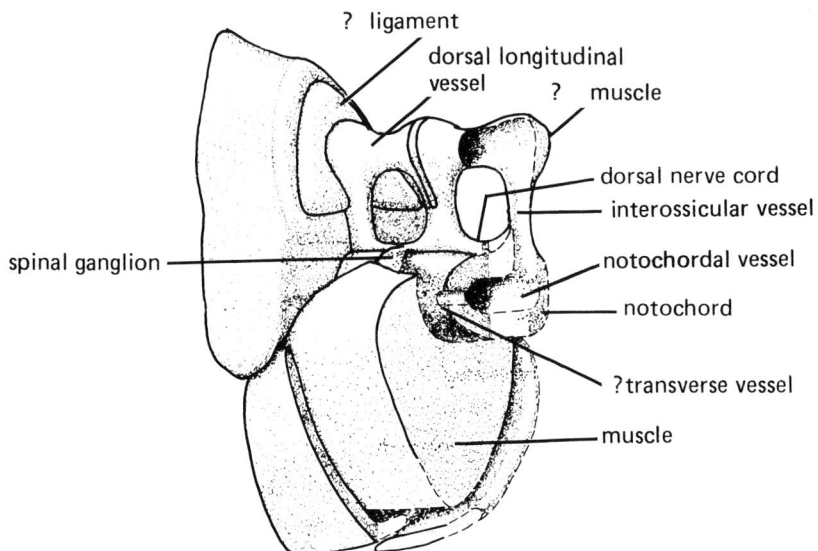

? ligament

dorsal longitudinal vessel

? muscle

dorsal nerve cord

interossicular vessel

spinal ganglion

notochordal vessel

notochord

?transverse vessel

muscle

8.5. Mitrocystites mitra. *Block diagram of the hind-tail skeleton with reconstructed soft parts. The position of the transverse vessel is conjectural; from Jefferies (1968, fig. 26, p. 318).*

on the ventral surface is caused in part by the nature of the skeleton. For three layers can be recognized in the ventral skeleton of mitrates, as shown in fig. 8.19b (*Placocystites*) — there is a thick outer layer of calcite; a thin, intermediate soft layer in which nerves seem to have run; and a thin inner layer of calcite. In the centres of plates, as in the top part of the figure in fig. 8.19b, there is no intermediate soft layer, for inner and outer calcitic layers are joined together in crystalline continuity. Now the inner calcitic layer never extends in mitrates onto the floor of the posterior coelom, which is formed of outer layer only. Instead the ventral part of the anterior wall of the posterior coelom is formed by the upturned posterior edges of the inner layer. When the skeleton was dissolved in a fossil of *Mitrocystites* to leave a natural mould, this upturned wall of inner calcitic layer disappeared to leave a narrow deep groove. The inner calcitic layer of the ventral skeleton is a specialization of mitrates which cornutes do not possess. As discussed later, the posterior coelom of mitrates is not exactly equivalent to that of cornutes, for it includes a right-hand portion which in cornutes could not have existed.

The left atrium of *Mitrocystites* would be situated in the posterior left angle of the head, and its anterior boundary is defined by a distinct groove on the natural mould. The rectum and gonoduct probably opened into the left atrium, by comparison with *Mitrocystella*. The right atrium presumably existed inside the right posterior angle of the head, and its dorsal edge is sometimes indicated by a groove on the specimens.

To sum up, the head chambers of *Mitrocystites* seem to be fundamentally the same as those of cornutes, with the important difference that a right pharynx existed, and caused several concomitant changes. Also there were left and right atria with left and right atrial openings.

The tail of *Mitrocystites*, as already mentioned, is divided into fore, mid and hind portions. The fore-tail skeleton consists of about six overlapping rings, each ring built of four plates (right, dorsal and ventral; left, dorsal and ventral). Right and left plates meet in the dorsal and ventral mid-lines at sutures, while at right and left the plates movably overlap each other. The fore-tail skeleton encloses a large lumen which in life was presumably almost filled with muscle and would have acted as the main motor of the tail. Such a fore tail, with its loose skeleton, would need an anti-compressional device so that flexion could happen without telescoping. This was presumably the notochord, situated in the principal mechanical axis which would neither lengthen nor shorten in flexion.

The hind-tail skeleton (fig. 8.4) consists of dorsal ossicles and ventral plates, with one pair of plates beneath each ossicle — left and right plates meet each other ventrally at a suture. Successive ossicles are hinged together by a pair of protuberances on the posterior surface that fit into a pair of depressions on the anterior surface of the next ossicle behind. These hinges would allow flexion in a vertical plane. On its dorsal surface each ossicle has a median posterior apex. A cutting edge runs forwards and downwards in the mid-line from the apex while the externally visible part of the posterior surface of the ossicle is approximately transverse to the tail and vertical when the tail was extended horizontally, but concave ('bearing surface' in fig. 8.8).

The mid-tail skeleton consists dorsally of a massive element, the styloid, which seems to be serially homologous with two hind-tail ossicles fused together (fig. 8.1d). The styloid is deeply excavate anteriorly and has two pairs of ventral plates attached to it. The ventral surface of the styloid has the sculpture appropriate to two hind-tail ossicles and there is a vertical interossicular canal between the two constituent ossicles which joins a horizontal dorsal longitudinal canal that passes through the styloid emerging anteriorly and posteriorly. (The meaning of these canals within the styloid will emerge when the hind-tail ossicles have been discussed.) Externally the styloid has two apices. The more posterior one is like that of a hind-tail ossicle, having a concave posterior surface beneath it and a cutting edge running forwards and downwards from it. The anterior apex of the styloid, representing the more anterior of the two constituent ossicles is knife-edged both anteriorly and posteriorly. The hind tail ends abruptly, as if a more distal portion had been broken off. So far as observed, it contains up to about nine ossicles.

The sculpture of the internal surfaces of the hind-tail ossicles is complicated (fig. 8.4a, c, d) and allows a reconstruction of the soft parts (fig. 8.5). There is a median groove in the ventral surface of the ossicle connected to a pair of pits in each ossicle. Between successive ossicles, on their anterior and posterior faces, dorsal to the hinge, are interossicular depressions each consisting of an outer shallower and an inner deeper part. The deeper parts of the depression of the anterior and posterior faces of the same ossicle are connected together by a horizontal dorsal longitudinal canal through the ossicle, and it is this canal that continues forward through the styloid. The deeper parts of the depression are also connected by vertical grooves on the posterior and anterior faces of the ossicle to the median groove on its ventral surface. The vertical grooves of contiguous ossicles together enclose vertical interossicular canals (containing interossicular vessels, fig. 8.5). A similar and serially homologous interossicular canal has already been mentioned within the styloid.

The median groove is composite, for, when natural moulds are examined, it seems to have been filled with two soft structures (fig. 8.5): there was a belt-shaped organ which overlay, in the mid-line, a roughly cylindrical organ. Moreover, the belt-shaped organ is connected in the natural moulds to a pair of rock-lumps in each segment, corresponding to the above-mentioned paired pits in each ossicle. Granted that mitrates are related to cornutes, and these, with their left-sided gill slits and post-anal tail, can be seen as chordates, then it is reasonable to interpret these organs by comparison with chordates. It is then natural to see the dorsal belt-like organ as the dorsal nerve cord, the roughly cylindrical organ as the notochord, and the paired rock-lumps as representing spinal ganglia connected to the dorsal nerve cord in each segment. The notochord would be a distal extension of the anti-compressional structure presumed to have existed in the fore tail. In the mid and hind tail, however, it would have no anti-compressional function, for compressional forces would be taken up by the dorsal ossicles and styloid. The notochord would probably help, however, to keep the ossicles in alignment, much like the string in a bead necklace.

How the tail functioned must be discussed before considering muscles and their likely antagonists in the tail. The head of *Mitrocystites* possesses, on the marginal plates just anterior to the postero–lateral angles, a small number of cuesta-shaped ribs (fig. 8.1b) which end dorsally as teeth in the peripheral flange (fig. 8.1a). A pair of similar ribs, or rather keels, extends, from right and left, part-way across the postero–ventral marginal plates (M_{1LV}, M_{1RV}) near the tail insertion. All these ribs are assymmetrical, steeper anteriorly than posteriorly. This direction of asymmetry is a universal rule in mitrates when cuesta-shaped ribs are present. In living animals, such as crabs and bivalves (figs 8.6, 8.7), cuesta-shaped ribs are used for gripping loose sediment, tending to hinder unwanted motion towards the steeper face of the rib. This suggests that these ribs in *Mitrocystites*, and in other mitrates where they are sometimes much better developed, were used in a similar way and would have favoured rearward motion, as also prevailed in cornutes. The same conclusion is suggested by the imbrication of the plates of the ventral surface of the head (fig. 8.1b). This is opposite in direction to that of the scales of a fish, for, when seen from outside, the plates overlap their anterior neighbours. This imbrication would produce a series of steps on the ventral surface, with the steep slopes of the steps anterior, as with the cuesta-shaped ribs. All this suggests that the head moved rearwards. The fact that the head of *Mitrocystites* is almost as wide as long, and the concentration of the ribs at right and left, suggests that locomotion was accompanied by yaw.

Now if the tail flexed ventralwards, as the shape of the hind-tail ossicles suggests, then in the position of greatest flexion the distal (morphologically posterior) external surfaces of the most distal dorsal ossicles would have faced forwards, while in more proximal ossicles they would have faced forwards and downwards (fig. 8.8). Such surfaces, being slightly concave, would be well adapted to push against sediment in the power stroke, thus pulling the head rearwards and upwards in the direction indicated by the cuesta-shaped ribs and the ventral imbrication of the head. The dorsal cutting edges of the ossicles, on the other hand, would have helped in extracting the tail from sediment

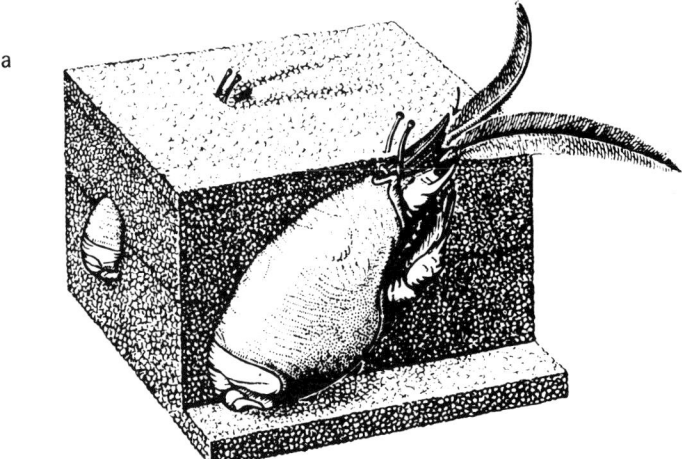

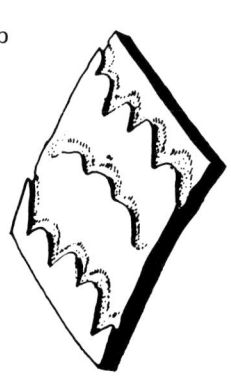

8.6. Cuesta-shaped ribs in the mole crab Emerita. *a) Two crabs in life-position, buried in the sand, b) detail of part of the surface of the carapace. The crab burrows into the sand, rear end first, by sweeping sand forwards with its walking legs. The cuesta-shaped ribs help to prevent the crab from slipping forwards (in the unwanted direction) during the return stroke of the legs. Their mode of action thus probably resembles that of the cuesta-shaped ribs of mitrates; from Seilacher (1959, fig. 1a, b, p. 258).*

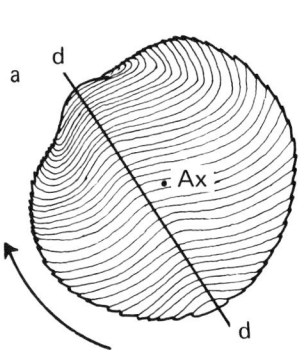

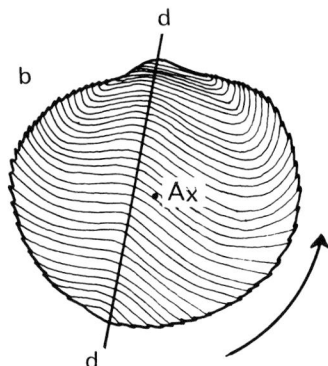

8.7. Cuesta-shaped ribs in the bivalve mollusc Divaricella. *This mollusc burrows by extending the foot into the soil approximately at the lower 'd' of the line d–d and rotating the shell alternately clockwise a) and anticlockwise b) about the axis Ax. During clockwise rotation the steep, dorsal slopes of the ribs left of d–d in the figure resist upward movement by pushing against sediment, while the ribs right of d–d slide downwards with the gentle slopes first. During anticlockwise rotation the process is reversed. The cuesta-shaped ribs of mitrates probably gripped sediment in a similar way.*

during the return stroke. Also, the dorsal cutting edge of the styloid, with its entirely knife-edged proximal apex, would have cut through the mud like the prow of a ship through water, helping the head to slide rearwards. The upwards component of motion of the head during the power stroke of the tail would probably help to keep the dorsal surface of the head free of mud. This is likely to have been the normal condition: (1) because no ribs are developed in this surface (though this is only a weak indication since ribless surfaces often were in contact with sediment), (2) because the openings of n_3 and n_{4+5}, whose exact nature is discussed below, can be interpreted as sensory openings serving to detect mud on the dorsal surface, (3) and because the downward and forward slope of the posterior surface of the head would tend to slide the head upwards as it moved rearwards against the bottom mud in the power stroke.

The power stroke of the tail therefore involved ventral

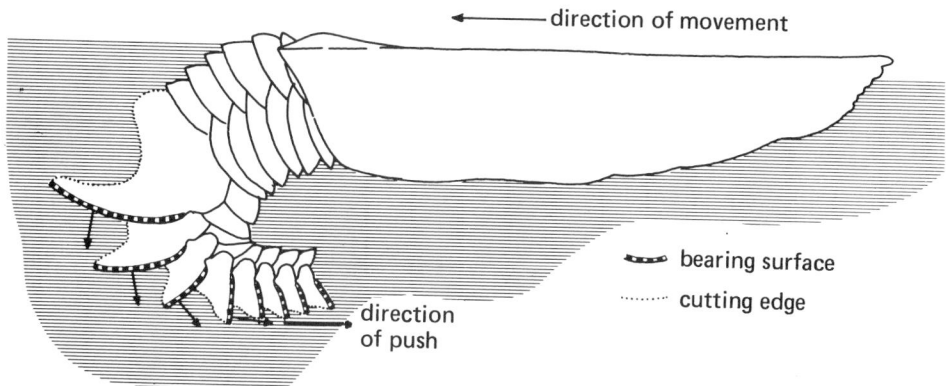

8.8. Mitrocystites mitra. *Suggested mode of action of the tail in locomotion. In the power stroke the tail would have pushed forwards and downwards on the mud, mainly by contraction of the muscles in the ventral part of the fore tail. The main bearing surfaces would have been the slightly concave distal surfaces of the dorsal hind-tail ossicles and stylocone. The proximal dorsal cutting edges of the ossicles and stylocone would have assisted the rearward and upward movement of the tail in the return stroke, diminishing the forward force then acting on the head. The cuesta-shaped ribs and the overlap of the ventral head plates would also have resisted forward movement during the return stroke. The width of the head (almost equal to the length) and the presence of ribs only at left and right (fig. 8.1) suggests that yaw was important in the animal's locomotion; from Jefferies (1981a, fig. 18, p. 521).*

flexion, and contraction of ventral muscles. The main source of power, however, would presumably lie in the large muscles of the fore tail which would move the mid and hind tail as a unit. The distal end of these muscles would insert in the anterior excavation of the styloid which would have functioned like the cup in the top of an old-fashioned wooden leg. The proximal insertion of these muscles is more difficult to reconstruct. It would have been situated in the tail insertion, where the tail joins the head. Within the tail insertion, the brain (fig. 8.9) probably coated only the anterior surface of the cerebral cup, for this surface is histologically smoother than the other surfaces of the tail insertion. Presumably the tail muscles pulled against whatever soft tissue covered the hind surface of the brain — especially they would pull against the posterior covering of the deuterencephalon (the part of the brain that occupied the peripheral parts of the cerebral basin). If so the deuterencephalon would need tensile strength. It was, therefore, probably thin, and solid rather than hollow.

In the hind tail the ventral muscles involved in the power stroke would presumably fill the large lumen between the dorsal ossicles and the ventral plates (fig. 8.5). Such ventral muscles would need antagonists dorsal to the interossicular hinges, and these could only have been located in the interossicular depressions. They would have comprised muscle or ligament or both. Perhaps the outer shallow depressions were filled with ligament and the inner deeper ones with muscle. For the inner depressions exist only in those mitrates which have interossicular canals, suggesting that, whatever tissue filled the inner

depressions required a blood supply that reached it through these canals. The more primitive condition, that lacks interossicular canals and has shallow interossicular depressions only, is represented in *Lagynocystis, Peltocystis* and *Chinianocarpos*. The depressions in these forms, were perhaps filled with ligament that needed no blood.

Mitrocystites, therefore, seems to have had blood vessels ascending to dorsal muscles from out of the notochord and passing up through the dorsal nerve cord. There is no living analogue for this situation among chordates, but it seems to have been a specialization within mitrates, confined to *Mitrocystites* and some related forms such as *Mitrocystella* and *Placocystites*. It need never have existed in the ancestry of living chordates. The vessel that gave rise to these dorsal interossicular blood vessels, on the other hand, probably followed the central axis of the notochord and probably existed in all mitrates and cornutes. If so, it has been lost or has migrated out of the notochord in all living chordates.

Thus the tail of *Mitrocystites* was a locomotory organ that pulled the head rearwards by ventral flexion. It was muscular and gives evidence of notochord, spinal ganglia and dorsal nerve cord. A more detailed discussion of locomotion in *Mitrocystites* and other mitrates can be found in Jefferies (1984).

The brain and cranial nerves of *Mitrocystites* were complicated (fig. 8.9) and can be reconstructed partly on direct evidence from *Mitrocystites* itself and partly by comparison with related forms such as *Mitrocystella* and *Placocystites*.

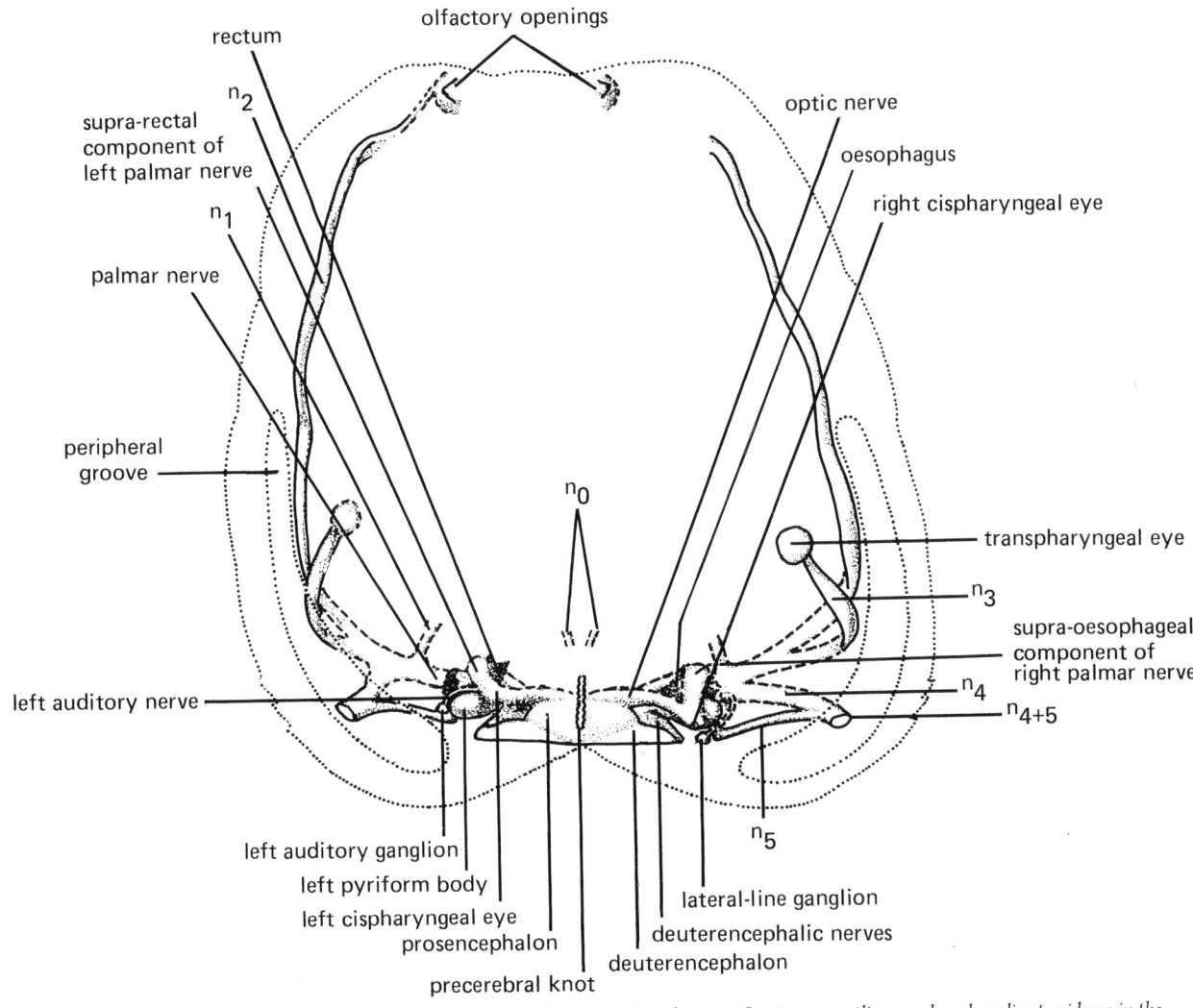

8.9. Mitrocystites mitra. *Reconstruction of the brain and cranial nerves in dorsal aspect. Continuous outlines are based on direct evidence in the fossils of* M. mitra. *Broken outlines are based on indirect evidence, mainly on comparison with* Mitrocystella; *from Jefferies (1981a, fig. 20, p. 524).*

At the anterior end of the notochord of *Mitrocystites*, which is also the anterior end of the tail, we should expect on chordate analogies to find the brain. And in an appropriate position, lodged in a depression in the head skeleton, the brain seems to have existed. Natural moulds show it to be bipartite with a more swollen central portion and a less swollen peripheral one. As pointed out in Chapter 5 (figs 5.13, 5.14), embryologists have argued that the primitive vertebrate brain was bipartite, and the two portions can therefore be identified provisionally as the central, more anterior prosencephalon and the peripheral, more posterior deuterencephalon (fig. 8.9). The supposed prosencephalon would presumably be anterior to the front end of the notochord while the deuterencephalon would

not be, or at least was in contact with the notochord. The prosencephalon has a large foramen (optic foramen in fig. 8.10) antero–ventral to it which is transversely elongate and constricted in the median plane, being shaped like a thick-waisted figure-of-eight lying on one side. The prosencephalon was supported beneath by a pair of skeletal processes (hypocerebral processes) which almost meet in the mid-line, so that a narrow median gap remains between them. If the prosencephalon is correctly identified then the foramen, by vertebrate analogies, would have given passage to the optic nerves, for in living vertebrates these always leave the prosencephalon antero–ventrally. The gap between the paired processes would then serve to connect the base of the prosencephalon with the

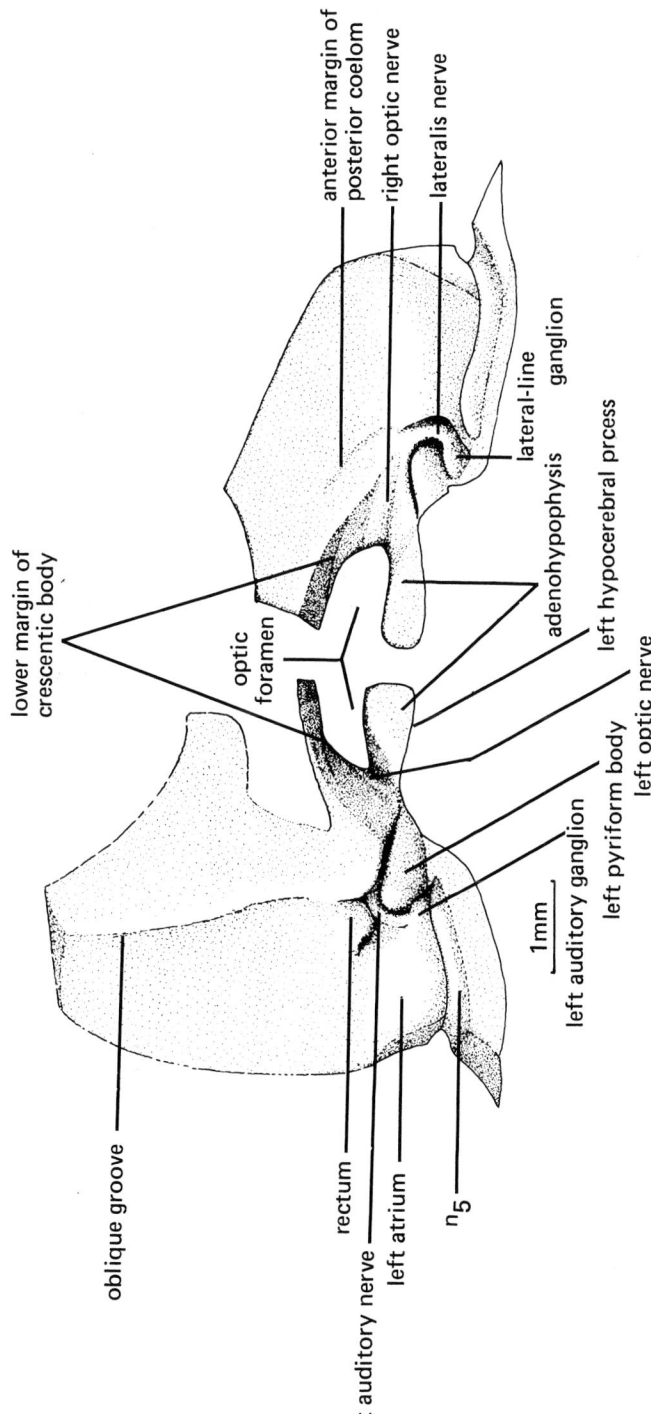

8.10. *Mitrocystites mitra. Camera-lucida drawing of internal moulds of plates* M_{1LV} *and* M_{1LD} *of a specimen from the Šárka shales of Osek near Rokycany, Czechoslovakia (BM(NH) E7517). The moulds simulate the soft parts of the rear part of the head, seen in postero-dorsal aspect. Note especially the left auditory ganglion and nerve, the bases of the optic nerves, the lateral-line ganglion on the right and, especially, the likely position of the adenohypophysis. (New.)*

adenohypophysis. One natural mould (fig. 8.10) shows a swelling beneath and on either side of the gap, which could represent the position of such an adenohypophysis. In the dorsal mid-line, just anterior to the presumed prosencephalon, in the vertical suture between the first two dorsal marginal plates (M_{1RD}, M_{1LD}), a characteristic pleating is developed (precerebral knot in fig. 8.9). The pleats are curved so as to be convex anteriorly, and, near their ventral ends, there are small points in the natural mould, which would correspond to outward-pointing pits in the two marginal plates. In life the pleats and points would represent a structure which can be called the precerebral knot and which would have resembled a thin corrugated curtain of soft tissue developed in the median plane with a series of right- and left-pointing cones near its ventral margin. These cones probably received soft-tissue fibres from the skeleton and correspond to the irregular conical projections of the paired dorsal lobes of *Ceratocystis* (fig. 7.26, p. 224). The position of this structure suggests that it represents the place where olfactory and terminalis fibres gathered before entering the brain, for such fibres in vertebrates join the brain antero–dorsal to the bases of the optic nerves. More exactly, the precerebral knot would represent the superficial layers of the optic bulbs of vertebrates. The presumed prosencephalon therefore has several features which, by vertebrate comparisons, confirm its identification. In *Mitrocystites* the prosencephalon was not divided by external constrictions into the olfactory telencephalon and the optic and hypophyseal diencephalon. Such a division, however, sometimes shows in *Placocystites*.

The front face of the presumed deuterencephalon carries flattened ridges that pass ventralwards and slightly medianwards. In view of their position, these probably represent the bases of deuterencephalic nerves (fig. 8.9), i.e. the trigeminal nerves, parts of the branchial nerves, the acustico-lateralis nerves and some others.

A pair of pear-shaped 'pyriform bodies' is developed just left and right of the deuterencephalon. In life those bodies were almost completely surrounded by skeleton in the form of a cup of calcite both above and beneath. Medianwards, however, the bodies were joined to the cavity of the head by a vertically elongate isthmus of tissue corresponding to a short length of rock in the natural mould and to a vertical slit in the two cups of the skeleton. The pyriform bodies suggest ganglia connected on their median faces with the nerves that came off the deuterencephalon. Their position with respect to the brain suggests, by vertebrate comparisons, that they represent the trigeminal ganglia, and they would therefore carry the cell bodies of the sensory constituents of the trigeminal complex.

Two ridges pass upwards on the posterior mould of the head cavity, in front of the brain from the presumed optic foramen. One goes off from the right end of the foramen and climbs directly up to the right dorsal angle of the posterior coelom. The other goes off from the left end of the foramen travels downwards for a short distance and then up again. These ridges, since they emerge from the optic foramen, probably represent the bases of optic nerves (fig. 8.10).

Peripheral to what has been so far described, the courses of nerves can be reconstructed partly on the basis of *Mitrocystites* itself, and partly by analogy with *Mitrocystella* which in some ways is more informative. The latter genus has on either side, extending from the anterior right and left ventral angles of the posterior coelom into the middle layer of the ventral skeleton, a palmar complex – so-called because it resembles a hand with five fingers (fig. 8.16). The central part of such a complex, corresponding to the 'palm' of the 'hand', can be called the palmar nerve and the five 'fingers' are labelled n_1, n_2, n_3, n_4 and n_5, reading in sequence outward from the mid-line. The nervous nature of these complexes in *Mitrocystella* is suggested by the way they radiate outward from the general position of the brain and by the fact that n_4 and n_5 finish just beneath the dorsal surface, consistent with some sensory function. In *Mitrocystites* the palmar nerves show in the natural moulds as short antero–lateral extensions of the cavity of the posterior coelom at right and left into the middle layer of the ventral skeleton (fig. 8.3). Presumably, by analogy with *Mitrocystella*, nerves n_1 to n_5 left the palmar nerves and radiated out in the middle layer, as shown in fig. 8.9, but there is no direct sign of this part of their course in *Mitrocystites*. If present, they had here no effect on the skeleton.

The more peripheral parts of the palmar nerves, except for n_1, have, by contrast, left obvious traces in *Mitrocystites*. The paired nerves n_2 can be followed in the form of ridges on either side of the natural mould which climb out of the ventral skeleton and run forward in the dorsal skeleton to the mouth region (figs 8.3, 8.9). The paired nerves n_3 can be traced for a short distance in the ventral skeleton and then pass into the dorsal skeleton (figs 8.3, 8.9) as obvious thick ridges on the natural mould which finally climb upwards as short rods of rock to the openings n_3 on the dorsal surface of the head, as already mentioned. The nerves n_3 change considerably in section, being belt-like in the ventral skeleton, and becoming inflated and circular in section in the dorsal skeleton. This change in section is what might be expected of a nerve where, so long as fibres remain continuous, their relative positions do not matter. The nerves n_4 run directly upward in the dorsal skeleton following the suture between two plates; they lead to the peripheral grooves of the dorsal surface. The nerves n_5, on the other hand, bend outwards to run almost horizontally shortly after entering the dorsal skeleton and then join the nerves n_4 to climb with them to the dorsal surface (figs. 8.3, 8.9).

As to vertebrate homologies, the nerves n_2, by their position and the fact that they end near the mouth, can probably be identified with the paired maxillary branches of the trigeminal. (The nerve n_1, for which there is evidence in *Mitrocystella* but not in *Mitrocystites*, may represent the mandibular branch of the trigeminal.) Nerves n_4 and n_5 may represent the dorsal touch-sensory branches of the trigeminal (respectively the profundus and superficialis branches).

The nerve n_3 is probably optic. This is suggested by its position dorsal to n_2 (the presumed maxillary trigeminal) and is functionally intelligible, since dorsally opening optic nerves make good sense. It is confirmed by a comparison with *Mitrocystella* (fig. 8.16) where the nerves n_3 do not open onto the dorsal surface of the head but stop short before entering the dorsal skeleton and where, probably in functional consequence: (1) the prosencephalon, which in living vertebrates is largely optic, is less inflated than in *Mitrocystites*; (2) the optic foramen in front of the brain is smaller; and (3) the bases of the optic nerves (or the ridges believed to represent these just above the optic foramen) are much slimmer (compare fig. 8.10 with 8.13b). If the nerves n_3 are optic, however, they are transpharyngeal optic nerves, situated on the far side of the pharynx from the brain.

The full course of the optic nerves can therefore be suggested. The right one, after leaving the brain through the optic foramen, would climb to the top right corner of the posterior coelom (right cispharyngeal eye of fig. 8.9), bend downwards, following the front wall of the posterior coelom, enter the right palmar nerve, run with the latter in the middle layer of the ventral skeleton, cross into the dorsal skeleton anterior to the right atrial opening and pass as nerve n_3 through a canal to open on the dorsal surface of the head. The opening of n_3 onto the dorsal surface sometimes lies at the bottom of a small basin, which suggests that the nerve ended in a small bulb or vesicle. The left optic nerve, after leaving the optic foramen would pass first downwards and then upwards as indicated by the ridge on the natural mould (fig. 8.10) to the left dorsal angle of the posterior coelom (left cispharyngeal eye of fig. 8.9), run down the front wall of the posterior coelom, enter the left palmar nerve, thus pass in the middle layer of skeleton to the canal for left n_3, and thence through the dorsal skeleton to the dorsal surface. The reconstruction of the course of the non-pharyngeal gut, based mainly on *Placocystites*, suggests that by this course the right optic nerve would arch in the posterior coelom over the opening of the oesophagus into the pharynx, while the left nerve would arch over the rectum. The proximal parts of

the optic nerves, on the same side of the pharynx as the brain, can be called the cispharyngeal optic nerves with their summits developed as cispharyngeal eyes.

The olfactory nerves are probably represented at their anterior ends by a pair of small cones on the natural mould, right and left of the mouth (olfactory openings in figs 8.3, 8.9). From these, nerve fibres would pass in the marginal plates rearwards to the cones of the precerebral knot (or the olfactory bulbs in vertebrate terms), and thus make contact with the antero–dorsal mid-line of the prosencephalon.

The acustico–lateralis system of *Mitrocystites* has already been mentioned when describing *Ceratocystis*. The natural internal moulds sometimes show a small lump in the left atrium (fig. 8.10), just left of and behind the left pyriform body (trigeminal ganglion). This lump corresponds in position to a a narrow groove on the external surface of *Ceratocystis* (fig. 7.17b,c, p. 214). The lump in *Mitrocystites* is connected to a ridge that reaches it by passing round the left pyriform body. Similarly the groove in *Ceratocystis* is connected to a canal that comes from just left of the left pyriform body (fig. 7.26b). The position of the lump in *Mitrocystites*, just behind the left trigeminal ganglion, recalls the acoustic ganglion of a vertebrate, in which case the ridge connected to it would represent the auditory (acoustic) nerve. This implies that the ear of *Mitrocystites* was situated in an atrium, but this is not surprising. For the cupular organs of *Ciona*, (fig. 4.21) which are likewise auditory and could well be homologous with the neuromast organs of vertebrates, are also situated in an atrium (Bone & Ryan, 1978). On this assumption the groove of *Ceratocystis* would represent the left ear in a primitive condition, functioning as lateral-line to detect displacements in the outside water. Invagination of such an organ in the left atrium of mitrates would produce an ear capable in a general way of detecting vibration, changes in orientation or acceleration.

The 'narrow groove' in the external surface of the right ventral marginal of *Mitrocystites* (figs 8.1b,c, 8.2) in some ways resembles an antimere of the groove left of the tail in *Ceratocystis* and therefore an antimere of the presumed left ear of *Mitrocystites*. Being external, however, it presumably functioned, not acoustically, but as lateral line. A carrot-shaped body in the natural mould is located just inside from it and could well represent the lateral-line ganglion. It is connected to a ridge of rock which comes to it from in front and above and probably represents the lateral-line nerve (fig. 8.10). The exact position of this presumed nerve suggests that it approached the lateralis ganglion from off the dorsal surface of the oesophagus (fig. 8.9). By analogy with *Mitrocystella* and *Placocystites* there may have been a right auditory ganglion situated in the right atrium, but there is no direct evidence of it in

Mitrocystites. The same could be said of the paired nerves n_0 (fig. 8.9), presumably supplying the endostyle in the ventral mid-line, which are well attested in *Mitrocystella* and *Placocystites*, but without trace in *Mitrocystites*.

Thus the nervous system of *Mitrocystites* probably comprised: a dorsal nerve cord with spinal ganglia; a bipartite brain, with prosencephalon and deuterencephalon; paired optic nerves that followed a complicated course, with cis- and transpharyngeal portions, running from the prosencephalon to the dorsal surface of the head; a trigeminal complex with paired ganglia and with maxillary and dorsal, presumably touch-sensory branches; an acustico–lateralis system in which both acoustic and lateral-line elements existed; and an olfactory system originating in the buccal cavity and connected with the dorsal mid-line of the prosencephalon. This complex system recurs, less or more reconstructable and with changes of detail, in the other mitrates that will be described.

The likely mode of life of *Mitrocystites* has already been mentioned in discussing the structure of the tail. The head was more strictly plano-convex than in *Mitrocystella* or *Placocystites*, for the dorsal surface was flatter, being totally invisible in posterior aspect (fig. 8.1c). This suggests that, in the normal course of locomotion, mud did not flow onto the surface. Indeed the transpharyngeal eyes and the presumably touch-sensory peripheral grooves supplied by nerves n_4 and n_5 may have functioned to detect mud that had spread accidentally onto the dorsal surface and thus threatened to bury the animal. Also the posterior surface of the head, in the middle of which the tail was inserted, sloped downwards and forwards, as seen from the side in fig. 8.1d. In consequence, when the tail pushed downwards and forwards in the power stroke, the head would slide rearwards and upwards against the bottom mud. The presence of a lateral line on the postero–ventral surface requires that this part of the head should sometimes emerge from the bottom mud. Perhaps, indeed, the lateral line served to detect such exposure. The atrial openings were postero–ventral in position and would have been forced against the mud by the weight of the animal above them. There is no reason to think, however, that they could not function in such a position, for, being outlets, they would open by the pressure of the water inside the head. They can thus be compared with the downward facing gill cover of a flatfish or the valve-guarded branchial openings on the ventral surface of a skate. All these outlets tend to be closed by the weight of the animal pressing against the underlying sediment, but they function nevertheless.

The convex-downwards shape of the head of *Mitrocystites*, like that of other mitrates and the advanced cornute *Reticulocarpos pissotensis*, can be seen as an adaptation for staying up on soft mud. As argued in Chapter 7, *Reticulocarpos hanusi*, which lived alongside *Mitrocystites*, was adapted to stay up on weak mud by resting lightly on the surface of the latter, without exceeding the bearing capacity of the mud (fig. 7.32, p. 233). This can be called the 'snow-shoe' method and is inherently unstable, since mud is thixotropic and likely to lose all its weak strength by stirring. The convex-downwards head of *Mitrocystites*, on the other hand, would sink down into and displace the mud, so compensating its own weight. If the mud had no strength whatever, the case becomes identical to staying up by buoyancy in a liquid. The method does not require that the mud has no strength, however. For it corresponds to the civil engineer's device of the compensated foundation, whereby a building weighing 1000 tonnes will just stand on ground of bearing capacity 500 tonnes if its base is inserted into a hole from which 500 tonnes of earth has been shifted. The snow-shoe method is merely uneconomic on very weak substrates while the compensated foundation is pointless on strong ones. Stanley (1970) has already noted the two methods of staying up on mud as illustrated by bivalve molluscs (see also Jefferies & Prokop, 1972, p. 100).

Mitrocystites thus appears as a more bilaterally symmetrical animal than the cornutes. In particular it had a right pharynx and presumed right gill slits. Moreover its nervous system was complicated, reconstructable and fish-like. Other mitrates will confirm and supplement the picture just given of its anatomy.

Mitrocystella incipiens miloni is the next mitrate to be considered (fig. 8.11). The subspecies *miloni* comes from the Lower Ordovician rocks of Brittany in France. It is preserved in siliceous nodules which can be picked up from the surfaces of ploughed fields at the locality of le Traveusot, north-west of Redon. It is found in the slightly sandy shales of the 'Formation de Traveusot' (formerly called 'schistes à calymènes') and thus belongs to the Llandeilo stage. It is therefore about 460 million years old and slightly younger than *Mitrocystites mitra*. It was first described and studied by the French palaeontologist Jean Chauvel in (1941) and named after the redoubtable Yves Milon who, besides being Professor of Geology at Rennes University, was also mayor of Rennes for several years. The typical subspecies *Mitrocystella incipiens incipiens*, however, which differs very little from *M. i. miloni*, is found in the Dobrotivá Beds (Llandeilo Stage) of Czechoslovakia and was originally described, like *Mitrocystites mitra*, by Joachim Barrande in (1887). As usual, I shall henceforth refer to *Mitrocystela incipiens miloni* by its generic name alone, although other species of the genus are known.

The general shape of *Mitrocystella* is like that of *Mitrocystites* – the head is almost flat dorsally and convex ventrally and the dorsal and ventral surfaces meet at a peripheral flange. Unlike *Mitrocystites*, however, the head is relatively narrow for its length and the dorsal surface slopes gently downwards posteriorly, being thus partly visible from behind (fig. 8.11c). The dorsal surface of the head is made up of 12 marginal plates and three centro–dorsal plates. The same marginal plates also form the peripheral parts of the ventral surface, but the rest of this surface is mostly composed of flexible integument (fig. 8.11b), the plates of which abut against each other posteriorly and overlap in the anterior region as in *Mitrocystites*. Near the posterior end of the ventral surface there are two ventral marginal plates (M_{1LV}, M_{1RV}). The ventral, more anterior, parts of these two plates are ornamented with transverse, cuesta-shaped ribs (fig. 8.11d) which, as usual, have the steeper slope anterior and the gentler slope posterior. Similar ribs, but somewhat broken up, are developed on the adjacent parts of the dorsal marginal plates, ventral to the peripheral flange.

The openings of the head are fewer in *Mitrocystella* than in *Mitrocystites*, since there are none on the dorsal surface. In ventral aspect, a large mouth is visible anteriorly, with a rigid upper lip and a flexible lower lip armed with long, spike-shaped, oral plates (fig. 8.11b). There were right and left atrial openings near the posterior, right and left angles of the head (fig. 8.12). The lateral line is developed on the posterior surface of the right ventral marginal plate, as in *Mitrocystites*. Sometimes it is not a straight groove but cruciform or with several branches.

Inside the head, *Mitrocystella* shows the same basic features as *Mitrocystites*, but is in some ways more informative and in others more specialized. A natural internal mould is divided in dorsal aspect (fig. 8.13a) into left and right fields by an oblique groove running from just right of the mouth to just left of the tail insertion. The anterior part of the groove, as in *Mitrocystites*, is arched to the rght and partly smudged out. This can be explained in the same way as in *Mitrocystites* – as suggesting that this part of the groove was deflected and partly weakened by the out-pouching of the right pharynx towards the posterior right corner of the head. There is likewise a mid-dorsal pit that extends posteriorly and rightwards from the groove. The subdivision of the right field is different from *Mitrocystites* however, for a straight ridge (cdl) passes rearwards in all specimens, not towards the right end of the posterior coelom, but more rightwards towards the right posterior angle of the natural mould. One specimen also has a slight groove (g) extending a short distance forward from the right dorsal corner of the posterior coelom. It is in the same position as the groove that crosses the right field of *Mitrocystites* (fig. 8.13) which probably represents the left boundary of the right pharynx in that genus. Probably these features, both the ridge and the groove, have to do,

in *Mitrocystella* as in *Mitrocystites*, with the subdivision of the right field into a right pharyngeal part and a right anterior coelomic part. The differences between the two mitrates can be explained by comparison with tunicate pharynxes, as explained later after dealing with *Placocystites*.

The posterior boundary of the buccal cavity is suggested by a weak transverse groove in some specimens in the anterior part of the left field. This groove would presumably mark the position of the velum (fig. 8.13a). (*Placocystites* has a much stronger groove in the same position.) The posterior part of the left field is longitudinally corrugated in several specimens. The corrugations end suddenly forwards, as if against an invisible boundary line, leaving the anterior part of the left pharynx smooth. As already mentioned, similar corrugations occur in one specimen in the right pharynx of *Mitrocystites* (fig. 8.14). All this suggests a functional similarity of the right pharynx with the corrugated part of the left pharynx and a functional contrast between these corrugated parts and the anterior, always smooth part of the left pharynx. Comparison with living primitive chordates, whether tunicates, acraniates or lamprey larvae, suggests that the contrast may have been connected with a mucous filter inside the pharynx. More exactly, the smooth part of the left pharynx could be anterior to the mucous filter while the corrugated parts were lateral to it (cf. the tunicate *Clavelina* in fig. 4.13, p. 96). Indeed the corrugations may represent pleats in the soft parts that served to keep the filter away from the pharyngeal wall proper, in the manner of the intrabranchial structures of tunicates. The anterior limits of the corrugated regions would thus represent the positions of the peripharyngeal bands (fig. 8.14c), whose function in living chordates is to hold the anterior edges of the mucous filter and pass them·dorsalwards. If so, the left peripharyngeal band would cross the left pharynx of *Mitrocystella* at the anterior ends of the corrugations, near the transverse level of the mid-dorsal pit, while the right peripharyngeal band would follow the anterior part of the oblique groove rearward to near the mid-dorsal pit. If these suggested positions are correct, they would imply that the right and left peripharyngeal bands in mitrates were not structural antimeres of each other, though they presumably functioned as such. As already discussed in Chapter 4, however, there are signs that the bands are not exactly antimeric in living tunicates (fig. 4.15, p. 97). As discussed later under *Placocystites*, the suggested position of the peripharyngeal bands in mitrates is a first step in comparing mitrate pharynxes with those of tunicates. Another peculiarity of the left field in *Mitrocystella* is the ridge *gd*. I shall discuss it later.

The posterior coelom of *Mitrocystella* is well defined on the natural moulds by grooves anterior to it on right and

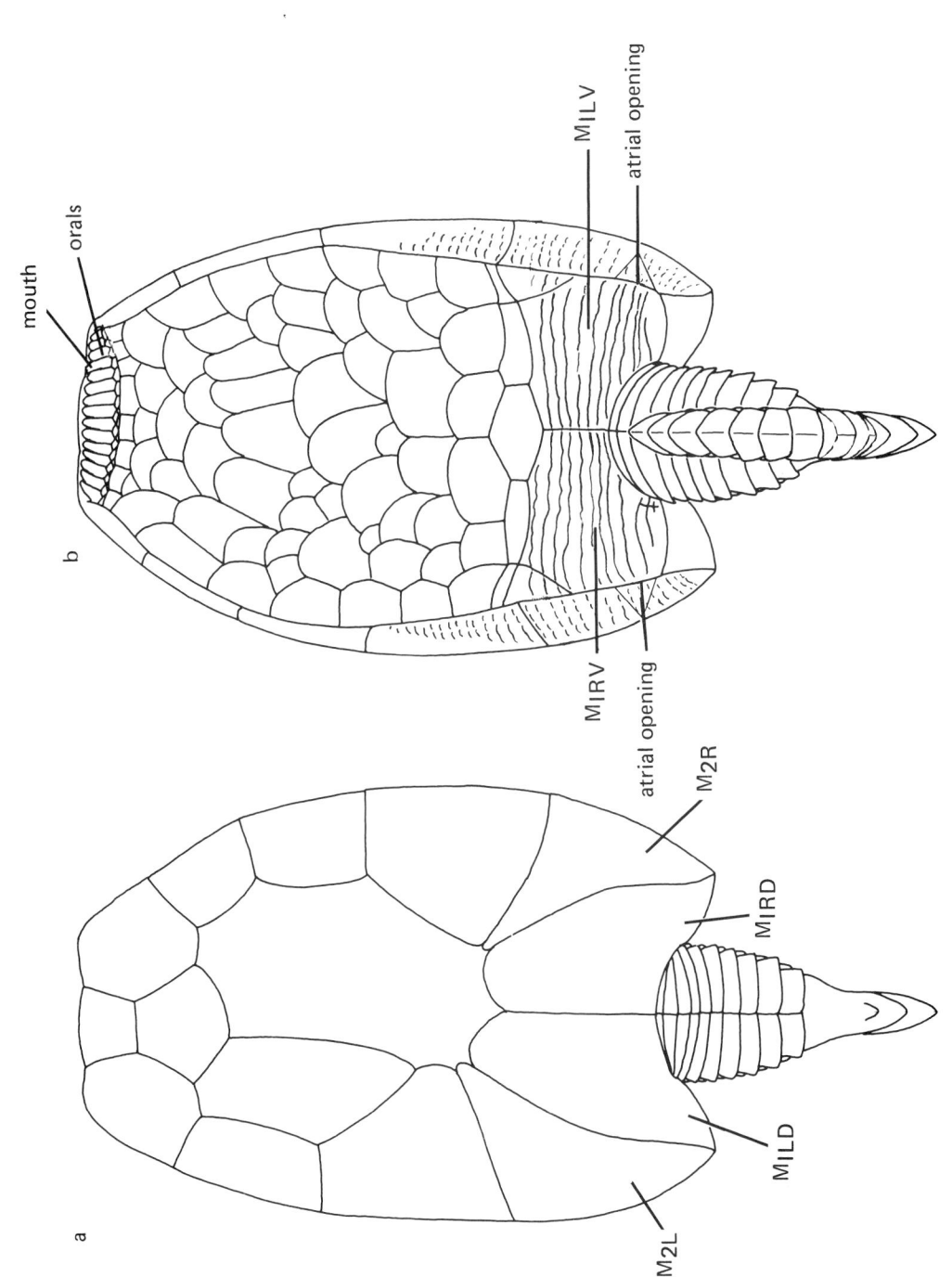

mouth

orals

M1LV

atrial opening

M1RV

atrial opening

M2R

M1RD

M1LD

M2L

b

a

Fig. 8.11.

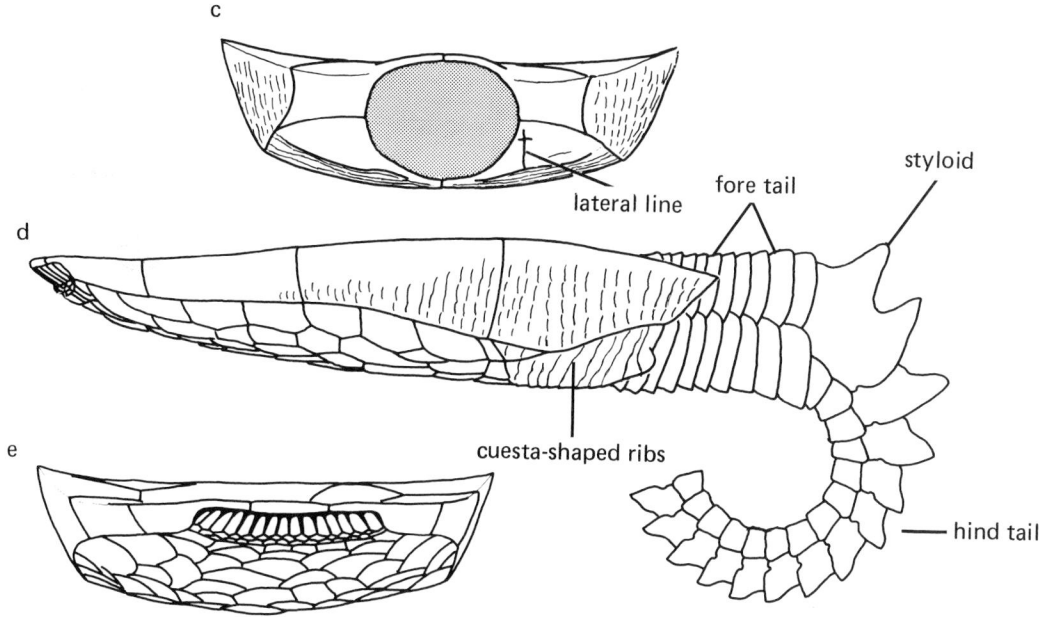

8.11. Mitrocystella incipiens miloni. *Reconstruction of external features in: a) dorsal aspect, b) ventral, c) posterior, d) left lateral, and e) anterior; from Jefferies & Lewis (1978, fig. 5).*

left and also dorsally and ventrally (figs 8.13, 8.14a). The left and right boundaries differ from each other (fig. 8.13b), for on the left there is an almost vertical hemi-cylinder of rock which is lacking on the right. At its ventral end the hemicylinder passes rightwards into the posterior coelom whilst at its dorsal end it passes leftward, presumably to open into the left atrium. It is comparable in position with the terminal portion of the rectum in *Cothurnocystis* (figs 7.4a, 7.7, pp. 198, 201) or *Reticulocarpos hanusi* (7.29a, p. 228), which runs vertically out of the posterior coelom to open to the outside, left of the tail insertion. It probably therefore represents the rectum of *Mitrocystella* which would open, not direct to the outside, but into the left atrium. Faeces would therefore leave the animal through the left atrial opening. As already noted (fig. 4.26), in tunicate tadpoles the rectum likewise opens into the left atrium. On one specimen of *Mitrocystella* a groove crosses the dorsal surface of the posterior coelom from posterior left to anterior right. This groove seems to divide the posterior coelom into two smaller chambers (fig. 8.13a between 'left and right epicardia'). Its significance will be further discussed under *Placocystites*.

The posterior face of the posterior coelom shows other complications (fig. 8.13b). There is a large ventral foramen which connects the posterior coelom with the cavity of the fore tail. Above this is the transversely elongate optic foramen, which, as in *Mitrocystites*, was underlain, and

separated from the ventral foramen, by the paired hypo-cerebral processes, which almost but not quite meet in the mid-line. Ridges on the natural mould that extend from the optic foramen at right and left probably represent the positions of the cispharyngeal parts of the optic nerves as in *Mitrocystites* (and as discussed later). Dorsal to the optic foramen, and medial to the cispharyngeal parts of the optic nerves, is a slight protuberance in the natural mould which can be called the crescentic body because of its shape. For its lower boundary is convex downwards, while its upper boundary, if coincident with the upper edge of the posterior coelom, would have been almost horizontal and straight. I shall discuss the nature of the crescentic body when considering the head segmentation of mitrates.

The ridge *gd* on the dorsal surface of the left pharynx can now be considered. Anteriorly it begins at the oblique groove and appears to be emerging from the cavity of the right anterior coelom. In passing rearwards it gradually widens and ends posteriorly by fading into the presumed position of the left atrium. However, the dorsal border of the latter is vague, unlike *Mitrocystites*, presumably being obscured by whatever organ filled the ridge *gd*. I have already argued in Chapter 7 that in cornutes the cavity of the right anterior coelom would have contained the gonad among other organs. Moreover, this gonad opened to the outside in cornutes either: primitively, by a direct

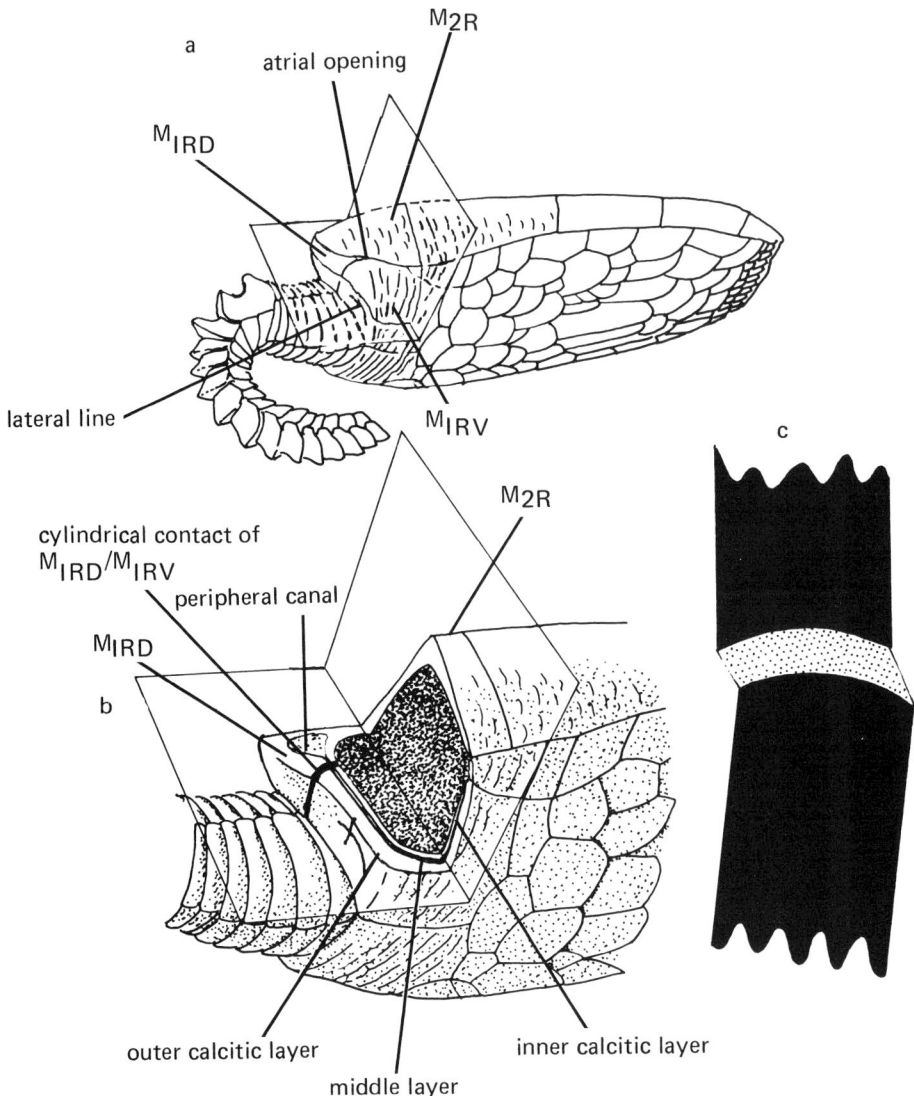

8.12. Mitrocystella incipiens miloni. *The right atrial opening. a) General view in left postero-ventral aspect, b) with the right posterior angle of the head cut away to show the plate contacts – the approximately horizontal junction of M_{1RD} and M_{1RV} is not plane but cylindrical, so as to allow M_{1RV} to rotate downwards and the atrial opening to gape. c) Suggested mode of action of the cylindrical junction (cf. the hind-tail ossicles of* Cothurnocystis *in Fig. 7.8b); from Jefferies (1981b, fig. 15, p. 378).*

opening right of the tail as in *Ceratocystis;* or, in the more advanced condition, by a gonoduct which accompanied the rectum to left of the tail, as in *Cothurnocystis* and presumably *Reticulocarpos;* or it opened into the gill slits as in *Scotiaecystis* (fig. 7.19). When, in the origin of the mitrates, the anus became included, along with the left gill slits, inside the left atrium, the opening of the gonoduct would have been taken into that atrium also. In the primitive mitrate condition the gonoduct would therefore follow

the rectum, as it presumably did in *Mitrocystites* and as it still does in primitive tunicates such as *Ciona* (fig. 4.2a,c, p. 89). In *Mitrocystella,* therefore, the ridge *gd* probably represents the gonoduct but straightened out in its course, so as to run directly from the gonad in the right anterior coelom across the dorsal surface of the left phrynx to the left atrium, without following the course of the gut.

The right atrium of *Mitrocystella* is defined on the natural moulds by a slight groove dorsal to it (fig. 8.13).

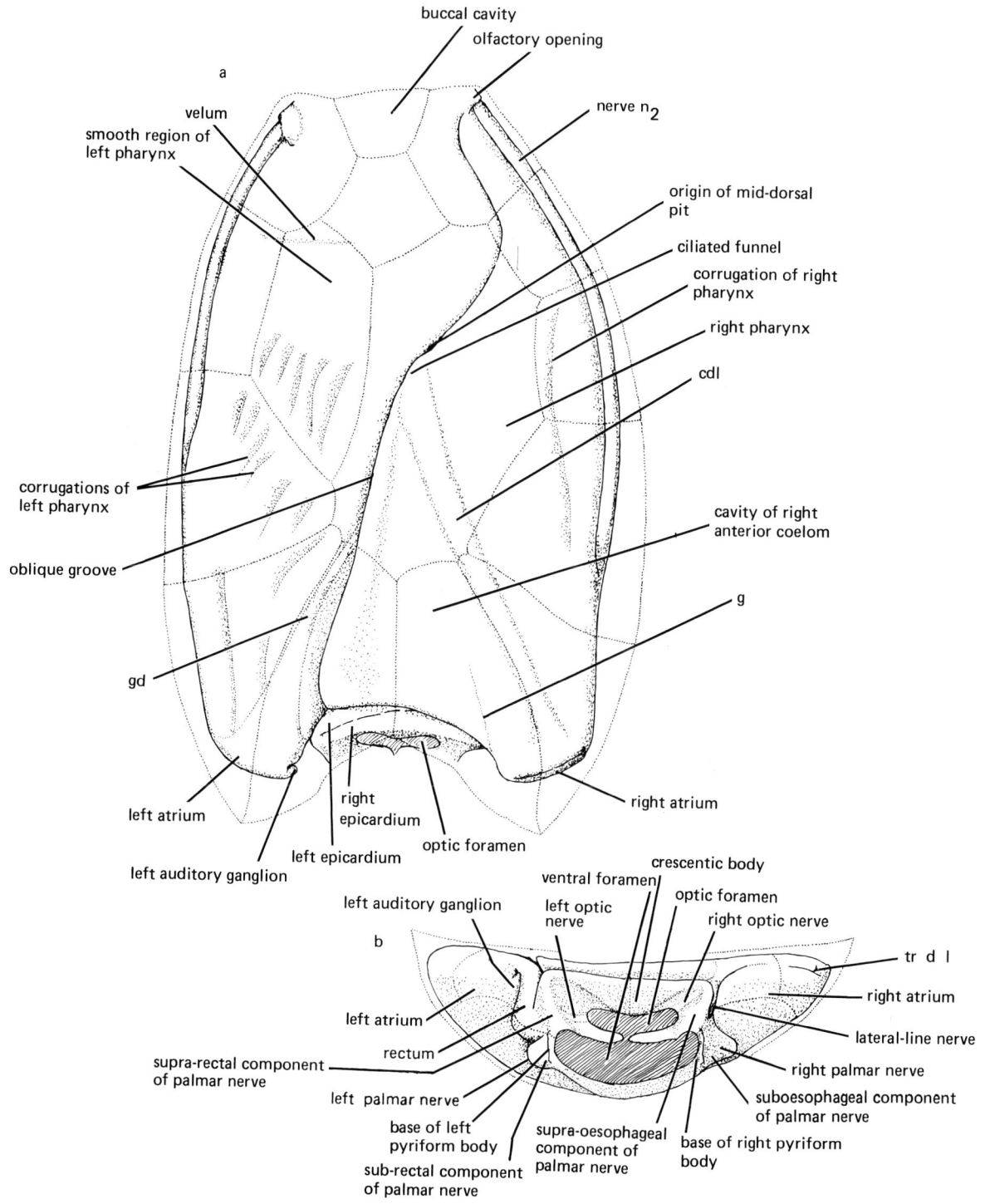

8.13. Mitrocystella incipiens miloni. *Reconstructed internal mould, simulating the soft parts. a) dorsal aspect, b) posterior aspect; from Jefferies & Lewis (1978, fig. 17, p. 251).*

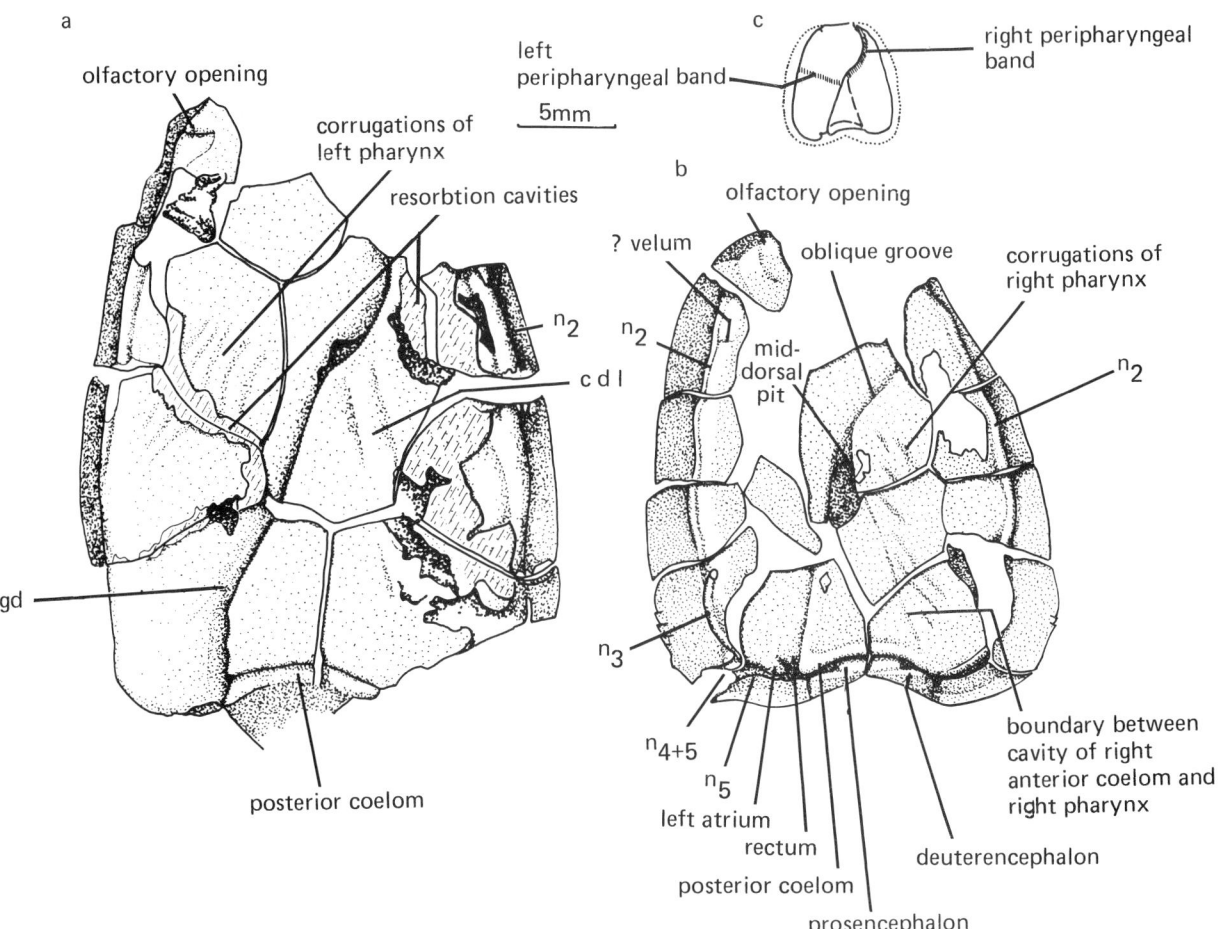

8.14. Mitrocystella incipiens miloni *a) and* Mitrocystites mitra *b). Camera-lucida drawings of internal moulds of the head in dorsal aspect. Note the corrugation of the right pharynx in* Mitrocystites *and of the left pharynx in* Mitrocystella; *from Jefferies (1981, fig. 10, p. 370).*

The left atrium, as already mentioned, was presumably in the same position as in *Mitrocystites*, but is not delimited on the natural moulds.

To sum up, the chambers of the head in *Mitrocystella* seem to have been similar to those of *Mitrocystites* in most ways. There are differences, however, and features that show in *Mitrocystella* but not *Mitrocystites*. The corrugation of the posterior part of the left pharynx, and its possible significance for the position of the left periphryngeal band, is the most important of these latter features, as should be apparent after *Placocystites* has been discussed.

The tail of *Mitrocystella* is much like that of *Mitrocystites*. However, it has about 11 rings of plates in the foretail, instead of about six. In the hind tail the spinal ganglia are sometimes more extensive than shown in the reconstruc-

tion of the soft parts (fig. 8.15) and in that case they leave impressions, not only on the ventral surfaces of the dorsal ossicles as shown but also on the internal surfaces of the ventral plates.

The brain and cranial nerves of *Mitrocystella* (figs. 8.16a,b,c) have already been mentioned while describing *Mitrocystites*. The total system is so complex that, in order to be intelligible, I shall describe the reconstructed nerves as though they were facts and mention the evidence for them as I proceed.

The brain lay at the anterior end of the tail and was divided, as always in mitrates, into prosencephalon and deuterencephalon. The prosencephalon was supported beneath and behind the optic foramen by paired hypocerebral processes which almost meet in the mid-line. All this resembles *Mitrocystites* except that the prosence-

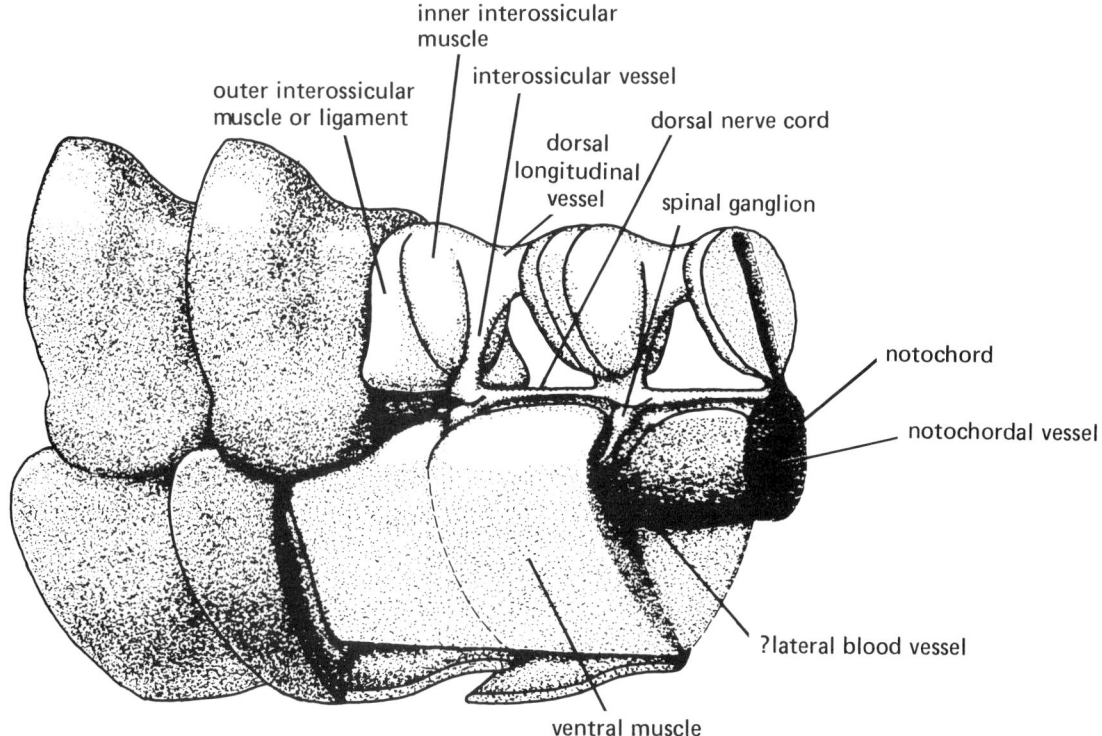

8.15. Mitrocystella incipiens miloni. *Block diagram of the hind tail with reconstructed soft parts. Hard outlines indicate direct evidence while broken outlines are reconstructional. In some tails, or in some regions of the tail, the spinal ganglia were probably placed more laterally, so as to be impressed on the insides of the ventral plates (cf. Jefferies & Lewis, 1978, pl. 2, fig. 61). The position of the lateral blood vessels is conjectural; from Jefferies (1968, fig. 18, p. 292).*

phalon was less inflated and the optic foramen smaller. As already said, this probably relates to the absence of trans-pharyngeal eyes in *Mitrocystella*. The part of the natural mould representing the precerebral knot was bigger than in *Mitrocystites* (compare figs 8.16a and 8.9). Possibly, however, this does not reflect differences in the nervous parts but is caused by resorbtion of the associated skeleton. For *Mitrocystella* has extensive irregular cavities in the dorsal skeleton probably caused by resorbtion (fig. 8.14a), whereas *Mitrocystites* has few or none such.

From the optic foramen, the optic nerves went off. The right optic nerve climbed from the right end of the foramen sideways and upwards to the dorsal right corner of the posterior coelom. The left optic nerve passed out of the left end of the optic foramen, ran downwards for a short distance and then up again (figs 8.13b, 8.16c). As already mentioned, the ridges on the natural mould of the posterior surface of the posterior coelom which indicate these proximal, cispharyngeal portions of the optic nerves are more slender than in *Mitrocystites*, which again probably relates to the lack of transpharyngeal eyes.

From the deuterencephalon, the deuterencephalic nerves, i.e. mainly the trigeminal complex, went off. On left and right a portion of these nerves climbed up the posterior face of the posterior coelom, as indicated by ridges on the natural mould of the latter (supra-rectal and supra-oesophageal components in fig. 8.13b). The right one of these ascending nerves joined the right optic nerve near the right dorsal corner of the posterior coelom. The left one joined the left optic nerve lower down than the right one, and passed over the rectum. Other, more ventral portions of the deuterencephalic nerves (sub-rectal and sub-oesophageal components in fig. 8.16b) ran forwards and sideways from the brain, passing, and communicating with, the median parts of the right and left pyriform bodies as they did so. The ascending parts of the deuterencephalar nerves, together with the optic nerves that they joined, passed in the posterior coelom over parts of the non-pharyngeal gut, i.e. over the oesophagus on the right and the rectum on the left, as appears by comparison with *Placocystites*. They can therefore be called supra-alimentary. The more ventral parts of the deuterencephalar nerves, on the other hand, passed beneath the gut and were subalimentary.

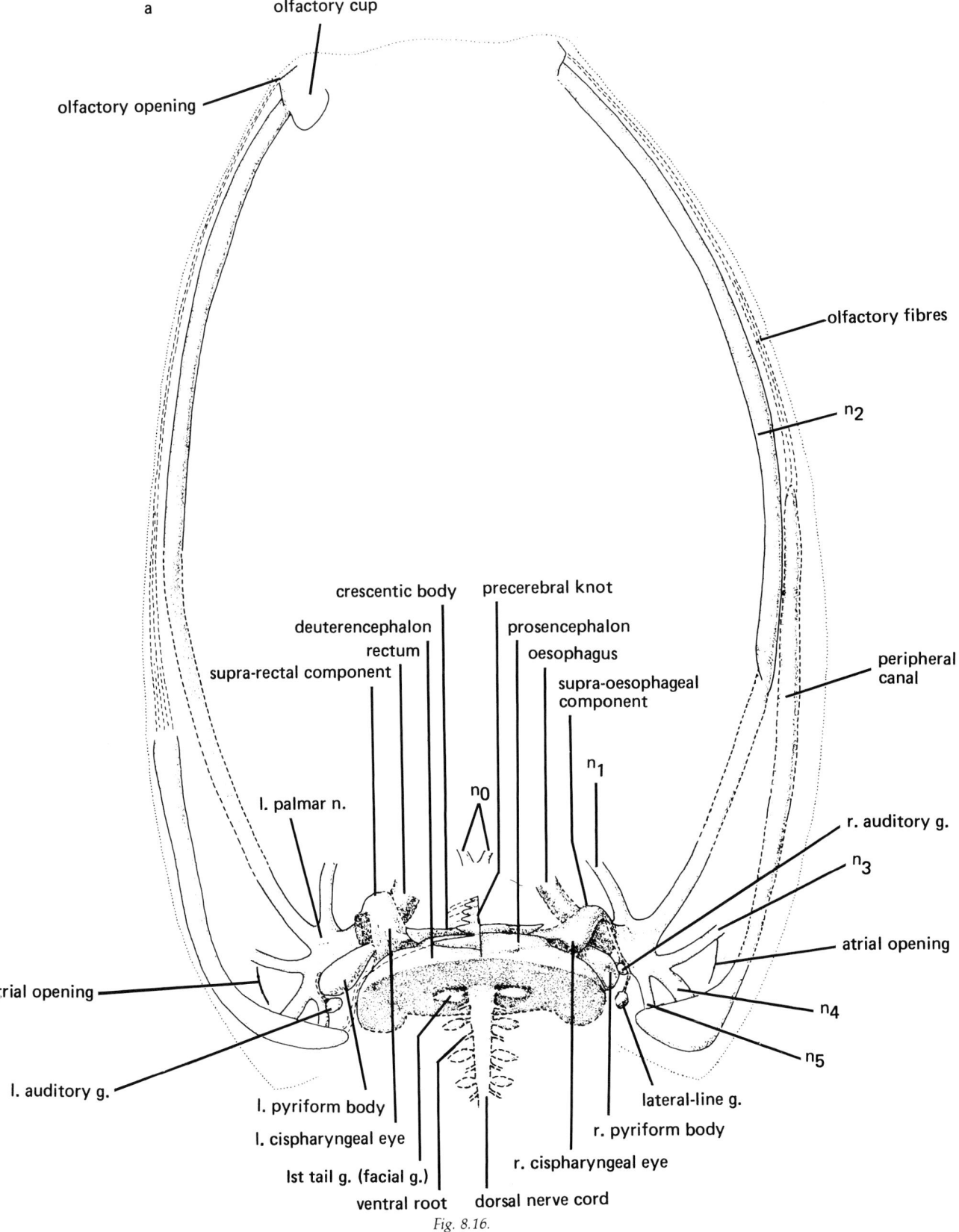

a

olfactory cup

olfactory opening

olfactory fibres

n_2

crescentic body   precerebral knot

deuterencephalon   prosencephalon

rectum   oesophagus

supra-rectal component   supra-oesophageal
component

peripheral
canal

n_1

l. palmar n.

n_0

r. auditory g.

n_3

atrial opening

rial opening

n_4

n_5

l. auditory g.

l. pyriform body

lateral-line g.

l. cispharyngeal eye

r. pyriform body

lst tail g. (facial g.)

r. cispharyngeal eye

ventral root   dorsal nerve cord

Fig. 8.16.

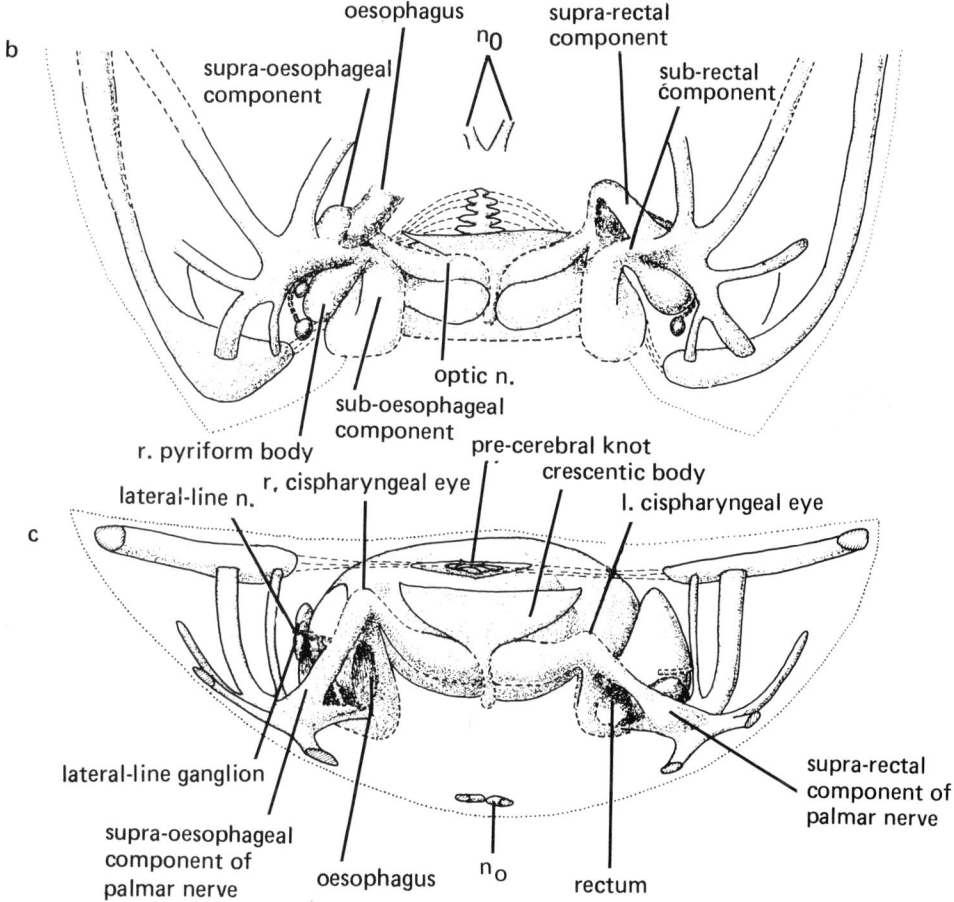

b

oesophagus

supra-oesophageal
component

supra-rectal
component

sub-rectal
component

n_0

optic n.

sub-oesophageal
component

r. pyriform body

lateral-line n.

r. cispharyngeal eye

pre-cerebral knot
crescentic body

l. cispharyngeal eye

c

lateral-line ganglion

supra-oesophageal
component of
palmar nerve

oesophagus

n_0

rectum

supra-rectal
component of
palmar nerve

8.16. Mitrocystella incipiens miloni. *Brain, cranial nerves and nerves of the fore tail. Solid outlines are based on direct evidence. Dashed outlines are reconstructional; a) dorsal aspect; b) ventral aspect of posterior part of head; c) anterior aspect of posterior part of head; from Jefferies & Lewis (1978, fig. 23, modified).*

The supra-alimentary nerves, both optic and deuterencephalic, descended in the front wall of the posterior coelom and ran to the right and left ventral anterior corners of this coelom. The thin inner calcitic layer present over most of the ventral skeleton (but not on the floor of the posterior coelom) turns upward posteriorly to form a low anterior rim to the posterior coelom. Special upward extensions of this rim, indicated by continuous lines in fig. 8.16c, covered the front faces of the descending supra-alimentary nerves so that the lower parts of these nerves can be traced directly in the fossils. At the anterior ventral corners of the posterior coelom the supra-alimentary nerves met and joined the subalimentary nerves whose distal courses are indicated by paired grooves in the floor of the posterior coelom. It follows that the sub- and supra-alimentary components totally encircled between them the oesophagus on the right and the rectum on the left.

The nerve trunk formed by the united sub- and supra-alimentary components is the palmar nerve and extends sideways into the middle, soft layer of the ventral skeleton.

A lucky feature of the ventral skeleton is here of great help in reconstruction – the soft middle layer was not uniform in thickness but formed a number of canals that can be traced in the fossils. Each canal is roofed by an upwardly convex fold of thin inner calcitic layer and is limited on either side by flat-topped ridges on the upper surface of the outer calcitic layer which reduce the middle layer above them to tenuous partings (fig. 8.22c). The canals allow the palmar nerves and their offshoots to be traced in some detail.

Each palmar nerve remained undivided for a short distance and then split into an anterior and a posterior furcation. The anterior furcation soon divided again into two nerves – n_1 and n_2, numbered outwards from the median line (fig. 8.16a). The nerve n_1 ran forwards and medianwards in the ventral skeleton but soon its canal ceases. This was not by blocking the canal but because the accompanying flat-topped ridges suddenly stop and the upward fold of inner layer gradually dies out. The nerve n_1 presumably passed forward beyond where the canal exists and would supply the floor of the head. Nerve n_2 passed forwards and lateralwards and then likewise ceased to have a discernible canal. The nerve itself, however, did not stop. For it can soon be traced again in the dorsal skeleton on right and left by means of a strong groove which climbs up from the ventral skeleton and then runs forward horizontally to end just beside the mouth. This nerve was presumably the main nervous supply of the mouth region.

The posterior furcation of the palmar nerve divided into three nerves – n_3, n_4, and n_5. Nerve n_3 passed sideways in the ventral skeleton in a distinct canal which runs to the junction with the dorsal skeleton and then stops. Its position, just in front of the atrial opening, suggests that it represents the proximal portion of the nerve n_3 of *Mitrocystites* (fig. 8.9) which in that animal ran in a canal through the dorsal skeleton to open on the dorsal surface of the head. Nerves n_4 and n_5 of *Mitrocystella* ran rearwards and upwards in the ventral skeleton and passed into the dorsal skeleton behind and medianwards of the atrial opening. Both nerves continued upwards in the dorsal skeleton in distinct canals which, unlike *Mitrocystites*, do not join nor open onto the surface. Instead they lose themselves in a large irregular peripheral canal which runs, almost horizontally, just beneath the dorsal surface, just internal to the peripheral flange and extends from near the tail insertion to half way along the length of the head. This peripheral canal corresponds in general position to the peripheral groove of *Mitrocystites* and receives the same nerves. It is homologous with the peripheral groove, except in probably containing extra nerves (olfactory and terminalis fibres) consonant with its internal position.

Homologies of the palmar nerves with the cranial nerves of living vertebrates can be proposed. In doing so, the central relations and the end organs of the nerves in question are more important than their intermediate courses. This is suggested by embryological evidence since a nerve will find its way in ontogeny to its correct end organ, bypassing obstacles that have been experimentally placed in its path (Balinsky, 1960). The optic nerve is a special case in that, being an outgrowing tract of the brain, it creates its own end organ. With it, only the origin counts in deciding homology – not its subsequent course, except from a functional viewpoint, and not its termination.

The nerve n_2, that ran forward to the mouth, was probably homologous with the maxillary branch of the trigeminal nerve which does the same thing in all living vertebrates. The nerve n_1 with its more median and ventral position, was probably the homologue of the mandibular trigeminal. For Lindström (1949) showed that separate mandibular and maxillary branches of the trigeminal exist in myxinoids and lampreys, and therefore preceded the phylogenetic origin of jaws. The dorsal branches n_4 and n_5 probably represented the dorsal touch-sensory branches of the trigeminal since they end superficially in *Mitrocystites* while the peripheral canal of *Mitrocystella*, into which n_4 and n_5 pass, runs only slightly below the dorsal surface. The relations in *Placocystites* show that n_5 supplied a region of the dorsal surface internal to that supplied by n_4, or more dorsal in terms of living vertebrates. It is therefore possible that n_4 is homologous with the profundus branch of the trigeminal and n_5 with the superficialis branch (cf. figs 5.48, p. 158, 6.32, p. 189 and Goodrich, 1930, fig. 728). The nerve n_3 which is much less well developed in *Mitrocystella* than *Mitrocystites* is probably optic; for this difference, as already said, can thus be related to the size of the prosencephalon in the two forms, and is consistent with its dorsal opening in *Mitrocystites*. It is likely that the presence of transpharyngeal eyes is more primitive in mitrates than their absence for they existed in the cornute *Reticulocarpos hanusi* (fig. 7.29, p. 228). If so, the nerve n_3 of *Mitrocystella* was a vestige. Optic function would depend on light falling on the optic nerve after passing through the presumably transparent skeleton. Indeed, the main optic function was probably located at the summits of the cispharyngeal portions of the optic nerves, at the right and left summits of these nerves, beneath the ceiling of the head. These cispharyngeal eyes could well be homologous with the paired eyes of vertebrates.

The olfactory end organs of *Mitrocystella*, by comparison with living vertebrates, would presumably arise from ectoderm. This suggests that they would be located in the buccal cavity and their likely position is indicated by a pair of pits (olfactory openings) in the dorsal skeleton over the buccal cavity on either side of the mouth (fig. 8.16a). These pits probably mark where olfactory fibres passed into the skeleton. These fibres would presumably pass rearwards parallel to the peripheral flange (as indicated in fig. 8.16a), either within the stereom mesh or, more posteriorly, inside the peripheral canal. Posteriorly they would enter the precerebral knot and thus make contact with the dorsal mid-line of the prosencephalon. Perhaps touch-sensory terminalis fibres accompanied them.

The acustico–lateralis system was probably much as in *Mitrocystites*. The lateral line had a ganglion just internal to it, as evidenced by an elongate carrot-shaped body situated in the natural moulds just behind and lateral to the right trigeminal ganglion (pyriform body). The right acoustic ganglion, presumably serving neuromasts in the

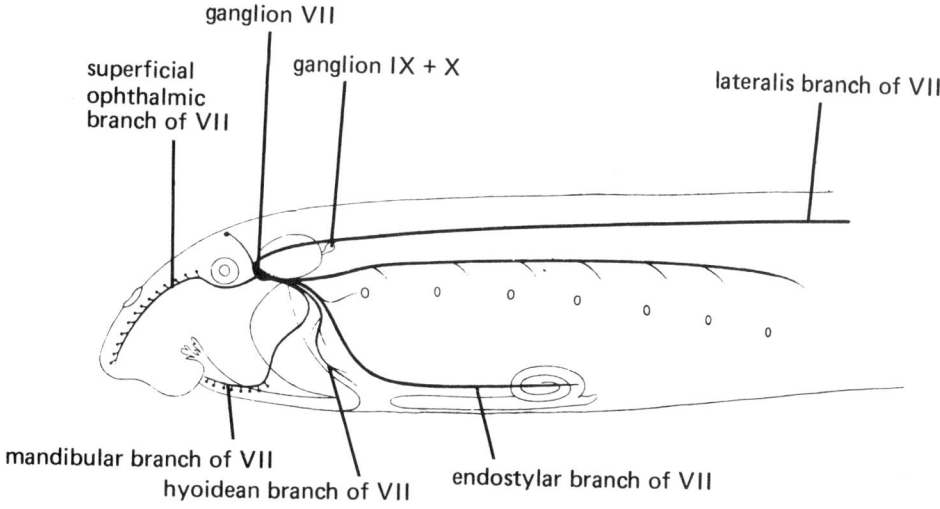

superficial
ophthalmic
branch of VII

ganglion VII

ganglion IX + X

lateralis branch of VII

mandibular branch of VII

hyoidean branch of VII

endostylar branch of VII

8.17. *The facial nerve of the ammocoete of* Lampetra planeri; *from Alcock (1898, pl. 2, fig. 2).*

right atrium, was situated just anterior to the lateral-line ganglion, and is evidenced by a body observed on the antero–lateral face of the right trigeminal ganglion in natural moulds. (This same body seems to have existed in *Placocystites* also, which had lost the lateral line and lateral-line ganglion.) The left acoustic ganglion would be located in the left atrium of *Mitrocystella* and is suggested in one specimen by a body in the left atrium (figs 8.13a,b), just behind the rectum and therefore in the same position as the supposed left auditory ganglion of *Mitrocystites*.

A pair of nerves, labelled n_0, passed from the posterior coelom near the mid-line being carried by short canals into the middle layer of the ventral skeleton. Judging by their position, these nerves probably supplied the endostyle since this is the only important organ in the ventral mid-line of the primitive chordate pharynx. They recur in *Placocystites*, as mentioned later, and are very important for a reconstruction of the pharynx. If they supplied the endostyle they are probably deuterencephalic in origin, for the endostyle of the larval lamprey is supplied by a branch of the facial nerve (fig. 8.17) (Alcock, 1898).

Thus the nervous system of *Mitrocystella* was much as in *Mitrocystites*. In the tail there were a dorsal nerve cord and a paired series of spinal ganglia. In the head there were: a trigeminal complex with paired ganglia and with maxillary, mandibular and dorsal touch-sensory branches; an optic system whose peripheral portions had partly disappeared and partly vestigialized; an acustico–lateralis system with paired auditory ganglia associated with the atria and with a lateral-line apparatus on the right; a pair of nerves supplying the endostyle; an olfactory system with end organs in the buccal cavity which were connected by intra-skeletal fibres with the prosencephalon; the olfactory

fibres were possibly accompanied by terminalis fibres; and there must have been other nerves such as those to the non-pharyngeal gut and gill slits, about which nothing can be said. As discussed later, the lateral line and the dorsal touch-sensory branches of the trigeminal (n_4, n_5) of *Mitrocystella*, *Mitrocystites* and some related mitrates were probably synapomorphies with vertebrates, so that these mitrates were stem vertebrates. The nervous system of *Mitrocystella* is therefore very important, as being almost an archetype for vertebrates as a whole, and even for man.

The mode of life of *Mitrocystella* will be dealt with when discussing *Placocystites*.

Placocystites forbesianus is the only known species of its genus and was first described by the Belgian L. de Koninck in (1869). It was collected in large numbers in the course of quarrying during the second half of the nineteenth century from the Middle Silurian Wenlock Limestone near the town of Dudley in western England. It has been studied by several workers after de Koninck, including Bather (1900), Ubaghs (1968) and Jefferies & Lewis (1978).

The Wenlock Limestone is a coral-reef limestone and contains marly partings in which the *Placocystites* occurred. It was probably laid down in water less than 30 m deep (Scoffin, 1972). Unlike the mitrates discussed already, *Placocystites* does not occur in siliceous nodules and it has the calcite skeleton preserved. To study its internal structure therefore, it was necessary to construct enlarged models made of expanded polystyrene, on the basis of serial sections. These were built by Mr David Lewis in the British Museum (Natural History), London. Two were made, on a scale of × 20, one based on longitudinal and the other on transverse sections.

In outside appearance (fig. 8.18) the head of *Placocystites*

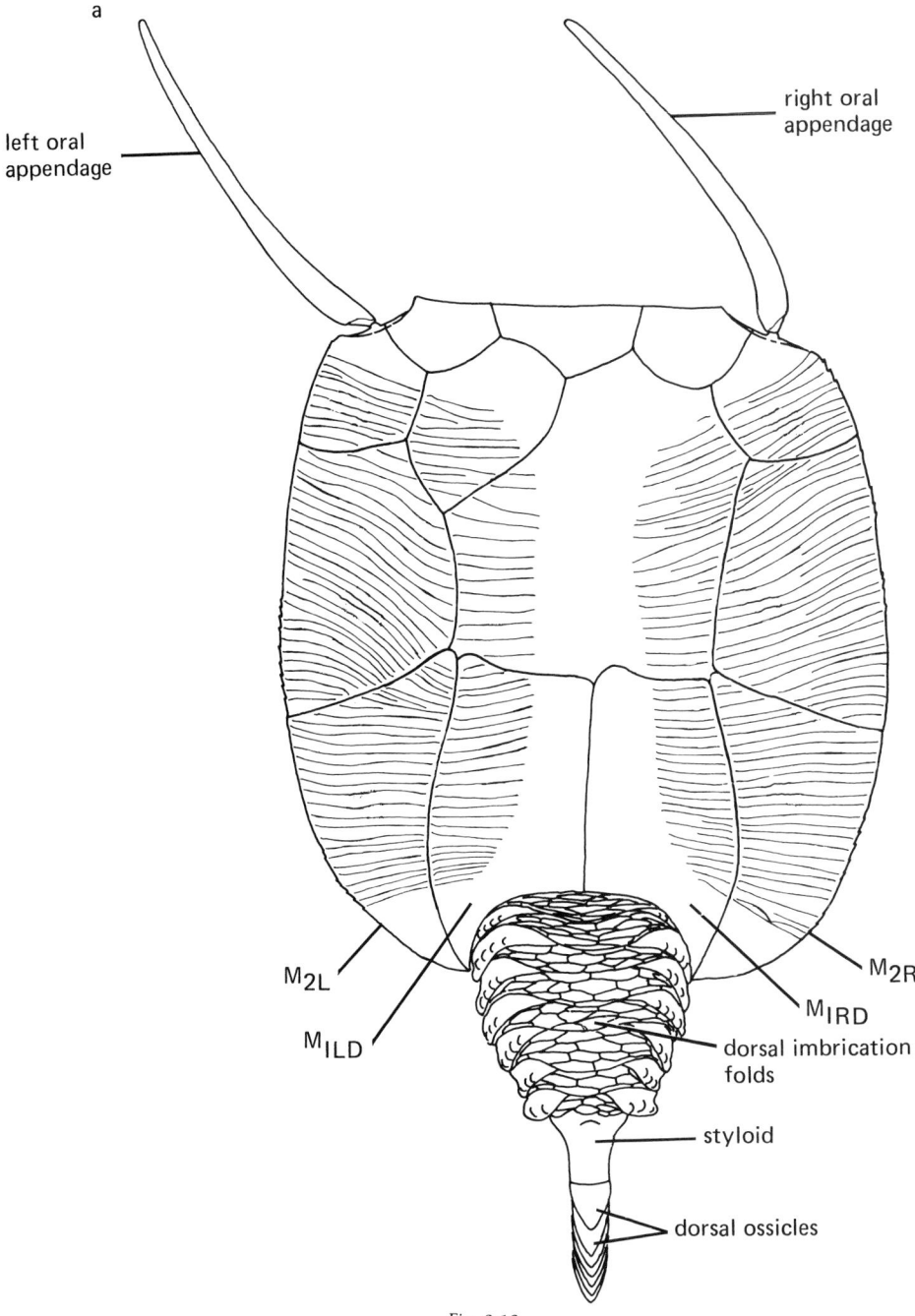

a

left oral
appendage

right oral
appendage

M₂ₗ

Mᵢₗᴅ

M₂ᵣ

Mᵢᵣᴅ

dorsal imbrication
folds

styloid

dorsal ossicles

Fig. 8.18.

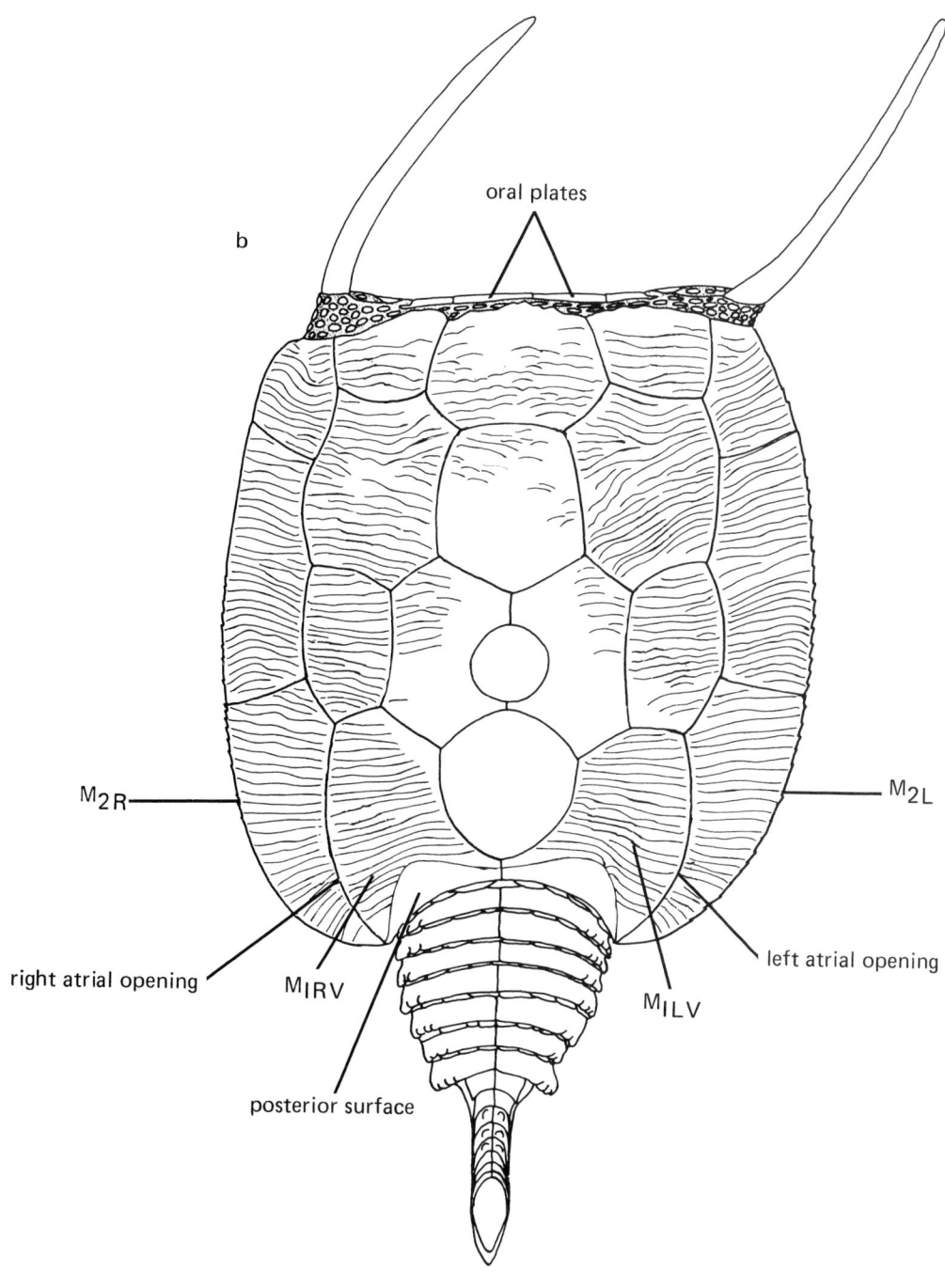

oral plates

b

M_{2R}

M_{2L}

right atrial opening

M_{lRV}

M_{lLV}

left atrial opening

posterior surface

Fig. 8.18.

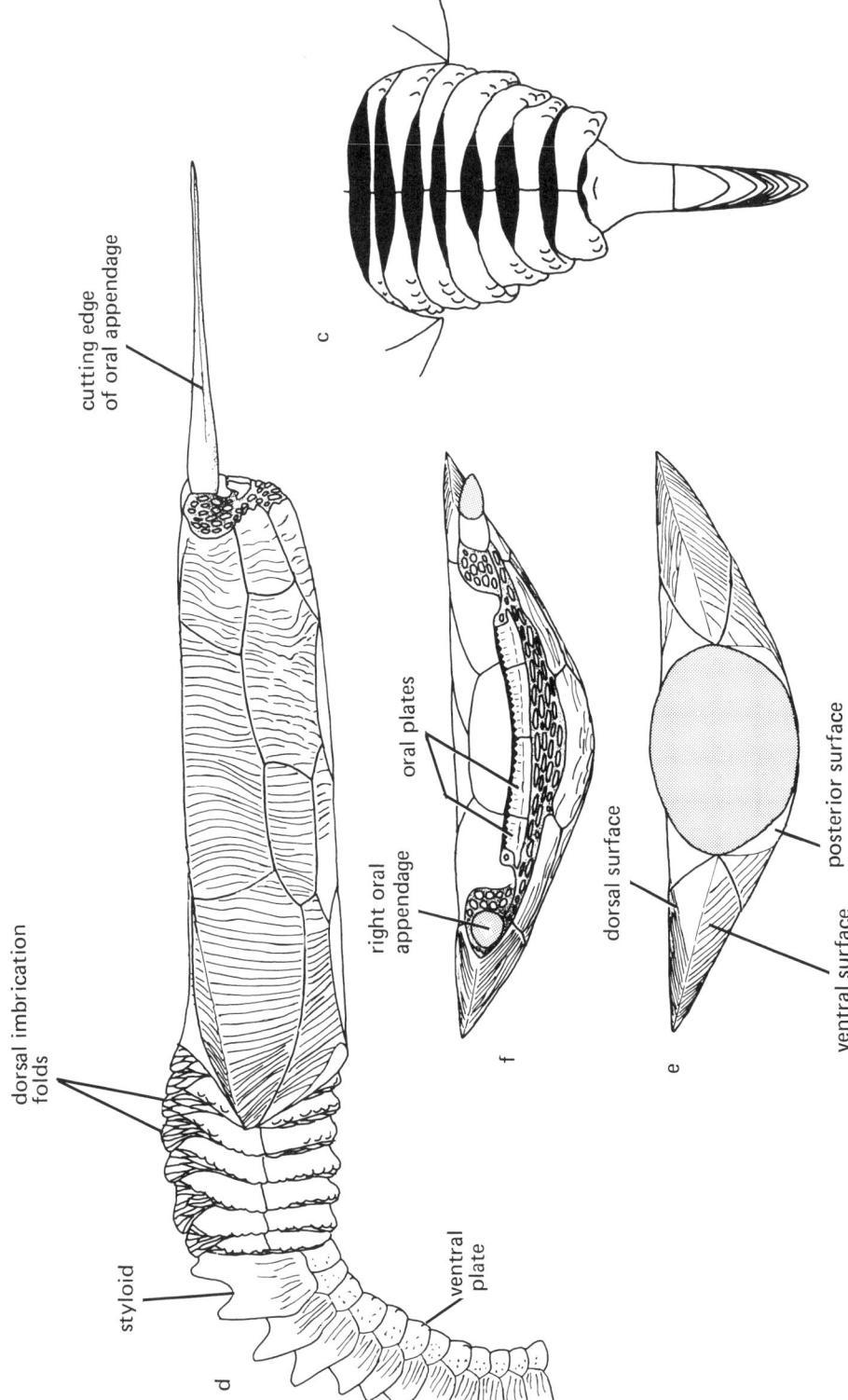

cutting edge
of oral appendage

c

oral plates

right oral
appendage

dorsal surface

posterior surface

ventral surface

e

f

dorsal imbrication
folds

styloid

ventral
plate

d

8.18.   Placocystites forbesianus. Reconstruction of external features. a) dorsal aspect, b) ventral, c) dorsal aspect of tail, with the imbrication folds removed from the fore tail to show gaps between major plates, d) right aspect, e) posterior and f) anterior.

is much like *Mitrocystella*, but there are several differences. Thus there is a pair of knife-like oral spines right and left of the mouth; the upper edge of the lower lip is guarded by four transversely elongate oral plates; the ventral surface of the head was rigid (except that posteriorly it would presumably flex to allow the atrial openings to gape) and made up of a fixed pattern of plates; there are cuesta-shaped ribs both on the dorsal and the ventral surface; the dorsal surface slopes down posteriorly more than in *Mitrocystella*, so as to be prominently visible in posterior aspect (fig. 8.18e); the posterior surface of the head is reduced to a smooth ventral patch just in front of the tail (fig. 8.18b); and there is no lateral line (a cladistic analysis shows that this organ has almost certainly been secondarily lost).

The outside of the tail of *Placocystites* is also much like *Mitrocystella*. However, the posterior end of the tail is unknown. And the rings of the fore tail are rigid, all four plates of a ring being joined at sutures while successive rings are connected together by plated imbrication membranes (fig. 8.18a,d) which dorsally form large recumbent folds covering big gaps between the rings. These folds suggest that the fore tail was capable of strong ventral flexion.

Most of these special features of *Placocystites* can be explained as adaptations for a shallow-burrowing mode of life. Thus the cuesta-shaped ribs on the dorsal surface show that sediment passed over that surface in sufficient quantity and with sufficient residual strength, to be gripped by the ribs. The strong downward posterior slope of the dorsal surface and the reduction of the posterior surface would 'streamline' the head, so that it would slide rearwards through mud more easily. The loss of the lateral line is to be expected if the head never emerged from mud. The rigid ventral wall of the head would prevent the head collapsing under the load of overlying mud. And the oral spines, whose shape and articulation show that they swept mainly from side to side, probably served to clear mud away from the mouth after it had fallen off the anterior, but trailing, edge of the head. The oral spines are sabre-shaped with a convexly curved cutting edge laterally and a blunt surface medianly. They are probably homologous with the leftmost and rightmost spike-shaped oral plates of the lower lip of *Mitrocystella* (fig. 8.11b). Their mode of functioning is discussed in Jefferies & Lewis (1978, 224 ff). The special features of the fore tail, especially the recumbent dorsal folds of the imbrication membranes, suggest that it flexed ventrally with great vigour and was the chief motor for this shallow-burrowing animal. The hind tail would have pushed against the sediment, probably using ventral bearing surfaces on the ventral plates in the early part of the power stroke and dorsal bearing surfaces, on the distal surfaces of the ossicles,

during the later part of the power stroke. As already mentioned, I have discussed locomotion in *Mitrocystites*, *Mitrocystella* and *Placocystites* in more detail elsewhere (Jefferies, 1984).

The morphological series from *Mitrocystites*, through *Mitrocystella*, to *Placocystites* probably represents increasing adaptation to a shallow-burrowing mode of life. For it coincides with an increase in cuesta-shaped ribbing, a covering-over of sensitive areas or organs on the dorsal surface of the head (between *Mitrocystites* and *Mitrocystella*), a progressive reduction in the eyes and the optic part of the brain, and a progressive tapering of the posterior leading edge of the head to help it penetrate mud.

Inside the head of *Placocystites* features (figs 8.19a–g) exist much like those of *Mitrocystites* and *Mitrocystella*. An oblique ridge runs across the ventral face of the dorsal skeleton from anterior right to posterior left (fig. 8.19a), dividing the skeleton into a right and a left field. The anterior part of the ridge is curved rightward and somewhat effaced. Behind this portion, where the ridge crosses the mid-line, a very long mid-dorsal process goes off to the posterior right. From near the origin of this process a low ridge stretches posteriorly across the right field of the dorsal skeleton to just right of the tail insertion (fig. 8.19a,d, ridge between r. ant. coelom and r. pharynx). The anterior part of the left field carries an arcuate roughly transverse groove (velar groove). There are irregular cavities, probably indicating resorbtion of the skeleton during life and surrounded by 'resorbtion cliffs' (fig. 8.18a,d), situated as in *Mitrocystella* in the posterior part of the left field and in the right field, right of the above-mentioned low ridge. Behind the origin of the mid-dorsal process the oblique ridge increases in strength, becomes asymmetrical with a steep or concave right side and shows other complexities discussed later. In the ventral skeleton (fig. 8.19b) the same three layers exist as in *Mitrocystella* or *Mitrocystites* (thin inner calcitic, middle soft, thick outer calcitic). The inner layer is developed only at the centres of plates anteriorly but becomes more widespread posteriorly except just in front of the tail where it is conspicuously absent from a semi-circular area floored by outer layer only (posterior coelom). The anterior limit of this semi-circular area is formed of an upturned rim of inner layer which extends upwards and forwards into several processes of calcite.

In terms of chambers, the internal features of the skeleton can be interpreted as in other mitrates. The right and left fields, separated by the oblique ridge, are the fields of the right and left anterior coeloms (the right being partly virtual and the left entirely so). The arcuate velar groove probably marks the posterior boundary of the buccal cavity, carrying the velum. The low ridge crossing the right field would separate the right pharynx from the cavity of the right anterior coelom, and would therefore coincide with the junction of the patent and virtual parts of the right anterior coelom, like the corresponding ridge (or groove on the natural mould) in *Mitrocystites*. The mid-dorsal process would be related to the same boundary. It is consistently right of the low ridge and is D-shaped in section, with the vertical stroke of the D median. Consequently, if we try to reconstruct the chambers in this part of the head, we are compelled to suppose a fold of flesh (fig. 8.20), bending downwards and rightwards in transverse section and with the mid-dorsal process stiffening its free edge. The areas of resorbtion of the dorsal skeleton are located over the right and left pharynxes. In the ventral skeleton, the conspicuous semi-circular area at the rear end without inner layer represents the floor of the posterior coelom. The dorsal boundary of this coelom is visible in the model based on longitudinal sections as a subtle but definite break in slope in the front surface of the posterior wall of the head in much the same position as the groove that defines this coelom dorsally in *Mitrocystites* or *Mitrocystella*. The crescentic body of *Placocystites* seems to have been situated as in *Mitrocystella*. It corresponds to a flattened area (fig. 8.19d) on the anterior face of the posterior wall of the head. This area is situated medianly above the optic foramen, is limited on right and left by upwardly diverging ridges and ends dorso–laterally in a pair of deep pits (fig. 8.19a).

A comparison with tunicates, however, reveals the true importance of *Placocystites*. As already mentioned in Chapter 4, tunicates feed by means of a mucous filter inside the pharynx. The mucus is produced by the endostyle which lies in the mid-ventral line of the pharynx. The front ends of the mucous sheets are grasped and carried dorsal-wards by the paired peripharyngeal bands which meet dorsally in the mid-line. The opening (ciliated funnel) of the duct of the neural gland is situated just in front of the dorsal meeting place of the peripharyngeal bands (fig. 4.7, p. 92). In the dorsal mid-line, the mucous sheets are rolled up together by the dorsal languets or dorsal lamina to form the mucous rope (figs 4.12, 4.13c, p. 95, 97) and pulled rearwards, along with entrapped food particles, into the opening of the oesophagus (fig. 4.14). There is a marginal band of cilia on the crests of the right and left endostylar folds (fig. 4.5, p. 91); and the right band, but not the left one, continues into the retropharyngeal ciliated band

which runs to the opening of the oesophagus (fig. 4.8b, p. 93). The dorsal languets or lamina slope in transverse section rightwards and ventralwards. The oesophageal opening is right of the mid-line, in order to receive the rear end of the mucous rope as it leaves the concavity of the dorsal languets or lamina. The right and left pharyngo–epicardial openings in adult *Ciona* and in the ontogeny of other tunicates, are situated just right and left of the retropharyngeal band (fig. 4.8a).

The peripharyngeal bands of *Placocystites* were probably situated as discussed in *Mitrocystella* and *Mitrocystites* (figs 8.14c), i.e. the left band crossed the left pharynx near the origin of the mid-dorsal process whereas the right one followed the oblique ridge rearwards to near the origin of the mid-dorsal process and there joined the left peripharyngeal band. These conclusions are based on the distribution of corrugations in the pharynges of *Mitrocystella* and *Mitrocystites*, which in *Placocystites* are not seen. However, all three genera are so similar that conclusions derived from two of them can be applied, with caution, to the third. The position of the left peripharyngeal band may be indicated in *Placocystites* by a very slight groove seen on one model and one specimen (fig. 8.19a).

The position of the ciliated funnel is suggested, in the model based on transverse sections, by the rounded-off edge of the oblique ridge immediately behind the origin of the mid-dorsal process (fig. 8.19a). For the shape of this portion of the ridge in itself suggests that some cylindrical organ came out of the right anterior coelom and entered the left pharynx here. And from what has already been said, this organ would be in the dorsal mid-line near the dorsal meeting place of the peripharyngeal bands.

The reconstructed fold of flesh with the mid-dorsal process along its free edge would therefore be a dorsal lamina. This follows from its position at the left boundary of the right pharynx, from the fact that it begins anteriorly near the ciliated funnel and the dorsal meeting place of the right and left peripharyngeal bands, and from its reconstructed transverse section, sloping ventralwards and rightwards. The mucous rope would be rolled up in the concavity above the dorsal lamina, rotating anti-clockwise as seen from behind.

The oesophageal opening would need to be located behind the dorsal lamina, where it could receive the mucous rope. Its likely position is indicated in the ventral skeleton of the model based on transverse sections by a

---

8.19. Placocysites forbesianus. *Photographs of polystyrene models. The individual sheets of polystyrene are 2 mm thick and represent 0·10 mm in the specimen. a–e, g represent the transversely sectioned model; f = longitudinally sectioned model. a) Ventral aspect of dorsal skeleton; b) dorsal aspect of ventral skeleton; c) detail of right side of posterior coelom to show especially the oesophageal opening; d) detail of posterior wall of the dorsal skeleton; e) posterior aspect of ventral skeleton to show especially the front margin of the posterior coelom with the retropharyngeal process situated over the right-hand canal n_0; f) view of longitudinally sectioned model corresponding to (e), to show the anterior margin of the posterior coelom with the elongate, rightward directed retropharyngeal process notched at the base; g) detail of left side of posterior coelom.*

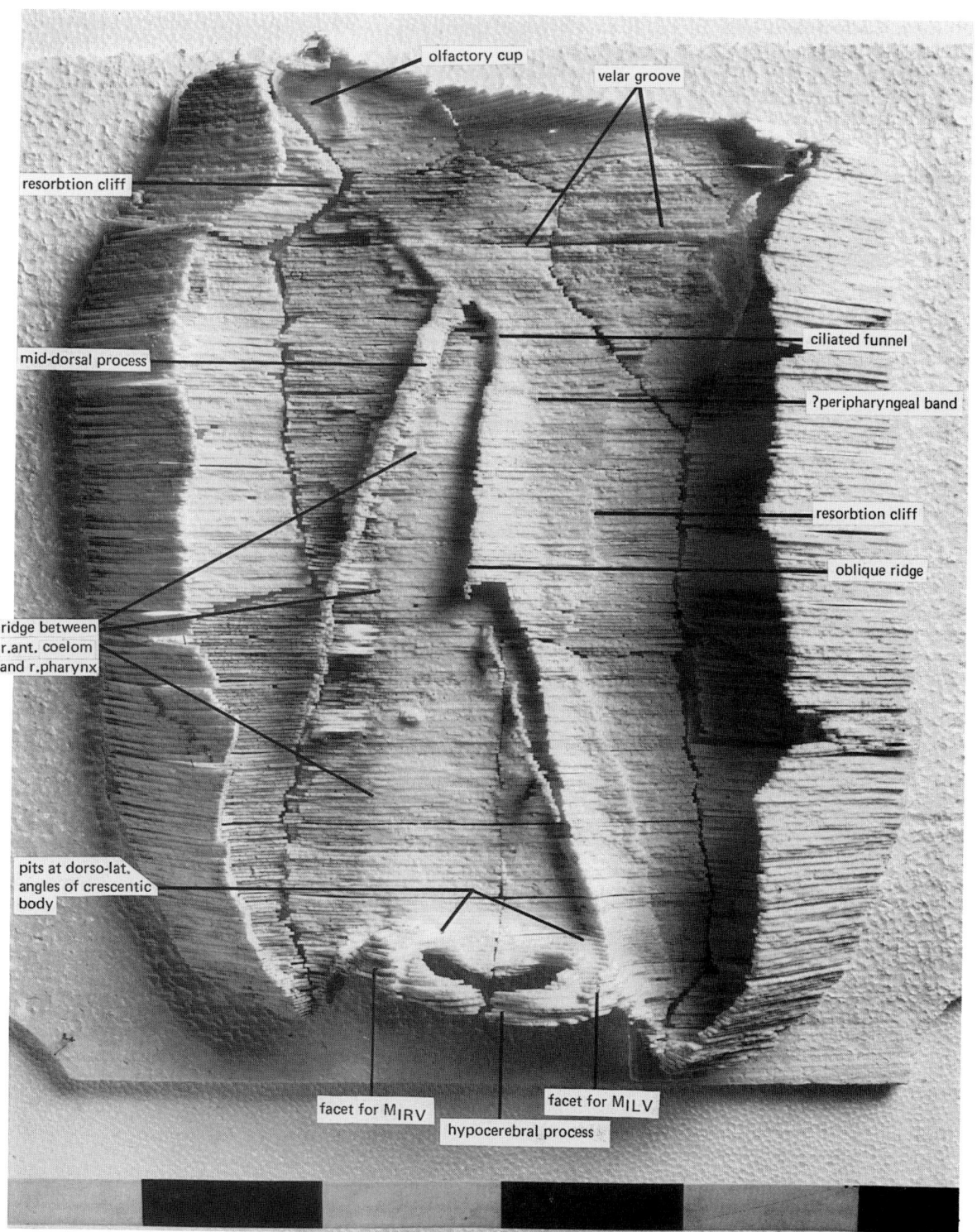

olfactory cup

velar groove

resorbtion cliff

mid-dorsal process

ciliated funnel

?peripharyngeal band

resorbtion cliff

oblique ridge

ridge between
r.ant. coelom
and r.pharynx

pits at dorso-lat.
angles of crescentic
body

facet for M$_{IRV}$

facet for M$_{ILV}$

hypocerebral process

Fig. 8.19a

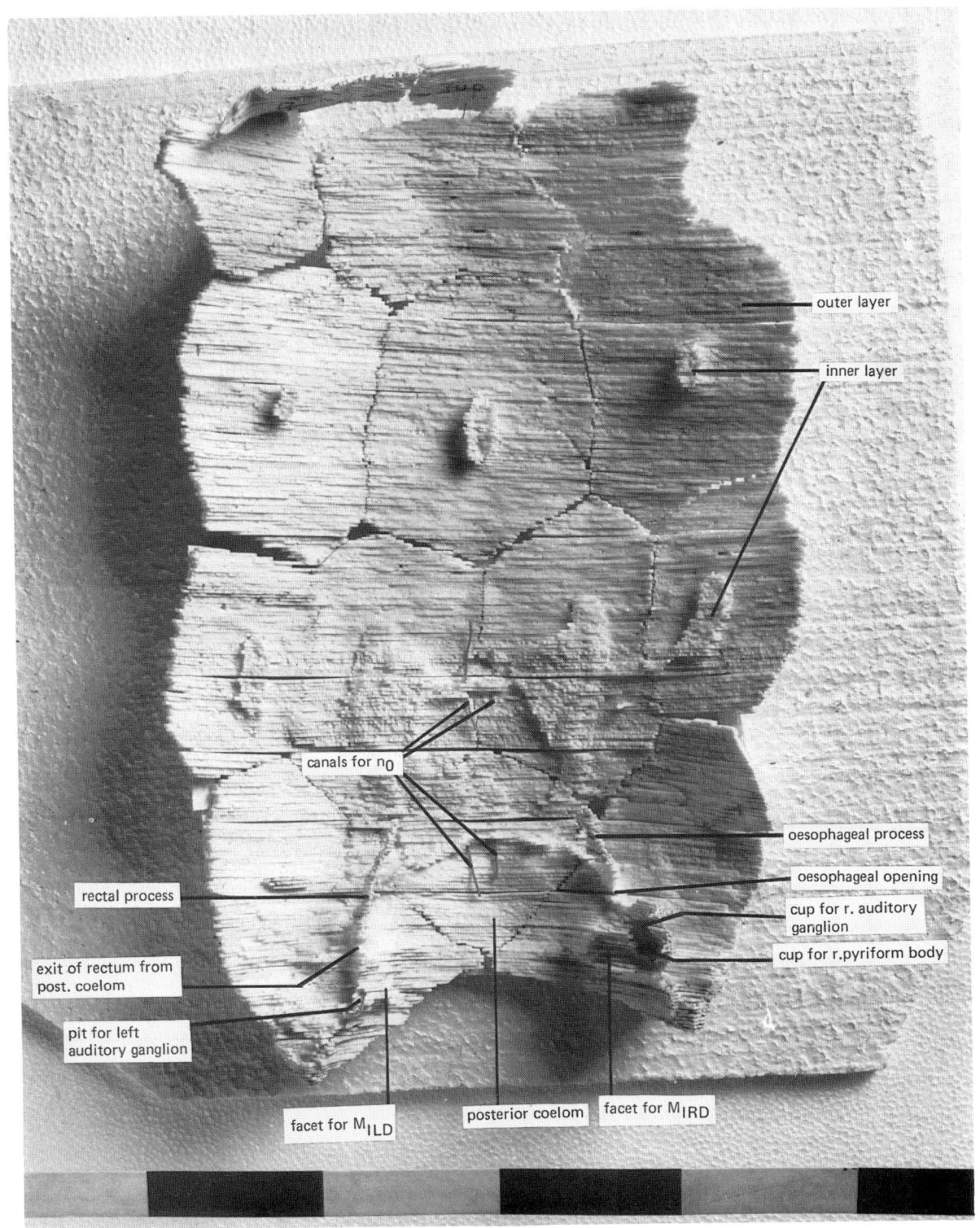

outer layer

inner layer

canals for n_0

oesophageal process

rectal process

oesophageal opening

cup for r. auditory ganglion

exit of rectum from post. coelom

cup for r. pyriform body

pit for left auditory ganglion

facet for M_{ILD}

posterior coelom

facet for M_{IRD}

Fig. 8.19b

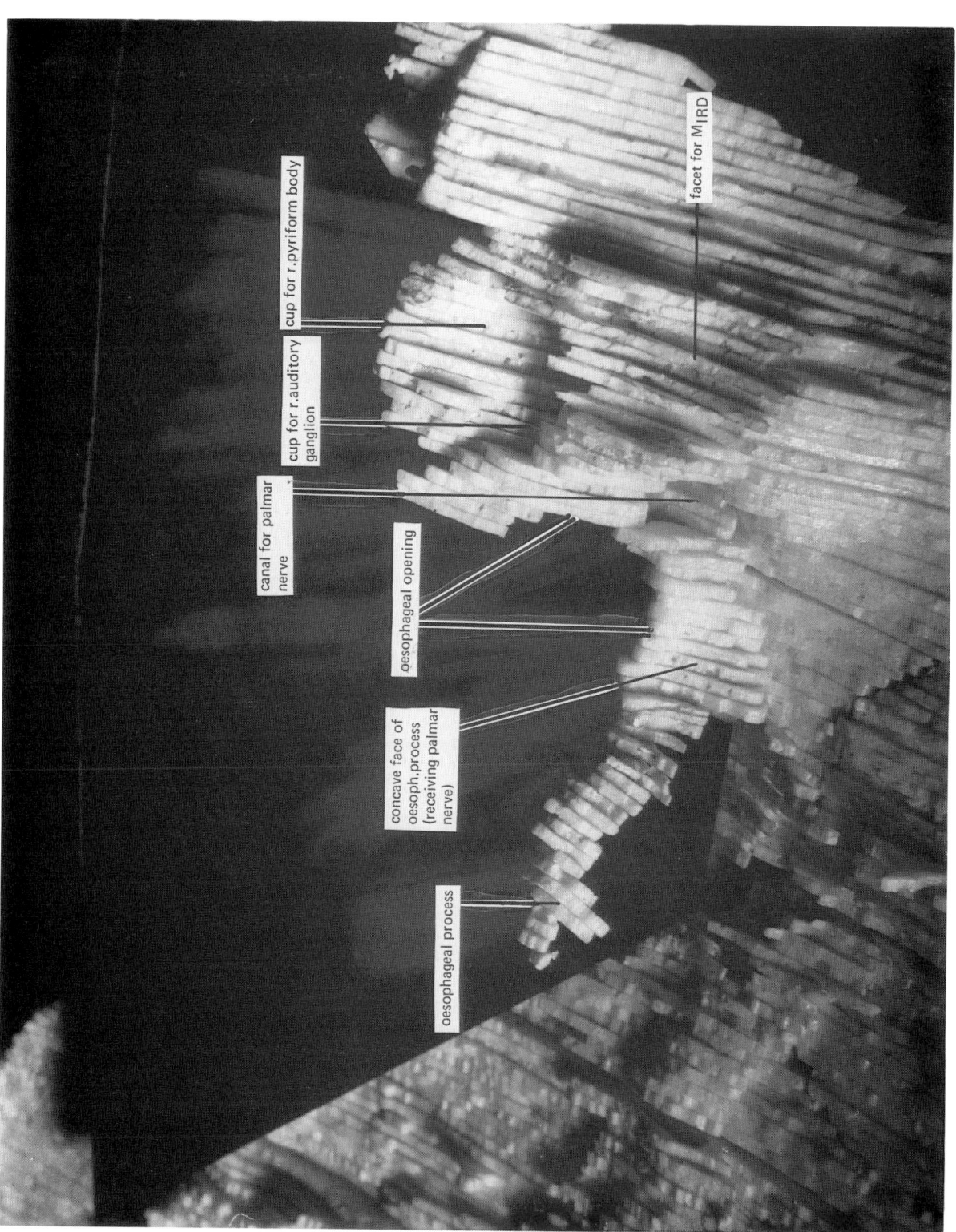

cup for r.pyriform body

cup for r.auditory ganglion

canal for palmar nerve

oesophageal opening

concave face of oesoph.process (receiving palmar nerve)

oesophageal process

facet for M$_{1}$RD

Fig. 8.19c

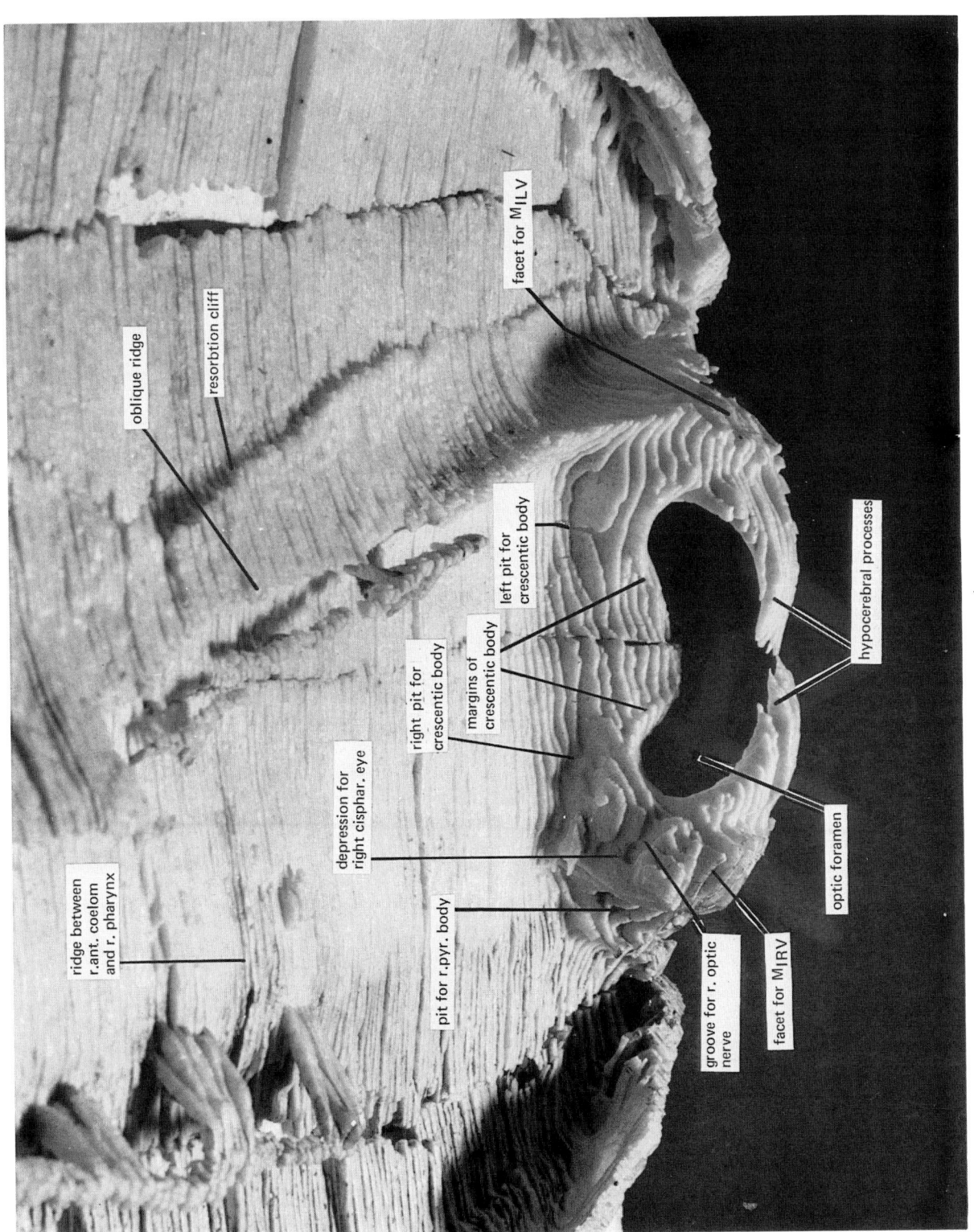

oblique ridge

resorbtion cliff

facet for M$_{ILV}$

left pit for crescentic body

ridge between r.ant. coelom and r. pharynx

right pit for crescentic body

margins of crescentic body

hypocerebral processes

depression for right cisphar. eye

pit for r.pyr. body

groove for r. optic nerve

facet for M$_{IRV}$

optic foramen

Fig. 8.19d

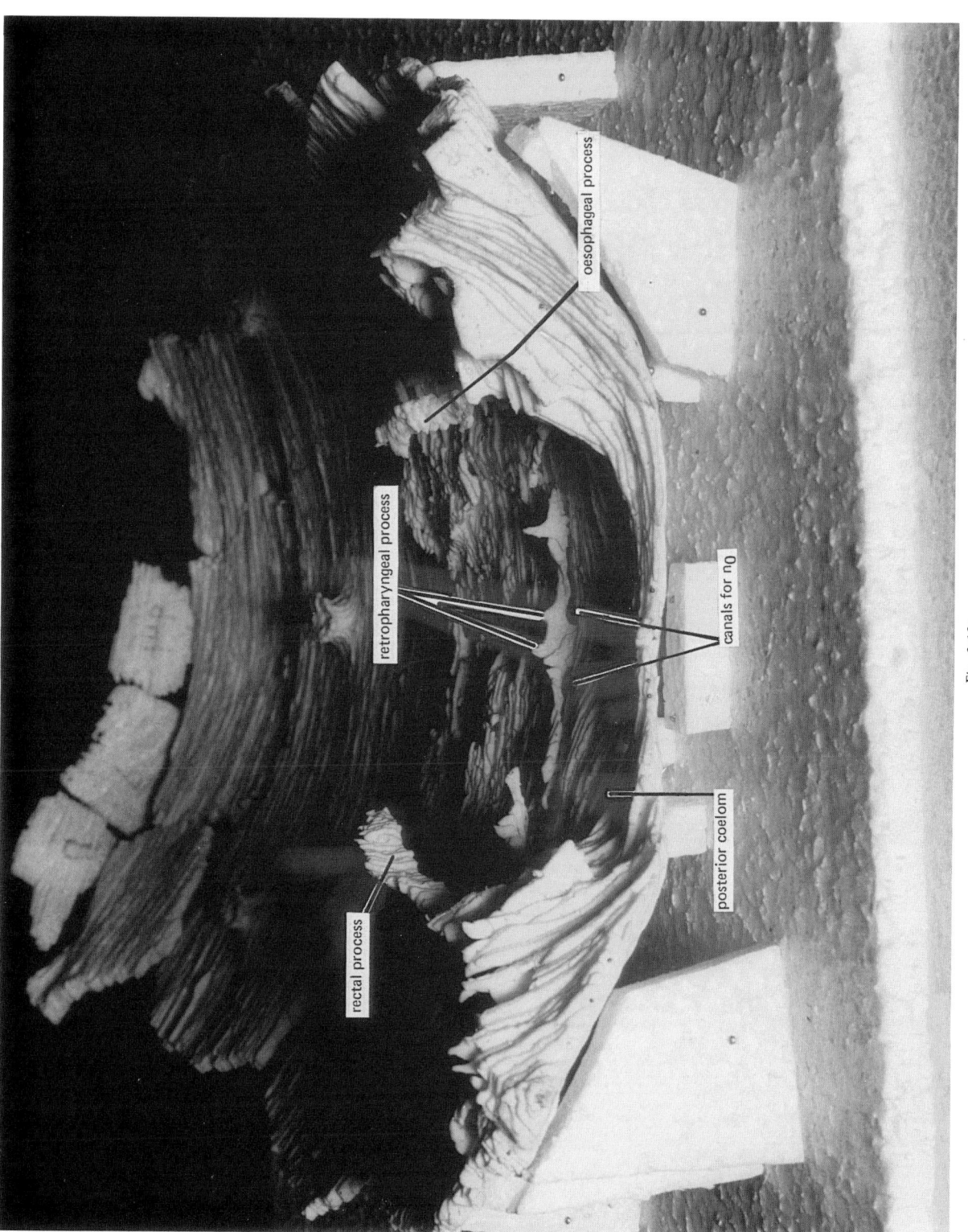

oesophageal process

retropharyngeal process

canals for n0

rectal process

posterior coelom

Fig. 8.19e

Fig. 8.19f

rectal process

exit of rectum from
posterior coelom

canal for n₅

pit for left
auditory ganglion

facet for M₁LD

canal for palmar
nerve

concave surface of
rectal process to
receive palmar nerve

ant. boundary of
posterior coelom

posterior coelom

Fig. 8.19g

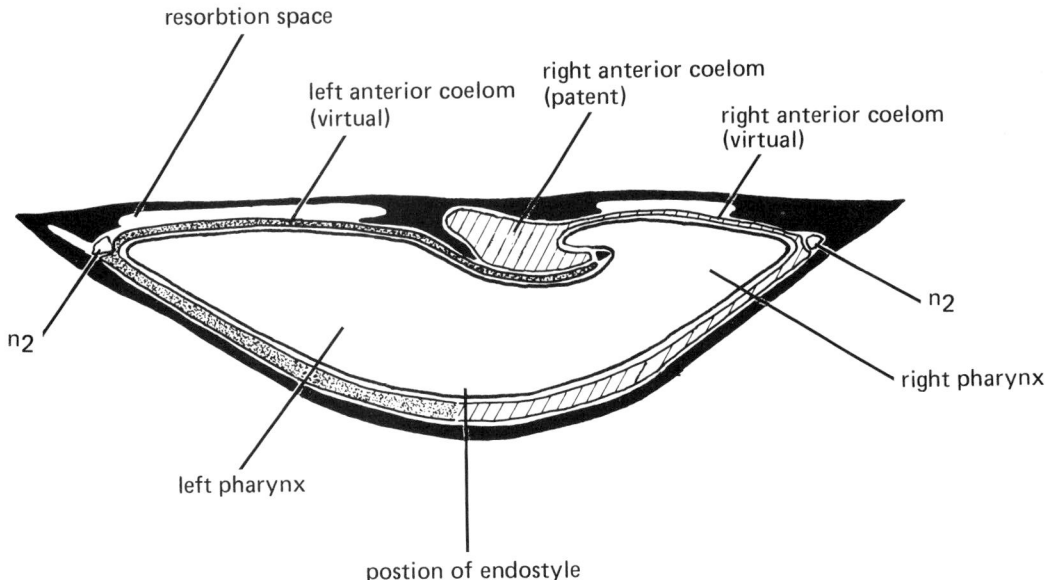

8.20. Placocystites forbesianus. *Transverse section of head at about half its length, with reconstructed soft parts; the animal's right is at the right. Note the reconstructed dorsal lamina, hooked downwards and rightwards; from Jefferies (1981a, fig. 30, p. 542).*

semi-circular notch, with rounded-off edges, in the right wall of the posterior coelom (fig. 8.19b,c). The position of this opening, well right of the mid-line, is comparable with that of tunicates (fig. 4.8a), being especially comparable with the oesophageal opening of the pelagic tunicates called salps.

A pair of canals, presumably for nerves n_0 like those of *Mitrocystella*, leaves the posterior coelom anteriorly (fig. 8.19b,e), just on either side of the mid-line, and passes forward into the middle layer of the ventral skeleton of the model based on transverse sections. These canals probably supplied the posterior end of the endostyle, for this is the only important organ in the mid-ventral line of the pharynx in tunicates, acraniates and ammocoetes. In the same model there is a U-shaped process (retropharyngeal process) in the front wall of the posterior coelom (fig. 8.19e). This is asymmetrical in position, being situated more over the right canal n_0 than over the left one. In the model based on longitudinal sections the canals n_0 cannot be found (fig. 8.19f). Presumably they existed in the living animal but after death became filled with secondary calcite and thus became untraceable. The retropharyngeal process in this second model is developed as a complicated blade, rooted to the floor of the head in the anterior wall of the posterior coelom but bending over strongly to the right. This blade has a notch in its tip, corresponding to the notch in the U-shaped process of the first model and also has a notch on either side near its base. Combining results

from both models, and assuming that the nerves n_0 supplied a bilaterally symmetrical endostyle, then the retropharyngeal process was rooted at the posterior right end of the endostyle and curved rightwards towards the presumed position of the oesophageal opening.

This suggests that the retropharyngeal process carried the retropharyngeal ciliated band on its front surface. For if so the band would have the same relations as in tunicates, running from the posterior right end of the endostyle towards the oesophageal opening. The notches of the base of the retropharyngeal process can then be interpreted as representing the pharyngo–epicardial openings, which in *Ciona* are situated on either side of the retropharyngeal band. And this implies that the posterior coelom of *Placocystites* is homologous with the right and left epicardia of tunicates, in agreement with the indications in *Mitrocystella* that it was divided obliquely into two chambers (fig. 8.13a). More precisely, the notches probably represent a more median pair of pharyngo–epicardial openings in *Placocystites*; the presence in *Placocystites* of a more lateral pair of such openings, not represented in tunicates, is suggested by details of the skeleton ventral to the rectal and oesophageal processes.

This homology, of the posterior coelom of mitrates with the right and left epicardia of tunicates, will prove important when discussing *Lagynocystis* and the acraniates below. It is also relevant in considering the posterior coelom of cornutes. For the epicardia of tunicates arise symmetrically, the left one pouching out in ontogeny from the left pharynx and the right one from the right pharynx (fig. 4.31). However, the cornutes had no right pharynx. It follows that the posterior coelom of cornutes can be homologous only with the left epicardium of tunicates and with only part of the posterior coelom of mitrates. For the right epicardium cannot have arisen in phylogeny before the right pharynx existed.

In the pharynx of *Placocystites* it is therefore possible to reconstruct (fig. 8.21a): right and left peripharyngeal bands; ciliated funnel; dorsal lamina, sloping downwards and rightwards as in tunicates; the oesophageal opening, right of the mid-line as in tunicates; the retropharyngeal band, coming off the right posterior corner of the endostyle and running to the right oesophageal opening as in tunicates; and the right and left pharyngo–epicardial openings, situated on either side of the tetropharyngeal band as in *Ciona*. All this shows that *Placocystites* fed like a tunicate and confirms the correctness of the basic interpretation.

The atria of *Placocystites* would have been situated at the right and left posterior corners of the head as in *Mitrocystella*. A notch in the left wall of the posterior coelom in the ventral skeleton (exit of rectum, fig. 8.19b,f) probably shows where the rectum left the posterior coelom to run to the left atrium, again in *Mitrocystella*.

A feature of *Mitrocystella* (fig. 8.22a,b,c) which remained unexplained above can now be interpreted. This is the low, straight ridge (cdl) on the dorsal surface of the natural moulds (fig. 8.13a) which runs from the oblique groove posteriorly across the right field to near the right-hand corner of the natural mould. This ridge, by analogy with *Placocystites*, probably represents the cavity of the dorsal lamina. In *Mitrocystella*, however, this lamina seems to have been longer than in *Placocystites* and to have been fixed at its posterior end to the posterior wall of the natural mould above the right atrium. A short curved groove on the natural mould (trdl, in fig. 8.13b) of one specimen probably marks the place of this attachment. On this interpretation the oesophagus would need to open into the pharynx directly into the cavity of the dorsal lamina, and therefore further rightwards than in *Placocystites*. A polystyrene model of *Mitrocystella* suggests that this was probably so.

The organs in the cavity of the right anterior coelom of *Placocystites* (fig. 8.21b) can be deduced partly from direct evidence and partly by comparison with tunicates and living vertebrates. The direct evidence comes from two calcitic processes, built from inner layer, that extend upwards and forwards from the rim that limits the posterior coelom anteriorly in the ventral skeleton. These two processes can be called the rectal process on the left and the oesophageal process on the right (fig. 8.19b). Their position suggests that they would have stiffened the left

and right ventral edges of the cavity of the right anterior coelom. Their surfaces facing the posterior coelom are especially significant, particularly near their bases which are situated just above the places where the left and right palmar nerves passed from the posterior coelom into the soft layer of the ventral skeleton (fig. 8.19c). For these surfaces are concave, presumably to receive the supra-alimentary components of the palmar nerves after these had passed over the rectum on the left and the oesophagus on the right. The right such concavity is anterior to the oesophageal opening (in the right wall of the posterior coelom) (fig. 8.19c) while the left concavity is anterior to the rectal opening (in the left wall of the posterior coelom) (fig. 8.19g). And this implies that the oesophagus passed forwards from the oesophageal opening into the cavity of the right anterior coelom, squashing the supra-oesophageal nerve component against the front wall of the posterior coelom on the right, while the rectum passed forwards from the rectal opening into the cavity of the right anterior coelom, squashing the supra-rectal nerve component against the front wall of the posterior coelom on the left. The loop of the non-pharyngeal gut (fig. 8.21b), therefore, was situated in the cavity of the right anterior coelom as it likewise presumably was in cornutes. The rectal process would support the rectum or intestine on the left, while the oesophageal process would support the oesophagus on the right.

The ciliated funnel of *Placocystites*, as already argued, probably opened into the left pharynx at the anterior end of the cavity of the right anterior coelom, just behind the origin of the mid-dorsal process. In tunicates this funnel is the opening of the duct of the neural gland and the gland itself is closely associated with the ganglion of the adult and with the brain of the tadpole (fig. 4.20). Indeed in the primitive tunicate condition the gland is ventral to the ganglion or brain as it is in *Ciona*. By analogy, the neural gland of *Placocystites* would likewise be ventral to the brain. It was presumably located just beneath the gap between the right and left hypocerebral processes. This, also, is where we should expect it to be on vertebrate comparisons if the neural gland is homologous to the adenohypophysis, for in living vertebrates this always lies ventral to the prosencephalon. From the neural gland the duct would run forwards, through the posterior coelom and the cavity of the right anterior coelom, to open at the ciliated funnel. Notably enough, the ciliated funnel of *Placocystites* apparently opened, as in tunicates, into the ceiling of the pharynx. Unlike Rathke's pouch, which in most vertebrates is the anlage of the adenohypophysis, it did not emerge in the buccal cavity. Its position is more like the endodermal, pharyngeal Seessel's pouch (pre-oral gut) of vertebrates, which represents the anlage of the adenohypophysis in myxinoids, and the endodermal

a

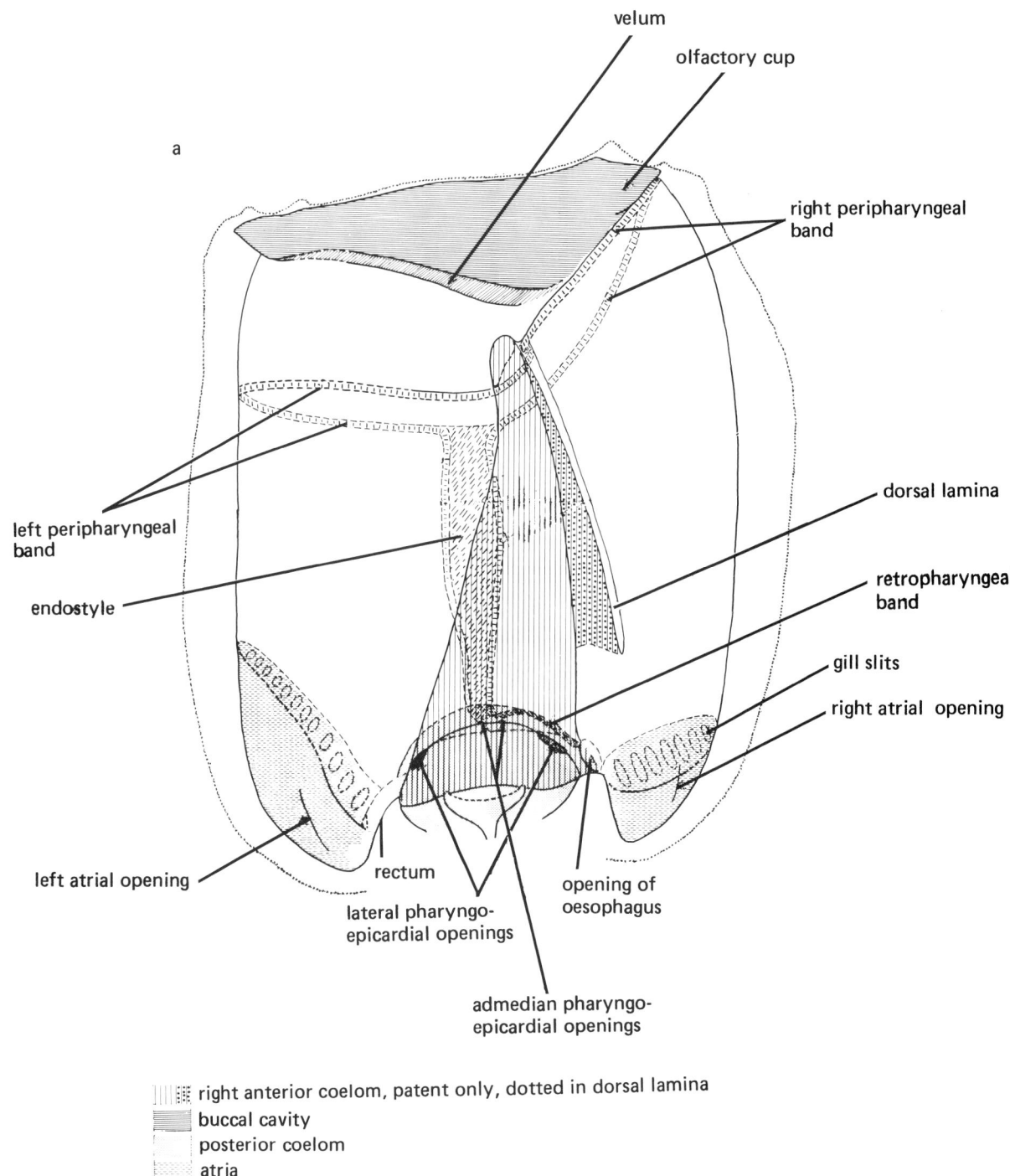

velum

olfactory cup

right peripharyngeal
band

left peripharyngeal
band

endostyle

dorsal lamina

retropharyngeal
band

gill slits

right atrial opening

left atrial opening

rectum

opening of
oesophagus

lateral pharyngo-
epicardial openings

admedian pharyngo-
epicardial openings

right anterior coelom, patent only, dotted in dorsal lamina
buccal cavity
posterior coelom
atria

Fig. 8.21.

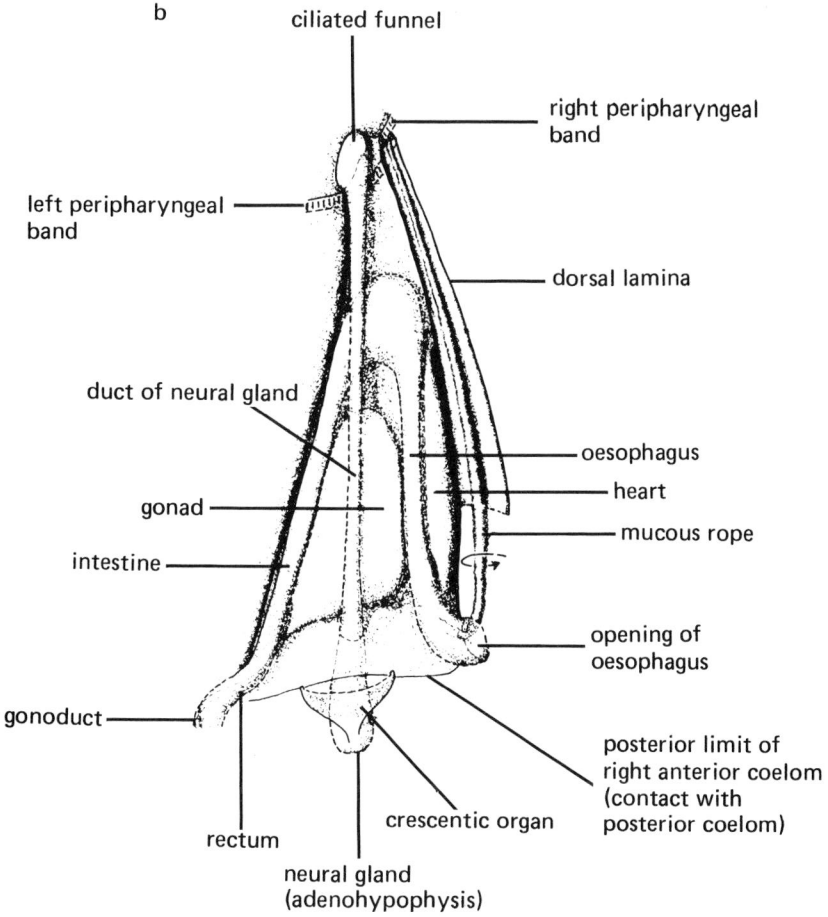

b

ciliated funnel

right peripharyngeal band

left peripharyngeal band

dorsal lamina

duct of neural gland

oesophagus

heart

gonad

mucous rope

intestine

opening of oesophagus

gonoduct

posterior limit of right anterior coelom (contact with posterior coelom)

rectum

crescentic organ

neural gland (adenohypophysis)

8.21. Placocystites forbesianus. *Reconstructed soft parts of head in dorsal aspect. a) head in general; b) contents of right anterior coelom; modified from Jefferies & Lewis (1978, fig. 18, p. 259).*

anterior gut diverticula of amphioxus. However, before concluding that the adenohypophysis of mitrates was endodermal it must be remembered that in tunicates the neural gland and its duct and the ciliated organ arise from brain anlage.

The gonad of *Placocystites* was probably also located in the cavity of the right anterior coelom, in the loop of the gut, for this is its primitive enterogonous position in tunicates (e.g. *Ciona*, fig. 4.17a, p. 100) – this location is confirmed by the gonorectal groove of *Mitrocystella* (if correctly identified), which emerges from the anterior part of the cavity of the anterior coelom, to run rearwards across the ceiling of the left pharynx to the left atrium (fig. 8.13). The heart probably lay just right of the oesophagus and gut loop, where it is in ascidians such as *Ciona* (fig. 4.18, p. 100).

Thus the cavity of the right anterior coelom of *Placocystites* probably contained, as deduced also for

cornutes, the non-pharyngeal gut, the gonad, the heart and the duct of the neural gland.

The relative positions of the organs in the right anterior coelom of cornutes can now be suggested, by mentally subtracting the right pharynx from the mitrate head (fig. 7.6, p. 200). This subtraction would allow the right anterior coelom to slide rightwards and fall to the floor of the head, into its cornute position, and the gut loop would be displaced rightwards, running around the 'heel' region of the boot-shaped head of such a cornute as *Cothurnocystis*. The opening of the oesophagus into the pharynx would not be moved by subtracting the right pharynx, but would remain at the right extremity of the posterior coelom. It follows that the oesophagus of cornutes would have run rightwards from this position, instead of forwards as in mitrates. If the heart and pericardium were right of the oesophagus in mitrates, then the 90° rotation implied by subtracting the right pharynx

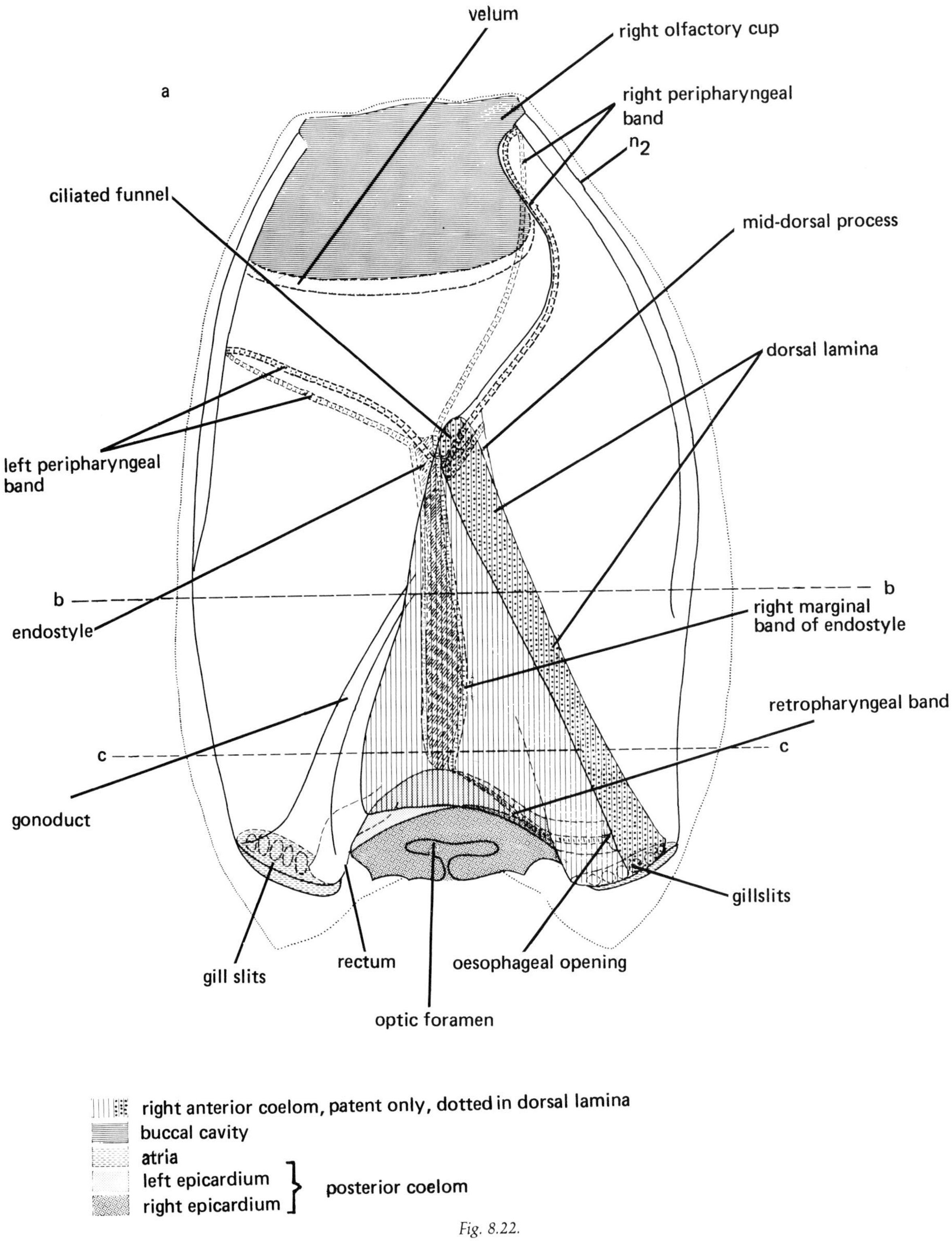

a

velum

right olfactory cup

right peripharyngeal band

n_2

ciliated funnel

mid-dorsal process

dorsal lamina

left peripharyngeal band

b ————————————————————— b

endostyle

right marginal band of endostyle

c ————————————————————— c

retropharyngeal band

gonoduct

gillslits

gill slits

rectum

oesophageal opening

optic foramen

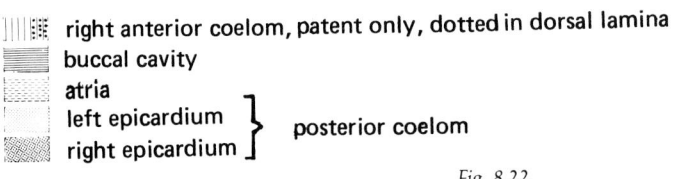

||| right anterior coelom, patent only, dotted in dorsal lamina
buccal cavity
atria
left epicardium }
right epicardium] posterior coelom

Fig. 8.22.

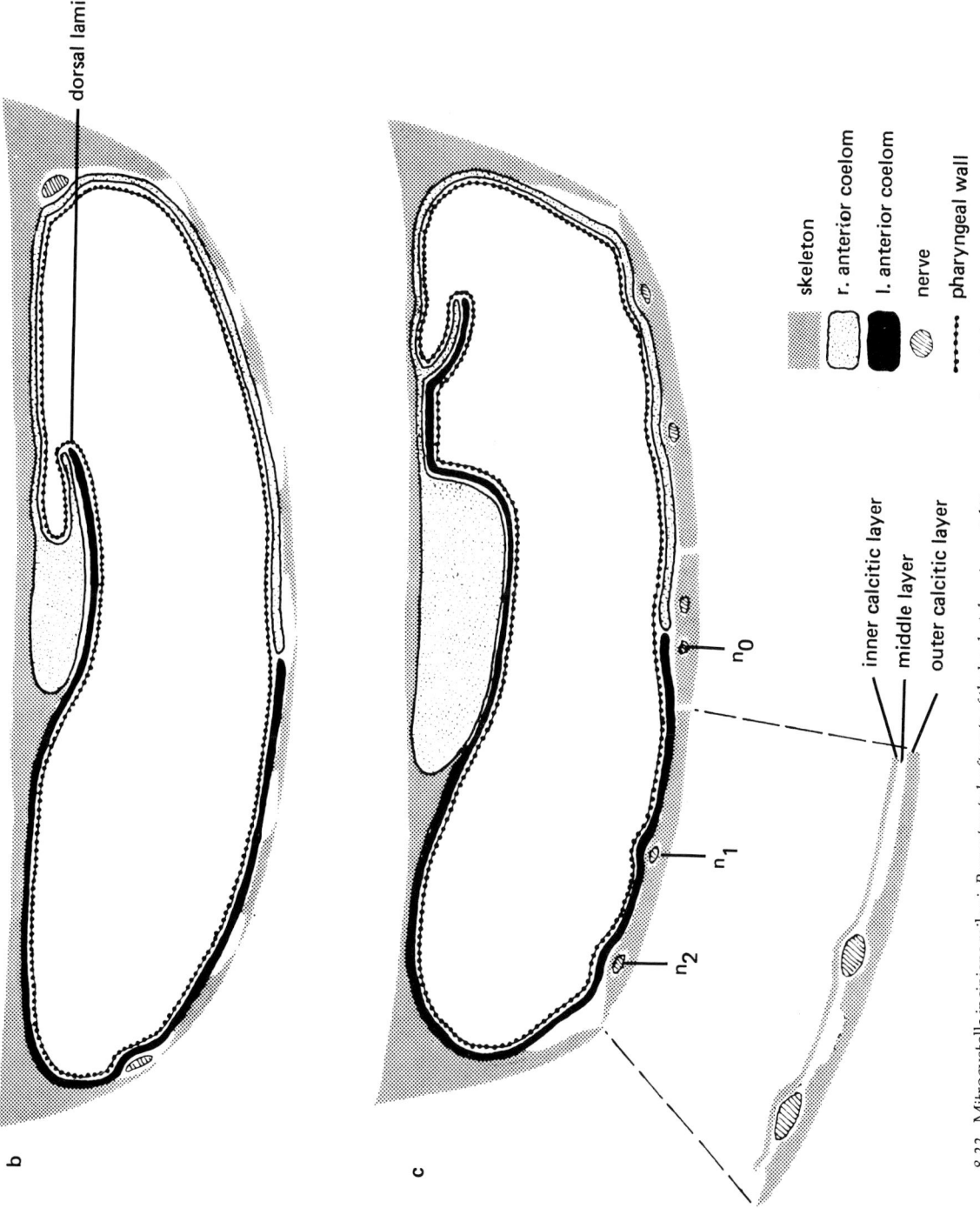

8.22. Mitrocystella incipiens miloni. *Reconstructed soft parts of the head; a) dorsal aspect; b) transverse section at b–b in a; c) transverse section at c–c in a; modified from Jefferies & Lewis (1978, fig. 20, p. 263).*

would bring it posterior to the oesophagus in cornutes (as suggested in figs 7.4, 7.7). Indeed, as already suggested in Chapter 7 heart and pericardium probably occupied a spindle-shaped hollow in the internal face of the marginals right of the tail, for the following reasons: (1) topologically, as just explained, this would be the same position as deduced for the heart of mitrates, (2) the pericardium and heart of tunicates is spindle-shaped, and (3) the hydropore of *Ceratocystis* penetrates the marginal frame in a comparable position right of the tail (fig. 7.17c) – and the hydropore of echinoderms is associated with the madreporic vesicle and head process of the axial gland (fig. 7.22, p. 220) which are homologous respectively with the pericardium and heart of hemichordates and chordates. The gonad of cornutes would probably be situated inside the gut loop and right of the rectum by comparison with its deduced position in mitrates (and its known primitive position in tunicates). This position for the gonad is somewhat confirmed by the position of the gonopore in *Ceratocystis*, right of the anus. These suggested positions of the contents of the right anterior coelom of cornutes are necessarily uncertain, but fit the known facts.

The position of the classical vertebrate head segments in mitrates can now be discussed (fig. 8.23). As already mentioned, the velum of *Placocystites*, where attached to the ceiling of the head, seems to have lain entirely in the field of the left anterior coelom. (There is no evidence as to its ventral relationships.) The velum would have been penetrated by the velar mouth. This asymmetry recalls the strange fact, mentioned in Chapter 3, that the larval mouth of amphioxus, which becomes the velar mouth of the adult, first penetrates the body wall in the ventral part of the left mandibular somite (fig. 3.30, p. 77). (The larval mouth of amphioxus later expands posteriorly into the territory of other left somites, but, if Haeckel's law applies, its initial penetration of the left mandibular wall will represent its primitive condition.) This asymmetry shared by *Placocystites* and amphioxus suggests that the left anterior coelom of *Placocystites* and other mitrates may be homologous with the left mandibular somite of amphioxus and therefore of living vertebrates. If so, then the right anterior coelom of mitrates would be homologous with the right mandibular somite. These suggested homologies are reasonable on direct vertebrate comparisons, for the muscles of the mouth and velum are formed in living vertebrates from the left and right mandibular somites, whereas in mitrates, to judge by the observed relations in the fossils, they were presumably formed mainly from the left one, as in amphioxus. To explain the differences of living vertebrates from mitrates it is only necessary to suppose that a process of symmetrization has occurred in them which did not happen in mitrates nor acraniates. Also in support of the suggested homologies is the fact that the

presumed mandibular somites of mitrates, like those of living vertebrates, were supplied by the paired trigeminal nerves.

The crescentic body of *Placocystites* and *Mitrocystella* was unpaired, situated immediately in front of the prosencephalon and therefore anterior to the notochord, dorsal to the likely position of Seessel's pouch (if this is equivalent to the duct of the neural gland) and median to the bases of the optic nerves (assuming that the left optic nerve in *Placocystites* behaved as in *Mitrocystella*). In all these respects it resembles the premandibular somite of the early embryos of living vertebrates and is therefore probably homologous with it. It may be objected that the premandibular somite of living vertebrates, at its first ontogenetic origin, is anterior to the mandibular pair of somites, not behind it like the crescentic body (figs 6.6a, 6.15, pp. 168, 175). Indeed, it is the only somite which arises in front of the notochord. This difference from the presumed mitrate condition, however, does not persist embryologically. For the mandibular pair of somites of living vertebrates, which is destined to form the muscles of the mouth and velum, grows forward on either side of the premandibular somite, which will give rise to most of the oculo-motor muscles (figs 6.9b, 6.16, pp. 171, 176). In the lamprey for example, the mandibular somites extends in front of the premandibular somite from the 20-somite stage onward, and this resembles the presumed mitrate condition.

The hyoidean pair of somites in living vertebrates is located immediately behind the mandibular somites. Its likely homologue in the mitrates was the first pair of muscle blocks in the tail, for these would be the first somites behind the left and right anterior coeloms. Moreover, they would be situated median to the right and left atria, which contained the ears and were therefore probably homologous with the optic capsules. In the same way the hyoidean somites of living vertebrates, together with the anterior part of the first pair of branchial somites behind them, are median to the otic capsules. Indeed the hyoidean somites and the anterior part of the first branchial somites are squashed out of existence by the median growth of these capsules as explained in Chapter 6 (figs 6.6a, 6.17, pp. 168, 176). The second pair of tail somites would be homologous with the first branchial somites of living vertebrates; the third pair with the second pair of branchial somites, and so on; and the more posterior tail somites of mitrates would be equivalent to the trunk and tail somites of living vertebrates.

There was no trunk region in mitrates, nor any notochordal part of the head. The gill slits were located entirely in the head, in mandibular territory. It follows that there could have been no alternation between somites and gill slits, unlike the embryos of living vertebrates. This

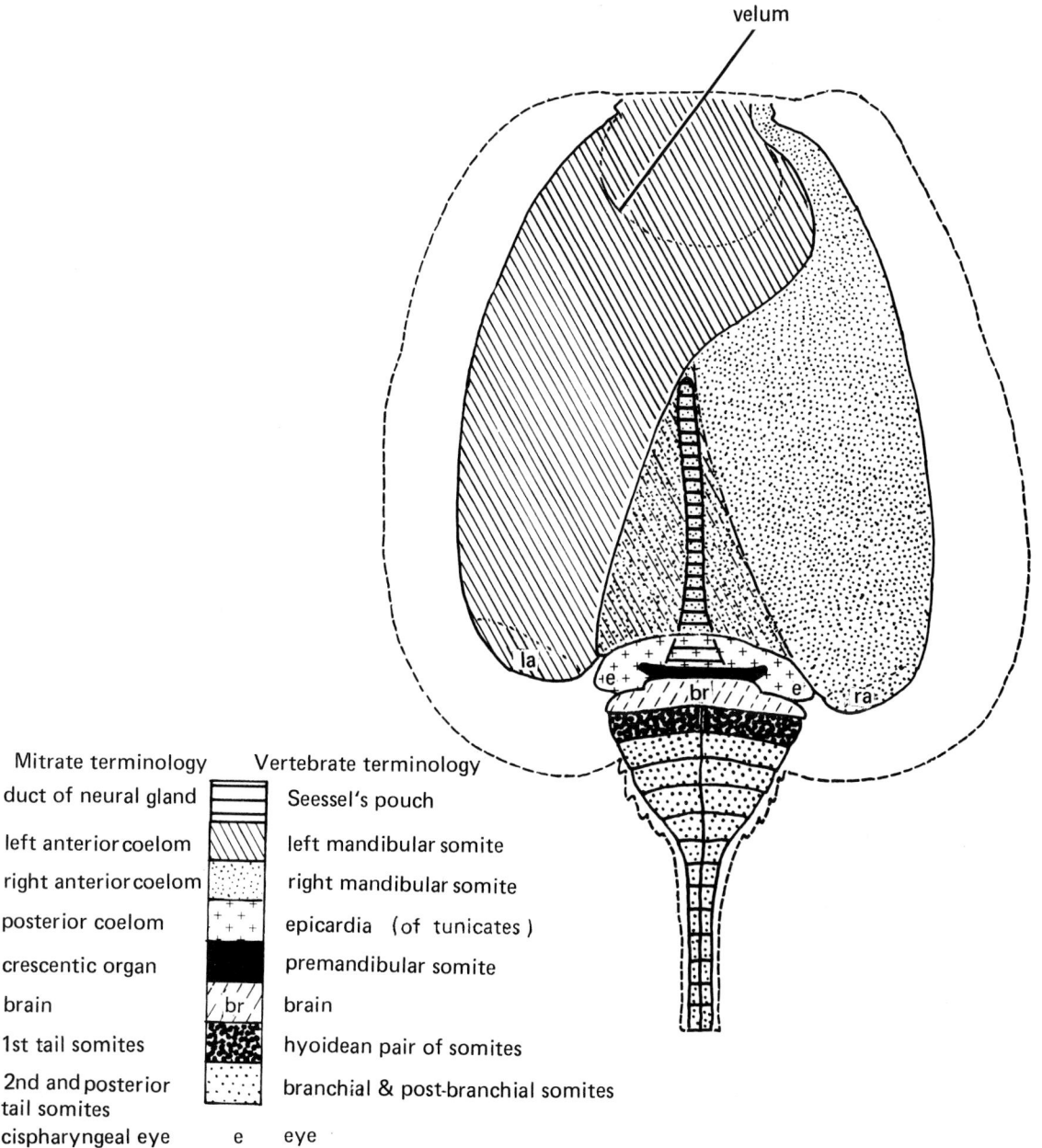

velum

Mitrate terminology		Vertebrate terminology
duct of neural gland	☰	Seessel's pouch
left anterior coelom	⧅	left mandibular somite
right anterior coelom	∴	right mandibular somite
posterior coelom	+ + +	epicardia (of tunicates)
crescentic organ	■	premandibular somite
brain	br	brain
1st tail somites	▨	hyoidean pair of somites
2nd and posterior tail somites	∴	branchial & post-branchial somites
cispharyngeal eye	e	eye

8.23. *The arrangement of vertebrate head segments in a mitrate, in dorsal aspect; la = left atrium, ra = right atrium; from Jefferies (1981a, fig. 33, p. 547).*

alternation could only have arisen, in post-mitrate stem vertebrates, by the rearward extension of the head beneath the anterior part of the tail, so that the somites in question could grow ventralwards each down a gill bar as the hyoidean somite does in *Lampetra* (fig. 6.16). The

precise nature of the transition from mitrates to crown vertebrates, however, will be discussed in the next chapter. Thus the old intuition, expressed in the quotation from Goodrich in Chapter 6 (p. 186), that vertebrates have an important landmark at the transverse level of the

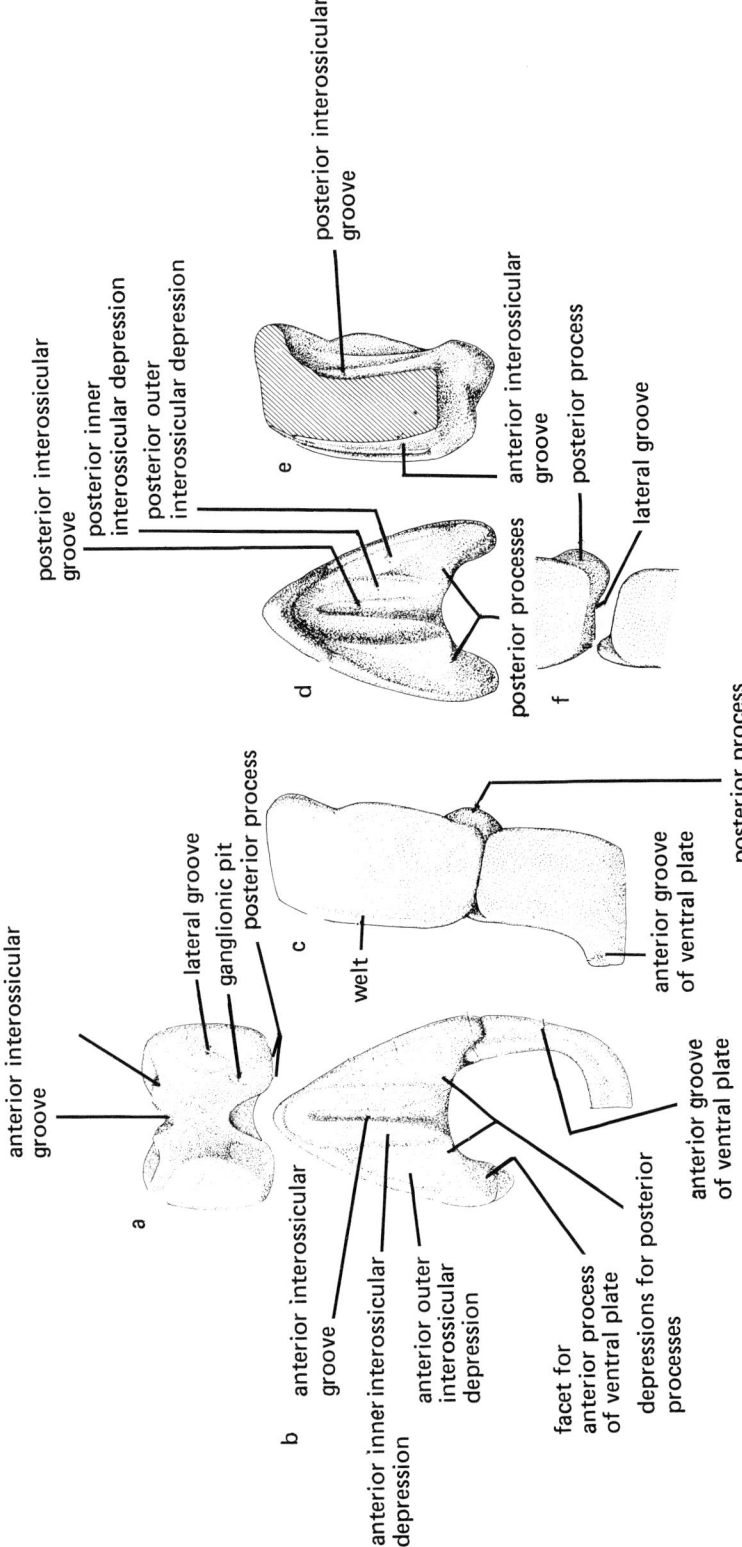

8.24. Placocystites forbesianus. *Reconstructed skeleton of hind tail (a dorsal ossicle and a left ventral plate). a) Ventral aspect of ossicle, b) anterior aspect, c) left lateral aspect of ossicle and plate, d) posterior aspect of ossicle, e) sagittal section of ossicle, and f) disarticulated junction of ossicle and plate; from Jefferies & Lewis (1978, fig. 6, p. 228).*

hypophysis and of the front end of the notochord, can now be explained phylogenetically. It corresponds to the junction between mitrate head and tail.

The eye relationships of the anterior mitrate somites are of interest. In gnathostomes and lampreys the mandibular somites give rise to the superior oblique oculo-motor muscle pair (called posterior oblique in lampreys), the hyoidean somites produce the posterior (external) rectus muscle pair, and the premandibular somites produce all the other oculo-motor muscles. The relationships of their mitrate equivalents with the proximal, cispharyngeal portions of the optic nerves and cispharyngeal eyes (*e* in fig. 8.23), which are probably homologous with the optic nerves and eyes of living vertebrates, suggest how this situation arose. For the mandibular somites (right and left anterior coeloms) are antero–lateral to the cispharyngeal parts of the optic nerves, as the superior obliques are to gnathostome eyes; the hyoidean somites (first pair of tail somites) are posterior to the cispharyngeal parts of the optic nerves, as the posterior (external) recti are to the eyes of gnathostomes; and the premandibular somite (crescentic body) is median to the cispharyngeal optic nerves (figs 8.16a, 8.25a), as the remaining oculo-motor muscles are to the eyes of gnathostomes. The mitrate condition, therefore, could easily be converted into that of living lampreys and gnathostomes.

The tail of *Placocystites* has already been described in its external features. Internally it does not much differ from *Mitrocystites* or *Mitrocystella* except that there is no dorsal longitudinal canal in the mid and hind tail (fig. 8.24). Vertical interossicular canals exist but they end blind in the inner deep, presumably muscular, parts of the interossicular depressions, instead of being connected together by a horizontal canal through the ossicles. This canal is likewise absent in the stem-acraniate mitrate *Lagynocystis*, the stem-tunicate mitrate *Peltocystis* and the primitive stem-vertebrate mitrate *Chinianocarpos*. Its absence in *Placocystites* is therefore probably a primitive feature. Another difference between *Placocystites* and *Mitrocystites* or *Mitrocystella* is the lack of a median groove in the ventral faces of the styloid and hind-tail ossicles (fig. 8.24a). Presumably the notochord and dorsal nerve cord were present but were not impressed in the skeleton. A pair of pits in each ossicle indicate the presence of paired segmental spinal ganglia.

The soft parts of the fore tail can partly be reconstructed. I have already argued that the mitrate fore, mid and hind tails are not homologous with their namesakes in cornutes. For the cornute mid and hind tails were probably lost in the early evolution of mitrates, lopped off by the autotomy which was a general feature of cornutes and mitrates, and the three mitrate tail regions were then regionated in the remaining stump. It follows that there

ought to be a continuous fundamental metamery extending throughout the mitrate tail, corresponding to, and derived from, the obvious metamery of the cornute fore tail. This supposition agrees with the observation that major plates alternate with plated imbrication membranes throughout the entire tail of *Chinianocarpos* (fig. 8.26). Soft parts deduced for the mitrate hind tail should therefore have existed in the fore tail also, if they did not arise during the mitrate tail regionation. In the mitrate fore tail we should therefore expect notochord, dorsal nerve cord and spinal ganglia alternating with muscle blocks. These blocks probably filled almost all of the lumen of the fore-tail skeleton and would have needed, as already said, the anti-compressional notochord to push against. The number of muscle blocks in the fore tail was probably equal to the number of rings of major plates in the skeleton (about seven in *Placocystites*) as it seems to have been in the hind tail. There would have been paired segmental ganglia between successive muscle blocks and also a pair anterior to the foremost, hyoidean pair of muscle blocks, between it and the brain. From in front rearwards, these would correspond to the facial, glossopharyngeal and the several pairs of vagal ganglia of living vertebrates insofar as these are derived from neural crest and not from epibranchial placode (as suggested for *Mitrocystella* in fig. 8.16a).

The brain and cranial nerves of *Placocystites* are also like those of *Mitrocystella* in most ways. They have been reconstructed partly (fig. 8.25a–d) from the polystyrene models, partly by dissection (of the cerebral basin only) and partly by means of glass-plate models of the nerves based on the serial sections.

In the brain, the prosencephalon was less inflated than in *Mitrocystella*, presumably because vision was less important to the continually burrowing animal. In one specimen the prosencephalon is divided into a flattened dorsal telencephalon and a more inflated diencephalon round the optic foramen (fig. 8.25d). There is no sign of the precerebral knot. Olfactory and terminalis fibres presumably gathered in the dorsal mid-line before entering the telencephalic part of the prosencephalon, but in *Placocystites* there is no direct evidence of this.

In front of the brain the right optic nerve can be followed latero-dorsally out of the optic foramen and leads to a swelling, presumably the optic vesicle of the right cispharyngeal eye, at the top right corner of the posterior coelom (corresponding to a groove and a depression in fig. 8.19d). There is no direct evidence of the course of the left optic nerve, though such presumably existed since the optic foramen is bifid and bilaterally symmetrical (fig. 8.19d). In the reconstructions it has been given the same course as in *Mitrocystella* and *Mitrocystites*. Also, there is no evidence, more peripherally, of transpharyngeal optic nerves, so it is uncertain whether or not the optic nerve

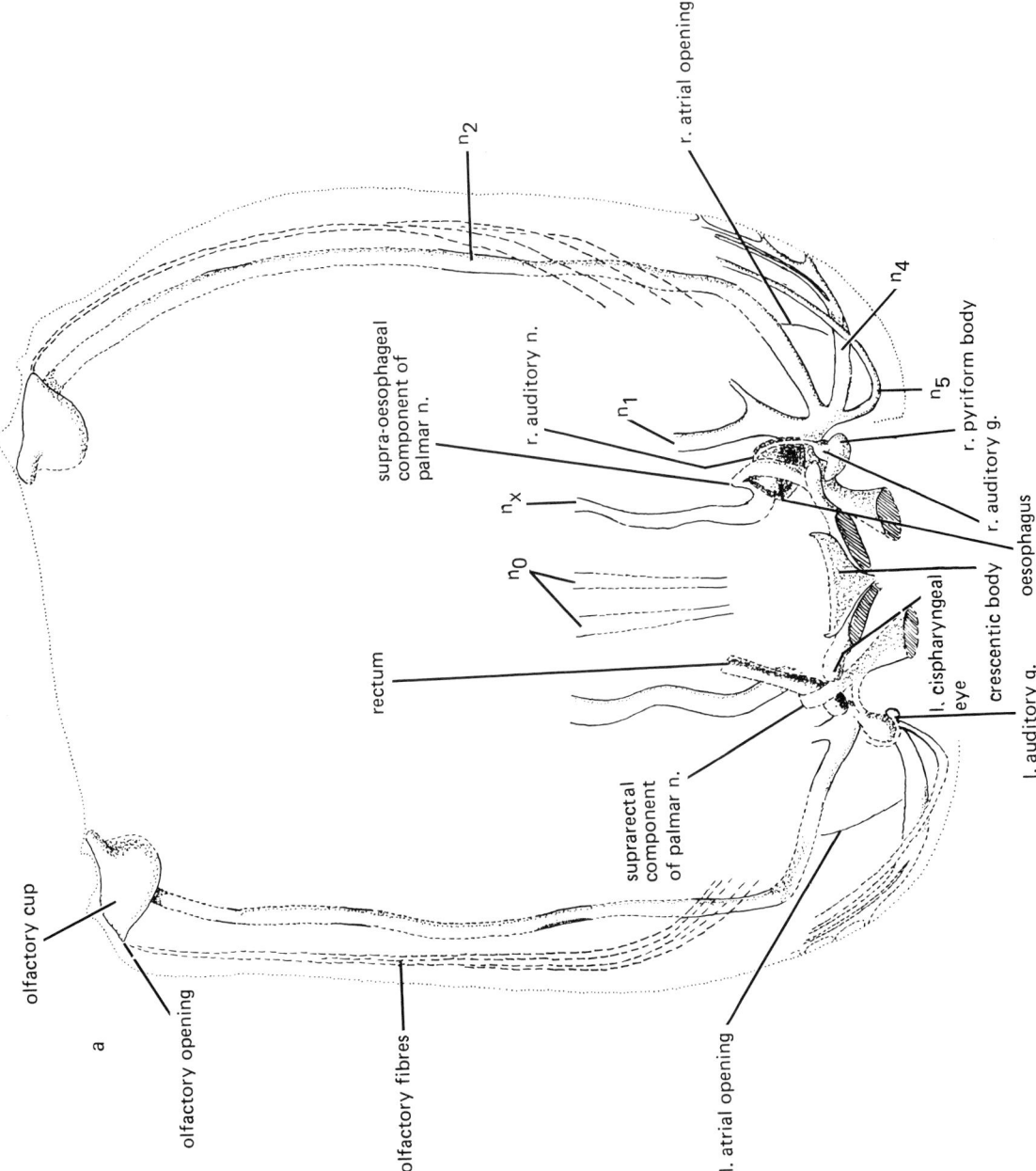

Fig. 8.25.

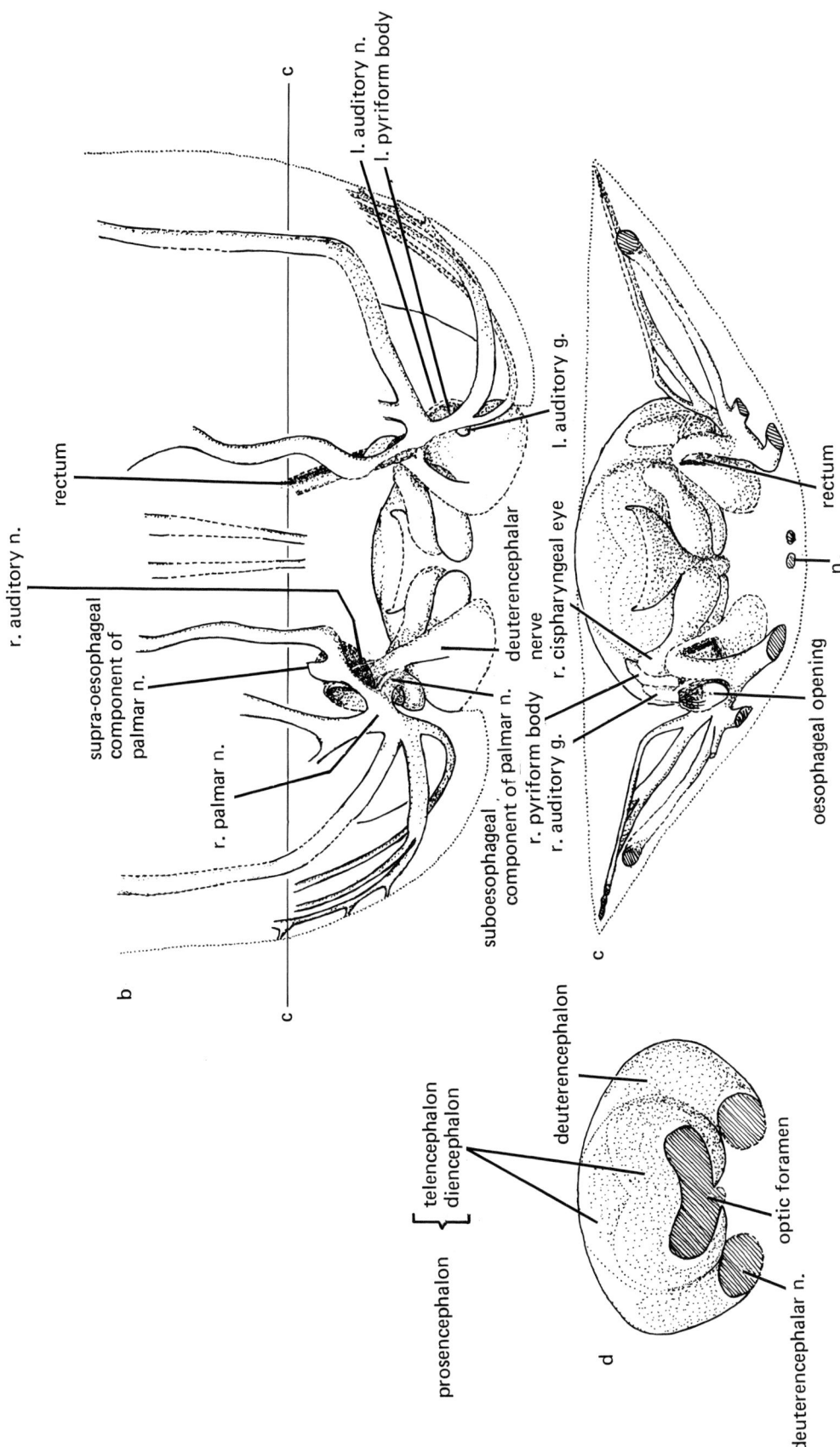

8.25. *Placocystites forbesianus. Reconstruction of brain and cranial nerves. a) Dorsal aspect, brain removed. b) ventral aspect of brain and posterior part of head, c) anterior aspect of brain and posterior part of head, and d) anterior aspect of brain only; from Jefferies & Lewis (1978, fig. 25, p. 279).*

descended the front wall of the posterior coelom (with the supra-alimentary components of the trigeminal) to contribute to the palmar nerves.

The palmar nerve complexes of *Placocystites* resemble those of *Mitrocystella* in having mandibular trigeminal (n_1), maxillary trigeminal (n_2) and dorsal, presumably touch-sensory branches (n_4, n_5). However, as just mentioned, there is no sign of the transpharyngeal optic nerves (n_3). Also there was a large additional canal presumed to be for a nerve trunk (n_x), which ran forwards as a zig-zag in the middle layer of skeleton (fig. 8.25a). The dorsal branches of the palmar complex ($n_4 + n_5$) run through a canal in the ventral skeleton and enter the dorsal skeleton behind the atrial opening. Here they separate into a thick lateral n_4 and a thin more medial n_5 which continue forward just beneath the surface of the skeleton. These nerves were presumably sensory to judge by their superficial position (at least, being enclosed in plates they could not have been, in the strict sense, motor). The nerve n_5, on the evidence of glass models and X-ray photographs, innervated a region of the dorsal surface more median (in terms of living vertebrates more dorsal) than n_4, and the latter was connected at several points with the peripheral flange. This suggests that n_5 may be equivalent to the ophthalmicus superficialis branch of the trigeminal of living vertebrates and n_4 to the ophthalmicus profundus.

The paired palmar nerves, as in *Mitrocystella*, received subalimentary and supra-alimentary components. The right trigeminal ganglion and right acoustic ganglion occupied a skeletal cup in the transversely sectioned specimen, but there was no corresponding cup on the left, while the longitudinally sectioned specimen had cups neither at left nor right. To judge by the skeletal cup, the right acoustic ganglion received its nerve supply from in front through a slit, probably from off the dorsal surface of the oesophagus. The left auditory ganglion is suggested in the transversely sectioned model by a vertically elongate cavity in the skeleton median to the left atrium and connected with that atrium by a skeletal slit. In the absence of a lateral line, there was neither lateral-line ganglion nor nerve.

The paired canals for n_0, presumably supplying the endostyle, have already been mentioned. Their central connections are obscure, though in ammocoetes the endostylar nerve is a branch of the facial (Alcock, 1898). There is some sign in the skeleton that they continued forward a short distance parallel to each other, though they are not continuously traceable.

The anterior ends of the olfactory nerves are associated with well-developed skeletal cups right and left of the mouth at the lateral extremities of the buccal cavity (figs 8.19a, 8.25a). Each cup has probable inhalent and exhalent openings. The cups terminate laterally in conical

depressions which probably mark where the olfactory fibres entered the skeleton as in *Mitrocystites* or *Mitrocystella*. From these depressions they would have passed rearwards to the brain through the stereom mesh, as already suggested.

Thus *Placocystites* was in most ways like *Mitrocystella* but supplies important additional evidence. In particular it complements the picture of the nervous system, makes possible a detailed comparison with the pharynges of tunicates and allows the classical vertebrate head segments to be recognized.

The next mitrate to be described is *Chinianocarpos*, only known from the type species *Chinianocarpos thorali* Ubaghs (fig. 8.26) from the Lower Ordovician (lowest Arenig stage) of the Montagne Noire in the south of France. It has been described by Ubaghs (1961a, 1961b, 1967, 1970, 1981) with his usual thoroughness and according to his preferred interpretation with the tail as an arm. I have restudied it myself and hope to publish a fuller account elsewhere, but here I shall discuss only its external features. I differ from Ubaghs in some factual details as well as in gross interpretation. Unfortunately only six specimens of this important species are known.

The head of *Chinianocarpos* is only about 10 mm long, and therefore smaller than in most mitrates. It has a flat dorsal and a convex ventral surface. It is framed by marginal plates which carry a denticulate, dorso–lateral peripheral flange, and the denticulations coincide with short cuesta-shaped ribs near the right and left margins of the dorsal surface. As always, the cuesta-shaped ribs are steeper anteriorly than posteriorly (fig. 8.26c).

As to the head openings, the lower lip of the mouth was rigid. The upper lip was largely formed of a big plate (n) which, being carried by plated integument, could presumably lift up from the mouth like a lid. Also on the dorsal surface were openings for paired nerves n_3 situated much as in *Mitrocystites* but leading to the posterior ends of elongate depressions on the dorsal surface which presumably carried large paired transpharyngeal optic vesicles (eyes). Unlike *Mitrocystites*, *Chinianocarpos* does not have paired openings on the dorsal surface for nerves n_{4+5} of the palmar complex. However, there are notches in the peripheral flange situated just in front of the right and left posterior angles of the head and straddling the sutures M_{1LD}/M_{2L} and M_{1RD}/M_{2R}, and thus resembling in their general situation the openings for n_{4+5} in *Mitrocystites*. Moreover, those notches are located at the lateral end of peripheral grooves in the dorsal surface and surmount vertical grooves, in the ventro–lateral surfaces, which climb up to the notches from the right and left atrial openings. These vertical grooves almost certainly carried homologues of the nerves n_{4+5} of *Mitrocystites*, while the peripheral grooves of *Chinianocarpos* would be

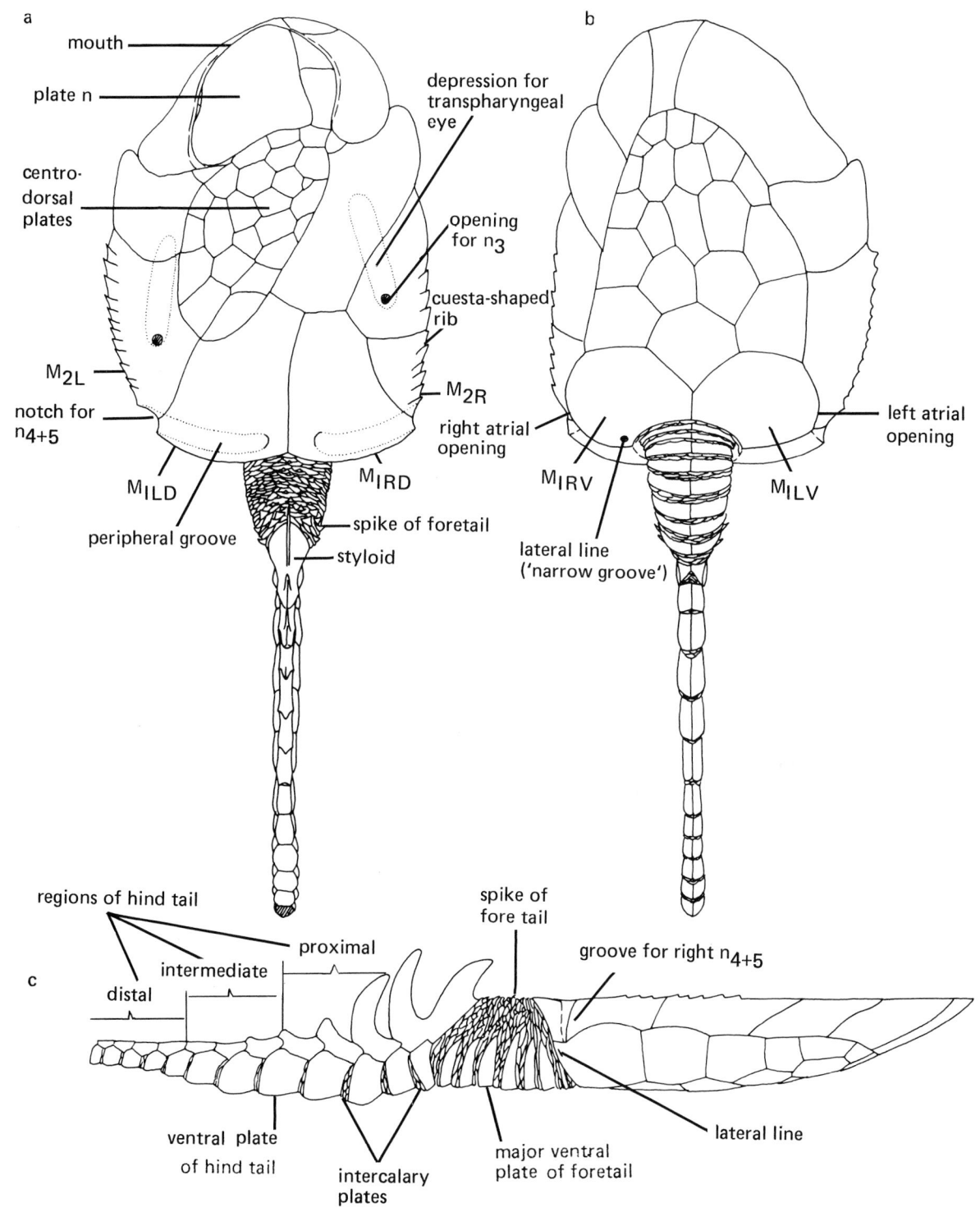

a

mouth

plate n

centro-
dorsal
plates

M_{2L}

notch for
n_{4+5}

M_{ILD}

peripheral groove

depression for
transpharyngeal
eye

opening
for n_3

cuesta-shaped
rib

M_{2R}

right atrial
opening

M_{IRD}

spike of foretail

styloid

b

left atrial
opening

M_{IRV}

lateral line
('narrow groove')

M_{ILV}

c

regions of hind tail

distal

intermediate

proximal

ventral plate
of hind tail

intercalary
plates

spike of
fore tail

groove for right n_{4+5}

lateral line

major ventral
plate of foretail

8.26. Chinianocarpos thorali. *Reconstruction of external features: a) Dorsal aspect, b) ventral, and c) right lateral. (New.)*

homologous with those parts of the peripheral grooves of *Mitrocystites* that lie medial to the openings of n_{4+5} (fig. 8.1a, p. 240). And the associated notches of the peripheral flange therefore correspond to the openings of n_{4+5} on to the dorsal surface of *Mitrocystites*. *Chinianocarpos* thus seems to have had dorsal touch-sensory branches of the trigeminal complex but in very simple form where the nerves n_4 are not separable from n_5.

On the ventral surface were the right and left atrial openings. They have the normal mitrate structure – the postero–ventral marginals (M_{1LV}, M_{1RV}) were hinged on either side to the first dorsal marginals (M_{1RV}, M_{1RD}) about a pair of approximately horizontal sutures and overlap the adjacent plates (M_{2L}, M_{2R}) on either side. When the postero–ventral marginals slightly rotated downwards and rearwards about the horizontal hinge, the atrial openings would gape beneath the overlaps, allowing water to escape. Then the postero–ventral marginals would rotate upwards and forwards and the gapes would shut. As already mentioned, the atrial openings would also give passage, by their dorsal edges to the nerves n_{4+5}. Another opening on the ventral surface is a small circular pit, about 70 μm across, situated on the right postero–ventral marginal (M_{1RV}) near the tail insertion. In position it corresponds to the dorsal end of the 'narrow groove' of *Mitrocystites* and *Mitrocystella* and therefore, as already argued for these forms, represents the lateral line. Its size and circular shape suggest that it contained only a single neuromast.

The dorsal surface of the head shows an elliptical area of flexible plated integument. This is situated left of an oblique ridge on the internal surface of the dorsal skeleton and therefore represents most of the roof of the left pharynx. Flexibility of the head roof is a primitive feature among mitrates, being cornute-like and reminiscent, in particular, of *Reticulocarpos hanusi* and *R. pissotensis*.

The ventral surface of the head is largely formed of irregularly plated, stiffish, ventral integument, as in *Mitrocystites*. Anteriorly, two of the marginal plates join to form an inverted arch beneath the mouth. And another plate, presumably by origin an anterior plate of the ventral integument, is held rigidly between them anteriorly. These three plates together form the rigid lower lip.

The tail of *Chinianocarpos* has already been mentioned in comparison with *Reticulocarpos hanusi*. The plating of the fore tail consists of major plates, intercalary plates and small spike-shaped plates. The major plates are in a right and left ventral row, joined their antimeres at median sutures, and number about eight pairs. Successive major plates are connected together by rather loose membranes covered with intercalary plates. Dorsally there are no major plates in the fore tail but, arranged among the intercalary plates there are small spike-shaped plates.

The plating of the fore tail is therefore much like that of *Lagynocystis* described later, except that the spike-shaped plates of *Chinianocarpos* are fewer and less prominent.

The mid-tail skeleton comprises a massive two-bladed styloid dorsally, to which two pairs of ventral plates are articulated. Both blades of the styloid are tall and recurved towards the head. The anterior edges of the blades are rounded while the dorsal parts of the posterior edges are cutting edges. The blades of the styloid are well suited to cut into the mud when rotated rearwards and downwards into the sea bed and would ease the passage of the animal as it crawled rearwards through the mud (fig. 8.27).

The hind tail contains about ten segments and seems to end abruptly. Its skeleton consists of dorsal ossicles and ventral plates and the plates alternate with the ossicles along the length of the hind tail, for each pair is articulated to the more proximal neighbouring ossicle and overlaps the more distal one. In lateral aspect, therefore, the junction between plates and ossicles is a zig-zag. This contact seems very ill-suited to behave as Ubaghs' interpretation would require, i.e. it is most unlikely that the plates could open outwards as cover plates. Successive ventral plates in the hind tail are connected together by plated imbrication membranes, as already mentioned under *Reticulocarpos hanusi* (fig. 7.31, p. 232). This suggests that the ventral plates of the hind tail are serially homologous with the major plates of the fore tail. If so, and seeing that the whole tail of *Chinianocarpos* was able to flex ventrally, it is likely that the whole tail of *Chinianocarpos* is homologous solely with the fore tail of *Reticulocarpos*, as already suggested.

The dorsal ossicles of the hind tail vary along its length (fig. 8.26c). The more proximal ones have recurved blades like the two blades of the styloid, and are similarly adapted, by dorsal posterior cutting edges, to cut into the mud when rotated downwards and rearwards, as would happen during the power stroke of the tail. More distal ossicles have successively shorter and smaller blades until, finally, the most posterior ossicles are without blades, have rounded dorsal surfaces, and also are wider than more proximal ossicles. Successive dorsal ossicles are connected by curved, transversely cylindrical articulations, allowing the hind tail to flex in a vertical plane. The dorsal ossicles of the mid and hind tail are probably serially homologous with the small spike-shaped plates on the dorsal surface of the fore tail – a suggestion that gains in likelihood when *Lagynocystis* is considered (cf. fig. 8.28c).

The functioning of the tail can be understood if it is reconstructed in the likely position of maximum flexion (fig. 8.27). For then the mid tail and the proximal well-bladed portion of the hind tail form an almost vertical limb, projecting downwards distally; the intermediate small-bladed portion forms an almost horizontal limb; and the distal, round-backed, wide ossicles form another vertical

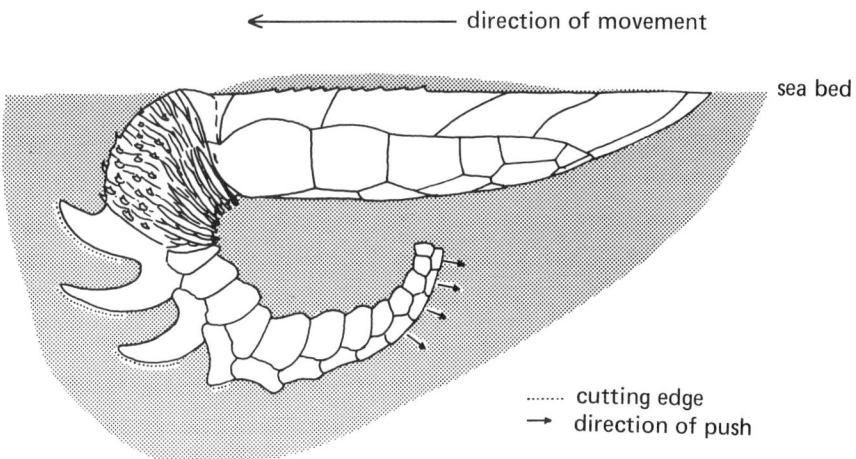

← —— direction of movement

sea bed

.... cutting edge
→ direction of push

8.27. Chinianocarpos thorali. *Suggested mode of action of the tail in locomotion. The hind tail is tripartite, much as in* Peltocystis, *with a terminal portion that would have acted as an anchor to push against the mud. (New.)*

limb, projecting upwards distally. *Chinianocarpos* probably moved rearwards by analogy with other calcichordates, and also as indicated by the forwards-pointing denticulations on the peripheral flanges and the cuesta-shaped ribs on the dorsal surface. If so, the cutting blades of the styloid and proximal hind tail would help to slice a way through the mud at the leading, morphologically posterior, end of the animal during the power stroke; the wide rounded backs of the terminal ossicles would be bearing surfaces so that the distal, upward projecting, part of the tail would be a terminal anchor, pushing forward by its dorsal surface against the mud; and the intermediate, narrow, short-bladed, presumably almost horizontal part of the hind tail would serve to transmit the force of the more proximal parts of the tail to this terminal anchor. Whilst crawling, mud would presumably slide up on the dorsal surface, at right and left, where the ribs were, and would be gripped by the ribs. Mud probably did not slide onto the more median parts of the head where the eyes and peripheral grooves would serve to give warning of its presence and so stop the animal burying itself completely.

Chinianocarpos was therefore a primitive mitrate, as shown in particular by the flexible head roof. It is shown to be a stem vertebrate by the lateral line and by the presence of dorsal, presumably touch-sensory, branches of the trigeminal (n_{4+5}), for these features can be seen as synapomorphies with living vertebrates which stem-tunicate and stem-vertebrate mitrates do not share (Chapter 9). Moreover, *Chinianocarpos* has specialized features of the tail which, as argued later, are shared with stem-tunicate mitrates but not with the only known stem-acraniate mitrate (*Lagynocystis*). These tail features suggest

that vertebrates are closer related to tunicates than to acraniates. *Chinianocarpos thorali* is, indeed, a most important animal, being, as I shall argue in the next chapter, the most primitive vertebrate known, though not quite the earliest.

Lagynocystis pyramidalis Barrande is another mitrate found in the siliceous nodules of the Lower Ordovician Šárka Shales (Llanvirn) of Bohemia, especially at Osek near Rokycany and Šárka near Prague. It is a common fossil, a few thousand having been collected – the largest and finest available collection is that of František Hanuš. This was made at Šárka in the early years of the twentieth century and is now preserved in the National Museum, Prague. The species was first described by Barrande in 1887, under the name of *Anomalocystites pyramidalis*. Jaekel (1918) made it the type species of a new genus *Lagynocystis* and a new family Lagynocystidae. Chauvel (1941) restudied it, noted several new internal features, and published the first sketches of the internal gill slits. Ubaghs (1867) described it in better detail and since then I have studied it myself (Jefferies, 1973). Chauvel & Nion (1977) have recently recorded it from Brittany. It is one of the best known mitrates. Henceforth I shall call it *Lagynocystis*.

In external appearance (fig. 8.28), the head of *Lagynocystis* is elongate, being about 30 mm long, 8 or 9 mm wide and about 6 mm deep. It tapers forward, ending anteriorly in a single knob-like plate (anterior appendage). The skeleton of the dorsal surface of the head, as usual in mitrates, consists mainly of a few large plates. These are divided into an anterior and a posterior group which touch, probably with a rocking articulation, on the

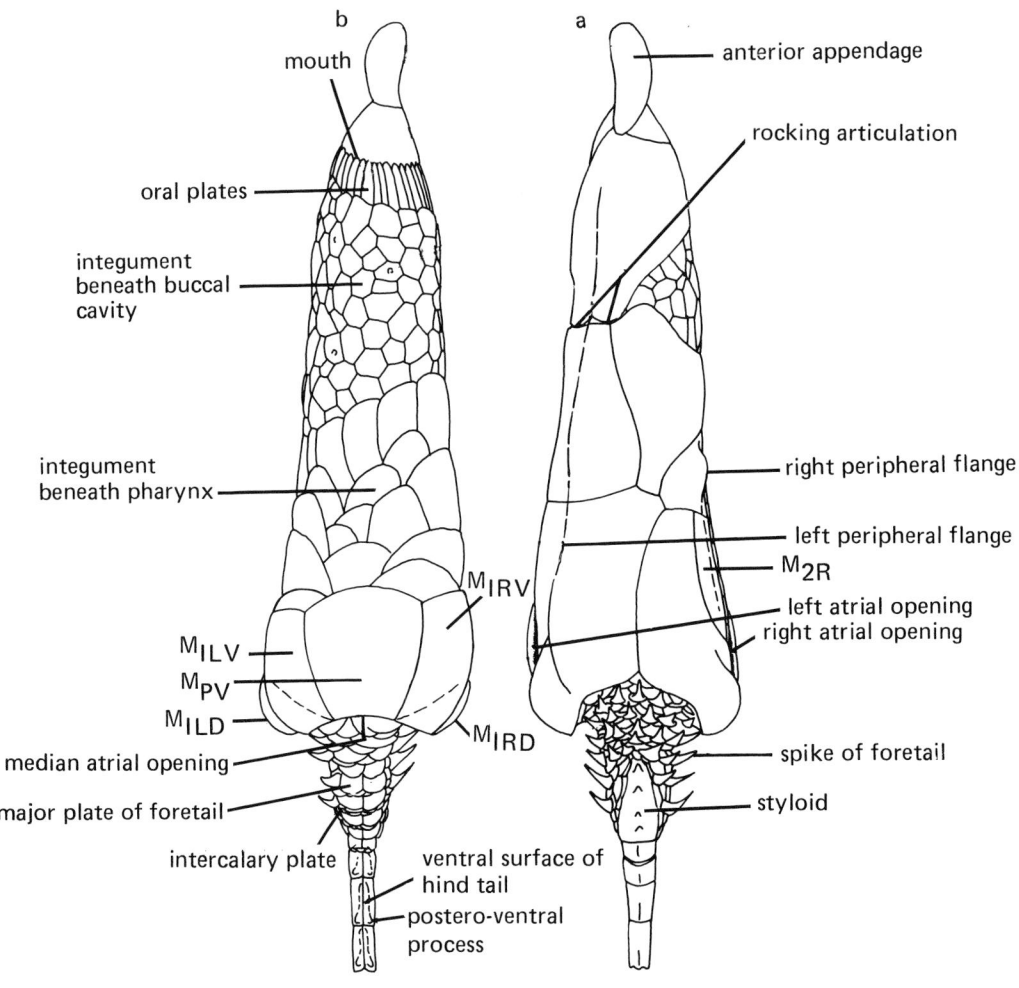

Fig. 8.28.

left, and are separated by a triangle of plated integument on the right. The left peripheral flange is much longer and stronger than the right one and the anterior part of the dorsal surface of the head slopes forwards and rightwards. The ventral surface, as usual, is mainly formed of plated integument. This is of two types – a smaller-plated, probably more flexible, anterior region is separated by a sudden obliquely transverse boundary from a larger-plated, probably stiffer, posterior region. The ventral surface ends behind in postero–ventral marginal plates. But there are three such plates (M_{1LV}, M_{PV}, M_{1RV}), not the usual two, since a median plate (M_{PV}) extends rearwards to the tail insertion. The tail was even longer than the head and was divided into the usual fore, mid and hind regions. The openings of the head comprise the mouth and three atrial openings (right, left and median). There is no

lateral line. The mouth has a flexible lower lip, guarded by spike-shaped oral plates, and a rigid upper lip formed of a single plate.

The right and left atrial openings have the same postero–lateral position as in other mitrates. The left and right postero–ventral marginal plates (M_{1LV}, M_{1RV}) are hinged to the posterior dorsal marginal plates (M_{1LD}, M_{1RD}) and overlap the marginal frame at the sides. The right overlap, within which the opening would have gaped, covers part of the first and second dorsal marginals (M_{1RD}, M_{2R}), while the left overlap covers part of the left first dorsal marginal only (M_{1LD}). Compared with the usual mitrate condition, therefore, the paired atrial openings are more rearward with respect to the plates – M_{2L} forms no part of the left opening and M_{2R} forms little of the right one. This arrangement may be more advanced

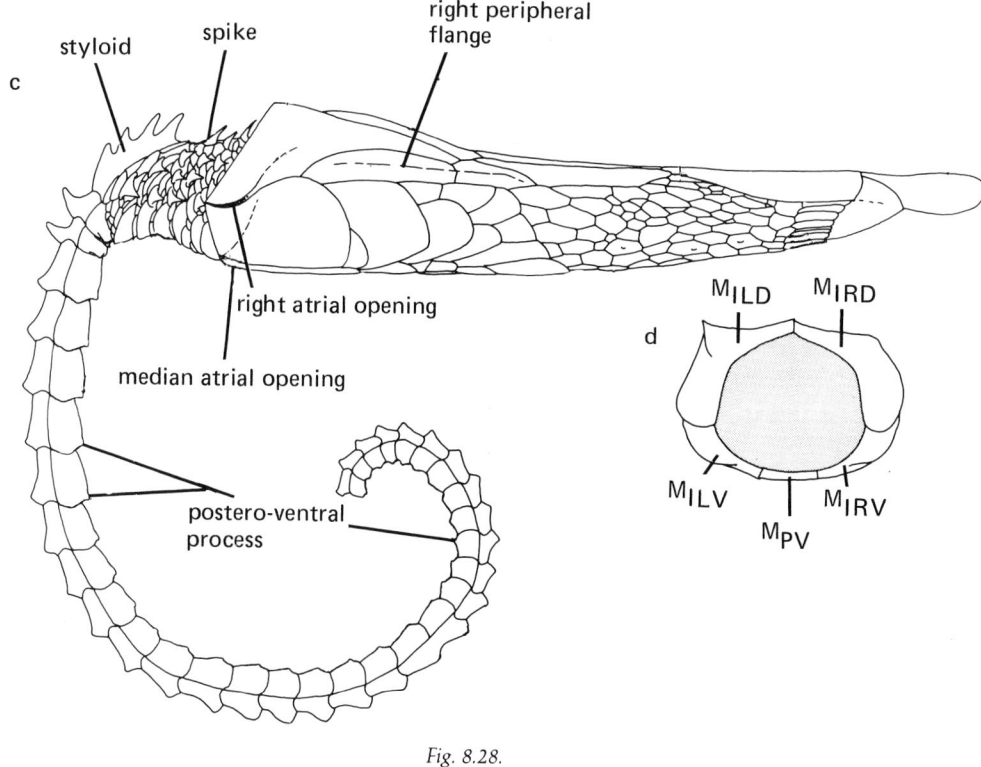

Fig. 8.28.

8.28. Lagynocystis pyramidalis. *Reconstruction of external features. a) Dorsal aspect, b) ventral, c) right lateral, and d) posterior aspect of head; from Jefferies (1973, fig. 5, p. 424).*

than the usual mitrate condition. It probably arose because in *Lagynocystis* the width of the head has been reduced and its posterior end sharpened, so as to move more easily through mud. These changes would have shortened the atrial hinges and shifted the atrial openings rearwards onto the first dorsal marginals.

The median atrial opening lay in the ventral mid-line where the head joined the tail, between the upper surface of the ventral head skeleton and the lower surface of the fore tail. The evidence for it comes from the internal structure, as discussed later.

The soft parts inside the head can be visualized from natural internal moulds (figs 8.29, 8.30). An oblique groove runs, as usual, from posterior left to anterior right on the dorsal surface of the natural mould (fig. 8.29a). It is very weak compared with other mitrates however, as if it had been smudged out, and varies in strength from specimen to specimen. Also its anterior end, where it reaches the right side of the head, is very posterior in position, located about halfway along the head length. As a result the left field, left of the oblique groove, is much larger than the right one. The posterior margin of the buccal cavity is indicated by a cleft at dorsal left of the natural moulds

situated slightly in front of the transverse level of the anterior end of the oblique groove. This cleft lies just dorsal to the boundary between the two types of ventral integument. This suggests that the anterior integument, which was probably more flexible because small-plated, floored the buccal cavity while the posterior, stiffer integument floored the pharynx. And the boundary between the two types of integument would be the ventral boundary between buccal cavity and pharynx. If so the partition, or velum, separating the two chambers would be an approximate plane, slightly more posterior on the right than the left. And in that case the dorsal surface of the buccal cavity would lie entirely in the left field, in the territory of the purely virtual left anterior coelom (as in *Placocystites*). There is no evidence concerning the ventral boundary between the right and left anterior coeloms, but from the dorsal relationships it seems likely that the buccal cavity as a whole, and the velum in particular, was predominantly in the field of the left anterior coelom. Consequently the arguments already used, in *Placocystites*, for homologizing the left anterior coelom with the left mandibular somite of amphioxus, apply also to *Lagynocystis*. The buccal cavity of *Lagynocystis*, like that of

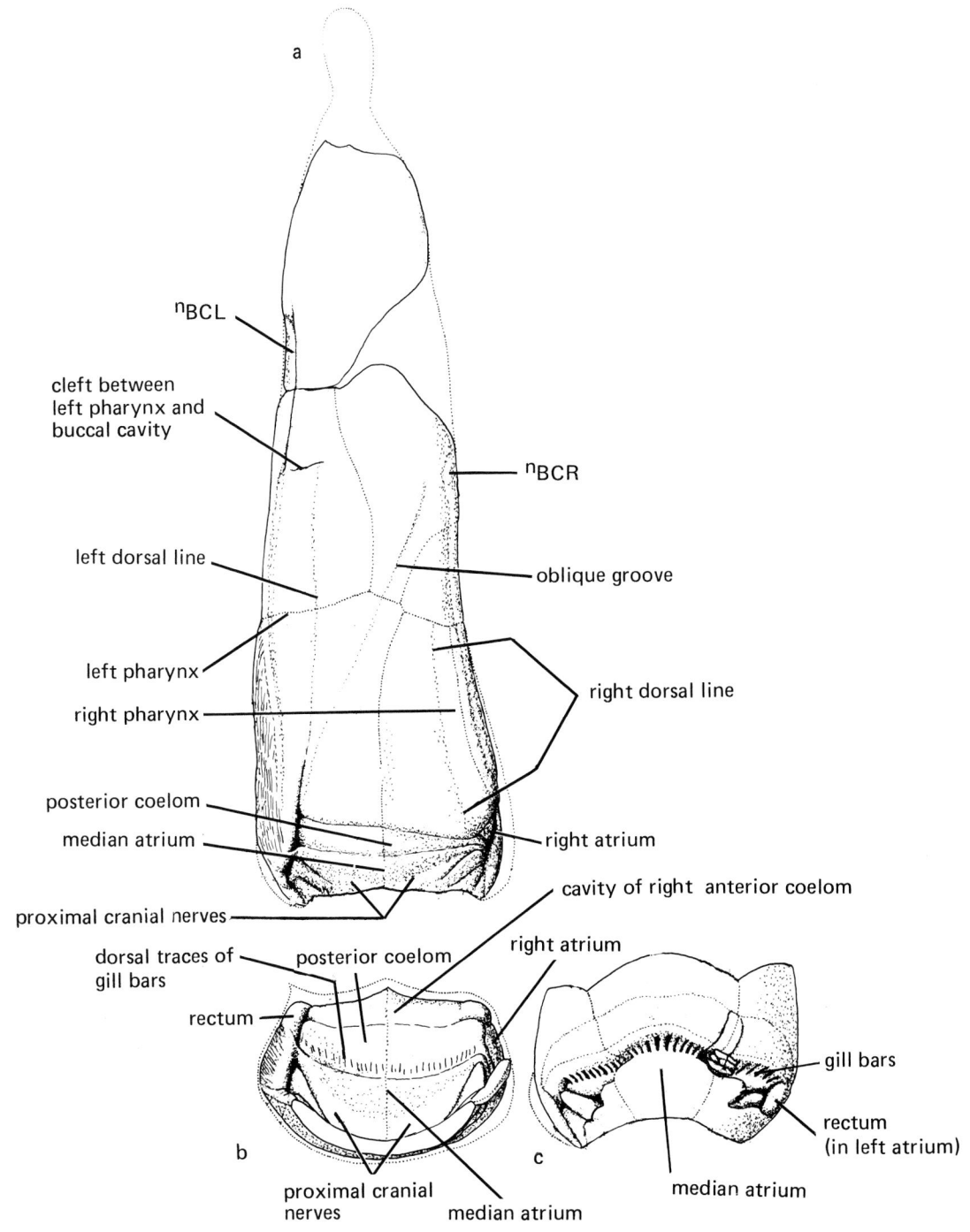

8.29. Lagynocystis pyramidalis. *Reconstructed internal mould of head, simulating the soft parts. a) Dorsal aspect, b) posterior, and c) ventral aspect of posterior part; from Jefferies (1973, fig. 6, p. 426).*

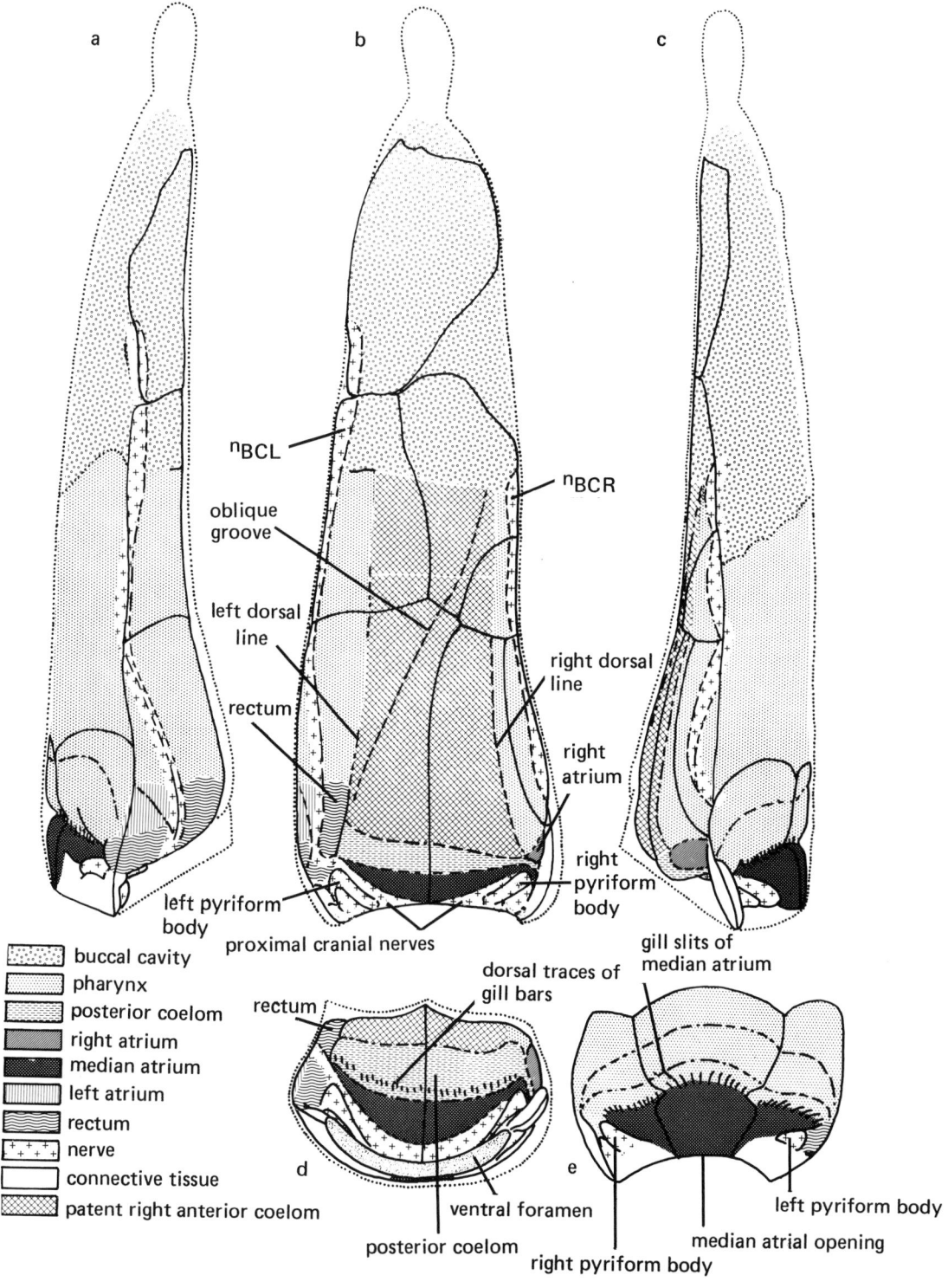

a

b

c

nBCL

nBCR

oblique
groove

left dorsal
line

right dorsal
line

rectum

right
atrium

right
pyriform
body

left pyriform
body

proximal cranial nerves

gill slits of
median atrium

buccal cavity

pharynx

posterior coelom

right atrium

median atrium

left atrium

rectum

nerve

connective tissue

patent right anterior coelom

rectum

dorsal traces of
gill bars

d

e

ventral foramen

posterior coelom

right pyriform body

left pyriform body

median atrial opening

8.30. Lagynocystis pyramidalis. *Head chambers as deduced from natural moulds (cf. fig. 8.31); from Jefferies (1981a, fig. 36, p. 552).*

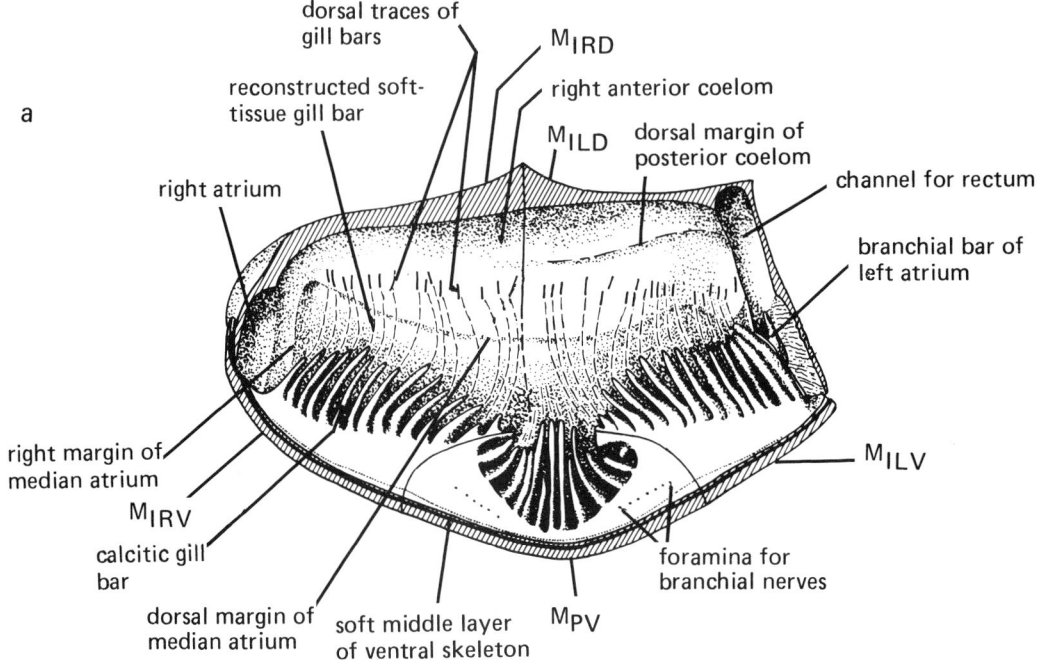

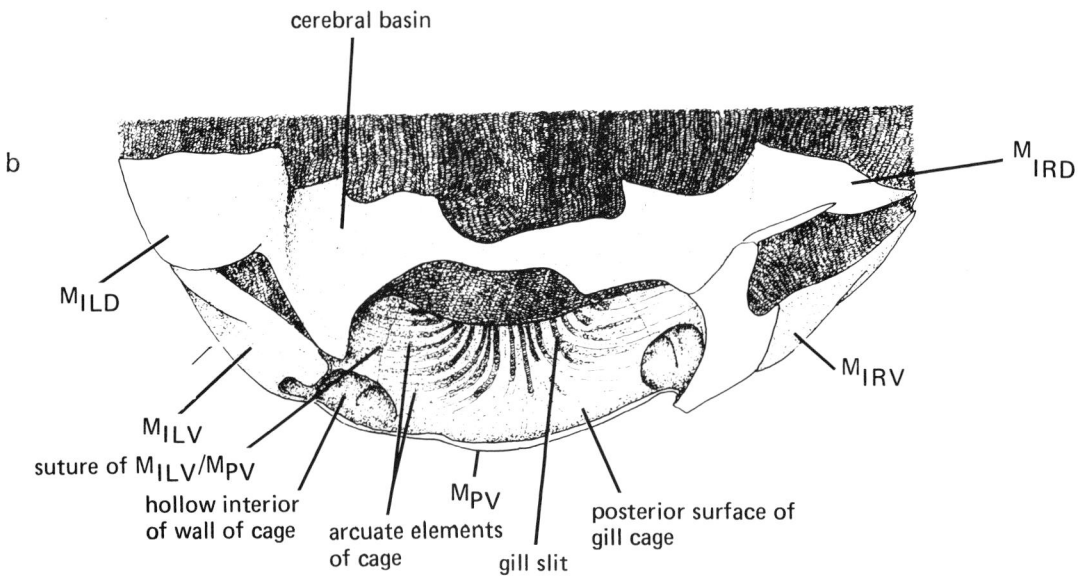

8.31. Lagynocystis pyramidalis. *a) Reconstruction of branchial cage and posterior wall of head in anterior aspect. b) Cameralucida drawing of posterior surface of branchial cage (anterior wall of median atrium); from Jefferies (1981, fig. 35, p. 550; 1973, fig. 11, p. 430).*

amphioxus, seems thus to have been an organ mainly or entirely of the left side. The antero–dorsal margin of the posterior coelom is indicated by a transverse groove near the posterior end of the internal natural mould of the head, as in other mitrates. The right atrium is visible as a well-defined little protuberance at the posterior right corner of the natural mould. Unlike other mitrates there is a pair of weak dorsal lines which run forwards, parallel to the longitudinal axes, on the dorsal surface of the natural moulds from near the posterior right and left angles. The left of these dorsal lines begins posteriorly with a short but deep longitudinal cleft, which in the skeleton represents a thin wall parallel to the sagittal plane. This cleft is situated left of the posterior coelom, like the posterior part of the oblique groove in other mitrates. However, it runs forward into the left dorsal line, rather than into the posterior end of the oblique groove. The right dorsal line has no such cleft posteriorly. The significance of those paired lines will be discussed later.

A strange cage near the posterior end of the head (fig. 8.31), in a median–ventral position, is the most striking internal feature of *Lagynocystis*. In terms of the skeleton, or of latex casts of the natural moulds, the cage consists of about 45 subvertical bars arranged in three subgroups, together with a transverse wall from the top of which the bars emerge. The three sub-groups are rooted respectively in the three postero–ventral marginal plates (M_{1LV}, M_{PV}, M_{1RV}) and are therefore symmetrical, being right, left and median. The wall beneath the bars has a pair of peaks which are also symmetrical, each bisected by a suture, and which separate the three subgroups from each other. The whole cage, both walls and bars, is formed from the inner calcitic layer of the ventral skeleton, and represents the anterior surface of a little median postero–ventral chamber (fig. 8.29c). The dorsal and right and left boundaries of this chamber, where they came in contact with the dorsal skeleton, are visible on the postero–dorsal faces of the natural moulds or the corresponding latex casts. The dorsal boundary of the chamber is approximately horizontal (fig 8.31a) and some distance below the dorsal boundary of the posterior coelom. Thus there was an area above the median postero–ventral chamber where the posterior coelom was in contact with skeleton, and posterior coelom also extended as a vertical strip, right of the chamber, between it and the right atrium. Where the posterior coelom was in contact with the dorsal skeleton, above the median postero–ventral chamber, a transverse series of about 35 short vertical striae is visible on the natural moulds. These probably mean that some of the calcitized bars of the cage continued upwards as threads of soft tissue whose dorsal ends were connected to the calcite skeleton, each one at a stria. The slits between the bars of the cage converge rearwards and medianwards into the median postero–ventral chamber. This convergence is especially obvious if natural moulds are viewed in ventral aspect (fig. 8.29c). The cage was called the ctenoid organ by Ubaghs (1967) while its discoverer, Chauvel, referred to it as 'crêtes aborales' (1941).

The bars of the cage are probably gill bars separated by gill slits. For such slits have already been postulated in the mitrates previously discussed, although at right and left rather than median in position. Moreover, the parallel-sided nature of the slits confirms this conclusion and suggests that they were lined in life with ciliated epithelium. For the slits of tunicates and amphioxus (figs 3.5, 4.3b, pp. 58, 90) are likewise parallel-sided so that the cilia on either side, which are constant in length, almost meet in the middle of the slit, leaving no dead water between them (Berrill, 1955, p. 88). The slits of *Lagynocystis* are 60–100 µm wide. This should be compared with tunicate slits, whose maximal width is about 60 µm and commonest width about 25 µm (figures in Herdman, 1882–8); and with the slits of amphioxus which are about 40 µm wide (Bone, 1961, fig. 20). If the skeletal slits of *Lagynocystis* were clothed on both sides with ciliated epithelium 20 µm thick, therefore, the living slit would be the same width as in tunicates or amphioxus. The fact that the slits converge rearwards indicates that water flowing through the slits would also converge on the median postero–ventral chamber. This makes no contact with left or right atrial openings and is symmetrical and constricted behind, on right and left, by the pyriform bodies and other soft paired structures and their skeletal housing. Water could therefore only issue from the chamber in the posterior mid-line, at the median ventral atrial opening already postulated. The chamber can therefore be called the medium atrium. Like the gill slits in its front wall, it did not exist in other mitrates.

The left atrium would have been situated left of the median atrium, just internal to the left atrial opening. It is limited on the right by a short, vertical wall parallel to the long axis of the head which is represented by a cleft in the natural mould, as already mentioned. From this wall, in one specimen, a single bar, similar to those of the cage of the median atrium, projects upwards and leftwards. It probably represents a single gill bar in the front wall of the left atrium.

The right atrium would have lain right of the median atrium, just internal to the right atrial opening. Its position is indicated by a little chamber excavated in the right wall of the head in the appropriate position. In the natural internal moulds, as already said, this chamber appears as a small protuberance. No skeletal trace of gill bars exists at its anterior margin. I assume such bars were present, but no calcite grew into them. In life there would therefore have been three groups of gill slits in

Lagynocystis associated with the left, right and median atria. And the gill slits of the median atrium were themselves divided into three subgroups, over the left, right and median postero–ventral marginal plates (M_{1LV}, M_{PV}, M_{1RV}).

The ontogeny of these gill slits can partly be deduced. The internal surface of the dorsal plates is marked by growth lines which show that they grew by accretion at the edges. The same mode of growth probably held for the postero–ventral marginals that carry the cage of the median atrium, though these plates lack growth lines. Calcitic gill bars, or the dorsal branchial traces in the posterior coelom, that are situated near sutures as in fig. 8.31a, therefore probably originated later than similar structures farther away from sutures. And this suggests that the dorsal branchial traces were connected only with the right and left subgroups of the gill bars of the median atrium. For if they are reconstructed with a link to the calcitic gill bars of the median subgroup (the ones rooted on plate M_{PV}) then inevitably old ventral gill bars would be connected with new dorsal traces, which is absurd. More important in reconstructing ontogeny is the fact that some of the gill traces are so near the median dorsal suture (M_{1LD}/M_{1RD}) that the attached gill bars could only have formed shortly before the animal died or stopped growing.

The same conclusion is confirmed by markings on the posterior surface of the cage (fig. 8.31b). For these show that the cage was built of successively superposed arcuate elements. Each arc is concave upwards so that it crosses the suture almost at right angles (and almost horizontally) but ends distally in a sub-vertical calcitic gill bar. Each arc would be generated at the suture and would follow a stratigraphical law of superposition there, by overlying pre-existing arcs. Consequently a gill bar which at death was far from the suture can be traced along its arc to a low-lying horizontal element at the suture. It would thus be older than a gill bar near the suture, which can be traced to a higher-lying and younger horizontal element at the suture. It also follows that new gill bars were still being added to the cage so long as the cage was increasing in size. If growth of the animal continued until death, therefore, then median gill bars and slits were added to the median atrium until death.

The left pharynx was probably formed earlier in ontogeny than the right one in mitrates other than *Lagynocystis* as already suggested. For the oblique groove in these forms, which separated the left from the right anterior coeloms and the left pharynx from the cavity of the right anterior coelom, has been partly smudged out anteriorly by the passage of the right pharynx across it. Like the pharynges therefore, it is probable that left gill slits arose ontogenetically before right ones. The same

conclusion applied to *Lagynocystis* which also had an oblique groove, though a weak one. Seeing that the median gill slits of *Lagynocystis*, as just shown, were still being produced when the animal died, the likely ontogenetic sequence for its gill slits is therefore: (1) slits of the left atrium, (2) slits of the right atrium, (3) slits of the median atrium. This is the same as the deduced phylogenetic sequence for the origin of those groups of slits revealed when *Lagynocystis* is placed on a cladogram. It is also comparable with the ontogenetic origin of the slits of amphioxus. For, as mentioned in Chapter 3, this animal has: (1) a primary group of slits, morphologically left, which appear in the larva (fig. 3.33); (2) a secondary group, morphologically right, which appear suddenly at the end of larval life (fig. 3.38); and (3) a tertiary group, which continue to be produced behind the others, on left and right, as the post-larva grows into the adult.

The gill slits of the median atrium of *Lagynocystis*, issuing by a median postero–ventral atrial opening, are therefore probably homologous with the tertiary slits of amphioxus, which issue by a median postero–ventral atriopore. There are differences from the amphioxus arrangement. In particular *Lagynocystis* had three atria, and three atrial openings, not one. This presumably means that the atria shown in *Lagynocystis* have become confluent in the ancestry of amphioxus, and the paired atrial openings have been lost. Nevertheless, *Lagynocystis* and living acraniates are the only animals known to have tertiary gill slits and a mid-ventral atriopore. This constitutes a synapomorphy between them which shows that *Lagynocystis* was an acraniate. If the fossil *Pikaia*, mentioned below, was likewise an acraniate, as seems probable, and was more advanced than *Lagynocystis* in lacking a calcite skeleton, then it too would probably have had tertiary gill slits and a mid-ventral atriopore, but there is no direct evidence of these organs.

The relationships of the posterior coelom to the median atrium are of interest. As already mentioned, the presence of branchial traces in the posterior wall of the posterior coelom above the median atrium indicates that the left and right subgroups of the median gill bars continued upwards, as threads of soft tissue which ended dorsally at the posterior wall of the posterior coelom. Now the anatomy of *Placocystites*, as explained above, suggests that the posterior coelom of mitrates is homologous with the combined right and left epicardia of a tunicate. Furthermore, as mentioned in Chapter 4, Berrill has made it probable (1955, p. 102) that the epicardia of tunicates were primitively excretory organs; and they are known to arise in ontogeny as outpouchings of the pharynx. Amphioxus, as mentioned in Chapter 3, has nephridia as excretory organs, each one of which overlies a primordial gill bar and probably arises in ontogeny as a proliferation of the

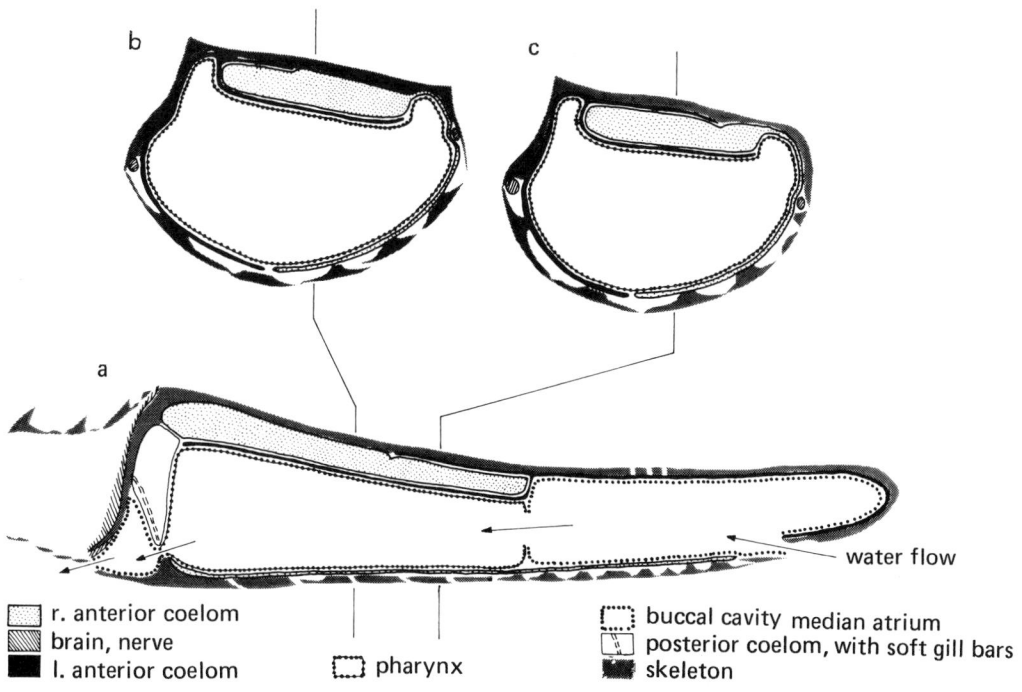

8.32. Lagynocystis pyramidalis. *Head chambers in longitudinal section a) and two transverse sections b) and c). (New.)*

pharyngeal wall (fig. 3.7). The nephridia of amphioxus could therefore have evolved from the *Lagynocystis* situation if, in phylogeny, the median (tertiary) gill slits elongated upwards, so dividing the posterior coelom (epicardia) into numerous pieces, each of which would overlie a gill bar. If so, the nephridia of amphioxus would collectively be homologous with tunicate epicardia. It is strange that this suggestion, which I made in (1973, p. 442), had never been put forward by neontologists. *Lagynocystis* gives no clue to the origin of the nephridial flame cells (cyrtopodocytes = solenocytes). However, the histological comparison of these cells with those of annelids, which Goodrich interpreted as homology, now seems less cogent than it did, following the ultra-structural work of Brandenberg & Kümmel (1961). I argued in Chapter 3 that the nephridia of amphioxus very probably arise in ontogeny from pharyngeal wall, like epicardia (fig. 3.40).

Peculiarities of the dorsal surface of the natural mould of *Lagynocystis* have already been mentioned. They are: the weakness of the oblique groove; the existence of a pair of dorsal lines running forwards parallel to the long axis at right and left; and the fact that the cleft just left of the posterior coelom continues forward, not onto the oblique groove which it narrowly misses, but into the left one of these dorsal lines. The existence of the median atrium, likewise peculiar to *Lagynocystis*, suggests how these

dorsal peculiarities arose. For this atrium could only function if its front wall, where the gill slits were, was in contact with pharynx. Probably, therefore, the median parts of the left and right pharynxes pushed rearwards to meet the median atrium, both in ontogeny and phylogeny. In so doing they would squash the cavity and contents of the right anterior coelom upwards and splay them out left of the oblique groove (fig. 8.32). The cavity would thus be transformed from an asymmetrical triangle entirely right of the oblique groove as in *Mitrocystites* (fig. 8.3) into a very shallow rectangle with symmetrical right and left edges and with a weakened oblique groove running diagonally across it from posterior left to anterior right. The paired dorsal lines would be the left and right boundaries of this widened chamber, where it met the left and right pharynges. The dorsal position of the oblique groove, like that of mitrates but unlike cornutes, indicates that left and right pharynges both existed in ontogeny before this symmetrization of the right anterior coelom, i.e. before the median parts of the pharynx pushed rearwards to meet the median atrium. This confirms the reconstructed ontogenetic sequence for the groups of gill slits – left, then right, then median.

The rectum in other mitrates, as already argued, went out from the posterior left corner of the right anterior coelom and debouched in the left atrium. The same

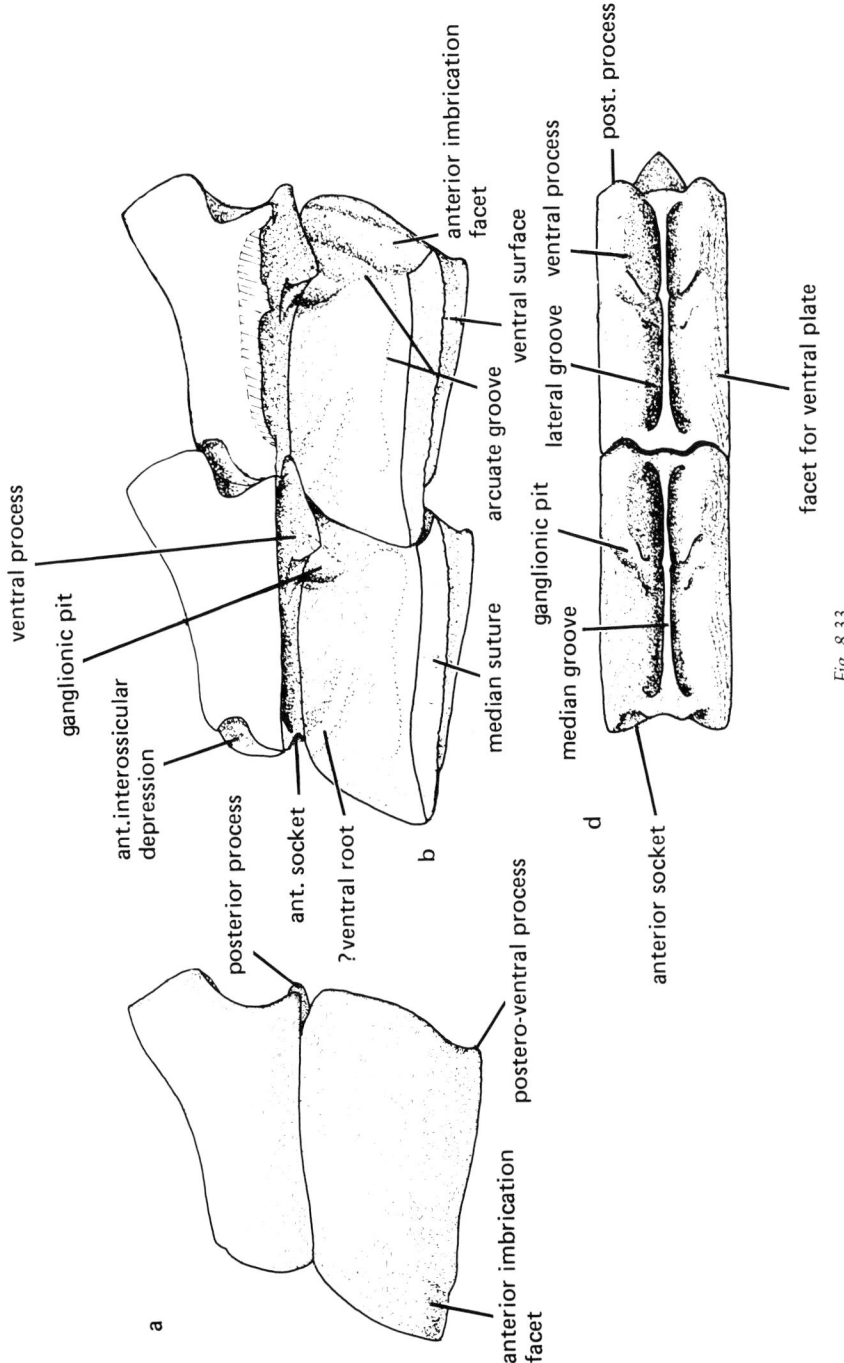

Fig. 8.33.

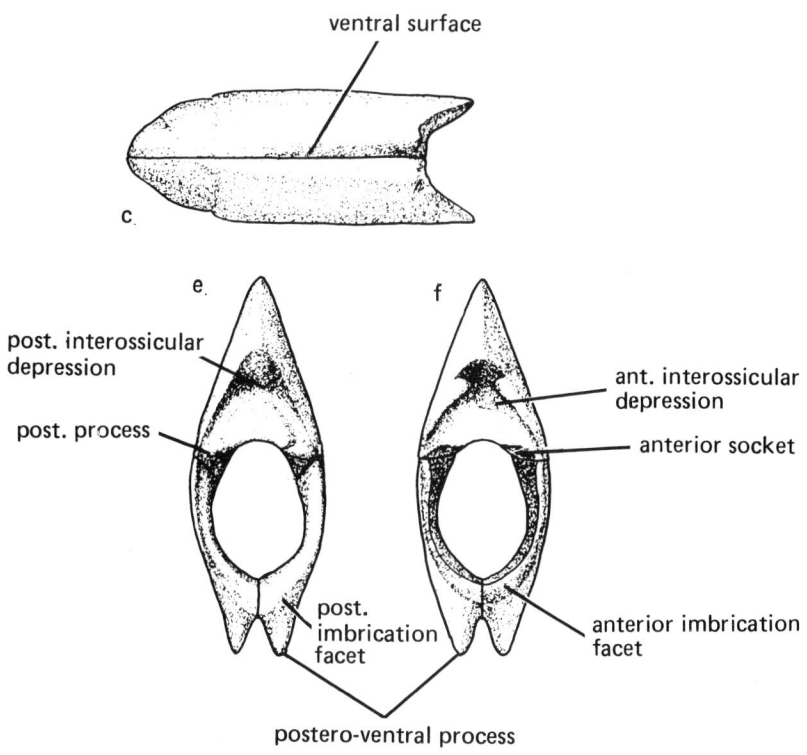

ventral surface

c

e

post. interossicular
depression

post. process

ant. interossicular
depression

anterior socket

post.
imbrication
facet

anterior imbrication
facet

f

postero-ventral process

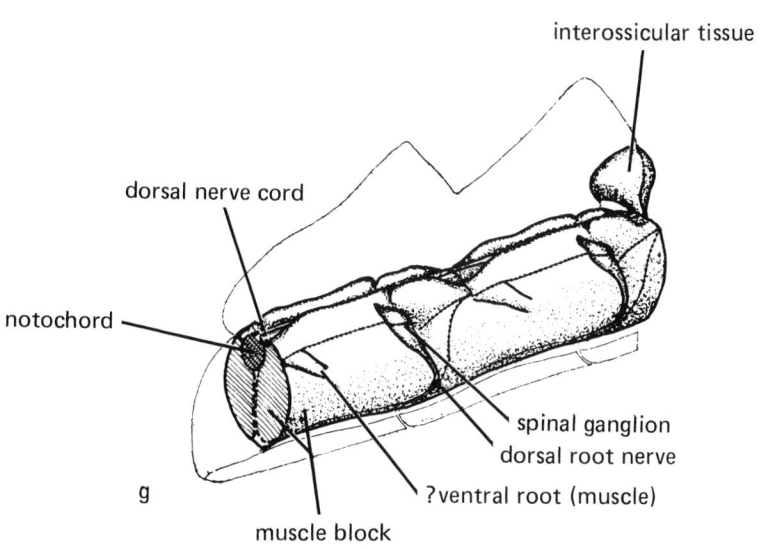

interossicular tissue

dorsal nerve cord

notochord

spinal ganglion

dorsal root nerve

?ventral root (muscle)

muscle block

g

8.33. Lagynocystis pyramidalis. *Skeleton and reconstructed soft parts of hind tail. a) Left aspect of one skeletal segment, b) left aspect of two skeletal segments in sagittal section – the stipple indicates the extent of the right lateral groove in one ossicle, c) ventral aspect of a pair of ossicles, d) ventral aspect of two ossicles, e) posterior aspect of a skeletal segment, f) anterior aspect of same, and g) reconstructed soft parts of two segments; from Jefferies (1973, fig. 12, p. 431).*

probably held for *Lagynocystis*, but the proximal end of the rectum would have been lifted up by the splaying out of the cavity of the right anterior coelom, so that the rectum ran distally downwards instead of upwards (figs 8.29, 8.30a). It probably followed the leftmost posterior portion of the natural mould, left of the posterior coelom. This is suggested by the course of a probable nerve, called the left proximal cranial nerve (fig. 8.29b) and corresponding to the suprarectal component of the left palmar nerve of other mitrates, which has been forced against the posterior wall of the head in this region, presumably by the pressure of the rectum in front of it. If the rectum opened into the left atrium, it recalls the curious fact that the anus of amphioxus is on the left side (fig. 3.1a), though, unlike *Lagynocystis*, there is no left atrium in amphioxus and the anus is very near to the posterior end of the animal.

The tail of *Lagynocystis* has the usual fore, mid and hind portions (figs 8.28, 8.34). The skeleton of the fore tail was loose, comprising major plates, intercalary plates and spike-shaped plates. The major plates are in a right and left ventral row and number about four pairs. Each pair is separated from succeeding pairs by imbrication membranes set with intercalary plates, and there are similar small intercalary plates in front of the foremost major plates and also on the sides and dorsal surface of the fore tail. The spike-shaped plates are developed on the sides and dorsal surface of the fore tail, surrounded by intercalary plates. They are curved forwards and are probably homologous with the spike-shaped plates of the fore tail of *Chinianocarpos* (fig. 8.26).

The mid tail is less well defined as a region than in other mitrates. The styloid has four or five spikes according to the individual, not two as with all other mitrates. Each spike resembles a spike-shaped plate of the fore tail, being thorn-like and curved forwards. Indeed the styloid gives the strong impression of being four or five such spike-shaped plates fused together. The styloid has the usual anterior cavity, though this is shallower than in other mitrates; and the posterior part of the styloid includes two sets of interossicular tissue, presumably separating the posterior three constituent ossicles. Lying as it does in the dorsal mid-line, the styloid can be seen from a different viewpoint as resembling four or five hind-tail ossicles fused together. It is likely, in fact, that the spike-shaped plates of the fore-tail, the constituent ossicles of the styloid and the hind-tail ossicles are serial homologues, each of the others. No paired ventral plates are articulated with the styloid. Presumably the two most posterior pairs of major fore-tail plates are homologous with the paired plates articulated to the styloid in *Chinianocarpos*, *Mitrocystites* or *Mitrocystella*.

The hind-tail skeleton (fig. 8.33) consists of dorsal ossicles and ventral plates, constituting 32 segments in the longest tail seen. The tail ends abruptly, as if a more distal portion had broken away by autotomy. The most anterior two segments are shorter than those behind them and their pairs of ventral plates are separated from each other by a plated imbrication membrane, as are the major plates of the fore tail (and the ventral plates of the hind tail of *Chinianocarpos*). This imbrication membrane confirms the suggestion, made originally on the basis of *Chinianocarpos*, that the ventral plates of the hind tail are serially homologous with the ventral major plates of the fore tail. Each ventral plate of the hind tail ends postero–laterally in a pointed process (fig. 8.33a,e) and has a flat ventral surface which slopes dorsalwards and inwards towards the median plane. In a pair of ventral plates, therefore, these flat surfaces combine to form a concave Λ-shaped longitudinally elongate surface that ends postero–laterally in a pair of points or spikes. This surface was probably a bearing surface and recalls the bearing surfaces of the fore tail of *Reticulocarpos hanusi*, which were likewise ventral and flanked by paired postero–lateral spikes (fig. 7.28b, p. 229, 8.34d,e). Each dorsal hind-tail ossicle has a peak in the dorsal mid-line near the posterior end of the ossicle, at the meeting place of an anterior and a posterior median cutting edge. In the two most anterior hind-tail ossicles these peaks are more prominent than more posteriorly, as if in morphological transition to the spikes of the styloid and the spike-shaped plates of the fore tail. This tends to confirm that the hind-tail ossicles are serially homologous with the styloid and with the spike-shaped plates of the fore tail. The hind-tail segments are not uniform in length – they are shortest in the first two segments of the hind tail and in the last five or six. These short-segmented regions were probably capable of particularly abrupt ventral flexion. In lateral aspect each dorsal hind-tail ossicle meets its proximal and distal neighbours at a transversely cylindrical surface, convex forwards. Such surfaces would allow flexion in a vertical plane. Hind-tail ossicles numbers 19 and 20 are fused together in the only specimen where they were seen, but this may be pathological.

As to the internal structure of the hind tail, the dorsal ossicles are attached to the ventral plates at a pair of wide flat sutures, and the ventral plates of a pair are joined by a flat, median suture. The presence of these broad flat surfaces argues against Ubaghs' view that the plates could open outwards as cover plates. Successive ossicles are separated from each other by interossicular cavities which, however, did not communicate, or scarcely communicated, with the general, ventral lumen of the hind tail. Each pair of ventral plates imbricates inside the pair next in front by a convex imbrication facet. The ventral surface of each dorsal ossicle bears a median groove flanked, except near the fore and hind end of each ossicle, by deep lateral

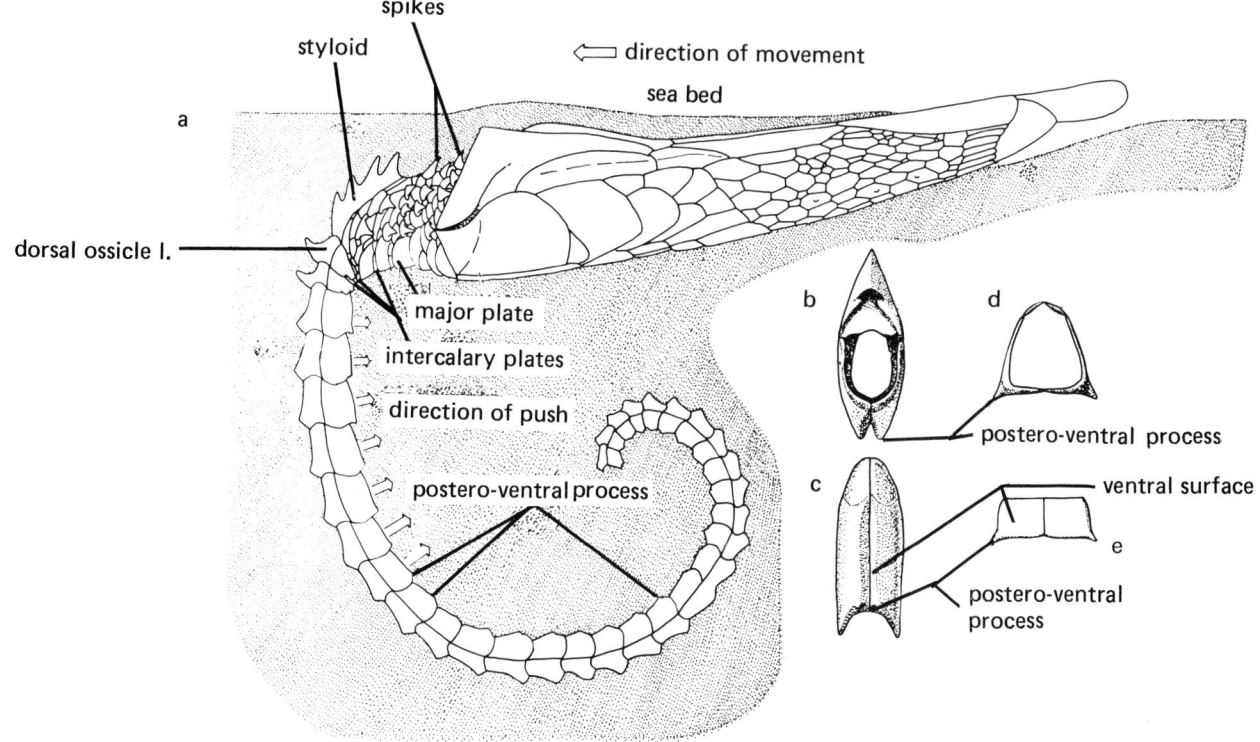

8.34. Lagynocystis pyramidalis. *Suggested mode of action of the tail in locomotion. The main bearing surfaces of the tail were ventral, as they were in the fore tail of* Reticulocarpos hanusi. *The spikes of the fore tail and styloid would hinder forward movement during the return stroke of the tail. a) Right aspect, b) anterior, and c) ventral aspect of hind-tail segment in* Lagynocystis; *d) and e) anterior and ventral aspects of a fore-tail segment of* Reticulocarpos hanusi.

grooves. Where these lateral grooves are thus absent, the ventral surface of an ossicle gives the impression of being moulded over a cylindrical notochord with a narrow, belt-like dorsal nerve cord on its back. A pair of pits in the ventral surface of the ossicle, excavated in the lateral surfaces of the lateral grooves, coincide in position with a widening of the median groove and are continuous with pits on the inside surfaces of the ventral plates (fig. 8.33a). These paired pits probably represent the positions of spinal ganglia. The pits continue ventralwards as grooves on the internal surfaces of the ventral plates. These grooves run vertically downwards and then curve to run horizontally forwards. They probably represent the positions of dorsal-root nerves coming out of the spinal ganglia. A wide, shallow groove in the internal surface runs rearwards and ventralwards from the antero–dorsal angle of each plate. These paired grooves could represent the positions of the 'ventral roots' of *Lagynocystis* which, by analogy with amphioxus, would be made of muscle, not nerve (fig. 3.11). Seeing that the hind tail is adapted for ventral flexion, the lumen between the dorsal ossicles and ventral plates was probably largely filled with segmented

muscle. The 'ventral roots' would pass upward from the muscle blocks to the dorsal nerve cord, while the spinal ganglia and dorsal-root nerves would be located between successive muscle blocks. Ligaments or muscles, or both, located in the dorsal interossicular spaces (fig. 8.33g), would act as antagonists to those ventral muscles.

Lagynocystis would use its tail to pull the head rearwards by ventral flexion, as was normal in mitrates, but would have pressed against the ventral bearing surfaces of the hind tail in the power stroke (fig. 8.34). These surfaces are a unique feature of *Lagynocystis* among known mitrates, but as already mentioned, they are comparable to, and probably homologous with, the ventral bearing surface of the fore tail of *Reticulocarpos hanusi*. The cutting edges of the dorsal ossicles would help the tail to move rearwards and upwards out of the sediment in the return stroke. The spike-shaped plates of the fore tail also probably acted during the return stroke – they would grip the overlying sediment and prevent the animal from slipping upward and forward. They would be specially effective in the early part of the return stroke, when the hind tail was rotating downward and rearward.

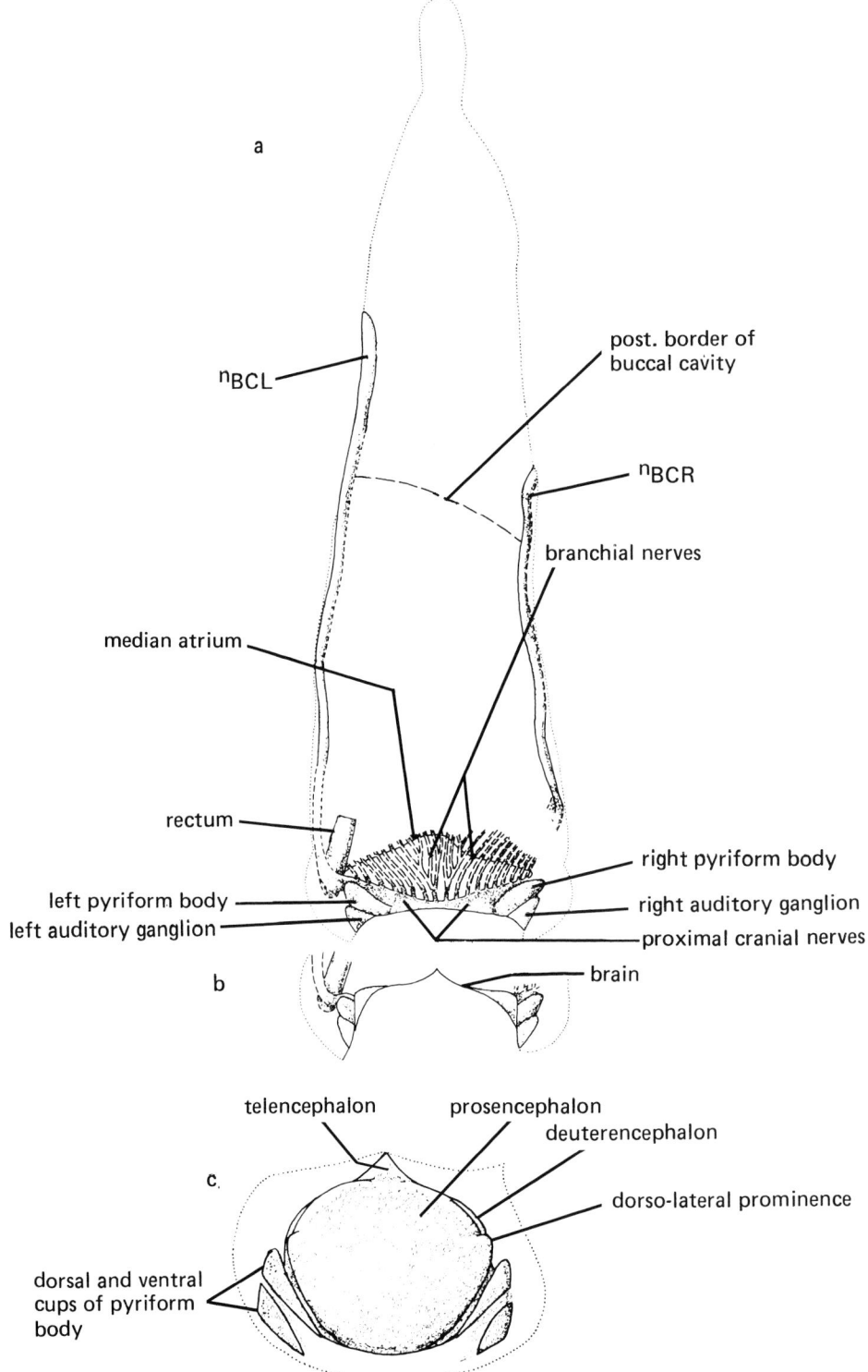

a

nBCL

post. border of
buccal cavity

nBCR

branchial nerves

median atrium

rectum

right pyriform body

left pyriform body
left auditory ganglion

right auditory ganglion
proximal cranial nerves

b

brain

telencephalon

prosencephalon

deuterencephalon

dorso-lateral prominence

c.

dorsal and ventral
cups of pyriform
body

8.35. Lagynocystis pyramidalis. *Reconstruction of brain and cranial nerves. a) Dorsal aspect of head with brain removed, b) dorsal aspect of posterior part of head with brain present, and c) anterior aspect of brain and pyriform bodies (trigeminal ganglia); from Jefferies (1973, fig. 13, p. 42).*

In phylogeny the tails of mitrates probably evolved from an unregionated, distally truncated ancestral tail which resembled the fore tail of *Lagynocystis* except that, along the whole extent of the tail, the ventral major plates would have had ventral bearing surfaces and postero–lateral spikes, as in the fore tail of *Reticulocarpos hanusi* and the hind tail of *Lagynocystis*. Later, by regionation, the more distal dorsal spike-shaped plates became dorsal hind-tail ossicles, which cut through the sediment in the return stroke of the tail, while the proximal ventral plates lost their spikes and bearing surfaces and become separated by plated membranes so as to make this region more flexible in a ventral direction. Also, some of the spikes in the evolving mid-tail fused to form the styloid.

In summary, the tail of *Lagynocystis* is more primitive than that of other known mitrates in having: (1) ventral bearing surfaces and paired spikes in the hind-tail plates, and no dorsal bearing surfaces, (2) large dorsal spike-shaped plates in the fore tail, of a shape that suggests serial homology with the styloid and the dorsal hind-tail ossicles (though it is important to note that *Chinianocarpos* has smaller but probably homologous dorsal spike-shaped plates), (3) a variable and large number (four or five, instead of two) of constituent plates in the styloid, and (4) no ventral plates articulated to the styloid (whereas all other mitrates have two such pairs).

The brain and cranial nerves of *Lagynocystis* (fig. 8.35) are basically as in other mitrates, but seem to have been modified by the presence of the median atrium. The brain was lodged, as usual, in a cerebral basin excavated in the posterior dorsal marginals (M_{1LD}, M_{1RD}) at the rear end of the head and was divided into prosencephalon and deuterencephalon (fig. 8.35c). There was no optic foramen, however, nor any hypocerebral processes separated by a median gap. Instead the line separating the prosencephalon from the deuterencephalon can be followed ventrally to the median plane, suggesting that deuterencephalon existed everywhere (ventrally, lateral and dorsally) behind the prosencephalon. Consequently there seems to be no place where optic nerves could have left the brain, nor would the brain have been in contact with the adenohypophysis (= Seessel's pouch = neural gland = ventral rostral coelom and Hatschek's pit (amphioxus)). Presumably the optic nerves and the contact with the adenohypohysis had been destroyed by the origin of the median atrium.

The prosencephalon extends dorso-laterally into a pair of prominences. Perhaps these functioned as eyes, detecting light that reached them through the skeleton and replacing the lost cis- and transpharyngeal eyes. The dorsalmost part of the prosencephalon, on either side of the median dorsal suture, is slightly swollen. This portion of the brain is immediately behind the precerebral knot which, as usual, is developed across the suture (M_{1LD}/M_{1RD}) and probably represents the meeting place of terminalis and olfactory fibres just before they entered the brain. If so, the swollen part of the brain would be the telencephalon. However, *Lagynocystis* shows no sign of specialized olfactory organs in the buccal cavity. The presumed olfactory sensory cells, and touch-sensory cells, were probably scattered over the surface of the head, as in echinoderms, and also in the buccal cavity.

In front of the brain there is a pair of pyriform bodies (trigeminal ganglia) which, as usual, were almost enclosed in cups above and beneath (fig. 8.35a,b). Behind these was another pair of ganglia (evidenced by lumps in the natural mould) which may have been auditory, to judge by their position behind the trigeminal ganglia, but which seem to have been directly connected to the deuterencephalon.

A pair of proximal cranial nerves, indicated by gentle ridges on the natural moulds, passed upwards and outwards from the ventral part of the brain in front of the trigeminal ganglia. They probably correspond to the deuterencephalic nerves of other mitrates and, as already said, probably lacked any optic component. The left one, as mentioned above, passed over the postero–dorsal face of the rectum and then ran forwards, traceable as a ridge on the natural mould of the dorsal skeleton, into the buccal cavity. (There is a small distance, just anterior to the rectum, where this nerve cannot be followed but only inferred). This left nerve to the buccal cavity (n_{BCL}) seems to be homologous with the left n_2 (maxillary trigeminal) of other mitrates (e.g. fig. 8.16a). It can be followed in the dorsal skeleton a long distance into the buccal cavity but finally passes downwards into the ventral skeleton where it is not traceable. There was a corresponding right nerve to the buccal cavity (n_{BCR}), as shown by a ridge in the dorsal skeleton on the right. The central connections of this nerve are obscure, though it comes out of the ventral skeleton anterior to the right atrial opening, like right n_2 in other mitrates. The right nerve seems to extend less far into the buccal cavity than the left nerve (so far as can be judged from the ridges in the dorsal skeleton). It is, therefore, likely that the buccal cavity, and in particular the mouth with its flexible lower lip, was innervated mainly from the left. The asymmetry is comparable with amphioxus, where the innervation of the buccal cavity is entirely from the left (fig. 3.4).

In the front wall of the median atrium, beneath the calcite gill bars, a series of small holes is present with about the same spacing as the bars (foramina for branchial nerves, fig. 8.31a). The gill bars were hollow and this suggests that small nerves passed from the proximal cranial nerves, over the median atrium, down the gill bars, and out to the floor of the pharynx through these holes. These nerves would be comparable to those supplying the endostyle of amphioxus which likewise reach it by passing over the atrium and down the gill bars.

Thus the nervous system in *Lagynocystis* was much more complex than in amphioxus. In particular it had spinal ganglia in the tail, a larger and bipartite brain, paired trigeminal and probably acoustic ganglia, and paired trigeminal nerve trunks to the mouth. It follows, if *Lagynocystis* is a stem acraniate, that most of the simple features of the amphioxus nervous system are secondary, not primitive. The presence of spinal ganglia probably implies the presence in the embryology of *Lagynocystis* of a neural crest which in amphioxus is lacking. At the same time, the brain and cranial nervous system of *Lagynocystis* are simpler than in other mitrates, especially in the absence of optic nerves and eyes. Since such existed in other mitrates and in *Reticulocarpos* it is fairly certain that their absence in *Lagynocystis* is secondary.

As to mode of life, much has already been said (fig. 8.34). *Lagynocystis* seems to be adapted to crawl rearwards, pulled by its tail, just below the surface of the bottom mud. This is suggested by the elongate shape of the head, sharpened posteriorly by the downward sloping dorsal surface, and the narrowness of the posterior part of the head. It is also suggested by the dorsal, spike-shaped plates of the fore tail which probably served to grip sediment dorsal to this region during the return stroke of the tail. However, *Lagynocystis* presumably fed in amphioxus fashion, using an endostylar mucous trap in the pharynx and drawing water into the mouth by means of cilia on the gill bars. If so the mouth would need to be free of sediment, so the animal probably did not bury itself deeply. The tail would have been used in primitive fashion by pressing on ventral bearing surfaces in the hind tail during the power stroke, unlike any other known mitrate but directly comparable with *Reticulocarpos*. A curious feature is the external asymmetry of the head with the right peripheral flange much shorter and weaker than the left one and the dorsal surface sloping downwards and rightwards anteriorly (fig. 8.28a). These asymmetries do not exist in other mitrates, nor in *Reticulocarpos*, which suggests that they must have been special adaptations of *Lagynocystis*.

To summarize, *Lagynocystis* has features which place it as a primitive acraniate: (1) tertiary gill slits debouching, by way of a median ventral atrium, though a median postero–ventral atriopore; and (2) a posterior coelom dorsal to the median atrium, suggesting a probable mode of origin for the suprabranchial nephridia of amphioxus, as homologues of the excretory epicardia of tunicates. These features form a complex of synapomorphies of *Lagynocystis* with living acraniates which other mitrates do not share. In addition it is likely that the buccal cavity of *Lagynocystis* was innervated mainly from the left and lay in the field of the left anterior coelom (left mandibular

somite). These buccal asymmetries, however, are probably primitive features of mitrates shared with stem-tunicate mitrates and stem-vertebrate mitrates, but which living tunicates and vertebrates have lost. They are not synapomorphies with living acraniates, therefore, though they constitute striking and probably homologous resemblances. Other phylogenetically important features of *Lagynocystis* are found in the tail, which in many ways is primitive for mitrates.

Pikaia gracilens Walcott from the Middle Cambrian Burgess Shale of British Columbia, is probably another fossil acraniate (fig. 8.36). It has not yet been properly described in print, but according to Conway Morris (1979a,b) it is an elongate, laterally compressed, roughly amphioxus-shaped animal with repeated myotomes and with a notochord confined to the posterior two-thirds of the animal. The gut seems to extend, median to the myotomes, to the hind end of the animal and the myotomes are thought to extend, anterior to the notochord, as far as the front end. Here *Pikaia* had a pair of sensory tentacles and behind them a row of small appendages which, in my opinion, could be homologous with the buccal cirri of amphioxus. *Pikaia* was naked, without trace of a calcite skeleton. If it is an acraniate, as seems likely, then it probably does not belong to the crown group since, unlike all living acraniates, the notochord seems to stop well short of the front end (though it is puzzling that myotomes extend in front of the notochord). As a stem acraniate, however, it is more crownward than *Lagynocystis* (i.e. closer related to the acraniate crown group) in having no calcite skeleton, no clear distinction between head and tail, and a gut that extends to, or almost to, the posterior end of the body. Although more crownward, it is older than *Lagynocystis*, and shows that crown chordates, and therefore the mitrates, existed before the Burgess Shales were laid down although fossils of them have not been found.

Two other points should be mentioned as concerns acraniates. Firstly, another possible fossil soft-bodied acraniate, *Palaeobranchiostoma haumatotergum*, has been described by Oelofsen & Loock (1981) from the Permian Whitehill Formation of South Africa. Secondly, Medawar (1951) argued that the asymmetries of the larva of amphioxus were caused by descent from a sessile tunicate-like adult. This cannot be correct, however, since the fossils imply that fixation is a feature special to tunicates.

A further group of mitrates is of high interest but cannot yet be discussed in detail, being still under study. This group includes the three genera *Peltocystis*, *Balanocystites* and *Anatifopsis* and constitutes the order Peltocystida. In what follows I shall mainly consider external features and what can be deduced from them.

Peltocystis cornuta Thoral (fig. 8.37) is the most primitive known peltocystidan. It comes from the lowest

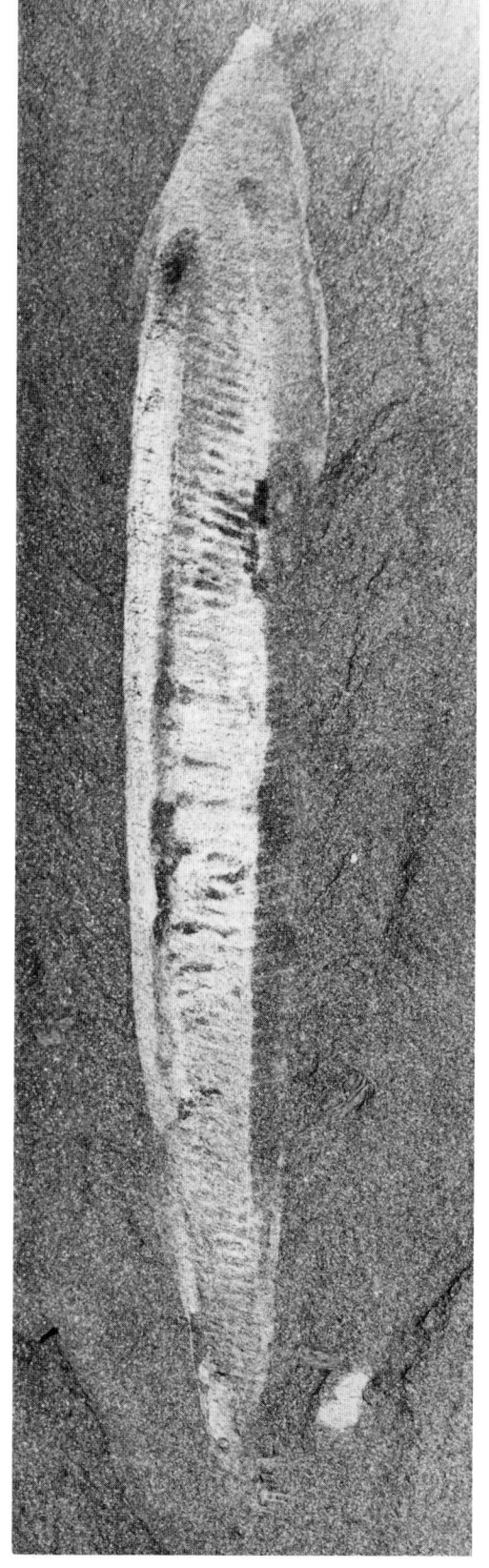

8.36. *Pikaia gracilens. Photograph of specimen in left aspect. Note the somites, fins and the absence of any distinctive head suggesting that* Pikaia *is* an acraniate.

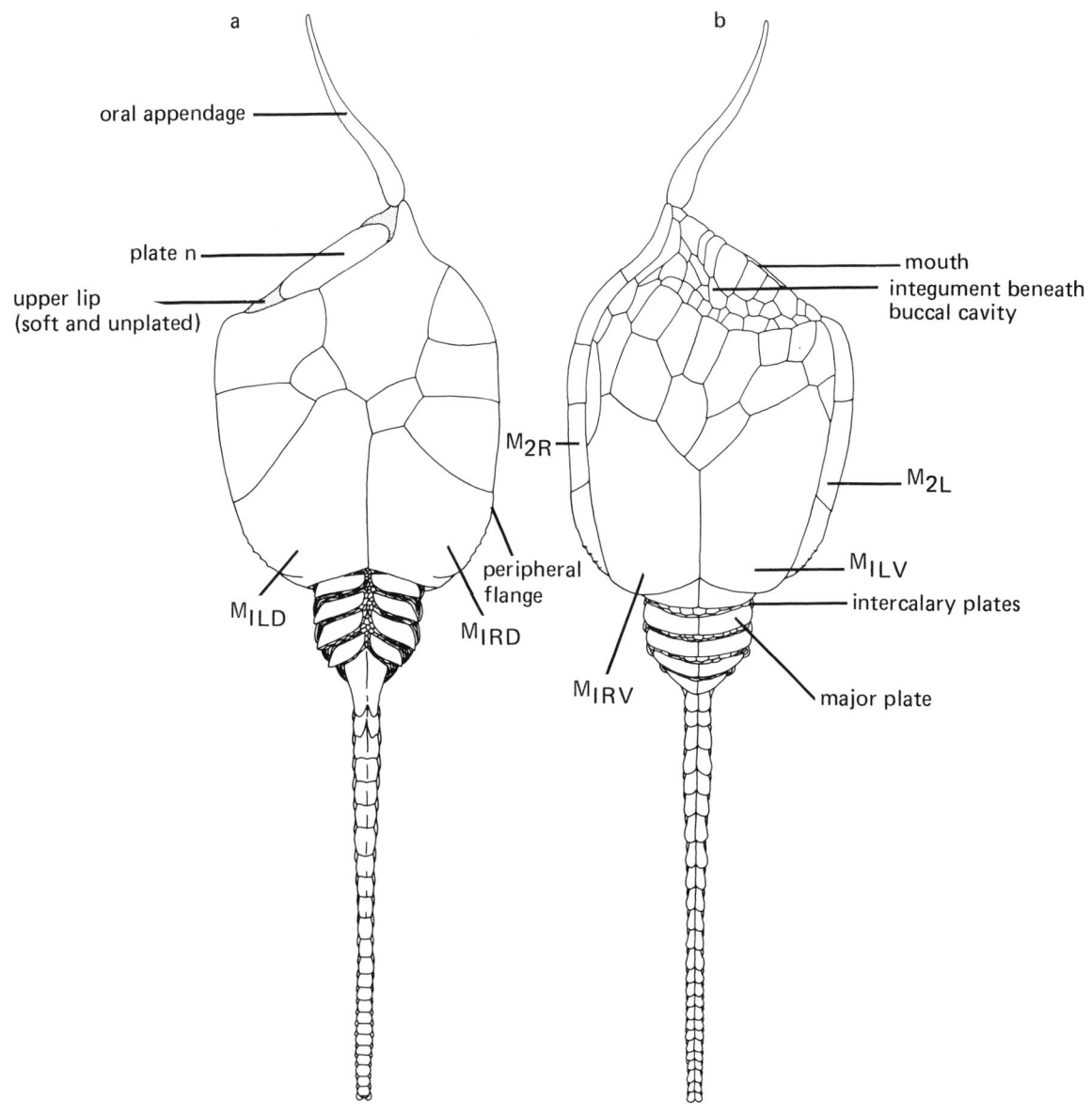

Fig. 8.37.

Ordovician of the Montagne Noire, together with *Chinianocarpos thorali*. The head has a flat dorsal and a convex ventral surface with a wide mouth anteriorly which is obliquely transverse, more posterior on the left than on the right, and has a single large plate above it referred to as plate *n*. Right of the mouth there is an oral appendage, formed of a single plate. The dorsal surface of the head is formed of large plates and is bounded by a peripheral flange. The ventral surface is formed anteriorly

of plated integument (fig. 8.37b). As in *Lagynocystis*, an anterior region of this integument is built of small plates, was probably relatively flexible, and presumably underlay the buccal cavity. While a more posterior region was formed of larger plates, was probably relatively stiff, and presumably underlay the pharynx (cf. fig. 8.28b). Approximately the posterior half of the ventral surface area is made up of two postero–ventral marginal plates (M_{1LV}, M_{1RV}). These are hinged to the posterior dorsal marginals (M_{1LD},

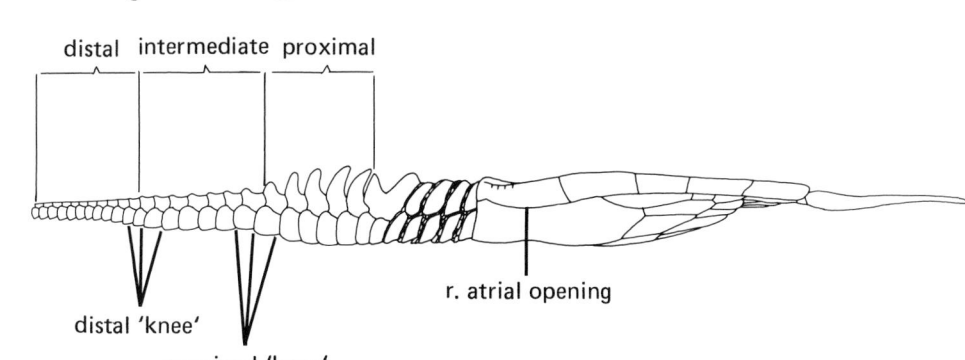

c

regions of hind tail

distal intermediate proximal

r. atrial opening

distal 'knee'

proximal 'knee'

8.37. Peltocystis cornuta. *Reconstruction of external features. a) Dorsal aspect, b) ventral, and c) right lateral; note the tripartition of the hind tail.* (New.)

M_{1RV}) on either side of the tail and are overlapped by them, in front of the hinge, while still further forward they contact the second dorsal marginals (M_{2L}, M_{2R}), again as in *Lagynocystis*. The great length of the postero–ventral marginals, when compared with *Mitrocystites*, *Chinianocarpos* (fig. 8.26), *Lagynocystis* and most other mitrates, suggests that the right and left atrial openings were especially long.

The fore-tail skeleton consists of four rows of major plates (dorsal, right and left; ventral, right and left) separated by imbrication membranes set with small intercalary plates. The ventral major plates meet at sutures in the mid-line and number about five pairs, whereas there are four pairs of dorsal major plates. The mid-tail skeleton consists of a dorsal styloid to which two pairs of ventral plates are articulated. The styloid is two-bladed – the anterior blade is short while the posterior one is long and recurved with a posterior cutting edge. The hind tail has dorsal ossicles alternating along its length with paired ventral plates. The junction between ossicles and plates is zig-zag in lateral aspect with each pair of plates overlapping the ossicle next behind. As with *Chinianocarpos*, this arrangement is hard to reconcile with Ubaghs' view that the plates were adapted to open outwards.

The hind tail is regionated (fig. 8.37c). The proximal four or five ossicles carry long, recurved dorsal blades with posterior convex cutting edges much like the posterior blade of the styloid. This long-bladed region is followed distally by a short-bladed region about seven ossicles long; and this by a terminal region, some 11 ossicles long, where the ossicles are bladeless and have a rounded dorsal surface. The hind tail ends abruptly, as usual, as if by autotomy. There is a corresponding regionation in the ventral plates of the hind tail. In particular, there are two short lengths where the ventral plates rapidly decrease distalwards in height. These places would act as 'knees' –

as sites of particularly strong ventral flexion. And they are situated where, in distal sequence, long blades on the dorsal ossicles give place to short blades, and short blades to no blades.

The functional significance of this regionation of the hind tail appears when the tail is reconstructed in the position of greatest flexion (fig. 8.38). For then, as in *Chinianocarpos* but more clearly (fig. 8.27), the hind tail divides itself into: (1) a proximal, long-bladed limb directed vertically downwards, (2) a horizontal, small-bladed limb, and (3) a distal bladeless limb pointing upwards. Indeed the regionation is clearer than in *Chinianocarpos* with sharper bends between the limbs and less gradation in the ossicles. The functional explanation is the same as in *Chinianocarpos*. The distal bladeless region was a terminal anchor whose rounded dorsal ossicles acted as a bearing surface in the power stroke. The horizontal, short-bladed limb transmitted force from the more proximal parts of the tail to this terminal anchor. And the proximal, long-bladed limb would carry force from the fore tail to the horizontal limb and also, with its curved cutting edges, would cut through the mud on being rotated downwards and rearwards in the power stroke. It would act like the prow of a boat cutting through water. The main motor of the tail as usual, would be the muscles of the fore tail, inserted distally in the proximal excavation of the styloid.

As regards phylogeny, *Peltocystis* therefore shares with *Chinianocarpos* several advanced features of the tail which *Lagynocystis* lacks. Such are: (1) the dorsal terminal bearing surfaces, (2) the absence of ventral bearing surfaces, (3) the concomitant subdivision of the hind tail into three regions with the proximal hind-tail ossicles as a mud-cutting prow, (4) the two-bladed styloid (as compared with the four- or five-pointed styloid of *Lagynocystis*), (5) articulation of two pairs of ventral plates to the styloid (none is articulated to

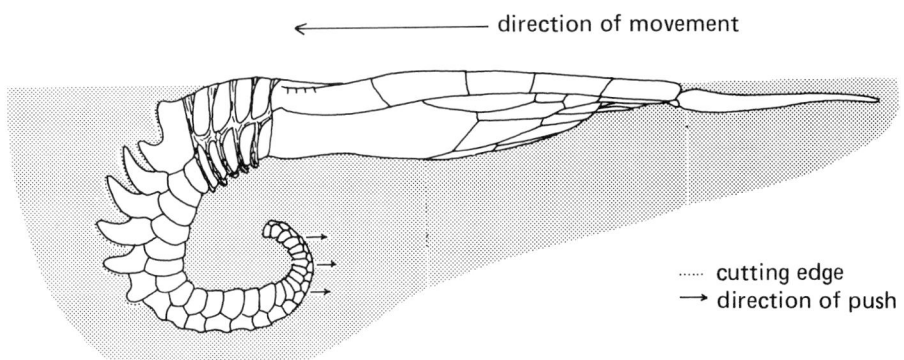

8.38. Peltocystis cornuta. *Suggested mode of action of the tail in locomotion. The tripartition of the hind tail probably had the same functional meaning as in* Chinianocarpos *(cf. fig. 8.27). (New.)*

it in *Lagynocystis*), and (6) reduction of dorsal spike-shaped plates in the fore tail (many large such plates in *Lagynocystis*, a lower number of smaller ones in *Chinianocarpos*, none in *Peltocystis*). These shared advanced features are probably synapomorphies and suggest that *Peltocystis* is closer related to *Chinianocarpos* than to *Lagynocystis*. This is an important result, for *Peltopcystis* is probably a stem tunicate (as will emerge when *Balanocystites* and *Anatifopsis* have been discussed), *Chinianocarpos* is a stem vertebrate and *Lagynocystis* a stem acraniate. If so, then tunicates are closer related to vertebrates than to acraniates.

Balanocystites primus is the next peltocystidan to be discussed (fig. 8.39). It is abundant in the siliceous nodules of Osek near Rokycany (Šárka Shales, Llanvirn) in Czechoslovakia. Its nomenclatural history is complicated, for the isolated large ventral plates of the head (M_{1LV}, M_{1RV}) were described first by Barrande (1872) under the name of *Anatifopsis prima*, in the belief that they probably belonged to a crustacean, perhaps a barnacle. The whole animal in associated state, complete with tail, was later described by Barrande in his posthumous work of (1887), under the name of *Balanocystites lagenula* and regarded as an echinoderm. In my view these names refer to one species which is a chordate and is correctly called *Balanocystites primus* (Barrande, 1872).

The head of *Balanocystites* was almost bilaterally symmetrical in outline with, as usual, a flattish dorsal and a convex ventral surface. A single oral appendage projected from the front end very nearly in the mid-line, but right of the mouth which was asymmetrical in position, opening leftwards. The mouth was largely covered dorsally, as in *Chinianocarpos* and *Peltocystis*, by a large plate *n* but plated integument was also present. The ventral surface of the head (fig. 8.39c) was almost entirely made up of two huge postero–ventral marginal plates (M_{1LV}, M_{1RV}) which join

each other at a suture at the mid-ventral line and extend on right and left onto the dorsal surface of the head as a pair of immovable opercula. (These two giant plates were the basis for Barrande's *Anatifopsis prima*.) These opercular parts of M_{1RV} and M_{1LV} overlap, and partly occlude, a dorsal skeleton (fig. 8.39b) which in other respects is much like that of *Peltocystis*. At the anterior end this dorsal skeleton is connected to the slightly mobile oral appendage.

The postero–ventral plates (M_{1LV}, M_{1RV}) were hinged posteriorly, at right and left, to the posterior dorsal plates M_{1LD} and M_{1RD}. More anteriorly they overlap onto these plates, completely occlude the second dorsal marginals (M_{2L}, M_{2R}) and mask parts of more anterior plates also. The atrial openings would have emerged, by comparison with other mitrates, on the dorsal surface of the head along the free edges of the postero–ventral plates, where these come in contact with the dorsal skeleton. In *Balanocystites* the openings were therefore very elongate and situated dorsally. Their length suggests that the atria were long also. The dorsal position of the openings is comparable with that of the atrial openings of a tunicate tadpole (fig. 4.26, p. 107) and can be seen as a synapomorphy with tunicates which suggests that *Balanocystites* was a stem tunicate. Forward-pointing spikes exist on the opercula of the dorsal surface of the head (fig. 8.39c), presumably for gripping sediment that passed over these parts during crawling. Such spikes are absent from the parts of the dorsal skeleton medial to the atrial opening, however, presumably because the branchial current would have removed any sediment from this region.

The fore tail of *Balanocystites* was shorter than that of *Peltocystis*. Its skeleton likewise consisted of four series of major plates separated by plated membranes but the plates are joined together to form rings in *Balanocystites* In the mid tail the styloid is two-bladed, with two pairs of ventral

8.39. Balanocystites primus. *Reconstruction of external features and relations of dorsal skeleton. Note that the plates* M_{1RV}, M_{1LV} *have encroached on the dorsal surface. a) Dorsal aspect, b) dorsal aspect of dorsal skeleton, c) ventral aspect, d) transverse section – the arrows indicate how water escaped through the atrial openings, and e) right lateral aspect; Jefferies (1981b, fig. 12, p. 374).*

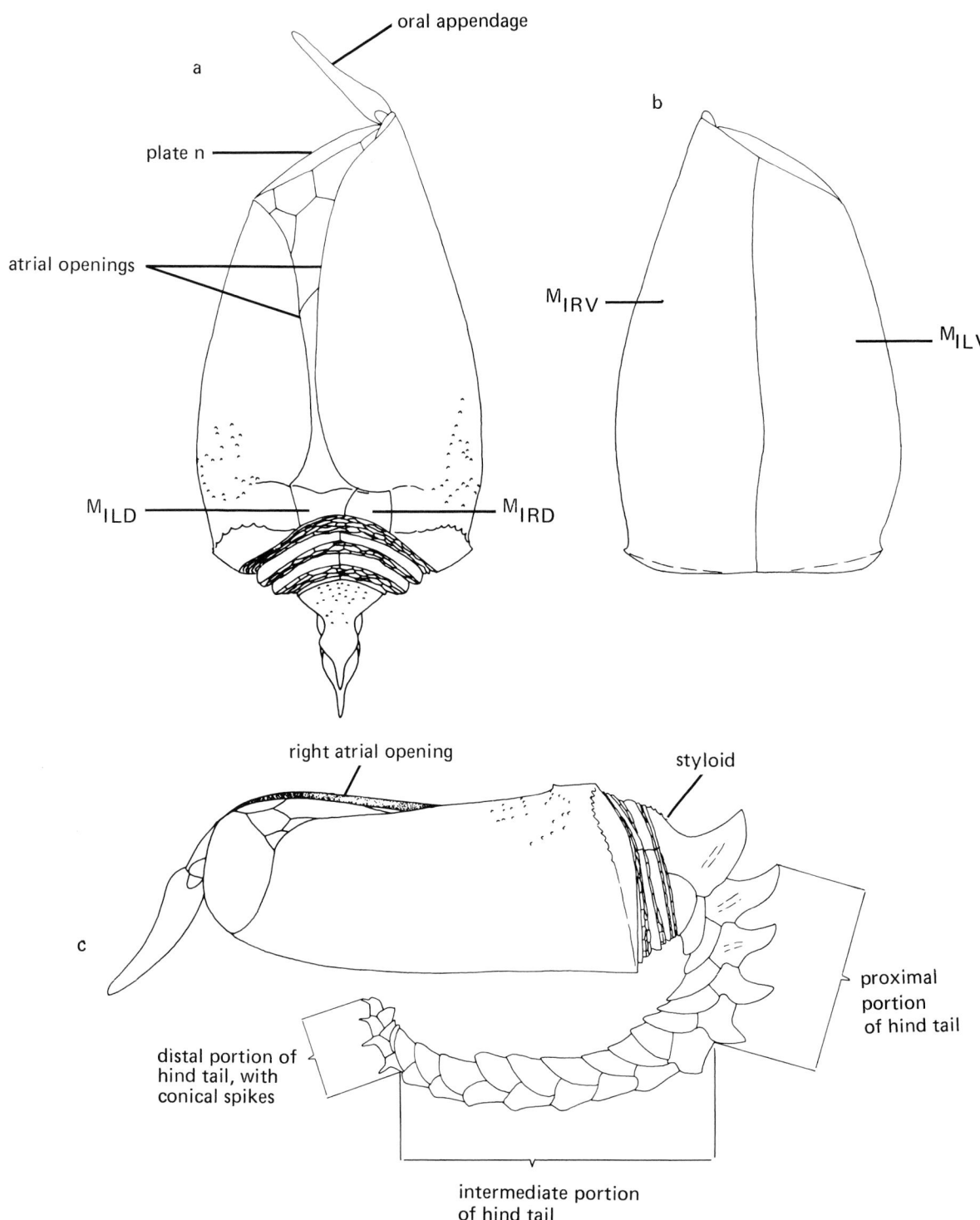

8.40. Anatifopsis barrandei. *Reconstruction of external features. a) Dorsal, and b) ventral aspect of head, c) left lateral aspect. Note that plates* M_{ILV}, M_{IRV} *almost meet in the dorsal mid line and the hind tail is tripartite as in* Peltocystis *and* Chinianocarpos *(figs 8.38, 8.27). (New.)*

plates articulated to it. The hind tail is truncated behind the tenth ossicle in the longest observed specimen, and the ossicles are long-bladed. If short-bladed and bladeless portions existed, as they did in *Peltocystis* and *Anatifopsis*, then they had broken away in this specimen. It seems likely that the hind tail of mitrates was fragile throughout life, and always liable to lose its distal part.

As to mode of life, *Balanocystites* probably crawled rearwards just beneath the sea bottom so that mud rode up over the lateral parts of the dorsal surface, where the opercula and dorsal spikes are. The oral appendage, almost exactly in the mid-line with respect to the dorsal outline although right of the mouth, was perhaps a steering device. It may have acted like a steering pole dragged in the snow behind a sledge.

Anatifopsis barrandei is the last peltocystidan to be discussed (fig. 8.40). It likewise occurs in the siliceous nodules of the Šárka Shale of Osek and Šárka. Its describer Chauvel (1941) recognized it as a mitrate but placed it in a new genus *Anatiferocystis* because it resembled, but in his belief was not the same as, *Anatifopsis* Barrande, 1872. I differ from Chauvel in this matter. The correct name for the species in my view is *Anatofopsis barrandei* (Chauvel, 1941). (As already implied, however, some of Barrande's '*Anatifopsis*' species ought to be referred to his genus *Balanocystites*.)

Anatifopsis is much like *Balanocystites* in most ways. The main differences are that: (1) the opercula extend further onto the dorsal surface of the head, almost meeting in the mid-line, and occluding an even large part of the dorsal skeleton, and (2) the mouth is different, for the big plate *n* slopes downwards almost vertically, instead of being nearly horizontal, and in consequence the oral appendage, right of the mouth, is well right of the mid-line. The extent of the opercula means that the atrial openings would be even more dorsal than in *Balanocystites* and very nearly median.

The tail is much as in *Balanocystites*, but more reconstructable, for one specimen shows 16 successive ossicles in the hind tail. The same regions are recognizable in the hind tail as in *Peltocystis*. Thus there is a proximal, long-bladed region, followed by a short-bladed region, followed by a terminal region that ends abruptly. The latter, however, has a short, conical spike on each ossicle instead of being smooth-backed (fig. 8.40c). If the tail is reconstructed in the likely position of greatest flexion, these hind-tail regions suggest the same functional significance as in *Peltocystis*. The spikes of the terminal region, however, would help it to grip the mud and confirm that it was an anchor. They would act, like the spikes on a running show, to prevent slip in the plane of the bearing surface.

Thus the peltocystidans form a morphological series. (1)

In *Peltocystis* the atrial openings have lengthened as compared with the primitive mitrate condition but still remain ventro–lateral, (2) in *Balanocystites* they have become still longer, and dorsal, and (3) in *Anatifopsis* they are even more dorsal and almost median in position. This morphological series is one of successive approach to the condition seen in living tunicates with their median dorsal atrium and probably represents successive degrees of relationship to crown tunicates within the tunicate stem group.

The phylogenetic significance of peltocystidans is threefold: (1) they are probably stem tunicates, (2) within the mitrates they seem to be closer related to stem vertebrates than to stem acraniates on the basis of synapomorphies in the tail (articulation of two pairs of ventral plates to the styloid, absence in the hind tail of ventral bearing surfaces comparable with those on the fore tail of *Reticulocarpos hanusi*, regionation of the hind tail into three

common atrial opening

mouth

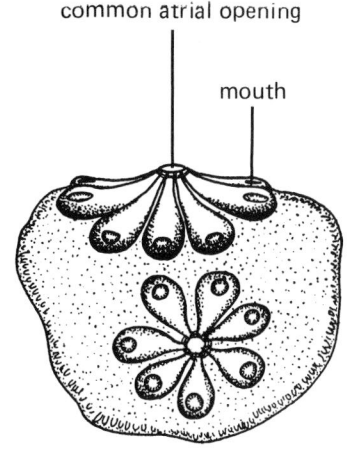

8.41. a) *The possible fossil tunicate Palaeobotryllus, from Müller (1977, fig. 2, p. 111); b) the Recent colonial tunicate Botryllus schlosseri – the radial arrangement of zooids in a colony is like that of Palaeobotryllus; from Berrill (1950, fig. 74b, p. 218).*

parts of which the most distal acted as a terminal anchor with dorsal ossicular bearing surfaces), and (3) if pelto-cystidans are stem tunicates they refute Garstang's pro-posal (1928) that the fixed, tailless adult tunicate stage existed in phylogeny before the tailed free-swimming tadpole. Instead, the tadpole stage, represented by cornutes and mitrates in general and peltocystidans in particular, is primitive, though it used its tail for crawling rather than swimming. The life cycle of modern tunicates is recapitulatory as Haeckel's law would imply, rather than vertebrates and acraniates being by origin paedogenetic.

Other tunicate fossils are sparse. They include spicules from Eocene rocks of France (Deflandre & Deflandre-Rigaud, 1956) which can plausibly be ascribed to the tunicate family Didemnidae but are of no help from the comparative anatomical viewpoint. More recently Müller (1977) has described a problematical fossil from the Upper Cambrian of Nevada under the name *Palaeobotryllus* (fig. 8.41a). This was a colonial animal whose zooids surrounded themselves in life with phos-phatic cups and were attached to a substrate. The zooidal cups of a colony radiate from a common centre, somewhat like the Recent tunicate *Botryllus*. In the latter, however, the radiality is due to the fact that the atrial siphons of the individual zooids open into a common cloaca (fig. 8.41b) with a common atrial opening. There is no sign of such a central cloaca in *Palaeobotryllus*, though conceivably it existed above the level of the preserved phosphate. The radial arrangement of the zooidal cups is, therefore, only weak evidence of affinity. Also, no living tunicates produce a phosphatic skeleton. If *Palaeobotryllus* is a tunicate, it would well be a crown tunicate since it was colonial and attached to a substrate. Possibly, however, it

is not a tunicate at all. Coming from the Upper Cambrian, it is somewhat older than the oldest known peltocystidan (*Peltocystis cornuta*, Lower Ordovician).

To summarize the argument of this chapter, the mitrates resembled giant calcite-plated tunicate tadpoles with a complex fish-like nervous system and an endostylar mucous filter in the pharynx comparable with that of amphioxus, tunicates and lamprey larvae. They have many internal asymmetries which recur in living primitive chordates. These include, among others: the rectum open-ing into the left atrium as in living tunicate tadpoles; a left-sided buccal cavity like amphioxus; sometimes inner-vation of the mouth mainly from the left, as in amphioxus; and the likelihood that the left gill slits preceded right gill slits in ontogeny as in amphioxus. Moreover, all known mitrates can be ascribed, on the basis of synapomorphies, to the stem groups of the living chordate subphyla – *Lagynocystis* was a stem acraniate; *Peltocystis, Balanocystites* and *Anatifopsis* were stem tunicates; and *Chinianocarpos, Mitrocystites, Mitrocystella* and *Placocystites*, along with some 15 genera not discussed here, were stem vertebrates. And tail synapomorphies suggest that the stem ver-tebrates were closer related to the stem tunicates than to the stem acraniate *Lagynocystis*. This means that, among extant forms, the tunicates, rather than the acraniates, are the sister group of the vertebrates. Inside the mitrate animal the classical vertebrate head segments (mandibular, premandibular, hyoidean etc.) can plausibly be identified. The mitrates had a locomotory tail which was used in ventral flexion for crawling rearwards. In the next chapter I shall discuss, among other matters, how the living chordates, and the living vertebrates in particular, evolved from their mitrate ancestors.

Chapter 9   The phylogeny of the deuterostomes

In this chapter I shall place on a cladogram the forms which I have so far discussed. I shall assume that the morphological interpretation of calcichordates already argued is correct, and their organs correctly identified. I shall often repeat arguments that have already been put, but here the sequence, so far as possible, will be strictly phylogenetic.

The deuterostomes include the chaetognaths (which I shall not discuss further), the hemichordates, the echinoderms and the chordates. These latter three groups are conventionally regarded as equal in rank and as separate phyla. However, as I shall argue below, and as already mentioned in Chapter 2 and under *Ceratocystis* in Chapter 7, the echinoderms and chordates are linked together in the monophyletic group Dexiothetica by the synapomorphies of dexiothetism and the calcitic skeleton of stereom mesh, dexiothetism being the condition of lying down on the primitive right side (fig. 2.55, p. 52). Consequently the hemichordates emerge merely as those deuterostomes which are not dexiothetes (nor chaetognaths). In embryological terms this means that the bilateral symmetry of hemichordate larvae becomes the bilateral symmetry of the adult, instead of going through a dexiothetic metamorphosis as with echinoderms. In terms of outgroup comparison, it means that the bilateral symmetry of hemichordates is probably homologous with that of the tentaculate protostomes (phoronids, brachiopods, bryozoans fig. 2.2, p. 18) but not with that of chordates. The lack of dexiothetism in hemichordates is therefore probably primitive. And being thus defined on a primitive condition, the hemichordates are suspect of being paraphyletic. This would mean that some of them were closer related to Dexiothetica than others are, where relationship is defined strictly in terms of recency of shared ancestry. However, it is not yet possible to say whether this suspicion is true.

The hemichordates are conventionally divided into the lophophorate, fixed, colonial pterobranchs such as *Cephalodiscus* (fig. 2.4, p. 20) and the worm-like, free-living, solitary enteropneusts (fig. 2.12, p. 25). The pterobranchs probably represent the primitive condition, for the lophophorate feeding habit is probably homologous with that of tentaculate prostostomes and also with the primitive feeding habit of the Dexiothetica as now seen in crinoids (fig. 2.43, p. 43). Indeed 'lophophorate' and 'tentaculate' are synonyms arbitrarily applied to different animals. In having lost this feeding habit and becoming worm-like, therefore, the enteropneusts are likely to be a monophyletic group within the hemichordates. The pterobranchs, however, being defined on a primitive retention of lophophorate feeding, are suspect of paraphyly. But once again, it is impossible to say whether the suspicion is true.

Living pterobranchs include *Cephalodiscus* and *Atubaria* (fig. 2.9, p. 23), with gill slits, and *Rhabdopleura* without them (figs 2.10, 2.11), as described in Chapter 2. It is difficult to tell which of these two conditions is the more primitive. Perhaps *Rhabdopleura* has never had slits and is closer related to the protostomes than *Cephalodiscus* and *Atubaria* are. Or perhaps *Rhabdopleura* once had slits and has lost them. The interrelationships between *Cephalodiscus* (and *Atubaria*), *Rhabdopleura*, the enteropneusts and the dexiothetes are unknown.

As concerns the origin of the dexiothetes, however, this uncertainty within the hemichordates is not crucial. For if the gill slits of chordates are homologous with those of hemichordates, then the latest common ancestor of living dexiothetes (corresponding to node w in fig. 9.2b) had gill slits (or a gill slit). And if the lophophorate mode of feeding seen in crinoids is homologous with that of pterobranchs, then this same latest common ancestor of living dexiothetes was lophophorate. In these respects it was more like *Cephalodiscus* (and *Atubaria*) than like *Rhabdopleura*, which lacks gill slits, or enteropneusts, which lack a lophophore. *Atubaria* need not be considered further because it is not well known and in most ways seems the same as *Cephalodiscus*. Consequently in discussing the origin of Dexiothetica, I shall begin with a *Cephalodiscus*-like ancestor. This is a safe beginning for, in constructing a cladogram to include only *Cephalodiscus*, the echinoderms and the chordates, it is fairly certain that *Cephalodiscus* is the sister group of the others.

Probable homologies between *Cephalodiscus* and echinoderms can be tabulated on the basis of features mentioned in Chapter 2, listed in sequence from the anterior to the posterior end of *Cephalodiscus*.

This table suggests that, if the condition seen in *Cephalodiscus* is taken as primitive for the reasons already given, then echinoderms are advanced in: (1) having the division between the right and left somatocoels horizontal, with the right somatocoel downward in crinoids and the left somatocoel upward (fig. 2.53b,c,d), (2) having lost the right hydrocoel (mesocoel), while the left hydrocoel became the water vascular system and, like the left somatocoel, faces upward in crinoids, (3) having lost the right hydropore (right protocoel pore) while the left hydropore became the hydropore of adult echinoderms – the numerous hydropores open upward in crinoids, (4) having established a secondary connection (stone canal) between the left hydrocoel and the axocoel which in *Cephalodiscus* does not exist and which in crinoids happens late in ontogeny, (5) having lost the right and left gill slits, and (6) in having lost the right gonad and gonopore.

The table suggests, in fact, that the *Cephalodiscus*-like ancestor of deuterostomes, after fixing itself to a substrate by the ventral surface of the head shield, fell over on the right side, as already proposed in Chapters 2 and 7. As a result the right somatocoel was downward, the left somatocoel and left hydrocoel upward and the right hydrocoel tentacles and the openings of the right side (right gill slit, right hydropore, right gonopore) became non-functional and were lost. This was the dexiothetic condition. At about this stage, the left hydrocoel pore was

Cephalodiscus	Echinoderms
protocoel (i.e. head shield coelom)	axocoel (i.e. axial sinus)
left protocoel pore	hydropore (upward in crinoids)
right protocoel pore	right hydropore (pathological only)
glomerulus	axial gland
pericardial vesicle	madreporic vesicle (i.e. dorsal sac)
central sinus of pericardial vesicle	head process of axial organ
ventral face of protosome	attachment disc of larva (in some sense = attachment of stalked crinoids)
left mesocoel	left hydrocoel (i.e. water vascular system)
left mesocoel pore	absent
right mesocoel	right hydrocoel (pathological only)
right mesocoel pore	absent
absent	stone canal (i.e. the ontogenetically secondary connection of left hydrocoel with axocoel)
stomochord	absent
left metacoel	left or oral somatocoael (upward in crinoids)
left gill slit	absent
right gill slit	absent
left gonad and gonopore	gonad and gonopore
right gonad and gonopore	absent
stem	absent (for the stem of crinoids is probably not homologous with that of *Cephalodiscus*)

lost, the stone canal was produced and a calcitic skeleton was acquired. Such an animal would be the latest common ancestor of chordates and echinoderms. Dexiothetism, the stone canal and the calcite skeleton were acquired in segment 1 of fig. 9.2b.

Later, in evolving towards crown echinoderms, life-long fixation by the primitive right side was adopted, so that the attachment area of crinoid larvae is homologous in some sense with the ventral surface of the head shield of *Cephalodiscus* and in some sense with its right body surface. (The chambered organ, which is the hydrostatic skeleton of the crinoid stem, pouches out from the right, aboral, definitively downward-facing somatocoel, but ends in the early larva at the attachment disc which is in the larval median plane.) The mouth, however, moved to the centre of the originally left, now upward, surface and the base of the hydrocoel grew around the mouth to form the circumoral water vascular ring. For in the ontogeny of all living echinoderms this ring is at first U-shaped (fig. 2.51b). It closes secondarily, but only when, as is normal, the right hydrocoel aborts or never appears. This onto-genetic loss of the right hydrocoel probably recapitulates phylogeny. At about the same time, in phylogeny, the remaining, originally left, gill slit was lost.

The radial symmetry of echinoderms, usually based on the number five and induced in embryology by the water vascular system, probably resulted from this fixation. For fixed organisms or organs, whether buttercups, oak trees or sponges, often develop radial symmetry as a way of being equally exposed on all sides to the environment. Stephenson (1979) has discussed why the radiality became five fold. In my opinion, which in this respect differs from his, the five-fold condition probably followed a three-fold condition, as suggested by Bather (1900) and confirmed by Bell's recent discovery (1976) of a triradiate stage in the ontogeny of some Ordovician and Devonian edrioasteroids (fig. 9.1). The acquisition of quinqueradi-ality and the loss of the left gill slit happened in segment 2 in fig. 19.2b, though they were probably not simultaneous.

Living echinoderms can be divided into: (1) the attached Pelmatozoa, all the surviving members of which are crinoids (e.g. *Antedon* in Chapter 2), and (2) the free-living Eleutherozoa (e.g. *Asterias* in Chapter 2). Pelmato-zoans have the mouth facing upwards and use the water vascular system in somewhat the primitive lophophorate manner, to catch food particles from the surrounding water. They retain the primitive mode of feeding among living echinoderms, but there is no doubt that living crin-oids make a monophyletic group. For they are character-ized by many advanced features which all other living echinoderms lack and have probably never had. Such include: the stem (present in *Antedon* only in the larva): the ring-shaped stem ossicles; the chambered organ of the stem; the massive aboral nervous system; the loss of the madreporic vesicle; the multiplication of hydropores and stone canals; the fact that hydropores and stone canals open secondarily into the body coelom; and the many-times-branched arms.

Eleutherozoans, on the other hand, are held together as a monophyletic group by the advanced features of the downward-facing mouth and the use of the water vascular system in locomotion. Consequently the larval left side, the left or oral somatocoel and the left hydrocoel (water vascular system) are downward rather than upward in them (fig. 2.38b). This indicates that, in their phylogeny, the eleutherozoans have turned over and fallen on what had previously been the upward surface. Their descent from a pre-existing overtly dexiothetic ancestor, however, is demonstrated by the embryological asymmetries already mentioned such as the lack of the right hydrocoel and right hydropore.

The early phylogeny within the phylum Echinodermata and the allocation of the known primitive fossils to their correct stem groups is an open field of study where great advances will happen in the next few years, now that the requisite methodology has been invented. Indeed, major discoveries have been made already (Paul & Smith, 1984; Smith, 1984). I shall not discuss it further except to say that the Middle and Lower Cambrian fossil *Stromatocystites*, mentioned and illustrated in Chapter 2 (fig. 2.56, p. 53), is probably not far distant from the latest common ancestor of living pelmatozoans and eleutherozoans, for it was stemless but lived mouth-upward on the sea floor. Also its quinqueradiate ambulacral system seems to retain traces of an earlier triradiality since it conforms to the pattern $(2 + 1 + 2)$ seen in many early echinoderms.

The cornutes share synapomorphies with living chordates which echinoderms do not show and probably never had. Such include: (1) the muscular, locomotory, post-anal tail with the notochord, and (2) the brain and paired trigeminal ganglia at the front end of the notochord. Both 1 and 2 were acquired in segment 3 in figs 9.2b, 9.3. These synapomorphies show that cornutes were chordates. Unlike mitrates, however, cornutes had no right gill slits which all living adult chordates possess, or used to possess. (In my view the right gill slits of mitrates and living chordates are new acquisitions, not homologous with the right gill slits of hemichordates which were lost during the overtly dexiothetic stage, i.e. within the stem group of the Dexiothetica (fig. 9.2a)). The lack of right gill slits shows that cornutes were stem chordates, not crown chordates like mitrates. The large number of gill slits penetrating the upper surface of cornutes may be another synapomorphy of cornutes with living chordates. For if the first crown dexiothete (the latest common ancestor of living echinoderms and chordates) resembled a

a

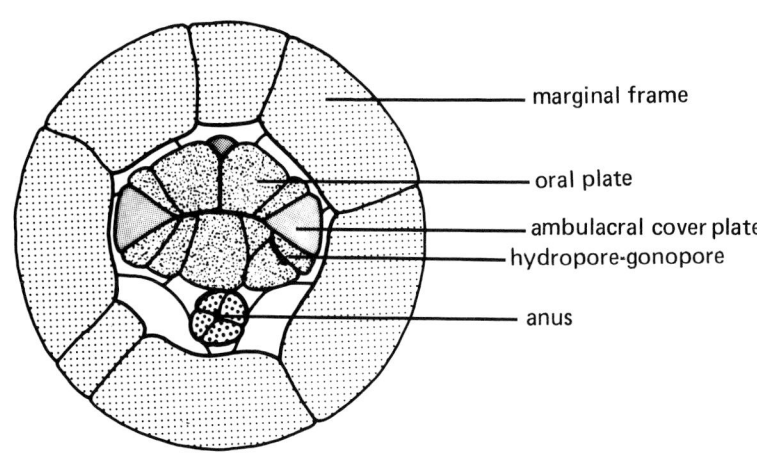

— marginal frame

— oral plate

— ambulacral cover plate
— hydropore-gonopore

— anus

b

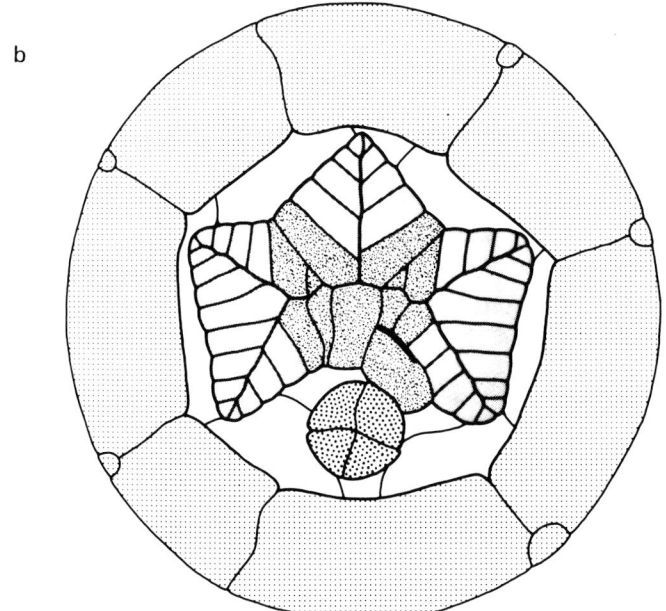

9.1. The triradiate stage in the early ontogeny of an edrioasteroid echinoderm (Timeischites, M. Devonian). a) Juvenile about 1·20 mm in diameter with three conjoined ambulacral grooves, b) larger juvenile (about 3 mm in diameter) with five ambulacral grooves but retaining an obvious 2 + 1 + 2 arrangement; the largest adults are about 7·75 mm in diameter; from Bell (1976, text fig. 4, e,h).

Cephalodiscus that had lain down on its right side, it would have had only one such slit. This conclusion is tentative, however, and implies that the numerous gill slits of cornutes and other chordates are a parallelism with those of enteropneusts.

The cornutes retain several resemblances to echinoderms which living chordates have lost. The most important of these are the skeleton built of stereom mesh with each plate a single crystal of calcite, and strong asymmetry inherited directly from the dexiothetic latest common ancestor of chordates with echinoderms (node *w* in fig. 9.2b). Thus the right anterior coelom underlay the major part of the gut, as does the homologous right somatocoel of crinoids or the right somatocoel reconstructed in *Stromatocystites* (fig. 2.56). Again, the position of the head openings as seen in the dorsal aspect of the cornute *Ceratocystis* can be compared directly with that of the left aspect of *Cephalodiscus*, as discussed in Chapter 7 (fig. 7.23, p. 221). This was presumably because the primitive dexiothetic arrangement had not been modified beyond recognition in *Ceratocystis*. Thus cornutes, as stem chordates, preserve synapomorphies with echinoderms of which living chordates have lost all trace. This is their importance from the point of view of systematics and shows that fossils, contrary to Løvtrup (1977), can sometimes decide high-level groupings.

Subdivision of the cornutes, viewed as stem chordates, involves assigning known cornutes to plesions distinguished from each other by different degrees of relationship to the crown chordates (figs 9.2b, 9.3), or more particularly to the mitrates as primitive crown chordates. I shall speak of a 'less crownward' or a 'more crownward' plesion to indicate a more distant or a closer degree of relationship to the crown group. Behind these expressions there is an evolutionary meaning. For within the chordate stem group there would have been a stem lineage ('Stammlinie' of Ax, 1984) the members of which were all directly ancestral to all more crownward members of the stem lineage and to the crown group. When I speak of species *a* being less crownward than species *b*, I mean that *a*'s latest common ancestor with the crown chordates – a member of the stem lineage – was older than *b*'s latest common ancestor with the crown chordates, which was likewise a member of the stem lineage. This is not exactly the same as saying that *a* is more primitive than *b*, for *a* may show advanced features which never existed in the stem lineage, and, therefore, never in the ancestry of *b*. Also, 'less crownward' does not mean 'stratigraphically older'. Without a perfect stratigraphical record, which for the calcichordates we certainly do not have, it is not possible to assign cornutes to the stem lineage of the crown chordates. On the other hand, it is often possible to say that a particular known form could *not* have been a member of the stem lineage, either because it is stratigraphically younger than more crownward members of the stem group or members of the crown group, or else because it shows advanced features which never existed in the stem lineage. A plesion, as here defined, includes all those members of a stem group which so far as can be discerned, are equally closely related to the crown group. As already pointed out in Chapter 1, a plesion, if completely known, is paraphyletic.

Among known cornutes, *Ceratocystis* (fig. 7.17, p. 215) is the least crownward (figs 9.2b, 9.3). This is shown by the fact that it has a hydropore, like echinoderms and like hemichordates (left protocoel pore). All other cornutes lack a hydropore, the loss of which is a synapomorphy of

them with crown chordates. Several other features confirm that *Ceratocystis* is less crownward than *Nevadaecystis* (fig. 9.3), but they are recognized as primitive on functional rather than comparative grounds, as explained in Chapter 7. Thus: (1) the gonopore-anus is right of the tail rather than left, which is probably primitive because it makes no use of the branchial current as a flushing mechanism, (2) the roof of the head is rigid rather than flexible (except for the beginnings of flexibility suggested by accessory gaps and rounded plate junctions), and this rigidiy is probably primitive because *Nevadaecystis* seems to have traces of it in functionless and vestigialized form, and (3) there is no strut in the floor of the head, which is probably primitive because the strut was mechanically unnecessary given a primitively rigid roof. *Ceratocystis* is the only known member of its plesion. The presence of only one hydropore and only one gonopore are synapomorphies of *Ceratocystis* with echinoderms as opposed to *Cephalodiscus*. They show that chordates are closer related to echinoderms than to *Cephalodiscus*.

The median eye seen in *Ceratocystis* is not special to the genus for it occurs in several cornutes not discussed in Chapter 7, as witnessed by a groove in the median dorsal plate M_{PD}. Such cornutes include Sprinkle's Middle Cambrian '*Cothurnocystis*' and also three French Lower Ordovician species *Phyllocystis crassimarginata* Thoral, *P. blayaci* Thoral (fig. 9.5) and *Cothurnocystis fellinensis* Ubaghs (see Ubaghs, 1969, especially figures 2 (1a, 1c), 3, 4, 5, 11, 19, 21 (1,2)). These forms have a flexible floor to the head but lack certain mitrate-like features of the more crownward cornutes *Galliaecystis*, *Amygdalotheca* and *Reticulocarpos*. They, therefore, belong to the plesion of *Cothurnocystis* which, as shown below, is more crownward than *Ceratocystis* or *Nevadaecystis* but less crownward than *Galliaecystis*. The taxonomic distribution of the median eye, therefore, suggests that it was a primitive feature of cornutes. It should exist in *Nevadaecystis* also, unless it has been secondarily and independently lost in that genus, but the only known specimen is not good enough to show whether it was present.

The acustico–lateralis sense organ ('narrow groove'), located just left of the tail in *Ceratocystis*, is likewise probably not peculiar to that genus, although it has never been found in any other cornute. For it probably existed just left of the tail in these forms but is conflated with the gonopore-anus in the fossils. Its occurrence in the left atrium of mitrates shows that it existed in the stem lineage from the *Ceratocystis* plesion crownwards. It was, therefore, probably general in cornutes, although usually invisible.

The stratigraphically early date of *Ceratocystis* cannot be used as evidence of primitiveness since Sprinkle's Middle Cambrian '*Cothurnocystis*' is slightly older,

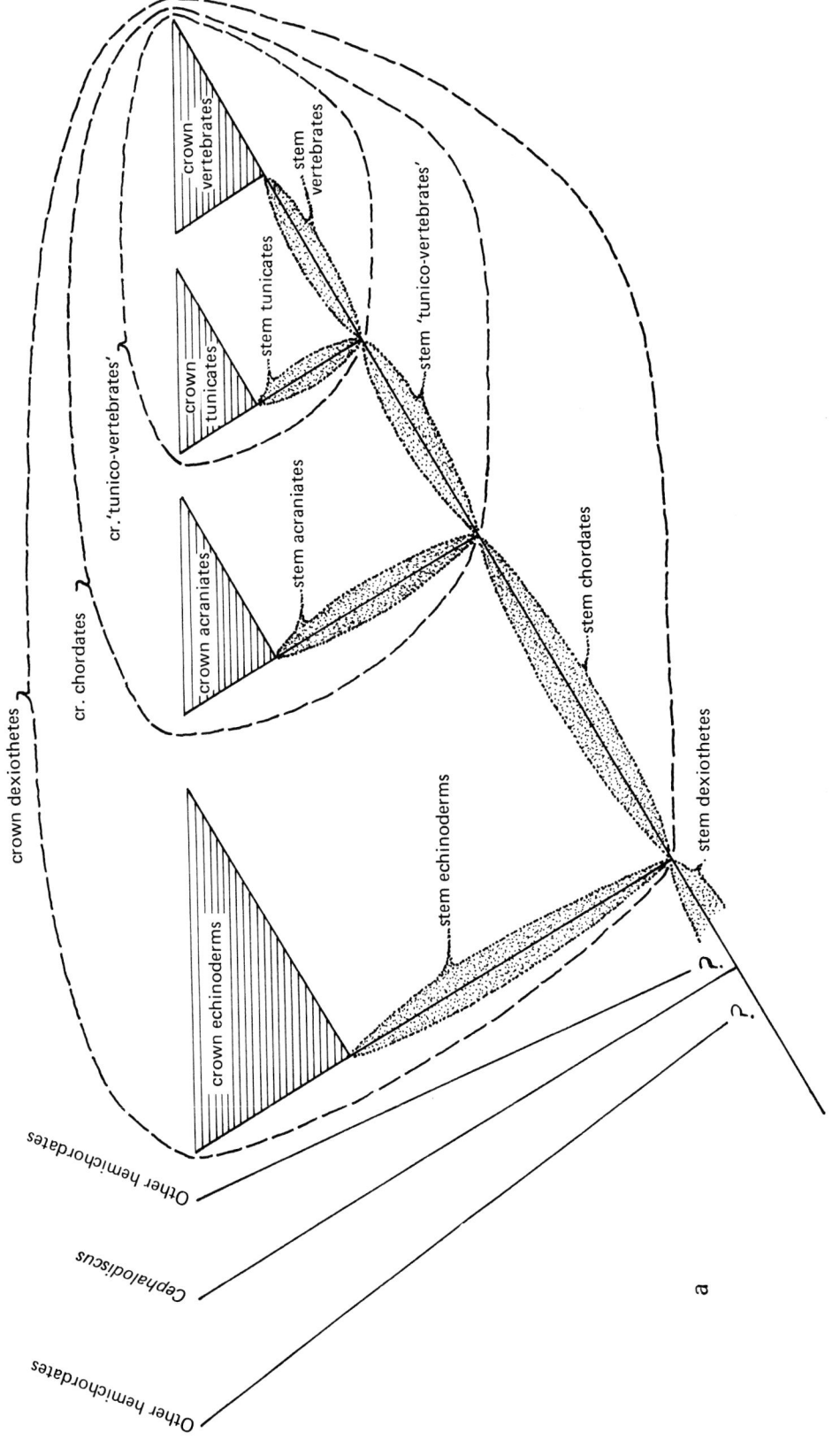

a

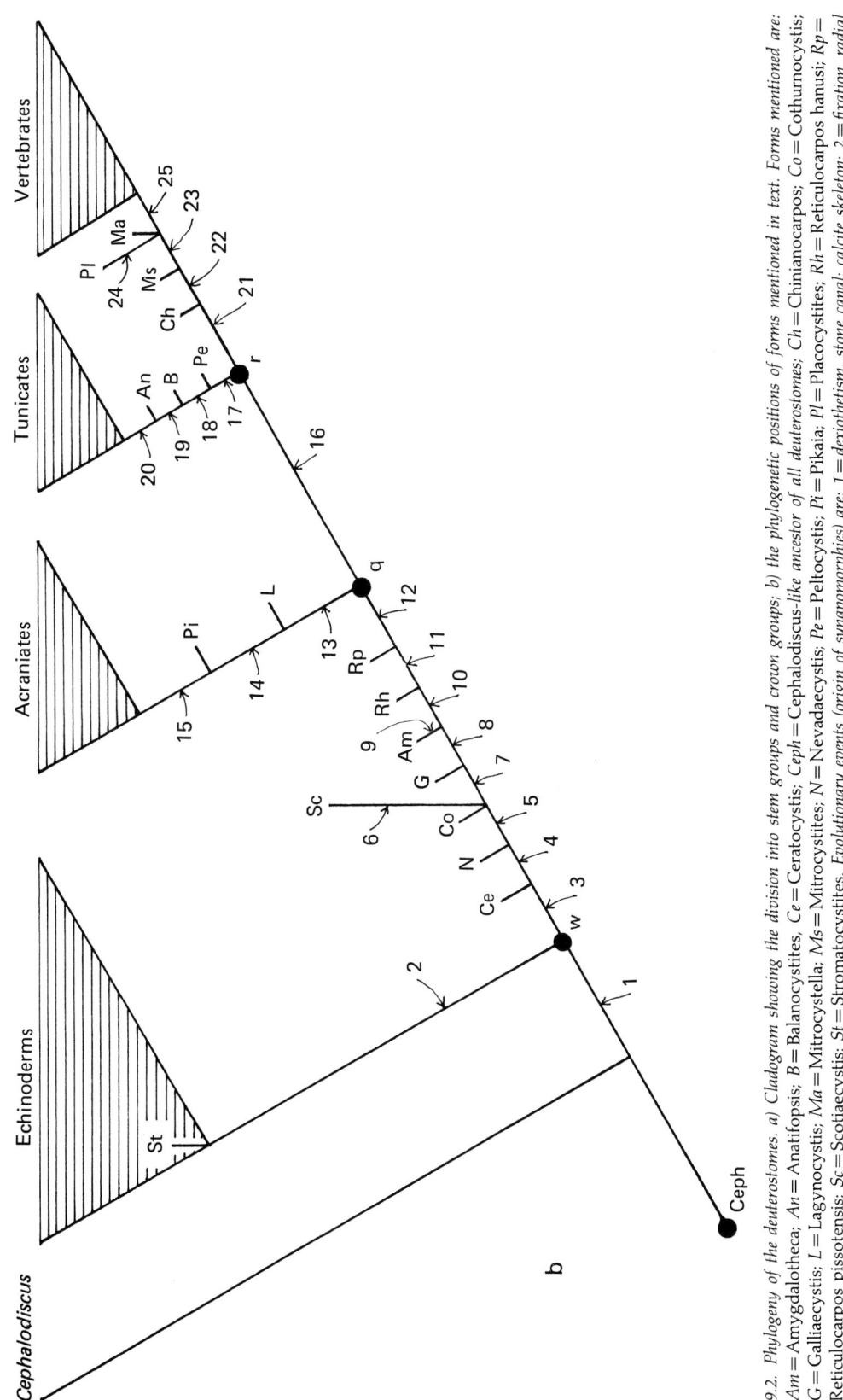

9.2. *Phylogeny of the deuterostomes. a) Cladogram showing the division into stem groups and crown groups; b) the phylogenetic positions of forms mentioned in text. Forms mentioned are: Am = Amygdalotheca; An = Anatifopsis; B = Balanocystites; Ce = Ceratocystis; Ceph = Cephalodiscus-like ancestor of all deuterostomes; Ch = Chinianocarpos; Co = Cothurnocystis; G = Galliaecystis; L = Lagynocystis; Ma = Mitrocystella; Ms = Mitrocystites; N = Nevadaecystis; Pe = Peltocystis; Pi = Pikaia; Pl = Placocystis; Rh = Reticulocarpos hanusi; Rp = Reticulocarpos pissotensis; Sc = Scotiaecystis; St = Stromatocystites. Evolutionary events (origin of synapomorphies) are: 1 = dexiothetism, stone canal, calcite skeleton; 2 = fixation, radial symmetry, loss of remaining gill slit; 3 = locomotory tail (with notochord, muscle blocks, dorsal nerve cord), brain, trigeminal ganglia; 4 = flexible head roof, strut (as thickening of floor), gonopore-anus left of tail; 5 = flexible head floor, loss of median eye; 6 = chevron-shaped branchial plates, dorsal mouth, gonopore-anus opening into gill slits; 7 = dorsal bar, 8 = symmetrical outline, peripheral flange, anterior strut plate separated from the frame by a suture; 9 = loss of dorsal bar, dorsal plates bobbin-shaped; 10 = small size, anterior strut plate does not touch frame; 11 = ventral surface convex; 12 = right gill slits and pharynx, left and right atria, loss of cornute mid and hind tail, regionation of remaining cornute fore tail into new fore, mid and hind parts; 13 = median atrium with tertiary gill slits and mid-ventral atriopore; 14 = loss of skeleton, loss of tail autonomy, blurring of distinction between head and tail; 15 = notochord extends to front end of animal; 16 = tripartition of hind tail; 17 = elongation of atrial openings; 18 = dorsal position of atrial openings; 19 = near median position of atrial openings; 20 = fixation (with torsion and loss of tail in adult), mantle, loss of calcite skeleton; 21 = lateral line (as a pit), dorsal branches of trigeminal (n_{4+5}); 22 = unbranched, groove-shaped lateral line, partial separation of n_4 and n_5, reduction of transpharyngeal eyes, epipharyngeal resorbtion of skeleton (but not much); 23 = sometimes branched lateral line, dorsal separation of n_4 and n_5 and burial beneath dorsal surface of head, loss of transpharyngeal eyes, extensive epipharyngeal resorbtion of skeleton; 24 = loss of lateral line, oral spines developed; 25 = total loss of skeleton, origin of notochordal head and of trunk region and other changes in the ancestry of the crown vertebrates. For further information see text. The nodes q and r are reconstructed in figs 9.6 and 9.8. The node w was the first crown dexiothete (latest common ancestor of living chordates and echinoderms).*

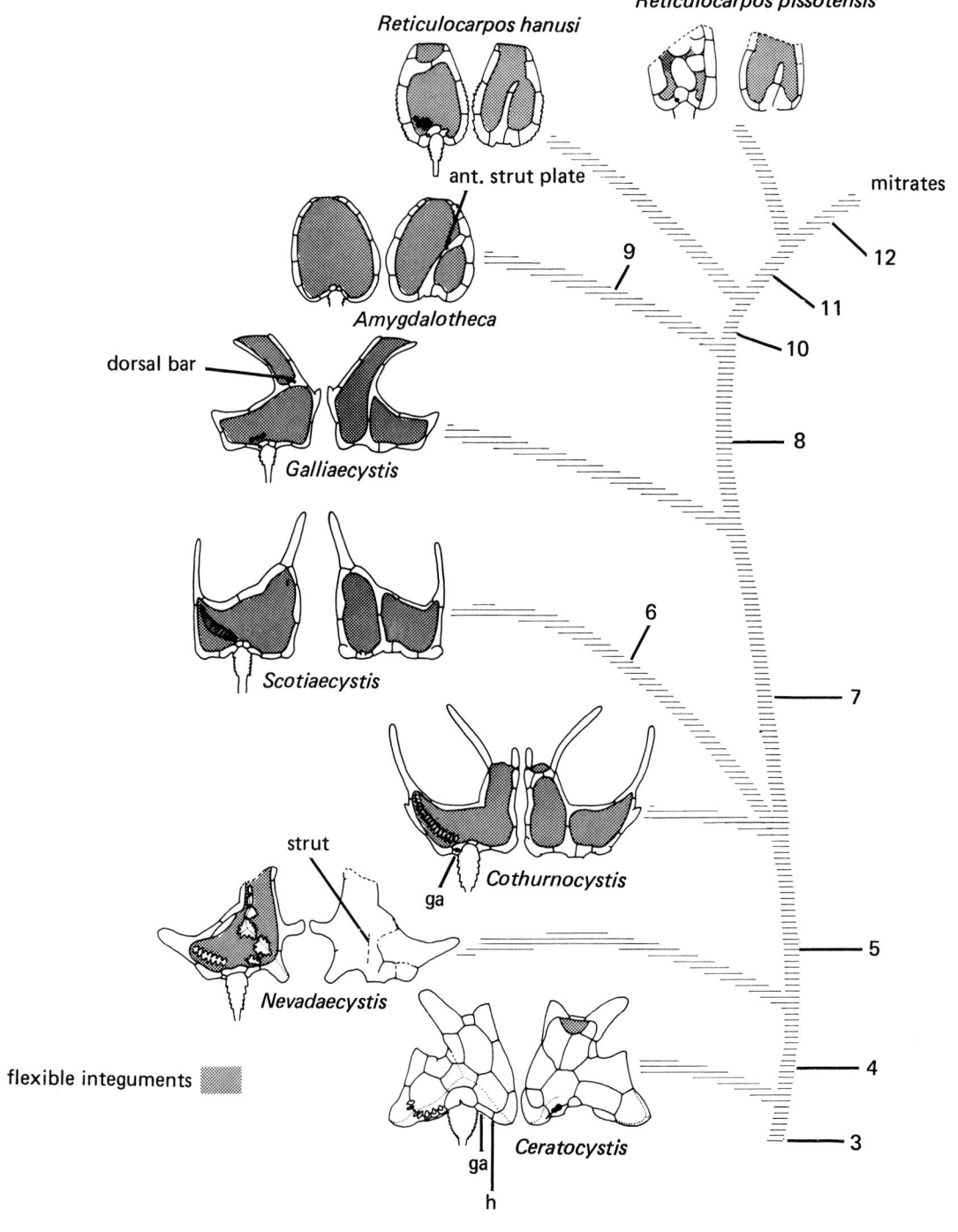

9.3.   *The phylogeny of stem chordates (cornutes). For each illustrated species dorsal aspect at left, ventral at right. Synapomorphies 1 to 12 explained in Fig. 9.2 and text. ga = gonopore-anus, h = hydropore.*

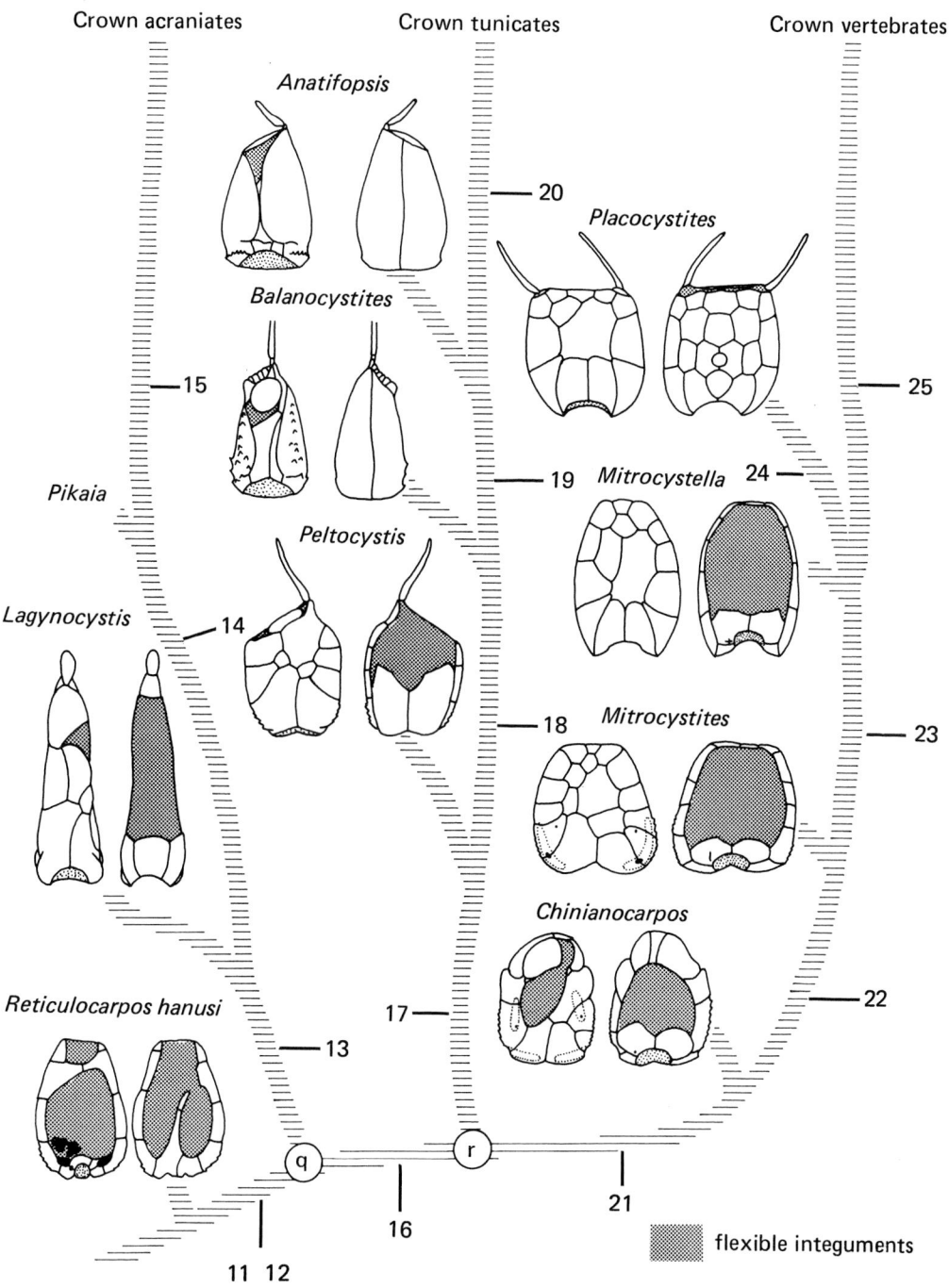

9.4. *The phylogeny of primitive crown chordates (mitrates). For each species, dorsal aspect at left, ventral at right. Synapomorphies 11 to 25 as explained in fig. 9.2 and text.*

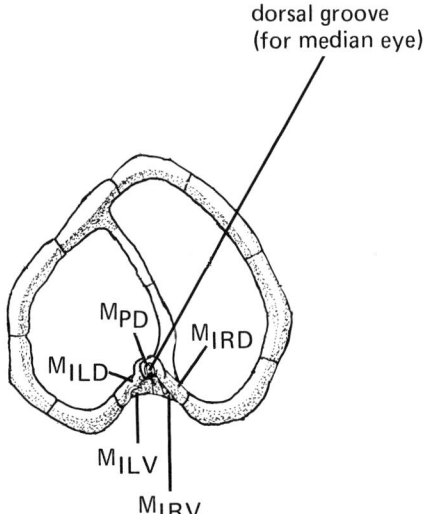

dorsal groove
(for median eye)

M_{PD} M_{IRD}

M_{ILD}

M_{ILV}

M_{IRV}

9.5. The cornute Phyllocystis blayaci *(Lower Ordovician, Montagne Noire). Marginal plates of the head in dorsal aspect with median eye in plate* M_{PD}; *from Ubaghs (1970, fig. 3/1).*

but closer to the crown group than *Ceratocystis* or *Nevadaecystis* in having a flexible floor to the head.

Nevadaecystis[*] (figs 7.27, p. 226, 9.3) is more crownward than *Ceratocystis*, but less crownward than any other known cornute. For it shares with still more crownward cornutes a number of advanced features which *Ceratocystis* lacked. These include: (1) the flexible roof of the head, (2) the likely position of the gonopore-anus left of the tail, (3) the *Cothurnocystis*-like gill slits framed by two U-shaped plates, and (4) the strut in the floor of the head, though this was represented in *Nevadaecystis* by a mere thickening. These can be seen as synapomorphies with these more crownward cornutes. On the other hand, *Nevadaecystis* has several *Ceratocystis*-like and presumably primitive features which other known cornutes lack. Such include: (1) the rigid floor of the head, (2) the facts that the plates of the dorsal integument are large, homologizable with those of the roof of *Ceratocystis* and crossed by keels homologous with those of *Ceratocystis*, and (3) the equality in size of the anterior and posterior U's of each gill slit, which is comparable with the situation in *Ceratocystis* where each slit is excavated about equally from the plate above and the plate beneath it. The keels of the dorsal integument make no functional sense in *Nevadaecystis* itself and are, therefore, probably vestiges of the keels seen in *Ceratocystis* which

[*]Recent work has shown that *Protocystites menevensis* Hicks, from the Middle Cambrian of Wales, represents a plesion intermediate between those of *Ceratocystis* and *Nevadaecystis* (Jefferies, Lewis & Donovan, in press).

would have served to stiffen the roof in that genus. Again, the equality of the U-shaped plates of a *Nevadaecystis* gill slit is hydrodynamically inferior to the inequality shown in *Cothurnocystis* since it would have blocked the outflow more – equal U-plates are therefore probably more primitive than unequal ones. These *Ceratocystis*-like features of *Nevadaecystis* have presumably been lost, for functional reasons, in all cornutes except *Nevadaecystis* and *Ceratocystis*. These losses are synapomorphies of these more crownward cornutes, therefore, and show that they are closer to the crown group than *Nevadaecystis* was.

Stratigraphically speaking, *Nevadaecystis* from the Upper Cambrian is younger than *Ceratocystis* from the Middle Cambrian, as a naive stratigrapher would expect. However, both are younger than Sprinkle's Middle Cambrian '*Cothurnocystis*', so the stratigraphical criterion of primitiveness fails again. *Nevadaecystis* is the only known member of its plesion.

Cothurnocystis elizae (figs 7.2, p. 194, 9.3) is closer to the crown group than *Nevadaecystis* is. This is shown by several features, presumed synapomorphies, which *Cothurnocystis elizae* shares with still more crownward cornutes, but which *Nevadaecystis* lacks. The more crownward cornutes in question, in sequence of increasing closeness to the crown group, are: *Galliaecystis*, *Amygdalotheca*, *Reticulocarpos hanusi* and *Reticulocarpos pissotensis*. And the presumed synapomorphies, lacking in *Nevadaecystis*, of *C. elizae* with these forms are: (1) the flexible floor of the head (except for the rigid strut), (2) the fact that the dorsal integument of the head is covered with uniform small plates without keels, and (3) the lack of a median eye (the relevant anatomy is unknown in *Galliaecystis* and *Amygdalotheca*, as also in *Nevadaecystis*, but such an eye was present in *Ceratocystis* and some other cornutes, as already pointed out, and was certainly absent in *Reticulocarpos hanusi*, as it was in the mitrates, and probably was in *R. pissotensis*). These features would have been acquired in segment 5 of figs 9.2b, 9.3.

On the other hand, *Cothurnocystis elizae* lacks synapomorphies which connect these more crownward cornutes with the crown group. Thus: (1) it has no dorsal bar connecting right and left parts of the marginal frame behind the buccal cavity, as seen in *Galliaecystis* and *R. hanusi*, (2) it has two U-shaped plates framing each gill slit, as in *Nevadaecystis*, while these plates are absent, presumed lost, in *Galliaecystis*, *Reticulocarpos hanusi*, *Amygdalotheca* and *R. pissotensis*, (3) the oral integument does not extend to the anterior end of the marginal frame in *C. elizae* – this situation resembles *Ceratocystis*, but is unlike *Galliaecystis*, *Amygdalotheca* or *Reticulocarpos*, and (4) there is a ventral mouth frame in *Cothurnocystis elizae*, formed of two plates homologizable with floor plates behind the mouth in *Ceratocystis*, whereas *Galliaecystis* and more crownward cornutes lack such a frame, presumably having lost it. On these grounds it seems certain that *Cothurnocystis elizae* is more crownward than *Nevadaecystis*, but less crownward than *Galliaecystis*. A plate in the frame, which exists in *Cothurnocystis elizae* (M_{3L} of fig. 7.2a) but not in *Ceratocystis*, *Nevadaecystis*, *Galliaecystis* nor *Reticulocarpos*, is probably a specialization of *Cothurnocystis elizae* and some other cornutes, and never existed in the stem lineage. *Cothurnocystis elizae*, which comes from the uppermost Ordovician, cannot be ancestral to more crownward cornutes such as *Galliaecystis*, *Amygdalotheca* and *Reticulocarpos* nor to the mitrates, for stratigraphical reasons.

Scotiaecystis (figs 7.12, p. 208, 9.3) is likewise more crownward than *Nevadaecystis* but less crownward than *Galliaecystis*. It is more crownward than *Nevadaecystis* in having a flexible floor to the head and in the absence of a median eye. It is probably less crownward than *Galliaecystis* in having a differentiated branchial skeleton. (The chevron-shaped branchial plates of *Scotiaecystis* were probably formed in phylogeny by the fusion of neighbouring U-shaped branchial plates of *Cothurnocystis* type, whereas *Galliaecystis* has lost such plates.) On the other hand, *Scotiaecystis* shows many advanced features which probably never existed in the stem lineage. Such include: (1) the chevron shape of the branchial plates, (2) the internal gonopore-anus opening into the gill slits, which is not homologous to the internal gonopore-anus of mitrates since the opening was external in the more crownward cornutes *Galliaecystis* and *Reticulocarpos*, (3) the dorsal mouth, (4) the presence of only one oral appendage, (5) the bobbin-shaped integument plates, (6) the upward convexity of the head in the mid-line, and (7) the presence of an extra plate (M_{5L} in fig. 7.12) in the left anterior part of the frame. These features were acquired in segment 6 of figs 9.2b, 9.3. Most of them are shared with two other genera – *Thoralicystis* from the lowest Ordovician of southern France and Morocco, and *Bohemiaecystis* from the Middle Ordovician (Llandeilo) of *Bohemia*. The three genera together constitute a well-defined monophyletic group named Scotiaecystidae by Caster & Ubaghs in Ubaghs (1967). On the other hand, nothing indicates whether the Scotiaecystidae are less or more crownward than *Cothurnocystis elizae*. The only solution at present is to include them both, along with some other forms, in the plesion of *Cothurnocystis elizae*.

The other forms which must be assigned to the plesion of *Cothurnocystis* were not described in Chapter 7, though in part they have already been mentioned in the present chapter. They are placed in the genera *Phyllocystis*, *Chauvelicystis* and *Cothurnocystis* and are described in the works of Thoral (1935), Gigout (1954), Chauvel (1966), Ubaghs (1968, 1969, 1983) and Sprinkle (1976). The 'Cothurnocystis' of Sprinkle (1976) is Middle Cambrian in age and comes from Utah, as already mentioned, whereas the other forms come from the lowest Ordovician of France and Morocco. It is to be hoped that future work will subdivide the *Cothurnocystis* plesion.

Galliaecystis (figs 7.33, 7.34, pp. 234, 235, 9.3), as already implied, is more crownward than *Cothurnocystis*, but less crownward than *Amygdalotheca* or *Reticulocarpos*. Its most striking mitrate-like feature is the dorsal bar, posterior to the buccal cavity, which connects the left and right sides of the marginal frame. This bar consists of extensions of two plates and is homologous with, and formed of the same plates as, the anterior margin of the dorsal head shield of the mitrate *Peltocystis* (fig. 8.37a, p. 310, 9.4). *Galliaecystis* resembles the still more crownward cornute *Reticulocarpos hanusi* in: (1) having this bar, (2) lacking the U plates of the branchial skeleton, (3) having the dorsal and ventral integument extend to the front end of the marginal frame, and (4) having postero–ventral points in the mid-line of the hind-tail ossicles, comparable with the spikes of *R. hanusi*. All these features are probably synapomorphies of *Galliaecystis* with *Reticulocarpos hanusi* which show that *Galliaecystis* was more crownward than the members of the *Cothurnocystis* plesion. They were acquired in segment 7 of figs 9.2b, 9.3.

On the other hand *Galliaecystis* was more primitive than *Amygdalotheca* (fig. 9.3), and less crownward, in retaining a boot-like outline of the head, in having no peripheral flange, and in having the strut of the ventral integument formed from two processes of the marginal frame. (In *Amygdalotheca*, *Reticulocarpos hanusi* and *R. pissotensis* the anterior part of the strut is individualized as a separate plate.) The species *Galliaecystis lignieresi* is the only described member of its genus and the only known member of its plesion. Coming from the lowest Ordovician of the south of France, it is contemporaneous with the earliest known mitrates and younger than the oldest known agnathan fishes and the oldest known acraniate *Pikaia*, although a member of the chordate stem group. Here again, stratigraphical sequence is no guide to primitiveness.

Amygdalotheca (figs 7.35, p. 236, 9.3), if compared with *Galliaecystis*, shows the additional mitrate-like features of: (1) an almost bilaterally symmetrical head outline, and (2) a peripheral flange. Also it is like the still more crownward cornutes *Reticulocarpos hanusi* and *R. pissotensis*, and unlike *Galliaecystis*, in having the anterior part of the ventral strut individualized as a separate plate (anterior strut plate in fig. 9.8). And it is like *R. hanusi*, and unlike *Galliaecystis*, in having a plane ventral surface to the head. All these features can be regarded as synapomorphies of *Amygdalotheca* with the mitrates or the respective more crownward cornutes, which *Galliaecystis* does not share.

They would have been acquired in segment 8 of figs 9.3b, 9.3. They show that *Amygdalotheca* is more crownward than *Galliaecystis*. On the other hand, *Amygdalotheca* shares two features with *Galliaecystis* which are probably primitive and not shown by still more crownward cornutes or mitrates. These are: (1) the ventral strut contacts the marginal frame anteriorly (although separated by a suture in *Amygdalotheca*), and (2) the head is large — about 20 mm long as against 10 mm in *Reticulocarpos hanusi* and *R. pissotensis*. This combination of features suggests that *Amygdalotheca* is more crownward than *Galliaecystis* but less crownward than *R. hanusi*. In addition *Amygdalotheca* has some features which are probably specializations never present in the stem lineage. These, which would have been acquired in segment 9 of figs 9.2b, 9.3, are: (1) the absence of a dorsal bar behind the buccal cavity, (2) bobbin-shaped plates in the dorsal integument – a parallelism to those of *Scotiaecystis*, and (3) extra plates in the marginal frame – making 16 plates in total (assuming that there were two posterior dorsal marginals M_{1LD}, M_{1RD}) as opposed to 12 in *Galliaecystis* and *Reticulocarpos hanusi* and probably 11 in the reconstructed first crown chordate, (see discussion of *q* below; the number of such plates in *Reticulocarpos hanusi* is unknown.) The only described species of *Amygdalotheca* is *A. griffei*. It is the only known member of its plesion. Coming from the lower Ordovician of France it is exactly contemporary with the oldest mitrates known.

Reticulocarpos hanusi (fig. 7.28, p. 228, 9.3, 9.4) is more crownward than *Amygdalotheca* because of: (1) the small size of the head (about 10 mm in length), in common with *R. pissotensis* and the primitive mitrates *Peltocystis cornuta* and *Chinianocarpos thorali*, (2) the separation of the front end of the strut from the marginal frame, foreshadowing the condition in mitrates where the strut has disappeared completely, (3) the presence of transpharyngeal eyes, in common with mitrates such as *Chinianocarpos* and *Mitrocystites*, but apparently lacking in cornutes less crownward than *R. hanusi*, (4) the presence of flat, ventral bearing surfaces on the major plates of the fore tail, which terminate postero–ventrally in a lateral spike – these surfaces are probably homologous with the ventral bearing surfaces and spikes of the hind tail of the mitrate *Lagynocystis* (fig. 8.34b–e, p. 305) and (5) the presence of intercalary plates in the fore tail probably homologous with the intercalary plates of the fore tail of *Placocystites* and of the fore, mid and hind tail of *Lagynocystis* and *Chinianocarpos* (Chauvel & Nion, 1977, however, report that such plates are absent in the still more crownward cornute *Reticulocarpos pissotensis*). All these features would have been acquired in segment 10 of figs 9.2b, 9.3.

As to more distal features of the tail, the spikes in the ventral mid-line of the mid and hind tail of *Reticulocarpos*

hanusi are probably homologous with those of *R. pissotensis*. They constitute a synapomorphy of *R. hanusi* with *R. pissotensis*, as compared with *Galliaecystis* which has short points in the corresponding positions. The shortness of the hind tail is likewise a synapomorphy of *Reticulocarpos hanusi* and *R. pissotensis* as compared with *Galliaecystis*. It foreshadows the total lack of the cornute mid and hind tail in mitrates. Unfortunately no comparison can be made in these respects with *Amydgalotheca*, for the only known specimens of this do not preserve the mid and hind tail. The mid-ventral spikes of *R. hanusi* and the smooth ventral surface of the stylocone indicate that the tail pulled the head rearwards by flexing downwards. This ventral flexion is another synapomorphy with mitrates which *Galliaecystis* probably did not share. Once again, however, the relevant parts are not known in *Amygdalotheca*. Taken together, these features are synapomorphies of *Reticulocarpos hanusi* with mitrates or with *Reticulocarpos pissotensis*, as compared with *Galliaecystis* and ignoring *Amygdalotheca*. They would have been acquired either in segment 8 or segment 10 of figs 9.2b, 9.3.

On the other hand, *R. hanusi* differs from *R. pissotensis* (fig. 9.3) and resembles *Amygdalotheca* in the flatness of the ventral surface of the head, which in *R. pissotensis* is convex, and in the fact that the marginal plates extend only a short distance onto the dorsal surface of the head (in common with all less crownward cornutes). In both respects *R. pissotensis* shows a more crownward condition which is probably a synapomorphy of it with mitrates. Thus, taking all its features together, *Reticulocarpos hanusi* is more crownward than *Amygdalotheca* but less crownward than *Reticulocarpos pissotensis*.

Reticulocarpos hanusi is not morphologically disqualified as a member of the stem lineage leading to the mitrates and other crown chordates, but it is stratigraphically too late. For it comes from the Llanvirn Stage of the Ordovician, whereas the earliest known mitrates belong to the immediately preceding Arenig Stage, while the earliest known fish is Upper Cambrian and the earliest known acraniate is Middle Cambrian.

Reticulocarpos pissotensis (figs 7.36, p. 237, 9.3) is closer to the crown chordates than is any other known cornute, because of features already mentioned (convex ventral surface of head; extension of marginals onto dorsal surface) which would have been acquired in segment 11 of figs 9.2b, 9.3. It does not seem to be morphologically disqualified as a member of the stem lineage except that it may (though perhaps not) lack the dorsal bar behind the buccal cavity. Unfortunately the known specimens are not good enough to show whether this bar is present. Like *R. hanusi*, however, *R. pissotensis* is stratigraphically too late to be a member of the stem lineage. For it comes from the

Llandeilo Stage (Middle Ordovician) of France and is, therefore, younger than *R. hanusi* and considerably younger than the first known mitrates, fishes and acraniates. *R. pissotensis* is the only known member of its plesion.

To sum up, the cornutes as stem chordates can be arranged in sequence of increasing relationship to the mitrates, and thus to the crown chordates, as follows: plesion *Ceratocystis*; plesion *Nevadaecystis*; plesion of *Cothurnocystis*, including *Cothurnocystis elizae*, the Scotiaecystidae and several other forms; plesion *Galliaecystis*; plesion *Amygdalotheca*; plesion *Reticulocarpos hanusi*; plesion *Reticulocarpos pissotensis*. The important open questions are: What were the stem chordates less crownward than *Ceratocystis*? What was between *Reticulocarpos pissotensis* and the mitrates? And how should the *Cothurnocystis* plesion be subdivided? Future work will probably answer some of these questions.

All known mitrates are primitive crown chordates, as already said in Chapter 8, for all can be assigned to the stem groups of one or other of the three living chordate subphyla as I shall argue in this chapter. The defining feature of the mitrates, separating the group at its 'lower' end from cornutes, is the origin of the right gill slits and right pharynx. This is a synapomorphy of mitrates with all living adult chordates, which cornutes do not share and it would have arisen in segment 12 of figs 9.2b, 9.3, 9.4. Mitrates, however, are a paraphyletic group which has three times given rise to non-mitrates, being ancestral separately to the living acraniates, tunicates and vertebrates as I shall seek to show. The 'upper' defining limit of the mitrates, therefore, is the loss of the calcite skeleton. It was passed three times.

Several other features, besides the right gill slits and pharynx, separate mitrates from all known cornutes, even from *Reticulocarpos pissotensis*. These include: (1) the lifting up of the viscera by the right pharynx and their resulting shift to a median position hanging, in the cavity of the right anterior coelom, from the ceiling of the head (fig. 7.6) – these changes would coincide with the origin of right pharynx and right gill slits, (2) the concomitant development of an oblique ridge, which would have arisen as a strut in the ceiling, serving to steady the viscera in their new position, (3) the probable development of a right epicardium as an outpouching of the new right pharynx, (4) the enclosure of the gill slits inside atria debouching by atrial openings, and the enclosure of the gonopore-anus in the left atrium, (5) the loss of the cornute mid and hind tail by autotomy and the regionation of the mitrate tail regions in the remaining stump, (6) the development of an inner calcitic layer in the ventral skeleton, (7) the fact that the brain became distinctly bipartite (prosencephalon and deuteroencephalon) and was enclosed in the postero-dorsal marginals (M_{1LD}, M_{1RD}), not in the postero-ventral marginals (M_{1LV}, M_{1RV}), (8) the fact that the trigeminal ganglia became enclosed in skeleton dorsally, as well as ventrally, unlike all cornutes except the least crownward (*Ceratocystis*), and (9) the loss of a marginal plate (M_{2L} or M_{3L} or *Reticulocarpos hanusi*) from the left side of the head. Again, these features would have arisen in segment 12 of figs 9.2b, 9.3, 9.4. It is most unlikely that they arose simultaneously but it is not yet possible to put them in phylogenetic sequence.

All known mitrates can be divided into three groups as follows: (1) *Lagynocystis*, (2) peltocystidans, including *Peltocystis*, *Balanocystites* and *Anatifopsis* among the forms described in Chapter 8, and (3) a group for which I shall use, for the present, the non-committal name of mitrocystitidans, including *Chinianocarpos*, *Mitrocystites*, *Mitrocystella* and *Placocystites* among the forms discussed in Chapter 8. (For the moment I shall not consider to which living subphyla of chordates these three groups are most closely related.) From the viewpoint of phylogenetic logic these three groups can be treated as monophyletic, if non-mitrates are disregarded, so there are three possible solutions to the problem of how they are interrelated, to wit: 1 is sister group to [2 + 3]; 2 is sister group to [1 + 3]; and 3 is sister group to [1 + 2].

The first of these solutions is almost certainly correct, i.e. among mitrates along, *Lagynocystis* is sister group to [peltocystidans + mitrocystitidans]. The arguments for this solution came from tail anatomy and depend on the deduction that *Lagynocystis* retains primitive features of the tail from which peltocystidans and mitrocystitidans have homologously departed. In making the relevant comparisons *Peltocystis*, as the most primitive peltocystidan, and *Chinianocarpos*, as the most primitive mitrocystitidan, are of special importance.

The mitrate tail arose, as argued in Chapters 7 and 8, by differentiation from a cornute fore tail, after the cornute mid and hind tail had been lost by autotomy. The tail in question would have had dorsal recurved spikes (not known in any cornute but acquired in segment 12 of fig. 9.2b), paired ventral major plates, and small intercalary plates interpolated between the other plates and in particular between successive pairs of major plates. The dorsal spikes would have given rise in mitrates to dorsal fore-tail spikes (such as occur in *Lagynocystis* and *Chinianocarpos*), dorsal hind-tail ossicles and to the constituent ossicles of the styloid. Observed mitrate tail features can, therefore, probably be regarded as primitive for mitrates: (1) if they resemble conditions seen in the fore tails of cornutes, particularly in the mitrate-like cornute *Reticulocarpos hanusi* (*R. pissotensis* is closer related to mitrates but disregarded here because it is less well known), and (2) more generally, if they show serial uniformity, seeing that the fore tail of cornutes is serially uniform. Bearing these considerations in mind, the following features of the tail of *Lagynocystis* (figs 8.28, p. 294, 8.33, 8.34, p. 302) are probably more primitive than those of peltocystidan or mitrocystitidan mitrates: (1) The hind-tail plates have concave ventral surfaces, interpreted as bearing surfaces, produced on each side into a postero–lateral spike; the surfaces and spikes resemble, despite differences in proportion, those on the fore tail of *Reticulocarpos hanusi* and would have acted similarly in locomotion by pressing downwards and forwards against the mud. (2) The hind tail of *Lagynocystis* is to a large degree serially uniform along its length except for signs that the distal portion was specially flexible ventrally; in this uniformity it differs from the hind tails of *Chinianocarpos* (fig. 8.26c, p. 291) and *Peltocystis* (fig. 8.37c, p. 310) which are differentiated into three parts, i.e. proximal long-bladed, medial short-bladed and distal bladeless. (Both in *Chinianocarpos* and in *Peltocystis* these three parts formed essentially the same locomotory device (figs 8.27, 8.38), the bladeless portion being a terminal anchor which pushed against the mud by means of a dorsal bearing surface on the ossicles). (3) The tail of *Lagynocystis* also approaches serial uniformity in that the spikes on the dorsal surface and sides of the fore tail are of about the same size and shape as the serially homologous con- stituent ossicles of the styloid and the proximal hind-tail ossicles (more distal hind-tail ossicles are larger, longer and of different shape); this contrasts with *Chinianocarpos*, where the fore-tail spikes are smaller and not much like the constituent ossicles of the styloid in size or shape, and differs still more from *Peltocystis*, which lacks fore-tail spikes. (4) *Lagynocystis* also approaches serial uniformity of the tail in that the two most distal pairs of major plates in the fore and mid tail are not articulated to the styloid, but independent of it, just as the major plates of the fore tail are independent of the dorsal spikes – *Lagynocystis* differs in this respect from all other known mitrates where two pairs of ventral plates are articulated to the styloid.

The large and variable number of ossicles in the styloid of *Lagynocystis* (four or five, as opposed to two in all other known mitrates) may also be a primitive feature. For it suggests that the styloid was less standardized and, there- fore, less 'well established' than in other mitrates. This conclusion, however, is not based on outgroup comparison and is therefore very tentative.

The tail anatomy thus suggests that *Chinianocarpos* (as a primitive mitrocystitidan) is closer related to *Peltocystis* (as a primitive peltocystidan) then to *Lagynocystis*. To summarize, this is on the basis of the following synapo- morphies: (1) reduced fore tail spikes (absent in *Peltocystis*), (2) two pairs of major plates (rather than none) articulated to the styloid, (3) absence of ventral bearing surfaces on the hind tail, (4) differentiation of the hind tail into three

regions with dorsal bearing surfaces on the ossicles of the distal region, and (5) perhaps standardization of the styloid to contain only two ossicles. These features would have arisen in segment 16 of figs 9.2b, 9.4. As mentioned later, some of these synapomorphies are not seen in mitrocystitidans more advanced than *Chinianocarpos*, presumably having been lost. But the tail structures show that the oldest dichotomy among known mitrates is between *Lagynocystis* and the rest.

Lagynocystis, as already argued in Chapter 8, is a primitive acraniate. The synapomorphies which it shares with living acraniates are: (1) the presence of tertiary gill slits, (2) the presence of a median ventral atriopore, (3) the situation of the posterior coelom (= nephridia of living acraniates = epicardia of tunicates) dorsal to the gill bars, so that upward extension of the gill slits in phylogeny could produce the supra-branchial nephridia of amphioxus, and perhaps (4) the absence of eyes and optic nerves. All these synapomorphies would have arisen in segment 13 of figs 9.2b, 9.4. In addition *Lagynocystis* seems to share with amphioxus the predominantly left-sided innervation of the buccal cavity. However, this probably represents a primitive condition that living tunicates and vertebrates have lost, rather than a synapomorphy of *Lagynocystis* with living acraniates.

Pikaia (fig. 8.36, p. 309) is probably another fossil acraniate. If its notochord is truly confined to the posterior two-thirds of the animal then this is probably a primitive resemblance to mitrates which distinguishes it from all living acraniates and suggests that it belongs to the stem group of acraniates, rather than to the crown groups. The extension of the notochord into the anterior part of the animal would, in that case, have happened in segment 15 of figs 9.2b, 9.4. On the other hand, *Pikaia* has many synapomorphies with living acraniates which *Lagynocystis* does not share. Such include: (1) the lack of clear distinction between head and tail with myotomes occurring up to the front end of the animal, (2) lack of a calcitic skeleton, (3) lack of autotomy at the end of the tail, (4) fins, and (5) the situation of the anus near the posterior end of the animal. All of these synapomorphies would have arisen in segment 14 of figs 9.2b, 9.4. As a stem acraniate, therefore, *Pikaia* is more crownward than *Lagynocystis*. Indeed, *Lagynocystis* and *Pikaia* constitute two successive plesions within the acraniate stem group. *Pikaia*, being Middle Cambrian in age, is older than *Lagynocystis*, but both are known from only a few horizons and localities. It is, there- fore, clear, to say the least, that no reliable stratigraphical range chart of the two forms can yet be drawn up. Their observed stratigraphical sequence is thus not relevant in deciding which is the more primitive.

In passing from a form like *Lagynocystis* to living acraniates some other changes must have occurred. (*Pikaia* is too poorly known to say whether they had already been completed in that form.) Such include: (1) loss of the right and left atria and atrial openings and expansion of the median atrium so that it covered the primary and secondary gill slits as well as the tertiary ones, and (2) further simplification of the brain with loss of the trigeminal ganglia, auditory ganglia and ears, and spinal ganglia – these spinal ganglia imply that neural crest existed in *Lagynocystis* but has been lost in living acraniates; these changes would have occurred in segments 14 or 15 of figs 9.2b, 9.4.

Other changes probably occurred in the stem lineage of the acraniates, but cannot be located in any particular segment whether 13, 14 or 15. Such include: (1) migration of the gonad to a new position lateral to the atrium, followed by its myotomic subdivision – it may be that *Asymmetron*, with gonads on the right side only, preserves a more primitive condition than amphioxus (*Branchiostoma*) with gonads both on right and left, (2) loss of the heart – this would have existed either in *Lagynocystis* or its ancestors since the heart or its homologues occurs in hemichordates (pericardial vesicle and central sinus), echinoderms (dorsal sac), tunicates and vertebrates, and (3) transformation of the crescentic organ, assuming that it existed in the ancestry of *Lagynocystis* or in *Lagynocystis* itself, into the anterior gut diverticula of larval amphioxus and thus into Hatschek's pit and the ventral rostral coelom of adult amphioxus – this implies that the mitrate ciliated organ migrated forward to open into the buccal cavity as Hatschek's pit.

Most of these various changes seem, from a human and subjective viewpoint, to be degenerative. In any case, contrary to some elementary textbooks, amphioxus is far from representing, almost unchanged, the latest common ancestor of the chordates. It is unfortunate that *Asymmetron* has never been properly described and that its embryology is almost unknown. For its asymmetrical gonads now suggest, contrary to what was thought previously, that it may be more primitive than amphioxus.

The next oldest dichotomy among mitrates is between peltocystidans and mitrocystitidans (figs 9.2b, 9.4). The peltocystidans include *Peltocystis* (fig. 8.37, p. 310), *Balanocystites* (fig. 8.39) and *Anatifopsis* (fig. 8.40) among the forms described in Chapter 8 and these three forms are connected together by the following synapomorphies: (1) expansion of the postero–ventral plates M_{1LV}, M_{1RV}, (2) elongation of the atrial openings, and (3) specialization of an anterior plate as a moveable oral appendage right of the mouth (this is probably not homologous with the static anterior appendage of *Lagynocystis* (fig. 8.28a, p. 294) which probably derived from a plate left of the mouth). These synapomorphies would have evolved in segment 17 of figs 9.2b, 9.4. In addition, *Balanocystites* and

Anatifopsis are connected to each other, and distinguished from *Peltocystis*, by the following synapomorphies: (1) the plates M_{1LV} and M_{1RV} have expanded even further so as to make up all of the ventral surface of the head and much of the dorsal surface and to occlude a large part of the ventral skeleton, and (2) the atrial openings have increased further in length and moved onto the dorsal surface of the head. These two synapomorphies would be acquired in segment 18 of figs 9.2b, 9.4. Moreover, in *Anatifopsis* the atrial openings are not merely dorsal but almost median – a condition that would have been acquired in segment 19 of figs 9.2b, 9.4.

The peltocystidans are probably stem tunicates as already argued in Chapter 8. For: (1) the elongation of the atrial openings probably implies elongation of the atria on either side of the pharynx as seen in tunicates but not in other living chordates, and (2) the dorsal atrial openings of *Balanocystites* and *Anatifopsis* can be compared with the dorsal atrial openings of tunicate tadpoles and the dorsal position of the atrium in adult tunicates (fig. 4.2a, p. 89).

Indeed, within the stem-tunicate mitrates, *Peltocystis*, *Balanocystites* and *Anatifopsis* represent successively more crownward plesions. *Peltocystis* is the least crownward of these, for its atrial openings, though more elongate than in non-tunicate mitrates, remain in the primitive ventro–lateral position. The only known species is *Peltocystis cornuta* from the lowest Ordovician of the south of France. The next more crownward plesion is that of *Balanocystites*, for the atrial openings are longer than in *Peltocystis*, have moved onto the dorsal surface of the head, but remain more lateral in position than in *Anatifopsis*. The *Balanocystites* plesion contains several species, for beside *Balanocystites primus* from the Llanvirn of Bohemia described in Chapter 8, it also contains the two species '*Anatifopsis*' *escandei* Thoral and '*Anatifopsis*' *trapeziiformis* Thoral from the lowest Ordovician of the south of France (*A. escandei* is known also from the same horizon in Morocco (Chauvel, 1966)). It may be that other species ascribed in the literature to '*Anatifopsis*' in fact belong to *Balanocystites*. Finally, the genus *Anatifopsis*, in the strict sense, is the most crownward plesion among stem-tunicate mitrates. For in it the atrial openings almost meet in the dorsal mid-line, foreshadowing the dorsally joined right and left atria of living tunicates. The oldest known definite member of the genus is *Anatifopsis barrandei* Chauvel from the Lower Ordovician (Llanvirn) of Czechoslovakia, as described in Chapter 8. *Palaeobotryllus*, from the Upper Cambrian of Nevada (fig. 8.41a, p. 315), if it is a tunicate at all, is more advanced than *Anatifopsis* in being colonial and attached to a substrate. Possibly, as already said in Chapter 8, it is a crown tunicate, although older than any known stem-tunicate mitrate. Once again there is no correlation between stratigraphical age and position with respect to the crown group.

The stem-tunicate mitrates are clearly not crown tunicates because they lack several advanced features which living tunicates possess. Such include: (1) Absence of a calcite skeleton – there is no reason to think that the aragonitic spicules of didemnid tunicates are homologous with the calcite plates of mitrates, for they are made of aragonite, not calcite, and are deposited, not in the mesoderm, but in the gelatinous tunic – a sort of soft exoskeleton which mitrates did not have. (2) Fixation – in modern tunicates this happens in almost all benthic forms (ascidians) and it involves attachment of the larva by its front end to a substrate, resorbtion of the tail and torsion of the body so that the mouth opens upward away from the substrate. Fixation does not occur in pelagic tunicates but presumably existed in their ancestors, assuming they are a monophyletic group within the tunicates. For the pelagic doliolids have a tadpole stage (fig. 4.35, p. 114) which, without fixation, loses its tail and becomes adult, and these two phases of the ontogeny are presumably homologous with those of benthic tunicates. The pelagic tunicates known as appendicularians or larvaceans (fig. 4.33) remain as tailed tadpoles throughout life. I agree with Garstang (1928) that this is almost certainly not primitive but results from neoteny. For within tunicates the doliolids are probably the sister group of the appendicularians, sharing with them the synapomorphy, not found elsewhere, of a frequently moulted test (= 'house' of appendicularians) without cellulose. (3) The absence of tail autotomy (the tail of a tunicate tadpole disappears, not by dropping off, as happened to the end of the tail in mitrates, but by resorbtion at its proximal end). (4) The junction of right and left atria to form a single atrium with right, left and dorsal portions. The formation of a single atrium from paired atria happens in the ontogeny of most benthic tunicates such as *Ciona*, which can be seen as a Haeckelian recapitulation. In some other benthic tunicates (stolidobranchs) the atrium arises medially which presumably represents a secondary shortening of ontogeny. (5) The cellulosic tunic. This is a gelatinous exoskeleton, stiffened with cellulose, that the tunicate animal secretes outside itself. These features would have originated in segment 20 of figs 9.2b, 9.4.

Coloniality is widespread among living tunicates but is probably not a primitive feature for crown tunicates. More probably the primitive condition is solitary, as seen in stem-tunicate mitrates and also in *Ciona* and many other benthic tunicates. From this solitary condition, coloniality has probably evolved three times, having arisen independently, with different styles of budding (Berrill, 1950), within the aplousobranchs, the stolidobranchs and the pelagic tunicates.

Hermaphroditism is universal in tunicates, except for some appendicularians. There is no direct evidence as to whether the sexes were separate in cornutes and mitrates but indirect evidence suggests that they were so, for they are separate in almost all living echinoderms, all acraniates, and almost all living vertebrates. Hermaphroditism is likely to be a specialization of tunicates following fixation, for the latter makes cross-fertilization difficult in sparse populations (Tomlinson, 1966). Interestingly enough, the only large hermaphroditic group among crustaceans is the barnacles which are attached like tunicates (Ghiselin, 1969). Hermaphroditism most probably evolved in segments 17, 18, 19 or 20 of figs 9.2b, 9.4.

Mitrocystitidan mitrates are connected together by two synapomorphies: (1) the presence of dorsal, presumably touch-sensory, branches of the palmar nerves ($n_4 + n_5$, homologous respectively to the profundus and superficialis branches of the trigeminal complex), and (2) the lateral-line, present in a pit or groove in the external surface of plate M_{1RV} just right of the tail insertion (of the forms discussed in Chapter 8, *Placocystites* has no lateral-line, but its absence is almost certainly secondary). Both these synapomorphies, which would have been acquired in segment 21 of figs 9.2b, 9.4, are also synapomorphies with living vertebrates. They indicate, as already implied in Chapter 8, that mitrocystitidans are stem vertebrates.

Among mitrocystitidan mitrates *Mitrocystites* (fig. 8.1, p. 240), *Mitrocystella* (fig. 8.11, p. 256) and *Placocystites* (fig. 8.18, p. 266) share several synapomorphies which *Chinianocarpos* (fig. 8.26, p. 291) does not have (limiting discussion to those forms described in Chapter 8). These synapomorphies include: (1) incorporation of plate *n* into the front margin of the dorsal head shield, (2) rigidification of the dorsal skeleton with loss of the dorsal integument and reduction in the number of dorsal plates, (3) posterior enclosure of the nerves n_4 and n_5 by skeleton, (4) individualization of nerves n_4 and n_5 from each other (in *Mitrocystites* the joint trunk of n_{4+5} which ascends to the dorsal surface (fig. 8.9) seems to be constituted at its base from a stout, lateral n_4 and a more slender n_5 which approaches the trunk from a median direction, while in *Mitrocystella* and *Placocystites* nerves n_4 and n_5 in the dorsal skeleton (figs 8.16a, 8.25a) are entirely separate – in *Chinianocarpos*, on the other hand, nerves n_4 and n_5 are simply indistinguishable from each other), (5) loss of the tripartition of the hind tail and its functional replacement by presumed bearing surfaces on the distal, posterior surfaces of the dorsal ossicles, (6) elongation of the lateral-line (except in *Placocystites* which has no lateral-line), and (7) flexibility of the lower lip and the acquisition of spike-shaped oral plates. These synapomorphies would have been acquired in segment 22 of figs 9.2b, 9.4 and indicate that *Mitrocystites, Mitrocystella* and *Placocystites* form a group (paraphyletic or monophyletic) more advanced than *Chinianocarpos*.

Mitrocystella and *Placocystites* share synapomorphies which *Mitrocystites* does not have. These include: (1) lack of a particular marginal plate from near the front left angle of the head (plate M_{5L} in fig. 8.1a), (2) existence of cuesta-shaped ribs on the postero–ventral surface of the head (*Placocystites* has these also over most of the ventral surface and dorsal surface), (3) absence of two centro–dorsal plates, (4) extensive resorbtion of the head skeleton dorsal to the right and left pharynges, as witnessed in the fossils by irregular rock-filled cavities in those areas (figs 8.14a, 8.19a, pp. 260, 271) (some resorbtion in these regions sometimes occurs in *Mitrocystites*, but is rare and never extensive), (5) posterior slope of the dorsal surface, (6) covering-over of the peripheral grooves to form peripheral canals, (7) clear separation of nerves n_4 and n_5 in the dorsal skeleton, and (8) absence of transpharyngeal eyes. These synapomorphies would have been acquired in segment 23 of figs 9.2b, 9.4.

They indicate that *Mitrocystella* and *Placocystites* are close related to each other than to *Mitrocystites*, and also less primitive than *Mitrocystites*. If the deduced relationships of these four genera are correct, then the absence of the lateral-line in *Placocystites* is secondary, for otherwise this organ must have evolved separately in *Chinianocarpos*, *Mitrocystites* and *Mitrocystella*.

As concerns relationship to the crown vertebrates, and again considering only the four genera discussed in Chapter 8, then the series: (1) *Chinianocarpos*, (2) *Mitrocystites*, and (3) [*Mitrocystella* + *Placocystites*] probably represents successive plesions of the vertebrate stem group, successively more closely related to the crown group. This conclusion is tentative but is based on the following considerations. Firstly, *Mitrocystites* shares with crown vertebrates the following likely synapomorphies which *Chinianocarpos* lacks: (1) reduction of transpharyngeal eyes (which in crown vertebrates are absent), (2) partial individualization of nerves n_4 and n_5 (the distal courses of the equivalent profundus and superficialis nerves are entirely separate in living vertebrates) and (3) extension of the lateral-line from being a pit in *Chinianocarpos*, probably containing a single neuromast, to a groove, presumably containing many neuromasts, in *Mitrocystites*. These synapomorphies with crown vertebrates are also synapomorphies, as already mentioned, with *Mitrocystella* and *Placocystites* (except that the lateral-line was secondarily lost in *Placocystites*). They would have arisen in segment 22 of figs 9.2b, 9.4. Secondly, *Mitrocystella* and *Placocystites* have synapomorphies with crown vertebrates which *Mitrocystites* does not show. Such include: (1) absence of transpharyngeal eyes, (2) further extension of the lateral-line to form a branched groove in *Mitrocystella* (the line is absent in *Placocystites*, but by secondary loss), (3) extensive resorbtion of the

dorsal skeleton above the left and right pharynges (a synapomorphy with crown vertebrates, for in these the calcitic skeleton is entirely absent), and (4) further separation of nerves n_4 and n_5. These synapomorphies with crown vertebrates, which have already been mentioned as separating [*Mitrocystella* + *Placocystites*] from *Mitrocystites*, were probably acquired in segment 23 of figs 9.2b, 9.4. It is impossible to say whether *Placocystites* and *Mitrocystella* are equally closely related to the crown vertebrates, or whether one is more closely related to them than the other, and if so which. In segment 24 of figs 9.2b, 9.4, on the way to *Placocystites*, other changes occurred as well as the loss of the lateral-line. Such include the development of oral spines from the rightmost and leftmost lower-lip plates, the rigidification of the ventral head skeleton and acquisition of cuesta-shaped ribs on the dorsal surface of the head. These features unite *Placocystites* with several other mitrates, none of which will be described here.

Chinianocarpos, which is only known from the species *Chinianocarpos thorali* Ubaghs, is the only known member of its plesion. *Mitrocystites mitra* is not the only known member of its plesion since there are other known species of the genus, and other similar mitrates, which, so far as can be discerned, are equally related to the vertebrate crown group. (Such include *Aspidocarpus bohemicus* Ubaghs, 1979 and *Mitrocystites* sp. 2 of Chauvel (1971), but I shall not describe these forms here). Again *Placocystites* and *Mitrocystella* are not the only members of their plesion, for a large number of somewhat similar mitrates are known, and it is not possible to say whether some of these are more closely related to crown vertebrates than others are.

None of the known stem-vertebrate mitrates can be part of the stem lineage of vertebrates. For the oldest known fish – *Anatolepis* cf. *heintzi* of Repetski (1978) – is connected to the crown vertebrates by the synapomorphy of its phosphatic skeleton which mitrates lack and, being Upper Cambrian, is older than the oldest known mitrocystitidan mitrate which is *Chinianocarpos thorali* from the Lower Ordovician.

I have already argued that mitrocystitidan mitrates are closer related to peltocystidan mitrates than to *Lagynocystis*, on the basis of synapomorphies in the tails. I have also argued that mitrocystitidans are stem vertebrates, peltocystidans are stem tunicates and that *Lagynocystis* is a stem acraniate. If all this is true then vertebrates are more closely related to tunicates than to acraniates.

I originally proposed such a grouping (Jefferies, 1973, p. 463) on the basis of an argument which now seems false. It concerned the innervation of the somatic musculature in amphioxus, tunicates (tail muscles of the tadpole) and vertebrates. Flood (1966) had shown, namely, that amphioxus innervates its somitic muscles by means of

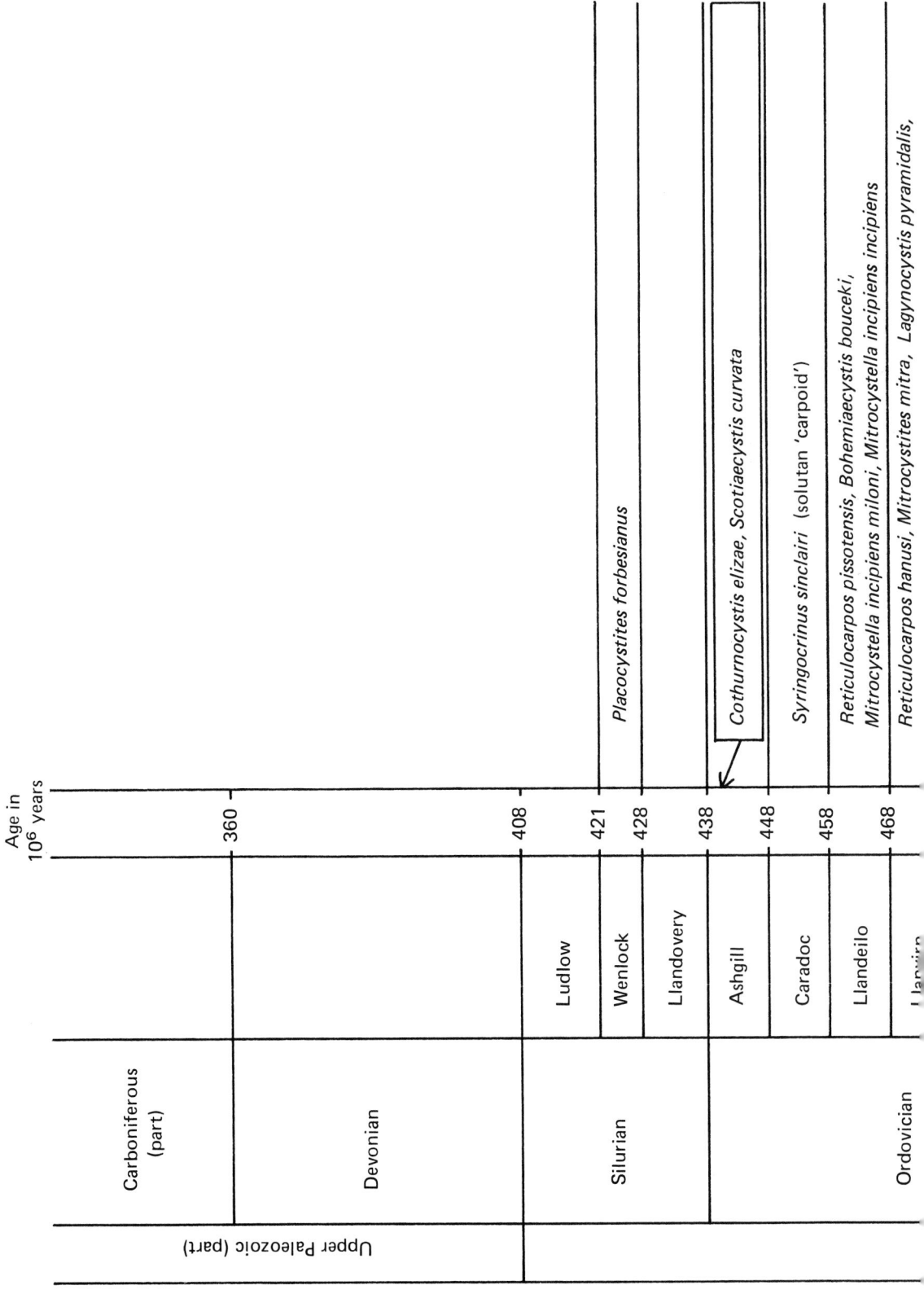

Thoralicystis spp., *Phyllocystis* spp. *Chauvelicystis* spp., *Peltocystis cornuta*, '*Anatifopsis*' *escandei*, '*Anatifopsis*' *trapezilformis*

Nevadaecystis americana, *Palaeobotryllus taylori* (? crown tunicate), *Anatolepis* cf. *heintzi* . (agnathan fish)

Ceratocystis perneri, '*Cothurnocystis*' sp. of Sprinkle (1976), *Pikaia gracilens* (soft-bodied acraniate), *Stromatocystites pentangularis* (echinoderm)

Age		
488	Tremadoc	Lower Palaeozoic
505	Upper Cambrian (Merioneth)	
525	Middle Cambrian (St David's)	Cambrian
540	Lower Cambrian (Caerfai)	
590	Pre-Cambrian (part)	Pre-Cambrian

9.6. *Stratigraphical distribution of forms mentioned in text. The subdivisions of the stratigraphical column and their ages in years after Harland et al. (1982).*

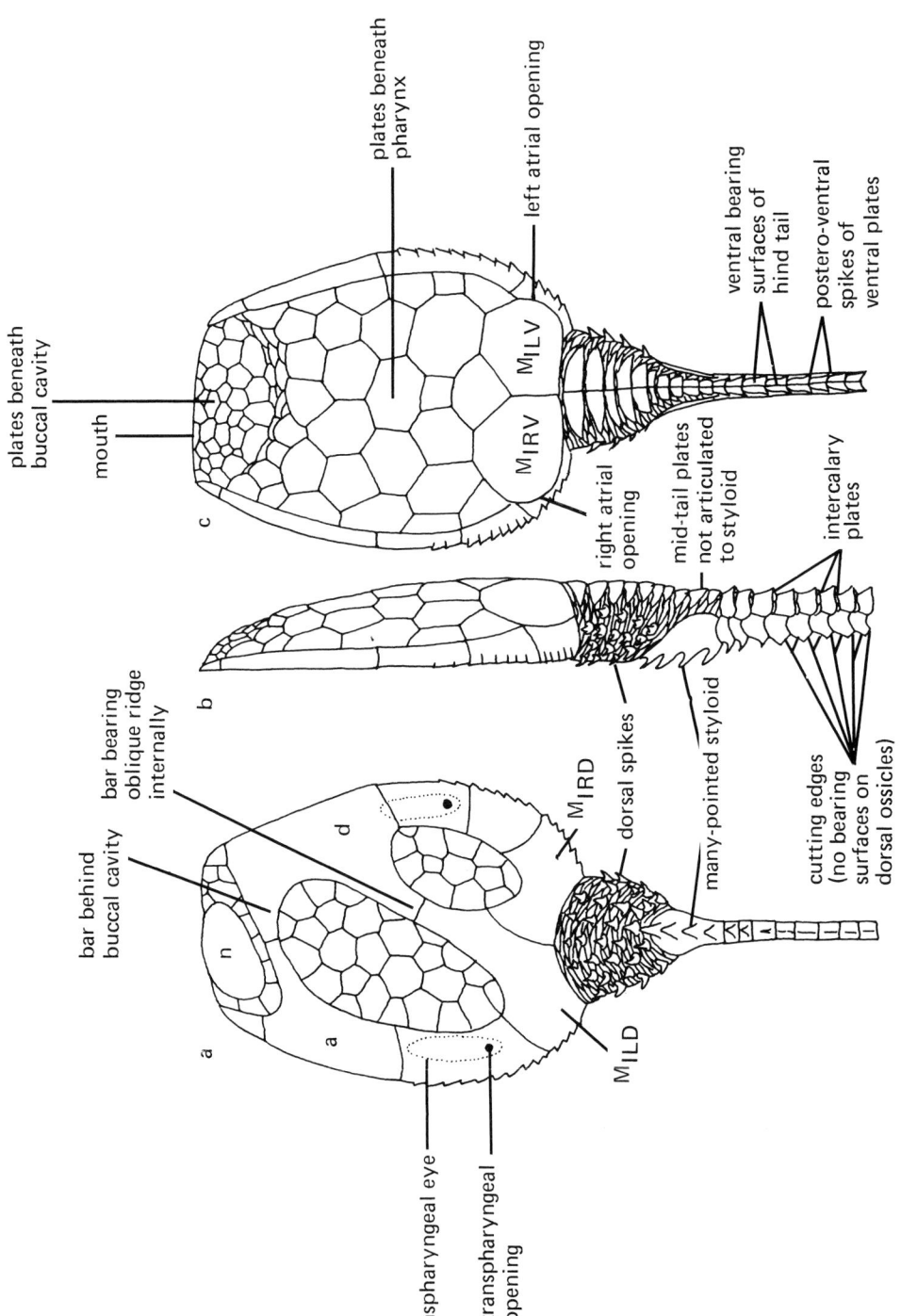

9.7. Reconstruction of the hypothetical animal q – the first crown chordate and the latest common ancestor of all living chordates.

muscle 'tails' and that the actual neuro-muscular junctions are at the ventro–lateral angles of the nerve cord (fig. 3.11, p. 62). Since echinoderms often innervate their muscles by means of muscle 'tails' (Cobb & Laverack, 1967) the amphioxus situation probably represents the primitive condition. Vertebrates, however, innervate their somitic muscles by means of ventral spinal nerves which go out to the muscles, so that the neuro-muscular junctions are at the ends of these spinal nerves. I had thought that, in this respect, tunicates were more like vertebrates than like amphioxus, but this was the result of misunderstanding a paper by Pucci-Minafra (1965). In fact the neuro-muscular junctions of the tunicate-tadpole tail are ventro–lateral or lateral to the dorsal nerve cord much as in amphioxus, as pointed out to me by Dr M. Katz and Dr Q. Bone (*in litt.*) (see also Mackie & Bone, 1976; Torrence & Cloney, 1982; Katz, 1983). The argument from neuro-muscular junctions therefore fails. However, the argument from the tail anatomy of the mitrates, viewed as members of the stem groups of chordate subphyla, is probably valid and is based on facts that I have ascertained personally.

The first crown chordate (fig. 9.7), being the latest common ancestor of all extant chordates, can now be reconstructed on the basis of fossils. It would be the latest common ancestor of all living chordates and can be referred to as *q*. The reconstruction depends on the fact that *q* would also be the latest common ancestor of *Lagynocystis*, *Peltocystis* and *Chinianocarpos*, as being respectively, among known forms, the least crownward stem acraniate, stem tunicate and stem vertebrate. The reconstruction also involves *Reticulocarpos hanusi* as a crownward stem chordate. (*Reticulocarpos pissotensis* is a still more crownward stem chordate, but is less well known than *R. hanusi*.) For the sake of brevity, I refer to the known animals in question by their initial letters as *R*, *L*, *P* and *C*. The likely cladogram of *q* and the four known fossils is shown in fig. 9.8.

A feature can be reconstructed in *q* with the highest probability if it occurs in all four of these fossils {*R, L, P, C*}. It can be reconstructed in *q*, but with a lower probability, if it occurs in only three out of the four known fossils, whether {*L, P, C*, not *R*} as a probable mitrate synapomorphy, or {*R, L, P*, not *C*}, {*R, P, C*, not *L*} or {*R, L, C*, not *P*}. It can be reconstructed in *q*, though with a still lower probability, if present in only one of the mitrates and *Reticulocarpos hanusi*, i.e. {*R, L*, not *P*, nor *C*}, {*R, P*, not *L*, nor *C*}, {*R, C*, not *L*, nor *P*}, or present in *Lagynocystis* and either *Peltocystis* or *Chinianocarpos*, i.e. {*L, P* not *R*, nor *C*} or {*L, C*, not *R*, nor *P*}. It cannot be reconstructed in *q* if present in only *Peltocystis* and *Chinianocarpos* {*P, C*, not *L* nor *R*}, since the most parsimonious assumption is then that it evolved in the exclusive common ancestry of *Peltocystis* and *Chinianocarpos*, which is also the line *qr* in fig. 9.8 and the stem lineage of the tunicate–vertebrate crown group. Most of the reconstructed features of *q* are independent of the assumed mutual relationships of the three mitrates in question. The only exceptions are these features shared by *Peltocystis* and *Chinianocarpos* alone.

As regards external features, *q* would have had a head and a tail {*R, L, P, C*}. The head was probably about 10 mm

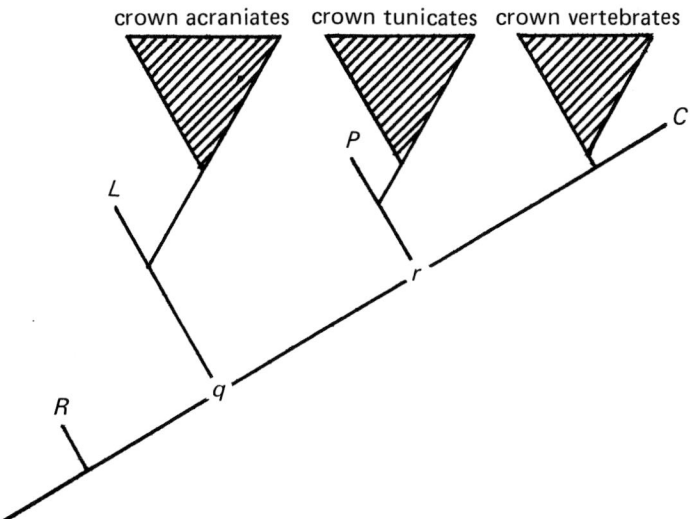

9.8. *Cladogram to show the logic of reconstructing the hypothetical animals* q *(fig. 9.7) and* r *(fig. 9.9).* C = Chinianocarpos *(the most primitive known stem-vertebrate mitrate);* L = Lagynocystis *(the only known stem-acraniate mitrate);* P = Peltocystis *(the most primitive known stem-tunicate mitrate);* R = Reticulocarpos hanusi *(one of the most crownward stem chordates).*

long {*R, P, C,* not *L,* but *L* is elongate rather than large}. There were peripheral flanges {*R, L, P, C*} and these were dorso–lateral rather than ventro–lateral {*L, P, C* and also *R. pissotensis* not *R*} which means that the head was flat dorsally and convex ventrally. The head was broad rather than elongate {*R, P, C* not *L*}. The right and left peripheral flanges were symmetrical {*R. P. C,* not *L* in which the right one is shorter and weaker}. There were right and left atrial openings between the postero–ventral marginals and the first or second dorsal marginals {*L, P, C,* not *R*}. A dorsal bar was present behind the buccal cavity {*R, P,* not *L* nor *C*}; plates M_{1LD} and a right anterior plate joined in the middle of the dorsal head shield and carried the oblique ridge internally {*L, P, C* not *R*}. The upper lip was flexible {*R, P,* not *C* nor *L*}. There was a large plate *n* dorsal to the mouth {*L, C, P* not *R*}. Plate *n,* assuming it is homologous with one of the dorsal integument plates of *Reticulocarpos hanusi,* was probably included in the dorsal buccal integument rather than sutured to the marginal frame {*C, P, R* (in a sense), not *L*}. The lower lip was flexible, not rigid {*R, L, P* not *C*}. There was flexible integument included in the dorsal surface behind the buccal cavity {*R, C,* not *L* nor *P*} – indeed there were probably two such patches of integument separated by a dorsal strut as suggested by Chauvel's *Mitrocystites* sp.2 (1971, see Jefferies & Lewis, 1978, fig. 29 under 'Moroccan *Chinianocarpos*') and the presence of two apparently homologous centro–dorsal plates in *Peltocystis.* A marginal plate was present left of the mouth {*R, L,* not *P,* nor *C*}. Another marginal plate was present right of the mouth {*R, P, C,* not *L*}. There was a pair of transpharyngeal eyes on the dorsal surface of the head {*R, C,* not *L,* nor *P*}. There were no dorsal touch-sensory branches of the trigeminal {*R, L, P,* not *C*}. There was no lateral-line {*R, L, P,* not *C*}. There was no median atrial opening {*R, P, C,* not *L*}.

Inside the head a right pharynx and right gill slits existed, the cavity of the right anterior coelom hung from the ceiling and there were right and left atria behind the gill slits, internal to the atrial openings {*L, P, C,* not *R*}. There were also a buccal cavity, left and right anterior coeloms and a posterior coelom {*R, L, P, C*}. The brain was bipartite with an optic foramen anteriorly, and paired hypocerebral processes beneath {*L, P, C,* not *R*}. There were paired trigeminal ganglia left and right of the brain {*R, L, P, C*}. There was no median atrium nor were there tertiary gill slits {*R, P, C* not *L*}.

In the tail, the cornute mid and hind tail had been lost {*L, P, C,* not *R*}. The tail was divided into mitrate fore, mid and hind tail with dorsal ossicles and ventral plates in the hind tail and a styloid in the mid tail {*L, P, C* not *R*}. There were dorsal spike-shaped plates in the fore tail {*L, C* not *P,* nor *R*}, serially homologous with the styloid and dorsal ossicles. The hind tail had ventral bearing surfaces, each of

which ended postero–laterally in a pair of spikes {*R, L,* not *P* nor *C*}.

By comparison with living animals it is fairly certain that *q* had a mucous pharyngeal trap produced by an endostyle and held anteriorly by peripharyngeal bands. This follows from the position of *q* as the latest common ancestor of amphioxus, tunicates and lamprey larvae. It is confirmed by the detailed anatomy of the reconstructed pharynxes of *Placocystites, Mitrocystites* and *Mitrocystella,* as discussed in Chapter 8.

The description of *q* might continue almost indefinitely for a vast number of relevant facts are known. It is a hypothesis, but not an empty one. It is the bauplan, groundplan or morphotype of chordates. It is not a mystical concept but an animal that once existed (reconstructional mistakes excepted). Even if we found it fossil, however, we could never prove that we had done so.

The latest common ancestor of living tunicates and vertebrates (fig. 9.9), which would be the first member and morphotype of the tunicate–vertebrate crown group, can be called *r.* Its phylogenetic position is shown in fig. 9.8. In most ways it would have been like *q.* The tail would differ, however. For the hind tail would have been tripartite with a terminal anchor provided with dorsal bearing surfaces and there would have been no ventral bearing surfaces {*P, C,* not *L,* nor *R*}. There would have been spike-shaped plates in the fore tail {*C, L,* not *P,* nor *R*}. The styloid would consist of two ossicles only {*P, C,* not *L*} which were developed as knife-like blades and would have two pairs of plates articulated to it {*P, C,* not *L*}.

The latest common ancestor of living vertebrates, being the first crown vertebrate, can be called *s* (fig. 9.10). In reconstructing it (fig. 9.11) I shall assume the correctness of the cladogram for living vertebrates proposed by Løvtrup (1977) and Janvier (1978, 1981b), i.e. that living petromyzonids are the sister group of the gnathostomes, while the myxinoids are the sister group of the petromyzonid–gnathostome group which Janvier has called the myopterygians. As end members of the cladogram I shall take *Ciona* (tunicate), *Mitrocystites* (stem-vertebrate mitrate), *Eptatretus* (myxinoid), *Lampetra* (petromyzonid) and *Squalus* (gnathostome). The logic involved is the same as in reconstructing *q* and *r* but the result is less certain because the end members differ more from each other, with a greater chance of parallelism, convergence and mistaken homologies. I shall refer to the end members of the cladogram by their first and second letters {*Ci, Mi, Ep, Sq*} except for *Lampetra* which I call *Lm.*

The first crown vertebrate *s* would have lived solely in the seas {*Ci, Mi* (presumably), *Ep,* not *Lm* nor *Sq* (which is probably only secondarily marine)}. It was probably eel-shaped {*Ep, Lm,* not *Ci,* nor *Sq,* nor *Mi*}. It swam by side-to-side undulation {*Ep, Lm, Sq,* probably not

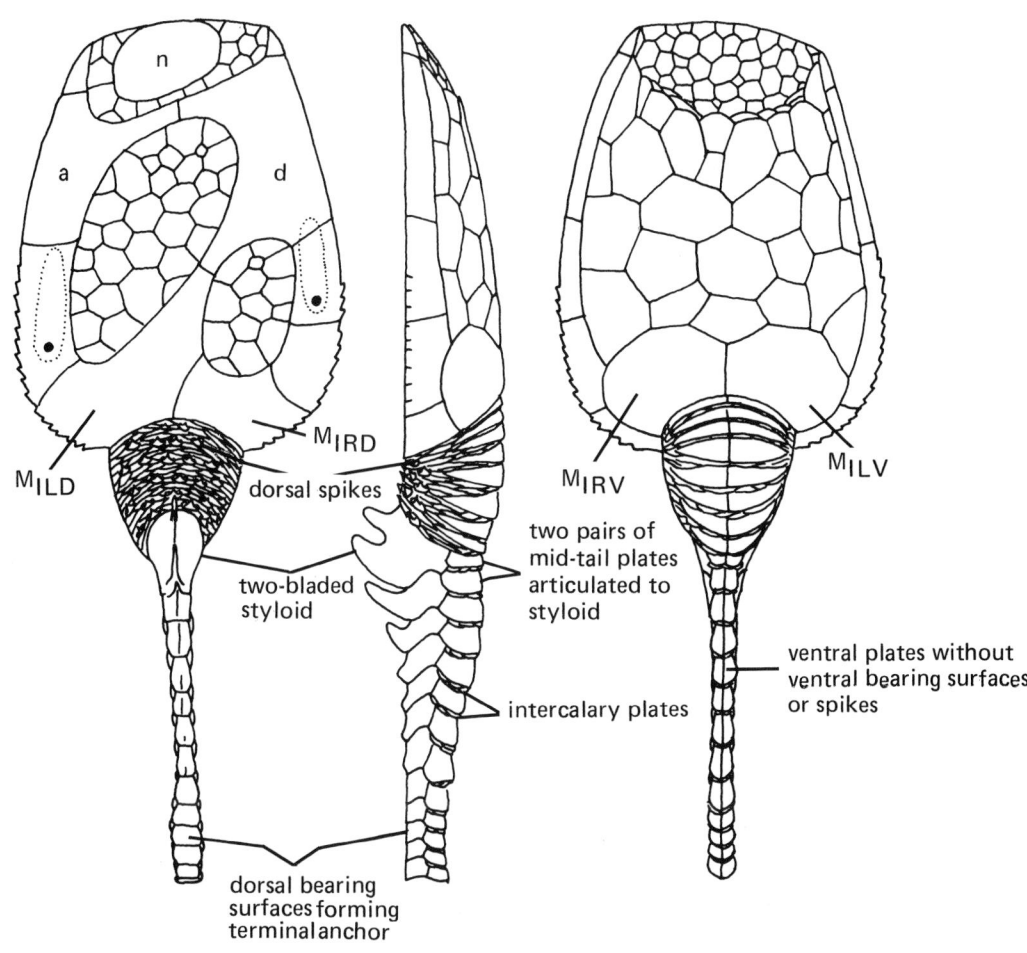

9.9. Reconstruction of the hypothetical animal r – *the latest common ancestor of living tunicates and vertebrates.*

Mi} – lateral undulation in the tadpole of *Ci* was probably acquired separately. It had a prenotochordal head region {*Ci* (tadpole), *Mi*, *Ep*, *Lm*, *Sq*}, a notochordal head region {*Ep*, *Lm*, *Sq*, not *Ci* (tadpole) nor *Mi*}, a trunk region {*Ep*, *Lm*, *Sq*, not *Ci* (tadpole) nor *Mi*} and a tail region {*Ci* (tadpole), *Mi*, *Ep*, *Lm*, *Sq*}. The skeleton was solely cartilaginous, with neither calcite nor bone {*Ep*, *Lm*, not *Sq* (which has scales and teeth) nor *Ci* (which has probably lost its calcite independently) nor *Mi*}. The tail was not truncated by autotomy {*Ep*, *Lm*, *Sq*, not *Mi*, nor *Ci* (which has probably lost tail autotomy independently)}. The mandibular somites, homologous with the left and right anterior coeloms of mitrates, were symmetrical {*Ep*, *Lm*, *Sq*, not *Mi* nor *Ci* (where the mandibular somites are secondarily absent)}. The ears were lateral to the hyoidean somites {*Mi*, probably *Ep*, *Lm*, *Sq*, not *Ci* (tadpole)} and squashed inwards onto them and onto the anterior part of

the fourth somites in ontogeny {probably *Ep* (where the ears are immediately lateral to the anterior part of the notochord), *Lm*, *Sq*, not *Mi* nor *Ci*}.

In the alimentary canal, the larva of *s* would have had an endostylar pharyngeal mucous filter held anteriorly and transported dorsalwards by peripharyngeal bands {*Ci*, *Mi*, *Lm* (larva) not *Ep* nor *Sq*} but this filter would be absent in the adult {*Ep*, *Lm*, *Sq*, not *Mi* nor *Ci*}. This larva would have resembled the ammocoete larva of a lamprey, and would have been followed by a non-ammocoete adult which fed by rasping horny teeth actuated by a great muscular body as in adult myxinoids and lampreys. The ammocoete larva was presumably lost independently in myxinoids and gnathostomes. The adenohypophysis was probably endodermal {*Ep*, ?*Mi*, perhaps hemichordates if the stomochord is homologous, amphioxus if Hatschek's pit and the ventral rostral coelom are homologous with

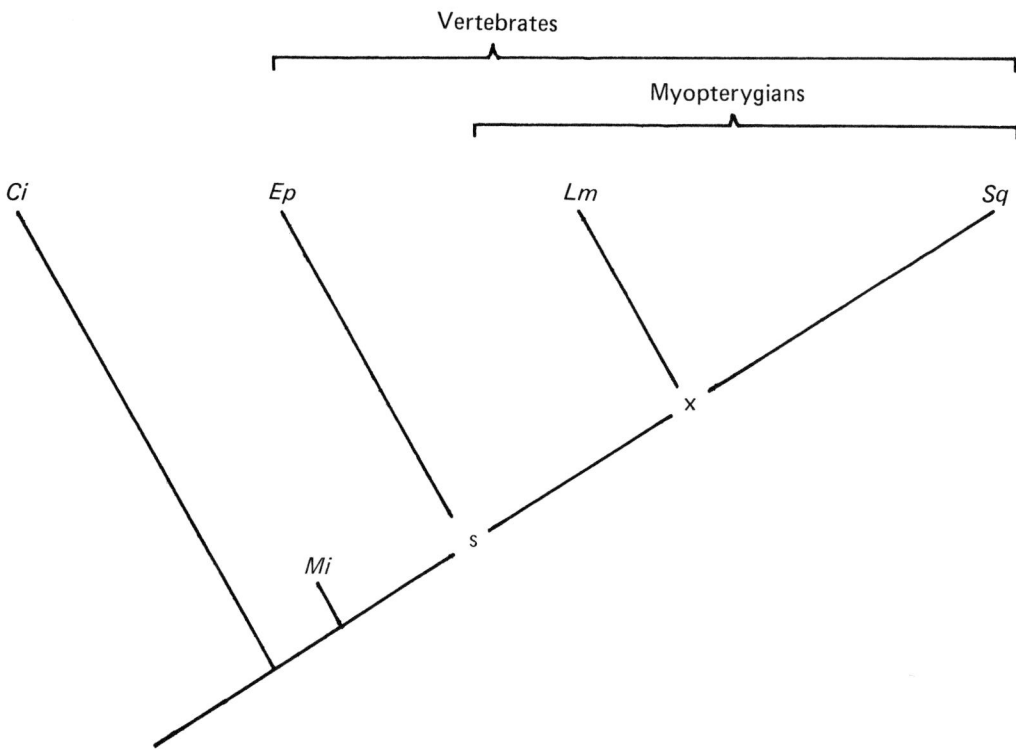

9.10. *Cladogram to show the logic of reconstructing the hypothetical animals* s *and* x. *Ci* = Ciona *(tunicate), Ep* = Eptatretus *(myxinoid), Lm* = Lampetra *(lamprey), Mi* = Mitrocystites *(mitrate), Sq* = Squalus *(gnathostome).* s *is shown in fig. 9.11.* x *is discussed in the text.*

the adenohypophysis, not *Lm*, nor *Sq*, nor *Ci* (where the neural gland develops from a nervous anlage)}. The non-pharyngeal gut would have included oesophagus, stomach and intestine {*Ci, Sq*, not *Ep*, nor *Lm* which both lack a stomach; no information for *Mi*}. A pancreas would have opened into the gut behind the stomach {*Ci* (pyloric gland), *Lm, Sq*, not *Ep* (no information for *Mi*}. The anus would have opened into a median ventral cloaca independent of the left atrium {*Ep, Lm, Sq*, not *Ci* nor *Mi*}.

The nervous system of *s* included a dorsal nerve cord and a brain {*Ci* (tadpole), *Mi, Ep, Lm, Sq*}. The brain was divided into prosencephalon and deuterencephalon {*Ci, Mi, Ep, Lm, Sq*}, and the prosencephalon would have been divided further into telencephalon and diencephalon {*Ep, Lm, Sq*, not *Mi* (but the two regions are sometimes distinguishable in *Placocystites*), nor *Ci*}. There was no pineal nor parapineal eye {*Ep, Mi, Ci* (tadpole), not *Lm*, nor *Sq*}. The paired eyes were lens-less and non-motile, without oculo-motor muscles {*Ep, Mi, Ci* (tadpole), not *Lm* nor *Sq*}. The ears contained one or more semi-circular canals and were not included in atria, or, rather, the atria had converted into auditory capsules and no longer included the anus or the gonopore {*Ep, Lm, Sq*, not *Ci* nor *Mi*}. An

external lateral-line system existed {*Mi*, probably *Ep, Lm, Sq*, not *Ci*}. Trigeminal, acoustic, lateral-line and spinal ganglia existed together with their associated nerves {*Mi, Ep, Lm, Sq*, not *Ci*}. Facial, glossopharyngeal and vagus nerves were present, formed in ontogeny by combining a placodal epibranchial constituent, and a dorsal-root constituent {*Ep, Lm, Sq*, not *Mi* (where the epibranchial constituent, if it existed, would be in the head and the dorsal-root constituent in the tail), nor *Ci*}. Olfactory receptors and nerves existed {*Mi, Ep, Lm, Sq*, nothing certainly homologous in *Ci*}, but they were connected directly to the exterior through a nostril rather than opening into the buccal cavity {*Ep, Lm, Sq*, not *Mi*, nothing comparable in *Ci*}. There were ventral-root nerves innervating the muscle blocks {*Ci* (tadpole), *Ep, Lm, Sq*, no information in *Mi*}. The first crown vertebrate *s*, unlike the mitrates, had a pair of kidneys and kidney ducts {*Ep, Lm, Sq*, not *Ci* nor *Mi*}. The statement that *Mitrocystites* and other mitrates had no kidneys is a deduction from embryology. For kidneys develop in ontogeny from nephrotomes which are formed where myotomes make contact ventrally with lateral plate. And *Mitrocystites*, since it lacked a trunk region and a notochordal head

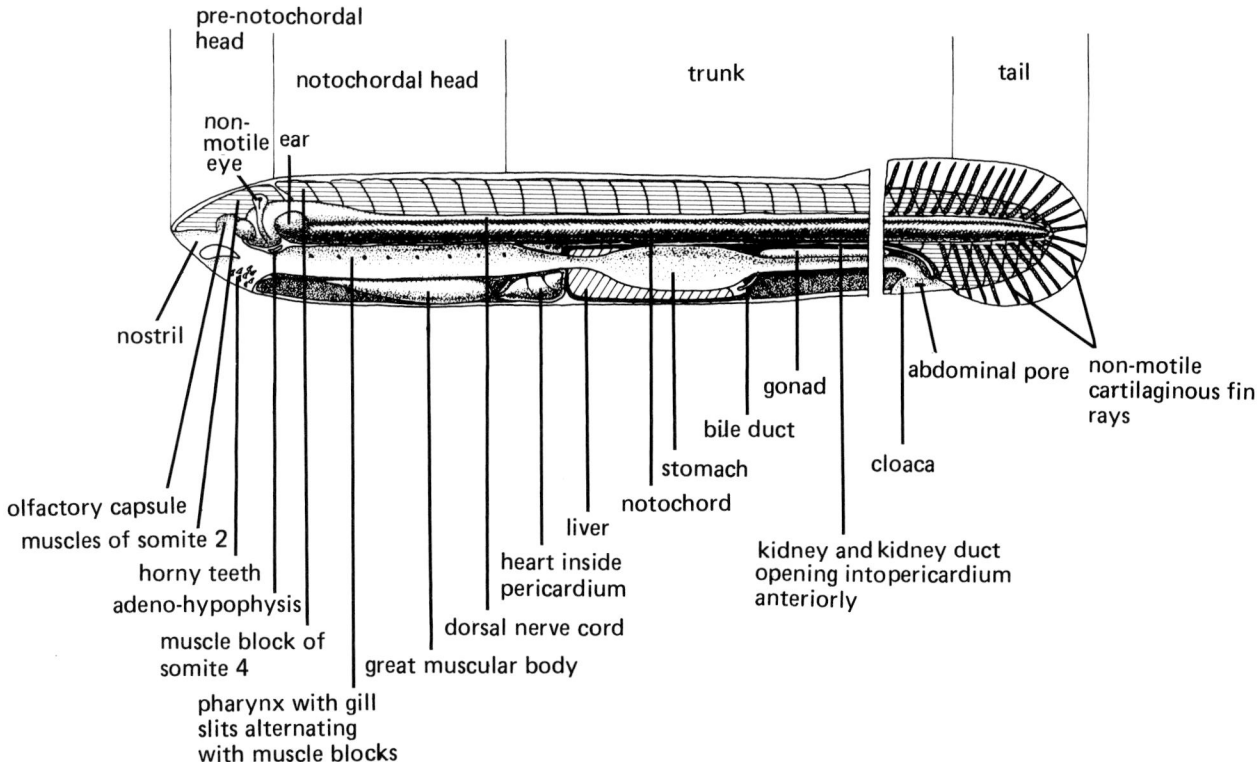

pre-notochordal
head

non-
motile ear
eye

notochordal head

trunk

tail

nostril

olfactory capsule
muscles of somite 2

horny teeth
adeno-hypophysis

muscle block of
somite 4

pharynx with gill
slits alternating
with muscle blocks

great muscular body

dorsal nerve cord

heart inside
pericardium

liver

notochord

stomach

bile duct

gonad

kidney and kidney duct
opening into pericardium
anteriorly

cloaca

abdominal pore

non-motile
cartilaginous fin
rays

9.11. Reconstruction of the hypothetical animal s – the latest common ancestor of living vertebrates and being the first crown vertebrate. The animal is cut through sagittally and the gut, cloaca and liver are shown in sagittal section. This animal resembles the Carboniferous conodont animal described by Briggs, Clarkson & Aldridge (1983).

region, had no lateral plate ventral to myotomes, therefore no nephrotomes, and therefore no kidneys. The blood was isotonic with sea water {*Ci* (and amphioxus and echinoderms) *Ep*, not *Lm* nor *Sq*; no information for *Mi*}. There were no epicardia {*Ep, Lm, Sq*, not *Ci* nor *Mi*}.

The circulatory system of *s* would have contained a heart and blood vessels {*Ci, Ep, Lm, Sq*; no information on *Mi*}. Flow would have been uni-directional as in a fish with a three-chambered heart (sinus venosus, auricle, ventricle) {*Ep, Lm, Sq*, not *Ci*; no information on *Mi*}. There would have been a ventral aorta with afferent vessels leading to the gills {*Ci, Ep, Lm, Sq*; no information on *Mi*}. There would have been respiratory capillaries in the gills {*Ep, Lm, Sq*, not *Ci*; no information on *Mi*}. Efferent branchial vessels would pass into a dorsal aorta, or a pair of such, and from there would have distributed blood to the organs {*Ci, Ep, Lm, Sq*; no information on *Mi*}. The organs would have been bathed in extensive blood sinuses, rather than supplied by capillary networks {*Ci, Ep*, not *Lm* (except for the branchial sinuses), nor *Sq*; no information on *Mi*}.

The reproductive system of *s* would probably involve separate sexes {*Ep, Lm, Sq* (and also acraniates and echino-

derms), not *Ci*; no information on *Mi*}. Eggs and sperm were probably released separately into the sea water and fertilized there {*Ci*, presumably *Ep, Lm*, not *Sq* (but there is little doubt that external fertilization is primitive for gnathostomes); no information on *Mi*}. The gametes probably left the body by way of the abdominal pores in the cloacal wall {*Ep* (presumably), *Lm*, not *Sq*, nor *Ci* (gonoducts), nor *Mi* (evidence exists for a gonoduct homologous with that of tunicates in *Mitrocystella*)}.

The respiratory system of *s* would involve branchial lamellae, with capillary networks, on the gill bars {*Ep, Lm, Sq*, but *Ci*; no information on *Mi*}. Most probably the gills opened direct to the outside, rather than into atria or opercular cavities {*Ep, Lm, Sq*, not *Ci* nor *Mi*}.

In the embryology of *s* all somites were at first probably separated from each other by complete dorso–ventral myocommata. Later, in the notochordal head region and trunk, the ventral parts of the myocommata would break down, so producing uninterrupted lateral plate in those regions {*Lm*, not *Sq* nor *Ep* (where lateral plate is from the first distinct from somites); not *Ci*, nor presumably *Mi* (both of which have no notochordal head nor trunk)}. This sequence of events is similar to amphioxus, where the lateral plate likewise arises by ventral fusion of somites, but this probably represents a convergence with vertebrates, in view of the stem acraniate *Lagynocystis* with its completely distinct head and tail. The gill slits probably alternated with the hyoidean and branchial somites in the notochord head region, as a result of somites growing ventralwards down gill bars, for there seem to be traces of such an alteration in *Lampetra* and sharks, as discussed in Chapter 6 (fig. 6.30, 6.31, pp. 187, 188). Neural crest would have existed in s {Mi (to judge by the presence of spinal ganglia which are always of crestal origin in living vertebrates), Lm, Sq, not Ci; no information on Ep}. The illustration of s in fig. 9.11 is necessarily too exact and over-explicit but may have some merit as a 'straw man'. It resembles the conodont animal from the Scottish Carboniferous recently described by Briggs, Clarkson & Aldridge (1983) and perhaps this resemblance is no accident.

Considerable changes were thus involved in passing from a form like *Mitrocystites* to the first crown vertebrate s. These same changes, which would have happened in segment 25 of figs 9.2b, 9.4, transformed the tunicate-tadpole-like 'antisegmentationist' protovertebrate (fig. 6.28, p. 185) (which was a mitrate) into the almost uniformly segmented protovertebrate of the segmentationists (somewhat like fig. 6.32, p. 189, but jawless). They thus hold the answer to many ancient riddles. Perhaps the greatest such change was the origin of the trunk and the notochordal region of the head. This meant that the posterior part of the mitrate head, where the gill slits were, grew rearwards beneath the anterior part of the mitrate tail. Also the viscera (non-pharyngeal gut, heart, gonads) which had been located in the head, in the cavity of the right anterior coelom, moved rearwards into the new trunk region and came to be hung in a perivisceral coelom formed, by fusion into lateral plate, from the ventral parts of the affected myotomes. As a consequence the rectum escaped from the left atrium. The more anterior somites of the mitrate tail grew ventralwards down the gill bars which had thus come to underlie them. This produced the alternation of somites and gill slits which the segmentationist school of classical anatomists emphasized. At about the same time the ears, located in the right and left atria (which became the right and left otic vesicles), pressed medianwards in ontogeny onto somites 3 and 4, destroying the dorsal part of 3 and the antero–dorsal part of 4 as in the ontogeny of living *Lampetra* and sharks. The branchial nerves (facial, glossopharyngeal, vagus) probably arose by the junction of a placodal, epibranchial component, which had formerly been in the head, with a crestal dorsal-root component, which had formerly been in the tail. (Indeed, all the placodal nerves of living

vertebrates probably evolved from sensory nerves located in the mitrate head.) The result of the origin of the notochordal head region and the trunk was to blur the clear mitrate distinction between head and tail and to give the animal a more or less eel-like shape, as in *Eptatretus* or *Lampetra*. At the same time the atria were lost, except for the portion that remained as otic vesicles. Also the gonads, in agreement with their new position in the trunk, lost the gonoduct and acquired a new exit to the exterior, in the shape of abdominal pores in the cloacal wall. These changes coincided with a change in the adult mode of feeding. The larva would have continued to feed by a pharyngeal mucous trap, as the ammocoete still does, but the adult either ceased to feed or more probably became a predator, perhaps using rasping horny teeth actuated by a great muscular body as in modern myxinoids and lampreys. The acquisition of an eel-like shape was probably an adaptation to swimming in a forwards direction, for if *Mitrocystella* attempted to swim forwards a crick would tend to happen where the head joined the tail. The origin of a trunk region with lateral plate ventral to somites made it possible for kidneys to evolve. And these would be necessary because of increased excretion consequent on swimming. Because of the origin of the notochordal head and trunk, the prenotochordal head decreased in size and the ears and eyes come to be situated near the front end of the animal, which was advantageous since it swam forwards. With a changed mode of feeding, no longer dependent on a branchial current and now requiring the location of prey, it became necessary to take chemical samples of the outside water and consequently the olfactory receptors became walled off from the buccal cavity, probably to open by a single nostril as in living myxinoids and lampreys. At the same time the ears improved as accelerometers by the development of semi-circular canals, and the lateral-line system extended over the surface of the body. The eyes remained lens-less and without oculo-motor muscles but they acquired optic cups. These made them to some extent sensitive to the direction of received light. They would probably be located beneath special pigment-free patches of skin as they are in *Eptatretus*. The calcite skeleton was lost during all these changes, probably as a swimming adaptation. The animal s was the first crown vertebrate. It probably existed already in the Cambrian period, earlier than the American fossils described by Repetski – if indeed these are crown vertebrates rather than stem vertebrates.

From this first crown vertebrate s, two living groups of animals descended – the myxinoids and the [lamprey + gnathostome] group of myopterygians (Janvier, 1978) (fig. 9.10). In the origin of myxinoids from s the main changes were: the loss of the ammocoete larva; the concomitant development of a large egg; and the evolution of the characteristically myxinoid system of rasping horny teeth- - though a recognizable precursor of the lamprey and myxinoid dental system probably existed in s.

In the evolution of myopterygians, whose first crown member is labelled x in fig. 9.10, from the first crown vertebrate s the changes were greater: (1) The paired eyes developed lenses, oculo-motor muscles and oculo-motor nerves – most of the oculo-motor muscles (all except the external rectus and superior oblique) evolved from the mitrate crescentic organ (= premandibular somite) (fig. 8.23, p. 285) and were supplied by a pair of branches from the pre-existing trigeminal complex which became the oculo-motor nerves in the strict sense (III); however, the superior oblique pair of muscles evolved from the dorsal posterior extremity of the mitrate left and right anterior coeloms (= mandibular pair of somites) and were supplied by a pair of branches of the trigeminal complex which became the trochlear nerves (IV); and the external rectus pair of muscles evolved from the foremost pair of somites in the mitrate tail (= hyoidean somites) and were supplied by the paired ventral roots of the hyoidean somite which became the abducens nerves (IV). (2) Pineal and parapineal eyes evolved. (3) The ears were modified to produce two vertical semi-circular canals at right angles to each other. (4) The anterior part of the kidney (pronephros) ceased to be used except in the ammocoete larva. (5) The kidney acquired osmotic control and, concomitantly, part or all of the life cycle came to be lived in fresh water. (6) The heart came under nervous control. (7) The cartilaginous fin skeleton acquired muscles (hence the name myopterygians). (8) The adenohypophysis became ectodermal in origin, probably because the diencephalon had moved forward to overlie the buccal cavity. Most of these changes, particular the improvement of the sense organs, can be regarded as predatory adaptations.

Two living groups of animals evolved from this first crown myopterygian x – the lampreys and the gnathos-

tomes. In the origin of the lampreys, the changes were relatively minor – in particular the horny dental apparatus was elaborated with a relatively static piston of lamprey type. In the origin of gnathostomes, on the other hand, there were great changes: (1) The ammocoete larval stage was lost. (2) Horny teeth, and the agnathan tongue system were lost. (3) Large cartilaginous gill bars of neural-crest origin were developed. (4) And the most anterior of these bars was modified into the mandibular arch with upper and lower jaws. (5) Teeth made of dentine arose on these jaws to make them effective in grasping prey. The animal so produced would be the first crown gnathostome – the latest common ancestor of all living vertebrates with jaws.

In a recent paper Janvier (1981) has placed the various groups of fossil jawless fishes on a cladogram and reached the tentative conclusion that all the known forms can be placed as stem myopterygians, or stem lampreys or stem gnathostomes. Bony tissue probably arose within the stem group of the myopterygians (which are the 'Vertebrata' in Janvier's new restricted sense). However, the diversification of the vertebrates, in the broad sense used here for the word 'vertebrate', is a huge and difficult subject which is beyond the scope of this book.

To sum up, a complex history can now be written of the origin of the vertebrates and their closest relatives. This history is self-consistent, agrees in most respects with classical theory, and fits the complicated morphological data of fossil and recent groups. Many questions are still unanswered, and some of the answers given here will surely prove to be wrong. However, it is likely that segmentationists and antisegmentationists were both, in their different ways, correct. Also the methodological clarity created by Willi Hennig is a necessary tool in unravelling the sequence of events. And finally, above all, as Gislén and Matsumoto first realized, the nub of the whole matter is the extraordinary group of fossils called calcichordates.

Chapter 10   Some other opinions

In this chapter I aim to refute a number of theories that differ from my own. It is impossible to consider all the interpretations that have been proposed. I shall therefore discuss only those views that have been argued in particular detail or have been widely accepted. They fall into two groups – those concerned with the origin of vertebrates or chordates, and those that interpret the fossils here called calcichordates.

As to the origin of vertebrates, the views of Løvtrup have recently attracted much attention. They are developed in a thought-provoking and original book, for which I have much respect, entitled *The Phylogeny of Vertebrata* (1977); it has four main parts, dealing respectively with: (1) an axiomatization of cladistic methodology, (2) the ancestry of vertebrates, (3) the divergence of the vertebrates, and (4) the mechanism of evolution.

Løvtrup's views on the divergence of the vertebrates, presented in the third part of this book, are of great importance. They include the suggestion that myxinoids are the sister group of all other living vertebrates which I have accepted in Chapters 5, 6 and 9 above. His methodological opinions are interesting, mainly because of his painstaking attempt to make them explicit. On the other hand his views on the ancestry of the vertebrates are wildly implausible, as I shall try to show.

Løvtrup's methodology is rigorously cladistic. Its special feature is that, in trying to reconstruct the phylogenetic relationships of groups, he uses a series of 'basic classifications'. These are cladograms with three terminal members (three extant monophyletic groups) which I shall call a, b and c. He sets up two-sided comparisons for each of the three groups, looking for features that are present in two members of the basic classification but not in the third. If he finds a large preponderance of such features shared, for example, by b and c but not a, he presumes that most of them are synapomorphies of b and c, and would mean that b and c are more closely related to each other than

to a. Closeness of relationship is here defined in the strict Hennigian manner in terms of recency of common ancestry. He does not first examine the features to judge whether they are likely to be homologous, since this would introduce subjectivity into the comparisons. It contrasts with Hennig's methodology, as used by Hennig (1969, 1981, 1983) and as applied also by me in Chapter 9, which, by implication or overtly, reconstructs a series of groundplans (latest common ancestors) for monophyletic groups. Løvtrup's method is perhaps more objective than Hennig's, for it requires no prior decision as to what is primitive and what is derived, but is weaker in ignoring loss of features. Indeed Løvtrup believes, or hopes, that such loss is unimportant for he says, as his Theorem 2.5 (1977, p. 22): 'Each animal is the bearer of the taxonomic characters defining all the taxa to which it belongs'.

Løvtrup believes that fossils are irrelevant to a phylogenetic classification. This is argued, in his quasi-Euclidean manner, as follows – Axiom 2.3: 'All actual and potential taxonomic characters possessed by an animal can be known only for extant ones'. This leads to Theorem 2.2: 'Only extant animals can be classified'. And this in turn leads to Theorem 2.3: 'The discovery of a new fossil has no impact on classification.'

From this he concludes (1977, p. 22) that the calcichordates are irrelevant to reconstructing the ancestry of vertebrates. Obviously, I am personally biased against this result, but this is unimportant. What matters is that 'Theorem 2.2' does not follow apodictically from 'Axiom 2.3'. In my view Løvtrup has omitted from his argument a sentence between the axiom and the theorem which should have read: 'An animal can be classified only if all its actual and potential characters can be known.' This sentence, however, would clearly be mistaken. All attempts yet made to classify animals, whether phylogenetic or not, have been based on only a small proportion of potentially usable characters. Fossils, it is true, give fewer relevant facts than still-living animals and this makes them difficult to use phylogenetically, as Løvtrup does right to stress. I contend, however, that, in lucky cases, particularly when soft parts are reconstructable, stem groups can preserve synapomorphies that extant forms have lost. Løvtrup's syllogistic arguments do not successfully exclude this possibility.

Løvtrup values non-morphological characters above morphological ones for purposes of phylogenetic reconstruction. But his argument here is in fact a naked assertion for it runs as follows: Axiom 2.10, 'The origination of new non-morphological characters is much less probable than is the origination of new morphological ones'. This leads, logically enough, to: Theorem 2.55, 'In phylogenetic classification non-morphological characters carry much more weight than morphological ones.' However, the axiom is not self-evident and therefore the theorem is worthless. No generalization about the relative stability of biochemical or histological *versus* morphological features will stand up, for both types of feature show a great range of stability. Thus, among biochemical features, the genetic code is very stable and so is the basic molecule of cytochrome *c*, but the nature of particular amino-acid residues in mammalian fibrino-peptides is not stable. Likewise, among morphological features, the heart or the notochord are stable features, present as homologues in *Ciona* and man (in my view though not in Løvtrup's), whereas hair length, as between a wild and an Angora rabbit, is highly unstable. It is truer to say that complex features, which could not have arisen instantaneously in evolution, which are held together by Riedl's burden relationships, and which represent complicated, convincingly recognizable gestalten, are more likely to be stable than simple ones. Such features, however, can be either non-morphological or morphological. At the same time, morphological features do at present have one great advantage over other features – their taxonomic distribution is much better known.

When analysing relationships between phyla, Løvtrup's preference for non-morphological features is most marked. For he asserts (1977, p. 55) that morphological characters are 'not stable enough to warrant the establishment of a satisfactory superphyletic classification'. And indeed (1977, p. 56): 'In phylogenetic classification of taxa above the phylum level it is necessary to employ non-morphological characters exclusively'. In Løvtrup's eyes the vertebrates count as a phylum, the relationships between vertebrates, tunicates and acraniates are interphyletic and therefore the traditional morphological features uniting the Chordata are irrelevant. The list of features common to vertebrates, tunicates and acraniates, which is thus swept out of the path, is long. It includes: the post-anal tail; notochord; muscle blocks (or the probably

homologous tiers of muscle cells in tunicate tadpoles); hollow dorsal nerve cord with Reissner's fibre; hollow brain; pharynx with gill slits; endostyle with subendostylar vessel and with gland strips, ciliated strips and paired iodine-binding strips dorsal to the gland strips; peripharyngeal bands and pharyngeal mucous filter; and adenohypophysis = neural gland of tunicates = Hatschek's organ of acraniates. Moreover, comparing tunicates and vertebrates alone, the heart and pericardium situated just behind the subendostylar vessel arbitrarily become irrelevant as does the presence of similar retinal cells in the vertebrate eye and the tunicate-tadpole brain. And in comparing acraniates and vertebrates alone, the homology of the anterior dorso-lateral pair of somites of amphioxus with the mandibular somites of vertebrates can be discounted, as also that of the anterior gut diverticula with the premandibular somite of vertebrates. Indeed Løvtrup's rejection of the phylum Chordata is a *reductio ad absurdum* of his belief that 'non-morphological' features are intrinsically better than morphological ones as phylogenetic indicators. His anti-morphological bias also rejects as irrelevant any morphological features that support the distinction between protostomes and deuterostomes. Similarly it can be applied, as a convenient pruning knife, to the detailed comparisons between *Cephalodiscus* and echinoderms drawn by classical authors and listed by me in Chapters 2 (p. 30) and 9 (p. 318).

The phylum Chordata is non-existent as a monophyletic group, according to Løvtrup, besides being invalidly based. Indeed he thinks that the huge, possibly monophyletic group of [Molluscs + Arthropoda + Annelida] is more closely related to vertebrates than the acraniates or tunicates are. But this result can likewise be seen as a *reductio ad absurdum*. It would mean either that the traditional chordate features just listed are not homologous between acraniates, tunicates and vertebrates, or else that they once existed in the ancestry of [Mollusca + Arthropoda + Annelida] but have since been lost. Both these propositions are, to my mind, grossly improbable. They indicate that, at the 'interphyletic' level, Løvtrup's methodology does not properly function. The likely reason is the vagueness and simplicity of the features used in comparison. Thus, on p. 126 (1977), echinoderms are said to lack 'nerves' while acraniates, vertebrates and molluscs possess then, but on p. 127 Løvtrup refers to the radial nerves of starfishes. The absence of 'ganglia' in echinoderms is similarly a matter of definition. If Løvtrup's methodology leads him to interpose the [Mollusca + Arthropoda + Annelida] in the middle of the phylum Chordata, we can reasonably disregard as baseless his view that vertebrates are more closely related to molluscs than to echinoderms in particular. I also believe that this view disagrees with the fossil evidence of the calcichordates, but, as already said, Løvtrup and I differ as to the value of fossil evidence.

Løvtrup proposes an epigenetic mechanism (1977, p. 148) to explain the difference between vertebrates and annelids. In my view such embryological sleight-of-hand, in transforming one phylum to another by way of the egg, is in no way more convincing than the 'credulous formanalytical strategems' which he condemns (1977, p. 146) when used by anatomists for the same purpose. Both types of supposed transformation, one affecting the egg and the other the adult, would produce huge changes and leave no evidence behind them.

When applied within the vertebrates, Løvtrup's methodology gives much more reasonable results. The reason for the difference is probably that there he is mainly, though not solely, using morphological features whose taxonomic distribution is well known and whose homology is long-established. To sum up, I have genuine respect for Løvtrup's book which is partly penetrating and partly incredible. The conclusions about the ancestry of vertebrates are, in my view, incredible.

Neotenous transformation of a tunicate tadpole is central to several theories of the origin of vertebrates which have been widely discussed and adopted in English-speaking countries and in Germany. The first of these theories was proposed by Garstang (1894, developed in detail in 1928). In Germany a similar theory was proposed by Veit (1924, 1947) and adopted by Starck in his influential writings (1944, 1963, 1975). In English-speaking countries Garstang's original suggestions have been developed and modified in different ways by Berrill (1955), Romer (1972), Tarlo (1960), Bone (1960, 1981) and Barrington (1965).

Garstang (1928) considered that budding and coloniality were homologous in all compound tunicates and also homologous with the same conditions in pterobranchs. Also the budding primitively evolved as a result of fixation to a substrate, so that pelagic tunicates had inherited their coloniality from a fixed benthic ancestor. But tunicates, to use modern terminology, shared with vertebrates and acraniates several synapomorphies which pterobranchs lacked (elaborated endostyle and peripharyngeal bands, notochord, dorsal nerve cord, muscular tail, brain) and most of these synapomorphies show, in tunicates, only in the tadpole. Garstang, therefore, supposed that the muscular, tailed, motile tadpole-phase was inserted into the life cycle of forms with fixed adults by modification of a pre-existing ciliated larva of the type seen in enteropneusts or echinoderms (tornaria or auricularia). He supposed that subsequently, in the exclusive common ancestry of vertebrates and acraniates, the previous fixed adult stage had been lopped off the life cycle by precocious sexual development (neoteny).

Consequently adult vertebrates and amphioxus were homologous with the larva of the ancestral tunicate. Garstang found confirmation of his views on the origin of acraniates and vertebrates in a powerfully argued case that appendicularians were not primitive tunicates, as previously believed by most, but close relatives of the doliolids (as accepted by me in Chapter 4). They differed from these mainly by neotenous loss of the primitive tail-less adult stage. I cannot decide by reading his paper, however, whether he regarded the appendicularians as the sister group of [acraniates + tunicates] or merely as having arisen in phylogeny by an analogous process. In the absence of any statement to the contrary I suppose he regarded these two situations as mere parallels. I stress that, in modern terminology, Garstang's argument for thinking that tunicates, and therefore chordates, were primitively attached, was an outgroup comparison with pterobranchs, with the assumption that budding primitively implied attachment to a substrate.

The lack of gill slits in echinoderms was primitive for Garstang, and he, therefore, thought that hemichordates were more closely related, in Hennig's sense, to chordates than echinoderms are. He also assumed that bilateral symmetry was primitive and homologous in hemichordates and chordates, as most other authors have done. He explained the asymmetry of larval amphioxus as a special adaptation for larval feeding, necessitated by the small-yolked egg and producing the benefit of an enlarged mouth. He considered that myomeric segmentation was incipient, rather than degenerate, in the tails of tunicate tadpoles and fully developed in acraniates. Full myomery was, therefore, a synapomorphy of acraniates and vertebrates that tunicates lacked. He homologized the U-shaped gill slits of post-larval and adult amphioxus with the U-shaped slits of post-larval tunicates and adult *Balanoglossus*. The cladogram in fig. 10.1 represents Garstang's views, so far as I can reconstruct them, assuming that the neotenous origin of appendicularians was a parallelism with that of [acraniates + vertebrates].

Parts of Garstang's theory are probably correct in my view. As explained in Chapter 4 (p. 113) the arguments for regarding appendicularians as the sister group of doliolids within the tunicates are compelling. They are based on several synapomorphies including: the dextral twist of the gut, the posterior position of the atrium or its appendicularian equivalent, simplicity of the endostyle, and, above all, the habit of moulting the test (appendicularian house) and the absence of cellulose in it. Also, given this sister-group relationship, then the doliolids are probably more primitive than appendicularians in retaining a post-tadpole phase like benthic tunicates (as originally suggested by Willey, 1894). and this implies that appendicularians have lost the post-tadpole

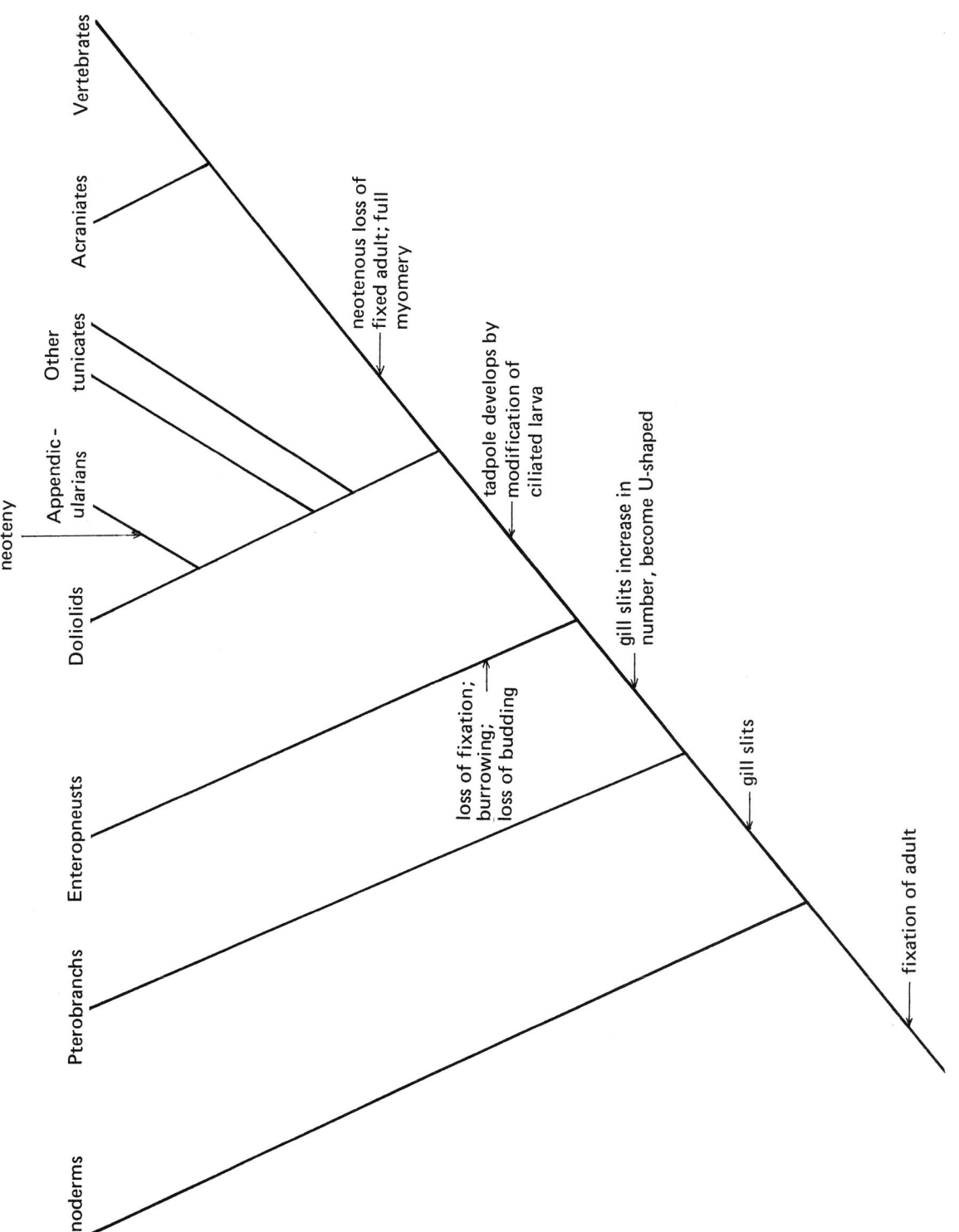

neoteny

Vertebrates

Acraniates

Other tunicates

Appendic-ularians

Dolioliids

Enteropneusts

Pterobranchs

Echinoderms

neotenous loss of fixed adult; full myomery

tadpole develops by modification of ciliated larva

loss of fixation; burrowing; loss of budding

gill slits increase in number, become U-shaped

gill slits

fixation of adult

10.1. *The phylogeny of the deuterostomes according to Garstang (1928) – a cladogram deduced from Garstang's paper.*

by neoteny. However, since Garstang nowhere states that this neoteny is homologous with the one that ostensibly gave rise to [acraniates + vertebrates], this strong part of the argument is irrelevant to the main point at issue.

The weakest part of Garstang's argument is his assumption (followed by Millar, 1966 but rejected by Berrill, 1936, 1955) that budding is homologous throughout tunicates and with pterobranchs, for there are great differences: *Rhabdopleura* has a stolon made of mesoderm and ectoderm; *Cephalodiscus* produces buds from the free end of a stalk and these later break away from the parent; pelagic tunicates have a stolon which, in the likely primitive state as seen in *Pyrosoma,* has separate strands, or pairs of strands, from the heart, gonads, pharynx and atria; aplousobranch tunicates have budding which always involves the epicardium; and stolidobranch tunicates only bud in the subfamily Botryllinae and the buds always start from the atrial wall. Moreover, if budding is primitive, there are several groups of solitary tunicates which must then have lost it independently. Garstang was the outstanding English-speaking expert on tunicates of his day, and well aware of those different modes of budding 'even in genera so closely related as *Clavelina* and *Distaplia*' (1928, p. 175). It is hard to discover why he thought they were all homologous. But if they are not, then the outgroup comparison with pterobranchs falls and the whole plausible web is broken, apart from some hair-thin arguments based on adaptation. If budding is not everywhere homologous in tunicates and hemichordates, then Garstang's reason for thinking that the latest common ancestor of tunicates and hemichordates was attached disappears, and with it the conclusion that the tadpole is a larval interpolation in phylogeny, rather than a Haeckelian recapitulation. Barrington (1965) and Bone (1981) have repeated and elaborated Garstang's proposal, but without giving any additional evidence, though evidence is what this hypothesis lacks.

Berrill (1955) proposed a hypothesis on the origin of vertebrates which is intellectually descended from Garstang's and resembles it in many ways. He believed that adult attachment to a substrate was primitive for tunicates but that budding was not. Indeed he argued, on good grounds, that budding had arisen several times within the tunicates. He was uncertain, however, whether hemichordates were closely related to chordates, and Garstang's outgroup comparison with pterobranchs plays no part in his argument. Berrill pointed out that the tunicate tadpole in present-day benthic tunicates serves mainly for habitat selection and he proposed that it had arisen, in a previously attached tailless group, for this purpose. In particular the tadpole, with its eye, is able to find the dark clean undersides of rocks. From these initially benthic forms the pelagic tunicates had arisen,

to exploit the vastness of the open sea. Berrill accepted Garstang's interpretation of the appendicularians as neotenous doliolids and strengthened it by pointing out that appendicularians have in their bodies about half as many cells as benthic tunicate tadpoles (1955, p. 137). The vertebrate–acraniate group is supposed to have arisen from early pelagic tunicates by a similar, or possibly the same, neoteny. Shortly after arising, the ancestors of the [acraniates + vertebrates] entered fresh water, during at least part of their life cycle. Then acraniates evolved, by returning again to the sea and losing their brains and sense organs. Modern vertebrates evolved from forms that remained in the rivers.

Berrill commands respect as a great student of tunicates. In particular, his prescient classification of tunicates, published in 1936, is almost cladistic in outlook and form. His account of the origin of vertebrates has literary charm, but is based not on outgroup comparison whether weak or strong but entirely on adaptational arguments. Thus his contention that the tunicate tadpole serves for site selection, particularly for finding dark overhangs (1955, p. 20), is cogently defended. But the conclusion drawn, that the tadpole evolved for this reason, rather than being retained for this reason, is thin. Adaptational arguments are weak as criteria of primitiveness because, like this one, they can usually be read in two opposite directions. Berrill admits that his case is weak. Indeed he describes it, in the last sentence of his book, as: 'such stuff as dreams are made on'.

Remaining theories that invoke a neotenous tunicate tadpole were put forward by vertebrate anatomists and strongly resemble each other. Such authors include principally two Germans (Veit, 1924, 1947; Starck, 1944, 1963, 1975) and the American Romer (1972). All these theorists point out, correctly, that gill slits are more widely distributed in the natural system than the post-anal tail with its notochord, dorsal nerve cord and muscle blocks. Thus gill slits occur in pterobranchs, enteropneusts and chordates, whereas the post-anal tail and its constituent organs are found only in chordates. This difference in distribution strongly suggests that gill slits arose earlier than muscle blocks and so could not, in the beginning, have alternated with them anatomically. This is seemingly contrary to the claims of the segmentationists. Many facts from the embryology of vertebrates were brought forward to support the claim that the tail was initially very distinct from the head, leading to the important conclusion that the vertebrates had, at some stage, a tunicate-tadpole-like ancestor (fig. 6.28, p. 185). Indeed, in embryological form the assertion that head and tail are fundamentally distinct goes back to Froriep (1882). These compelling arguments of the antisegmentationists have already been summarized in Chapter 6.

However, Veit, Starck and Romer go on to assert, with

Garstang and Berrill, that the tunicate tadpole arose as a larval adaptation in forms with a fixed adult stage, and that this fixed adult was lost by neoteny. Romer's ideas descended from Garstang's, but Veit's perhaps did not since his first statement of them, published in 1924, precedes Garstang's major paper on the subject (1928), though later than Garstang's preliminary note (1894). None of these authors considered the possibility, advocated here, that the segmentationist vertebrate ancestor, in which muscle blocks alternated with gill slits, might be descended from the tadpole-like ancestor. None of them gave a good reason why neoteny of a tadpole larva in the ancestry of vertebrates is more likely than the older view that tunicate tadpoles recapitulate a free-living adult. Neoteny is introduced by all of them as a graceful hypothesis in the subjunctive mood, which gradually becomes solider, not by bringing evidence but by mere repetition.

The suspicion that vertebrates had a tunicate-tadpole-like ancestor is well founded but neoteny of such a tadpole in the origin of vertebrates has the scientific status of a creation myth. It is an explanatory fable that flourished in the absence of evidence, helped by the authority of its advocates. In English- and German-speaking countries it largely replaced, by these means, an earlier view which Garstang summarized as follows (1928, p. 58): 'The development of Ascidians has hitherto been interpreted as an example of adult degeneration consequent upon the assumption of sessile habits by a free-swimming ancestor. The organization of this ancestor is generally assumed to have been more or less like an *Amphioxus* or an Ammocoete, sharing an open endostyle and ciliated gill slits with the former and cerebral sense organs ... with the latter, together with the neuromuscular metamerism of both. The tailed larva is supposed to 'represent' such an ancestor, which, having fixed itself to a rock, was eventually superseded by an adult of the modern sessile creature.' This is the theory, grounded on Haeckel's law, which Willey (1894) propounded and which baseless neoteny drove out. The fossils show that, unlike its successor, it was very near the truth. People who believe in the neotenous-tadpole theory of the origin of vertebrates should ask themselves, and try to explain to others, *why* they believe it.

A new theory of vertebrate origins, not invoking neoteny, has been proposed by Northcutt & Gans (1983) and summarized by Gans & Northcutt (1983). These authors note, like Romer (1972), the many contrasts between the head and trunk region of a vertebrate – in particular that placodes and paired sense organs are especially associated with the head and that neural crest gives rise to cartilage in the head but not in the trunk nor tail. They also list a number of 'shared-derived' features of vertebrates which, as they assert, characterize vertebrates in the Recent fauna but not other chordates (Gans & Northcutt, 1983, p. 270). Their list comprises: cranial nerves with sensory ganglia; trunk nerves with sensory ganglia; peripheral motor ganglia; prosencephalon; nose; eyes; ears; lateral-line mechanoreceptors; lateral-line electroreceptors; taste-sensory organs; cartilaginous pharyngeal bars; branchiomeric muscle; smooth muscle of the gut; calcitonin cells; chromaffin cells; gill capillaries; muscularized aortic arches; muscular heart; anterior neurocranium and sensory capsules; cephalic armour and its derivatives. They note in particular that these features are lacking in modern acraniates which they regard as the closest living relatives of vertebrates. They then suggest that the crucial event in the origin of vertebrates was the acquisition of a head which acraniates lack. It would have arisen anterior to the notochord which latter would have extended to the anterior end of the ancestor, as in amphioxus. They relate the origin of this head, with its sense organs, placodes and cartilage of crestal origin, to the adoption by the adult of a predatory mode of life, instead of the older habit of filter feeding. This predatory mode of life required the origin of respiratory gills with gill capillaries and pharyngeal pumping muscles and the acquisition of sense organs for detection of prey. Ultimately it led to the origin of biting jaws in the gnathostomes. This is a brief summary of a long narrative, presented with great embryological learning, in a paper of 28 pages (Northcutt & Gans, 1983).

A reader can only despair. For the contrasts between head and post-head suggest to Northcutt & Gans that the head is a new invention of the vertebrates which acraniates have never had. Whereas the same contrasts led Romer (1972) and Starck (1975) to see the head–tail distinction as an ancient feature, originating with the tadpole of the supposed tunicate-like ancestor of all living chordates and lost by the acraniates. When equally eminent workers, starting from the same data, reach mutually contradictory conclusions, it might seem that all phylogenetic reconstruction is vain. Indeed, I believe that it is so, when based entirely on extant organisms without benefit of cladistic methodology.

Taxonomic vagueness permeates the work of Northcutt & Gans. For example, they recognize 'protochordates' as a group but write of it, according to context, sometimes as equivalent to [pogonophores + hemichordates + tunicates + acraniates] and sometimes as equivalent to acraniates only (Gans & Northcutt, 1983, p. 270), though their commonest usage seems to make it equivalent to [tunicates + acraniates]. They do not recognize the logical consequences of the fact that the group 'protochordata' in their widest sense is polyphyletic (for the pogonophores are allied to the annelids (Southward, 1975)) and in their commonest sense is probably paraphyletic, nor do they warn the reader that it swells and shrinks from paragraph to paragraph. Again they see the agnathans as a group, which they assume to be monophyletic, so that features common to lampreys and gnathostomes, such as an electrical sense located in the lateral-line organs, are ascribed also to the earliest vertebrates (Gans & Northcutt, 1983, p. 271). This does not follow if the agnathans are paraphyletic with the myxinoids as sister group to the other vertebrates (myopterygians) as argued in Chapter 5 and by Løvtrup (1977) and Janvier (1981). For in that case the latest common ancestor of myxinoids and myopterygians was older than the latest common ancestor of the myopterygians. Taxonomic vagueness vitiates a phylogenetic argument, since only by arranging organisms into monophyletic groups within monophyletic groups is it possible to discern the sequence in which their respective synapomorphies arose.

An aspect of this vagueness is that tunicates are largely ignored. This leads Northcutt & Gans to assume that living amphioxus accurately represents an ancestral condition from which vertebrates have arisen. Thus their list of 'shared-derived' vertebrate features comes to include some which also occur in tunicates, to wit: the muscular heart; the prosencephalon (sensory vesicle of tunicate tadpole (Katz, 1983)); ears – probably equivalent to the cupular organs of adult tunicates (Bone & Ryan, 1978); eyes – if the eye of a tunicate tadpole is homologous with the eye of a vertebrate (Dilly, 1964), or particularly with the right eye (Froriep, 1906). Indeed the head-tail distinction, whose importance is rightly stressed by Northcutt & Gans, already exists in a tunicate tadpole, as Starck and Romer have emphasized. This gravely weakens the view that the head is a vertebrate invention.

A still graver objection to the views of Northcutt & Gans is that they ignore the fossil evidence. Thus, of their 'shared-derived' vertebrate characteristics, the following existed in mitrates: cranial nerves with sensory ganglia; trunk (tail) nerves with sensory ganglia; prosencephalon; nose; eyes; ears; lateral-line (in stem-vertebrate mitrates only); muscular heart (present in mitrates by deduction). The presence in the mitrate head of olfactory organs, ears, lateral-line and dorsal branches of the trigeminal (superficialis and profundus, or n_5 and n_4) indicates that placodes existed in the head embryologically. Moreover, the presence of trigeminal ganglia in forms that do not have dorsal (placodal) branches of the trigeminal (*Lagynocystis* and the peltocystidans) suggests that the head neural crest existed in mitrates, just as the spinal ganglia of mitrates suggest tail neural crest. Also the presence of a clear distinction between head and tail in all cornutes and mitrates, and especially in the stem acraniate *Lagynocystis*, shows that the extension of the notochord

to the front end of the animal in living acraniates is a specialization that vertebrates have never had in their ancestry. Indeed, the mitrates confirm that the tunicate tadpole gives a better picture of the lay-out of the first crown chordates than does amphioxus.

Northcutt & Gans (1983, p. 2) reject the calcichordate theory without discussion. They quote no paper of mine more recent than (1975) and mention the critique by Philip (1979) without citing my reply (1981b). It is clear that they have never understood the calcichordate theory. For they ascribe to me the opinion (Northcutt & Gans, 1983, p. 2) that calcichordates are the sister group of chordates, rather than being a paraphyletic grouping ancestral to the living chordates. To sum up, their narrative is less plausible than those of Romer or Starck, because it disregards the tunicates, and is basically unsound because it uses vague and variable groups and ignores the fossil evidence.

Concerning the fossils central to this book, three rival interpretations are now current which can be called the calcichordate, aulacophore and stele theories. The calcichordate theory is the one advocated here. The aulacophore theory holds that the appendage of cornutes and mitrates, here called the tail, was a feeding arm. The stele theory holds that it was not a feeding arm, was sometimes propulsive like a tail, was homologous with the appendage (stele) of other 'carpoid echinoderms', but was not homologous with the tail of chordates.

The aulacophore theory was proposed by Ubaghs (1961), was applied by him in the *Treatise of Invertebrate Paleontology* (1967) and has been assumed by him and others in several later works (Ubaghs, 1969, 1971, 1975, 1979, 1981, 1983; Caster, 1983; Kolata & Guensburg, 1979; Chauvel, 1981; Chauvel & Nion, 1977; Sprinkle, 1976, 1983). The stele theory was proposed by Philip (1979) in a review paper where he considered and rejected the calcichordate and aulacophore theories. It has since been applied by him (1981) in describing an Australian Silurian mitrate and by Kolata & Jollie (1982) in describing some American Ordovician mitrates. Ubaghs (1981) and I (Jefferies, 1981b; 1982) have both answered Philip from our different angles. According to the stele and aulacophore theories, the cornutes and mitrates are echinoderms. They are referred to collectively as stylophorans, in reference to the massive ossicle (stylocone or styloid) thought to be homologous in the middle part of the appendage.

The aulacophore theory, as mentioned, holds that the appendage of cornutes and mitrates was a feeding arm. It uses the name aulacophore ('groove-bearer') for this appendage, alluding to the median groove in the inner face of the ossicles of its distal parts. It holds that the proximal, mesial and distal parts of the appendage (fore, mid and hind tail) are homologous as between cornutes and mitrates. Moreover, the ossicles of the distal and mesial

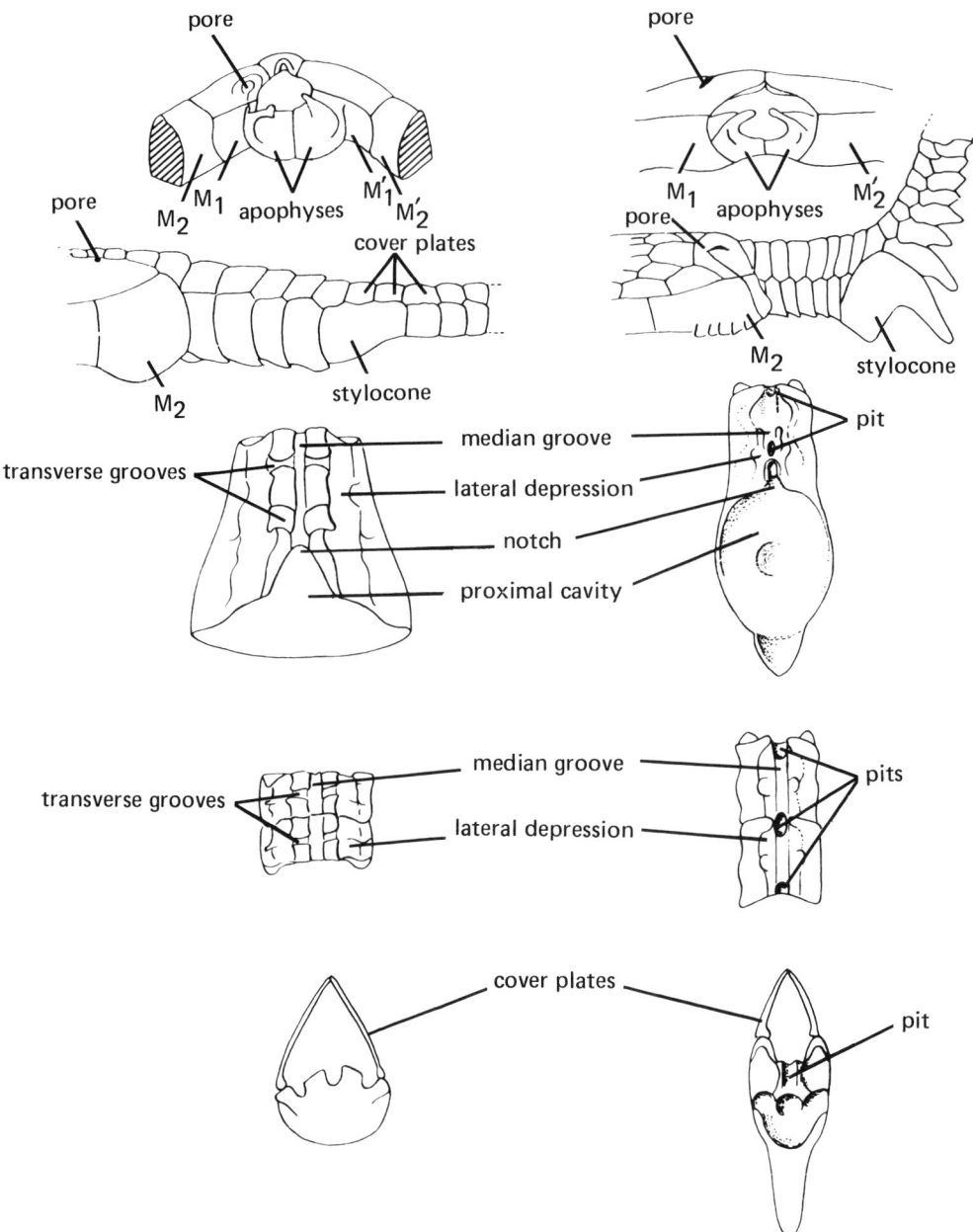

10.2. Ubaghs' views on the homologies between cornutes (represented by Phyllocystis crassimarginata *Thoral) and mitrates (*Mitrocystites mitra *Barrande); from Ubaghs (1981, fig. 3, p. 10) with translated labels.*

parts (what I call the hind-tail ossicles, and the cornute stylocone and mitrate styloid) were also homologous between cornutes and mitrates and were downwards (ventral) in life in mitrates as well as in cornutes. These views imply, since everyone agrees which face of the cornutes was downward, that what I call the ventral face

of mitrates was upward in life. Workers who accept the aulacophore theory, therefore, refer to the mitrate styloid as the stylocone, by presumed homology with cornutes. The paired plates of the distal and mesial parts of the aulacophore (hind and mid tail) are also seen as homologous between cornutes and mitrates and regarded as cover plates, able to open outwards. The median groove on the inner face of the ossicles (by the aulacophore interpretation always the upper face) is thought to have carried a radial water vessel from which lateral vessels, housed in the transverse grooves, went off to tube feet, seated in the lateral pits (fig. 7.8a, p. 203) of cornutes or in what I have called the ganglionar pits of mitrates (fig. 8.4c, p. 245). Ubaghs' proposed homologies between the 'aulacophores' of cornutes and mitrates are shown in fig. 10.2, taken from Ubaghs (1981).

Ubaghs (1981) has recently defended the aulacophore theory against Philip's criticisms, and also against mine, in a short paper which brings the chief issues into focus. He presents five main arguments, as follows:

(1) The echinoderm-type skeleton of cornutes and mitrates favours an anatomical interpretation in echinoderm terms, whence the appendage is more likely to be a food-gathering arm than a chordate tail. (This is an argument against the calcichordate theory but not against the stele theory which agrees with Ubaghs that the cornutes and mitrates are echinoderms.)

(2) The appendage of cornutes and mitrates is in several ways unlike that of other 'carpoid echinoderms' (often called the stele) and therefore probably not homologous with it (cf. fig. 10.3). (This is important because the stele of these other forms, since it lacks 'cover plates', and is completely enclosed in immovable skeleton at the distal end, cannot in any way be interpreted as a feeding arm.)

(3) The 'theca' (head) of cornutes was probably incapable of locomotion because of its frequently asymmetrical outline, the downward pointing spikes on its lower surface, and occasionally the presence of spines all about it, spread out in the horizontal plane (the genus *Chauvelicystis*; fig. 10.4). A sessile mode of life is, therefore, likely, so that a locomotory appendage, as required by the calcichordate and stele theories, would be redundant.

(4) The plates of the distal part of the 'aulacophore' were able to open outwards, as indicated particularly by their mode of preservation in cornutes.

(5) The proposed homologies between the skeletons of the 'aulacophores' of cornutes and mitrates are very detailed (fig. 10.2). It follows that the calcichordate theory places the mitrates upside-down and therefore that the whole chordate interpretation is mistaken. (This, again, argues against the calcichordate theory, but not against the stele theory, since Philip agrees with Ubaghs that what I call the ventral face of mitrates was upward.)

I shall answer Ubaghs' arguments in the same sequence. As to the first one, stereom mesh certainly implies an echinoderm in the Recent fauna, but we can reasonably expect a phylum to have been less distinct from related

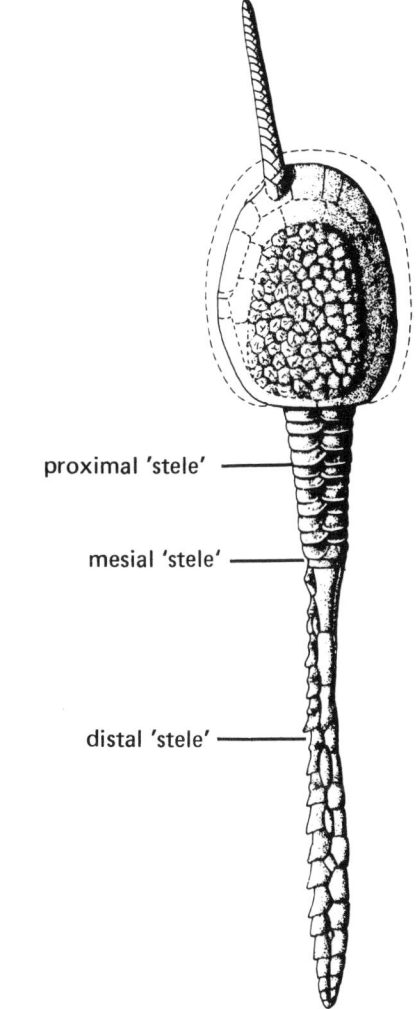

proximal 'stele' ——

mesial 'stele' ——

distal 'stele' ——

10.3. *The solutan 'carpoid'* Syringocrinus sinclairi *Parsley & Caster in ventral aspect to show the non-autotomized distal end of the 'stele' (tail) (Middle Trenton Group (Middle Ordovician), Ontario); from Parsley & Caster (1965, text-fig. 1, 2, p. 122).*

phyla in the remote past. The presence of an echinoderm-type skeleton in early chordates is not surprising since the two phyla have long been seen as related and cornutes have convincing gill slits.

Ubaghs' second argument implies the non-homology of the aulacophore with the stele of other 'carpoid echinoderms'. The most relevant fossils here are the solute 'carpoids' (Cambrian to Devonian) where the stele consists of three parts – proximal, mesial and distal (fig. 10.3). The proximal part, which was evidently highly mobile, has a large lumen and a loose skeleton formed of quadripartite rings. The distal part has two series of plates, one dorsal

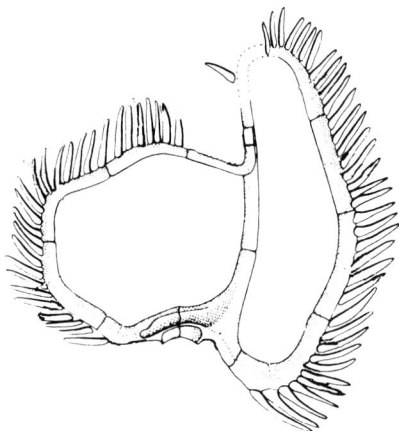

10.4. *The cornute* Chauvelicystis spinosa *Ubaghs with its fringe of spines – the marginal frame and spines in dorsal aspect; from Ubaghs (1969, fig. 23, p. 54).*

and one ventral, and was relatively stiff. The medial part is transitional between the other two portions. Ubaghs agrees that the proximal part of the stele is like the proximal part (fore tail) of the 'aulacophore' of cornutes and mitrates. He correctly points out, however, that the mesial and distal parts are not very similar to the positionally corresponding regions of the appendages of cornutes and mitrates. In my opinion the resemblance of the proximal parts suggests that the appendage of cornutes and mitrates are probably homologous, as wholes, with the stele of solutes. On the other hand, it is quite likely that the three regions of the appendages are not homologous between the three groups (as indeed they were not between cornutes and mitrates, in my view). For in solutes, unlike cornutes or mitrates, there is no sign of terminal autotomy. It follows that the solute stele includes a distal portion which in cornutes and mitrates has been lopped off. The whole question of homologies between solutes and cornutes, and of the systematic position of the solutes, requires further study, however.

Ubaghs' third argument suggests the non-motility of cornutes. I think it is mistaken, because by their shape the spikes on the ventral surface and the anterior downward-sloping appendages of *Cothurnocystis* or *Scotiaecystis*, for example, would hinder forward locomotion, but would favour rearward locomotion pulled by the tail (figs 7.2, 7.12, pp. 194, 208). The same direction of locomotion is required by the cuesta-shaped ribs of several mitrates (figs 8.1, p. 240, 8.11, 8.18, 8.26). For these ribs are always steeper anteriorly than posteriorly and almost certainly served to push against sediment, by analogy with similar ribs in still-living animals (figs 8.6, 8.7). These ribs strongly

suggest that rearward movement was habitual, and it could only have happened if the animal was pulled by the appendage (tail). Moreover, in mitrates the appendage is often equipped with a series of knifes on the ossicles whose likely function was to cleave the mud morphologically posterior to the animal, making rearward locomotion easier (figs 8.8, 8.27, 8.38). As to the spines around the cornute *Chauvelicystis* (fig. 10.4), they are absent from the rearmost part of the head, the posterior ones are transversely placed and the more anterior ones tend to run longitudinally. Contrary to Ubaghs, therefore, the fringe of spines would probably have favoured rearward movement, as was normal in calcichordates. In addition the spines were articulated at their bases, so that they could presumably be lifted at will so as not to impede locomotion.

Ubaghs' fourth argument, that the plates of the distal part of the 'aulacophore' (hind tail) could open as cover plates, is suggested to him in particular by the mode of preservation sometimes seen in cornutes (fig. 10.5). He means by this that the plates are often found right and left of the ossicles, still connected with each other in a right and a left series, and adjacent to the particular ossicles to which they were attached in life. He says (1981, p. 7, my translation): 'If they had not been able to open and constituted the skeletal cover of the underlying tissues, as supposed by those who see the aulacophore as a stele ["pedoncule"] or a tail, they would have fallen down on one another after the death of the animal and the disappearance of its soft parts, but they would not be separated into two well ordered columns while maintaining, in their fall, along the flanks of the aulacophore, their original arrangement' . . . 'It is necessary to suppose that the cover plates were open when the organism was buried by the sediment which preserved it. For putrefaction would not have been able to separate them at this place [along the dorsal mid-line] without disturbing their relations with each other and with the ossicles with which they are connected' ('dont elles dépendent').

However, it is risky to deduce the life habits of an animal from the precise arrangement of its parts after death. Thus, one of the specimens which Ubaghs illustrates shows the hind tail of *Ceratocystis* in dorsal aspect (fig. 10.5b, taken from Ubaghs, 1981, pl. 1, fig. 2). The plates of a column have to some degree kept their mutual relations, as Ubaghs points out. But the columns have been broken into several portions containing a few plates each. The plates within each portion are correctly arranged, which shows that they were linked together, probably by ligaments between them, when they were displaced. The individual portions of columns, however, are not in alignment, having been translated sideways from their original position to differing distances. Sometimes, indeed, the

a   b

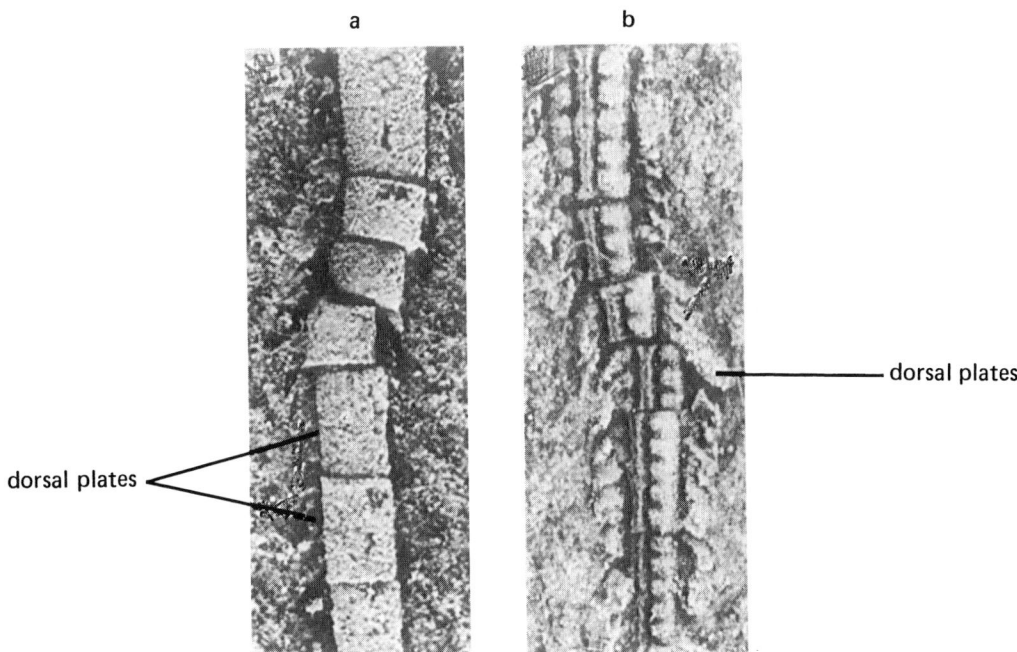

dorsal plates

dorsal plates

10.5. Ceratocystis perneri. *The hind tail in a) ventral, and b) dorsal aspects to illustrate Ubaghs' view that the dorsal plates ('cover plates') could open in life on the basis of their positions in the fossils; from Ubaghs (1981, pl. 1, fig. 2).*

lateral edges of the formerly underlying ossicles are visible *medial* to the plates. This cannot have been caused by pure outward rotation of the plates in life, since the plates and ossicles no longer meet at all at the ostensible surface of articulation. Something must have moved the plates sideways in small groups after death. The likely cause, suggested by me but explicitly rejected by Ubaghs, is the explosive release of gases by rotting of the soft parts in the lumen of the hind tail, after the death by burial of the animal. The same cause has probably operated in all the other specimens with opened-out plates that Ubaghs figures in support of his contention. The mid-dorsal junctions between the plates were presumably weaker than the junctions between plates and ossicles – they were certainly smaller in area. This weakness would explain why the usual effect of exploding gases of putrefaction was to break the mid-dorsal junction and to blow the plates sideways with some rotation.

The hind tail of *Scotiaecystis curvata* is instructive in this connection. As shown in fig. 7.16, each ossicle of *Scotiaecystis curvata* has a transverse hemicylindrical buttress. The plates were locked against this in a complex manner which would perhaps allow a slight rocking motion about a transverse axis but would not have permitted, without dislocation, a rotation outwards from the mid-line (see photographs in Jefferies, 1968a, pl. 4, figs

1,4). To summarize, the preservational argument used by Ubaghs to show that the plates of the cornute 'aulaco-phore' could rotate outwards on the ossicles in life, is not valid. The specimens cited show dislocation probably by explosive release of gases after burial and death.

As to the 'cover plates' of the mitrates, Ubaghs admits (1981, p. 8), that the preservational argument does not apply, for the plates are 'normally closed or ajar' ('entrou-vertes'). I have previously said that, in some mitrates such as *Mitrocystella barrandei, Peltocystis* and *Chinianocarpos* the plates and ossicles overlap each other in a complicated manner – each pair of plates imbricates inside an ossicle proximally and overlaps the neighbouring ossicle distally. Such an arrangement, though conceivably it *could* allow the plates to open, is exceedingly ill-adapted to do so, as Ubaghs agrees – '... ce qui, bien entendu, ne leur permettait pas de s'ouvrir' (1981, p. 8). However, he states that in examining the two known hind tails of *Chinianocarpos thorali* he does not see this alternation of imbrication (1981, p. 8). I can only say that both he and I have to make allowance for slight post-mortem shifting of the plates. If that is done, the alternation seems clear enough to my eyes (photograph in Jefferies & Prokop, 1972, pl. 6c). Ubaghs asserts that, because the preser-vational argument allegedly shows that plates of cornutes could open, then those of mitrates could have done so too

(1981, p. 9). However, this result depends on his next argument. (I note that Parsley, who is one of Ubaghs' leading American supporters, has recently admitted that the hind-tail plates of mitrates could not open at the mid-line (1982).)

Ubaghs' fifth and last argument is that the homologies proposed by him between the mitrate and cornute 'aulacophores' are so detailed and complex as to be beyond doubt. The mitrates were, therefore, oriented in life the opposite way up to what I suppose, and, therefore, the calcichordate theory falls. These suggested homologies can be presented most fairly by quoting a drawing from Ubaghs 1981 paper, which I do (fig. 10.2). As already explained in Chapters 7 and 8 I do not regard the three regions of the cornute appendage (fore, mid and hind tail of this book) as being homologous with those of mitrates. Rather, in the early evolution of mitrates, the cornute mid and hind tail was lopped off by the autotomy which was normal in calcichordate tails, and the mitrate fore, mid and hind tails were then regionated in the remaining stump. The mitrate mid- and hind-tail ossicles were probably serial homologues of the fore-tail spikes of *Lagynocystis* and *Chinianocarpos*. The similarities which Ubaghs notes between cornute and mitrate tails are probably functional analogies, rather than homologies. The hind tail of mitrates is adapted for flexing downwards, whence it has plates ventrally and ossicles dorsally. The hind tail of cornutes, on the other hand, is fundamentally adapted for flexing upwards. By this means, in *Cothurnocystis elizae* for example, the down-curved distal tip was lifted out of the sea floor. To produce this dorsal flexion, the plates are dorsal and the ossicles ventral. The styloid of mitrates and stylocone of cornutes, with their excavate proximal surfaces, probably served to receive the muscles of the fore tail, so that these muscles could move the mid and hind tail as a unit. The sculpture of the inner surfaces of the ossicles of cornutes and mitrates resemble each other largely because they represent impressions of the same longitudinal structures from above in mitrates and from below in cornutes. In particular, the median groove is a ventral impression of the notochord in cornutes and a dorsal impression of dorsal nerve cord and notochord in mitrates.

The reconstruction of the head chambers in cornutes and mitrates, and the fact that the reconstructed mitrate pharynx is in several points like that of tunicates, are my main reasons for believing that the surfaces that I have called ventral are homologous in cornutes and mitrates, and were downward. I agree with Ubaghs that '... the cavities are obviously hypothetical, since they are not preserved' (1981, p. 12). There is, however, a great deal of evidence for them, as detailed in Chapters 7 and 8. The water vascular structures in the 'aulacophore', which are so important in Ubaghs' theory, are also hypothetical 'since they are not preserved'.

In summary, the calcichordate theory is superior to the aulacophore theory because it interprets all the known skeletal details, both of the head and the tail, in mitrates and cornutes. It does so in functional terms and by point-by-point comparison with living animals. The aulacophore theory seriously attempts to explain only the tail, fits that of cornutes more easily than that of mitrates, but does not exactly fit even the cornutes.

The stele theory is much more tentative than the calcichordate or aulacophore theories. It was put forward by Philip (1979) in a paper which is mainly important for its criticisms of the other two theories, rather than for its positive proposals. Philip's reasons for rejecting the aulacophore theory largely coincide with mine, and have been discussed already. His reasons for rejecting the calcichordate theory have been summarized by him as follows (1979, p. 468).

'(a) Identification of the various thecal pores [head openings] is highly contentious. This applies not only to the so-called gill slits, but also to the mouth and anus. In particular the interpreted rectum passes through solid calcite in certain cornutes.

(b) Interpretation of the stele as [containing] a notochord is conceptually difficult. It is impossible to see how such a structure could have functioned, enclosed as it was in calcite plates, or why it should have been necessary.

(c) The vertebrate-like brain is based on major reconstruction. There would have been no room for such an enlarged ganglion among the stele muscles at the base of the theca [tail muscles near the posterior end of the head].

(d) The theory demands loss of calcite skeleton in three independent lines of descent. Again it is difficult to see how this could have been brought about. Indeed the evolution of carpoids [cornutes and mitrates] shows a progressive development of a more rigid and box-like theca [head].

(e) The fossil record suggests that Ordovician mitrates could not have given rise to late Cambrian vertebrates and Middle Cambrian acraniates.

(f) The homology of the stele [tail] in cornutes and mitrates indicates that the thecal faces are to be identified in some other way than required by the theory. This leads to a complete break-down in the proposed homology of the internal thecal chambers of cornutes and mitrates.

(g) The calcichordate theory requires there to be two posterior gill openings [atrial openings] under the theca of mitrates. Such openings do not exist in later mitrates such as the anomalocystitids.'

The words in square brackets were added by me, by way of explanation.

I have answered Philip's criticisms of the calcichordate theory at length elsewhere (Jefferies, 1981b) so here I shall deal with them briefly. I take them in the sequence used in Philip's summary.

'(a)' As to the head openings of cornutes, the identifications which I have given make good functional sense. They explain: why *Ceratocystis* had a hydropore and why the gonopore-anus was right of the tail (both as representing the primitive condition); why in all other cornutes the gonopore-anus was left of the tail – as lying in the outwash of the gill slits; and why the gonopore-anus of *Scotiaecystis* opened forward into the gill slits whose peculiar structure would produce a vertical rather than horizontal branchial current (fig. 7.19, p. 218). They also make good phylogenetic sense, since the left-sided gill slits can be compared with the morphologically left gill slits of larval amphioxus. (Philip always ignores the many cases where asymmetries in calcichordates match those in tunicates and amphioxus.) Also the distribution of gill slits in the dorsal integument of *Cothurnocystis elizae* coincides with the limits of the pharyngeal sculpture on the internal surfaces of the marginals (fig. 7.4, p. 198). The rectum passes through solid calcite only in *Scotiaecystis* (fig. 7.14b), for in *Ceratocystis* the anus is on a suture which would readily allow growth (fig. 7.17c). If it enlarged during the growth of *Scotiaecystis* then resorbtion of calcite would be necessary, but there seems no reason why this should not have happened.

'(b)' Concerning the notochord in the tail ('stele'), Philip is wrong to say that it would have been entirely surrounded by calcite. The fore tail of cornutes and mitrates had a large lumen and usually a loose skeleton (except in *Cothurnocystis*). The notochord in the fore tail would thus act as an anti-compressional structure in locomotion, very much as classical theory requires. In the hind tail it would not be anti-compressional, since compressional stresses would be taken up by the ossicles, but would presumably help to keep the ossicles in alignment.

'(c)' The vertebrate-like brain, for which there was allegedly no room in mitrates, is situated at the anterior end of the tail, in the same general position as the aboral ganglion presumed to exist by Philip. In his fig. 7, Philip gives two diagrams of my reconstruction of the brain and cranial nerves of *Mitrocystites*, purporting to contrast the objective part with the reconstructional. These diagrams are in some ways misleading – thus there is good direct evidence in *Mitrocystites* of the dorsal surface of the brain (fig. 8.9, p. 250), the optic foramen and the peripheral grooves all of which are left out of Philip's 'objective' figure (his fig. 7b). The rest of the reconstruction, as emphasized above (Chapter 8) is largely based on direct evidence from *Mitrocystella*. Contrary to Philip (1979, p.

459) I have never held that cispharyngeal eyes opened on the dorsal surface of the head, whether in *Mitrocystites* or elsewhere.

'(d)' Loss of the calcitic skeleton in three independent lines of descent is not unlikely for there are signs of resorbtion of calcite in the stem groups of acraniates, tunicates and vertebrates as I define them. These signs take the form of irregular rock-filled cavities in the dorsal skeleton. Philip's alleged trend of increasing rigidity of the head skeleton, which he discerns in the stratigraphical record of cornutes and mitrates, is ill-based and highly selective. In any case it would not forbid the loss of skeleton in forms that did not follow the 'trend'.

'(e)' The stratigraphical sequence of fossils is not a useful guide to phylogeny when the record is patchy, as it clearly is for the calcichordates, for the earliest known agnathan fishes and for the Middle Cambrian acraniate *Pikaia* (which is known from only one horizon and one locality).

'(f)' As already discussed in dealing with the aulacophore theory, the attempt to homologize the 'ossicular' face of the tail in cornutes with that of mitrates is weakly based. Comparison of the tail of *Lagynocystis* or *Chinianocarpos* with that of *Reticulocarpos hanusi* suggests that the whole tail of these two mitrates is homologous only with the fore tail of the cornutes. For (1) ventral flexibility existed throughout the tail of *Chinianocarpos*, *Lagynocystis* and other mitrates but was confined to the fore tail of *Reticulocarpos hanusi*, (2) ventral plates alternate with intercalary plates throughout the tail of *Chinianocarpos*, but are confined to the fore tail of *Reticulocarpos hanusi*, and (3) ventral bearing surfaces with postero–ventral spikes are found in the hind tail only of *Lagynocystis* (presumably having been lost in the fore tail), but are confined to the fore tail of *Reticulocarpos hanusi*. Also autotomy of the end of the tail seems to have been a real phenomenon, and it acted more proximally in *Reticulocarpos hanusi* than in less crownward cornutes. To this extent the total loss of the cornute mid and hind tail in the early phylogeny of mitrates is not surprising. As already mentioned, the dorsal ossicles and styloid of all mitrates were probably serial homologues of the spike-shaped ossicles of the fore tail of *Lagynocystis* and *Chinianocarpos*.

'(g)' As to the atrial openings, which I formerly called 'gill openings', I consider that they always existed in mitrates. There can be little doubt of their presence in some forms, given the broad overlap of the ventral onto the dorsal skeleton in *Balanocystites* (fig. 8.39, p. 313) and *Anatifopsis*. This affects the same plates as are involved in the deduced atrial openings of other mitrates (M_{1LV} and M_{1RV} articulate in all mitrates respectively with M_{1LD} and M_{1RD}; they overlap onto M_{2L} and M_{2R} and sometimes, in

Lagynocystis, onto parts of M_{1LD} and M_{1RD} also). The existence of gill slits inside the head of mitrates, implying the existence of atrial openings at the surface of the head, is based on direct evidence in the case of *Lagynocystis* (fig. 8.31), though most of the gill slits so preserved are tertiary gill slits, not exactly homologous with gill slits of other mitrates, nor of tunicates and vertebrates. In the stem-vertebrate mitrates *Placocystites, Mitrocystites* and *Mitrocystella* the evidence for internal gill slits is circumstantial but strong – it takes the form of a close comparison of the mitrate pharynx with that of tunicates.

Chauvel (1981), Jollie (1982) and Sprinkle (1983) have also criticized the calcichordate theory, but on the same grounds as Philip, so their contributions do not need to be separately rebutted.

Philip's stele theory can correctly be described as residual – it is what remains to him when the aulacophore and calcichordate theories are discarded. It is also highly changeable at present, which makes it difficult to outline even its main features beyond what I have done above. Thus, throughout the main text of his 1979 paper Philip assumes that what I call the ventral surface of mitrates is homologous with the ventral surface of cornutes, but he changed his mind in an addendum (1979, p. 471). Similarly, in 1979 he thought that the mitrates moved forward, but in 1981 decided that they moved rearward. The stele theory is stated to be 'traditional and simplistic' (1979, p. 458). This is presented as a virtue but in fact means that Philip, like Ubaghs, refuses to interpret the complex evidence inside the heads of cornutes and mitrates.

In trying to de-code the script of an extinct and forgotten language no interpretation would be acceptable that did not produce coherent sentences, or better still, complete and intelligible texts. It would not be thought convincing to give plausible readings of a few details only. Similarly with the calcichordates, failure to interpret the vast majority of observed features, in terms of soft parts, comparative anatomy and phylogenetic position, is not praiseworthy caution but a refusal to grapple with the facts. The calcichordate theory, unlike its rivals, explains all the details in a coherent and logical way. It must therefore be largely right.

Literature cited

Afzelius, B.A. & Olsson, R. 1957. The fine structure of the subcommissural cells and of Reissner's fibre in *Myxine*. *Z. Zellforsch. mikrosk. Anat.* **46**: 672–685.

Alcock, A. 1898. The peripheral distribution of the cranial nerves of ammocoetes. *J. Ant. Physiol. Lond.* **33**: 131–153.

Andersson, K.A. 1907. Die Pterobranchier der Schwedischen Südpolarexpedition 1901–1903, nebst Bemerkungen über *Rhabdopleura normani* Allman. *Wiss. Ergebn. schwed. Südpolar-exped.* **5**(1) Zoologie: 1–122.

Anonymous [Hanström, B.] 1954. Torsten Richard Emanuel Gislén. *K. fysiogr. Sälsk. Lund. Förh.* **24**: 61–65.

Ayers, H. & Worthington, J. 1907. The skin end organs of the trigeminus and lateralis nerves of *Bdellostoma dombeyi*. *Am. J. Anat.* **7**: 327–336.

Azariah, J.A. 1965. On the reversal of heart beat in *Branchiostoma lanceolatum*. *J. mar. biol. Ass. India* **7** (1965): 58–60.

Ax, P. 1984. *Das Phylogenetische System.* pp. 347. Stuttgart.

Ax, P. (in press). *The phylogenetic system.* Chichester and New York.

Azariah, J.A. 1969. Studies on the cephalochordates of the Madras Coast. V. The effect of the concentration of particulate matter and the oxygen tension in sea water on the filtration rates of amphioxus (*Branchiostoma lanceolatum*). *Proc. Indian Acad. Sci.* **49**: 259–268.

Baatrup, E. 1981. Primary sensory cells in the skin of amphioxus (*Branchiostoma lanceolatum* (D)). *Acta zool., Stockh.* **62**: 147–157.

Babin, C. 1971. *Elements de Paléontologie.* Paris.

Babin, C. (Tr. Orriss, N.) 1980. *Elements of Palaeontology.* 446 pp. Chichester.

Balabai, P.P. 1951. Nablyudeniya nad pitaniyem pyeskoroiki. [Observations on the feeding of the ammocoete]. *Dokl. Akad. Nauk. SSSR* **77**: 341–344.

Balabai, P.P. 1956. Morfologiya i filogeneticheskoye razvitiye gruppy bezchelyustnykh. [*Morphology and phylogenetic development of the agnatha*]. 139 pp. Kiev.

Balfour, F.M. 1878. *A monograph of the development of elasmobranch fishes.* 295 pp. London.

Balfour, F.M. 1880–1881. *Comparative embryology.* Vol. 1, 492 pp. Vol. 2, 655 pp. London.

Balinsky, B.I. 1960. *An introduction to embryology.* 562 pp. Philadelphia.

Barrande, J. 1872. *Système Silurien du centre de la Bohême.* Supplement au Vol. I. *Trilobites, crustacés divers, et poissons.* 647 pp. Prague.

Barrande, J. 1887. *Système Silurien du centre de la Bohême.* Vol. VII. *Classe des échinodermes, ordre des Cystidées.* 233 pp. Prague.

Barrington, E.J.W. 1937. The digestive system of *Amphioxus* (*Branchiostoma*) *lanceolatus*. *Phil. Trans. R. Soc.* (B) **228**: 269–312.

Barrington, E.J.W. 1945. The supposed pancreatic organs of *Petromyzon fluviatilis* and *Myxine glutinosa*. *Q. Jl. microsc. Sci.* **85**: 391–417.

Barrington, E.J.W. 1965. *The biology of Hemichordata and Protochordata.* 176 pp. Edinburgh.

Barrington, E.J.W. 1968. Phylogenetic perspectives in vertebrate endocrinology. pp. 1–46 *In* Barrington, E.J.W. & Jørgensen, C.B. (Eds) *Perspectives in endocrinology.* Academic Press. London and New York.

Barrington, E.J.W. 1972. The pancreas and intestine. Chapter 14: pp. 135–169 *In* Vol. 2 of Hardisty, M.W. & Potter, I.C. (Eds). 1971–83. 466 pp. London.

Barrington, E.J.W. & Sage, M. 1972. The endostyle and thyroid gland. Chapter 13: pp. 105–134 *In* Vol. 2 of Hardisty, M.W. & Potter, I.C. (Eds). 1971–83. 466 pp. London.

Bassler, R.S. 1943. New Ordovician cystidean echinoderms from Oklahoma. *Am. J. Sci.* **241**: 694–703.

Bateson, W. 1884. The early stages in the development of *Balanoglossus* (sp. incert.). *Q. Jl microsc. Sci.* **24**: 208–236.

Bateson, W. 1885. The later stages in the development of *Balanoglossus kowalevskii*, with a suggestion as to the affinities of the Enteropneusta. *Q. Jl microsc. Sci.* **25** Supplement: 81–122.

Bateson, W. 1886. The ancestry of the Chordata. *Q. Jl microsc. Sci.* **26**: 535–571.

Bather, F.H. 1900. Chapters 8–12 (pp. 1–216) *in* Lankester, E.R. *A treatise on zoology.* Part III Echinoderma. 344 pp. London.

Bather, F.A. 1913. Caradocian Cystidea from Girvan. *Trans. R. Soc. Edinb.* **49**: 359–529.

Bell, B.M. 1976. Phylogenetic implications of ontogenetic development in the class Edrioasteroidea (Echinodermata). *J. Paleont.* **50**: 1001–1009.

Berrill, N.J. 1935a. Studies in tunicate development. 3. Differential retardation and acceleration. *Phil. Trans. R. Soc.* (B) **225**: 255–326.

Berrill, N.J. 1935b. Studies in tunicate development. 4. Asexual reproduction. *Phil. Trans. R. Soc.* (B) **225**: 327–379.

Berrill, N.J. 1938. Studies in tunicate development. 5. Evolution and classification. *Phil. Trans. R. Soc.* (B) **226**: 43–70.

Berrill, N.J. 1947a. Metamorphosis in ascidians. *J. Morph.* **81**: 249–267.

Berrill, N.J. 1947b. The development and growth of *Ciona. J. mar. biol. Ass. U.K.* **26**: 616–625.

Berrill, N.J. 1950a. Budding and development in *Salpa. J. Morph.* **87**: 553–606.

Berrill, N.J. 1950b. *The Tunicata.* 354 pp. London.

Berrill, N.J. 1955. *The origin of vertebrates.* 257 pp. Oxford.

Berrill, N.J. & Sheldon, H. 1964. The fine structure of the connections between muscle cells in ascidian tadpole larvae. *J. Cell. Biol.* **23**: 664–669.

Billings, E. 1858. On the Cystideae of the Lower Silurian rocks of Canada. Canada Geological Survey. *Figures and descriptions of Canadian organic remains*, Dec. **3**: 9–74.

Bjerring, H.C. 1967(?) The second somite with special reference to the evolution of its myotomic derivatives. pp. 341–357 *in* Ørvig, T. (Ed.) Mobel Symposium 4. *Current problems of lower vertebrate phylogeny.* 539 pp. Stockholm.

Bjerring, H. 1977. A contribution to structural analysis of the head of craniate animals. *Zoologica Scr.* **6**: 127–183.

Bockelie, T. & Fortey, R.A. 1976. An early Ordovician vertebrate. *Nature, Lond.* **260**: 36–38.

Bone, Q. 1958. Observations upon the living larva of amphioxus. *Pubbl. Staz. zool. Napoli* **30**: 458–471.

Bone, Q. 1959. The central nervous system in larval acraniates. *Q. Jl microsc. Sci.* **100**: 509–527.

Bone, Q. 1960a. The origin of the Chordates. *J. Linn. Soc.* (Zoology) **44**: 252–269.

Bone, Q. 1960b. The central nervous system of amphioxus. *J. comp. Neurol.* **115**: 27–64.

Bone, Q. 1961. The organization of the atrial nervous system of amphioxus (*Branchiostoma lanceolatum* (Pallas)). *Phil. Trans. R. Soc.* (B), **243**: 241–269.

Bone, Q. 1963a. Some observations upon the peripheral nervous system of the hagfish, *Myxine glutinosa. J. mar. biol. Ass. U.K.* **43**: 31–47.

Bone, Q. 1963b. The central nervous system. pp. 50–91 *In* Brodal, A. & Fänge, R. (Eds). *The biology of* Myxine. 588 pp. Oslo.

Bone, Q. 1981. The neotenic origin of chordates. pp. 465–486 *In* Ranzi, L. (Ed.) Origine dei grandi Phyla dei Metazoi. *Atti conv. Lincei* **49.**

Bone, Q. & Best, A.C.G. 1978. Ciliated sensory cells in amphioxus (*Branchiostoma*). *J. mar. biol. Ass. UK.* **58**: 479–486.

Bone, Q. & Ryan, K.P. 1978. Cupular sense organs in *Ciona* (Tunicata: Ascideacea). *J. Zool., Lond.* **186**: 417–429.

Bonik, K., Gutmann, W.F. & Haude, R. 1978. Stachelhäuter mit Kiemen-Apparat: Der Beleg für die Ableitung der Echino-dermen von Chordatieren. *Natur Mus. Frankf.* **108**: 211–214.

Boolootian, R.A. & Campbell, J.L. 1964. A primitive heart in the echinoid *Strongylocentrotus purparatus. Science N.Y.* **145** (3628): 173–175.

Borcéa, J. 1906. Système urogenitale des elasmobranches. *Archs Zool. exp. gén.* (4) **4**: 199–484.

Boveri, T. 1890. Über die Niere des Amphioxus. *Sber. Ges. Morph. Physiol. Münch.* **6**: 65–77.

Boveri, T. 1892. Die Nierenkanälchen des Amphioxus. Ein Beitrag zur Phylogenie des Urogenitalsystems der Wirbeltiere. *Zool. Jb. (Anatomie)* **5**: 429–510.

Brandenburg, J. & Kümmel, G. 1961. Die Feinstruktur der Solenocyten. *J. Ultrastruct. Res.* **5**: 437–452.

Brien, P. 1948. Embranchement des Tuniciers. Morphologie et Reproduction. pp. 553–894 *In* Grassé, P.P. *Traité de zoologie.* Vol. 11. Paris.

Briggs, D.E., Clarkson, E.N.K. & Aldridge, R.J. 1983. The conodont animal. *Lethaia* **16**: 1–14.

Brodal, A. & Fänge, R. (Eds) 1963. *The biology of* Myxine. 588 pp. Oslo.

Burdon-Jones, C. 1952. Development and biology of the larva of *Saccoglossus horsti* (Enteropneusta). *Phil. Trans. R. Soc.* (B) **236**: 553–590.

Burn, O.M. (Ed.) 1980. *The complete encyclopaedia of the animal world.* 400 pp. London.

Bury, H. 1888. The early stages in the development of *Antedon. Phil. Trans. R. Soc.* (B) **179**: 257–301.

Cahn, P.H. 1967. *Lateral line detectors.* 496 pp. Bloomington, Indiana.

Carlisle, D.B. 1953. Origin of the pituitary body of chordates. *Nature, London.* **172**: 1098.

Caster, K.E. 1952. Concerning *Enoploura* of the Upper Ordovician and its relation to other carpoid Echinodermata. *Bull. Am. Paleont.* **34**(141): 1–47.

Caster, K.E. 1954. A new carpoid from the Parana Devonian. *Anais Acad. bras. Cienc.* **26**: 123–147.

Caster, K.E. 1956. A Devonian Placycystoid echinoderm from Parana Brazil. *Paleontologie de Parana* (Centennial volume): 137–148. Curitiba.

Caster, K.E. 1967. Contributions on pp. S550, S561–4 in Ubaghs, G. (1967b).

Caster, K.E. 1983. A new Silurian carpoid echinoderm from Tasmania and a revision of the Allanicytidiidae. *Alcheringa* **7**: 321–336.

Caster, K.E. & Eaton, J.H. 1956. Microstructure of the plates in the carpoid echinoderm *Paranacystis. J. Paleont.* **30**: 611–614.

Castle, W.E. 1896. The early embryology of *Ciona intestinalis* Fleming (L). *Bull. Mus. comp. Zool. Harv.* **27**(7): 203–280.

Chadwick, H.C. 1907. *Antedon. L.M.B.C. memoirs on typical British marine plants and animals* **15**: 1–55.

Chadwick, H.C. 1923. *Asterias. L.M.B.C. memoirs on typical British marine plants and animals* **25**: 1–63.

Chauvel, J. 1941. Recherches sur les cystoides et les carpoïdes armoricains. *Mém. Soc. géol. minér. Bretagne,* 286 pp. 7 pls.

Chauvel, J. 1966. *Echinodermes de l'ordovicien de Maroc.* 120 pp. Paris. (Cahiers de Paléontologie).

Chauvel, J. 1971. Les echinodermes carpoides du paléozoique inférieur marocain. *Notes Serv. géol. Maroc.* **31**: 49–60.

Chauvel, J. 1981. Etude critique de quelques echinodermes stylophores du Massif Armoricain. *Bull. Soc. géol. minér. Bretagne* (C) **13**: 67–101.

Chauvel, J. & Nion, J. 1977. Echinodermes (Homalozoa: Cornuta et Mitrata) nouveaux pour l'ordovicien du Massif Armoricain et conséquences paléogéographiques. *Géobios,* **10**: 35–49.

Cobb, J.L.S. 1970. The significance of the radial nerve ends in asteroids and echinoids. *Z. Zellforsch. mikrosk. Anat.* **108**: 457–474.

Cobb, J.L.S. & Laverack, M.S. 1967. Neuromuscular systems in echinoderms. *Symp. zool. Soc. Lond.* **20**: 25–51.

Cole, F.J. 1912. A monograph on the general morphology of the myxinoid fishes, based on a study of *Myxine*. Part IV. On some peculiarities of the afferent and efferent branchial arteries of *Myxine*. *Trans. R. Soc. Edinb.* **48**: 215–230.

Cole, F.J. 1926. A monograph on the general morphology of the myxinoid fishes based on a study of *Myxine*. Part VI. The morphology of the vascular system. *Trans. R. Soc. Edinb.* **54**: 309–342.

Conel, J.L. 1929. The development of the brain of *Bdellostoma stouti*. I. External growth changes. *J. comp. Neurol.* **47**: 343–344.

Conel, J.L. 1931. The development of the brain of *Bdellostoma stouti*. II. Internal growth changes. *J. comp. Neurol.* **52**: 365–499.

Conklin, E. 1905. The organization and cell-lineage of the ascidian egg. *J. Acad. nat. Sci. Philad.* **13**: 1–119.

Conklin, E. 1932. The embryology of *Amphioxus*. *J. Morph.* **54**: 69–151.

Conway-Morris, S. 1979a. The Burgess Shale (Middle Cambrian) fauna. *Ann. Rev. Ecol. Syst.* **10**: 327–349.

Conway-Morris, S. 1979b. The animals of the Burgess Shale. *Scient. Am.* **241**: 122–133.

Crowson, R.A. 1982. Computers versus imagination in the reconstruction of phylogeny. pp. 245–255 *In* Joysey, K.A. & Friday, A.E. Problems of phylogenetic reconstruction. *Syst. Ass. Spec. Vol. 21,* 442 pp. London.

Crowther, P.R. 1981. The fine structure of graptolite periderm. *Spec. Pap. Palaeont.* **26**: 1–119.

Damas, D. 1900. Les formations epicardiques chez *Ciona intestinalis* (L). *Archs Biol., Paris* **16**: 1–25.

Damas, H. 1944. Recherches sur le développement de *Lampetra fluviatilis* L. – contribution à l'étude de la cephalogénèse des vertébrés. *Archs. Biol. Paris* **55**: 1–289.

Dawydoff, C. 1928. *Traité d'embryologie comparée des invertébrés:* 930 pp. Paris.

Dean, B. 1899. On the embryology of *Bdellostoma stouti*. A general account of myxinoid development from the egg and segmentation to hatching. pp. 220–276 *In Festschrift zum 70ten Geburtstag Carl von Kupffer,* Jena.

De Beer, G.R. 1922. The segmentation of the head in *Squalus acanthias*. *Q. Jl microsc. Sci.* **66**: 457–474.

De Beer, G.R. 1958. *Embryos and ancestors.* 3rd edition. 197 pp. Oxford.

Deflandre, G. & Deflandre-Rigaud, M. 1956. *Micrascidites* manip. nov., sclerites de Didemnidés (Ascidies, Tuniciers) fossiles du Lutétien du Bassin parisien et du Balcombian d'Australie. *C.r. somm. Séanc. Soc. géol. Fr.* **(1956)**: 47–49.

Dehm, R. 1933. Cystoideen aus dem rheinischen Unterdevon. *Neues Jb. Miner. Geol. Paläont.* BeilBd, Abt. B **69**: 63–93.

Dehm, R. 1934. Untersuchungen über Cystoideen des rheinischen Unterdevons. *Sber. bayer. Akad. Wiss.* **(1934)**: 19–43.

Denison, R.H. 1971. The origin of vertebrates: a critical evaluation of current theories. *Proc. North American Paleontological Convention* (H): 1132–1146.

Dijkgraaf, S. 1967. Biological significance of the lateral line organs. pp. 83–95 *In* Cahn, P.H. (Ed.).

Dilly, P.N. 1962. Studies on the receptors in the cerebral vesicle of the ascidian tadpole. 1. The otolith. *Q. Jl microsc. Sci.* **103**: 393–398.

Dilly, P.N. 1964. Studies on the receptors in the cerebral vesicle of the ascidian tadpole. 2. The ocellus. *Q. Jl microsc. Sci.* **105**: 13–20.

Dilly, P.N. 1969. Studies on the receptors in *Ciona intestinalis*. 3. A second type of photoreceptor in the tadpole larva of *Ciona intestinalis*. *Z. Zellforsch. mikrosk. Anat.* **96**: 63–65.

Dilly, P.N. 1973. The larva of *Rhabdopleura compacta*, (Hemichordata). *Mar. Biol. Berlin* **18**: 69–86.

Dilly, P.N. 1975a. The dormant buds of *Rhabdopleura compacta* (Hemichordata). *Cell Tissue Research* **159**: 387–397.

Dilly, P.N. 1975b. The pterobranch *Rhabdopleura compacta*: its nervous system and phylogenetic position. *Symp. zool. Soc. Lond.* **36**: 1–16.

Dodd, J.M. & Dodd, M.H.I. 1966. An experimental investigation of the supposed pituitary affinities of the ascidian neural complex. pp. 233–252 *In* Barnes, H. (Ed.) *Some contemporary studies in marine science.* London.

Dohrn, A. 1886. Studien zur Urgeschichte des Wirbelthierkörpers. *Mitt. zool. Stn. Neapel.* **6**: 1–92.

Drach, P. 1948. Embranchement des céphalochordés. pp. 931–1037 *In* Grassé, P.P. (Ed.) *Traité de Zoologie. Tome XI, Echinodermes, Stomochordés, Prochordés.* 1007 pp. Paris.

Dücker, M. 1924. Über die Augen der Zyklostomen. *Jena. Z. Med. Naturw.* **60**: 471–530.

Dupuis, C. 1978. La 'Systématique phylogénétique' de W. Hennig (Historique, discussion, choix de references). *Cah. Nat.* **34:** 1–72.

Eakin, R.M. 1963. Lines of evolution of photoreceptors. *In* Mazia, D. & Tyler, A. (Eds) *General physiology of cell specialization.* New York.

Eakin, R.M. 1968. Evolution of photoreceptors. *Evolut. Biol.* **2:** 194–242.

Eakin,. R.M. & Kuda, A. 1971. Ultrastructure of sensory receptors in ascidian tadpoles. *Z. Zellforsch. mikrosk. Anat.* **112:** 287–312

Eakin, R.M. & Kuda, A. (1972). Glycogen in the lens of the tunicate tadpole (Chordata, Ascidiacea). *J. exp. Zool.* **180:** 267–276.

Eakin, R.M. & Westfall, J.A. 1962. Fine structure of photo-receptors in amphioxus. *J. Ultrastruct. Res.* **6:** 531–539.

Eaton, T.H. 1970. The stem-tail problem and the ancestry of chordates. *J. Paleont.* **44:** 969–979.

Eldredge, N. & Cracraft, J. 1980. *Phylogenetic patterns and the evolutionary process.* 349 pp. New York.

Eldredge, N. & Gould, S.J. 1972. Punctuated equilibria: an alternative to phylogenetic gradualism. pp. 82–115 *In* Schopf, T.J.M. (Ed.) *Models in Paleontology.* San Francisco.

Elwyn, A. 1937. Some stages in the development of the neural complex in *Ecteinascidia turbinata. Bull. nurol. Inst. N.Y.* **6:** 163–177.

Fänge, R. 1972. The circulatory system. Chapter 17, pp. 241–259 *In* Vol. 2 of Hardisty, M.W. & Potter, I.C. 1971–83.

Fänge, R. (Ed.) 1973. *Myxine glutinosa,* biochemistry, physiology and structure. *Acta R. Soc. scient. litt. gothoburg.* (Zoology) **8:** 1–104.

Fechter, H. 1965. Über die Funktion der Madreporenplatte der Echinoidea. *Z. vergl. Physiol.* **51:** 227–257.

Fernholm, B. & Holmberg, K. 1974. The eyes in three genera of hagfish (*Epratretus, Paramyxine* and *Myxine*) – a case of degenerative evolution. *Vision Res.* **15:** 253–259.

Flock, A. 1967. Ultrastructure and function in the lateral line organs. pp. 163–197 *In* Cahn, P.H. (Ed.).

Flood, P.R. 1966. A peculiar mode of muscular innervation in amphioxus. *J. comp. Neurol.* **126:** 181–218.

Flood, P.R. 1975. Fine structure of the notochord of amphioxus. *Symp. zoo. Soc. Lond.* **36:** 81–104.

Flood, P.R. & Fiola-Medioni, A. 1979. Filter characteristics of ascidian food trapping mucous films. *Acta zool. Stockh.* **60:** 271–272.

Flood, P.R., Guthrie, D.M. & Banks, J.R. 1969. Paramyosin muscles in the notochord of amphioxus. *Nature, Lond.* **222:** 87–88.

Fortey, R.A. & Jefferies, R.P.S. 1982. Fossils and phylogeny – a compromise approach. pp. 197–234 *In* Joysey, K.A. & Friday, A.E. (Eds) *Problems of phylogenetic reconstruction. Systematics Assoc. Spec. Vol. 21,* 440 pp. London.

Franz, V. 1923. Hautsinnesorgane und Nervensysteme der Akranier. *Jena Z. Naturw.* **59:** 401–526.

Franz, V. 1927. Morphologie der Akranier. *Ergebn. Anat. Entwickl.* **27:** 464–692.

Franz. V. 1933. Das Gefäßsystem der Akranier. *Handb. Vergl. Anat.* **6:** 451–466.

Froriep, A. 1882. Über ein Ganglion des Hypoglossus und Wirbelanlagen in der Occipitalregion. *Arch. Anat. EntwGesch.* (1882): 279–302.

Froriep, A. 1906. Über die Herleitung des Wirbeltierauges vom Auge der Aszidienlarve. *Anat. Anz.* 29 *(Verh. anat. Ges. 20. Vers. in Rostock):* 145–151.

Gans, C. & Northcutt, R.G. 1983. NeuraL crest and the origin of vertebrates: a new head. *Science N.Y.* **220:** 268–274.

Garstang, W. 1894. Preliminary note on a new theory of the phylogeny of the Chordata. *Zool. Anz.* **17:** 122.

Garstang, W. 1928. The morphology of the Tunicata and its bearing on the phylogeny of the Chordata. *Q. Jl microsc. Soc.* **72:** 51–187.

Garstang, W. 1929. On the dextricolic condition in tunicates. *Proc. Leeds phil. lit. Soc.* **1:** 506–515.

Gegenbaur, C. 1871. Über die Kopfnerven von *Hexanchus* und ihr Verhältnis zur Wirbeltheorie des Schädels. *Jena Z. Med. Naturw.* **6:** 497–559.

Gemmill, J.F. 1914. The development and certain points in the adult structure of the starfish *Asterias rubens* L. *Phil. Trans. R. Soc.* (B) **205:** 213–294.

Gemmill, J.F. 1915. Double hydrocoel in the development and metamorphosis of the larva of *Asterias rubens* L. *Q. Jl microsc. Soc.* **61:** 51–60.

Gemmill, J.F. 1919. Rhythmic pulsations in the madreporic vesicle. *Q. Jl microsc. Soc.* **63:** 537–540.

Ghiselin, M.T. 1969. *The triumph of the Darwinian method.* 287 pp. Berkeley.

Gigout, F. 1954. Sur un hétérostélé de l'ordovicien marocain. *Bull. Soc. Sci. nat. phys. Maroc* **37:** 3–7.

Gilchrist, J.D.F. 1915. Observations on the Cape *Cephalodiscus* (*C. gilchristi* Ridewood) and some of its early stages, with an appendix by S.F. Harmer. *Ann. Mag. nat. Hist.* (8).**16:** 233–246.

Gilchrist, J.D.F. 1917. On the development of the Cape *Cephalodiscus* (*C. gilchristi* Ridewood). *Q. Jl microsc. Sci.* **62:** 189–211.

Gill, E.D. & Caster, K.E. 1960. Carpoid echinoderms from the Silurian and Devonian of Australia. *Bull. Am. Paleont.* **41**(185): 1–71.

Gislén, T. 1930. Affinities between the Echinodermata, Entero-pneusta and Chordonia. *Zool. Bidr. Upps.* **12:** 199–304.

Glaesner, L. 1910. Studien zur Entwicklungsgeschichte von *Petromyzon fluviatilis* 1. Furchung und Gastrulation. *Zool. Jb.* (Anatomie) **29:** 139–180.

Goldring, R. & Stephenson, D.G. 1972. The depositional environment of three starfish beds. *Neues Jb. Geol. Palaeont. Mh.* (1972): 611–624.

Godeaux, J. 1964. Que connaissons-nous de la glande neurale des tuniciers? *Publs Univ. Elisabethville* **7**: 51–64.

Goodbody, I. 1974. The physiology of ascidians. *Adv. mar. Biol.* **12**: 1–149.

Goodrich, E.S. 1902. On the structure of the excretory organs of amphioxus. *Q. Jl microsc. Sci.* **45**: 493–501.

Goodrich, E.S. 1909a. On the structure of the excretory organs of amphioxus. Part 2 – The nephridium in the adult. Part 3 – Hatschek's nephridium. Part 4 – The nephridium in the larva. *Q. Jl microsc. Sci.* **54**: 185–205.

Goodrich, E.S. 1909b. Vertebrata craniata (first fascicle: cyclostomes and fishes) 528 pp. *In* Lankester, E.R. (Ed.) *A treatise on zoology.* London.

Goodrich, E.S. 1917. 'Proboscis pores' in craniate vertebrates, a suggestion concerning the praemandibular somites and hypophysis. *Q. Jl microsc. Sci.* **62**: 539–554.

Goodrich, E.S. 1918a. Development of the pericardio–peritoneal canals in selachians. *J. Anat.* **53**: 1–13.

Goodrich, E.S. 1918b. On the development of the segments of the head in *Scyllium. Q. Jl microsc. Sci.* **63**: 1–30.

Goodrich, E.S. 1930a. *Studies on the structure and development of vertebrates.* London. Reprinted 1958, 837 pp. London & New York.

Goodrich, E.S. 1930b. The development of the club-shaped gland in Amphioxus. *Q. Jl microsc. Sci.* **74**: 155–164.

Goodrich, E.S. 1933. The nephridia of *Asymmetron* and *Branchiostoma* compared. *Q. Jl microsc. Sci.* **75**: 723–734.

Goodrich, E.S. 1934a & b. The early development of nephridia in amphioxus. Introduction and Part I. Hatschek's nephridium. II. the paired nephridia. *Q. Jl microsc. Sci.* **76**: 493–510, 655–674.

Goodrich, E.S. 1937. On the spinal nerves of Myxinoidea. *Q. Jl microsc. Sci.* **80**: 153–158.

Gorbman, A. 1963. The myxinoid thyroid gland. pp. 477–480 *In* Brodal, A. & Fänge, R. (Eds).

Gosselck, F. & Kuehner, E. 1973. Investigations of the biology of *Branchiostoma senegalense* larvae off the West African coast. *Mar. Biol. Berlin* **22**: 67–73.

Gould, S.J. 1977. *Ontogeny and phylogeny.* 501 pp. Cambridge, Mass.

Gould, S.J. & Eldredge, N. 1977. Punctuated equilibria: the tempo and mode of evolution reconsidered. *Paleobiol.* **3**: 115–151.

Grave, C. 1921. *Amaroucium constellatum* (Verrill). II. The structure and organization of the tadpole larva. *J. Morph.* **36**: 71–91.

Gregory, W.K. 1935. Reduplication in evolution. *Q. Rev. Biol.* **10**: 272–290.

Grobben, K. 1908. Die systematische Einteilung des Tierreiches. *Verh. Zool. bot. Ges. Wien* **58**: 491–511.

Grobben, K. 1924. Theoretische Erörterungen betreffend die phylogenetische Ableitung der Echinodermen. *Sber. Akad. Wiss. Wien.* (Mat. Nat. Klasse) **132**: 262–290.

Grove, A.J. & Newell, G.E. 1961. *Animal biology.* 6th edition, 820 pp. London.

Guthrie, D.M. 1975. The physiology and structure of the nervous system of amphioxus (the lancelet), *Branchiostoma lanceolatum* Pallas. *Symp. Zool. Soc. Lond.* **36**: 43–80.

Haeckel, E. 1866. *Generelle Morphologie der Organismen.* 1. Band, 574 pp. 2. Band, 462 pp. Berlin.

Haeckel, E. 1874a. *Anthropogenie oder Entwicklungsgeschichte des Menschen.* 732 pp. Leipzig.

Haeckel, E. 1874b. Die Gastraea – Theorie, die phylogenetische Klassifikation des Tierreiches und die Homologie der Keimblätter. *Jena Z. Naturw.* **8**: 1–55.

Hall, J. 1859–1861. *Natural history of New York. Paleontology: vol. III, containing descriptions and figures of the organic remains of the Lower Helderberg Group and the Oriskany Sandstone.* Text 532 pp. 120 pls.

Hardisty, M.W. 1979. *Biology of the cyclostomes.* 428 pp. London.

Hardisty, M.W. 1982. Lampreys and hagfishes: analysis of cyclostome relationships. Chap. 36: 165–259 *In* vol. 4a of Hardisty, M.W. & Potter, I.C. 1971–82.

Hardisty, M.W. & Potter, I.C. (Eds) 1971–82. *The biology of lampreys.* 5 vols, London.

Harland, W.B., Cox, A.V., Llewellyn, P.G., Pickton, C.A.G., Smith, A.G. & Walters, R. 1982. *A geologic time scale.* 131 pp. Cambridge.

Harper, C. 1976. Phylogenetic inference in paleontology. *J. Paleont.* **50**: 180–193.

Hatschek, B. 1881. Studien über die Entwicklung des Amphioxus. *Arb. zool. Inst. Univ. Wien* **4**: 1–88.

Hatschek, B. 1906. Das Acromerit des *Amphioxus. Morph. Jb.* **35**: 1–16.

Hehn, G. von 1970. Über den Feinbau des hyponeuralen Nervensystems des Seesternes (*Asterias rubens* L.) *Z. Zellforsch. mikrosk. Anat.* **105**: 137–154.

Heider, K. 1912. Über Organverlagerungen bei der Echinodermen – Metamorphose. *Verh. dt. zool. Ges.* **22**: 239–251.

Hennig, W. 1950. *Grundzüge einer Theorie der phylogenetischen Systematik.* 270 pp. Berlin.

Hennig, W. 1966. *Phylogenetic systematics.* 263 pp. Urbana.

Hennig, W. 1969. *Die Stammesgeschichte der Insekten.* 436 pp. Frankfurt-am-Main.

Hennig, W. 1975. 'Cladistic analysis' or 'Cladistic classification': a reply to Ernst Mayr. *Syst. Zool.* **24**: 244–256.

Hennig, W. 1981. (tr. Pont, A.C.). *Insect phylogeny.* 514 pp. Chichester.

Hennig, W. 1983. Stammesgeschichte der Chordaten. *Fortschr. zool. Syst. Evolutionsforsch.* **2**: 1–208.

Hennig, W. 1984. *Aufgaben und Probleme stammesgeschichtlicher Forschung.* Pareys Studientexte 35), 65 pp. Hamburg.

Hicks, H. 1872. On some undescribed fossils from the Menevian group. *Q. Jl geol. Soc. Lond.* **28**: 173–185.

Holmberg, K. 1971. The hagfish retina: electron microscopic study comparing receptor and epithelial cells in the Pacific

Hagfish *Palistotrema stouti* with those in the Atlantic Hagfish *Myxine glutinosa*. *Z. Zellforsch. mikrosk. Anat.* **121**: 249–269.

Holmgren, N. 1940. Studies on the head in fishes – Embryological, morphological and phylogenetic researches. Part I. Development of the skull in sharks and rays. *Acta zool., Stockh.* **21**; 51–267.

Holmgren, N. 1946. On the embryos of *Myxine glutinosa*. *Acta zool. Stockh.* **27**: 1–90.

Holmgren, N. 1950. On the pronephros and the blood in *Myxine glutinosa. Acta zool., Stockh.* **31**: 233–248.

Horst, C.J. van der 1927–1939. Hemichordata in *Bronn's Klassen und Ordnungen des Tierreichs.* Bd. 4, Abt. 4, Buch 2 Teil 2, 737 pp. Leipzig.

Hoyle, G. 1953. Spontaneous squirting of an ascidian, *Phallusia mammillata* Cuvier. *J. mar. biol. Ass. U.K.* **31**: 541–562.

Hull, D.L. 1979. The limits of cladism. *Syst. Zool.* **28**: 416–440.

Hûus, J. 1924. Genitalorgane und Ganglio-Genital-Strang bei *Corella parallelogramma* O.F.M. *Skr. VidenskSelsk. Christiania* **19** (**1923**): 1–50.

Hûus, J., Ihle, I.E.W., Lohmann, H. & Neumann, G. 1933–1940. Tunicata. *In* Kükenthal, W. & Krumbach, T. (eds) *Handbuch der Zoologie* 5(2. Hälfte). 771 pp. Berlin & Leipzig.

Hyman, L.H. 1955. *The invertebrates: Echinodermata*. Vol. IV, 763 pp. New York.

Hyman, L.H. 1959. *The invertebrates: smaller coelomate groups, Chaetognatha, Hemichordata, Pogonophora, Phoronida, Ectoprocta, Brachiopoda, Sipunculida the coelomate Bilateria.* Vol. V. 783 pp. New York.

Jaekel, O. 1899. *Stammesgeschichte der Pelmatozoen.* 441 pp. Berlin.

Jaekel, O. 1900. Über Carpoideen; eine neue Klasse von Pelmatozoen. *Z. dt. geol. Ges.* **52**: 661–677.

Jaekel, O. 1918a. Über fragliche Tunikaten aus dem Perm Siziliens. *Paläont. Z.* **2**: 66–74.

Jaekel, O. 1918b. Phylogenie und System der Pelmatozoen. *Palaeont. Z.* **3**: 1–128.

Janvier, P. 1978. Les nageoires paires des Ostéostracés et la position systématique des Céphalaspidomorphes. *Annls. Paléont.* (Vertebrés) **64**: 1–30.

Janvier, P. 1980. *Les Osteostraci de la Formation de Wood Bay (Devonien inférieur, Spitzberg) et le problème des relations phylogénétiques entre Agnathes et Gnathostomes.* Thesis. Université P. & M. Curie (Paris VI), 404 pp.

Janvier, P. 1981a. *Norselaspis glacialis* n.g., n.sp., et les relations phylogénétiques entre les kiaeraspidiens (Osteostraci) du Devonien inférieur du Spitzberg. *Palaeovertebrata* **11**: 19–130.

Janvier, P. 1981b. The phylogeny of the Craniata, with particular reference to the significance of fossil 'agnathans'. *J. vert. Paleont.* **1**: 121–159.

Jefferies, R.P.S. 1967. Some fossil chordates with echinoderm affinities. *Symp. zool. Soc. Lond.* **20**: 163–208.

Jefferies, R.P.S. 1968a. Fossil chordates with echinoderm affinities. *Proc. geol. Soc.* **1649**: 128–140.

Jefferies, R.P.S. 1968b. The subphylum Calcichordata (Jefferies 1967) – primitive fossil chordates with echinoderm affinities. *Bull. Br. Mus. nat. Hist. Geol.* **16**: 243–339.

Jefferies, R.P.S. 1969. *Ceratocystis perneri* – a Middle Cambrian chordate with echinoderm affinities. *Palaeontology* **12**: 494–535.

Jefferies, R.P.S. 1971. Some comments on the origin of chordates. *J. Paleont.* **45**: 910–912.

Jefferies, R.P.S. 1973. The Ordovician fossil *Lagynocystis pyramidalis* (Barrande) and the ancestry of amphioxus. *Phil. Trans. R. Soc.* (B) **265**: 409–469.

Jefferies, R.P.S. 1975. Fossil evidence concerning the origin of the chordates. *Symp. Zool. Soc. Lond.* **36**: 253–318.

Jefferies, R.P.S. 1979a. The origin of chordates – a methodological essay. pp. 443–477 *In* House, M.R. (Ed.) *The origin of major invertebrate groups. Systematics Association Special Volume* **12**: 1–515.

Jefferies, R.P.S. Jefferies, R.P.S. 1979b. Calcichordates: pp. 161–167 *In* Fairbridge, R.W. & Jablonski, D. *The encyclopedia of palaeontology.* 886 pp. Stroudsberg, Penn.

Jefferies, R.P.S. 1980. Zur Fossilgeschichte des Ursprungs der Chordaten und der Echinodermen. *Zool. Jb. (Anatomie)* **103**: 285–353.

Jefferies, R.P.S. 1981a. Fossil evidence on the origin of the chordates and echinoderms. pp. 487–561 *In* Ranzi, L. (Ed.) Origine dei grandi phyla dei metazoa. *Atti conv. Lincei* **49**: 1–565.

Jefferies, R.P.S. 1981b. In defence of the calcichordates. *Zool. J. Linn. Soc.* **73**: 351–396.

Jefferies, R.P.S. 1982. The calcichordate controversy – comments on *Notocarpos garratti* Philip. *Alcheringa* **6**: 78.

Jefferies, R.P.S. 1984. Locomotion, shape, ornament and external ontogeny in some mitrate calcichordates. *J. Vert. Paleont.* **4**: 292–319.

Jefferies, R.P.S. & Lewis, D.N. 1978. The English Silurian fossil *Placocystites forbesianus* and the ancestry of the vertebrates. *Phil. Trans. R. Soc.* (B) **282** (990): 205–323.

Jefferies, R.P.S., Lewis, M. & Donovan, S.K. (in press). *Protocystites menevensis* Hicks 1872 – a stem-group chordate (Cornuta) from the Middle Cambrian of South Wales. *Palaeontology.*

Jefferies, R.P.S. & Prokop, R.J. 1972. A new calcichordate from the Ordovician of Bohemia and its anatomy, adaptations and relationships. *Biol. J. Linn. Soc.* **4**: 69–115.

Johansen, K. 1960. Circulation in the hagfish *Myxine glutinosa* L. *Biol. Bull. mar. biol. Lab. Woods Hole* **118**: 289–295.

Jollie, M. 1973. The origin of chordates. *Acta zool., Stockh.* **54**: 81–100.

Jollie, M. 1977. The origin of the vertebrate brain. *Ann. N.Y. Acad. Sci.* **299**: 74–86.

Jollie, M. 1982. What are the 'calcichordata'? and the larger question of the origin of chordates. *Zool. J. Linn. Soc.* **75**: 167–188.

Jørgensen, C.B. 1952. On the relation between water transport and food requirements in some marine filter-feeding invertebrates. *Biol. Bull. mar. biol. Lab. Woods Hole* **103**: 356–363.

Julin, C. 1899. Contribution a l'histoire phylogénétique des tuniciers. Recherches sur le développement du péricarde, du coeur, et les transformations de l'epicarde chez les ascidies simples. *Trav. Stn zool. Wimereux* **7**: 311–366.

Julin, C. 1904. Recherches sur la phylogénèse des Tuniciers. Développement de l'appareil branchial. *Z. wiss. Zool.* **76**: 544–611.

Katz, M.J. 1983. Comparative anatomy of the tunicate tadpole, *Ciona intestinalis. Biol. Bull.* **164**: 1–27.

Keibel, F. 1928. Beiträge zur Anatomie, zur Entwicklungsgeschichte und zur Stammesgeschichte der Sehorgane der Cyklostomen. *Z. mikrosk. -anat. Forsch.* **12**: 391–456.

Kennedy, M.C. & Rubinson, K. 1977. Retinal projections in larval, transforming and adult sea lamprey, *Petromyzon marinus. J. comp. Neurol.* **171**: 465–480.

Kieckebusch, H.H. von 1928. Beiträge zur Kentniss des Baues und der Entwicklung der Schilddruse bei den Neunaugenlarven (*Lampetra fluviatilis* L. und *Lampetra planeri* Bl.). *Z. Morph. Okol. Tiere* **11**: 247–360.

Kleerekoper, H. 1969. *Olfaction in fishes.* 222 pp. Bloomington.

Kleerekoper, H. 1972. The sense organs. pp. 373–404 *In* Hardisty, M.W. & Potter, I.C. (Eds).

Knight-Jones, E.W. 1952. On the nervous system of *Saccoglossus cambrensis* (Enteropneusta). *Phil. Trans. R. Soc.* (B) **236**: 315–354.

Kobayashi, H. 1964. On the photo-perceptive function in the eye of the hagfish *Myxine garmani* Jordan & Snyder. *J. Shimonoseki Univ. Fish.* **13**: 141–157.

Koch, J. 1881. Die neuesten Veröffentlichungen der 'Chaucer Society' und die Überlieferung der 'Minor Poems'. *Anglia Z. Englische Philologie Anz.* **4**: 93–117.

Kolata, D.R. & Guensburg, T.E. 1979. *Diamphidiocystis,* a new mitrate carpoid from the Cincinnattian (Upper Ordovician) Maquoketa Group in southern Illinois. *J. Paleont.* **53**: 1121–1135.

Kolata, D.R. & Jollie, M. 1982. Anomalocystitid mitrates (Stylophora-Echinodermata) from the Champlainian (Middle Ordovician) Guttenberg Formation of the Upper Mississippi Valley Region. *J. Paleont.* **56**: 531–565.

Koltzoff, N.K. 1901. Entwickelungsgeschichte des Kopfes von *Petromyzon planeri. Bull. Soc. Nat. Moscou* **15**: 259–289.

Komai, T. 1949. Internal structure of the pterobranch *Atubaria heterolopha* Sato with an appendix on the homology of the 'notochord'. *Proc. Japan Acad.* **25**: 19–24.

Koninck, M.L. de 1869. Sur quelques echinodermes remarquables des terrains paléozoiques. *Bull. Acad. r. Belg. Cl. Sci.* (Z) **28**: 544–552.

Koninck, M.L. de 1870. On some new and remarkable echinoderms from the British Palaeozoic rocks. *Geol. Mag.* (1) **7**: 258–263.

Kowalevsky, A.O. 1867a. Entwicklungsgeschichte der einfachen Ascidien. *Mém. Acad. Sci. St. Petersb.* (7)**10** (15): 1–19.

Kowalevsky, A.O. 1867b. Entwicklungsgeschichte des *Amphioxus lanceolatus. Mém. Acad. Sci. St. Petersb.* (7) **11** (4): 1–17.

Kozłowski, R. 1949. Les graptolithes et quelques nouveaux groupes d'animaux du Tremadoc de la Pologne. *Palaeont. pol.* **3**: 1–235.

Kükenthal, W. & Krumbach, T. (Eds) 1933–40. *Handbuch der Zoologie* 5. Band 2. Hälfte. Tunicata. 1771 pp. Berlin & Leipzig.

Kümmel, G. 1967. Die Podocyten. *Zool. Beitr.* (N.F.) **13**: 245–263.

Kupffer, C. von 1870. Die Stammverwandtschaft zwischen Ascidien und Wirbeltieren. *Arch. mikrosk. Anat.* **6**: 115–172.

Kupffer, C. von 1894. *Studien zur vergleichenden Entwicklungsgeschichte des Kopfes der Kranioten. 2. Heft. Die Entwicklung des Kopfes von* Ammocoetes planeri. 79 pp. München & Leipzig.

Kupffer, C. von 1895. *Studien zur vergleichenden Entwicklungsgeschichte des Kopfes der Kranioten. 3 Heft. Die Entwicklung der Kopfnerven von* Ammocoetes planeri. 80 pp. München & Leipzig.

Kupffer, C. von 1900. *Studien zur vergleichenden Entwicklungsgeschichte des Kopfes der Kranioten. 4 Heft. Zur Kopfentwicklung von* Bdellostoma. 86 pp. München & Leipzig.

Kupffer, C. von 1906. Die Morphogenie des Centralnervensystems. 1–272 *In* Buch 2, Teil 3 (Vol. 4) of Hertwig, O. *Handbuch der Entwicklungslehre der Wirbeltiere.* 6 vols. Jena.

Lamarck, J. 1816. Histoire naturelle des animaux sans vertèbres. Vol. 2, 568 pp. Paris.

Lankester, E.R. & Willey, A. 1890. The development of the atrial chamber of amphioxus. *Q. Jl microsc. Sci.* **31**: 445–466.

Larsell, O. 1947. The cerebellum of myxinoids and petromyzonids including developmental stages in the lampreys. *J. comp. Neurol.* **86**: 395–445.

Legros, R. 1898. Développement de la cavité buccale de l'*Amphioxus lanceolatus*. Contribution a l'étude de la morphologie de la tête. *Archs Anat. microsc.* **1**: 497–542.

Legros, R. (Anon) 1909. Développement des fentes branchiales et des canalicules de Weiss-Boveri chez l'*Amphioxus. Anat. Anz.* **34**: 126–151.

Legros, R. 1910. Sur quelques points de l'anatomie et du développement de l'*Amphioxus. Anat. Anz.* **35**: 561–587.

Leuckardt, R. & Pagenstecher, A. 1858. Untersuchungen über niedere Seethiere. *Arch. Anat. Physiol. Wiss. Med.* **(1858)**: 558–610.

Lindström, T. 1949. On the cranial nerves of the cyclostomes, with special reference to n. trigeminus. *Acta zool. Stockh.* **30**: 314–458.

Lohmann, H. 1899. Das Gehäuse der Appendicularien, sein Bau, seine Funktion und Entstehung. *Schr. naturw. Ver. Schlesw.-Holst.* **11**: 345–407.

Lohmann, H. 1922. *Oesia disjuncta* Walcott, eine Appendicularie aus dem Kambrium. *Mitt. zool. St Inst. Hamb.* **38(1920)**: 69–75.

Lohmann, H. 1933. Tunicata. 5(2): 1–202 *In* Kükenthal-Krumbach, *Handbuch der Zoologie.*

Lönnberg, E. 1901–13. Pisces in *Bronns Klassen und Ordnungen des Tierreiches,* Band 6, Teil 1, 582 pp. Leipzig.

Løvtrup, S. 1977. *The phylogeny of the Vertebrata.* 330 pp. Chichester.

Lowenstein, O.E. 1967. The concept of the acousticolateral system. pp. 3–12 *In* Cahn, P.H. (Ed.) *Lateral-line detectors*. 496 pp. Bloomington.

Lowenstein, O.E. 1970. The electrophysiological study of the responses of the isolated labyrinth of the lamprey (*Lampetra fluviatilis*) to angular acceleration, tilting and mechanical vibration. *Proc. R. Soc.* (B) **174**: 419–434.

Lowenstein, O.E. & Osborne, M.P. 1964. Ultrastructure of the sensory hair-cells in the labyrinth of the ammocoete larva of the lamprey, *Lampetra fluviatilis. Nature, Lond.* **204**: 197–198.

Lowenstein, O.E., Osborne, M.P. & Thornhill, R.A. 1968. The anatomy and ultrastructure of the labyrinth of the lamprey (*Lampetra fluviatilis* L.). *Proc. R. Soc.* (B) **170**: 113–134.

Lowenstein, O.E. & Thornhill, R.H. 1970. The labyrinth of *Myxine*. Anatomy, ultrastructure and electro-physiology. *Proc. R. Soc.* (B) **176**: 21–42.

Lowenstein, O.E. & Wersäll, J. 1959. A functional interpretation of the electron-microscopic structures of the sensory hairs in the cristae of the elasmobranch *Raja clavata* in terms of directional sensitivity. *Nature, Lond.* **184**: 1807–1808.

MacBride, E.W. 1909. The formation of the layers in amphioxus and its bearing on the interpretation of the early ontogenetic processes in other vertebrates. *Q. Jl microsc. Sci.* **54**: 279–345.

MacBride, E.W. 1914. *Text book of embryology. Vol. I Invertebrata.*

MacBride, E.W. 1926. The recapitulation theory. *Sci. Progr., Lond.* **20**(2): 461–474.

Mackie, G.O. & Bone, Q. 1976. Skin impulses and locomotion in an ascidian tadpole. *J. mar. biol. Ass. U.K.* **56**: 751–768.

Mallatt, J. 1981. The suspension feeding mechanism of the larval lamprey *Petromyzon marinus. J. Zool., Lond.* **194**: 103–142.

Marcus, E. 1958. On the evolution of the animal phyla. *Q. Rev. Biol.* **33**: 24–58.

Marinelli, W. & Strenger, A. 1954. *Vergleichende Anatomie und Morphologie der Wirbeltiere. I Lieferung.* Lampetra fluviatilis *(L)*: 1–80, Vienna.

Marinelli, W. & Strenger, A. 1956. *Vergleichende Anatomie und Morphologie der Wirbeltiere. II Lieferung.* Myxine glutinosa *(L)*: 81–172, Vienna.

Marinelli, W. & Strenger, A. 1959. *Vergleichende Anatomie und Morphologie der Wirbeltiere. III Lieferung.* Squalus acanthias *L*: 173–308, Vienna.

Matsumoto, H. 1929. Outline of a classification of Echinodermata. *Sci. Rep. Tohoku Univ.* (2, Geology) **13**: 27–33.

Mayr, E. 1942. *Systematics and the origin of species from the viewpoint of a zoologist.* 334 pp. New York.

Mayr, E. 1974. Cladistic analysis or cladistic classification? *Z. Zool. Syst. Evol. Forsch.* **12**: 95–128.

Mayser, P. 1882. Vergleichend anatomische Studien über das Gehirn der Knochenfische mit besonderer Berücksichtigung der Cyprinoiden. *Z. wiss. Zool.* **36**: 259–364.

Medawar, P.B. 1951. Asymmetry of larval *Amphioxus. Nature, Lond.* **167**: 852–853.

Meves, A. 1973. Elektronenmikroskopische Untersuchungen über die Zytoarchitektur des Gehirns von *Branchiostoma lanceolatum. Z. Zellforsch. mikrosk. Anat.* **139**: 511–532.

Millar, R.H. 1953. *Ciona. L.M.B.C. memoirs on typical British marine plants and animals* **35**: 1–84.

Millar, R.H. 1966. Evolution in Ascidians. pp. 519–534 *In* Barnes, H. (Ed) *Some contemporary studies in marine science.* 716 pp. London.

Millott, N. 1966. A possible function for the axial organ of echinoids. *Nature, Lond.* **209**; 594–596.

Millott, N. 1967. The axial organ of echinoids – reinterpretation of its structure and function. *Symp. zool. Soc. Lond.* **20**: 53–63.

Millott, N. & Vevers, H.G. 1968. The morphology and histochemistry of the echinoid axial organ. *Phil. Trans. R. Soc.* (B) **253**: 201–230.

Moller, P.C. & Philpott, C.W. 1973. The circulatory system of amphioxus (*Branchiostoma floridae*), 2. Uptake of exogenous proteins by endothelial cells. *Z. Zellforsch. mikrosk. Anat.* **143**: 135–141.

Morris, R. 1972. Osmoregulation. pp. 193–239 *In* Hardisty, M.W. & Potter, I.C. *The biology of lampreys.* Vol. 1.

Müller, A. 1856. Über die Entwicklung der Neunaugen. *Arch. Anat. Physiol. wiss. Med.* (1856): 323–339. *Arch. Anat. Phys.* **(1856)**: 323–339.

Müller, J. 1851. Über die Jugendzustände einiger Seethiere. *Monatsb. K. Akad. Wiss. Berlin* **(1851)**: 468–474.

Müller, K.J. 1977. *Palaeobotryllus* from the Upper Cambrian of Nevada – a probable ascidian. *Lethaia* **10**: 107–118.

Müller, W. 1873. Über die Hypobranchialrinne der Tunikaten und deren Vorhandensein bei Amphioxus und den Cyklostomen. *Jena Z. Med. Naturw.* **7**: 327–332.

Nakao, T. 1965. The excretory organ of *Amphioxus* (*Branchiostoma*) *belcheri. J. Ultrastruct. Res.* **12**: 1–12.

Nansen, F. 1888. A protandric hermaphrodite (*Myxine glutinosa* L.) amongst the vertebrates. *Bergens Mus. Årsberetn.* **(1887) (7)**: 1–34.

Narasimhamurti, N. 1931. The development and function of the heart and pericardium in Echinodermata. *Proc. R. Soc.* (B) **109**: 471–486.

Narasimhamurti, N. 1933. Development of *Ophiocoma nigra. Q. Jl microsc. Sci.* **76**: 63–88.

Neal, H.V. 1898. The segmentation of the nervous system in *Squalus acanthias. Bull. Mus. comp. Zool. Harv.* **31**: 147–279.

Neal, H.V. 1914. Morphology of the eye muscle nerves. *J. Morph.* **25**: 1–186.

Nelsen, O.E. 1953. *Comparative embryology of the vertebrates.* 982 pp. New York.

Newman, H.H. 1921. On the occurrence of paired madreporic pores and pore canals in the advanced bipinnaria larva of *Asterina* (*Patiria*) *miniata* together with a discussion of the significance of similar structures in other echinoderm larvae. *Biol. Bull. mar. biol. Lab. Woods Hole* **40**: 118–125.

Newman, H.H. 1925. An experimental analysis of asymmetry in the starfish *Patiria miniata. Biol. Bull. mar. biol. Lab. Woods Hole* **49**: 111–138.

Newth, H.G. 1930. The feeding of ammocoetes. *Nature, Lond.* **126**: 94–95.

Newth, D.R. 1951. Experiments on the neural crest of the lamprey embryo. *J. exp. Biol.* **28**; 247–266.

Newth, D.R. 1956. On the neural crest of the lamprey embryo. *J. Embryol. exp. Morph.* **4**: 358–375.

Newth, D.R. & Ross, D.M. 1955. On the reaction to light of *Myxine glutinosa. J. exp. Biol.* **32**: 4–21.

Nichols, D. 1960. The histology of the tube-feet of *Antedon bifida. Q. Jl microsc. Sci.* **101**: 105–117.

Nieuwenhuys, R. 1977. The brain of the lamprey in a comparative perspective. pp. 97–145 *In* Dimond, S.J. & Blizard, D.A. (Eds) *Evolution and lateralisation of the brain. Ann. N.Y. Acad. Sci.* **299**: 1–480.

Northcutt, R.G. & Gans, C. 1983. The genesis of neural crest and epidermal placodes: a reinterpretation of vertebrate origins. *Q. Rev. Biol.* **58**: 1–28.

Nursall, J.R. 1956. The lateral musculature and the swimming of fish. *Proc. zool. Soc. Lond.* **126**: 127–143.

Oelofsen, B.W. & Loock, J.C. 1981. A fossil cephalochordate from the Early Permian Whitehill Formation of South Africa. *S. Afr. J. Sci.* **77**: 178–180.

Ohshima, H. 1922. The occurrence of situs inversus among artificially reared echinoid larvae. *Q. Jl microsc. Sci.* **66**: 105–148.

Olsson, R. 1955. Structure and development of Reissner's fibre in the caudal end of amphioxus and some lower vertebrates. *Acta zool. Scotkh.* **36**: 166–198.

Olsson, R. 1956. The development of Reissner's fibre in the brain of the salmon. *Acta zool. Stockh.* **37**: 235–250.

Olsson, R. 1962. The 'infundibular' cells of amphioxus and the question of fibre-forming secretions. *Ark. Zool.* **15**: 347–355.

Olsson, R. 1963. Endostyles and endostylar secretions: a comparative histological study. *Acta zool. Stockh.* **44**: 299–328.

Olsson, R. 1969a. General review of the Protochordata and Myxinoidea. *Gen. comp. Endocr.* (Supplement 2): 485–499.

Olsson, R. 1969b. Phylogeny of the ventricle system. pp. 291–299 *In* Sterba, G. (Ed.) *Zirkumventrikuläre Organe und Liquor.* Jena.

Olsson, R. 1972. Reissner's fibre in ascidian tadpole larvae. *Acta zool. Stockh.* **53**: 17–21.

Olsson, R. & Wingstrand, K.G. 1954. Reissner's fibre and the infundibular organ in amphioxus – results obtained with Gomeri's chrome alum haematoxylin. *Univ. Bergen Arb.* (Publ. Biol. Stat.) **1954(14)**: 1–14.

Orton, J.H. 1913. The ciliary mechanisms of the gill and the mode of feeding in amphioxus, ascidians and *Solenomya togata. J. mar. biol. Ass. U.K.* **10**: 19–49.

Owsjannikow, P. 1867. Über das Zentralnervensystem des *Amphioxus lanceolatus. Bull. Acad. Sci. St. Petersb.* **12**: 287–302.

Packard, A. 1968. Asexual reproduction in *Balanoglossus* (Stomochordata). *Proc. R. Soc.* (B) **171**: 261–272.

Parker, G.H. 1908. The sensory reactions of amphioxus. *Proc. Am. Acad. Arts. Sci.* **43**: 416–455.

Parsley, R.L. 1982. Functional morphology of mitrate homalozoans (Echinodermata). *Abstr. Prog. Geol. Soc. America* **14(7)**: 583.

Parsley, R.L. & Caster, K.E. 1965. North American Soluta (Carpoidea, Echinodermata). *Bull. Amer. Paleont.* **49(221)**: 109–174.

Patterson, C. 1980. Cladistics. *Biologist* **27**: 234–240.

Patterson, C. 1981. Significance of fossils in determining evolutionary relationships. *A. Rev. Ecol. Syst.* **12**: 195–223.

Patterson, C. & Rosen, D.E. 1977. Review of the ichthyodectiform and other Mesozoic teleost fishes and the theory and practice of classifying fossils. *Bull. Am. Mus. nat. Hist.* **158**: 85–172.

Paul, C.R.C. & Smith, A.B. 1984. The early radiation and phylogeny of echinoderms. *Biol. Rev.* **59**: 443–481.

Pentreath, V.W. & Cobb, J.L.S. 1972. Neurobiology of the Echinodermata. *Biol. Rev.* **47**; 363–391.

Pérès, J.M. 1979. Recherches sur le sang et les organes neuraux des tuniciers. *Annls. Inst. Océanogr. Monaco* **21**: 229–359.

Philip, G.M. 1979. Carpoids – echinoderms or chordates? *Biol. Rev.* **54**: 439–471.

Philip, G.M. 1981. *Notocarpos garatti* gen. et sp. nov., a new Silurian mitrate carpoid from Victoria. *Alcheringa* **5**: 29–38.

Platnick, N.I. 1980. Philosophy and the transformation of cladistics. *Syst. Zool.* **28**: 537–546.

Platnick, N.I. & Cameron, H.D. 1977. Cladistic methods in textual, linguistic and phylogenetic analyses. *Syst. Zool.* **26**: 380–385.

Pompeckj, J.F. 1896. Die fauna des Cambrium von Tejřovic und Skrej in Böhmen. *Jb. geol. Staatsanst. Wien* (1895) **45**: 495–614.

Prokop, R.J. 1963. *Dalejocystis*, n. gen., the first representative of the Carpoidea in the Devonian of Bohemia. *J. paleont.* **37**: 648–650.

Prouho, H. 1887. Recherches sur le *Dorocidaris papillata. Archs. Zool. exp. gén.* (2) **5**: 213–280.

Pucci-Minafra, I. 1965. Ultrastructure of muscle cells in *Ciona intestinalis* tadpoles. *Acta Embryol. Morph. exp.* **8**: 289–305.

Pumphrey, R.J. 1950. Hearing. pp. 1–18 *In* Dainelli, J.F. and Brown, R. (Eds) *Physiological mechanisms in animal behaviour. Symp. Soc. exp. Biol.* **4**: 1–482.

Rähr, H. 1979. The circulatory system of amphioxus (*Branchiostoma lanceolatum* (Pallas)). A light-microscope investigation based on intravascular injection technique. *Acta zool. Stockh.* **60**: 1–18.

Reed, F.R.C. 1925. Revision of the fauna of the Bokkeveld Beds. *Ann. South Afr. Mus.* **22**: 27–226.

Reese, A.M. 1902. Structure and development of the thyroid gland in Petromyzon. *Proc. Acad. nat. Sci. Philad.* **54**: 85–112.

Reichensperger, A. 1905. Zur anatomie von *Pentacrinus decorus*. *Z. Wiss. Zool.* **80:** 22–55.

Reissner, E. 1860. Beiträge zur Kentniss vom Bau des Rückenmarks von *Petromyzon fluviatilis* L. *Arch. Anat. Physiol. wiss. Med.* **(1860):** 545–588.

Remane, A. 1943. Die Geschichte der Tiere. pp. 589–677 *In* Heberer, G. *Die Evolution der Organismen, Ergebniße und Probleme der Abstammungslehre.* Band 1, 774 pp. Jena.

Remane, A., Storch, V. & Welsch, V. 1972. *Kurzes Lehrbuch der Zoologie.* 459 pp. Stuttgart.

Rennie, J.V.L. 1936. On *Placocystella*, a new genus of cystids from the Lower Devonian of South Africa. *Ann. S. Afr. Mus.* **31:** 269–275.

Repetski, J.E. 1978. A fish from the Upper Cambrian of North America. *Science, N.Y.* **200(4341):** 529–531.

Reynolds, F.E. 1931. Hydrostatics of the suctorial mouth of the lamprey. *Univ. Calif. Publs Zool.* **37;** 15–34.

Riedl, R. 1975. *Die Ordnung des Lebendigen – Systembedingungen der Evolution.* 372 pp. Hamburg.

Riedl, R. 1979. *Order in living organisms.* 313 pp. Chichester.

Ritchie, A. & Gilbert-Tomlinson, J. 1977. First Ordovician vertebrates from the southern hemisphere. *Alcheringa* **1:** 351–368.

Robertson, J.D. 1957. The habitat of the early vertebrates. *Biol. Rev.* **32:** 156–187.

Rohde, E. 1888. Histologische Untersuchungen über das Nervensystem von Amphioxus. *Zool. Beitr.* **2:** 169–210.

Romer. A.S. 1972. The vertebrate as a dual animal – somatic and visceral. *Evolut. Biol.* **6:** 121–156.

Roule, L. 1884. Recherches sur les ascidies simples des côtes de Provence. Phallusiadées. *Annls Mus. Hist. nat. Marseille* (Zoology) **2:** 1–270.

Rovainen, C.M. 1967. Physiological and anatomical studies on large neurons of the central nervous system of the sea lamprey (*Petromyzon marinus*). II Dorsal cells and giant interneurons. *J. Neurophysiol.* **30:** 1024–1042.

Rudwick, M.J.S. 1961. The feeding mechanism of the Permian brachiopod *Prorichthofenia. Palaeontology* **3:** 450–471.

Rudwick, M.J.S. 1964. The inference of function from structure in fossils. *Br. J. Phil. Sci.* **15:** 27–40.

Sargent, P.E. 1904. The optic reflex apparatus of vertebrates for short-circuit transmission of motor reflexes through Reissner's fibre; its morphology, ontogeny, phylogeny and function – the fish-like vertebrates. *Bull. Mus. comp. Zool. Harv.* **45:** 127–258.

Sato, T. 1936. Vorläufige Mitteilung über *Atubaria heterolopha* gen. nov. et sp. nov. einen in freiem Zustand aufgefundenem Pterobranchier aus dem Stillen Ozean. *Zool. Anz.* **1154:** 97–106.

Scammon, R.E. 1911. Normal plates of the development of *Squalus acanthias. Keibel's Normaltafeln zur Entwicklungsgeschichte der Wirbeltiere* **12:** 1–140.

Schaeffer, B. & Thomson, K.S. 1980. Reflections on agnathan-gnathostome relationships. pp. 19–33 *In* Jacobs, L.L. (Ed.) *Aspects of vertebrate history. Essays in honor of Edwin Harris Colbert.* 407 pp. Flagstaff.

Schuchert, C. 1913. Echinodermata cystoidea. pp. 227–248 in *Maryland Geological Survey Gen. Series Devonian* (Vol. 1). Lower Devonian 1–560 and (Vol. 3) Devonian Plates.

Schneider, A. 1879. *Beiträge zur vergleichenden Anatomie und Entwicklungsgeschichte der Wirbeltiere.* 164 pp. Berlin.

Schultze, M. 1852. Beobachtungen junger Exemplare von Amphioxus. *Z. wiss. Zool.* **3:** 416–419.

Scoffin, T.P. 1972. The conditions of growth of the Wenlock reefs of Shropshire, England. *Sedimentology* **17:** 173–219.

Scott, F.M. 1952. The developmental history of *Amaroecium constellatum* III. Metamorphosis. *Biol. Bull. mar. biol. Lab. Woods Hole* **103:** 226–241.

Seeliger, O. 1885. Die Entwicklungsgeschichte der socialen Ascidien. *Jena Z. Naturw.* **18:** 45–120.

Seeliger, O. 1892. Studien zur Entwicklungsgeschichte der Crinoiden. *Zool. Jb.* (Abt. Anatomie.) **6:** 161–444.

Seeliger, O. 1893. Über die Entstehung des Peribranchialraumes in den Embryonen der Ascidien. *Z. Wiss. Zool.* **56:** 365–401.

Seeliger, O. & Hartmeyer, R. 1893–1911. Tunicata (Manteltiere). 1. Abteilung. Die Appendicularien undAscidien. Vol. 3 (Suppl) of *Bronns Klassen und Ordnungen des Tier-Reichs.* 1280 pp. Leipzig.

Selys-Longchamps, M. de 1900. Développement du coeur, du péricarde et des épicardes chez *Ciona intestinalis. Archs Biol., Paris* **17:** 499–542.

Silén, L. 1954. Reflections concerning the 'stomochord' of the Enteropneusta. *Proc. zool. Soc. Lond.* **124:** 63–67.

Skramlik, E. von 1938. Über den Blutumlauf bei *Amphioxus lanceolatus* Y. *Pubbl. Staz. zool. Napoli* **17:** 130–157.

Smith, A.B. 1984. Classification of the Echinodermata. *Palaeontology* **27:** 431–459.

Smith, J.E. 1937. On the nervous system of *Marthasterias glacialis. Phil. Trans. R. Soc.* (B) **227:** 111–173.

Smith, K.M. & Newth, H.G. 1917. A note concerning the collar cavities of the larval amphioxus. *Q. Jl microsc. Sci.* **62:** 243–251.

Southward, E.C. 1975. Fine structure and phylogeny of the Pogonophora. *Symp. Zool. Soc. Lond.* **36:** 235–251.

Sprinkle, J. 1973. *Morphology and evolution of blastozoan echinoderms.* 284 pp. Cambridge, Mass.

Sprinkle, J. 1983. Patterns and problems in echinoderm evolution. *Echinoderm Stud.* **1:** 1–18.

Sprinkle, J. 1976. Biostratigraphy and paleoecology of Cambrian echinoderms from the Rocky Mountains. *Geology Stud. Brigham Young Univ.* **23:** 61–73.

Sprinkle, J. 1983. Patterns and problems in echinoderm evolution. *Echinoderm Studies* **1:** 1–18.

Stanley, S.M. 1970. Relation of shell form to life habits in the Bivalvia (Mollusca). *Mem. geol. Soc. Am.* **125:** 1–296.

Starck, D. 1944. Die Bedeutung der Entwicklungsphysiologie für die vergleichende Anatomie, erläutert am Beispiel des Wirbeltierkopfes. *Biologia gen.* **17:** 481–510.

Starck, D. 1963. Die Metamerie des Kopfes der Wirbeltiere. *Zool. Anz.* **170:** 393–429.

Starck, D. 1975. *Embryologie. Ein Lehrbuch auf allgemein biologischer Grundlage.* 3. Auflage. 704 pp. Stuttgart.

Stebbing, A.R.D. 1970. The status and ecology of *Rhabdopleura compacta* (Hemichordata) from Plymouth. *J. mar. biol. Ass. U.K.* **50:** 209–221.

Stebbing, A.R.D. & Dilly, P.N. 1972. Some observations on living *Rhabdopleura compacta* (Hemichordata). *J. mar. biol. Ass. U.K.* **52:** 443–448.

Stephenson, D.G. 1979. The trimerous stage in echinoderm evolution: an unnecessary hypothesis. *J. Paleont.* **53:** 44–48.

Sterba, G. 1961. Zur Phylogenese des Kiemendarmes der Chordaten. *Int. Revue ges. Hydrobiol.* **46:** 105–114.

Sterba, G. 1962. O znachenii Petromyzonidae v evolyutsii khordovykh. [On the significance of Petromyzonidae in the evolution of the Chordata]. *Vest. Leningr. gos. Univ.* **15:** 58–65.

Sterba, G. 1962. Die Neunaugen (Petromyzonida). pp. 282–352 in vol. 3(4) of Demoll, R., Maier, H.N. & Wundsch, H.H. (Eds) 1924–62. *Handbuch der Binnenfischerei Mitteleuropas,* 3 vols. Stuttgart.

Stiasny, G. 1914a. Studien zur Entwicklung von *Balanoglossus clavigerus.* I. Die Entwicklung der Tornaria. *Z. wiss. Zool.* **110:** 36–75.

Stiasny, G. 1914b. Studien über die Entwicklung von *Balanoglossus clavigerus.* II. Darstellung der weiteren Entwicklung bis zur Metamorphose. *Mitt. zool. StnNeapel.* **22;** 255–290.

Stockard, C.R. 1906. The development of the thyroid gland in *Bdellostoma stouti. Anat. Anz.* **29:** 91–99.

Studnička, F.K. 1912. Über die Entwicklung und die Bedeutung der Seitenaugen von Ammocoetes. *Anat. Anz.* **41:** 561–578.

Tarlo, L.B.H. 1960. The invertebrate origins of the vertebrates. *21st International Geological Congress* **22:** 113–123.

Tester, A.L. & Kendall, J.I. 1967. Innervation of free and canal neuromasts in the sharks *Carcarhinus menisorrah* and *Sphyrna levini.* pp. 53–59 In Cahn, P.H. (Ed.).

Thoral, M. 1935. *Contribution à l'étude paléontologique de l'ordovicien inférieur de la Montagne Noire et révision sommaire de la faune cambrienne de la Montagne Noire.* 362 pp. Montpellier.

Thornhill, R.A. 1972. The development of the labyrinth of the lamprey (*Lampetra fluviatilis* Linn). *Proc. R. Soc.* (B) **181:** 175–198.

Thorpe, A. & Thorndyke, M.C. 1975. The endostyle in relation to iodine binding. *Symp. zool. Soc. Lond.* **36:** 159–177.

Tomlinson, J. 1966. The advantages of hermaphroditism and parthenogenesis. *J. theor. Biol.* **11:** 54–58.

Torrence, S.A. & Cloney, R.A. 1982. The nervous system of ascidian larvae: primary sensory neurons in the tail. *Zoomorphology* **99:** 103–115.

Ubaghs, G. 1961a. Un echinoderme nouveau de la classe des carpoides dans l'ordovicien inférieur du département de l'Hérault (France). *C.R. Séanc. Soc. Biol.* **253:** 2565–2567.

Ubaghs, G. 1961b. Sur la nature de l'organe appelé tige ou pedoncule chez les carpoides Cornuta et Mitrata. *C.R. Séanc. Soc. Biol.* **253:** 2738–2740.

Ubaghs, G. 1963. *Cothurnocystis* Bather, *Phyllocystis* Thoral and an undetermined member of the order Soluta (Echinodermata, Carpoidea) in the uppermost Cambrian of Nevada. *J. Paleont.* **37:** 1133–1142.

Ubaghs, G. 1967a. Le genre *Ceratocystis* Jaekel (Echinodermata, Stylophora). *Paleont. Contr. Univ. Kans.* **22:** 1–16.

Ubaghs, G. 1967b. Stylophora: S495–565 In Moore, R.C. (Ed.) *Treatise on invertebrate paleontology. Part S. Echinodermata* 1(2): S297–650. New York.

Ubaghs, G. 1969. Les echinodermes carpoides de l'Ordovicien inférieur de la Montagne Noire. *Cahiers de Paléontologie* 112 pp. Paris.

Ubaghs, G. 1971. Diversité et specialisation des plus anciens echinodermes que l'on connaisse. *Biol. Rev.* **46:** 157–200.

Ubaghs, G. 1975. Early Paleozoic echinoderms. *Ann. Rev. Earth planet. Sci.* **3:** 79–98.

Ubaghs, G. 1978. Classification of the echinoderms. T359–T371. In Moore, R.C. & Teichert, C. (Eds) *Treatise on invertebrate paleontology. Part T Echinodermata* 2. Vol. 1, 401 pp. Boulder & Lawrence.

Ubaghs, G. 1979. Trois Mitrata (Echinodermata: Stylophora) nouveaux de l'ordovicien de Tchécoslovaquie. *Paläont. Z.* **53:** 98–119.

Ubaghs, G. 1918. Reflexions sur la nature et la fonction de l'appendice articulé des carpoides Stylophora (Echinodermata). *Annls. Paléont.* (Invertebrés) **67:** 33–48.

Ubaghs, G. 1983. Echinodermata. Notes sur les echinodermes de l'ordovicien inférieur de la Montagne Noire (France). Chap. 3: 33–55, pl. 8–10 in: Courtessole, R., Marek, L., Pillet, J., Ubaghs, G. & Vizcaino, D. Calymenina, Echinodermata et Hyolitha de l'Ordovicien de la Montagne Noire (France Meridionale). *Mem. Soc. d'Etudes Scientif. Aude* 62 pp. Carcassonne.

Ubisch, L. von 1929. Über Lage, Entwicklung, Induktionswirkung und Funktion von Chorda und Hydrocöl. *Verh. dt. zool. Ges.* **33:** 83–85.

Ubisch, L. von 1957. Doppelbildungen bei Seeigeln (*Psammechinus multituberculatus, Paracentrotus lividus*). *Zool. Anz.* **159:** 1–12.

Ubisch, L. von 1958. Die phylogenetischen Symmetrieveränderungen bei den Seeigeln. *Forschungsberichte des Wirtschafts und Verkehrsministeriums von Nordrhein-Westfalen* **539:** 1–56.

Uljanin, B. 1884. Die Arten der Gattung *Doliolum* im Golfe von Neapel. *Fauna und Flora des Golfes von Neapel, Monogr.* **10:** 1–140.

Veit, K. 1957. Einige Beobachtungen über die ersten Furchungsschritte bei *Petromyzon planeri. Gegenbaurs morph. Jb.* **98:** 1–34.

Veit, O. 1924. Beiträge zur Kenntniss des Kopfes der Wirbeltiere II. *Gegenbaurs morph. Jb.* **53:** 320–390.

Veit, O. 1947. *Über das Problem Wirbeltierkopf.* 102 pp. Kempen.

Walls, G.L. 1942. The vertebrate eye and its adaptive radiation. *Bull. Cranbrook Inst. Sci.* **19:** 1–785. Bloomfield Hill, Mich.

Webb, J.E. 1969. On the feeding and behaviour of the larva of *Branchiostoma lanceolatum. Mar. Biol. Berlin* **3:** 58–72.

Webb, J.E. 1973. The role of the notochord in forward and reverse swimming and burrowing in the amphioxus *Branchiostoma lanceolatum. J. Zool., Lond.* **170**: 325–338.

Webb, J.E. 1975. The distribution of amphioxus. *Symp. zool. Soc. Lond.* **36**: 179–212.

Weichert, C.K. 1965. *Anatomy of the chordates.* 758 pp. New York.

Weiss, F.E. 1890. Excretory tubules in amphioxus. *Q. Jl microsc. Sci.* **31**: 489–497.

Weissenberg, R. 1934. Untersuchungen über den Anlageplan beim Neunaugenkeim: Mesoderm, Rumpfdarmbildung und Übersicht der centralen Anlagezonen. *Anat. Anz.* **79**: 177–199.

Welsch, U. 1968a. Die Feinstruktur der Josephschen Zellen im Gehirn von Amphioxus. *Z. Zellforsch. mikrosk. Anat.* **86**: 252–261.

Welsch, U. 1968b. Über den Feinbau der Chorda dorsalis von *Branchiostoma lanceolatum. Z. Zellforsch. mikrosk. Anat.* **87**: 69–81.

Welsch, U. 1975. The fine structure of the pharynx, cyrtopodo-cytes and digestive caecum of amphioxus (*Branchiostoma lanceolatum*). *Symp. zool. Soc. Lond.* **36**: 17–41.

White, M.J.D. 1978. *Modes of speciation.* 455 pp. San Francisco.

Wickstead, J.H. 1967. *Branchiostoma lanceolatum* larvae: some experiments on the effect of thiouracil on metamorphosis. *J. mar. biol. Ass. U.K.* **47**: 49–59.

Wijhe, J.W. van 1883. Mesodermsegmente und die Entwicklung der Nerven des Selachierkopfes. *Verh. K. Akad. Wet., Amst.* **22**(5): 1–50.

Wijhe, J.W. van 1893. Über Amphioxus. *Anat. Anz.* **8**: 152–172.

Wijhe, J.W. van 1902. Beiträge zur Anatomie der Kopfregionen des *Amphioxus lanceolatus. Petrus Camper* **1**: 100–194.

Wijhe, J.W. van 1914. Studien über Amphioxus. I Mund und Darmkanal während der Metamorphose. *Verh. K. Akad. Wet., Amst.* (2. Sect.) **18**: 1–84.

Wiley, E.O. 1981. *Phylogenetics – the theory and practice of phylogenetic systematics.* 439 pp. New York.

Willey, A. 1891. The later larval development of amphioxus. *Q. Jl microsc. Sci.* **32**: 183–234.

Willey, A. 1893a. Studies on the Protochordata. No. 1. On the origin of the branchial stigmata, preoral lobe, endostyle, atrial cavities etc. in *Ciona intestinalis* L., with remarks on *Clavelina lepadiformis. Q. Jl microsc. Sci.* (2) **34**: 317–360.

Willey, A. 1893b. Studies on the Protochordata. No. 2. The development of the neuro-hypophyseal system in *Ciona intestinalis* and *Clavelina lepadiformis* with an account of the origin of the sense organs in *Ascidiamentula. Q. Jl microsc. Sci.* (2) **35**: 295–333.

Willey, A. 1894. *Amphioxus and the ancestry of the vertebrates.* 316 pp. New York.

Willmer, E.N. 1975. The possible contribution of the nemertines to the problems of the phylogeny of the protochordates. *Symp. zool. Soc. Lond.* **36**: 319–345.

Wolf, H. 1940. Über die Genauigkeit der Herztätigkeit in der Tierreihe. *Pflügers Arch. ges. Physiol.* **244**: 181–203.

Wolf, H. 1941. Über die Beeinflüßung der Kreislauftätigkeit bei *Amphioxus lanceolatus* Y. *Pflügers Arch. ges. Physiol.* **244**: 736–748.

Yalden, D.W. 1985. Feeding mechanisms as evidence for cyclostome monophyly. *Zool. J. Linn. Soc.* **84**: 291–300.

Young, J.Z. 1935. The photoreceptors of lampreys III. Control of colour change by the pineal and the pituitary. *J. exp. Biol.* **12**: 258–270.

Young, J.Z. 1981. *The life of vertebrates.* 3rd edition. 643 pp. Oxford.

Ziegler, H.E. 1908. Die phylogenetische Entstehung des Kopfes der Wirbeltiere. *Jena. Z. Naturw.* **43**: 653–684.

Zimmer, R.L. 1973. Morphological and developmental affinities of the lophophorates. pp. 593–606 *In* Larwood, G.P. (Ed.) *Living and fossil bryozoa.* 634 pp. London and New York.

Zimmer, R.L. 1980. Mesoderm proliferation and function of the protocoel and metacoel in early embryos of *Phoronis vancouverensis* Phoronida). *Zool. Jb. Anat.* **103**: 219–233.

Index

acorn worms *see* enteropneusts
acraniates 300
    classification as vertebrates 87
    loss of heart 331
    migration of gonad 331
    transformation of crescentic organ 331
acustico-lateralis system 146
agnathans 117
Amaroucium tadpole 105, 111
Ambulacralia 54
amnion 9
amphioxus (*Branchiostoma lanceolatum*) 56–86
    alimentary canal 56
        cilia 60
    asymmetries 2
    atrioporal sphincter 62, 64
    atrium 57
    brain 67–9
        Joseph cell 67–8
        nerves from 69–70
        pigment spot 68–9
    club-shaped gland 78, 81
    coeloms 62–4, 85
        epipterygial 64
        fin-ray 64
        general 62
        gonocoels 64
        inner lip 64
        metapterygial 64
        pterygial 62–3
        rostral 64
        velar 64
    Cuvierian duct 72
    dorsal nerve cord 64–7
        function 66
        giant-fibre system 66–7
        Hesse cells 67
        histology 66
        photoreception 67
        Rohde cells 66–7
    egg 73
    embryology 73–86
        anterior gut diverticula 75
        atrial floor 82 (fig.)
        atrial folds 79
        atriopore 79
        club-shaped gland 78, 81
        coeloms 84–5
        dorso-lateral somitic ridges 74
        endostyle 78, 81–3, 153 (fig.)
        enterocoely 74
        epipharyngeal band 78
        gastrulation 73–4
        gill bars 79
        gill slits 77–8, 81, 86
        metapleural fold 79
        mid-gut diverticulum 84
        mouth 77, 834–4
        nephridia 85
        neurula 74, 76 (figs.)
        notochord 75
        one-gill-slit stage 78–9, 80 (fig.)

        schizocoely 74
        somites 75–6
        subatrial fold 79
    endostyle 57
    epipharyngeal groove 56, 60
    filter-feeding mechanism 61
    fin rays 56
    gill bars 57–8
    gill slits 56–9, 77
    gonads 56, 71
    Hatschek's pit 56
    ilio-colonic ring 61
    infundibular organ 69
    innervation of somitic muscles 333–7
    Kölliker's pit 69
    larval mouth 77, 284
    male/female 56
    mode of life 56
    movements 62–2
    muscle-tails 61, 62
    muscular system 62
        innervation 62
        muscle-blocks 55–6, 62
    myocommata (myosepta) 56
    nephridia 70, 300–1
        solenocytes (cyrtopodocytes) 70–1
    nervous system 64–70
    notochord 61, 75
    notochordal plates 61
        muscles 61–2
    peripharyngeal bands 78
    peripharyngeal grooves 57, 60
    pharynx 56, 60–1
    primordial bars 58–60
    Reissner's fibre 69
    tongue bars 58–60
    vascular system 71–3
        arterial 71–2
        branchial 72
        plexi 72–3
        venous 72
    wheel organ 56
Amygdalotheca 233, 321, 326
Amygdalotheca griffei 239–9, 327–8
    gill slits 236
    head 235
    stratigraphical distribution 335 (fig.)
    tail 236
Anatifopsis 329
Anatifopsis barrandei 314 (fig.), 315, 331
    stratigraphical distribution 334 (fig.)
    '*Anatifopsis*' *escandei* 331, 335 (fig.)
'*Anatifopsis*' *trapeziiformis* 331, 335 (fig.)
Anatolepis cf. *heintzi* 333, 335 (fig.)
Antedon bifida 17, 41–51
    comparison with *Cephalodiscus* 50–1
        axocoel 50
        circumoral ectoneural nerve ring 51
        fixation sucker 50–1
        larval axocoel 50
        somatocoels 51
        water-vascular system 51

    embryology 47–50
        gut 44–5
        larva 48–50
            ciliated 48–50
            fixed 50
            stemmed 50
        nervous system 45–6
        perihaemal system 46
        water-vascular system 46, 51
anti-segmentationists 183–4
appendicularians (larvaceans) 113–15, 332
Archaeopteryx teeth 8
archencephalon 126
Ascidia nigra tadpole, ocellus 108
Aspidocarpus bohemicus 333
Asterias forbesi 29
Asterias rubens (common starfish) 17, 29–42
    alimentary canal 32 (fig.)
    ambulacral grooves 39
    auricularia larva 37–40
    axial organ 33–4, 51
    comparison with *Cephalodiscus* 50–1
        circumoral ectoneural nerve ring 51
        fixation sucker 50–1
        hydrocoel 51
        madreporic vesicle 51
        somatocoels 51
        water-vascular system 51
    dorsal sac (madreporic vesicle) 34, 51
    double hydrocoel 40, 42 (figs.)
    double hydropore 40, 41, 43 (fig.), 51
    embryology 36–7
    gonads 33
    haemal system 33–4
    inverse larvae 40, 41
    Lange's nerves 36
    larva 37–40
    nervous system 34–6
        apical (endoneural) 34
        ectoneural 34–6
        hyponeural 34
    optic cushion 35–6
    organs around mouth 31 (fig.)
    ovum 36
    pedicellariae 32
    perihaemal system 33–4
    peristome 29
    radial nerves 346
    skeleton 29–33
    stereom mesh 30, 32
    tube feet 29, 33
    young 41 (fig.)
Asterias vulgaris 29
Asymmetron 56, 71, 331
Asymmetron bassanum larva 79
Ateleocystites huxleyi 2
Atubaria 23–4, 317
auditory capsule in vertebrate embryos 189
aulacophore 191, 351–3
Aulechinus 54
Australopithecus 13
axial organ (gland) 34, 331

Balanocystites 329
Balanocystites primus 312–15, 331
  mode of life 315
  stratigraphical distribution 334 (fig.)
Balanoglossus 18, 25
Balanoglossus australiensis 28
Balanoglossus capensis 28
Balanoglossus clavigerus, tornaria larva 28 (fig.)
Bather, Francis, quoted 1–2
Bathochordaeus 115
Berrill's hypothesis 349
Bilateria 17
Bohemiaecystis 238
Bohemiaecystis bouceki 334 (fig.)
brachiopods 17, 317
branchiomery, in enteropneusts 187
Branchiostoma 56
Branchiostoma lanceolatum see amphioxus
bryozoans 17, 317
calcichordates 1, 2, 191–2
  appendage 191–2
  skeletons 15
  subphylum of Chordata (as formerly thought) 14
  tail ('stele') 205
  theory 356, 358
    answers to 357–8
  use of term 14
'carpoids' 353–4
Cephalodiscus 17, 20–3, 50, 317
  alimentary canal 21, 23
  blood system 22, 23 (fig.)
  body openings 21, 23
  coeloms 23
  comparison with *Asterias/Antedon* 50–1
    circumoral nerve ring 51
    mesocoels 51
    metacoels 51
    pericardial/heart vesicle 51
    protocoel 50
    protocoel pores 51
    protosome (head shield) 50–1
  embryology 23
  glomerulus 22
  heart vesicle 22, 51
  homologies with echinoderms 318
  muscles of protosome (head shield) 22
  nervous system 21–2
  sexual reproduction 23
  stomochord 21
Cephalodiscus-like animal 52
Ceratocystis 54
  gonopore-anus 197
Ceratocystis perneri 2, 213–25
  acustico-lateralis organ (narrow groove) 224–5, 253, 321
  anus 218
  axial organ 218–21
  axial sinus (axocoel) 218
  brain 223, 224 (fig.)
  branchial currents 218 (fig.)
  buccal cavity 222
  coeloms 222
  comparison with *Cephalodiscus* 221–2
  gonopore 218, 284

habits 225
head 216
  chambers 217, 222
  openings 216–17
hydropore 218, 284
least crownward chordate known 321
median eye 223, 321
pharynx 222
pyriform bodies (trigeminal ganglia) 223
ridge from left face of brain 225
stratigraphical distribution 335 (fig.)
tail 222–3, 335 (fig.)
voiding of faeces and gametes 218 (fig.)
water-vascular system absent 218–21
Chaetognatha 17–18, 317
chambered organ 319
Chaucer, Geoffrey (1340?–1400?) 14, 15 (fig.)
Chauvelicystis 327, 335 (fig.)
Chauvelicystis spinosa 354
Chelyosoma 198, 199 (fig.)
chick, notochord development 8
  transverse section of dorsal nerve cord 36
Chiniancarpos 329
Chinianocarpos thorali 290–3, 332–3
  atrial openings 292
  external features 291 (fig.)
  head 290–2
    openings 290
  interossicular canal absent 287
  mouth 290
  stratigraphical distribution 335 (fig.)
  tail 231–2, 287, 292
    functions 292–3
    hind 231, 355
chordates (Chordata) 317
  crown 321
  first use of word 87
  less crownward/more crownward plesion 321
  non-existence as monophyletic group (Løvtrup) 346
Chordonia 87
Ciona intestinalis 87–115
  alimentary canal 88
  alimentary ciliated loop 97–8
  asymmetric gland 102
  blood 100
  body 88
  crossed reflex 101
  cupular organ 101, 103 (fig.), 253
  development 103–13
    gastrulation 104
    neurulation 194–5
    post-larval 110 (fig.)
    see also below tadpole
  distribution 87
  dorsal languets 92–3
  dorsal strand 101
  egg 104
  endostylar groove 93
  endostylar hood 92
  endostyle 90–2
  epicardia 98
  external appearance 87–8
  feeding 95–7
  fertilization of ova 103

head of tadpole 106 (fig.)
heart 99–100
interepicardial mesenteries 98, 99 (fig.)
intestine 95
intrabranchial transverse bar 92
morula cells 100
movements 88
mucous rope 95, 96 (fig.), 97 (fig.)
mucous sheets 95
muscles 88, 91 (fig.)
nervous system 101
neural gland 101, 102–3
non-pharyngeal gut 94–5
oesophagus 94
ovary 98–9
oviduct 98–9
pericardium 100
peripharyngeal bands 92
pharyngo-epicardial openings 94
pharynx 88–90, 92 (fig.), 93–4
  assymmetries 93–4
posterior endostylar region 93, 94 (fig.)
pyloric gland 95
rectum 95
reproductive system 98–9, 100 (fig.)
retropharyngeal band 93, 97
retropharyngeal groove 93
sensory receptors 101
stigmata 90
stomach 95
tadpole 105–11
  alimentary canal 108
  atria 112
  cerebral ganglion 105
  dorsal nerve cord 105, 106 (fig.)
  dorsal strand 112
  epicardia 112
  fixation 111
  ganglion 112
  gonads 113
  head 105
  heart 108–11
  intrabranchial vessels 112
  muscle cells 105
  neural gland 112
  notochord 105
  ocellus 107
  otolith 107, 108
  pharynx 108
  pressure receptors 107
  protostigmata 111–12
  pyloric gland 108
  Reissner's fibre 105
  retropharyngeal groove 108
  sensory vesicle 105–7
  stigmata (gill slits) 111
  tail 105, 111
  test (tunic) 105
  torsion 111
test (tunic) 88
testis 98
vascular system 99–101
vas deferens 98
ventral groove 94
cladists 3, 6

cladograms 3–5, 10
  lampreys 157–60
  Metazoa 19 (fig.)
  myxinoids 157–60
  phylogenetic trees 9 (fig.)
  phylogeny of deuterostomes 323 (fig.)
    Garstang 348 (fig.)
  reconstruction of hypothetical animals 336
    (fig.), 337 (fig.), 340 (fig.)
  three-taxon 5
  with empty segment 6 (fig.)
Clavelina lepadiformis
  feeding 96–7 (figs.)
  tadpole 107
coelenterates, gastrula stage 17
common starfish *see Asterias rubens Corella*
  endostyle 153 (fig.)
  young individual 112 (fig.)
cornutes 2, 191, 192, 238
  blood vessel in notochord 204 (fig.)
  gill slits 192, 319
  head chambers 200 (fig.)
  heart 281–4
  marginal plates near tail 192
  pericardium 281–4
  relationship to mitrates 329
  resemblances to echinoderms lost by living
    chordates 320–1
  subdivision 321
  synapomorphies with living chordates
    319
Cothurnocystis 327
'*Cothurnocystis*' (Sprinkle's) 321, 335 (fig.)
Cothurnocystis elizae 192–207
  brain 205
  chambers 198–202
  closeness to crown group 326
  feeding 205
  gill slits 197, 198
  gonopore-anus 197
  head 192–3
    openings 193–8
    skeleton 193
  notochord 202
  pyramid of plates around mouth 197
  pyriform bodies (trigeminal ganglia) 205
  stratigraphical distribution 334 (fig.)
  tail 193, 202–5, 207
    function 204–5
    stylocone of mid-tail 202, 204
Cothurnocystis fellinensis 321
'*Cothurnocystis*' *melchiori* 238
crinoids 41
  stem 205–6
  see also Antedon bifida
crocodile 8
crown chordates, phylogeny 325 (fig.)
crown dexiothete 319–20
crown group (*group) 10
crown vertebrates 338–9
cystoids, openings in 217
Dendrodoa tadpole 105
'Dethe of Blanche the Duchess, The' (G.
  Chaucer) 14, 15 (fig.)
deuterostomes 17, 17–18, 317–43

embryological distinction from protostomes
  18 (fig.)
Dexiothetica 222, 317
dexiothetic orientation (dexiothetism) 52–4, 317
Didemnidae 316
Distaplia tadpole 107–8, 109 (fig.)
Divaricella, cuesta-shaped ribs 248 (fig.)
Doliolids 113–115
Doliolum denticulatum tadpole 114 (fig.)
Doliolum resistibile 114 (fig.)
double hydrocoel 40, 42 (figs.)
double hydropore 40, 41, 43 (fig.)
Echinodermata 1, 17, 29, 317, 319
  axial gland 221
  edrioasteroid, triradiate stage 320 (fig.)
  filter-feeding stage 52
  fixation by right side 319
  gonads 54
  homologies with *Cephalodiscus* 318
  hydropore 284
  innervation of muscles 337
  larval plane of symmetry 51
  nervous system 34
  radial symmetry 319
Ecteinascidia 112
edrioasteroids 319
elasmobranchs, blood isosmotic with sea water
  124
Eleutherozoa 29, 319
Emerita, cuesta-shaped ribs 248 (fig.)
empty segments 6
enterocoely 74
enteropneusts (acorn worms) 18, 25, 317
  alimentary canal 26–7
  blood system 28
  body openings 25–6
  excretion 28
  glomerulus 28
  mesosome (collar) 25
  metasome (trunk) 25
  nervous system 27–8
  protosome (proboscis) 25
  reproduction 28–9
Eocephalodiscus 25
Eptatretus burgeri (*Myxine garmani*) 132
Eptatretus stouti 132–3
  brain 128 (fig.)
  eggs 179
  embryology 179–85
    auditory capsules 179, 185
    blastoderm 179
    brain 179
    coelom 179
    cranial-nerve placodes 181
    head 179–81
    heart 179
    hepatic vein 179
    infundibulum 179–80
    kidney 179, 181 (fig.)
    mesodermal somites 179
    nasohypophyseal duct 181, 185
    nerve cord 179
    optic vesicles 179
    somatopleur 179
    splanchnopleur 179

    stomodaeal membranes 180
    thyroid gland 181–5
    velum 181
    visceral pouches 179
  eyes 129, 132, 133 (fig.), 160
  fertilization of egg 179
  lateral-line system 133
flying vertebrates 8
fossils
  arrangement in sequence of plesions 13
  methodology of functional morphology 14
Froriep's theory of eye in tunicate tadpole
  107–8, 109 (fig.)
Galliaecystis 223, 321, 326
Galliaecystis lignieresi 233–5, 327
  gill slits 234
  gonopore-anus 234
  gonorectal groove 234
  head 233–4
  integuments 234
  internal structure 235
  resemblances to *C. elizae* 235
  stratigraphical distribution 335 (fig.)
  tail 234–5
Garstang's views 113, 115, 347–9, 350
  cladogram 348 (fig.)
Girvan Starfish Bed 192, 207
Glyptosphaerites leuchtenbergi 216–17, 334 (fig.)
gnathostomes 117, 343
  absence of ammocoete larva 160
  cladogram 157–60
  classification within vertebrates 160
  features in common with myxinoids/
    lamphreys 158–61
  mandibular somites 287
  Müller cells 125
  oculomotor muscles 140
  ova pass through coelom 138
  spinal nerves 130 (fig.)
graptolites 25
Haeckel's law 8–9
hagfishes *see* myxinoids
hemichordates 1, 17, 317
  bilateral symmetry 317
  more primitive than echinoderms 51
  perhaps paraphyletic 54
Heterostelea 191
holothurians 54, 217
homology 6
hypothetical animals reconstructed 337–42
inverse larvae 40, 41
Lagynocystis 329
Lagynocystis pyramidalis 205, 293–308
  as acraniate 300
  atria 299–300
  atrial openings 294–5
  brain 306 (fig.), 307
  branchial cage (ctenoid organ) 298 (fig.),
    299–300
  buccal cavity 295–9
  cranial nerves 306 (fig.), 307
  external appearance 293–5
  gill bars 299–300
  gill slits 299–300
  head 293–5

openings 294
  soft parts inside 295–301
interossicular canal absent 287
mode of life 308
mouth 294
nervous system 306 (fig.), 307–8
oblique groove 295, 300, 301
ossicles in styloid process 304, 330
pyriform bodies (trigeminal ganglia) 66, 307
rectum 301–2
spinal ganglion 66
stratigraphical distribution 334 (fig.)
synapomorphies with recent acraniates 330
tail 223–3, 294, 304–7
  movements 305
Lampetra fluviatilis 134–48
abducens nerve 143
acoustic nerve 143
adenohypophysis 142
alimentary canal 137
ammocoete 134, 178
  branchial current 135
  buccal cavity 148
  duct, pharyngo-endostylar 149
  endostyle 149, 154
  kidney 141
  pharynx 148–53
aorta, dorsal/ventral 141
arcualia (vertebrae) 136
blood vascular system 140–1
brain 141–3
  diencephalon 142
  fibrous connections inside 143 (fig.)
  medulla oblongata 143
  mesencephalon 142–3
  rhombencephalon 143
  telencephalon 141–2
branchial current 135
branchial system 137–8
buccal cavity 120
cardinal veins 141
cloaca 120
coeloms 138
ear 146, 147 (fig.), 148 (fig.)
embryology 173–9
  brain 174
  cartilages 179
  dorsal nerve cord 174
  eyes 143–4
  gastrulation 174
  homologies with amphioxus 179
  hypobranchial musculature 178
  naso-pharyngeal duct 178
  neurulation 174
  oculomotor muscles 178
  pro-ammocoete 178
  skeleton 178–9
  somites 174–8
  trigeminal placodes 175
  visceral ('branchial') pouches 174–5, 178
external appearance 134–5
eyes 143–6
  compared with eyes of *Eptatretus stouti* 144–5
facial nerve 138, 143

fin rays 136–7
glossopharyngeal nerve 138, 143
heart 140
hyoidean segment 285
intestine 120
kidney 140
larval metamorphosis into adult 178
life cycle 134
muscles 138–40
  oculomotor 138–40
  somitic 138
    nerve supply 138
myomery/branchiomery, numerical
    correspondence 188 (fig.)
naso-hypophyseal duct 120
neurohypophysis 142
oesophagus 137
olfactory capsule 148
olfactory nerve 141–2, 148
olfactory organ 148
pancreas, exocrine 137
pharynx 120
pro-ammocoetes 134
Reissner's nerve 141, 142
sinus venosus 141
skeleton 135–7
  branchial 136
  tongue 135–6
spinal nerves 138
spino-occipital nerve 138
subcommisural organ 142
thyroid gland 15
tongue compared with that of *M. glutinosa*
    136 (note)
trigeminal nerve 138, 143
trochlear nerve 143
urogenital papilla ('penis') 135
vagus nerve 138, 143
velum 135
venous sinuses 141
venous system 141
Lampetra mariae ammocoete 152
Lampetra planeri
  ammocoete 153
    facial nerve 256 (fig.)
    optic cup 145 (fig.)
  endostyle cells 152 (fig.)
lampreys (Petromyzonids) 117, 133–4, 343
  brain 127 (fig.)
  choroid body 157
  cladogram 157–60
  classification within vertebrates 160
  dorsal nerve cord 141
  features shared with myxinoids/primitive
    gnathostomes 157–60
  mandibular somites 284, 287
  mesencephalon 142
  monophyletic group with myxinoids 160
  Müller cells 125
  oculomotor nerve 140
  osmoregulation system 140
  prosencephalon 141
  spinal nerves 130 (fig.), 141
  three-taxon problems in phylogeny 157–61
larvaceans (appendicularians) 113–15, 332

'less crownward', definition of 13
Loto vulgaris. lateral line 147 (fig.)
Løvtrup's views 345–7
  Chordata non-existent as monophyletic
    group 346
  non-morphological characters 346
Marthasterias glacialis, nervous system 34, 35 (fig.)
mesencephalon 126
mesodermalizing factor 186
Metazoa
  cladogram 19 (fig.)
  earliest phylogeny 17
  gastrula stage 17
  non-coelenterate (Bilateria) 17
metazoan phyla 17
mitrates 191, 192, 239–326
  blood vessel in notochord 204 (fig.)
  brain 126
  gill slits 192, 284
  gonoduct 258
  groups 329
  head chambers 200 (fig.)
  heart 284
  interossicular canals 249
  interossicular depressions 249
  left pharynx 300
  marginal plates near tail 192
  mitrocystitidan 332
  muscles of mouth and velum, formation of 284
  notochord of hind tail 202
  ossicles of hind tail 202
  peripharyngeal bands 255
  separation from cornutes 329
  stem-tunicate 332
  tail 329–30
  vertebrate head segments 284, 285 (fig.)
Mitrocystella 197, 329
  dorsal longitudinal canal 202
Mitrocystella barrandei 205
Mitrocystella incipiens incipiens 254, 334 (fig.)
Mitrocystella incipiens miloni 254–65, 332–3
  acustico-lateralis system 264–5
  atria, left 260
    right 258–60
  brain 260–3
  cranial nerves 260–3
  crescentic body 284
  head 254–60
    interior 255
    openings 255
  mouth 255
  nervous system 260–5
  olfactory end organs 264
  optic foramen 257
  palmar nerves 263–4
  pharynx 255
  posterior coelom 255–7
  ridge *cdl* (dorsal lamina) 279
  ridge *gd* (gonoduct) 257
  stratigraphical distribution 344 (fig.)
  tail 260
Mitrocystites 329
Mitrocystites sp. 2 (Chauvel) 333
Mitrocystites mitra 239–54, 332–3
  acustico-lateralis system 253

atria 239
atrial openings 239, 242 (fig.), 243 (fig.)
aulacophore 243
belt-shaped organ (dorsal nerve cord) 247
blood vessels 249
brain 249–52
calcite layers 246
carrot-shaped body (lateral-line ganglion) 253
coeloms, left/right, fields of 243
cranial nerves 249, 250 (fig.)
cylindrical organ (notochord) 247
gonoduct 246
head 239
   chambers 243–6, 260 (fig.)
homologies 352 (fig.)
hydropore 239
lateral-line nerve 253
lateripores 239
median groove of hind tail 247
mode of life 247–9, 254
mouth 239
narrow groove (lateral line) 239, 253
nervous system 249–54
notochord 247
oblique groove 243–4
olfactory nerves 253
optic nerves 252, 253
palmar complex 239, 252
palmar nerves 252–3
paripores 239
pharynx 244
pyriform bodies (trigeminal ganglia) 252
rectum 246
ribs 247
stratigraphical distribution 334 (fig.)
tail 239, 246–7
   function 247–9
   hind-tail ossicles 247
   Ubaghs' views 239–43
mitrocystitidans 329, 331
mollusc-arthropod group 1
[Mollusc + Arthropod + Annelida] group 346
monophyletic groups, categorial rank 14
monophyly 6, 7 (fig.)
'more crownward', definition of 13
Müller cells 125–6
Mustelus canis, Reissner's fibre in 157
myomery 187
myopterygians 137, 160, 342, 343
Myxine garmani (*Eptatretus burgeri*) 132
Myxine glutinosa 117–32
   abdominal coelom 120
   alimentary canal 120–1, 122 (fig.)
      associated organs 122 (fig.)
   auditory capsules 131
      ganglia associated with 129–30
   blood salts 124
   blood-sinus ('red lymphatic' system) 124, 125
   blood system 124–5
      'red lymphatic' 124, 125
      'white lymphatic' 124, 125
   brain 126–30
   branchial apparatus 120
   buccal ganglion/nerve 129
   cartilages 119

cloaca 121
coelomic pore 121
diet 117
egg 124
external appearance 117–18
giant cells in dorsal nerve cord 125–6
glossopharyngeal nerve 130
heart, accessory 125
   caudal 125
   systemic 124–5
kidney 121–4
lateral-line system absent 132
liver 121
medulla oblongata 129
membranous labyrinth 131–2
mesonephros 121, 124 (fig.)
movements 117
mucus expelled 118
muscles 121
   anterior part of head 123
   'lingual' 119–20, 121
naso-hypophyseal duct 118
nervous system 125–32
neurocranium 119
notochord 119
optic cups 128
optic nerves 128–9
ovary 124
pancreas, endocrine/exocrine 121
pericardium 120
pharyngo-cutaneous duct 120
photo-sensitive skin 129
pronephros 121, 122, 123 (fig.)
renal system 121–4
reproductive system 124
retina 128
skeleton 119
   'lingual' 119
sperm 124
spinal nerves 130
subcommissural organ 129
testis 124
tongue compared with that of *L. fluviatilis* 136
trigeminal nerves 129
vagus nerves 130
velum 118–19
myxinoids (hagfishes) 117, 342–3
   absence of ammocoete larva 160
   adenohypophysis 180
   blood salts 124
   cladogram 157–60
   embryology 185; *see also under E. stouti*
   features shared with lampreys/primitive
      gnathostomes 157–60
   living sister group of [lampreys +
      gnathostomes] 160
   monophyletic group with lampreys 160
   Müller-like fibres 126
   renal tubules short/absent 140
   spinal nerves 130 (fig.)
   three-taxon problem in phylogeny 157–61
   thyroid gland 181
nemertines 1
nephridium(a) 85
   Hatschek's 85

neural gland 221
neuralizing factor 186
neuro-muscular junctions 337
Nevadaecystis americana 13, 225–7, 326
   Ceratocystis-like features 326
   gill slits 225
   gonopore-anus 226
   head 225
   integumental plates 225
   keels 225
   mouth 225–6
   stratigraphical distribution 335 (fig.)
   strut 226
   tail 226
new species, origin of 6
Northcutt and Gans' theory 350–1
Oikopleura 113, 115 (fig.)
olfactory epithelium, vertebrate (frog) 149
paedogenesis 9
Palaeobranchiostoma haumatotergum 309
Palaeobotryllus taylori 315 (fig.), 316, 335 (fig.)
paraphyly 7
Patiria miniata larva, reversed asymmetry 43 (fig.)
pelmatozoans 29, 51–2, 319
Peltocystida 308, 315, 329, 331
Peltocystis 329
Peltocystis cornuta 308–12, 331
   interossicular canal absent 287
   stratigraphical distribution 335 (fig.)
   tail 311
'*Pentacrinus*' *decorus* 206 (fig.)
petromyzonids *see* lampreys
Petromyzon marinus 148
Philip's stele theory 351, 356, 358
phoronids 317
Phyllocystis 238, 327
   stratigraphical distribution 335 (fig.)
Phyllocystis blayaci 321, 326 (fig.)
Phyllocystis crassimarginata 321, 352 (fig.)
Pikaia gracilens 308, 309 (fig.), 330
   stratigraphical distribution 335 (fig.)
Placocystites forbesianus 2, 265–90, 323
   atria 279
   brain 287–90
   calcitic layers 269
   canals 278
   ciliated funnel 270, 279
   coeloms 279
      left anterior 284
      right, organs in 281–4
   cranial nerves 287–90
   crescentic body 284
   cylindrical organ 270
   fields, right/left 269
   gonad 281
   head 265–9
      interior 269
   interossicular canal 287
   mode of life 269
   oesophageal opening 270–8
   olfactory nerves 290
   opening in head 197
   palmar nerves 290
   peripharyngeal bands 270
   pharyngo-epicardial openings 278

pharynx 279
retropharyngeal process 278
skeleton 265, 269
    resorbtion 269
stratigraphical distribution 334 (fig.)
tail 269, 286 (fig.), 287
velum 269, 284
plesions 12–13
arrangement of fossils 13
monophyletic concept 13
paraphyletic concept 13
pogonophores 17, 350
polyphyly 7 (fig.), 8
primitiveness, criteria of 8
adaptational 10
embryological 8
outgroup comparison 8
stratigraphical 9–10
prosencephalon 126
Protocystites menevenis 227 (note), 326 (note)
Protostomia 17
embryological distinction from
    deuterostomes 18 (fig.)
tentaculate 17, 317
Psolus 198, 199 (fig.)
pterobranchs 18, 317
pterodactyl's wing 14
punctuated equilibria 10
radial symmetry 52, 319
Rathke's pouch 279
reciprocal illumination 15
reptiles 7–8
Reticulocarpos hanusi 227–33, 254, 326, 328
coeloms 227
differences from *R. pissotensis* 328
gill slits 227
gonopore-anus 227
habits 231
head 227
    chambers 227
locomotion 231
mouth 227
pharynx 227
stratigraphical distribution 334 (fig.)
tail 227–31
    comparisons with tails of *Chinianocarpos*
        thorali and *Lagynocystis pyramidalis*
        231–3
    ossicles 230, 230–1
    stylocone 230
    ventral spikes 328
Reticulocarpos pissotensis 238, 254, 328
stratigraphical distribution 334 (fig.)
reversed asymmetry 43 (fig.)
Rhabdopleura 24–5, 317
Rhabdopleurides 25
Rhabdopleurites 25
rhombencephalon 126
Romer's theory of vertebrate origins 349–50
Saccoglossus 25
Saccoglossus cambriensis 25 (fig.), 26 (fig.), 27
    (figs.)
Saccoglossus horsti larva 29 (figs.)
Saccoglossus kowalevskii 26 (figs.)
Scotiaecystis curvata 192, 205, 207–13, 327

brain 213
branchial current 207–10
branchial plates 207, 210 (fig.), 327
buccal cavity 211
feeding 210
gonopore-anus 197, 207, 210, 326
gonorectal groove 211–12
head 207
    appendages 207, 213
    openings 207
    skeleton 207
integument plates 210–11
    mouth 210
stratigraphical distribution 334 (fig.)
tail 212–13
    hind 355
Scyllium canicula (common dogfish)
blastomeres 164
early embryo 165 (fig.)
segmentation of head 189 (fig.)
sea lilies 47
sea squirts *see* tunicates
segmentationists 185–7
'somatico-visceral animal', Romer's 185
Seessel's pouch 279, 284
somites
hyoidean pair in living vertebrates 284
mandibular pair in living vertebrates 284
relationship to gill bars in vertebrate embryo
    187 (fig.)
speciational tree for group of fossil species
    11 (fig.)
Spiralia 17
Squalus acanthias 154–7
branchial current 154
cerebellum 157
egg 163
embryology 163–73
    auditory capsules 164–5
    blood vascular system 169
    cloaca 173
    dorso-lateral placodes 164
    gastrulation 163–4
    intestine 173
    liver 173
    neurulation 166 (fig.)
    neural crest 164, 169–72
    oculomotor muscles 169, 171 (fig.)
    optic vesicles 164
    pancreas 173
    Platt's vesicles 164
    precaudal plate 164
    premandibular somites 164
    skull 172
    somite differentiation 164, 166 (fig.), 167
        (fig.)
    trigeminal placodes 172 (fig.)
    visceral pouches 165–9
external appearance 154
hypoglossal nerve 157
mouth 154
nervous system 157, 158 (fig.)
oculomotor muscles/nerves 139 (fig.)
skeleton 154–6
terminalis nerve 157

urogenital system 157 (fig.)
Starck's theory of vertebrate origins 349–50
stele theory (Philip) 351, 356, 358
stem chordates 191, 321, 324 (fig.)
stem groups 10–14
    proposed abolition of (Ax) 13
    subdivision into plesions 13
stem lineage ('Stammlinie') 10–12
stemmatics (textual analysis) 14
stem species 6–7
Stolonica socialis 115
stratigraphical succession of species 9–10,
    354–5 (fig.)
Stromatocystites pentangularis 53, 319
    stratigraphical distribution 335 (fig.)
Stylophora 2, 191
symplesiomorphy 5
synapomorphies 4
Syringocrinus sinclairi 353 (fig.)
    stratigraphical distribution 334 (fig.)
tadpole, paedogenetic tunicate 1
Tentaculata 17, 18 (fig.)
textual analysis (stemmatics) 14
Thoralicystis 238, 335 (fig.)
Timeischites (edrioasteroid) 320 (fig.)
tornaria larva 28
tunicates (sea squirts) 87–115
alimentary ciliated loop 98 (fig.)
ciliated funnel 279
classification 113
coloniality 332
epicardia 278–9, 300
feeding 95, 270
heart 100–1, 284
hermaphroditism 332
neural gland 279–81
pelagic 332
pericardium 284
periodic reversal of heart beat 100
phylogenetic position 115
phylogenetic significance 115
salps 278
tadpole tail, neuro-muscular junctions 337
Ubaghs' aulacophore theory 351, 353
    answers to 353–6
Urochordata 87
Veit's theory of vertebrate origins 349–50
vertebrates
body, interplay of anterior neuralising/
    posterior mesodermalising factors 186 (fig.)
brain 126
flying 8
head problem 185–90
    anti-segmentationist view 185
    segmentationist view 186–7
innervation of somitic muscles 337
living 117–61
    agnathans 117
    gnathostomes 117; *see also* gnathostomes
olfactory epithelium (frog) 149
use of word 1
von Baer's law 8
Wenlock Limestone 265
wood warbler, relationship to willow warbler
    and chiff-chaff 5